Handbook of Formal Argumentation
Volume 3

Handbook of Formal Argumentation
Volume 3

Edited by

Dov Gabbay
Gabriele Kern-Isberner
Guillermo R. Simari
Matthias Thimm

© Individual authors and College Publications 2024. All rights reserved.

ISBN 978-1-84890-481-1

College Publications
Scientific Director: Dov Gabbay
Managing Director: Jane Spurr

http://www.collegepublications.co.uk

Cover produced by Laraine Welch

All rights reserved. No part of this publication may be reproduced, stored in a retrieval system or transmitted in any form, or by any means, electronic, mechanical, photocopying, recording or otherwise without prior permission, in writing, from the publisher.

Dedicated to Trevor, in grateful memory

Preface

After two successful volumes of the *Handbook of Formal Argumentation*, we now proudly present another volume to which experts of the field contributed chapters summarizing background and recent work of their research focus in formal argumentation. While the first two volumes aimed at providing introductory articles and surveys on a diversity of argumentation formalisms and their relations in particular to dialogues, dynamics, and preferences, highlighting computational aspects and formal properties, this third volume contains research works broadening the borders of formal argumentation by elaborating on applications and links to other areas of research, and also by addressing general challenges in argumentation theory nowadays. Accordingly, the structure of this volume consists of three parts, focussing on *Applications*, *Cross-field Connections*, and *Perspectives and Future Directions*, respectively. Each paper had been peer-reviewed by two reviewers before it was accepted for publication here. In the following, we summarize the contributions to each part in the following.

Part I Applications starts with a paper on *Social Argumentation Systems* by Antonis Bikakis, Giorgos Flouris, Joao Leite, and Theodore Patkos. The authors present frameworks and semantics that are particularly suitable for argumentation in the social web and in online communities where more complex interactions among arguments than just attack and defense occur.

The next paper on *Searching, Navigating, and Querying Arguments and Debates: Tools, Languages and Methodologies* by Giorgos Flouris, John Lawrence, and Dimitra Zografistou also addresses the Web as a place of dialogical exchange and debate. The authors provide a survey of tools and languages for searching, exploring and querying arguments in the Web, aiming at identifying relevant arguments for a given context.

Argumentation in legal contexts is the topic of the paper *Computational Models of Legal Argument* by Trevor Bench-Capon, Katie Atkinson, Floris Bex, Henry Prakken, and Bart Verheij. Indeed, a legal dispute at court is quite a prototypical example for abstract argumentation where two parties take the clear roles of proponent and opponent, arguing with one another to come up with most conclusive arguments attacking the arguments of the respective other party. The authors organize their survey along a chronological perspective.

The paper *Applications of Argumentation-based Dialogues* by Andreas Xydis, Federico Castagna, Elizabeth I Sklar, and Simon Parsons deals with formal argumentation for modelling dialogues. The authors present formal models of dialogues and then discuss software tools to construct argumentation-based dialogue systems, reaching out to recent work on chatbots from natural language processing. Finally, they give a comprehensive survey on selected applications and point out future directions.

Agent-based models are a central paradigm of research work in artificial intelligence. The paper *Argumentative Agent-based Models* by Louise Dupuis, Matteo Michelini, Dunja Šešelja, and Christian Straßer puts such models on an argumentative base, defining argumentative agent-based models (AABMs). The authors provide a systematic overview of AABMs, pointing out the various roles that argumentation plays in these models, and elaborating on the underlying methodology to bridge gaps between different disciplines.

Finally, Srdjan Vesic and Bruno Yun present their paper *Argumentation-based Applications for Decision-Making* as the last contribution to Part I. This paper considers decision making within the context of formal argumentation and gives a comprehensive survey on existing frameworks and systems, pointing out tools for decision making and visualization tools for argumentation.

Part II Cross-field Connections contains papers that link formal argumentation to the fields of logic programming, conditionals, machine learning, causation, epistemic planning, and modal logic. The paper *Syntactic and Semantic Connections between Logic Programming and Argumentation Systems* by Samy Sá, Wolfgang Dvořák, and Martin Caminada examines the especially strong (both historically and methodologically) connections between logic programming and argumentation. The authors consider translations between these two formal frameworks and investigate if and how semantic equivalence can be proven or restored, where various argumentation formalisms and common semantics are taken into regards. They also give an overview on approaches that make use of these connections to implement argumentation via answer set programming.

In the paper *The Semantical Structure of Conditionals, and its Relation to Formal Argumentation*, Jesse Heyninck, Gabriele Kern-Isberner, Tjitze Rienstra, Kenneth Skiba, and Matthias Thimm investigate connections between conditionals and abstract dialectical frameworks (ADFs) by interpreting the links between nodes and their acceptance conditions in ADFs as conditionals. This allows for lifting well-known techniques for conditional (and nonmonotonic) reasoning and belief revision to ADFs. Moreover, the paper gives a general survey on works where ideas from conditional logics are applied in formal argumentation.

Regarding *Argumentation and Machine Learning*, Antonio Rago, Kristijonas Čyras, Jack Mumford, and Oana Cocarascu present a review of the relevant literature from which two major types of interactions between these two fields become clear: on the one hand, argumentation is used to improve and/or explain machine learning models; on the other hand, machine learning is applied to support, analyze, or replace argumentation. The authors investigate these interactions along various dimensions, including the type of machine learning, the specific argumentation framework, and the degree of integration between learning and argumentation.

In their paper *Causation and Argumentation*, Alexander Bochman, Federico Cerutti, and Tjitze Rienstra elaborate on the crucial link between these two fields that is given by the explanatory power which is associated with both. This makes integrative frameworks connecting argumentation and causal reasoning strong candidates for explainable AI. The authors present and discuss several such integrative approaches.

The paper *Defeasible Argumentation-based Epistemic Planning with Preferences* by Juan C. L. Teze, Lluis Godo, and Gerardo I. Simari studies applications of formal argumentation to epistemic planning. The authors first give a brief overview on the field and then present a general architecture for defeasible argumentation-based epistemic planning. This architecture is then realized by an approach that is based on defeasible logic programming and possibility theory.

Carlos Iván Chesñevar, Jürgen Dix, Beishui Liao, Jieting Luo, Carlo Proietti, and Antonio Yuste-Ginel investigate different connections between *Formal Argumentation and Modal Logic*. They introduce three different combinations of modal operators and argumentation systems, namely via dynamic operators, temporal operators, and epistemic operators. Argumentation and modal logic may also take different positions in these combined frameworks, respectively.

Finally, in Part III Perspectives and Future Directions, Liuwen Yu, Leendert van der Torre, and Réka Markovich conclude this volume with *Thirteen Challenges in Formal and Computational Argumentation*. The authors consider Dung's attack-defense paradigm shift as an origin of their challenges, taking three perspectives with respect to this reference point: regarding its context, investigating the paradigm shift itself, and considering its consequences. Examples from many disciplines illustrate these challenges.

We would like to thank all authors for their valuable contributions to this handbook, and also all reviewers who helped maintain the high quality of these handbooks. We are also grateful for any suggestion, critique, and other form of feedback that we received while preparing this handbook. As usual, special thanks go to College Publications for realizing this project, in particular to Jane Spurr for her helpful and reliable support.

Dov M. Gabbay
Gabriele Kern-Isberner
Guillermo R. Simari
Matthias Thimm

Contents

Articles

Applications

1 Social Argumentation Systems 15
Antonis Bikakis, Giorgos Flouris, Joao Leite, Theodore Patkos

2 Searching, Navigating, and Querying Arguments and Debates: Tools, Languages and Methodologies 67
Giorgos Flouris, John Lawrence, Dimitra Zografistou

3 Computational Models of Legal Argument 101
Trevor Bench-Capon, Katie Atkinson, Floris Bex, Henry Prakken, Bart Verheij

4 Applications of Argumentation-based Dialogues 225
Andreas Xydis, Federico Castagna, Elizabeth I Sklar, Simon Parsons

5 Argumentative Agent-based Models 305
Louise Dupuis, Matteo Michelini, Dunja Šešelja, Christian Straßer

6 Argumentation-based Applications for Decision-Making 377
Srdjan Vesic, Bruno Yun

II Cross-field Connections

7 Syntactic and Semantic Connections between Logic Programming and Argumentation Systems 42
Samy Sá, Wolfgang Dvořák, Martin Caminada

8 The Semantical Structure of Conditionals, and its Relation to Formal Argumentation 51
Jesse Heyninck, Gabriele Kern-Isberner, Tjitze Rienstra, Kenneth Skiba, Matthias Thimm

9 Argumentation and Machine Learning 57
Antonio Rago, Kristijonas Čyras, Jack Mumford, Oana Cocarascu

10 Causation and Argumentation 62
Alexander Bochman, Federico Cerutti, Tjitze Rienstra

11 Defeasible Argumentation-based Epistemic Planning with Preferences 71
Juan C. L. Teze, Lluis Godo, Gerardo I. Simari

12 Formal Argumentation and Modal Logic 76
Carlos Iván Chesñevar, Jürgen Dix, Beishui Liao, Jieting Luo, Carlo Proietti, Antonio Yuste-Ginel

III Perspectives and future directions

13 Thirteen Challenges in Formal and Computational Argumentation ... 84
Liuwen Yu, Leendert van der Torre, Réka Markovich

Part I

Applications

CHAPTER 1

SOCIAL ARGUMENTATION SYSTEMS

ANTONIS BIKAKIS
Department of Information Studies, University College London, UK
a.bikakis@ucl.ac.uk

GIORGOS FLOURIS
Institute of Computer Science, FORTH, Heraklion, Greece.
fgeo@ics.forth.gr

JOAO LEITE
NOVA LINCS, NOVA University Lisbon, Portugal
jleite@fct.unl.pt

THEODORE PATKOS
Institute of Computer Science, FORTH, Heraklion, Greece.
patkos@ics.forth.gr

1 Introduction

The use of the Web is constantly evolving. Although users were originally expected to be merely consumers of Web information, in recent years we experienced a proliferation of portals and online systems allowing users to become also producers of information. As a result, the social aspect of the Web has been flourishing, allowing users to post opinions, comments and reviews populating a wide range of online systems, from social media and online discussion forums to news sites and product review sites [Bikakis *et al.*, 2023]. The impact of this user-generated information is clearly evident, especially on consumer behaviour and businesses [Luca, 2011]. Due to the vast number of comments that users need to search through to locate the most helpful ones, their proper ranking, filtering and recommendation become critical functionalities. Apart from the open-ended discussions on the Web, more goal-oriented debates are also becoming popular, e.g., in debate portals for active citizenship[1], or in decision support systems, such as issue-based information systems (IBIS) [Kunz and Rittel, 1970], [Baroni et al., 2013], [Baroni et al., 2015].

In this setting, it is not enough for applications to provide functionality for opinion or argument exchange. In order to help users reach informed, well-justified and sensible conclusions or decisions, such applications also share the need to *evaluate arguments* based on quantitative methods. Towards this goal, various methods have been used to rate user arguments or comments, which vary from voting mechanisms, such as like/dislike counters, and expert ratings, to combinations of these with user responses in the form of counter (attacking) or follow-up (supporting) arguments. Methodologies from computational argumentation have been proposed as a powerful tool for a more accurate evaluation of an argument's acceptance, and a number of formal frameworks have emerged that properly adapt argumentation algorithms to the needs of the Social Web (e.g., [Leite and Martins, 2011], [Patkos et al., 2016c], [Evripidou and Toni, 2014]) or decision support systems (e.g., [Baroni et al., 2015], [Rago et al., 2016]).

This chapter focuses mainly on argumentation frameworks that treat two different types of reaction to an argument: verbal responses, which are commonly modelled as arguments that are related to the original one (e.g., via an attack or support relationship); and (positive or negative) votes, which express someone's approval or disapproval of the argument. We call such frameworks *social argumentation systems*. Their main aim is to provide a way to evaluate the social acceptance of an argument (which in most of these frameworks is referred to as *strength*, or *score*) taking into account the responses and the votes it has received and the strengths of such responses. Also, in many cases, the computed strength is also affected by a *base score*, which reflects the strength of an argument before any reactions are considered. The base score may reflect different

---

[1] https://www.kialo.com, https://www.kialo-edu.com, https://www.createdebate.com

intuitions, such as an a priori assessment by experts, argument popularity, or other features that are supported by the underlying application. The strength of an argument, as computed by a social argumentation system, usually quantifies the degree of acceptance of the argument, although some social argumentation systems may also evaluate other aspects of arguments, such as their acceptability, quality, relevance, objectivity and others (see, for example, Section 4 in this chapter and the related papers [Patkos et al., 2016c], [Patkos et al., 2016a]).

Note that, under this viewpoint, the argument evaluation process in a social argumentation system differs from the one in the standard Abstract Argumentation Frameworks (AAFs) of Dung [Dung, 1995] and most of their extensions. In AAFs and similar frameworks, an argument is either in the extension, or not. On the other hand, social argumentation systems employ *gradual semantics*, i.e., the strength of an argument is expressed in terms of a numerical value, enabling a more fine-grained evaluation compared to the two- or three-valued acceptability semantics of most argumentation frameworks. Having said that, one can view extensions as a special case of a numerical assignment, where the only values allowed are 0 (not in the extension) and 1 (in the extension).

The aim of this chapter is *to provide an overview of applications of formal argumentation to the Social Web, in the form of Social Argumentation Systems*. Towards this, we start by presenting a number of relevant principles for such systems (Section 2). Then, we present two examples of social argumentation systems, namely Social Argumentation Frameworks (Section 3) and s-mDiCE (Section 4). We then briefly survey other social argumentation systems, including some examples of extensions or applications of existing argumentation frameworks designed or used to model other aspects of discussions in social media, such as the semantic relations among posts, the social relevance of a post, the influence of users and multi-topic discussions (Section 5). We conclude in Section 6.

2 Principles and properties

As will become obvious by our presentation in the following sections of this chapter, methodologies for computing different notions of argument strength in a social context abound. Thus, a question naturally arises: which one is best? The answer, of course, depends on the needs of the application, but even so, how can we evaluate and compare these methodologies in the context of any given application?

To address this question, many authors have proposed several different *principles*, i.e., logical constraints, or axioms, that formalise certain intuitive properties that a "good" social argumentation system should satisfy. In this section, we study some of these principles, with emphasis on the ones that are most relevant to social argumentation.

2.1 The principle-based approach in argumentation

The *principle-based approach* in argumentation, sometimes called the *axiomatic approach* [van der Torre and Vesic, 2018], is a methodology for developing principles for assessing the "quality" or "relevance" of different semantics for specific applications. Importantly, such principles are not necessarily ubiquitous or indisputable, and different principles may be relevant to different applications: as a matter of fact, for any given application, some principles may be desirable, others irrelevant, and others even undesirable. But this is exactly the strength of the principle-based approach: by choosing the principles that are (un)desirable for a certain application, the designer of the application can immediately identify the semantics that are (in)appropriate for the application at hand, and therefore make an informed choice.

Thus, the principle-based approach can be viewed as a methodology for choosing the most appropriate semantics to use for a particular application. Also, it has been argued that this approach provides a systematic way of viewing semantics and their properties, guiding the search for novel interesting argumentation semantics [van der Torre and Vesic, 2018], and allowing the identification and definition of new relevant principles [Baroni *et al.*, 2018].

The principle-based approach has been used for many different types of argumentation frameworks. As expected, the bulk of the related work deals with the development of principles for the standard Abstract Argumentation Frameworks of Dung [Dung, 1995], and, thus, are not entirely relevant to the gradual argumentation semantics that interest us here. We start by briefly presenting some of the studies that adopt the principle-based approach in settings other than gradual argumentation semantics, before analysing in more detail the principles that are relevant to gradual argumentation semantics.

2.2 Proposed principles

In [Baroni and Giacomin, 2007], the authors present the principle-based approach and apply it to abstract argumentation. The work is comprehensive, evaluating 13 different principles against 11 different semantics. A very similar discussion appears in [van der Torre and Vesic, 2018], which evaluates 8 principles against 15 proposed semantics. Note that there is some overlap in the considered principles and semantics (and thus in some of the results as well) between these two studies. Importantly, both approaches apply to the standard semantics of abstract argumentation frameworks, and not for gradual argumentation, and thus their relevance to the present chapter is limited.

A similar work appears in [Yu et al., 2021] for abstract agent argumentation frameworks, i.e., frameworks that extend Dung's AAFs with agents. The idea in this setting is that each argument is associated with certain agents, and this association may affect the semantics of the framework and the extensions that it has. Obviously, it also affects the potentially desirable principles that such semantics should satisfy. In that work, the authors examine 52 agent semantics and 17 principles under this setting, all of them considering the standard AAF semantics.

A work that is more closely related to this chapter is [Bonzon et al., 2016], which examines the principle-based approach for *ranking-based semantics*, in which the argument evaluation process results to a ranking determining whether an argument is "more acceptable" than another. That work evaluates 7 ranking-based semantics against 18 principles.

In [Caminada, 2018], the author examines a set of properties related to how argumentation frameworks (and their associated semantics) behave when used for reasoning. Thus, the principles presented in this paper apply to the output of the argumentation process (i.e., the conclusions), rather than the (accepted) arguments themselves.

In [Gonzalez et al., 2021], the authors propose a set of principles for bipolar argumentation. Their approach is based on the use of labels that are defined in an abstract manner based on an appropriate algebra.

Depending on how the labels and the respective algebra are defined, they could be used to represent different things, including the "strength" of an argument, under various different notions of what "strength" may mean. Thus, this work can be considered to fall within the scope of gradual and/or social argumentation. In fact, the authors of [Gonzalez et al., 2021] explicitly mention the possible application of this work to social platforms.

Other studies define principles that are specific to the formalism at hand. For example, [Flouris et al., 2023] propose 4 principles, most of which are adaptations of AAF principles for the formalism proposed in that paper (Abstract Argumentation Frameworks with Domain Assignments – AAFDs), and show which AAFD semantics satisfy such principles.

More in relation to the present chapter, numerous principles for gradual and social argumentation have appeared in various papers [Gonzalez et al., 2021], [Amgoud and Ben-Naim, 2016b], [Amgoud and Ben-Naim, 2017], [Bonzon et al., 2016], [Matt and Toni, 2008], [Rago et al., 2016], [Amgoud and Ben-Naim, 2013], [Amgoud et al., 2016], [Baroni et al., 2015], [Leite and Martins, 2011], [Thimm and Kern-Isberner, 2014], [Amgoud and Ben-Naim, 2016a], [Amgoud et al., 2017]. Although these principles often use different notation and terminology, they capture similar intuitions. Still, their comparison is difficult. This problem was identified and addressed by [Baroni et al., 2019], resulting in an organisation of the principles that will be the focus of the following subsection.

2.3 Organising the gradual argumentation principles

Perhaps the most prominent work presenting principles for gradual argumentation appeared in a series of papers [Rago et al., 2018], [Baroni et al., 2018], [Baroni et al., 2019], where the authors presented an effort to organise the numerous principles proposed in other studies under a single unifying umbrella of flexible definitions. The focus of this chapter on the work of [Baroni et al., 2019] is justified by the fact that it essentially unifies many previously-expressed principles in other papers (such as those described above).

In particular, [Baroni et al., 2019] collected several principles that have appeared in the literature, noticing that they have common conceptual roots and are based on a small set of common patterns. Then, they formalised these patterns into 11 *group properties* (GPs), which are essentially generic principles that correspond to most of the principles that have appeared in previous work. Further, these 11 GPs can be viewed as instantiations of 4 novel parametric principles.

The above organisation of principles has many benefits. First, it provides a systematic way to view principles relevant to gradual argumentation, thereby revealing the existence of novel principles that have so far not been studied in the literature. Secondly, it provides a simplifying and unifying terminology to study and discuss the relevant principles, ending the polyphony that hindered the comparison of principles appearing in previous papers. In other words, it provided a unifying substrate with the use of which most of the principles in the gradual argumentation literature can be expressed, compared against, and studied.

From a formal perspective, the authors of [Baroni et al., 2019] base their analysis on an argumentation model called *Quantitative Bipolar Argumentation Framework (QBAF)*, which is a tuple $(X, \mathcal{R}^-, \mathcal{R}^+, \tau)$, where X is a finite set of arguments, $\mathcal{R}^-, \mathcal{R}^+$ is the attack and support relation (respectively) between arguments, and $\tau : X \mapsto \mathbb{I}$ is a total function mapping each argument with a *base score* in $\mathbb{I}$, where $\mathbb{I}$ is a set equipped with a preorder (usually $\mathbb{I} = [0, 1]$, i.e., a real number in the $0 \dots 1$ range, but other options are also possible). The base score corresponds to the argument's evaluation before considering its relationships (attack, support) with other arguments, and is a common feature in social argumentation systems. In [Baroni et al., 2019], the use of τ is overloaded, and exploited as a convenient abstraction to hide a possibly complex method of computing the base score, based, e.g., on experts' assessment, argument popularity, votes, various types of non-verbal reactions on arguments, or other features that are supported by the underlying application. Note also that either $\mathcal{R}^-$ or $\mathcal{R}^+$ could be empty, leading to special cases of QBAFs where only support (or only attack, respectively) relations are allowed.

The *strength* (or *score*) of an argument is computed using another function, $\sigma : X \mapsto \mathbb{I}$. The function σ corresponds to the final arguments' assessment, after taking into account both the base score and all relationships (attack, support) between arguments. The principles associated with gradual argumentation frameworks (i.e., QBAFs, in the terminology of [Baroni et al., 2019]) are all meant to restrict the behaviour of σ in ways that make intuitive sense.

Using this formalisation, [Baroni et al., 2019] presented 11 intuitive group properties, and their formal formulation as a GP, as well as 4 principles, which were shown to imply the GPs and thus provide a more general intuition behind them. These 4 principles are called *balance*, *strict balance*, *monotonicity*, and *strict monotonicity*. Balance and strict balance capture the intuition that any difference between the score of an argument ($\sigma(a)$) and its base score ($\tau(a)$) must correspond to some imbalance between the scores of its attackers and supporters. Monotonicity and strict monotonicity capture the intuition that each of the factors that affect an argument's score (base score, attackers, supporters) has a monotonic effect on the argument's score.

Further, the authors of [Baroni et al., 2019] parameterise the QBAF model using the following five features:

- Whether it is required that $\mathcal{R}^- = \emptyset$ or $\mathcal{R}^+ = \emptyset$.

- The exact definition of $\mathbb{I}$ and its associated preorder $\leq$ (as well as its strict counterpart, $<$).

- A special relation $\ll$ between elements of $\mathbb{I}$, such that $<\,\subseteq\,\ll\,\subseteq\,\leq$.

- Whether $\mathbb{I}$ has a bottom element and whether arguments whose strength equals that bottom element are (or should be) considered in the evaluation or not.

- The definition of τ.

This parameterisation essentially allows different frameworks that have appeared in the literature to be recast as a QBAF. Equally importantly, it allows the parameterisation of principles in order to capture different intuitions.

A comprehensive evaluation of the principles that different argumentation frameworks and their semantics satisfy, appears in [Baroni et al., 2019] (see Table 5 in that paper). In particular, the authors recast 19 different argumentation frameworks in the QBAF terminology, using the parameterisation options described above, and then showed which of the 4 main principles are satisfied by each of these 19 approaches. As a corollary (and combined with other results in the same paper), one can easily identify the GPs and other properties that are satisfied by each of these approaches.

2.4 Principles for dynamic frameworks

The aforementioned principles all deal with static frameworks, i.e., given a framework, they suggest how the scoring function σ should behave. An interesting exception is the principle of *smoothness*, which was informally defined in [Patkos et al., 2016a] and formally in the extended version of that paper [Patkos et al., 2016b]. Unlike the other principles, smoothness deals with how the scoring function behaves under *changes in the framework*. For this reason, it was not considered by [Baroni et al., 2019], despite the generality and comprehensiveness of the principles presented there.

In particular, smoothness, as the name implies, guarantees that the scoring function of a gradual argumentation framework will behave "smoothly", i.e., small changes in the framework (e.g., a new relationship, or a small change in the base score of an argument) cannot have large effects on the overall evaluation of arguments. This is an essential feature for any rating framework, in order to be adopted by the public, as the effect of an action on the framework should be commensurate with the importance of the action itself; big leaps that are not justified by the underlying changes may seem counterintuitive to users, causing them to lose their trust in the objectivity of the rating algorithms.

The mathematical notion that is closest to the intuition presented above is the notion of the derivative of a function: in functions over real numbers, the derivative determines the rate at which the function changes at each point. However, for the considered setting this notion must be adapted to apply over more complex (non-continuous) domains, because changes in our case are not necessarily continuous (e.g., the addition of a new relationship between arguments). Thus, to define smoothness over arbitrary functions and sets, *semi-metrics* were used to determine the "rate" of change:

Definition 2.1. *Given a set S, a function $d_S : S \times S \mapsto \mathbb{R}$ is called a semi-metric for S iff for all $x, y \in S$: $d_S(x, y) \geq 0$, $d_S(x, y) = d_S(y, x)$ and $d_S(x, y) = 0$ iff $x = y$.*

Definition 2.2. *Consider two sets S, T equipped with semi-metrics d_S, d_T. A function $f : S \mapsto T$ is called ℓ-smooth (for d_S, d_T) iff*

$$d_T(f(x), f(y)) \leq \ell \cdot d_S(x, y)$$

for all $x, y \in S$.

The value of ℓ in Definition 2.2 determines the "smoothness" of the function: a large ℓ implies that the function has at least some "abrupt" points, i.e., in our setting, that there are cases where simple (small) actions by the users would lead to major changes in the assessment result of the related arguments. On the other hand, small ℓ guarantees that a large number of changes are required to achieve a large effect on the assessment results, thus making the function more reluctant to change. Given a function f, when there is no ℓ such that f is ℓ-smooth, we will say that f is ∞-smooth. Moreover, we will say that f is *exactly ℓ-smooth* when it is ℓ-smooth and there is no $\ell' < \ell$ such that f is ℓ'-smooth.

In the considered context, smoothness should be applied over the argument rating function (σ in the [Baroni et al., 2019] terminology), to determine how quickly the ratings change when the framework changes. Note that the input to σ is an argument, therefore, to apply smoothness we technically need to also include the QBAF itself as an input to σ, i.e., define σ as $\sigma_F : X \mapsto \mathbb{I}$, where F is the QBAF under consideration. Now smoothness studies "how much" a change in F affects the output $\sigma_F(a)$ for $a \in X$. To apply smoothness here, one should use the semi-metric d_S to quantify the effect of each possible change in F (e.g., the addition/deletion of arguments, the addition/deletion of attack/support relationships, and the modification of one or more base scores), and d_T to quantify the effect on the actual score (given that $\mathbb{I}$ is usually a numerical domain, such as $[0, 1]$, a reasonable choice for d_T is the simple difference, i.e., $d_T(x, y) = |x - y|$).

Note that different applications may have different requirements regarding smoothness. As an example, using functions with high sensitivity to input (i.e., less smooth) will allow applications that attract few users to lay more emphasis on maintaining the liveness of discussions by having users' feedback cause reasonable, yet evident, effects in the course of the discussions. On the contrary, applications that lay emphasis on the reliability of the outcome of a debate, such as product/services rating sites, probably want to disallow small changes to significantly impact the outcome, in order to secure reliable results, therefore requiring functions that are less sensitive to user input. In this respect, smoothness is no different than other principles, in that it is application-dependent, and parameterisable (using ℓ, as well as the definitions of d_S, d_T). For applications of this principle, see Section 4.

3 Social argumentation frameworks

Justified by the fact that a growing percentage of users were giving up on the social web for lack of intellectually stimulating discussions, [Leite and Martins, 2011] argued that a Social Network should facilitate:

- More *open participation* where users with different levels of expertise are able to easily express their arguments, even without knowing formal argumentation and any formal rules of debate.

- More *flexible participation* where debates are not restricted to a pair of users arguing for antagonistic sides, but where there may be more than just two sides, more users can propose arguments for each side, and each user is allowed to contribute with arguments for more than one side of the debate.

- More *detailed participation* where users are allowed to express their opinions by voting on individual arguments and on argument relations, instead of just on the overall debate's outcome.

- Appropriate *feedback* to users so that they can easily assess the strength of each argument, taking into account not only the logical consequences of the debate, but also the popular opinion and all its subjectiveness.

To that end, they envisioned a self-managing online debating system capable of accommodating two archetypal levels of participation.

On the one hand, experts, or enthusiasts, would be provided with simple mechanisms to specify their arguments and also a way to specify which arguments attack which other arguments. When engaging in a debate, users always propose arguments for specific purposes, like making a claim central to the issue being discussed, or defeating arguments supporting an opposing claim. Thus, the envisioned system should allow users to describe an abstract argument, capable of attacking other arguments, simultaneously with its natural language (or image, video, link, etc.) representation. Therefore, the formal specification of arguments and attacks becomes a natural by-product of the users' intent when proposing new arguments. To make this process as painless and easy as possible, and enable more people to participate, no particularly deep knowledge (such as logic) can be required. It is natural that a new argument might attack a previously proposed argument - indeed, that was likely the object of its creation. However, it is also possible that an older argument attacks the new argument as well. Therefore, the system should allow users to add this new attack relation to the system.

On the other hand, less expert users who prefer to take a more observational role, and do not wish to engage in proposing arguments or attacks, should also be accommodated in the system through a less complex participation scheme. These users may simply read the arguments in natural language (or image, video, link, etc.) and state whether they agree with them. This induces a voting mechanism similar to what is found in current social networks. There are alternatives, such as having argument's social trustworthiness be based on people's opinions of who proposed it. Voting on arguments seems to offer the path of least resistance for being the closest to current social networks. Additionally, it is apparent that not all attacks bear the same weight. Some attacks might have an obvious logical foundation (e.g., undercuts or rebuts), thus gaining trust from the more perceptive users. Other attacks might be less obvious or downright senseless, especially in open online contexts, making users doubt or wish to discard them. Thus, extending the ability to vote to attacks becomes eminently desirable. Not only does voting on attacks more accurately represent a crowd's opinion in a variety of situations, but it also allows the system to self-regulate by letting troll-attacks be "down-voted" to irrelevance.

The system should also be able to autonomously and continuously provide an up to date view of the outcome of the debate, e.g., by assigning a value to each argument that somehow represents its social strength, taking the structure of the argumentation framework (arguments and attacks) and the votes into account. A nice GUI, e.g., depicting arguments with a size and/or colour proportional to these values would make the debate easier to follow, bringing forward relevant (socially) winning arguments, while downgrading unsound, unfounded (even troll) arguments. So that users may understand and follow a debate, small changes in the underlying argumentation framework and its social feedback (i.e., votes) should result in small changes to the formal outcome of the debate. If a single new vote entirely changes the outcome of a debate, users cannot gauge its evolution and trends, and are likely to lose interest.

In addition, any debating system as the one envisioned should also ensure, as argued in [Leite and Martins, 2011], that a few crucial properties be satisfied.

- *There should always be at least one solution to a debate.* From a purely logical standpoint, one may consider that some debates simply contain inconsistencies that make it impossible to assign them meaningful semantics. However, we are dealing with the Social Web, where inconsistency is the *norm*. If the system is incapable of providing solutions to every debate, then there is too much risk involved in using it. We believe that most of its users would prefer a system that would, nevertheless, provide them some valuation of the arguments that is somehow justifiable, instead of telling them that the debate is inconsistent.

- *There should always be at most one solution to a debate.* Logicians and mathematicians find it perfectly natural for there to be multiple, or even infinite, solutions to a given problem. However, in a social context as far-reaching as the Internet, it is disingenuous to assume that the general user-base, which likely covers a large portion of the educational spectrum, shares these views with the same ease. It is very hard for someone who has invested personal effort into a debate to accept that all arguments are in fact true (in a multitude of models)!

- *Argument outcomes should thus be represented very flexibly.* In particular, to accurately represent the opinions of thousands of voting users, arguments should be evaluated using degrees of acceptability, or gradual acceptability. Two-valued or three-valued semantics risk grossly under-representing much of the user-base.

- *Formal arguments and attacks must be easy to specify.* For example, assuming knowledge of first-order logic for specifying structured arguments would alienate many potential users when the present goal is to include as many as possible. Moreover, simpler frameworks would make implementing and deploying such a system in different contexts (web forums, blogs, social networks, etc) much easier.

- *Argument strength should be limited by popular opinion, and every vote should count.* In a true social system, there should be no arguments of authority, nor votes without effect. Argument strength can be weaker than its direct support base, since arguments may be attacked by other arguments, but the direct opinion expressed by the votes should act as an upper bound on the strength of the argument. Also, positive (resp. negative) votes should increase (resp. decrease) the strength of the argument/attack on which they are cast (how much can depend on many factors).

- *Computing and updating debate outcomes should be highly efficient.* With the increasing speed of social interactions on the Web 2.0, users would grow impatient if new arguments and votes would not have an almost immediate effect on the debate outcome.

In the remainder of this section, we describe Social Argumentation Frameworks [Leite and Martins, 2011; Egilmez et al., 2013], which can serve as the underlying formal backbone of an online debating system as the one described above.

Social Argumentation Frameworks use abstract arguments in the sense of Dung [Dung, 1995][2], but add the possibility of associating pro

---

[2]Abstract Argumentation [Dung, 1995] and Argumentation Theory in general grounds debates in solid logical foundations and has in fact been shown to be applicable in a multitude of real-life situations (c.f. [Modgil et al., 2013]).

and con votes on both arguments and attacks, and a (family of) semantics that goes beyond the classical accepted/defeated valuations assigning each argument a value from an ordered set of values (e.g. the [0, 1] real interval, a set of colours, textures, etc.). One particular semantics – based on the popular product T-norm and probabilistic sum T-co-norm and assigning arguments values from the [0, 1] real interval – deserves particular attention because of its formal properties, namely guaranteeing the existence and uniqueness of a model, and also because of the existence of an algorithm that can effectively and efficiently compute the debate outcome. Social Argumentation Frameworks provide the theoretical foundations on which to build interaction tools that will provide more robust, flexible, pervasive and interesting social debates than those currently available.

In the following, we start by describing Social Argumentation Frameworks (SAF) and their semantics, which we subsequently illustrate with a very simple example. We then discuss some important formal properties of SAF. Finally, we present an efficient algorithm for computing debate outcomes.

3.1 Framework and Semantics

We start by describing a Social Argumentation Framework. First introduced in [Leite and Martins, 2011] and later extended in [Egilmez et al., 2013], it is an extension of Dung's AAF, composed of arguments and an attack relation to which we add an assignment of votes to each argument and each attack.

Definition 3.1 (Social Argumentation Framework). *A social argumentation framework is a triple* $F = \langle \mathcal{A}, \mathcal{R}, V \rangle$, *where*

- $\mathcal{A}$ *is a set of arguments,*

- $\mathcal{R} \subseteq \mathcal{A} \times \mathcal{A}$ *is a binary attack relation between arguments,*

- $V : \mathcal{A} \cup \mathcal{R} \to \mathbb{N} \times \mathbb{N}$ *is a total function mapping each argument and each attack to its number of positive and negative votes.*

The notion of a semantic framework is used to aggregate operators representing the several parametrisable components of a semantics:

Definition 3.2 (Semantic Framework). *A social argumentation semantic framework is a 6-tuple* $\langle L, \curlywedge_{\mathcal{A}}, \curlywedge_{\mathcal{R}}, \Upsilon, \neg, \tau \rangle$ *where:*

- L is a totally ordered set with top element $\top$ and bottom element $\bot$, containing all possible valuations of an argument.

- $\curlywedge_{\mathcal{A}}, \curlywedge_{\mathcal{R}} : L \times L \to L$ are two binary algebraic operation on argument valuations used to determine the valuation of an argument based on its valuation given by the votes and how weak its attackers are ($\curlywedge_{\mathcal{A}}$), and to determine the strength of an attack given the votes on the attack and the valuation of the attacking argument ($\curlywedge_{\mathcal{R}}$).

- $\Upsilon : L \times L \to L$ is a binary algebraic operation on argument valuations used to determine the valuation of a combined attack;

- $\neg : L \to L$ is a unary algebraic operation on argument valuations used to determine how weak an attack is.

- $\tau : \mathbb{N} \times \mathbb{N} \to L$ is a vote aggregation function which produces a valuation of an argument based on its votes.

The definition of a semantic framework imposes very little on the behaviour of the operators. As such, many specific semantic frameworks could result in systems whose behaviour would be far from intuitive – a semantic framework where an increase in the strength of the attacking arguments would result in an increase in the strength of the attacked argument would make little sense. There are several basic properties that the operators should obey so that the resulting semantics is adequate for its purpose. For example, $\neg$ should be antimonotonic, continuous, $\neg\bot = \top$, $\neg\top = \bot$ and $\neg\neg a = a$; $\curlywedge_{\mathcal{A}}, \curlywedge_{\mathcal{R}}$ should be continuous, commutative, associative, monotonic w.r.t. both arguments and have $\top$ as their identity element; Υ should be continuous, commutative, associative, monotonic w.r.t. both arguments and have $\bot$ as its identity element; and τ should be monotonic w.r.t. the first argument and antimonotonic w.r.t. the second argument.

Continuity of operators guarantees small changes in the social inputs result in small changes in the models. Were this not the case, outcomes of debates would be very unstable, hard to follow and more easily exploited by trolls. The remaining algebraic properties simply state that the order in which arguments are attacked makes no difference; that an argument's valuation is proportional to its crowd support; that aggregated attacks are proportional to the attacking arguments; and so forth.

Notice also that the valuation set L of arguments (often denoted as $\mathbb{I}$ in other frameworks, as mentioned in the previous section) is parametrisable. L could be $[0,1] \subseteq \mathbb{R}$, but it could also be any finite, countable or uncountable set of values such as booleans, colours, textures, or any other set that is deemed appropriate for users of the final application, so long as it is totally ordered.

One particular semantic framework has received great attention because of its properties. It uses a simple vote aggregation function and is based on the well known product T-norm and probabilistic sum T-conorm, which combine the desirable properties discussed above, i.e. they are continuous, commutative, associative and monotonic[3]. It is dubbed the Simple Product Semantics and is defined as follows:

Definition 3.3 (Simple Product Semantics). *A simple product semantic framework is* $\mathcal{S}_\epsilon^\cdot = \langle [0,1], \curlywedge^\cdot, \curlywedge^\cdot, \curlyvee^\cdot, \neg, \tau_\epsilon \rangle$ *where*

- $x \curlywedge^\cdot y = x \cdot y$, *i.e. the product T-norm,*
- $x \curlyvee^\cdot y = x + y - x \cdot y$, *i.e. the T-conorm dual to the product T-norm,*
- $\neg x = 1 - x$,
- $\tau_\varepsilon\left(v^+, v^-\right) = \frac{v^+}{v^+ + v^- + \varepsilon}$, *with* $\epsilon > 0$.[4]

The heart of the semantics is in the definition of a model, which combines the operators of a semantic framework $\mathcal{S}$ into a system of equations, one for each argument, that must be satisfied.

Definition 3.4 (Social Model). *Let* $F = \langle \mathcal{A}, \mathcal{R}, V \rangle$ *be a social argumentation framework and* $\mathcal{S} = \langle L, \curlywedge_\mathcal{A}, \curlywedge_\mathcal{R}, \curlyvee, \neg, \tau \rangle$ *a semantic framework.*

---

[3] Besides other uses, product t-norm and its dual probabilistic sum t-conorm are the standard semantics for conjunction and disjunction, respectively, in Fuzzy Logic [Wang et al., 2024]

[4] The meaning of ϵ is explained in [Leite and Martins, 2011] and, in practice, it should be a sufficiently small value with no significant influence on the result of the voting aggregation function.

A total mapping $M : \mathcal{A} \to L$ is a social model of F under semantics $\mathcal{S}$, or $\mathcal{S}$-model of F, if

$$M(a) = \tau(a) \curlywedge_{\mathcal{A}} \neg \bigvee_{a_i \in \mathcal{R}^-(a)} (\tau((a_i, a)) \curlywedge_{\mathcal{R}} M(a_i)) \quad \forall a \in \mathcal{A}$$

where

$$\mathcal{R}^-(a) \triangleq \{a_i : (a_i, a) \in \mathcal{R}\}$$
$$\curlyvee\{x_1, x_2, ..., x_n\} \triangleq ((x_1 \curlyvee x_2) \curlyvee ... \curlyvee x_n)$$
$$\tau(x) \triangleq \tau(v^+, v^-) \quad (whenever\ V(x) = (v^+, v^-))$$

We refer to $M(a)$ as the social strength, or value, of a in M, dropping the reference to M whenever unambiguous.

Each equation encodes the contribution of votes and attacks to the social strength of an argument, for which we now proceed to provide further intuition.

Whenever an argument a_i attacks another argument a, then the strength of the attack is the valuation of the attacking argument a_i reduced by the social support of the attack: no argument's attack is stronger than either its own valuation or the social support of the attack itself. We use $\curlywedge_{\mathcal{R}}$ to restrict these values.

$$\tau((a_i, a)) \curlywedge_{\mathcal{R}} M(a_i)$$

As an argument may have multiple attackers, all of their attack strengths must be aggregated to form a stronger combined attack value, using operator $\curlyvee$.

$$\bigvee_{a_i \in \mathcal{R}^-(a)} (\tau((a_i, a)) \curlywedge_{\mathcal{R}} M(a_i))$$

The above equation results in a combined attack strength that must be turned into a restricting value, representing how permissive or weak the attack is, using the $\neg$ operator.

$$\neg \bigvee_{a_i \in \mathcal{R}^-(a)} (\tau((a_i, a)) \curlywedge_{\mathcal{R}} M(a_i))$$

1 - Social Argumentation Systems

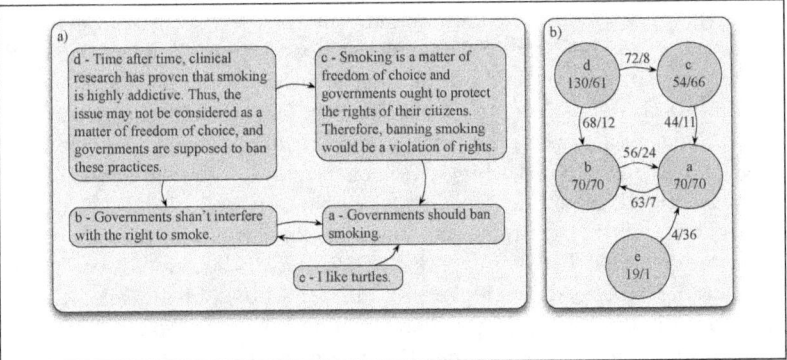

Figure 1: Social Argumentation Framework: a) arguments and attacks; b) votes

In a social context where the crowd has given its direct opinion on argument a through the votes, it seems clear that a's valuation should never turn out higher than a's social support $\tau(a)$. Thus, an argument's valuation is given by restricting $\tau(a)$ with the value of the aggregated attack using the final operator $\curlywedge_{\mathcal{A}}$.

$$\tau(a) \curlywedge_{\mathcal{A}} \neg \bigvee_{a_i \in \mathcal{R}^-(a)} \left(\tau\left((a_i, a)\right) \curlywedge_{\mathcal{R}} M\left(a_i\right) \right)$$

Throughout the remainder of the section, $\mathcal{S}$ stands for the Simple Product Semantics.

3.2 Illustrative Example

Consider a social interaction inspired by [Walton, 2009] where several participants, while arguing about the role of the government in what banning smoking is concerned, set forth the arguments and attack relations depicted in Fig. 1 a).

Note that these arguments are structurally different: a and b are unsupported claims, c and d contain multiple premises and a conclusion, while e, despite being rather consensual (who doesn't like turtles?), seems to be totally out of context and can hardly be seen as an attack on a (here, the attack by e on a is meant to represent a troll attack). Our goal is to show that SAFs' level of abstraction allows meaningful arguments to be construed out of most forms of participation – in fact, with suitable GUIs, arguments could even be built from videos, pictures, links, etc. – while the participation through voting will help deal with mitigating the disturbing effect of unsound arguments and poorly specified (troll) attacks.

After a while, the arguments and attacks garner the pro/con votes depicted in Fig. 1 b). Arguments a and b obtain the same direct social support as expressed by the 70 *pro* and *con* votes. Meanwhile, a's attack on b is deemed stronger than its counterpart, judging from their votes. One might speculate that this is a consequence of a delivering a more direct message. Whereas argument c does not get much love from the crowd (a vote ratio of 54/66), its attack on a is still supported by the community (44/11). Perhaps initially there was a better sentiment towards c but the introduction of d, which amassed a decent amount of support itself (130/61), turned the odds against c. Both of d's attacks on b and c materialise to be strong enough, the former being slightly weaker (72/8 versus 68/12). Lastly, argument e received just a mere number of votes, most being positive (19/1). However, there seems to have been a significant effort from the users to discredit the attack on a by e (4/36). Note that e is a perfectly legitimate argument. Indeed the crowd endorses the fondness for turtles – it's the attack, not the argument, that is not logically well-founded.

With the abstract argumentation framework and the votes on arguments and attacks in hand, we can turn our attention to the valuation of the arguments.

If we consider the social support of each argument, i.e. its value considering only the votes it obtained while ignoring attack relations, we obtain the following values:[5] $\tau(a) = 0.50$, $\tau(b) = 0.50$, $\tau(c) = 0.45$, $\tau(d) = 0.68$ and $\tau(e) = 0.95$, as depicted in Fig. 2 a) (where the size of each node is proportional to its value).

The original SAF semantics [Leite and Martins, 2011], which considers attacks between arguments but not the votes on attacks, assigns the following values to arguments: $M(a) = 0,02$, $M(b) = 0,16$,

---

[5]We will consider the Product Semantics as in Def.3.3, with a neglectable low ϵ

1 - Social Argumentation Systems

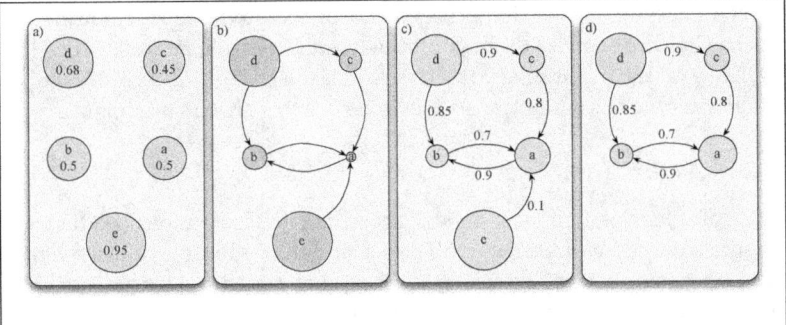

Figure 2: Model of the Social Abstract Argumentation Framework: a) considering social support only; b) considering attacks but not their strength; c) considering attack strength; d) considering attack strength, without argument e.

$M(c) = 0,14$, $M(d) = 0,68$ and $M(e) = 0,95$, as depicted in Fig. 2 b). As expected, d and e retain their initial social support values, since they are not attacked, while the remaining arguments see a decrease in their social support value. Argument a decreases the most while b and c maintain a reasonable fraction of their initial strength. Since two of a's attackers – b and c – are attacked by d, which is a non-attacked argument with strong social support, their value is weakened, so their effect on a is lessened. Thus, we can conclude that the main cause for the downfall in a's value is e's attack.

We can now turn our attention to the model that also takes votes on attacks into consideration, which assigns the following values to arguments: $M(a) = 0,35$, $M(b) = 0,14$, $M(c) = 0,17$, $M(d) = 0,68$ and $M(e) = 0,95$, as depicted in Fig. 2 c). The value assigned to a by the model increases from 0.02 to the more plausible level of 0.34, mostly due to e's weakened capability to attack a. Indeed, the crowd's overwhelming con votes on the (troll) attack of e on a essentially neutralised it. To confirm, we compare it with the model obtained if argument e was simply removed, depicted in Fig. 2 d), whose valuations of $M(a) = 0,39$, $M(b) = 0,14$, $M(c) = 0,17$ and $M(d) = 0,68$ are very similar to those obtained in the presence of e but with a very weakened attack on a, which allows us to conclude the success of the model in discounting attacks that are socially deemed unsound, such as troll attacks. Since the weights of the remaining attacks are relatively high and also close to each other at the same time, their impact is somewhat minimal.

3.3 Algorithms

The problem of finding a model according to the simple product semantics can be cast to the problem of finding a solution to a nonlinear system where variables represent the arguments and equations encode their attacks, with the following generic form:

Definition 3.5. *A Social Abstract Argumentation System is a square nonlinear system with n variables $\{x_1, \ldots, x_n\}$ and n equations:*

$$x_i = \tau_i \prod_{j \in A_i} (1 - \tau_{ji} x_j) \qquad 1 \leq i \leq n \qquad (1)$$

where $\tau_i, \tau_{ji} \in\,]0,1[$ and $A_i \subseteq \{1, \ldots, n\}$.[6]

Contrary to the linear case, systems of nonlinear equations cannot be solved exactly using a finite number of elementary operations. Instead, iterative algorithms are usually used to generate a sequence $(\mathbf{x}^{(k)})_{k \in N_0}$ of approximate solutions. These algorithms start with an initial guess $\mathbf{x}^{(0)}$ and, to generate the approximating sequence, follow an iteration scheme of the form $\mathbf{x}^{(k+1)} = \mathbf{g}(\mathbf{x}^{(k)})$ where the fixed-points for $\mathbf{g}$ are solutions $\mathbf{x}^*$ of the nonlinear system.

The success of iterative algorithms depends on their convergence properties. Given a domain of interest, an iterative method that converges for any arbitrary initial guess is called globally convergent. If convergence is only guaranteed when the initial approximation is already close enough to the solution, the algorithm is called locally convergent. In the case of Social Abstract Argumentation Systems the domain of interest is $]0,1[^n$ thus the iterative algorithm must converge to a solution $\mathbf{x}^* = (x_1^*, \ldots, x_n^*) \in\,]0,1[^n$.

---

[6]We can exclude $\tau_i, \tau_{ji} \in \{0,1\}$ because arguments and attacks (x) with $\tau(x) = 0$ have no effect in the system while $\tau(x) = 1$ can never occur, according to the simple product semantics in Def. 3.3, because $\epsilon > 0$.

Two classical algorithms that can be used to approximate the solution of such a system are the Iterative Fixed-Point Algorithm (IFP) where the iteration scheme is directly obtained from the equations (1), and the Iterative Newton-Raphson Algorithm (INR).[7]

Unfortunately, IFP is only locally convergent and often divergent, even for systems with a reduced number of variables, while INR, also only locally convergent, requires the computation of a Jacobian matrix at each iteration, which is prohibitive for large systems.

Based on the Iterative Successive Substitutions Algorithm (ISS) previously proposed for Social Argumentation Frameworks without votes on attacks [Correia et al., 2014] – itself an adaptation of the Gauss-Seidel method for systems of nonlinear equations –, here we present an adaptation to also admit votes on attacks.

Definition 3.6 (ISS). *The ISS algorithm uses the iteration rule:*

$$x_i^{(k+1)} = \tau_i \prod_{j<i, j \in A_i} \left(1 - \tau_{ji} x_j^{(k+1)}\right) \prod_{j \geq i, j \in A_i} \left(1 - \tau_{ji} x_j^{(k)}\right) \qquad (2)$$

From the initial guess $\mathbf{x}^{(0)}$, elements of $\mathbf{x}^{(k+1)}$ are computed sequentially using forward substitution until the stopping criterion is attained.

Following a similar strategy as [Correia et al., 2014], we can prove the global convergence of the algorithm.

The algorithm performs as well as its original version [Correia et al., 2014]. For example, it is able to approximate the model of debates with 5000 arguments and an attack density of 0.1 (i.e. 10% of all pairs of arguments are related through an attack) in well under 1 second.[8] A thorough analysis of the original ISS algorithm can be found in [Correia et al., 2014]. Additionally, just as with the original ISS, we can exploit the structure of the debate to obtain considerable gains in efficiency.

---

[7] A comprehensive treatment of methods for solving nonlinear systems of equations with some recent developments on iterative methods can be found in [Ortega and Rheinboldt, 2000; Argyros and Szidarovszky, 1993].

[8] The higher the attack density, the slower the convergence of the algorithm. However, an attack density of 0.1 seems to be a rather high value.

4 A multi-aspect comment evaluation framework

Another generic framework that formalises the most commonly used features found in online debate platforms is s-mDiCE. This approach also introduces a set of novel features, which serve diverse purposes of debate platforms.

As already discussed, almost all online debate platforms implement some form of voting mechanism, such as positive/negative votes, like/dislike counters, star-based rating etc. s-mDiCE formalises votes, which is a generic enough mechanism that enables other types of rating to be transformed to votes rather easily. In addition, it also incorporates the notion of a *base score* (or intrinsic strength) BS, which is often used in decision-making systems (e.g., in [Rago et al., 2016] or in [Baroni et al., 2019], where it is denoted as τ, as explained in Section 2). The base score offers an one-time, prior evaluation of an argument; as the dialogue evolves other users may influence this initial evaluation, positively or negatively, through their arguments or votes. It can be used to represent various notions, such as to capture an expert's initial rating over an opinion, before any debate has taken place. In some platforms, the base score may also obtain a more personalised flavor, representing for instance the trust that a user attributes to the person who issues an argument, regardless of its content.

In addition to voting mechanisms, many platforms, especially those intending to implement structured debates under the ranking-based semantics, appoint a characterisation of users' opinions as being in favour of or against other opinions or topics. According to such semantics, the strength of an opinion, or more accurately an argument, as is often referred to in these platforms, is determined by the type, number and strength of the arguments that respond to it, which are taken into account by s-mDiCE.

One novel feature of s-mDiCE is the fact that it treats positive and negative votes as arguments. The idea is that a positive vote signifies a person that is happy to submit exactly the same opinion as the one stated in the original argument; this means that all arguments that support or attack the original argument also place a support or attack

1 - Social Argumentation Systems

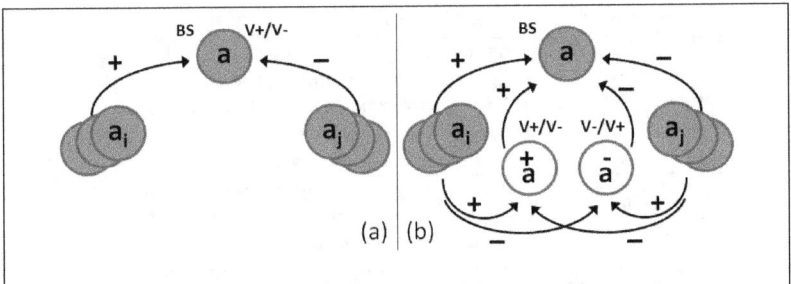

Figure 3: (a) A debate graph with votes, base score and user-generated supporting and attacking arguments, (b) The same debate graph as transformed with blank nodes.

to that vote (see Fig. 3). Since the arguments that represent the votes do not carry any content of their own, these are called *blank arguments*, and their strength is associated with the degree with which people identified themselves with the original opinion. The symmetricity exhibited among positive and negative blank arguments may seem redundant, but it helps promote the intuitiveness of the model, as explained later on.

Another novelty is the utilisation of a set of *aspects* (dimensions), upon which an argumentative opinion is evaluated. For example, and depending on the domain of interest, such aspects may concern how relevant an argument is to the topic of discussion, how complete its justification is, how objective it is, or others. This way, users can choose to specify the aspects of an opinion they agree on and the ones they disagree, affecting the final strength accordingly. Such a scheme also assists in better understanding the intentions of users, which is difficult to capture in many of the existing platforms (for instance, often it is very difficult to interpret a down vote as representing an objection to the position stated or to the explanation given).

Finally, the framework enables a debate platform to correlate all the aforementioned features, in order to calculate two distinct scores to characterize an argument, namely *acceptance* and *quality* scores. The former aims to represent how strongly the position expressed by the argument is supported by the community, whereas the latter represents how well the position is justified; in the vast majority of existing frameworks, those

Functions	Description
$g^{vot} : \mathbb{N}^0 \times \mathbb{N}^0 \to \mathbb{I}$	Aggregates positive and negative votes
$g^{set} : (\mathbb{N}^0)^{\mathbb{I}} \to \mathbb{I}$	Aggregates arguments of the same polarity
$g^{diff} : \mathbb{I} \times \mathbb{I} \to \mathbb{I}'$	Aggregates arguments of different polarity
$g^{dlc} : \mathbb{I} \times \mathbb{I}' \to \mathbb{I}$	Aggregates the base score, and the strength of support and attack arguments
$g^{\mathcal{ACC}} : \mathbb{I}^N \to \mathbb{I}$	Aggregates the acceptance score of all aspects
$g^{\mathcal{QUA}} : \mathbb{I}^N \to \mathbb{I}$	Aggregates the quality score of all aspects
$\mathcal{ACC} : \mathcal{A} \to \mathbb{I}$	Returns the acceptance score of an argument
$\mathcal{QUA} : \mathcal{A} \to \mathbb{I}$	Returns the quality score of an argument

Table 1: Overview of the s-mDiCE Generic Functions

two metrics are unified leading to misconceptions about what a user's reaction to an opinion signifies, e.g., when one agrees with a comment that is irrelevant to a given discussion.

In the remainder of this section, a formal definition of the main functions that transform features into comparable strength scores is given, along with a study of some of their properties. An algorithm that can apply these generic functions is then presented, with the goal of clarifying how s-mDiCE works.

4.1 Formalization

s-mDiCE is a generic formal framework that enables the evaluation of the strengths of arguments considering one or more aspects. To do so, it relies on a number of functions that help quantify and aggregate the strength of the various features. These functions are shown in Table 1 and are formally defined next.

Definition 4.1. *An s-mDiCE (symmetric multi-Dimensional Comment Evaluation) framework is an (N+1)-tuple $\langle \mathcal{A}, \mathcal{D}_{d1}^*, \ldots, \mathcal{D}_{dN}^* \rangle$, where $\mathcal{A}$ is a finite set of arguments and $\mathcal{D}_{d1}^*, \ldots, \mathcal{D}_{dN}^*$ are aspects (dimensions), under which an argument is evaluated.*

Definition 4.2. *An aspect $\mathcal{D}_x^*$ corresponding to an argument set $\mathcal{A}$ is a 5-tuple $\langle \mathcal{R}_x^{supp}, \mathcal{R}_x^{att}, BS_x, V_x^+, V_x^- \rangle$, where $\mathcal{R}_x^{supp} \subseteq \mathcal{A} \times \mathcal{A}$ is a binary acyclic support relation on $\mathcal{A}$, $\mathcal{R}_x^{att} \subseteq \mathcal{A} \times \mathcal{A}$ is a binary acyclic attack relation on $\mathcal{A}$, and $BS_x : \mathcal{A} \to \mathbb{I}$, $V_x^+ : \mathcal{A} \to \mathbb{N}^0$ and $V_x^- : \mathcal{A} \to \mathbb{N}^0$*

are total functions mapping each argument to a base score ($\mathbb{I}$ can be any totally ordered set), a number of positive and a number of negative votes relative to this aspect, respectively.

The attack and support relations are acyclic, since a new argument can only support/attack previously added comments. The $\mathbb{I}$ set is parameterisable, similar to the L set discussed in the previous section; in the sequel, it is assumed that $\mathbb{I} = [0,1]$, which is most frequently used range in similar frameworks. The base score is a fixed value assigned to each argument before any computation takes place. If no value is given, the default value can be set to a value that neutralises its effect.

Votes as blank arguments. The set of votes on any argument in an s-mDiCE framework is transformed into a pair of supporting and attacking blank arguments (Fig. 3). Before formally defining blank arguments, some convenient notation is needed. Let $\widetilde{\mathcal{A}}$ denote the set of user-generated arguments and $\mathring{\mathcal{A}}$ refer to the set of blank arguments of an s-mDiCE framework $\mathcal{F}$, such that $\mathcal{A} = \mathring{\mathcal{A}} \cup \widetilde{\mathcal{A}}$. Given an aspect $\mathcal{D}_x^* = \langle \mathcal{R}_x^{supp}, \mathcal{R}_x^{att}, BS_x, V_x^+, V_x^- \rangle$, the set of direct supporters of an argument $a \in \mathcal{A}$ is defined as $\mathcal{R}_x^+(a) = \{a_i : (a_i, a) \in \mathcal{R}_x^{supp}\}$. Similarly, the set of direct attackers of a is defined as $\mathcal{R}_x^-(a) = \{a_i : (a_i, a) \in \mathcal{R}_x^{att}\}$.

Definition 4.3. *Let $\mathcal{F}$ be an s-mDiCE framework and $\mathcal{D}_x^* = \langle \mathcal{R}_x^{supp}, \mathcal{R}_x^{att}, BS_x, V_x^+, V_x^- \rangle$ be an aspect of $\mathcal{F}$. For each argument $a \in \widetilde{\mathcal{A}}$, we define two new arguments $\overset{+}{a}$ and $\overline{a}$, called the supporting blank and attacking blank argument of a respectively, such that*

- $(\overset{+}{a}, a) \in \mathcal{R}_x^{supp}$,

- $V_x^+(\overset{+}{a}) = V_x^+(a)$, $V_x^-(\overset{+}{a}) = V_x^-(a)$,

- *for all $(a_i, a) \in \mathcal{R}_x^{supp}$ it also holds that $(a_i, \overset{+}{a}) \in \mathcal{R}_x^{supp}$,*

- *for all $(a_j, a) \in \mathcal{R}_x^{att}$ it also holds that $(a_j, \overset{+}{a}) \in \mathcal{R}_x^{att}$*

Similarly,

- $(\overline{a}, a) \in \mathcal{R}_x^{att}$,

- $V_x^+(\bar{a}) = V_x^-(a)$, $V_x^-(\bar{a}) = V_x^+(a)$,
- for all $(a_i, a) \in \mathcal{R}_x^{supp}$ it also holds that $(a_i, \bar{a}) \in \mathcal{R}_x^{att}$,
- for all $(a_j, a) \in \mathcal{R}_x^{att}$ it also holds that $(a_j, \bar{a}) \in \mathcal{R}_x^{supp}$.

Aggregation of Votes. The positive and negative votes that each argument receives over time need to be aggregated into a single value.

Function Definition 4.1. *The generic score function* $g^{vot} : \mathbb{N}^0 \times \mathbb{N}^0 \to \mathbb{I}$ *aggregates positive and negative votes into a single strength score.*

There are many different ways to instantiate g^{vot}, in order to represent an estimate of the community's stance towards that argument. Some are rather simplistic, e.g., averaging their population, while others provide more insights. Many frameworks often rely on the mean of the Wilson Score Interval [Wilson, 1927], which is a popular choice for systems that need more accurate estimations, as it assesses the probability that the next vote will be of a certain polarity:

$$g^{vot}(v^+, v^-) = \frac{2 \cdot v^+ + z^2}{2 \cdot (v^+ + v^- + z^2)} \tag{3}$$

where $z = 1.96$ for a confidence level of 95%. When no votes are placed, the initial score is 0.5. One can implement other instantiations instead, considering the particular requirements of the domain, in order to control for instance the rate of convergence as an argument is populated by more votes. Clearly, the above definition of g^{vot} has the desirable property of being increasing with respect to the number of positive votes, and decreasing with respect to the number of negative votes.

To determine the smoothness of the g^{vot} function, d_S is defined as the number of votes that were added or deleted, i.e., $d_S(\langle v_1^+, v_1^- \rangle, \langle v_2^+, v_2^- \rangle) = |v_1^+ - v_2^+| + |v_1^- - v_2^-|$ (essentially, the Manhattan distance for 2-dimensional vectors), and d_T as the difference in the output, i.e., $d_T(x, y) = |x - y|$ (also the Manhattan distance, for 1-dimensional vectors). Under these definitions, it can be shown that g^{vot} is exactly $\frac{1}{2 \cdot (1+z^2)}$-smooth, and that this extreme is reached only for the first positive vote placed; all subsequent votes have strictly smaller effects. Note how the parameter z can be used to enforce different smoothness properties on g^{vot}.

1 - SOCIAL ARGUMENTATION SYSTEMS

Aggregation of the strength of arguments with the same polarity. Supporting and attacking arguments form a set that collectively affects the strength of the target argument. This combined support or attack can take into account, for instance, the strongest argument in the set or it can average the strength of all members in the set. Such schemes can be captured by the $g^{set} : (\mathbb{N}^0)^{\mathbb{I}} \to \mathbb{I}$ function[9] in s-mDiCE.

Function Definition 4.2. *The generic score function $g^{set} : (\mathbb{N}^0)^{\mathbb{I}} \to \mathbb{I}$ aggregates the strength of a set of (supporting or attacking) arguments into a single strength score.*

For most purposes, the T-CoNorm function $\perp_{sum}: \mathbb{I} \times \mathbb{I} \to \mathbb{I}$, also known as the probabilistic sum, is a convenient choice, as it satisfies a number of useful properties, especially when $\mathbb{I} = [0, 1]$ [Klement et al., 2000]:

$$\perp_{sum}(x_1, x_2) = x_1 + x_2 - x_1 \cdot x_2 \qquad (4)$$

For a multiset S of natural numbers, it is defined:

$$\perp_{sum}^*(S) = \begin{cases} 0, \text{ if } S = \emptyset \\ \perp_{sum}(x_1, \perp_{sum}^*(\{x_2, ..., x_n\})), \\ \quad \text{if } S = \{x_1, x_2..., x_n\} \text{ with } n > 0 \end{cases} \qquad (5)$$

Consequently, the g^{set} function for the multiset of argument strengths can be instantiated as follows:

$$g^{set}(S) = \perp_{sum}^*(\{x_i : x_i \in S \text{ and } x_i \geq \vartheta\}) \qquad (6)$$

Here, it is assumed that the inputs x_i represent the strength (score) of each supporting (or attacking) argument. Constant ϑ can be used to discard arguments that fall below a given threshold, rendering them ineffective in changing the strength score of other arguments. This way, irrelevant or troll arguments can easily be neutralized.

The following monotonicity properties can be easily shown for g^{set}:

- If $A, B \in (\mathbb{N}^0)^{\mathbb{I}}$, $A \subseteq B$, then $g^{set}(A) \leq g^{set}(B)$, where $\subseteq$ should be interpreted as the subset relationship for multisets.

---

[9] g^{set} is meant to take as input a multiset over elements of $\mathbb{I}$.

- If $A, B, C \in (\mathbb{N}^0)^\mathbb{I}$, and $g^{set}(A) \leq g^{set}(B)$, then $g^{set}(C \uplus A) \leq g^{set}(C \uplus B)$, where $\uplus$ stands for "union" for multisets.

The first condition guarantees that the addition of arguments cannot decrease the combined strength of a set of arguments. The second condition ensures that the addition of "stronger" arguments has a more powerful effect than the addition of "weaker" ones. It also becomes clear from the above that when an argument's acceptance score increases, this has a negative effect on the acceptance score of all the arguments it attacks, and a positive effect on the acceptance score of all the arguments it supports. This effect propagates along the tree of arguments using this pattern.

To determine the smoothness of g^{set}, we need to define a semi-metric for $(\mathbb{N}^0)^\mathbb{I}$; our notion of distance will be based on the strength (based on g^{set}) of the symmetric difference between the sets compared, in particular: $d_{(\mathbb{N}^0)^\mathbb{I}}(X, Y) = g^{set}(X \setminus Y \uplus Y \setminus X)$. This can be viewed as a special type of edit distance, where the importance of the "edits" $(X \setminus Y, Y \setminus X)$ is judged by the g^{set} function itself. For the range of g^{set}, we will use, as usual, the semi-metric $d_\mathbb{T}(x, y) = |x - y|$. Under these assumptions, it can be shown that g^{set} is exactly 1-smooth.

Aggregation of the strength of arguments with opposite polarity. In addition to the cumulative effect of arguments that jointly support or attack another argument, s-mDiCE needs to calculate how to balance the antagonistic effect of the supporting and attacking sets.

Function Definition 4.3. *The generic score function $g^{diff} : \mathbb{I} \times \mathbb{I} \to \mathbb{I}'$ aggregates the strength of supporting and attacking arguments into a single strength score.*

There are different instantiations that combine these sets to specify the overall attitude. The polynomial $g^{diff} : \mathbb{I} \times \mathbb{I} \to [-1, 1]$ is often used, which offers a convenient solution for many domains:

$$g^{diff}(x_s, x_a) = x_s^n - x_a^n - x_a \cdot x_s^n + x_s \cdot x_a^n \qquad (7)$$

Apparently, this function is increasing with respect to x_s and decreasing with respect to x_a, for any $n \geq 1$ (under the assumption that

$\mathbb{I} = [0, 1]$). For $n = 1$ in particular, this equation results in the difference between the two values (as suggested, e.g., in [Rago et al., 2016]). In decision-making systems, where reaching reliable conclusions becomes critical, the effect of arguments should begin to matter only when they obtain a substantial strength. This behavior can be obtained for larger values of n, which force the system to react very slowly initially, but when some clear tendency for/against an opinion has appeared, the system quickly achieves a steeper increase in confidence.

In particular, it is n that determines the smoothness of g^{diff}, as g^{diff} is exactly n-smooth. The maximum effect of a change in the inputs of g^{diff} is reached when the current sentiment is very positive (close to 1) or very negative (close to 0) and someone casts an opposing argument. This means that it is easier to cast doubts on the strength of a strong argument, than to quickly trust a doubtful one. This aims at promoting the liveness of the discussion without damaging the credibility of conclusions. One can easily adapt this behavior by changing the degree of the polynomial in Eq. (7) or by applying a different function altogether. Note that lower degrees (n) in g^{diff} lead to more smooth functions; this is due to the fact that higher-degree polynomials tend to change very fast in the limit cases (i.e., when the current sentiment is close to 0 or 1), and more slowly in the intermediate points, whereas lower-degree polynomials are more uniform in their behaviour.

Of course, other schemes may also be applied according to the domain requirements. For example, one may decide to assess the informative quality of an argument by applying a scheme which increases with the strength of votes and decreases with the strength of both positive and negative arguments. This is based on the rationale that the "ideal" comment would attract only positive votes and no supporting arguments. In other words, in an ideal setting, supporting arguments are only asserted to add material or to explain better the opinion stated, thus giving a sense of discomfort related to the quality of the target argument.[10]

---

[10] Of course, in practice we often see arguments, such as "I totally agree", "True!", which offer support but no valuable content. The use of aspects in s-mDiCE can help identify such arguments, reducing their quality, without affecting the supporting or attacking effect they have on the target argument.

Dialectical Strength. The formalization so far incrementally builds the strength of an argument taking into consideration the strength of the support and opposition it attracts, including the votes, which are represented as blank arguments. These parameters are finally aggregated with the base score in the $g^{dlc} : \mathbb{I} \times [-1, 1] \to \mathbb{I}$ function, in order to provide the overall dialectical strength of an argument.

Function Definition 4.4. *The generic score function $g^{dlc} : \mathbb{I} \times [-1, 1] \to \mathbb{I}$ valuates the dialectical strength of an argument, considering the aggregation of the base score, and the strength of the supporting and/or attacking arguments.*

As with the previous functions, different instantiations of the dialectical strength can be devised. A popular choice is to trust the base score more when the other scores do not converge to a positive or a negative value, e.g.,:

$$g^{dlc}(x_b, d) = x_b \cdot (1 - |d|) + \frac{d + |d|}{2}$$
$$\text{where } d = g^{diff}(x_s, x_a) \tag{8}$$

In the equations above, it is assumed that x_b is the base score and d is the strength of the combined support and/or opposition that it has attracted. Yet another example could be to give more credit to the base score initially, and, as the argument attracts more votes and/or arguments, let the strength of the latter start weighing more in the final score. This way, rather than the supporting and attacking sets balancing each other out when they have equal strength, such a scheme will manage to capture the increasing confidence obtained as the discussion progresses.

The function defined in equation 8 is apparently increasing with respect to each of its inputs, i.e., both the base score x_b and the strength of the combined support and/or opposition that it has attracted (d). For smoothness, using the Manhattan distance for both the input and the output of the function, we can show that g^{dlc} is exactly 1-smooth. This value applies in two cases. The first is when the aggregated strength of the argument's responses is equal (or close) to 0, in which case modifications to the base score have a linear effect on the result of the function.

The second is when the base score is in one of the two extremes (very low, i.e., close to 0 or very high, i.e., close to 1), and only when the aggregated score d indicates an opposite stance by the community. For example, for an argument with a base score of 1, an overall negative stance by the community will more quickly lower its score, compared to the case where we had a lower base score, or a positive stance by the community. This applies also symmetrically to the opposite case.

Acceptance and Quality Scores. The aforementioned aggregation functions compute various metrics regarding the strength of an argument, considering a single aspect. In addition, s-mDiCE allows for more refined valuations of a single argument. For instance, an argument may have a different set of votes regarding its relevance, another set for its objectivity and a third one for its informativeness. As a result, the final score of the argument needs to be calculated by taking into consideration the scores obtained on each individual aspect.

Function Definition 4.5. *Let $\mathcal{F} = \langle \mathcal{A}, \mathcal{D}^*_{d1}, ..., \mathcal{D}^*_{dN} \rangle$ be an s-mDiCE framework. The generic score functions $g^{ACC} : \mathbb{I}^N \to \mathbb{I}$ and $g^{QUA} : \mathbb{I}^N \to \mathbb{I}$ can be used to aggregate the dialectical strength of each individual aspect. Eventually, the functions $\mathcal{ACC} : \mathcal{A} \to \mathbb{I}, \mathcal{QUA} : \mathcal{A} \to \mathbb{I}$ are used to denote the acceptance and quality scores of an argument $a \in \mathcal{A}$, respectively.*

A simple, weight-based solution that determines the influence of each aspect is often a sufficient solution. In the following equation, w_i quantifies the weight assigned by the system moderator on aspect $\mathcal{D}^*_{di}$ based on other metrics or by experience (where x_i will be the dialectical strength for the given aspect):

$$g^{QUA}(x_1, ..., x_n) = \sum_{i=1}^{n} w_i^{QUA} \cdot x_i \text{ , with } \sum_{i=1}^{n} w_i^{QUA} = 1, w_i^{QUA} \geq 0 \quad (9)$$

$$g^{ACC}(x_1, ..., x_n) = \sum_{i=1}^{n} w_i^{ACC} \cdot x_i \text{ , with } \sum_{i=1}^{n} w_i^{ACC} = 1, w_i^{ACC} \geq 0 \quad (10)$$

Given that $w_i^{QUA} \geq 0$ and $w_i^{ACC} \geq 0$ for all i, g^{QUA} and g^{ACC} are both increasing with respect to each of its inputs. Similarly, the smoothness of $g^{\mathcal{ACC}}$ and $g^{\mathcal{QUA}}$ is determined by (i.e., is equal to) the weight with the maximum value. The smoothness with respect to each aspect in particular (i.e., if we assume that the scores of all other aspects remain constant) is determined by its respective weight. An interesting conclusion from this is that "balanced" functions (where w_i are close to each other, or equal) are smoother, i.e., less sensitive to input changes. Moreover, as expected, the largest effects appear when the changed aspects are those that have the largest weights.

In the following section, we demonstrate how $\mathcal{ACC}, \mathcal{QUA}$ can be computed for a given argument, by successively applying all aggregate functions mentioned above. Given the monotonicity properties of their constituent functions, we observe that the effects of each aspect on the argument's acceptance and/or quality score depend on its respective weight: all aspects have a monotonically increasing effect on the outcome of the functions $\mathcal{ACC}$ and $\mathcal{QUA}$, whereas their smoothness is also determined by these weights (just like in the case of $g^{\mathcal{ACC}}$, $g^{\mathcal{QUA}}$ above).

4.2 Computation Loop in s-mDiCE

Given the aforementioned aggregation functions, Algorithm 1 below presents the steps that can be followed to calculate the acceptance and quality scores of an argument in a debate graph, as the one shown in Fig. 4 (a). Despite the procedural presentation of the algorithm, not all steps need to be executed in a sequential manner.

Algorithm 1 takes as input an s-mDiCE framework consisting of one or more aspects, and an argument, whose acceptance and quality scores are to be calculated. The majority of computations are executed for each aspect individually. The first step is to generate the blank nodes of a given aspect (Fig. 4 (b) and line 3 in Algorithm 1), followed by the computation of the vote strength of each (Fig. 4 (c) and lines 6, 7 in Algorithm 1).

Before calculating their overall strength, one needs to determine the support and attack they receive (Fig. 4 (d) and lines 10, 11 in Algorithm 1). Notice that the supporting (resp. attacking) set of a blank argument includes only the user-generated arguments that exist in the supporting (resp. attacking) set of the input argument. Their strength is then aggregated (Fig. 4 (e) and lines 14, 15 in Algorithm 1), producing all

1 - Social Argumentation Systems

Algorithm 1 $CalcScores(\mathcal{F}, a)$: Calculate Acceptance / Quality Score

INPUT: an s-mDiCE framework $\mathcal{F} = \langle \mathcal{A}, \mathcal{D}^*_{d1}, \ldots, \mathcal{D}^*_{dN} \rangle$ and an argument $a \in \mathcal{A}$
OUTPUT: All strength scores of a

1: **for** each aspect $\mathcal{D}^*_x = \langle \mathcal{R}^{supp}_x, \mathcal{R}^{att}_x, BS_x, V^+_x, V^-_x \rangle$ of $\mathcal{F}$ **do**
2:   % Generate blank arguments
3:   Create $\overset{+}{a}$ and $\overline{a}$, according to Definition 4.3
4:
5:   % Calculate vote strength of blank arguments
6:   **let** $g^{vot}_x(\overset{+}{a}) = g^{vot}(V^+_x(\overset{+}{a}), V^-_x(\overset{+}{a}))$
7:   **let** $g^{vot}_x(\overline{a}) = g^{vot}(V^+_x(\overline{a}), V^-_x(\overline{a}))$
8:
9:   % Calculate strength without blank arguments
10:   **let** $s^{blSuppSet}_x(a) = g^{set}(\{g^{dlc}_x(a_i) : a_i \in \mathcal{R}^+_x(a) \cap \widetilde{\mathcal{A}}\})$
11:   **let** $s^{blAttSet}_x(a) = g^{set}(\{g^{dlc}_x(a_i) : a_i \in \mathcal{R}^-_x(a) \cap \widetilde{\mathcal{A}}\})$
12:
13:   % Calculate combined strength for blank arguments
14:   **let** $g^{diff}_x(\overset{+}{a}) = g^{diff}(s^{blSuppSet}_x(a), s^{blAttSet}_x(a))$
15:   **let** $g^{diff}_x(\overline{a}) = g^{diff}(s^{blAttSet}_x(a), s^{blSuppSet}_x(a))$
16:
17:   % Calculate blank arguments strength
18:   **let** $g^{dlc}_x(\overset{+}{a}) = g^{dlc}(BS_x(\overset{+}{a}), g^{diff}_x(\overset{+}{a}))$
19:   **let** $g^{dlc}_x(\overline{a}) = g^{dlc}(BS_x(\overline{a}), g^{diff}_x(\overline{a}))$
20:
21:   % Calculate strength with blank arguments
22:   **let** $s^{suppSet}_x(a) = g^{set}(\{g^{dlc}_x(a_i) : a_i \in \mathcal{R}^+_x(a)\})$
23:   **let** $s^{attSet}_x(a) = g^{set}(\{g^{dlc}_x(a_i) : a_i \in \mathcal{R}^-_x(a)\})$
24:
25:   % Calculate combined support/attack strength
26:   **let** $g^{diff}_x(a) = g^{diff}(s^{suppSet}_x(a), s^{attSet}_x(a))$
27:
28:   % Calculate acceptance/quality for one aspect
29:   **let** $g^{dlc}_{acc,x}(a) = g^{dlc}(BS_x(a), g^{diff}_x(a))$
30:   **let** $g^{dlc}_{qua,x}(a) = g^{dlc}(BS_x(a), g^{diff}_x(a))$
31: **end for**
32: % Calculate overall acceptance/quality
33: **let** $\mathcal{ACC}(a) = g^{ACC}(g^{dlc}_{d1}(a), ..., g^{dlc}_{dN}(a))$
34: **let** $\mathcal{QUA}(a) = g^{QUA}(g^{dlc}_{d1}(a), ..., g^{dlc}_{dN}(a))$

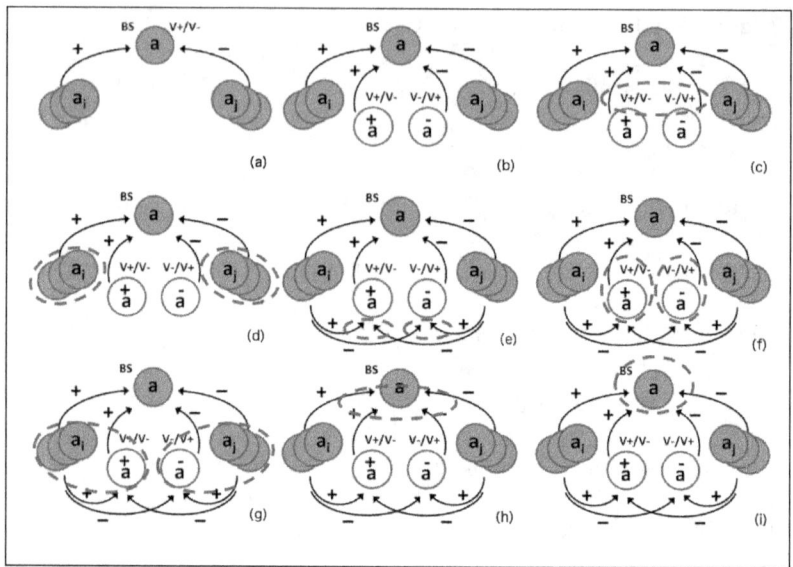

Figure 4: Computation steps of Algorithm 1.

values needed to calculate the dialectical strength of the blank arguments (Fig. 4 (f) and lines 18, 19 in Algorithm 1).

The rest of the algorithm continues in a similar style, in order to compute the corresponding scores for the input argument. It starts by calculating the strength of the supporting and attacking sets, which now include the corresponding blank arguments (Fig. 4 (g) and lines 22, 23 in Algorithm 1), and then their aggregated strength (Fig. 4 (h) and line 26 in Algorithm 1).

These values are then aggregated to compute the dialectical strength of the target argument for the given aspect (Fig. 4 (i)). Notice that the algorithm computes two different values, the acceptance score (line 29 in Algorithm 1) and the quality score (line 30 in Algorithm 1). As discussed in the previous section, a feature may weigh differently in each case, which can lead to different instantiations of the dialectical aggregation function.

Finally, the overall acceptance and quality scores are computed by aggregating all corresponding scores of each individual aspect (lines 33, 34 in Algorithm 1)

As can be seen from the above, the algorithm is recursive, triggering the computation of the dialectical strength of the arguments that exist one level below the input argument in the debate graph (see lines 10, 11, 22, 23), Based on the assumption of having a debate graph without cycles, the recursion is guaranteed to terminate.

To conclude, a note is needed to justify the symmetry in computations for the supporting and the attacking blank arguments, which is evident in Algorithm 1. In particular, one can see that having computed the dialectical strength of the one, one can easily compute the strength of the other (that is, $g_x^{dlc}(\overset{+}{a}) = 1 - g_x^{dlc}(\overset{-}{a})$), which raises the question of whether it is necessary to have both blank arguments in the framework. Indeed, it is possible to substitute this pair with a single blank argument, which according to its strength, it is assigned either to the supporting or the attacking set (following a pendulum pattern). A problematic situation would arise though when the strength of this argument is divided exactly in the middle, denoting for instance that the votes are shared among the participants in the debate. Omitting the argument altogether in this case would be counterintuitive, as its score should affect the supporting/attacking set to capture this dichotomy of opinions. For such cases, and for promoting clarity in the presentation of the model, s-mDiCE relies on a symmetric modeling of blank arguments.

4.3 Application to Discussion Platforms

The s-mDiCE framework was first deployed in the Apopsis platform[11] (Figure 5), a web-based debating platform that aims to motivate online users to participate in well-structured discussions by raising issues and posting ideas or comments that support or attack other opinions [Ymeralli et al., 2017]. The main goal of the system is to offer an automated opinion analysis that determines and extracts the most useful and strongest opinions expressed in a dialogue, eventually assisting decision-makers in understanding the opinion exchange process. The aspects chosen to determine the acceptance and quality scores of arguments within

---

[11] https://demos.isl.ics.forth.gr/apopsis/

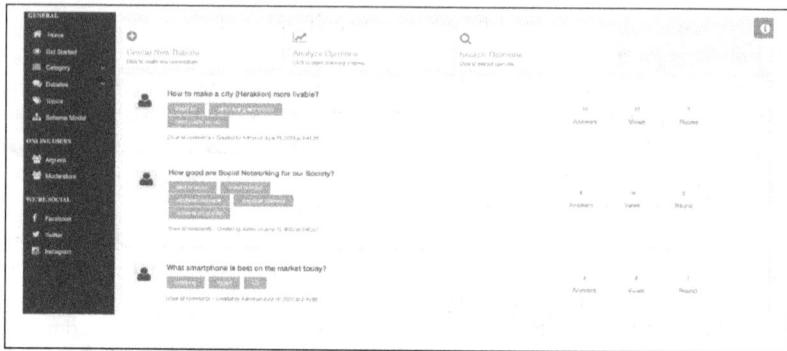

Figure 5: The Apopsis Debate Platform.

Apopsis are correctness, relevance, and sufficiency of evidence (Figure 6). Users can specify which of these aspects they consider inadequate when voting against a particular opinion, optionally adding a counterargument to support their claim. The s-mDiCE framework calculates the different opinion scores as the dialogue progresses, in order to pinpoint the strongest opinions in a debate, but also in the generation of different types of analytics, such as clustering users with similar opinions or preferences.

Recently, the framework was also used in the Argument Navigator[12] tool, developed in the context of the DebateLab project [Ymeralli *et al.*, 2022]. The project developed a suite of tools and services that can assist the work of the professional journalist in accomplishing everyday tasks (e.g., writing, archiving, retrieving articles), as well as the activity of the ordinary Web user (reader) who wishes to be well-informed about topics or entities of interest. The Article Navigator in particular is a search engine that can be used to explore, visualise and rank arguments on the Web. Among other functionalities, the user can vote for or against an argument with respect to certain aspects, such as informativeness, validity, and relevance. The argument mining process is accomplished automatically, based on a token classification / sequence labeling approach for extracting segments of argumentative discourse units, while

---

[12] https://debatelab.ics.forth.gr/tools/

1 - SOCIAL ARGUMENTATION SYSTEMS

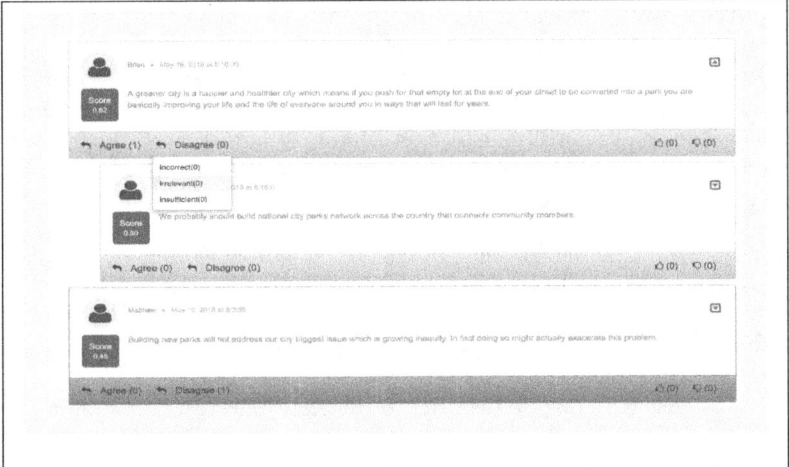

Figure 6: Rating opinions in the Apopsis platform.

diverse Deep Learning techniques and gradient-based modeling are applied to perform argument relation and stance classification.

5 Other approaches to social argumentation

In this section we review other approaches to social argumentation. We classify them into two categories: (a) other social argumentation systems, which like SAF and s-mDiCE integrate arguments with social votes aiming to model and reason with online debates; (b) extensions or applications of existing argumentation frameworks aiming to model other aspects of discussions in social media, such as the semantic relations among posts, the social relevance of a post, controversy and multi-topic discussions.

5.1 Frameworks integrating arguments and votes

Starting with the first category, *Quantitative Argumentation Debate for Voting* (QuAD-V) frameworks were developed to support collaborative debates and deliberation within e-Democracy [Rago and Toni, 2017].

QuAD-V evolved from the *Quantitative Argumentation Debate* (QuAD) frameworks [Baroni et al., 2015; Rago et al., 2016], which incorporate attack and support argument relations and intrinsic strengths of arguments[13]. QuAD-V extend QuAD frameworks with a set of users and their votes on arguments, based on which the base score (which in the rest of this section we refer to as *social support*) of an argument can be computed using the following formula:

$$\tau_v(a) = \begin{cases} 0.5 & if \ |\mathcal{U}| = 0 \\ 0.5 + (0.5 \times \frac{V^+(a) - V^-(a)}{|\mathcal{U}|}) & if \ |\mathcal{U}| \neq 0 \end{cases} \quad (11)$$

where $\mathcal{U}$ denotes the set of users, $V^+(a)$ the number of positive votes, and $V^-(a)$ the number of negative votes for argument a.

The following function is used to calculate the strength of an argument:

$$v_a = \begin{cases} \tau_v(a) \cdot (1 - |g(\mathcal{R}^+(a)) - g(\mathcal{R}^-(a))|) \\ \qquad if \ g(\mathcal{R}^-(a)) \geq g(\mathcal{R}^+(a))) \\ \tau_v(a) + (1 - \tau_v(a)) \cdot |g(\mathcal{R}^+(a)) - g(\mathcal{R}^-(a))| \\ \qquad otherwise \end{cases} \quad (12)$$

where g is a function that calculates the aggregated strength of a set of arguments given the strength of each argument in the set:

$$g(\{a_1, \cdots, a_n\}) = 1 - \prod_{i=1}^{n}(1 - v_{a_i}) \quad (13)$$

The QuAD-V frameworks exhibit the following properties. (i) $\tau_v(a)$ (the social support of a) is monotonically increasing (decreasing) with respect to the number of positive (negative) votes for a; (ii) v_a (the strength of a) is monotonically non-decreasing (non-increasing) with respect to the aggregated strength of the supporters (attackers) of a and the number of positive (negative) votes for a; (iii) an argument with stronger (weaker) attackers than supporters has a strength lower (higher) than the argument's social support, provided that the social support is not already minimal (maximal); (iv) for an argument to have

---

[13]Note that QBAFs (studied in section 2.3) are yet another extension/evolution of QuADs

the minimum (maximum) strength, either the supporters (attackers) have the minimum value and the attackers (supporters) the maximum or all votes for the argument are negative (positive) with its attackers (supporters) at least as strong as its supporters (attackers); (v) v_a is continuous with respect to $g(\mathcal{R}^+(a))$ and $g(\mathcal{R}^+(a))$, i.e. the aggregated strength of the attackers (resp., supporters) of a.

A novelty of this work is the characterisation of users as rational/irrational taking into account their votes on each argument, its attackers and its supporters. A user is considered irrational in the following two cases: (a) (s)he agrees with (votes positively for) an argument, agrees also with one of its attackers but does not agree with any of its supporters; (b) (s)he disagrees with (votes negatively for) an argument, agrees with one of its supporters but does not agree with any of its attackers. Based on the concept of rationality, they also introduce a methodology (*QuAD-V opinion polling*) for evolving polls, which aims at highlighting and eradicating irrationalities in user's opinions through a series of dynamic questions to irrational users, making the polls more informative to the pollster.

A similar approach to integrating votes and arguments was developed and implemented in [Evripidou and Toni, 2014] to support *Quaestio-it*, a web-based Q&A debating platform. The main underlying idea was the same: the strength of an argument is determined by the (positive and negative) votes it receives and the strength of its attacking and supporting arguments. Specifically, they define two functions (f_{att}, f_{supp}) that calculate the strength of an argument taking into account its social support (i.e. positive and negative votes it has received) and the aggregated strength of its attackers (resp. supporters):

$$f_{att}(a) = \tau_v(a) \cdot (1 - g(\mathcal{R}^-(a))) \qquad (14)$$

$$f_{supp}(a) = \tau_v(a) \cdot (1 + g(\mathcal{R}^+(a))) \qquad (15)$$

The strength of a set of arguments is calculated recursively using the following formula:

$$g(\{a_1, a_2, \ldots, a_n\}) = v_{a_1} + (1 - v_{a_1}) \cdot g(\{a_2, \ldots, a_n\}) \qquad (16)$$

where v_{a_1} is the strength of argument a_1 and $g(\emptyset) = 0$.

The social support for an argument is calculated using the lower bound of the Wilson Score Interval [Wilson, 1927]:

$$ws(x, y) = \frac{n}{n + z^2}\left[\hat{p} + \frac{z^2}{2n} - z\sqrt{\frac{\hat{p}(1-\hat{p})}{n} + \frac{z^2}{4n^2}}\right] \qquad (17)$$

where $n = x + y$, $\hat{p} = x/n$ and $z = 1.96$ for a confidence level of 95%. The social support of an argument a is given by:

$$\tau_v(a) = ws(V^+(a), V^-(a)) \qquad (18)$$

where $V^+(a)$ is the number of positive votes, and $V^-(a)$ the number of negative votes for argument a.

The following function is used to calculate the strength of a:

$$v_a = \begin{cases} \tau_v(a) & if\ \mathcal{R}^-(a) = \mathcal{R}^+(a) = \emptyset \\ f_{supp}(a) & if\ \mathcal{R}^-(a) = \emptyset \\ f_{att}(a) & if\ \mathcal{R}^+(a) = \emptyset \\ (f_{supp}(a) + f_{att}(a))/2 & otherwise \end{cases} \qquad (19)$$

Similarly with s-mDiCE, this framework is symmetric with respect to supporting and attacking arguments, i.e., a supporting argument increases the value of an argument's strength by the same amount by which an equivalent attack would decrease it. However, the aggregated strength of a set of arguments is defined in a way that induces discontinuity in certain cases.

5.2 Modelling other aspects of social web debates

In this section we review studies that apply models and methods from formal argumentation to model and/or reason with various aspects of social media discussions, such as the semantic relations among posts, the social relevance of a post, controversy and multi-topic discussions. One study of this type is presented in [Alsinet et al., 2017]; its main aim is to analyse discussions in Twitter, specifically to identify the social accepted tweets and measure the controversy between the users participating in a discussion. To achieve this aim, they model a discussion in Twitter as a Value-based Argumentation Framework $\mathcal{F} =$

$\langle T, attacks, R, W, Valpref \rangle$, where T is the set of tweets, $attacks = \{(t_1, t_2) \mid t_1, t_2 \in T \text{ and } t_1 \text{ criticises } t_2\}$, R is a non-empty set of ordered values used to model the social relevance of tweets, $W : A \to R$ is a function that assigns a value from R to each tweet and $Valpref \subseteq R \times R$ is the ordering relation over R. The ideal extension of $\mathcal{F}$ is the *accepted set of tweets* in the discussion. They consider three different ways to define W (i.e., to quantify the social relevance of a tweet), each of which takes into account a different type of information: the number of followers of the author of the tweet; the number of retweets of the tweet; and the number of favourites for the tweet. They also present an analysis discussion system, which consists of two components: the Discussion Retrieval component, which retrieves relevant information from a discussion, i.e. the tweets, their semantic relations, and the number of followers, retweets and favourites; and the Reasoning component, which computes the accepted set of tweets. Finally, they present two measures for controversy in a discussion: the *controversy degree*, which is the number of rejected tweets (i.e., tweets that are not in the ideal extension of the corresponding VAF) that criticise an accepted tweet, and the *controversy depth*, which is the length of the longest controversial path (sequence of tweets connected via the *attacks* relation) in a discussion.

A similar approach is used in [Alsinet et al., 2020] to model and reason with debates in Reddit. They model a Reddit debate Γ with root comment r as a *Debate Tree* $\mathcal{T} = \langle C, r, E, L \rangle$ such that for every comment in Γ there is a node in C, r is the root of $\mathcal{T}$, if c_1 answers c_2 in Γ then there is a directed edge (c_1, c_2) in E, and $L : E \to [-2, 2]$ assigns a value to each edge denoting the sentiment of the corresponding answer, from highly negative (-2) to very positive (2). They then prune the Debate Tree by discarding neutral comments (based on their sentiment values) and their subtrees. To find the accepted comments in a debate, they map the corresponding pruned debate tree to a Value-based Argumentation Framework: each comment is mapped to an argument, each answer to a comment is mapped to an attack from the answer to the comment if the sentiment of the answer is negative, and the score of each comment is mapped to a natural number that represents its social acceptance. The set of accepted comments in a Reddit debate is the ideal extension of the corresponding VAF. They also propose measures for quantifying the users' influence, the controversy that they generate throughout a debate, their contribution to the polarisation of the debate and their social acceptance. In all such measures, they use the notion of *filtered tree*, which results from a pruned debate tree by removing the comments of a given user (excluding the root comment) and all the comments in the subtrees rooted by the user's comments. Some of these measures are: (i) the *debate engaging degree* of a user u, which is used to quantify the interactions of the user; (ii) the *influence degree* of a user u, which is used to quantify the comments that change their status (from accepted to rejected or vice versa) if we disregard the comments from u; (iii) the *polarisation degree* of a solution S, which is a measure of the bias of S towards comments in favour of the root comment and comments against the root comment; (iv) the *rebalancing degree* of a user u, which quantifies the influence of the user on the polarisation of a debate solution; and (v) the *social acceptance* of a user u, which sums up the scores of the user's comments;

An argumentation framework that integrates the notion of *topic tags* or *hashtags* used in social media applications, such as Facebook and Twitter, was proposed in [Budán et al., 2020]. They introduce the notion of *hashtagged argument*, which they define as a pair $\langle a, \mathcal{H}_a \rangle$, where $\mathcal{H}_a$ denotes a set of hashtags associated with argument a. To model the relations among topics, they define *hashtag graphs*; a vertex in such a graph denotes a hashtag, and an undirected edge between two vertices denotes some relationship between the corresponding hashtags. The distance between two vertices in a hashtag graph is the number of edges in the shortest path connecting them. The distance between two hashtagged arguments can be defined in several ways, for example as the minimum or the maximum or the average distance between the hashtags of the two arguments. A *hashtagged argumentation framework* Ω is defined as a pair $\langle \Phi, \mathcal{G} \rangle$, where Φ is an AAF consisting of hashtagged arguments and $\mathcal{G}$ is a hashtag graph. To reason with such frameworks they adjust the standard acceptability semantics of AAFs to take into account the hashtags of arguments and their relations. Specifically, they redefine acceptability as follows: a hashtagged argument a is ϵ-*acceptable* w.r.t. a set of hashtagged arguments S when for every hashtagged argument b that attacks a there is a hashtagged argument $c \in S$ that attacks b and

$d_\Omega(a,c) \leq \epsilon$, where d_Ω is a distance function for Ω and ϵ a user-defined threshold. Admissible, complete, grounded and preferred semantics are then defined in the standard way. They also provide an alternative definition for acceptability semantics, which takes into account both the distance between an argument and its defenders but also the distance between the arguments in an extension.

An earlier study explored the use of Bipolar Argumentation Frameworks for modelling and reasoning with online debates [Cabrio and Villata, 2013]. Specifically, they proposed the use of textual entailment for classifying the relation between two sentences in one of the following types: *entailment*, i.e. the meaning of one of the two sentences can be inferred from the other; *contradiction*, i.e., the two sentences cannot be simultaneously true; and *unknown*, i.e., the truth of one sentence cannot be verified on the basis of the other. Using an empirical study they found a high correlation between entailment and support, i.e., in most (61.6%) of the cases where annotators identified that a sentence a supports another sentence b, they also identified that a entails b, and an even higher correlation between contradiction and attack, i.e., in most (71.4%) of the cases where the annotators identified that a sentence a attacks another sentence b, they also identified that a contradicts b. They also verified with another experiment that the correlation between attack and contradiction also holds for other types of attacks that can be deduced from a bipolar argumentation framework, i.e., *supported*, *secondary*, *mediated* and *extended* attacks.

Finally, as mentioned in Section 2, Labeled Bipolar Argumentation Frameworks [Gonzalez et al., 2021] is another formalism that can be applied to social argumentation systems. Similar to s-mDiCE, it enables the valuation of an argument with respect to different dimensions (argument features) taking into account the strength of its attackers and supporters. A distinctive characteristic of this formalism is that it allows assigning ranges of values to an argument, which is useful when there is uncertainty in the original valuation of an argument. On the other hand, it does not explicitly handle social votes; in their running example from the domain of social platforms, they represent the social rating of an argument as one of its features and assume that the original argument valuation for this feature are given. A common limitation of both approaches compared to Social Argumentation Frameworks (discussed in Section 3) is their inability to treat cycles in the argument graph.

6 Conclusion

The recent trend in Web usage has elevated users from pure consumers of information to "prosumers", i.e., both consumers and producers of information. A large part of this trend is attributed to websites that allow users to contribute their opinions on any conceivable topic, reviews on physical or digital products and services, as well as commentaries on events, people, ideas, or things. This trend often has the form of a discussion, with arguments that support one's opinions, as well as responses or other reactions to such opinions by other users. In such cases, making sense out of a (possibly long and complex) debate is important for users, and part of the sense-making process is the ability to automatically evaluate arguments.

The original argumentation theory is not fully suitable to cope with this evaluation procedure because the arguments appearing in these kinds of debates are rarely totally accepted or totally rejected; instead, a numerical assessment under the so-called *gradual semantics* is more suitable. Furthermore, the original argumentation theory has no support for the temporal dimension (i.e., the order in which the arguments are presented), or for other types of reactions (such as votes) that one can typically use in such systems.

In this chapter, we surveyed *social argumentation systems*, i.e., various systems and frameworks specifically designed to model, analyse or enable these kinds of debates. This includes theories and principles that such systems should satisfy, as well as specific technical solutions that address these issues and the properties that such solutions satisfy. Our aim is to provide an overview for interested researchers and practitioners in choosing the most suitable solution for their purposes, and/or in developing alternative methods for argument analysis or evaluation in such settings.

Future avenues of research in this area include (i) the further development or extension of social argumentation systems to take into account the characteristics of users (user profiles, expertise or popularity of users, etc,); (ii) the evaluation of social argumentation systems using data from online debates in social networks or debate websites; and (iii) the development of social web applications that fully exploit the capabilities of such frameworks to facilitate and analyse online debates (some examples of such applications are already available and are described in another chapter of this volume).

To the best of the authors' knowledge, social argumentation systems focus on abstract arguments. Recasting those ideas in the context of structured argumentation (e.g., ASPIC+ [Prakken, 2010], [Modgil and Prakken, 2017] or ABA [Cyras et al., 2018]) is another future work direction with significant potential. In particular, the additional information provided by the arguments' structure may be exploited both to allow more specific user input (e.g., votes relating only to a particular premise of an argument, or to the argument's reasoning), and a more fine-grained evaluation of the argument that takes into account its structure.

References

[Alsinet et al., 2017] Teresa Alsinet, Josep Argelich, Ramón Béjar, Cèsar Fernández, Carles Mateu, and Jordi Planes. Weighted argumentation for analysis of discussions in twitter. *Int. J. Approx. Reason.*, 85:21–35, 2017.

[Alsinet et al., 2020] Teresa Alsinet, Josep Argelich, Ramón Béjar, and Santi Martínez. Measuring user relevance in online debates through an argumentative model. *Pattern Recognit. Lett.*, 133:41–47, 2020.

[Amgoud and Ben-Naim, 2013] L. Amgoud and J. Ben-Naim. Ranking-based semantics for argumentation frameworks. In *Proceedings of the Seventh International Conference on Scalable Uncertainty Management (SUM)*, pages 134–147, 2013.

[Amgoud and Ben-Naim, 2016a] L. Amgoud and J. Ben-Naim. Axiomatic foundations of acceptability semantics. In *Proceedings of the Fifteenth International Conference on Principles of Knowledge Representation and Reasoning (KR-16)*, pages 2–11, 2016.

[Amgoud and Ben-Naim, 2016b] L. Amgoud and J. Ben-Naim. Evaluation of arguments from support relations: Axioms and semantics. In *Proceedings of the Twenty-Fifth International Joint Conference on Artificial Intelligence (IJCAI-16)*, pages 900–906, 2016.

[Amgoud and Ben-Naim, 2017] L. Amgoud and J. Ben-Naim. Evaluation of arguments in weighted bipolar graphs. In *Proceedings of the Fourteenth*

European Conference on Symbolic and Quantitative Approaches to Reasoning with Uncertainty (ECSQARU-17), pages 25–35, 2017.

[Amgoud et al., 2016] L. Amgoud, J. Ben-Naim, D. Doder, and S. Vesic. Ranking arguments with compensation-based semantics. In *Proceedings of the Fifteenth International Conference on Principles of Knowledge Representation and Reasoning (KR-16)*, pages 12–21, 2016.

[Amgoud et al., 2017] L. Amgoud, J. Ben-Naim, D. Doder, and S. Vesic. Acceptability semantics for weighted argumentation frameworks. In *Proceedings of the Twenty-Sixth International Joint Conference on Artificial Intelligence (IJCAI-17)*, pages 56–62, 2017.

[Argyros and Szidarovszky, 1993] I. K. Argyros and F. Szidarovszky. *The Theory and Applications of Iteration Methods*. Systems Engineering. Taylor & Francis, 1993.

[Baroni and Giacomin, 2007] Pietro Baroni and Massimiliano Giacomin. On principle-based evaluation of extension-based argumentation semantics. *Artificial Intelligence*, 171(10):675–700, 2007.

[Baroni et al., 2013] P. Baroni, M. Romano, F. Toni, M. Aurisicchio, and G. Bertanza. An argumentation-based approach for automatic evaluation of design debates. In *CLIMA XIV: Proceedings of the 14th International Workshop on Computational Logic in Multi-Agent Systems - Volume 8143*, pages 340–256, 2013.

[Baroni et al., 2015] Pietro Baroni, Marco Romano, Francesca Toni, Marco Aurisicchio, and Giorgio Bertanza. Automatic evaluation of design alternatives with quantitative argumentation. *Argument and Computation*, 6(1):24–49, 2015.

[Baroni et al., 2018] Pietro Baroni, Antonio Rago, and Francesca Toni. How many properties do we need for gradual argumentation? In *Proceedings of the 32nd AAAI Conference on Artificial Intelligence*, 2018.

[Baroni et al., 2019] Pietro Baroni, Antonio Rago, and Francesca Toni. From fine-grained properties to broad principles for gradual argumentation: A principled spectrum. *Int. J. Approx. Reason.*, 105:252–286, 2019.

[Bikakis et al., 2023] Antonis Bikakis, Giorgos Flouris, Theodore Patkos, and Dimitris Plexousakis. Sketching the vision of the web of debates. *Journal of the Frontiers of AI, research topic on Computational Argumentation: a foundation for Human-centric AI*, 6, 2023.

[Bonzon et al., 2016] Elise Bonzon, Jérôme Delobelle, Sébastien Konieczny, and Nicolas Maudet. A comparative study of ranking-based semantics for abstract argumentation. In *Proceedings of the 30th AAAI Conference on Artificial Intelligence*, 2016.

[Budán et al., 2020] Maximiliano Celmo David Budán, Maria Laura Cobo, Diego C. Martínez, and Guillermo Ricardo Simari. Proximity semantics for topic-based abstract argumentation. *Inf. Sci.*, 508:135–153, 2020.

[Cabrio and Villata, 2013] Elena Cabrio and Serena Villata. A natural language bipolar argumentation approach to support users in online debate interactions†. *Argument Comput.*, 4(3):209–230, 2013.

[Caminada, 2018] Martin Caminada. Rationality postulates: Applying argumentation theory for non-monotonic reasoning. In *Handbook of Formal Argumentation*. College Publications, 2018.

[Correia et al., 2014] Marco Correia, Jorge Cruz, and João Leite. On the efficient implementation of social abstract argumentation. In Torsten Schaub, Gerhard Friedrich, and Barry O'Sullivan, editors, *ECAI 2014 - 21st European Conference on Artificial Intelligence, 18-22 August 2014, Prague, Czech Republic - Including Prestigious Applications of Intelligent Systems (PAIS 2014)*, volume 263 of *Frontiers in Artificial Intelligence and Applications*, pages 225–230. IOS Press, 2014.

[Cyras et al., 2018] K. Cyras, X. Fan, C. Schulz, and F. Toni. Assumption-based argumentation: Disputes, explanations, preferences. In *Handbook of Formal Argumentation*. College Publications, 2018.

[Dung, 1995] Phan Minh Dung. On the acceptability of arguments and its fundamental role in nonmonotonic reasoning, logic programming and n-person games. *Artificial Intelligence*, 77(2):321–357, September 1995.

[Egilmez et al., 2013] Sinan Egilmez, João G. Martins, and João Leite. Extending social abstract argumentation with votes on attacks. In Elizabeth Black, Sanjay Modgil, and Nir Oren, editors, *Theory and Applications of Formal Argumentation - Second International Workshop, TAFA 2013, Beijing, China, August 3-5, 2013, Revised Selected papers*, volume 8306 of *Lecture Notes in Computer Science*, pages 16–31. Springer, 2013.

[Evripidou and Toni, 2014] Valentinos Evripidou and Francesca Toni. Quaestio-it.com: a social intelligent debating platform. *Journal of Decision Systems*, 23(3):333–349, 2014.

[Flouris et al., 2023] Giorgos Flouris, Theodore Patkos, Antonis Bikakis, Alexandros Vassliades, Nick Bassiliades, and Dimitris Plexousakis. Theoretical analysis and implementation of abstract argumentation frameworks with domain assignments. *International Journal of Approximate Reasoning (IJAR)*, 161C, 2023.

[Gonzalez et al., 2021] Melisa G. Escañuela Gonzalez, Maximiliano C. D. Budán, Gerardo I. Simari, and Guillermo R. Simari. Labeled bipolar argumentation frameworks. *Journal of Artificial Intelligence Research (JAIR)*, 70, 2021.

[Klement et al., 2000] Erich Peter Klement, Radko Mesiar, and Endre Pap. *Triangular Norms*. Springer, 1 edition, 2000.

[Kunz and Rittel, 1970] W. Kunz and H. Rittel. *Issues as elements of information systems. Working Paper 131*. Institute of Urban and Regional Development, University of California, Berkeley, California, 1970.

[Leite and Martins, 2011] João Leite and João G. Martins. Social abstract argumentation. In Toby Walsh, editor, *IJCAI 2011, Proceedings of the 22nd International Joint Conference on Artificial Intelligence, Barcelona, Catalonia, Spain, July 16-22, 2011*, pages 2287–2292. IJCAI/AAAI, 2011.

[Luca, 2011] Michael Luca. Reviews, reputation, and revenue: The case of yelp.com. Technical report, Technical Report 12-016, Harvard Business School, 2011.

[Matt and Toni, 2008] P. Matt and F. Toni. A game-theoretic measure of argument strength for abstract argumentation. In *Proceedings of the Eleventh European Conference on Logics in Artificial Intelligence (JELIA-08)*, pages 285–297, 2008.

[Modgil and Prakken, 2017] Sanjay Modgil and Henry Prakken. Abstract rule-based argumentation. *Journal of Applied Logics - IfCoLog Journal of Logics and their Applications (FLAP)*, 4, 2017.

[Modgil et al., 2013] S. Modgil, F. Toni, F. Bex, I. Bratko, C. I. Ches nevar, W. Dvořák, M. A. Falappa, X. Fan, S. A Gaggl, A. J. García, M. P. González, T. F. Gordon, J. Leite, M. Možina, C. Reed, G. R. Simari, S. Szeider, P. Torroni, and S. Woltran. The added value of argumentation. In S. Ossowski, editor, *Agreement Technologies*, volume 8 of *Law, Governance and Tech. Series*, pages 357–403. Springer, 2013.

[Ortega and Rheinboldt, 2000] J. M. Ortega and W.C. Rheinboldt. *Iterative Solution of Nonlinear Equations in Several Variables*. Society for Industrial and Applied Mathematics, Philadelphia, PA, USA, 2000.

[Patkos et al., 2016a] Theodore Patkos, Antonis Bikakis, and Giorgos Flouris. A multi-aspect evaluation framework for comments on the social web. In Chitta Baral, James P. Delgrande, and Frank Wolter, editors, *Principles of Knowledge Representation and Reasoning: Proceedings of the Fifteenth International Conference, KR 2016, Cape Town, South Africa, April 25-29, 2016*, pages 593–596. AAAI Press, 2016.

[Patkos et al., 2016b] Theodore Patkos, Antonis Bikakis, and Giorgos Flouris. Rating comments on the socialweb using a multi-aspect evaluation framework. Technical report, TR-463, Institute of Computer Science, Foundation for Research and Technology - Hellas, 2016.

[Patkos et al., 2016c] Theodore Patkos, Giorgos Flouris, and Antonis Bikakis. Symmetric multi-aspect evaluation of comments - extended abstract. In

Gal A. Kaminka, Maria Fox, Paolo Bouquet, Eyke Hüllermeier, Virginia Dignum, Frank Dignum, and Frank van Harmelen, editors, *ECAI 2016 - 22nd European Conference on Artificial Intelligence, 29 August-2 September 2016, The Hague, The Netherlands - Including Prestigious Applications of Artificial Intelligence (PAIS 2016)*, volume 285 of *Frontiers in Artificial Intelligence and Applications*, pages 1672–1673. IOS Press, 2016.

[Prakken, 2010] Henry Prakken. An abstract framework for argumentation with structured arguments. *Argument and Computation*, 1:93–124, 2010.

[Rago and Toni, 2017] Antonio Rago and Francesca Toni. Quantitative argumentation debates with votes for opinion polling. In Bo An, Ana L. C. Bazzan, João Leite, Serena Villata, and Leendert W. N. van der Torre, editors, *PRIMA 2017: Principles and Practice of Multi-Agent Systems - 20th International Conference, Nice, France, October 30 - November 3, 2017, Proceedings*, volume 10621 of *Lecture Notes in Computer Science*, pages 369–385. Springer, 2017.

[Rago et al., 2016] Antonio Rago, Francesca Toni, Marco Aurisicchio, and Pietro Baroni. Discontinuity-free decision support with quantitative argumentation debates. In Chitta Baral, James P. Delgrande, and Frank Wolter, editors, *Principles of Knowledge Representation and Reasoning: Proceedings of the Fifteenth International Conference, KR 2016, Cape Town, South Africa, April 25-29, 2016*, pages 63–73. AAAI Press, 2016.

[Rago et al., 2018] Antonio Rago, Pietro Baroni, and Francesca Toni. On instantiating generalised properties of gradual argumentation frameworks. In *Scalable Uncertainty Management*, 2018.

[Thimm and Kern-Isberner, 2014] M. Thimm and G. Kern-Isberner. On controversiality of arguments and stratified labelings. In *Proceedings of Computational Models of Argument (COMMA-14)*, 2014.

[van der Torre and Vesic, 2018] Leendert van der Torre and Srdjan Vesic. The principle-based approach to abstract argumentation semantics. In *Handbook of Formal Argumentation*. College Publications, 2018.

[Walton, 2009] D. Walton. Argumentation theory: A very short introduction. In I. Rahwan and G. R. Simari, editors, *Argumentation in Artificial Intelligence*, pages 1–22. Springer, 2009.

[Wang et al., 2024] Haohao Wang, Wei Li, and Bin Yang. An extension of several properties for fuzzy t-norm and vague t-norm. *Journal of Intelligent & Fuzzy Systems*, 46(3):6881–6891, 2024.

[Wilson, 1927] Edwin B. Wilson. Probable Inference, the Law of Succession, and Statistical Inference. *Journal of the American Statistical Association*, 22(158):209–212, 1927.

[Ymeralli et al., 2017] Elisjana Ymeralli, Giorgos Flouris, Theodore Patkos, and Dimitris Plexousakis. Apopsis: A web-based platform for the analysis of structured dialogues. In *On the Move to Meaningful Internet Systems. OTM 2017 Conferences: Confederated International Conferences: CoopIS, C&TC, and ODBASE 2017, Rhodes, Greece, October 23-27, 2017, Proceedings, Part II*, page 224–241, 2017.

[Ymeralli *et al.*, 2022] E. Ymeralli, G. Flouris, V. Efthymiou, K. Papantoniou, T. Patkos, G. Petasis, N. Pittaras, G. Roussakis, and E Tzortzakakis. Representing online debates in the context of e-journalism. In *The Sixteenth International Conference on Advances in Semantic Processing (SEMAPRO 2022)*, 2022.

[Yu *et al.*, 2021] Liuwen Yu, Dongheng Chen, Lisha Qiao, Yiqi Shen, and Leendert van der Torre. A principle-based analysis of abstract agent argumentation semantics. In *Proceedings of the 18th International Conference on Principles of Knowledge Representation and Reasoning*, pages 629–639, 2021.

CHAPTER 2

SEARCHING, NAVIGATING, AND QUERYING ARGUMENTS AND DEBATES: TOOLS, LANGUAGES AND METHODOLOGIES

GIORGOS FLOURIS
Institute of Computer Science, FORTH, Heraklion, Greece.
fgeo@ics.forth.gr

JOHN LAWRENCE
Centre for Argument Technology, University of Dundee, UK.
j.lawrence@dundee.ac.uk

DIMITRA ZOGRAFISTOU
Centre for Argument Technology, University of Dundee, UK.
dzografistou@gmail.com

1 Introduction

The Web has long now ceased to be a purely one-directional means of communication: Web users are no longer just passive consumers of information, but can also contribute actively to the content of the Web. As a matter of fact, modern Web allows users to create and post different forms of digital information, such as the uploading of images and videos in social media channels; the posting of reviews for all types of products or services; the commentary of articles, political ideas or other viewpoints; the expression of ideas in chat rooms; the debating for hot controversial issues in specialised websites; and others.

One of the important consequences of this transition is that users often see the Web as a means to enable dialogical exchange, debating, and commenting, as it allows their ideas to reach people in all corners of the world. As a result, a significant portion of Web content is of argumentative form, containing users' opinions on any conceivable topic, often with well-articulated arguments. Depending on the platform in which these arguments are expressed, they can be unstructured (e.g., free-text), or have some kind of structure, which is imposed by the tool. This structure could be imposed at the debate level, e.g., through replies that end up generating debates that have a tree-like or forest-like structure, or at the argument level, by requiring the users to formulate their arguments in a specific way, e.g., explicitly specifying (or annotating) the premises and conclusions of the arguments.

The plethora of arguments online are useful only with the support of appropriate tools for identifying relevant arguments for any given information need. Note that keyword-based searching is useful but not sufficient for supporting the information needs of users when it comes to argumentative content. Arguments have an inherent structure (whether explicitly specified or not) and they are related to each other (although such relationships may not be always obvious and/or explicitly recorded); these properties are often critical and should be queryable, so whatever method we use for searching arguments should include the ability to perform more complex searches and queries that involve the arguments' (or debates') structure and interrelationships. Moreover, depending on the form of the debate, the context, and the application at hand, argument searching should also allow navigational or explorational capabilities as well as the ability to perform analytical queries.

In this chapter, we provide a short survey of various approaches for searching, exploring and querying arguments, aiming to highlight the main advances in this area. We classify the relevant approaches into three main categories, namely, *query languages*, *argument retrieval systems*, and *closed debating systems*, which we explain below and in their respective sections.

Section 2 deals with the first category, and discusses structured query languages that one could use to query arguments, in a manner similar to other structured languages such as SQL or SPARQL. This approach is mainly applicable to arguments whose structure and/or interrelationships are explicitly recorded using an appropriate knowledge representation formalism, such as an ontology. Towards this direction, we present the only (to the best of our knowledge) structured query language that was designed explicitly for argumentative content and addresses these issues, namely *Argumentation Query Language (ArgQL)* [Zografistou et al., 2017], [Zografistou et al., 2018], [Roussakis et al., 2022]. ArgQL abstracts argument/debate retrieval operations from the implementation details of the underlying repository, allowing the user to express queries in a more "natural" way (using argumentative terminology), leading to queries that are short, easy to formulate, read, understand and maintain, as well as efficient.

The second approach (argument retrieval systems) is discussed in Section 3. Argument retrieval systems are used to search for arguments in a manner analogous to how popular search engines crawl and search the Web for textual content. Argument retrieval systems can retrieve arguments that answer questions like "is coffee good for you?", by looking at the argumentative content of various websites that have been previously analysed. To do so, such systems often use argument mining techniques for identifying the argumentative content of websites, and the structure/interrelationships of the contained arguments. In other cases, the arguments' structure and interrelationships may have become known to the argument retrieval system because this structure is somehow exposed by the respective website, or because it has been identified through manual annotation.

The third approach (closed debating systems, analysed in Section 4) includes tools whose main functionality is the management and/or analysis of debates that have been created *within* the tool. In this respect the system is "closed", because it provides a complete set of functionalities for both entering and retrieving arguments. Note how this contrasts with the argument retrieval systems, which provide searching, analytics and exploration capabilities for arguments expressed elsewhere. The fact that closed debating systems fully control the way in which the arguments are entered and stored in the underlying repository allows them to provide sophisticated searching and visualisation functionalities not easily attainable with other methods. On the downside, the pool of arguments over which searching is allowed is limited to the ones that have been ingested through the tool's interface, which often limits their applicability to well-focused, small-scale debates within well-defined groups (e.g., among people involved in the management of a company pondering on a specific business decision).

The rest of this chapter presents further details on various representatives of these three approaches to argument searching (in their respective sections). We conclude in Section 5 with a general discussion and outlook for future research directions on the topic.

2 Languages for querying arguments

As mentioned above, searching and navigating within dialogues introduce specific challenges and requirements. This is mainly due to the fact that, at the core of debating and argumentation, particular structures are created, which are common in any kind of discussion regardless of its topic, the structure of which is queryable. This sets new ground in the area of information retrieval, in which new theories and models can be developed.

One such direction is the identification of the different kinds of queries related to the problem of dialogue searching and the exploration of the different ways in which they can be expressed, evaluated or executed more efficiently. The most obvious and straightforward approach is to adopt some of the standard storage schemes to represent dialogical data, and then use the associated query language to express requests for particular fragments of the data [Cabrio et al., 2013], [Cimiano et al., 2013]. Such storage schemes include relational databases (MySQL), semantic databases (RDF/SPARQL), graph databases (GraphQL, Neo4j) etc. For all those frameworks, dialogical searching constitutes an application domain, using an appropriate knowledge representation formalism, such as an ontology. The most popular approach in this respect is the use of the *Argument Interchange Format (AIF)* [Chesñevar et al., 2006] a data model that was designed to bridge the various models of argumentation into a common ontological pattern of representation. Although the existence of this de-facto representation standard allows the use of a general-purpose query language (such as SPARQL) for querying the argumentative data, this approach has several disadvantages. First, it requires the user to deeply understand AIF in order to properly formulate the correct query. Second, the resulting queries are often long and complex, thus error-prone and hard to maintain. Third, this approach creates an undesirable bonding between the query formulation process and the implementation details of the argumentation repository, reducing robustness and interoperability, as the application logic is bound to the specific implementation and thus cannot be easily migrated to alternative implementations.

The existence of a high-level language that uses terminology and semantics related to the argumentation domain, would offer to the community a more familiar way to express queries, saving them from dealing with all the technical details of using standard technologies. Furthermore, there is a large volume of dialogical and argumentative data distributed across various platforms and represented in different formats, and the development of centralised mechanisms to search through all this data, facilitating their integration, would enable the development of techniques for the automatic analysis of dialogues and debates. The only known language that carries the momentum of becoming such a language is *ArgQL (Argumentation Query Language)* [Zografistou et al., 2017], [Zografistou et al., 2018], [Roussakis et al., 2022].

2.1 Argumentation Query Language (ArgQL)

ArgQL is a high-level, declarative query language, designed to express queries particularly related to the problem of searching within dialogues. Its specification is designed on top of a data model that captures the fundamental principles of structured argumentation, and thus, data comprise graphs of interlinked, structured arguments. ArgQL enables querying both the internal structure of arguments and their relations, but also the structure of the debate that the arguments are part of. The debate structure can be seen in terms of Dung's argument frameworks [Dung, 1995], where arguments and relations are abstracted away from their structure, and define graphs where nodes represent arguments and edges the relations between them. By allowing querying such graphs, ArgQL makes itself compatible with those frameworks, and in future extensions, it could incorporate semantics of abstract frameworks in the search mechanisms.

ArgQL offers a simple and clear way to express queries that fall into four main categories (or their combinations): a) Locating individual arguments, e.g. *"Find arguments with conclusion p"* b) Identifying commonalities between arguments' structure, e.g. *"Find arguments with common premises"* c) Extracting argument relations, e.g. *"Find arguments that attack or support each other"* and d) Navigating within the argument graph e.g. *"Find arguments that attack the attackers of argument a"*. In all of these categories, special emphasis is given in the factor of rephrasing, or otherwise propositional equivalence, namely the fact that two propositions might be saying the same thing (be equivalent), but expressed in a different way. In particular, assuming that arguments consist of propositions, a query searches for propositions, that either themselves or any of their equivalent ones are satisfying a particular expression.

The current implementation of ArgQL is built on top of the RDF storage scheme. In particular, in order for a query to be executed, it is first translated into SPARQL, which is then executed against an RDF database, to which the data model of ArgQL has been mapped. The RDF schema that is used is specified by the AIF ontology [Chesñevar et al., 2006]. In a similar way, more implementations of ArgQL can be developed, that will translate into other query languages, allowing that way for ArgQL to become a centralised language that integrates the results of querying different databases and heterogeneous argumentative data.

2.1.1 Data Model

The main concept in the data model of ArgQL is the *argument*, which consists of *propositions*. There are three kinds of relations between propositions, *equivalence*, *conflict* and *inference*. Equivalence and conflict relations define equivalence and contrariness in the content among propositions and are defined in a symmetric way. An argument is defined as a tuple $\langle pr, c \rangle$, where pr is a set of propositions, called *premise*, c is a single proposition, called *conclusion* and it holds that the premise logically infers the conclusion. A conflict relation between propositions defines two types of *attack* between arguments: *rebut* (conflict between two conclusions) and *undercut* (conflict between a conclusion and a premise). An equivalence relation between propositions defines two types of *support* between arguments: *endorse* (equivalent conclusions) and *backing* (equivalence between a conclusion and a premise). Overall, an argument base forms a graph, in which nodes are structured arguments, connected via four types of relations.

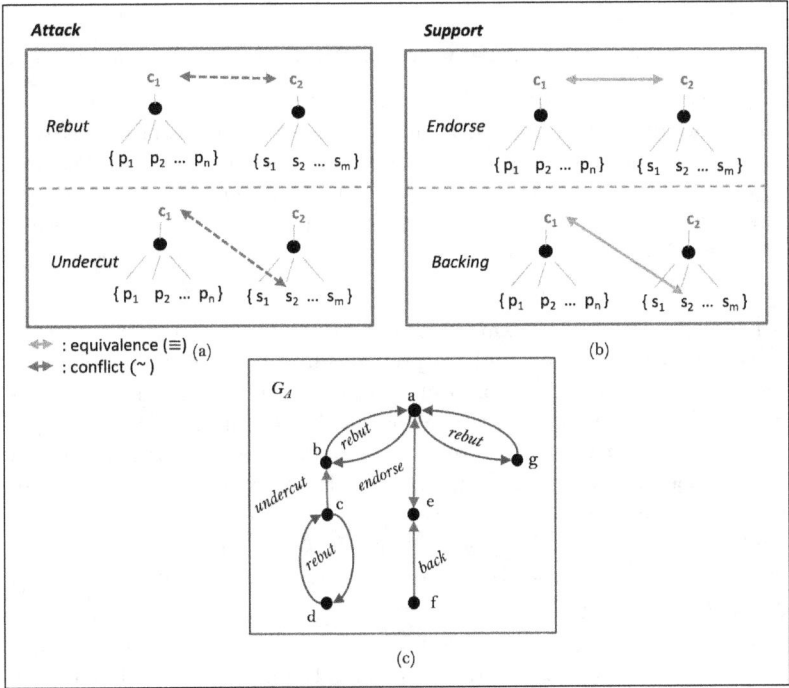

Figure 1: Example of data model

Figure 1 gives an example of the data model. Subfigures (a) and (b) show arguments' and relations' internal structure, while (c) depicts the structure of the debate. The nodes in this last graph view (c) represent an abstract version of arguments, the internal structure of which has the general form $\langle pr, c \rangle$, where pr infers c.

2.1.2 ArgQL specification

The general form of an ArgQL query is:

$$q \leftarrow \textbf{match } dialogue\_pattern \text{ (',' } dialogue\_pattern)*$$
$$\textbf{return } varlist \mid \textbf{path}(v_1, v_2)$$

where $varlist = (v_1, v_2, \ldots, v_n)$ is a list of variables. A *dialogue pattern* may have one of the following two forms:

```
dialogue_pattern ::= argpattern |
                    argpattern  pathpattern  dialogue_pattern
```

Argument patterns are the primary units in the language and are used to match arguments' internal structure. Syntactically, an argument pattern can either be a single variable v_a, or have the form v_a:⟨*premisePattern, conclusionPattern*⟩. *PremisePattern* and *conclusionPattern* specify the premise and conclusion part of arguments, respectively. More precisely, the second form of an argument pattern may be one of the following:

$$v_a : \langle \{p_1, .., p_n\}, c \rangle \quad \text{or} \quad v_a : \langle v_p[f], c \rangle$$

where $p_1, .., p_n$ are constant propositional values, c is a proposition or variable, v_p is a variable and f a premise filter. Variable $\mathbf{v_p}$ matches the premise part of arguments and in particular, it matches sets of propositions, whereas c matches the conclusion. The occurrence of the expression $[f]$ is optional. When it exists, the premise part is restricted based on a propositional set pattern, let s, which can either be a set of propositions (strings) or a variable that takes as value a set of propositions and, so, there can be 3 types of filters: *inclusion, join* and *disjointness* written as [/s] , [.s] and [!s], respectively. Below, are some examples of argument patterns:

- ⟨?v[/{"p_1"}], ?c⟩ : match arguments the premise of which include some proposition equivalent to "p_1".

- ⟨?v, "c"⟩ : match arguments with conclusion any proposition equivalent to "c"

- ⟨?v[.{"p_1", "p_2"}], ?c⟩ : match arguments whose premise intersects or is equivalent to the set {"p_1", "p_2"}

○ $\langle\{"p_1","p_2"\},"c"\rangle$: instantiated arguments are also argument patterns

Path patterns are expressions that match complete paths and allow for navigation in the graph. They are identified by sequences of relations separated by the character '/' (e.g. attack/support/support). Note that a *relation* can either be one of the sub-relations (rebut, undercut, endorse, backing) or one of the general ones (attack, support). In the second case, any of the corresponding sub-relations will match the pattern. The expression *$*n$* is a syntactic sugar to express the "n repetitions of a path pattern". For example rebut*2, is an alternative notation of rebut/rebut. In addition, we can express the case "up to n repetitions", by using the notation '$+n$'. In particular, attack+3 defines three different patterns: {attack, attack/attack, attack/attack/attack}. The existence of multiple '+' indicators in the same pattern defines a maximum number of combinations, equal to the proliferation of the number of '+'s. For example, the expression ((attack)+2)+3 will give $2 \times 3 = 6$ alternative path patterns: the '+2' indicator defines 2 path patterns, and for each of them the '+3' indicator will give 3 different patterns. Note the *path* clause in the *return* statement allows to return the whole parts of the graph that match the path patterns.

Next, we show some examples of complete queries in ArgQL:

Q1. Find arguments which are in a maximum distance of 3 "defend" (attack the attackers) relational steps, from arguments that have conclusion equivalent to *"Cloning includes ethical risks"*, and return the complete matching paths.

```
match ?a1 (attack/attack)+3
?a2:<?pr, "Cloning includes ethical risks">
return path(?a1, ?a2)
```

Q2. Find and return arguments which attack other arguments connected via a sequence of three support relations, to an argument, for which, one of the premises is equivalent to the proposition *"cloning contributes positively in artificial insemination"*.

```
match ?arg attack/(support)*3
<?pr[/{"cloning contributes ... insemination"}],?c>
return ?arg
```

Q3. Find pairs of arguments whose premises intersect and return them.

```
match ?a1:<?pr1, ?c1>, ?a2:<?pr2[.?pr1], ?c2>
return ?a1, ?a2
```

3 Argument retrieval systems

In this section we explore tools for searching and retrieving argumentative data. In particular, we consider tools which move beyond a standard web search, where a user may be able to ask questions like "is coffee good for you?" and receive in return a number of articles discussing this topic. Instead we look at tools which provide the ability for more nuanced exploration of the arguments; such as "give me arguments pro and con why coffee is good for you" or "give me an argument from expert opinion supporting coffee being good for you". The ability to search for arguments in this way has a wealth of potential applications, from assisting users in reaching decisions and forming opinions, to providing on-demand knowledge bases for dialogical agents.

Implementations of such argument search tools can be split into two broad categories: those for searching structured data, and those for searching unstructured data. In the former case, an interface is provided for searching an existing corpus of known argumentation structures gathered, for example, through manual annotation, guided argument construction, or as the record of a structured dialogue (see Section 4). In the latter case, searching in unstructured data, these tools generally combine a specific query with argument mining techniques to determine the arguments contained in unstructured text. The argument mining approach allows for results to be found in a broader range of material that has not been previously analysed, though often with somewhat less accurate results. This categorisation corresponds to the two shared tasks introduced in Touché 2020 [Bondarenko et al., 2020] which firstly looked at retrieval of arguments on socially important topics from a pre-existing set of structured arguments, and secondly, using argument mining techniques to retrieve documents with relevant arguments from a generic web crawl.

3.1 Tools for searching structured data

One of the first dedicated argument search engines to be developed was the *args.me* prototype [Wachsmuth *et al.*, 2017], which retrieves relevant arguments on a given query, ranks these according to their relevance, and presents them as lists of points pro and con (see Figure 2). In describing the development of args.me the authors noted that, at the time, no automatic argument mining approach seemed "robust enough, yet, to obtain arguments reliably from the web". As such, args.me instead returns results using an index of structured arguments crawled from a number of online debate portals, specifically: idebate.org, debatepedia.org, debatewise.org, debate.org, and forandagainst.com. These portals all allow users to contribute pro and con arguments for given controversial issues, where the stance is explicitly indicated by the user, and therefore offers a source of data which can be reliably classified by this stance. In total 291,440 arguments were collected from these sources. While the args.me prototype only offers a simple interface for retrieving lists of pro and con arguments, it offers a framework for future expansion, improving the individual steps of retrieving, ranking and displaying arguments.

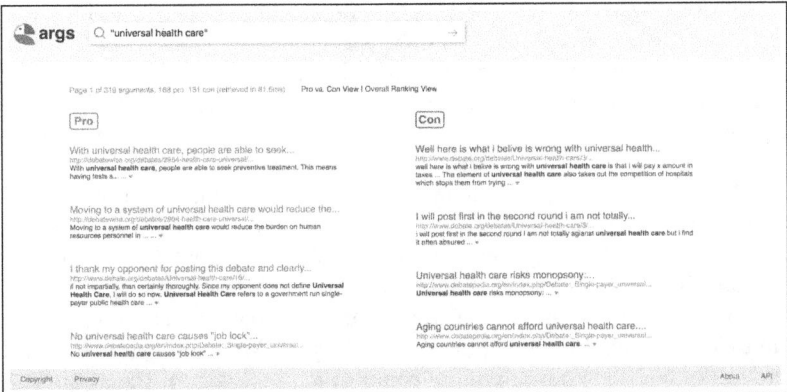

Figure 2: Args.me search results showing pro and con arguments on the topic of "universal health care".

Of similar size to the args.me index is the Argument Interchange Format (AIF) database, AIFdb [Lawrence *et al.*, 2012b], which contains over 28,000 argument maps, with over 3.3m words and 270,000 claims in more than twenty different languages. However, being based on AIF, AIFdb contains a significantly richer and more fine-grained representation of argument structure, where individual propositions are connected by specific argumentative relationships (e.g. support or conflict) with these relationships able to be further specified as instances of a given argumentation scheme. AIFdb offers three distinct search methods: a basic search functionality which matches the given text to propositions in the database; an advanced search functionality which offers the ability to narrow results by speaker, date, or argumentation scheme; and a web service interface allowing for direct queries to be performed on the underlying relational database. For the basic and advanced searches, results are returned as a list of matching elements with the ability to select any of these and view them in the context of their argument maps (see Figure 3). For the web service interface, results are returned in the JavaScript Object Notation (JSON) format, allowing for processing by software tools. AIFdb's native search interfaces do not perform ranking of the obtained results, however these have been combined with AIF *argument analytics* to give rankings by a number of measures including *Centrality*, which can be viewed as how important an issue is to the argument as a whole (calculated through eigenvector centrality, used in the Google Pagerank algorithm [Brin and Page, 1998]), and *Divisiveness*, which is used to assess how much an issue splits opinion (calculated based on how many other issues are in conflict with it and the amount of support which the two sides have) [Lawrence et al., 2017].

In addition to those tools mentioned above, a number of online debate portals (e.g. CreateDebate[1], Kialo[2] and PerspectroScope [Chen et al., 2019]) as well as argument mapping tools (e.g. DebateGraph[3] or Rationale Online[4]) offer users some minimal search functionality, however in each case this is limited to text searches showing results for either individual claims, or an argument map containing the specified text, and does not allow for any more complicated querying of the argument structure.

---

[1] https://www.createdebate.com
[2] https://www.kialo-edu.com
[3] http://debategraph.org
[4] https://www.rationaleonline.com/

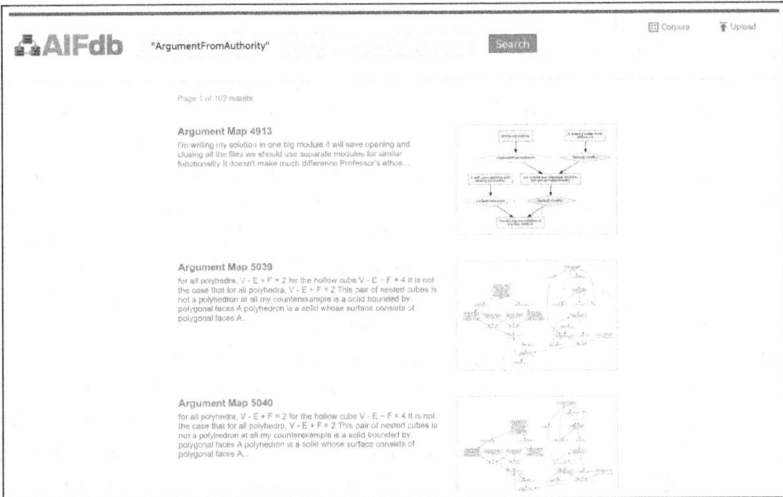

Figure 3: AIFdb advanced search results listing all matching instances of the *Argument from Authority* argumentation scheme.

3.2 Tools for searching unstructured data

Argument mining is the automatic identification and extraction of the structure of inference and reasoning expressed as arguments presented in natural language [Lawrence and Reed, 2020]. Whilst all applications of argument mining could be viewed as a form of argument search in that they find the argument structure contained within a given text, we constrain the discussion here to applications where either the argument components being mined match specific pre-determined search criteria, or where argument mining is applied in order to identify the full structure and some search technique is then used to return results based on this.

A prominent example in this first category can be found in much of IBM's work on Project Debater[5]. Debater can respond to a given topic by automatically constructing a set of relevant pro/con arguments phrased in natural language. For example, when asked for responses to the topic "The sale of violent video games to minors should

---

[5]https://www.research.ibm.com/artificial-intelligence/project-debater/

be banned", an early prototype of Debater scanned approximately 4 million Wikipedia articles and determined the ten most relevant articles, scanned all 3,000 sentences in those articles, detected sentences which contain candidate claims, assessed their pro and con polarity and then presented three relevant pro and con arguments[6], with more recent developments also working towards ranking and selecting the most convincing of these arguments [Gleize et al., 2019], expanding the topic of the debate [Bar-Haim et al., 2019], and providing "first principle" debate points, commonplace arguments which are relevant to many topics, where specific data is lacking [Bilu et al., 2019]. In [Levy et al., 2014], the challenge of searching for *Context Dependent Claims (CDCs)* in Wikipedia articles was first addressed, showing how, given a topic and a selection of relevant articles, a selection of "general, concise statements that directly support or contest the given topic" can be found. This work was followed in [Rinott et al., 2015] where finding supporting evidence from Wikipedia data for a given CDC was addressed. [Bar-Haim et al., 2017] introduced the task of claim stance classification, that is, detecting the target of a given CDC, and determining the stance towards that target. [Levy et al., 2017] further developed CDC identification, removing the need for pre-selected relevant articles, by first deriving a *claim sentence query* to retrieve CDCs from a large unlabelled corpus. Such large volumes of CDCs can be used both as potential points to be made by the debater system as well as to aid in the interpretation of spoken material containing breaks, repetitions, or other irregularities [Lavee et al., 2019].

Tools where argument mining is instead used as a component of argument search include ArgumenText [Stab et al., 2018], an argument search system capable of retrieving pro and con arguments relevant to a given topic from the English part of the CommonCrawl Web corpus [Patel, 2020]. ArgumenText first retrieves a list of documents relevant to a given topic (where a topic is considered as "some matter of controversy that can be concisely expressed through keywords") and then applies an argument mining model to identify the argument structure of the top-ranked documents; first classifying each document sentence as 'argument' or 'no-argument' with respect to the topic and then de-

---

[6] http://www.kurzweilai.net/introducing-a-new-feature-of-ibms-watson-the-debater

termining the stance (pro or con) of each topic-relevant argument. The results of a search on the topic of "self-driving cars" can be seen in Figure 4.

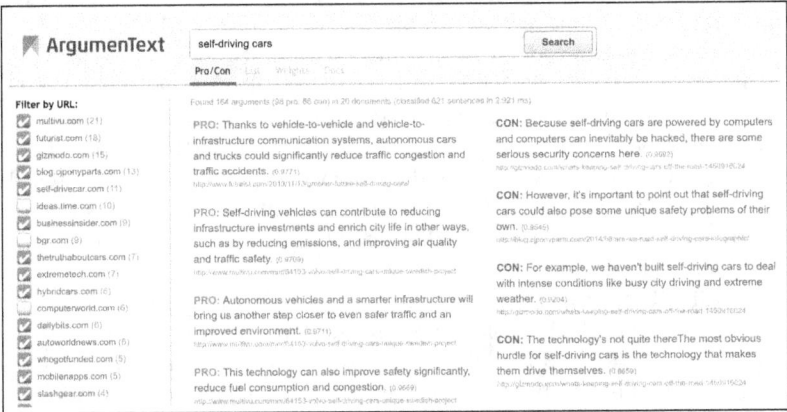

Figure 4: Argumentext search results showing arguments pro and con for the topic of "self-driving cars".

Similarly, DebateLab[7] was an HFRI-funded project that intended to pioneer research towards developing the theoretical machinery that can be used across diverse domains for representing, mining and reasoning with online arguments [Ymeralli et al., 2022]. It focused on journalistic articles written in the Greek language, and developed a suite of tools for article management[8], most importantly a portal where users can view, search, visualise and analyse journalistic articles, the arguments they contain and their relationships. The articles and their arguments are automatically retrieved using a crawler and an argument mining module, while a Named Entity Recognition (NER) tool allows connecting entities mentioned in articles and arguments with online resources (e.g., wikipedia pages) that describe these entities. DebateLab supports various advanced searching functionalities, that employ, in the backend, the ArgQL language (see [Zografistou et al., 2017], [Zografistou et al., 2018],

---

[7]https://debatelab.ics.forth.gr/
[8]https://isl.ics.forth.gr/debatelab_portal/

[Roussakis *et al.*, 2022] and also Section 2 in this Chapter), as well as advanced visualisations such as sunburst or tree-like views.

TARGER [Chernodub *et al.*, 2019] is an open source system combining argument mining techniques for tagging arguments in free text, with retrieval of arguments matching a given query. TARGER uses a pre-processing step to identify argumentative units and classify them as claims or premises. This pre-processing is carried out on the DepCC corpus [Panchenko *et al.*, 2018] to tag and store argument unit information in a web-scale index. The search component of TARGER allows the user to enter a keyword query and choose whether it should be matched in claims, premises, etc. Every retrieved result is rendered as a text fragment, with color-coded highlighting of each component's role in the overall argument structure.

DISPUTool [Haddadan *et al.*, 2019], [Goffredo *et al.*, 2023] allows the exploration and identification of argumentative components over political debates, in an automated manner, using argument mining technologies. It has been applied to analyse political debates from the presidential campaigns of the USA since 1960, and is framed as a tool to support humanities' scholars in exploring and analysing textual political debates. The argument mining component serves as a backend to a tool[9] that provides different functionalities on the analysed debates, including, among others, the (visual) exploration of claims, premises and their relationships, as well as the identification of named entities and fallacies.

4 Closed debating systems

In this section, we present several tools that have been used for debating, i.e., tools which allow the creation, management and/or analysis of debates. As with the rest of this chapter, the emphasis here is on the argument exploration and searching functionalities of the presented tools, many of which also allow various sophisticated visualisations and analytics. Our focus in this section lies on tools whose main functionality is the management and/or analysis of debates that have been created

---

[9]https://disputool.uni.lu/, newer version in: https://3ia-demos.inria.fr/disputool/.

within the tool; this is in contrast to Section 3, which focused on systems that provide searching, analytics and exploration capabilities for arguments expressed elsewhere and searched by the tool, or retrieved by the tool for further analysis.

The term "debating" is often used in a very broad sense to include various types of textual interaction among users. Therefore, we decided to provide some criteria (requirements) as to which tools are candidates for inclusion in this section. These requirements are the following:

Frontend The tool should provide a frontend, allowing users to create, manage, search, explore and/or interact with arguments.

Argument creation The tool should provide functionalities allowing the expression of opinions or arguments, either as free text, or using some semi-structured format.

Reactions to arguments The tool should allow reactions (replies) to other people's opinions in the form of new opinions that attack or support the original. We are not interested in other types of replies (e.g., responses to reviews that are used in standard reviewing tools, or the type of responses used in question-answering platforms like stackoverflow[10]). Votes and other types of structured reactions are possible, but not mandatory.

Argument search and navigation The tool should allow the user to search arguments, navigate through them and explore their contents. As this is the main focus of this chapter (and section), we prefer tools that provide advanced argument searching and navigation/exploration capabilities, including functionalities that allow analytics, advanced visualisations, and other sense-making features.

It should be noted that, even under this restrictive understanding, there are still numerous tools that fit this description and are relevant for this section. Thus, we had to make a selection, and only consider the most popular or significant ones. Importantly, some of the older systems that are presented here are no longer supported, and their URLs

---
[10] https://stackoverflow.com/

are non-functional; still, we decided to include them, on the basis of their importance in shaping the current state-of-the-art in the area. All the chosen systems are presented in Subsection 4.2 below. Before that, in Subsection 4.1, we present a number of concepts that are employed by the presented systems and may be useful for understanding their functionality.

4.1 Debating models

In this section, we present three models for debating in the context of decision-making that are employed by some of the tools to be presented later (in Subsection 4.2).

The *Issue-Based Information Systems* (*IBIS*) model, proposed in [Kunz and Rittel, 1970], is employed by many of the tools below. IBIS is a model, based on argumentation, that is used to help users obtain a better understanding of so-called "wicked problems", i.e., problems that are complex, ill-defined and involve multiple stakeholders. IBIS, as the name implies, is based on *issues*, which are questions that need to be answered by *positions*. Positions can be elaborated upon using *pro and con arguments*, i.e., arguments that support or object the respective position. Arguments themselves can be associated with other pro and con arguments that support or object the argument. Eventually, this creates an acyclic graph of arguments, rooted on an issue and including one or more positions and their pro/con arguments (and their pro/con arguments recursively). This graphical representation can be used to support the dialectical process and to improve sense-making in wicked problems.

The *Quantitative Argumentation Debate* (*QuAD*) framework [Baroni et al., 2013; Baroni et al., 2015] can be seen as the formal computational counterpart of the IBIS model. A QuAD framework supports issues, positions and arguments, in the same manner as IBIS, the only exception being that, in QuAD, issues cannot be directly linked to arguments. Moreover, QuAD supports a numerical score assigned to each argument, which represents its strength, or importance, according to the domain experts. This score can be used for an automated quantitative evaluation of the debate, and the determination of the "winning" positions or arguments. For this, different evaluation methods have been proposed in the literature, including, indicatively, [Leite and Martins, 2011; Egilmez *et al.*, 2013; Patkos *et al.*, 2016b; Patkos *et al.*, 2016a] and others.

Decision matrices [Pugh, 1991] is another way to visualise wicked problems in order to make well-informed decisions. The model is based on a matrix, whose rows correspond to evaluation criteria, and whose columns correspond to different options. The possible options are evaluated against the criteria, resulting in a positive, negative or indifferent evaluation. The options are then assessed against the criteria, taking into account weights that represent the importance of each different criterion. Note that this allows an automated quantitative evaluation of the alternative options.

4.2 List of tools

In this subsection we present some of the most important debating tools that have appeared online or in the relevant literature. The tools are organised in groups based on their features, functionality and similarities. Recall that, according to our desiderata set forth in the beginning of this section, we are listing tools providing an appropriate *frontend* that supports *argument creation*, *reactions to arguments* and *argument search and navigation* capabilities.

Tools based on the IBIS model. Many of the tools that support debating are based on the IBIS model, or its computational counterpart, QuAD, described in Subsection 4.1. As a matter of fact, almost all of the academic tools that will be presented here are based on this model. Below, we present some of these tools.

APOPSIS[11] provides an IBIS-based platform to support decision-making dialogues [Ymeralli *et al.*, 2017]. APOPSIS allows users to enter arguments in a structured form, as well as to react, in different ways, to previously-submitted arguments. It proposes a two-phase discussion process: in the first phase, all possible ideas for solutions (called *positions*) to the *issue* under discussion are gathered, and the most promising of them (based on users' reactions) are selected; in

---

[11]https://demos.isl.ics.forth.gr/apopsis/

the second phase, no further positions can be introduced, and discussion focuses on those selected positions, in order to understand better their pros and cons. The process is supported by an automated argument evaluation procedure, based on sm-dice [Patkos *et al.*, 2016a; Patkos *et al.*, 2016b], as well as by sophisticated analytics and visualisations for identifying trends and patterns related to users, their characteristics, and their expressed positions and arguments.

DesignVUE [Aurisicchio and Bracewell, 2013; Baroni *et al.*, 2013; Baroni *et al.*, 2015] was a decision-support system based on the IBIS and QuAD models. It supported the creation of IBIS debates and their automated evaluation. Unfortunately, its online version is no longer available.

A very similar tool developed by the same group was Quaestio-it [Evripidou and Toni, 2014], which used to be a popular system for creating, browsing, analysing and visualising arguments. The website of quaestio-it[12] is no longer maintained, and, to the best of our knowledge, the tool is no longer available. Quaestio-it allowed both attacking and supporting arguments.

Arg&Dec [Aurisicchio *et al.*, 2015] (standing for "Argue & Decide") was the third tool by the same group (developed after DesignVUE and Quaestio-it), with similar, but more advanced, functionalities. Unlike the other two, it is still supported and available as a web application[13]. Arg&Dec aims to assist collaborative decision-making through debating and argumentation. It is free to use (sign-up required), and supports collaborative work. Emphasis is placed on the mode of interaction (and decision-making) and two modes are supported, one based on QuAD frameworks and one based on Decision Matrices (see Subsection 4.1 for details on those decision-making models). Under the QuAD mode of interaction, users can create and edit different node types and (pro/con) links between them, as provisioned by the QuAD model. Under the Decision Matrices model, the users can create the matrix's rows and columns (evaluation criteria and options respectively), edit the weights of rows and the generated cells. The two models are interchangeable, in the sense that a Decision Matrix can be transformed to a QuAD and vice-

---

[12] www.quaestion-it.com
[13] http://arganddec.com/

versa. Under both modes, the system provides a visual representation of the input, and is able to automatically evaluate the debate and propose prevailing options, along with a numerical score. To further support the decision-maker, natural language explanations regarding option ranking are provided, including a speech synthesis functionality.

COLLAGREE [Ito *et al.*, 2015] is an open web-based forum system aiming to allow large-scale discussion for the purposes of agreeing on a certain matter (consensus-building). The system employs facilitators, whose aim is to coordinate, lead, integrate, classify, and summarise discussions towards consensus. Emphasis is placed on helping facilitators moderate discussions among many people. The discussions are based on published issues (in the sense of IBIS), but no strict structure on the discussion is imposed, and people can submit their opinions as plain text. Gamification techniques are employed to encourage participation. Keyword-based search facilities are provided, as well as some functionalities for sense-making (e.g., sentiment analysis, automated keyword extraction) and simple visualisations. The system has been employed in large-scale field experiments, e.g., for an internet-based town meeting in Nagoya, Japan [Ito *et al.*, 2014] and in the Aichi prefecture, Japan [Ito, 2021].

D-Agree [Ito *et al.*, 2019], [Ito *et al.*, 2022], [Ito, 2021] is a system very similar to COLLAGREE, and can be viewed as its continuation in many aspects. Just like COLAGREE, D-Agree uses the IBIS discussion model, gamification techniques, keyword-based searching facilities and visualisations based on a tag cloud. However, D-Agree adds a critical component compared to COLLAGREE: acknowledging the difficulty of the moderation task in large-scale discussions, D-Agree employs an automated facilitation agent that uses deep learning and natural language processing techniques to support the discussion. In particular, the facilitator agent is used to capture meaningful sentences and extract the discussion structure from the texts posted by users in discussions (i.e., identifying issues, positions and arguments in the posted free text), so that users' interaction is more natural, while at the same time adhering to the IBIS model. Importantly, the automated facilitator also prompts the users for additional information, keeping the discussion alive. D-Agree was also tested in a field test experiment in Nagoya, Japan, as well as in small-scale controlled experiments. The authors' experimental results show that the use of the automated facilitator agent led to a more lively discussion compared to the scenario of using human facilitators.

Deliberatorium[14] is a rather old, but still active web-based system for enabling people have productive discussions about complex (so-called wicked) problems [Gurkan et al., 2010]. As in other tools in this subsection, Deliberatorium follows the IBIS model. It organises contributions by topic, rather than submission time, using argumentation maps, and ensures that cycles and repetitions are kept to a minimum. Towards this, authors of arguments and other contributions need to follow certain rules, while moderators are employed to ensure that these rules are adhered to. In terms of argument search, the functionalities are really basic: some standard keyword-based searching facilities are supported, and the discussions are organised in a tree-based structure inspired by the IBIS model, facilitating sense-making.

Other online debating platforms. Various online debating platforms have emerged from a non-academic setting (e.g., Kialo, CreateDebate). These platforms aim to allow online communities to engage in discussions and express their opinions. Their business models vary, as will be seen below.

Kialo[15] is a website run by Kialo Inc., with the aim to "cut through the noise typically associated with social and online media, making it easy to engage in focused discussion". It supports the creation of and participation in debates on any conceivable topic. It provides numerous features for organising the ensuing dialogues and the involved arguments, allowing easy sense-making of lengthy dialogues, as well as different visualisations of the debate in the form of an argument tree or sunburst diagram. After login, users can create their own arguments (or debates), as well as respond and/or react to other users' arguments in different ways. Importantly, Kialo allows claims to be associated with external sources (e.g., scientific reports or articles) that back them up. A custom version of this platform for educators[16] has been used for class discus-

---

[14] https://deliberatorium.mit.edu/
[15] https://www.kialo.com/
[16] https://www.kialo-edu.com

sion and critical thinking and reasoning, teaching logical fallacies etc. Kialo is a very interesting success story, as it contains thousands of debates and arguments. The website is free to use, and does not display advertisements or sell users' data. Therefore, it produces no revenue for its maintainer. Instead, the website is used as a "demonstrator" of Kialo's capabilities, aiming to sell those functionalities to companies as a deliberation and decision-making tool.

CreateDebate[17] is an online platform that allows users to post a dilemma, in the form of a statement, and to provoke a debate over the truth of this statement. The debates consist of textual arguments that people post as a response to the original statement. An argument can be further supported or disputed by other users using their own arguments, and users can also ask for clarifications over an argument. The debate is organised into pro and con "sides", and the system automatically computes a score for each "side", encouraging participation in order to "win" the argument. It also provides some limited forms of visualisation allowing users to better understand the flow of the debate. The system's revenue model seems to be ad-based, and, in fact, the system is often being misused for ad-related postings that are irrelevant to its stated purpose. Nevertheless, it constitutes one of the first systems that attempted to create a global debating forum of users, aiming at a providing a podium for publicly expressing one's opinion on important matters.

A similar tool allowing online debates is the web-based discussion software *Arvina* [Lawrence *et al.*, 2012a]. Unlike the tools mentioned earlier in this categoty, Arvina emerged from academic work. Arvina allows participants to debate a range of topics in real-time in a way that is structured but at the same time unobtrusive. Arvina maintains flexibility in dialogue structure by using protocols written using the Dialogue Game Description Language (DGDL) [Bex *et al.*, 2014] to structure the discussion between participants. Such protocols determine which types of moves can be made (e.g. questioning, claiming, etc.), when these moves can be made (e.g. a dialogue starts with a claim; question moves can only made in the turn directly following a claim; etc.), and describe how each move updates the argument structure of the discussion taking

---

[17]https://www.createdebate.com/

place.

Arvina can support multiple human users interacting in the same dialogue, as well as incorporating software agents representing (the arguments of) specific authors who have their opinions stored in AIFdb [Lawrence et al., 2012b]. So, for example, say that a user has constructed a complex, multi-layered argument using the OVA argument analysis tool [Janier et al., 2014], concerning the use of nuclear weapons. An agent representing this user can then be added to an Arvina discussion and questioned about these opinions, with the agent answering by giving the user's pre-annotated opinions.

Tools employing structured arguments. All the debating tools presented above support free-text, unstructured arguments. Thus, the structure they impose lies on the debate level, rather than the argument level. Two debating tools, Gorgias-B [Spanoudakis et al., 2021; Spanoudakis et al., 2022], and Carneades [Walton and Gordon, 2012], [Walton and Gordon, 2017] stand out from the rest in the sense that they employ structured arguments, which allow them to provide more sophisticated analysis and reasoning functionalities. We present them below.

Gorgias-B [Spanoudakis et al., 2021; Spanoudakis et al., 2022] is a reasoning and argumentation tool[18], developed through a series of publications, and based on an earlier tool called Gorgias[19]. The stated purpose of Gorgias-B is to allow users to make informed decisions over a set of options, based on a thorough modelling of the underlying situation. Gorgias-B does not take as input free-text arguments; instead, it prompts registered users to define facts (predicates) and to connect them to form structured arguments (logical rules) and counter-arguments (or exceptions). Preferences, priorities and default options can also be defined graphically. In this sense, it can be seen as a no-code programming tool for declarative programming, with argumentation semantics. Once the domain has been modelled, specific scenarios (i.e., situations in which certain facts are true and others are false) can be input to the system in order for that scenario to be evaluated (reasoned upon) automatically

---

[18]http://gorgiasb.tuc.gr/
[19]http://www.cs.ucy.ac.cy/~nkd/gorgias/

by the system based on the domain knowledge. Explanations over the system's output are also provided.

Carneades [Walton and Gordon, 2012], [Walton and Gordon, 2017] is a tool[20] for representing and evaluating arguments, currently in its fourth version. Carneades uses a graph-based representation of arguments, where each node of the graph can be either a statement or an argument. Statements represent propositions that can be true or false, and function as premises or conclusions of arguments. The premises and conclusions of a given argument are determined via links (edges) between the respective argument node and the statements that represent its premises and conclusions, thereby explicitly representing the structure of an argument. Carneades is heavily based on previous work by Walton on argumentation schemes [Macagno et al., 2018], which are used for reasoning, argument evaluation, as well as for argument invention, i.e., the construction of arguments from statements known to be true. In terms of interaction with the user, Carneades supports a visual display of arguments, which is based on the user's specification of the statements considered true and on the ensuing evaluation which determines the (non-)acceptable arguments.

5 Conclusion

The evolution of the Web from a unidirectional information conduit to an interactive platform for global discourse has redefined the nature of user engagement, with the capacity for active contribution, debate, and exchange of ideas permeating every facet of online communication. Arguments, discussions, and viewpoints proliferate across diverse platforms, from social media to specialised forums, embodying a rich collection of perspectives and insights. Yet, amidst this expansive realm of user-generated content lies the challenge of navigating and accessing pertinent arguments efficiently.

As the volume of opinions and arguments grows, the significance of tools for discerning relevant content becomes paramount. Traditional keyword-based searches, while helpful, fall short when confronted with the intricacies of argumentative content. Arguments possess inherent

---

[20]https://carneades.github.io/

structure and interconnectedness, demanding nuanced search capabilities beyond simplistic keyword queries. The quest for suitable tools intertwines with the necessity to encompass the structural complexities inherent in debates, prompting the exploration of specific tools for creating, navigating and searching online debates. In this chapter we have explored three fundamental areas: structured argument query languages, argument retrieval systems, and closed debating systems. Although substantial advances have been made in each of these, it is evident that there is still much to be done to improve their utility and reach.

5.1 Future directions

Existing argument search tools are often limited in both their scope and functionality, relying on pre-structured material and limiting the results returned to either related documents, or lists of reasons pro and con. Advances in Argument Mining may help to address the first of these issues, opening up unstructured text to the same search techniques currently available for structured argument data. In parallel to broadening the scope of data which can be searched, improvements can also be made in the types of search available, from simple keyword search identifying related arguments, to the ability for a user to provide a (partially specified) argument graph as input and return arguments whose structure corresponds to the input graph.

Argument search can be broken down into a number of individual component steps: retrieval of relevant documents, identification of target argument components within these documents, ranking of components, and display of the identified results. While the first of these steps, retrieval of relevant documents, is an essential task, it largely relies on improvements in standard information retrieval techniques, and does not strictly depend upon any argument structure. The remaining three tasks on the other hand are all areas in which argument search tools could be directly improved. In terms of identifying target argument components within documents, not only can accuracy be improved by ongoing advances in argument mining, but as these techniques mature, the possibilities for more detailed queries grow. For example, a user could specify a particular graph as input and the search tool could return arguments whose structure corresponds to the input graph. Similar advances are also possible in improving the ranking of results. For example, incorporating elements of Argument Analytics [Lawrence et al., 2016] to supplement relevance with factors such as criticality or divisiveness of returned arguments. Finally, although some search tools offer visualisations of results, many just display lists of pro and con points. In these cases, there is considerable potential for incorporating richer visualisations which place arguments in their broader context and allow for fluid exploration of related points.

Whilst current debating systems allow for creation, management and analysis of debates, these tools almost exclusively represent isolated platforms lacking in interoperability and wider adoption. Future work in this area needs to address this issue, developing shared underlying concepts and a unified data fabric, as well as providing opportunities for integration with existing online platforms. The ability for a user to highlight a span of text anywhere on the Web, see reasons for and against from a broad range of sources, and provide their own structured response, would move such tools beyond niche communities and open them up to a global audience.

References

[Aurisicchio and Bracewell, 2013] Marco Aurisicchio and Robert Bracewell. Capturing an integrated design information space with a diagram-based approach. *Journal of Engineering Design*, 24(6):397–428, 2013.

[Aurisicchio et al., 2015] Marco Aurisicchio, Pietro Baroni, Dario Pellegrini, and Francesca Toni. Comparing and integrating argumentation-based with matrix-based decision support in arg&dec. In Elizabeth Black, Sanjay Modgil, and Nir Oren, editors, *Theory and Applications of Formal Argumentation*, pages 1–20, Cham, 2015. Springer International Publishing.

[Bar-Haim et al., 2017] Roy Bar-Haim, Indrajit Bhattacharya, Francesco Dinuzzo, Amrita Saha, and Noam Slonim. Stance classification of context-dependent claims. In *Proceedings of the 15th Conference of the European Chapter of the Association for Computational Linguistics: Volume 1, Long Papers*, volume 1, pages 251–261, Valencia, Spain, 2017.

[Bar-Haim et al., 2019] Roy Bar-Haim, Dalia Krieger, Orith Toledo-Ronen, Lilach Edelstein, Yonatan Bilu, Alon Halfon, Yoav Katz, Amir Menczel, Ranit Aharonov, and Noam Slonim. From surrogacy to adoption; from bitcoin to cryptocurrency: Debate topic expansion. In *Proceedings of the 57th*

Annual Meeting of the Association for Computational Linguistics, pages 977–990, Florence, Italy, 2019. Association for Computational Linguistics.

[Baroni et al., 2013] P. Baroni, M. Romano, F. Toni, M. Aurisicchio, and G. Bertanza. An argumentation-based approach for automatic evaluation of design debates. In *CLIMA XIV: Proceedings of the 14th International Workshop on Computational Logic in Multi-Agent Systems - Volume 8143*, pages 340–256, 2013.

[Baroni et al., 2015] P. Baroni, M. Romano, F. Toni, M. Aurisicchio, and G. Bertanza. Automatic evaluation of design alternatives with quantitative argumentation. *Argument and Computation*, 6:24–29, 2015.

[Bex et al., 2014] Floris Bex, John Lawrence, and Chris Reed. Generalising argument dialogue with the dialogue game execution platform. In S. Parsons, N. Oren, C. Reed, and F. Cerutti, editors, *Proceedings of the Fifth International Conference on Computational Models of Argument (COMMA 2014)*, pages 141–152, Pitlochry, Scotland, 2014. IOS Press.

[Bilu et al., 2019] Yonatan Bilu, Ariel Gera, Daniel Hershcovich, Benjamin Sznajder, Dan Lahav, Guy Moshkowich, Anael Malet, Assaf Gavron, and Noam Slonim. Argument invention from first principles. In *Proceedings of the 57th Annual Meeting of the Association for Computational Linguistics*, pages 1013–1026, Florence, Italy, 2019. Association for Computational Linguistics.

[Bondarenko et al., 2020] Alexander Bondarenko, Maik Fröbe, Meriem Beloucif, Lukas Gienapp, Yamen Ajjour, Alexander Panchenko, Chris Biemann, Benno Stein, Henning Wachsmuth, Martin Potthast, and Matthias Hagen. Overview of Touché 2020: Argument Retrieval. In *Experimental IR Meets Multilinguality, Multimodality, and Interaction. 11th International Conference of the CLEF Association (CLEF 2020)*, volume 12260 of *Lecture Notes in Computer Science*, pages 384–395, Berlin Heidelberg New York, September 2020. Springer.

[Brin and Page, 1998] Sergey Brin and Lawrence Page. The anatomy of a large-scale hypertextual web search engine. *Computer networks and ISDN systems*, 30(1-7):107–117, 1998.

[Cabrio et al., 2013] Elena Cabrio, Serena Villata, and Fabien Gandon. A support framework for argumentative discussions management in the web. In *The Semantic Web: Semantics and Big Data: 10th International Conference, ESWC 2013, Montpellier, France, May 26-30, 2013. Proceedings 10*, pages 412–426. Springer, 2013.

[Chen et al., 2019] Sihao Chen, Daniel Khashabi, Chris Callison-Burch, and Dan Roth. PerspectroScope: A window to the world of diverse perspectives. In Marta R. Costa-jussà and Enrique Alfonseca, editors, *Proceedings of the*

57th Annual Meeting of the Association for Computational Linguistics: System Demonstrations, pages 129–134, Florence, Italy, July 2019. Association for Computational Linguistics.

[Chernodub et al., 2019] Artem Chernodub, Oleksiy Oliynyk, Philipp Heidenreich, Alexander Bondarenko, Matthias Hagen, Chris Biemann, and Alexander Panchenko. Targer: Neural argument mining at your fingertips. In *Proceedings of the 57th Annual Meeting of the Association for Computational Linguistics: System Demonstrations*, pages 195–200, 2019.

[Chesñevar et al., 2006] Carlos Chesñevar, Jarred McGinnis, Sanjay Modgil, Iyad Rahwan, Chris Reed, Guillermo Simari, Matthew South, Gerard Vreeswijk, and Steven Willmott. Towards an Argument Interchange Format. *The Knowledge Engineering Review*, 21(04):293–316, 2006.

[Cimiano et al., 2013] Philipp Cimiano, Oscar Corcho, Valentina Presutti, Laura Hollink, and Sebastian Rudolph. *The Semantic Web: Semantics and Big Data: 10th International Conference, ESWC 2013, Montpellier, France, May 26-30, 2013. Proceedings*, volume 7882. Springer, 2013.

[Dung, 1995] P. M. Dung. On the acceptability of arguments and its fundamental role in nonmonotonic reasoning, logic programming and n-person games. *Artificial Intelligence*, 77(2):321–357, 1995.

[Egilmez et al., 2013] Sinan Egilmez, João G. Martins, and João Leite. Extending social abstract argumentation with votes on attacks. In Elizabeth Black, Sanjay Modgil, and Nir Oren, editors, *Theory and Applications of Formal Argumentation - Second International Workshop, TAFA 2013, Beijing, China, August 3-5, 2013, Revised Selected papers*, volume 8306 of *Lecture Notes in Computer Science*, pages 16–31. Springer, 2013.

[Evripidou and Toni, 2014] Valentinos Evripidou and Francesca Toni. Quaestio-it.com: a social intelligent debating platform. *Journal of Decision Systems*, 23(3):333–349, 2014.

[Gleize et al., 2019] Martin Gleize, Eyal Shnarch, Leshem Choshen, Lena Dankin, Guy Moshkowich, Ranit Aharonov, and Noam Slonim. Are you convinced? choosing the more convincing evidence with a Siamese network. In *Proceedings of the 57th Annual Meeting of the Association for Computational Linguistics*, pages 967–976, Florence, Italy, 2019. Association for Computational Linguistics.

[Goffredo et al., 2023] Pierpaolo Goffredo, Elena Cabrio, Serena Villata, Shohreh Haddadan, and Jhonatan Sanchez. Disputool 2.0: A modular architecture for multi-layer argumentative analysis of political debates. *Proceedings of the AAAI Conference on Artificial Intelligence*, 37:16431–16433, 06 2023.

[Gurkan et al., 2010] Ali Gurkan, Luca Iandoli, Mark Klein, and Giuseppe

Zollo. Mediating debate through on-line large-scale argumentation: Evidence from the field. *Information Sciences*, 180:3686–3702, 2010.

[Haddadan et al., 2019] Shohreh Haddadan, Elena Cabrio, and Serena Villata. Disputool – a tool for the argumentative analysis of political debates. In *Proceedings of the Twenty-Eighth International Joint Conference on Artificial Intelligence, IJCAI-19*, pages 6524–6526. International Joint Conferences on Artificial Intelligence Organization, 7 2019.

[Ito et al., 2014] Takayuki Ito, Yuma Imi, Takanori Ito, and Eizo Hideshima. COLLAGREE: Facilitator-mediated large-scale consensus support system. In *Proceedings of the ACM Collective Intelligence Conference Series*, 2014.

[Ito et al., 2015] Takayuki Ito, M. Okumura, and E. Hideshima. Implementation of a large-scale discussion support system collagree - large-scale discussion support based on a weakly structured discussion process. *Journal of Japan Industrial Management Association*, 66:83–108, 01 2015.

[Ito et al., 2019] T. Ito, D. Shibata, S. Suzuki, N. Yamaguchi, T. Nishida, K. Hiraishi, and K. Yoshino. Agent that facilitates crowd discussion. In *Proceedings of the ACM Collective Intelligence Conference Series*, 2019.

[Ito et al., 2022] Takayuki Ito, Rafik Hadfi, and Shota Suzuki. An agent that facilitates crowd discussion. *Group Decision and Negotiation*, 31:621–647, 2022.

[Ito, 2021] Takayuki Ito. *Discussion and Negotiation Support for Crowd-Scale Consensus*, pages 371–393. Springer International Publishing, Cham, 2021.

[Janier et al., 2014] Mathilde Janier, John Lawrence, and Chris Reed. OVA+: An argument analysis interface. In S. Parsons, N. Oren, C. Reed, and F. Cerutti, editors, *Proceedings of the Fifth International Conference on Computational Models of Argument (COMMA 2014)*, pages 463–464, Pitlochry, Scotland, 2014. IOS Press.

[Kunz and Rittel, 1970] W. Kunz and H. Rittel. *Issues as elements of information systems. Working Paper 131*. Institute of Urban and Regional Development, University of California, Berkeley, California, 1970.

[Lavee et al., 2019] Tamar Lavee, Matan Orbach, Lili Kotlerman, Yoav Kantor, Shai Gretz, Lena Dankin, Michal Jacovi, Yonatan Bilu, Ranit Aharonov, and Noam Slonim. Towards effective rebuttal: Listening comprehension using corpus-wide claim mining. In *Proceedings of the 6th Workshop on Argument Mining*, pages 58–66, Florence, Italy, 2019. Association for Computational Linguistics.

[Lawrence and Reed, 2020] John Lawrence and Chris Reed. Argument mining: A survey. *Computational Linguistics*, 45(4):765–818, 2020.

[Lawrence et al., 2012a] John Lawrence, Floris Bex, and Chris Reed. Dialogues on the argument web: Mixed initiative argumentation with arvina. In *Proceedings of the 4th International Conference on Computational Models of Argument (COMMA 2012)*, pages 513–514, Vienna, Austria, 2012. IOS Press.

[Lawrence et al., 2012b] John Lawrence, Floris Bex, Chris Reed, and Mark Snaith. AIFdb: Infrastructure for the argument web. In *Proceedings of the Fourth International Conference on Computational Models of Argument (COMMA 2012)*, pages 515–516, Vienna, Austria, 2012. IOS Press.

[Lawrence et al., 2016] J. Lawrence, R. Duthie, K. Budzysnka, and C.A. Reed. Argument analytics. In P. Baroni, M. Stede, and T. Gordon, editors, *Proceedings of the Sixth International Conference on Computational Models of Argument (COMMA 2016)*, Berlin, 2016. IOS Press.

[Lawrence et al., 2017] John Lawrence, Mark Snaith, Barbara Konat, Katarzyna Budzynska, and Chris Reed. Debating technology for dialogical argument: Sensemaking, engagement, and analytics. *ACM Transactions on Internet Technology*, 17(3):24:1–24:23, 2017.

[Leite and Martins, 2011] João Leite and João G. Martins. Social abstract argumentation. In Toby Walsh, editor, *IJCAI 2011, Proceedings of the 22nd International Joint Conference on Artificial Intelligence, Barcelona, Catalonia, Spain, July 16-22, 2011*, pages 2287–2292. IJCAI/AAAI, 2011.

[Levy et al., 2014] Ran Levy, Yonatan Bilu, Daniel Hershcovich, Ehud Aharoni, and Noam Slonim. Context dependent claim detection. In *Proceedings of the 25th International Conference on Computational Linguistics*, pages 1489–1500, Dublin, Ireland, 2014.

[Levy et al., 2017] Ran Levy, Shai Gretz, Benjamin Sznajder, Shay Hummel, Ranit Aharonov, and Noam Slonim. Unsupervised corpus–wide claim detection. In *Proceedings of the 4th Workshop on Argument Mining*, pages 79–84, Copenhagen, Denmark, 2017. Association for Computational Linguistics.

[Macagno et al., 2018] Fabrizio Macagno, Douglas Walton, and Chris Reed. *Argumentation Schemes*, pages 517–574. 08 2018.

[Panchenko et al., 2018] Alexander Panchenko, Eugen Ruppert, Stefano Faralli, Simone P. Ponzetto, and Chris Biemann. Building a web-scale dependency-parsed corpus from CommonCrawl. In *Proceedings of the Eleventh International Conference on Language Resources and Evaluation (LREC 2018)*, Miyazaki, Japan, May 2018. European Language Resources Association (ELRA).

[Patel, 2020] Jay M. Patel. *Introduction to Common Crawl Datasets*, pages 277–324. Apress, Berkeley, CA, 2020.

[Patkos et al., 2016a] Theodore Patkos, Antonis Bikakis, and Giorgos Flouris. A multi-aspect evaluation framework for comments on the social web. In *Proceedings of the 15th International Conference on Principles of Knowledge*

Representation and Reasoning (KR-16), Short Paper, 2016.

[Patkos *et al.*, 2016b] Theodore Patkos, Giorgos Flouris, and Antonis Bikakis. Symmetric multi-aspect evaluation of comments (extended abstract). In *Proceedings of the 22nd European Conference on Artificial Intelligence (ECAI-16), Short Paper*, 2016.

[Pugh, 1991] S. Pugh. *Total Design: Integrated Methods for Successful Product Engineering*. Addison-Wesley, Wokingham, 1991.

[Rinott *et al.*, 2015] Ruty Rinott, Lena Dankin, Carlos Alzate Perez, Mitesh M Khapra, Ehud Aharoni, and Noam Slonim. Show me your evidence-an automatic method for context dependent evidence detection. In *Proceedings of the 2015 Conference on Empirical Methods in Natural Language Processing*, pages 440–450, Lisbon, Portugal, 2015.

[Roussakis *et al.*, 2022] Yannis Roussakis, Giorgos Flouris, Dimitra Zografistou, and Elisjana Ymeralli. Extending the ArgQL specification. In *Proceedings of the 6th International Joint Conference on Rules and Reasoning (RuleML+RR-22), 16th International Rule Challenge session*, 2022.

[Spanoudakis *et al.*, 2021] N. Spanoudakis, K. Kostis, and K. Mania. "webgorgias-b": Argumentation for all. In *Proceedings of the 13th International Conference on Agents and Artificial Intelligence (ICAART 2021)*, 2021.

[Spanoudakis *et al.*, 2022] N. Spanoudakis, G. Gligoris, A.C. Kakas, and A. Koumi. Gorgias cloud: On-line explainable argumentation (system demonstration). In *9th International Conference on Computational Models of Argument (COMMA 2022)*, 2022.

[Stab *et al.*, 2018] Christian Stab, Johannes Daxenberger, Chris Stahlhut, Tristan Miller, Benjamin Schiller, Christopher Tauchmann, Steffen Eger, and Iryna Gurevych. Argumentext: Searching for arguments in heterogeneous sources. In *Proceedings of the 2018 conference of the North American chapter of the association for computational linguistics: demonstrations*, pages 21–25, 2018.

[Wachsmuth *et al.*, 2017] Henning Wachsmuth, Martin Potthast, Khalid Al Khatib, Yamen Ajjour, Jana Puschmann, Jiani Qu, Jonas Dorsch, Viorel Morari, Janek Bevendorff, and Benno Stein. Building an argument search engine for the web. In *Proceedings of the 4th Workshop on Argument Mining*, pages 49–59, 2017.

[Walton and Gordon, 2012] Douglas Walton and Thomas F. Gordon. The carneades model of argument invention. *Pragmatics & Cognition*, 20:1–31, 2012.

[Walton and Gordon, 2017] Douglas Walton and Thomas F. Gordon. Argument invention with the carneades argumentation system. *SCRIPTed*, 14, 2017.

[Ymeralli et al., 2017] Elisjana Ymeralli, Giorgos Flouris, Theodore Patkos, and Dimitris Plexousakis. APOPSIS: A web-based platform for the analysis of structured dialogues. In *Proceedings of the 16th International Conference on Ontologies, DataBases, and Applications of Semantics (ODBASE-17)*, 2017.

[Ymeralli et al., 2022] Elisjana Ymeralli, Giorgos Flouris, Vasilis Efthymiou, Katerina Papantoniou, Theodore Patkos, Georgios Petasis, Nikiforos Pittaras, Giannis Roussakis, and Elias Tzortzakakis. Representing online debates in the context of e-journalism. In *Proceedings of the 16th International Conference on Advances in Semantic Processing (SEMAPRO-22), Best Paper Award*, 2022.

[Zografistou et al., 2017] Dimitra Zografistou, Giorgos Flouris, and Dimitris Plexousakis. ArgQL: A declarative language for querying argumentative dialogues. In *Proceedings of the International Joint Conference on Rules and Reasoning 2017 (RuleML+RR), as the Best Paper of the Doctoral Consortium*, 2017.

[Zografistou et al., 2018] Dimitra Zografistou, Giorgos Flouris, Theodore Patkos, and Dimitris Plexousakis. Implementing the ArgQL query language. In *Proceedings of the 7th International Conference on Computational Models of Argument (COMMA-18), short paper*, 2018.

Chapter 3

Computational Models of Legal Argument

Trevor Bench-Capon*
University of Liverpool
tbc@liverpool.ac.uk

Katie Atkinson
University of Liverpool
katie@liverpool.ac.uk

Floris Bex
Utrecht University & Tilburg University
f.j.bex@uu.nl

Henry Prakken
Utrecht University
h.prakken@uu.nl

Bart Verheij
University of Groningen
bart.verheij@rug.nl

---

*During the finalization of this chapter, its initiator Trevor Bench-Capon passed away. He was a driving force behind its production, and added contributions until his final weeks.

1 Introduction

Argumentation is central to law. Consider for instance the following debate about the situation that Mary's bike is stolen and was bought by John:

A: Mary is the bike's owner.

B: Why?

A: She is the original owner.

B: I disagree. John is owner.

A: Why?

B: He is the buyer.

A: I disagree. He was not bona fide.

B: Why?

A: He bought the bike for €20.

B: I disagree. He bought the bike for €25.

A: You are right. That is still a reason he was not bona fide

In this brief argumentative dialogue, we already see several relevant phenomena. Initially a *conflict of opinions* is encountered, here about who is the owner of the bike. Also *claims* made are supported by *reasons*, here for instance about why there is ownership. Reasons can be *supporting or attacking*. For instance, Mary's original ownership supports her current ownership, and the fact that John bought the bike for €25 attacks that he paid €20. Furthermore, sometimes reasons are not about a claim, but about the *relation between a reason and a claim*. For instance, here John not being bona fide attacks the support relation between him being the buyer and being the owner. The example shows how argumentation can proceed in a *dialogue*, here between participants A and B. Apart from making claims and providing reasons, also *questions* are asked ('Why?').

3 - COMPUTATIONAL MODELS OF LEGAL ARGUMENT

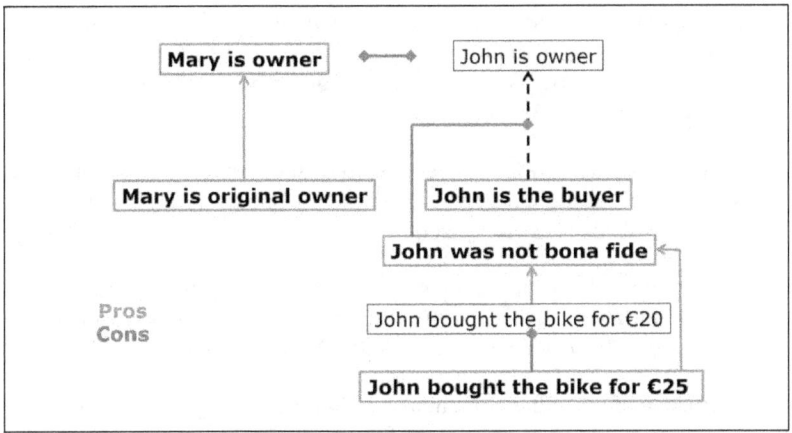

Figure 1: An example argument

In this dialogue, participants also make a *disagreement* explicit ('I disagree') which leads to a concession ('You are right').

Figure 1 provides a graphical representation of the arguments in the dialogue and their evaluation. Sentences in bold are accepted, either since they are undisputed claims (e.g. 'Mary is original owner') or since there is a successfully supporting reason for them (e.g. 'Mary is owner' supported by 'Mary is original owner'). Other sentences are not accepted, in fact they are rejected since there is a reason successfully attacking them (e.g. 'John is owner' attacked by 'Mary is owner'). Note that the figure shows that 'John was not bona fide' attacks the connection between him being the buyer and the owner. (A formalization is discussed in Section 5.3.4.)

Hence, since argumentation is so central, the topic of argumentation is prominent in research in AI and Law. For instance, the topics related to argumentation as they are discussed in the field of AI & Law include the following.

1. Legal cases have been studied from early on as the source of hypothetical arguments [Rissland and Ashley, 1987a; Ashley, 1990; Aleven and Ashley, 1995].

2. The dialogical use of legal rules, cases and values has been reconstructed as argumentation [Atkinson *et al.*, 2005; Bench-Capon and Sartor, 2001; Gordon, 1993b; Hage *et al.*, 1993; Loui and Norman, 1995; Prakken and Sartor, 1996, 1998].

3. Argumentation research has inspired schemes for decision-making and fact finding [Atkinson *et al.*, 2005; Bench-Capon *et al.*, 2000; Bex *et al.*, 2003; Verheij, 2003c; Walton, 1996; Gordon *et al.*, 2007].

4. Argument diagrams have been studied in the context of legal sense making [Bench-Capon, 1998; Bex *et al.*, 2010; Gordon and Karacapilidis, 1997; Gordon *et al.*, 2007; Verheij, 2003b].

5. Burden of proof has been analyzed in terms of argumentation [Gordon *et al.*, 2007; Prakken and Sartor, 2007b; Prakken *et al.*, 2005].

6. Legal decisions have been studied in early argument mining research [Mochales and Moens, 2011].

This correspondence between law and argumentation was also considered by philosophers, in particular Toulmin [Toulmin, 1958] and Perelman [Perelman and Olbrechts-Tyteca, 1969]. As a consequence, historically, AI and Law research has influenced computational argumentation research significantly, and vice versa. This in turn has led to the existence of various existing overview resources [Rissland *et al.*, 2003; Reed and Norman, 2004; Bench-Capon *et al.*, 2009; van Eemeren *et al.*, 2014; Prakken and Sartor, 2015a; Bench-Capon, 2020; Verheij, 2020b; Atkinson and Bench-Capon, 2021].

In this chapter, we aim to add to these resources by giving a chronological presentation of the development of various techniques for computational argumentation in AI and Law, with a section devoted to each decade. We will organise our discussion around three generic tasks:

- Argument Generation

- Evaluation of Arguments

- Use of Arguments

2 Early days: Semi-formal approaches at the start of AI

In the 1950s, Toulmin [1958] suggested to radically change the analysis and assessment of reasoning in purely formal logic and probability theory by looking at the situated, concrete context of debate in law. Whereas his work was primarily philosophical in nature, work on argumentation in AI applied to the law started in the 1970s with McCarty's TAXMAN [1976], an early contribution to the reconstruction of legal argument in a formal-computational style. Rissland [1983] initiated the idea of hypothetical cases as examples guiding argument in the 1980s, taking inspiration from the use of examples in a mathematical discovery dialogue as suggested by Lakatos [1963].

2.1 Argument Generation

2.1.1 Prototypes and Deformations

Perhaps the first project to address argumentation in AI and Law was TAXMAN [McCarty, 1976]. In this project McCarty attempted to reconstruct the arguments in a number of leading US Tax Law cases. One particular case was *Eisner v Macomber* and in [McCarty and Sridharan, 1981] McCarty attempted to reconstruct the arguments of both the majority (justice Pitney) and dissenting (justice Brandeis) opinions, using the mechanism of *prototypes and deformations*.

The idea was that both start with a case representing a paradigmatic instance of their position (the prototype), and then map this into the current case through one or more mapping operations (the deformations). The issue in *Macomber* was whether payment of a dividend in the form of the distribution of additional shares in the same stock was taxable as income. The distribution of a corporation's cash, as in *Lynch v Hornby*, and the distribution of the stock of an unrelated corporation, as in *Peabody v. Eisner*, were situations that all parties agreed should be taxable. On the other hand, the appreciation in the value of a stock without the actual transfer of stock certificates, a purely hypothetical case, was a situation that all parties agreed should be non-taxable. Pitney's argument was the construction of a mapping between the Macomber case and the unrealized appreciation case: the taxpayer in Macomber now owns 3300 shares of common stock out of 75.000.000 outstanding, but, Pitney claimed, that is the same as owning 2200 shares out of 50.000.000 outstanding, which is the situation that would have existed had there been no actual transfer of stock certificates. A more difficult mapping to represent is the one constructed by Justice Brandeis to demonstrate that it is possible to find a coherent path between the stock distribution of *Eisner v. Macomber* and the cash distribution of *Lynch v. Hornby*. In his argument Justice Brandeis posits a sequence of hypothetical cases: first the distribution of common stock, then preferred stock, then bonds; then the distribution of long-term notes, then short-term notes; and finally the distribution of cash.

The mechanism is to represent cases as frames[1] and then starting from a precedent or a clear case (prototype), change various attributes (deformations) to map through a sequence of precedents and hypotheticals to reach the target case. Although there was not a full implementation of this process, since the search procedure to find a suitable sequence of mappings was not yet finalised, this was an important step in the computational modelling of legal argument.

2.1.2 Hypotheticals

Another early attempt to model legal reasoning was [Rissland, 1983]. Here the idea was to take a "seed case" and, by modifying various features of that case, generate a series of hypothetical cases to explore doctrines and approaches, and to uncover assumptions and biases. Although Rissland's original inspiration was mathematics [Lakatos, 1963] and [McCarty and Sridharan, 1981] is not given as a reference, there are similarities between this proposal and that of [McCarty and Sridharan, 1981]. If we take the current case as the 'seed case", we can see the process as an attempt to produce a series of hypotheticals leading to a prototype with the desired outcome.

---

[1] Frames [Minsky, 1975] were a then standard form of knowledge representation, in which entities were represented as frames, which contained a number of 'slots' corresponding to attributes of that kind of entity. Individuals were represented by instantiating the frame and filling the slots with the values of the attributes appropriate to that individual.

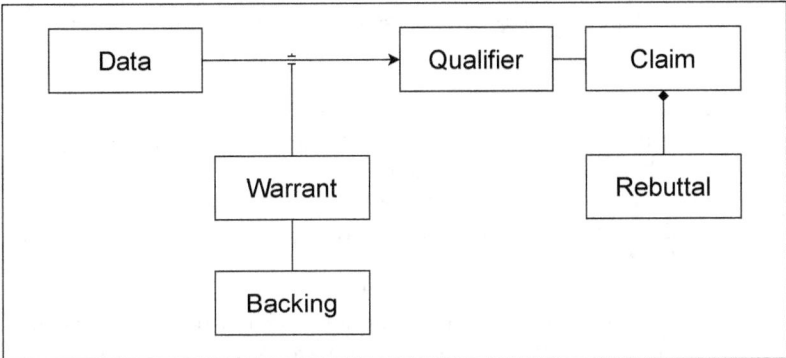

Figure 2: Toulmin's Argument Schema

Rissland's idea of using hypotheticals was more fully realised in the HYPO project with her then PhD student, Kevin Ashley, and we will give a more detailed discussion in Section 3.1.1.

2.2 Evaluation of Arguments

2.2.1 Toulmin

Toulmin was interested in encouraging critical thinking and as such in the defeasibility of most rules of inference. Very rarely does a set of premises entail its conclusion absolutely, in all circumstances. He therefore thought of argumentation in terms of justification rather than inference to a conclusion. His ideas are expressed in his argumentation schemes [Toulmin, 1958], shown in Figure 2.

Toulmin's scheme has six elements:

- **Claim**: This is the conclusion of the argument: note that Toulmin calls it a 'claim' to emphasise that the argument is intended to justify rather than establish it, and that it remains defeasible.

- **Warrant**: This is the rule used to justify the claim.

- **Backing**: This is the justification for the warrant. If the warrant is a legal rule, the warrant will be the statute or precedent case from which it derives.

- **Data** This is the basic premises needed to establish the antecedent of the rule.

- **Qualifier**: This expresses the degree of confidence in the claim, recognising that the warrants are rarely universally applicable, but often permit of exceptions. The qualifier will have different strengths depending on the nature of the warrant. Examples of qualifiers are 'certainly', 'probably', 'possibly', 'typically', 'usually'.

- **Rebuttal**: This represents exceptional circumstances under which the rule does not apply, expressed with an 'Unless' clause.

Toulmin's scheme thus adds the elements of Backing, Qualifier and Rebuttal to the standard *modus ponens* schemes of Premise (Data), Rule (Warrant) and Conclusion (Claim). This setup emphasises several ways in which an argument may be defeated: because the rule is unfounded in general, because it is inapplicable in the specific circumstances or because there is a stronger counter argument.

As we shall see later in the chapter, Toulmin's scheme had considerable influence in AI and Law, both for presentation (e.g. [Lutomski, 1989] and [Marshall, 1989]) and as a driver of dialogues [Bench-Capon, 1998]. This notion of an argumentation scheme was popular for a while, although it became replaced by Walton's more flexible notion of schemes [Walton, 1996].

Defeasibility was also an important aspect of the formalisms for argumentation developed in these early years. In recent structured accounts of argumentation (e.g. [Prakken, 2010]) three kinds of attack are identified. Two of these correspond to elements of Toulmin's scheme: the qualifier expresses that a warrant is defeasible, which allows for rebutting attacks on the claim, that is, arguments for a contradictory claim. Toulmin's rebuttals express explicit exceptions to warrants and are thus related to Pollock's undercutting counter arguments. Structured approaches to argumentation, however, do not require a backing for its rules: instead they allow undermining attacks, which are arguments claiming that the data is false, which did not arise for Toulmin, since there is no notion of chaining arguments in his scheme.

3 - Computational Models of Legal Argument

Toulmin's scheme underwent several adaptations by those interested in making it computable. This involved chaining schemes, so that the claim of one scheme became the data of another, leaving out the qualifier and/or the backing, redirecting the rebuttal to the rule rather than the claim, to represent undercut rather than rebuttal, and adding a Presumption element, justifying the rule by limiting the type of things to which it applied (e.g. [Bench-Capon, 1998]). Such adaptations will be discussed later in the context of particular systems which used them.

2.3 Use of Arguments

2.3.1 Logic Based Dialogue Games

[Hamblin, 1970] proposed an approach to the analysis of logical fallacies in terms of formal dialogues. The idea was that a dialogue would be formally specified as a set of rules, in such a way that the rules would prohibit the fallacy. The nature of the rules that would be broken if the fallacy is committed gives insight into the nature of the fallacy. These formal dialogue specifications became known in computational argumentation as 'dialogue games' (e.g. [Bench-Capon et al., 1991], [Gordon, 1993b], [Lodder and Herczog, 1995]).

[Hamblin, 1970] presented the dialogue game H, but it was Mackenzie's game DC [Mackenzie, 1979] that was the inspiration for several computational implementations including [Bench-Capon et al., 1991] and [Yuan et al., 2003]. The use of dialogue games both for providing interactive explanations, and as a means of modelling legal procedures became very popular in AI and law in the 90s, led by [Gordon, 1993b].

3   1980s: Rule-based and case-based knowledge representation

In the 1980s, adapting logical methods seems to be the way to go, since that is the language of computers. AI-wide this is the peak of nonmonotonic logic and logic programming [Gabbay et al., 1994]. Meanwhile, especially in the US, Case Based Reasoning (e.g. [Kolodner et al., 1985]) remained a widespread approach. In AI and Law, these developments give by the end of the decade prominent examples of a rule-based (British

Nationality Act, [Sergot *et al.*, 1986]) and a case-based (HYPO, [Rissland and Ashley, 1987a]) approach.

3.1 Argument Generation

3.1.1 Dimensions in HYPO

In order to explore the generation and use of hypotheticals, identified as important in sections 2.1.1 and 2.1.2, Edwina Rissland and her PhD student, Kevin Ashley, conducted the HYPO project [Rissland and Ashley, 1987a; Ashley, 1989, 1990]. HYPO is perhaps the most influential project in AI and law and has inspired work in case based reasoning ever since [Bench-Capon, 2017]. HYPO contained several important ideas, but was very firm in its conception of case based reasoning as adversarial argumentation.

Three Ply Structure

In HYPO argumentation was modelled as a *three ply* activity, described in Section 3.3.1.

- **Citation**: In the first ply the proponent cites a precedent case with similarities to the current case and an outcome for the desired side.

- **Response**: In the second ply the opponent responds by citing a *counter example*, a precedent case with similarities by the opposite outcome, or by pointing to a *distinction*: a difference between the current case and the cited precedent which makes the current case stronger for the opponent; or by using a hypothetical to question one of the cited similarities.

- **Rebuttal**: in the third ply the proponent attempts to counter the argument of the opponent by distinguishing the counter examples. downplaying the distinctions and hypotheticals, and citing cases which show any weakness not to be fatal, or provide additional reasons for the desired outcome.

Similarity of Cases

In HYPO cases are represented as a collection of facts. These facts are then used to identify which dimensions are applicable to a case, and to assess the case in terms of these dimensions. A dimension is an aspect of a case which may have legal significance by presenting a reason to decide for one party or the other. The aspect takes a range of values, with one end representing the extreme pro-plaintiff value and the other the extreme pro-defendant value.

As originally conceived [Rissland *et al.*, 1984], dimensions did not favour either party in particular, but could favour either depending on where on the range a particular case fell. An example would be the dimension of *SecurityMeasures*. At one extreme the plaintiff may have taken no security measures at all, which would be a reason to find for the defendant. At the other the extreme would be that the plaintiff had taken vert strict measures and this would be a reason to find for the plaintiff. In between which side is favoured is a matter for dispute, and courts will need to decide which, if any, side is favoured (moderate security measures may provide a reason for neither side). These decisions will become precedents, establishing how the dimension is used in future cases.

During the development of HYPO, however, Ashley's view of dimensions changed[2], and in [Ashley, 1990] he says that dimensions can be grouped "into those favoring the plaintiff generally and those favoring the defendant" ([Ashley, 1990], page 113)[3]. This shift was probably influenced by the nature of the dimensions implemented in HYPO. Ten of the thirteen are Boolean, with one value providing a reason, and the other not. For example, having a non disclosure agreement is a reason to find for the plaintiff, but the lack of one is not in itself a reason to find for the defendant, so it seems reasonable to describe this as a pro-plaintiff dimension. The three dimensions which do have a range, however, are less clear cut. *DisclosuresToOutsiders* was considered effectively binary since any disclosures at all were treated as a pro-defendant reason while no disclosures was not considered a reason for the plaintiff. With Security Measures, however, this breaks down: security measures is a dimension which can, and does, provide a reason for either side, and so cannot be classified as either pro-plaintiff or pro-defendant. The third

---

[2]For fuller discussions of the various different takes on dimensions and their evolution see [Rissland and Ashley, 2002] and [Bench-Capon and Rissland, 2001].

[3]By this time Ashley was already thinking in terms of factors, which would become the basis of CATO [Aleven, 1997], as shown by [Ashley, 1991].

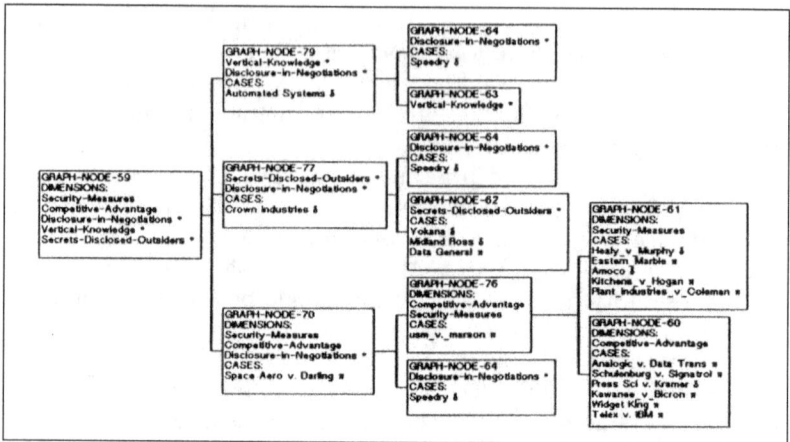

Figure 3: Claim Lattice used in HYPO taken from [Ashley, 1990]

numeric dimension, Competitive advantage, was considered by Ashley a pro-plaintiff dimension[4].

To determine the similarity between cases, the current case and the precedents are organised into a *claim lattice*. An example claim lattice, for USM[5] is shown in Figure 3[6]. The current case is shown as the root node, and its dimensions are listed. In Figure 3, USM has five dimensions. In the next level the nodes represent precedent cases with dimensions in common with the current case, where the dimensions in common are not a subset of any other precedent. For USM, there are three such cases, one with three dimensions in common and two with two dimensions in common. The next layer contains precedents with subsets

---

[4]This was criticised in [Bench-Capon and Gordon, 2022], where it was pointed out that the lack of competitive advantage could be seen as a reason to regard the information as not valuable and so to find for the defendant. This argument is in fact made in several precedent cases, whereas competitive advantage is rarely, if ever, given as a reason for the plaintiff, suggesting that the dimension is, if anything, pro-defendant.

[5]USM Corp. v Marson Fastener Corp. 379 Mass 90 (1979)

[6]Originally HYPO included "near miss" dimensions in the claim lattice [Rissland and Ashley, 1987a]. Rissland continued to use dimensions in their original sense and "near misses" play an important role in CABARET [Skalak and Rissland, 1992].

of these dimensions. Layers are added until we reach the leaves, which will have only a single dimension in common. Where precedents have the same dimensions in common they are represented in the same node. All the precedents in the lattice have some similarity to the current case: the closer to the root, the more similar they are.

Using the claim lattice we can construct the arguments to deploy in our three ply framework.

- We can cite a case closest to the root with the required outcome as a precedent. Thus in Figure 3, the plaintiff can cite *Space Aero*, and the defendant can cite either *Automated Systems* or *Crown Industries*.

- In the second ply, the respondent can cite a counterexample, such as a case supporting the respondent's side. Moreover it can distinguish the cited case by pointing to dimensions favouring the other side present in the root but not in the cited case, or to dimensions favouring the same side present in the cited case but not in the root, or to dimensions favoring the other side to a lesser degree in the root than in the precedent. Thus the defendant could respond to *Space Aero* by distinguishing with *Vertical-Knowledge* or *Secrets-Disclosed-Outsiders*. Finally, the defendant can also distinguish by saying that *SecurityMeasures* were less stringent or that the *CompetiveAdvantage* was less.

- In the third ply, counterexamples can be distinguished in the same way. The plaintiff could distinguish *Crown Industries* with *Security-Measures*. Distinctions can be rebutted by pointing to cases which also lacked the distinguishing feature: thus *Vertical-Knowledge* could be countered by pointing out that it was also not present in *USM* which was never the less found for the Plaintiff.

Thus the claim lattice can be used to generate arguments and counterexamples for both sides. The user is left to choose which arguments should be accepted.

Hypotheticals

Dimensions can also be used to generate hypothetical arguments, as discussed in [Rissland, 1989]. In the US Supreme Court, such arguments are typically used at the Oral Hearing stage to probe whether a particular dimension does indeed provide a reason to decide for the side mentioned. Here the idea is to consider a hypothetical case with a different point on the dimension. For example, suppose in Figure 3, the information had been disclosed to only seven outsiders, whereas in the precedent by the defendant, Crown Industries, the information had been disclosed to 150. Now one could argue that the current case is much weaker than *Crown Industries* on this dimension, and that in a hypothetical version of Crown Industries where it had been disclosed to only 50 outsiders, the plaintiff would have won. In this way Crown Industries is distinguished, since the current case is too weak on this dimension.

In [Ashley, 1990], four other ways of generating hypothetical cases are given (p. 148f.).

3.1.2 Logic Programs

In the 1980s the representation of legislation as logic programs was popularised with [Sergot *et al.*, 1986]'s work on the British Nationality Act as a well-known example. Given such a logic program, it could be deployed as a legal expert system by adding a facility for the user to supply information as to the status of the leaf predicates. As an example consider US Trade Secrets Law[7] as discussed for HYPO in section 3.1.1. We have

```
TradesSecretsMisappropriation:- TradeSecret,
                                Misappropriated.
TradeSecret:- InfoValuable, SecrecyMaintained.
Misappropriated if InfoUsed, Wrongdoing.
Wrongdoing:- BreachOfConfidence.
Wrongdoing:- IllegalMeans.
```

Such programs could explain their answer in the manner of the traditional *how?* explanations used in rule based systems since MYCIN [Buchanan and Shortliffe, 1984].

The problem is that the questions posed to the user are based on the terms of the legislation such as InfoValuable and BreachOfConfidence.

---

[7]This domain was not used by the original logic programmers, but we use it here to offer a direct comparison with HYPO.

But these terms are subject to interpretation, and need the clarification provide by case law. So reliable answers can only be given by a user expert in the case law of the domain.

To resolve this, the logic program was augmented with the reasons for applying these predicates established in case law. Thus, for example, that the information was disclosed in negotiations is a reason to find for the defendant, but that the defendant knew the information was confidential is a reason to find for the plaintiff. Thus we can add the clauses:

```
BreachofConfidence(d):-DisclosureInNegotiations.
BreachofConfidence(p):-KnewInfoConfidential.
```

Of course, both these things can be true in the same case, and so it is unclear in which way the issue should be resolved. In [Bench-Capon and Sergot, 1988] it was suggested that how? explanations of the set of answers could be seen as *arguments* for the two sides of the issue. Thus, here: *find for plaintiff since defendant knew the information was confidential* and *find for defendant since the plaintiff disclosed the information in negotiations*. The issue could then be resolved by choosing the better argument.

Generating arguments from a set of rules in this way became central to many current accounts of legal argumentation (e.g. [Prakken and Sartor, 1996, 1997], [Bench-Capon and Sartor, 2003], systems based on ASPIC+ [Prakken, 2010] and many more).

3.2 Evaluation of Arguments

3.2.1 Assessment by Users

In this period the emphasis was wholly on the generation of arguments. While both HYPO [Rissland, 1983] and the logic programming approach could generate arguments for both sides, they offered little support for choosing between them. The idea was that the users would evaluate the arguments on the basis of their knowledge and context.

Both systems were indeed often seen as being used before a trial by one of the parties to the case. In this scenario, the arguments for would be possible arguments to deploy in the trial, while the arguments against alerted the user to the potential counter arguments that might require rebuttal. In this scenario evaluation is unnecessary: the judge will have the ultimate decision.

Support for evaluation was left for future work. In [Bench-Capon and Sergot, 1988] the authors wrote:

> In the longer term, we hope to pursue what we have identified as a critical requirement: a representation in computer-intelligible terms of what it is that makes a legal argument persuasive.

Work on this was undertaken in the 1990s, as described in Section 4.2.

3.3 Use of Arguments

3.3.1 Three Ply

In HYPO [Rissland and Ashley, 1987b], arguments were deployed in the three ply structure described in Section 3.1.1. This three ply structure is common in Anglo Saxon law. The specific inspiration was the Oral Hearing stage of Supreme Court cases, in which the plaintiff makes a case, the defendant responds and the plaintiff rebuts, but there are other instances, such as witness examination in which, after the testimony has been elicited, there is a cross examination and a redirect.

The output from the program was a series of points relating to the three plies. Thus, using the claim lattice in Figure 3, we would get:

```
Point for Defendant as Side 1:
    Where plaintiff disclosed the information to outsiders
    and to the defendant in negotiations,
    defendant should win a claim for
    TradeSecretsMisappropriation.
    Cite:   Automated Systems.

Response for Plaintiff as Side 2;
    Automated Systems is Distinguishable.
    The plaintiff took security measures.
    Not so in Automated Stystems.
```

```
    Space Aero provides a counter example in which
    Plaintiff took security measures,
    there was competitive advantage
    and plaintiff disclosed information in negotiations

Rebuttal for Defendant as Side 1;
    Space Aero is distinquishable
    The infomation concerned constitutes vertical knowledge
    Not so in Space Aero.
```

This structure is also represented in the argumentation schemes of [Walton, 1996], in which an instantiation of the scheme by a proponent is challenged by the opponent using characteristic critical questions, which the proponent must then attempt to answer. Indeed the reasoning of an immediate successor to HYPO, CATO[8] [Aleven, 1997], was modelled as a set of Walton style argumentation schemes in [Prakken *et al.*, 2015].

3.3.2 How? and Why?

In the logic programming approach (e.g [Sergot *et al.*, 1986]), the arguments were deployed using the explanation facilities commonly found in the expert systems of the time, modelled on [Buchanan and Shortliffe, 1984]. Thus when presented with a conclusion, the user could ask *how?* and be presented with the sequence of inferences which led from the entered facts to the conclusion. Thus, using the example program given in Section 3.1.2, suppose the user had said that the information was known to be confidential and the system had responded that there had been a trade secrets misappropriation. The *how?* query now yields:

```
I can show TradesSecretsMisappropriation
    because I can show  TradeSecret and Misappropriated.

I can show TradeSecret
    because I can show  InfoValuable and SecrecyMaintained.
```

---

[8]CATO is discussed in Section 4.1.5.

```
I can show Misappropriated
   because I can show InfoUsed and  Wrongdoing.

I can show Wrongdoing
   because I can show BreachOfConfidence.

I can show BreachOfConfidence
   because I can show KnowInformationConfidential.
```

Here InfoValuable, SecrecyMaintained and InfoUsed are all taken to default to the plaintiff: i.e. the burden of proof is on the defendant.

The *why?* explanation was used when the user was asked a question in the interaction. For example, the system might ask *Was the Information known to be confidential by the Defendant?*. If the user wants to know why this question is asked the why? query can be used, and will yield the following response.

```
If I know that the Information is Known Confidential
   I can show Wrongdoing.
```

Reiterated use of the *why?* query enabled the user to move up the proof tree and see why the goals were significant.

These two queries enable a fairly primitive dialectical dialogue between user and machine. This dialogical interaction underwent a great deal of development in the 1990s.

3.3.3 Toulmin presentation

Arguments can be presented as text, in dialogue, and also visually, in particular following the diagrammatic nature of the Toulmin argument scheme described in Section 2.2.1. An early example of diagrammatic presentation is provided by [Marshall, 1989], who made two adaptations to the original scheme, dropping the qualifier and allowing the chaining of arguments so that the claim of one argument became the data of the next. Basically this gave a visual presentation of the how? explanation described in Section 3.3.2, but with the useful addition of the backing for each step, which provided the source of the rules used. Additionally

3 - COMPUTATIONAL MODELS OF LEGAL ARGUMENT

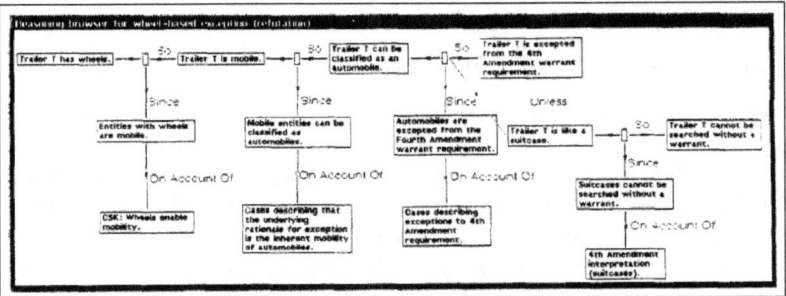

Figure 4: Marshall's presentation of an argument in *Carney*, taken from [Marshall, 1989]

it was possible to provide a counter example using the rebuttal link. The nature of the exception was explained by making the rebuttal node the data of an argument with a contrary conclusion as claim. The approach was illustrated with the case of *California v Carney*[9], also used in [Rissland, 1989], and in subsequent AI and Law research on the automobile exception to the Fourth Amendment (e.g.[Bench-Capon and Prakken, 2010]).

Diagrammatic presentation of arguments was to become very popular both in computational argumentation in general (e.g. Araucaria [Reed and Rowe, 2004]) and in AI and Law in particular (e.g. Carneades [Gordon *et al.*, 2007]).

4 1990s: Argumentation as an AI approach

In the 1990s, it became accepted that logic and logic programming do not suffice for the natural representation of debate. The focus turned to defeasibility, dialogue and procedure. Inspired by philosophy, argumentation takes center stage in AI ([Pollock, 1995; Dung, 1995]), and is immediately prominent in AI and Law. From the start, attempts are made to connect rules, cases, arguments in models of debate (in particular in the works of Bench-Capon, Prakken, Sartor, Hage, Gordon).

---

[9]California v. Carney :: 471 U.S. 386 (1985)

4.1 Argument Generation

4.1.1 Logic + Knowledge Base

That the *why?* explanation of logic programs could be seen as an argument comprising a series of *modus ponens* steps had been noted in [Bench-Capon and Sergot, 1988]. This idea was made more rigorous in [Prakken, 1993], where a formalisation of arguments and subarguments, conflicts between arguments and defeat was offered. The idea was that given a theory comprising facts and defeasible rules, arguments could be generated for and against a given statement. Particularly important was the idea of reinstatement, so that an argument which would otherwise be defeated by a preferred attacker could be reinstated if there was an argument to defeat that attacker.

This approach, generating arguments from an underlying knowledge base, was to become widespread, and is still used today in frameworks such as ASPIC+ [Prakken, 2010; Modgil and Prakken, 2014]. [Kowalski and Toni, 1996] advocated the use of assumption-based argumentation [Bondarenko *et al.*, 1997] for the same purposes.

4.1.2 Rationales

Loui and Norman [1995] discuss rationales in legal decision making, addressing the formal explication of various kinds of argument moves that use rationales. We here follow the discussion by [Governatori *et al.*, 2022] and [Bench-Capon and Verheij, 2022].

A key idea in the paper is that the rationales used in an argumentative dialogue can be interpreted as the summaries ('compilations') of extended rationales with more structure. By unpacking such summary rationales, new argument moves are possible. The paper distinguishes rationales for rules and rationales for decisions. In the authors' terminology, rule rationales express mechanisms for adopting a rule, while decision rationales express mechanisms for forming an opinion about the outcome of a case.

Here is an example of a small dialogue in which a compression rationale is unpacked, subsequently attacked and then defended against. The unpacking here has the form of adding an intermediate step, thereby interpreting a one step argument as a two step argument. We use a legal

3 - COMPUTATIONAL MODELS OF LEGAL ARGUMENT

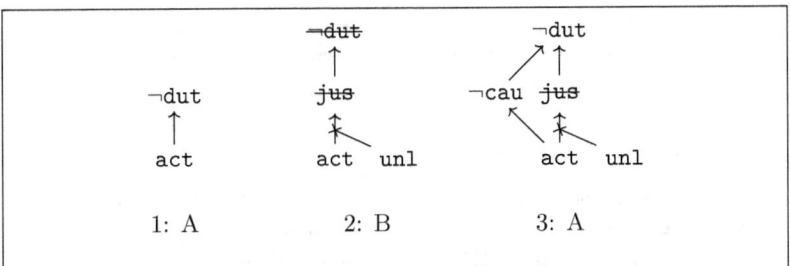

Figure 5: Unpacking a compression rationale

example (in the context of Dutch tort law), noting that the original paper focuses on abstract examples involving propositional constants a, b etc.

> A: I claim that there is no duty to pay the damages (¬dut) because of the act that resulted in damages (act).
> B: Unpacking your reasoning, you seem to claim ¬dut because of act using the additional intermediate reason that there is a ground of justification (jus). I disagree with jus, because the act was unlawful (unl), so there is no support for jus. Hence there is also no support for ¬dut.
> A: I agree with your reason for unl and that hence there is no support for jus. But I was not using jus as an intermediate step supporting ¬dut. Instead I used the intermediate step that there was no causal connection between the act and the damages (¬cau), hence my claim ¬dut because of act.

A graphical summary of the 3-step dialogue is shown in Figure 5. Normal arrows indicate a supporting reason and arrows ending in a cross indicate an attacking reason. All abbreviated statements are considered to be successfully supported, except those that are struck-through. Writing the first argument by A as act → ¬dut, B replies in the second move by interpreting the argument as actually having two steps act → jus → ¬dut, and then attacks the unpacked argument in the middle by the argument unl, making that jus and ¬dut are not successfully supported. But then at the third step A concedes that unl, while denying the unpacking via `jus`, instead claiming the unpacking act $\rightarrow$ ¬cau $\rightarrow$ ¬dut, providing an alternative way to support ¬dut, thereby still maintaining act $\rightarrow$ ¬dut.

4.1.3 Argument Moves

Deducing the consequences of a knowledge base provided a way of generating arguments for rule based approaches, but what of case based approaches, deriving from HYPO [Rissland, 1983]? Developments from HYPO took two distinct paths: Rissland worked with David Skalak on CABARET [Skalak and Rissland, 1992], while Ashley worked with Vincent Aleven on CATO [Aleven, 1997]. Both of them addressed argument generation through the use of *argument moves*.

4.1.4 CABARET

Arguments are generated in [Skalak and Rissland, 1992] with a three-tiered approach in terms of *argument strategies*, realised using *argument moves*, which are implemented using *argument primitives*. The appropriate strategy is selected by reference to the rule governing the case and the point of view. The move is determined by the precedents available and their dispositions. If the rule conditions are met and the point of view is positive, the hit must be confirmed but if the point of view is negative, the rule must be discredited. If the rule conditions are not met, the miss must be confirmed for a negative point of view, or the rule broadened for a positive point of view.

Once the strategy has been selected, the precedents are used to select a move. Depending on the outcome in the precedent and the strategy being employed the precedent must be analogised to or distinguished from the current case. These moves are then implemented through detailed comparison of the features of the current case and the precedent to determine the degree and nature of the matches and mismatches between the two[10]. For instance, when broadening a rule, citing a precedent with the desired outcome that also failed to satisfy a rule antecedent, and so

---

[10]These primitives play the role of the factor partitions in [Wyner and Bench-Capon, 2007] and the functions in [Prakken et al., 2015] used in the instantiation of their argument schemes.

can be used to argue that since the missed condition was not necessary in that case, it is not needed in this case either.

In [Skalak and Rissland, 1992], the argument moves are limited to those which can be produced using the form of argument the authors term a *straightforward argument*, in which the facts of a current case are compared with a precedent case with the desired outcome. The paper, however, gives a taxonomy of argument forms used in legal argumentation, which includes a variety of additional forms of argument.

4.1.5 CATO

CATO [Aleven and Ashley, 1995, 1997; Aleven, 1997, 2003] was designed to help law students to distinguish cases effectively, and hence its emphasis was on distinguishing. The key point was that not every difference in the case could serve as a significant distinction. CATO replaced the dimensions of HYPO with *factors*. Factors are boolean and can be seen as ranges on dimensions favouring a particular party to the case and so providing a reason to decide for that party. Thus if a factor present in a precedent was absent from the new case, this would only provide a distinction if it favoured the winning side: if it favoured the losing side it would make the new case stronger than the precedent. But even so not all possible distinctions are considered significant: it may be that the difference does not weaken the case sufficiently to change the outcome.

To model this in CATO factors were organised into a *factor hierarchy*, (or rather five factor hierarchies, one for each issue). The issue would be at the root, with abstract factors coming between the issues and the base level factors (the factors corresponding to ranges on the dimensions) and serving to group them together according to whether their reasons was related. The children are reasons to think that their parent is present or absent. The factor hierarchy in CATO for the Issue of whether or not there was a confidential relationship is shown in Figure 6. Here we have two abstract factors (or 'intermediate legal concerns' as they are termed in [Aleven, 1997]), *NoticeOfConfidentiality* and *ExpressConfidentialityAgreement*, each of which is associated with a variety of base level factors, some favouring the plaintiff and some the defendant.

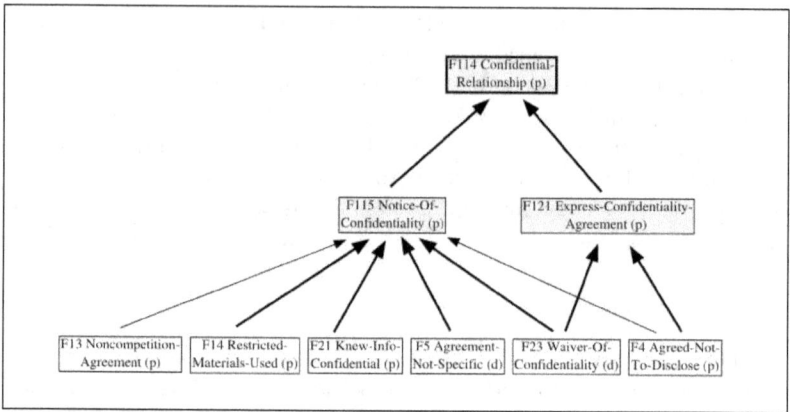

Figure 6: Factor Hierarchy for Confidential Relationship, taken from [Aleven, 1997]

CATO uses the standard moves of citing a precedent, citing a counterexample and distinguishing a precedent, as found in HYPO and described in Section 3.1.1. But with the factor hierarchy, CATO can add additional moves to argue about the significance of a distinction.

Consider Figure 6. Suppose that we have a precedent with F14 Restricted Materials and F13 Non Competition Agreement. We might cite a new case which had F14 but lacked F13. If we did this our opponent could distinguish the case by pointing to the absence of F13. We may, however, there are other factors present in the cases which enable us to *downplay* the distinction, to argue that it is not significant.

This can be done in two ways

- If there is a factor with the same polarity in the current, we can argue that this factor can be substituted for this missing factor. For example if F21 Knew Information Confidential was in the new case;

- If there is a factor with a different polarity in the precedent case we can argue that that factor cancels the missing factor. For example, if F5 Agreement Not Specific had been in the precedent.

If, however, we can neither substitute nor cancel the distinguishing factor, our opponent can emphasise the significance of the distinction.

For a further discussion of CATO's argumentation moves see [Bench-Capon, 1997]. For a formal treatment of these moves in terms of argumentation schemes see [Wyner and Bench-Capon, 2007] and [Prakken *et al.*, 2015].

4.1.6 Protoypes and Deformations

The idea of prototypes and deformations introduced in Section 2.1.1 was revived in [McCarty, 1995]. This paper claimed that knowledge representation languages available previously had been too inexpressive to implement the idea properly, in particular to formalise the notion of a prototype. Hence this paper presented an implementation, taking advantage of subsequent developments, in particular Language for Legal Discourse (LLD) [McCarty, 1989].

In [McCarty, 1995] we have a formalisation of the basic idea, illustrated with Prolog code, and a detailed computational reconstruction of the arguments of Justices Pitney and Brandeis in terms of the theory. But as noted in the discussion, while it was possible to generate the arguments, it was not possible to evaluate them: arguments that one was stronger than the other were not available. McCarty's suggestion was that this might in future be possible by considering the *coherence* of the competing arguments.

The notion of prototypes and deformation did not receive much subsequent take up, and rather CATO's factor based reasoning became the mainstream way of handling reasoning with precedent cases. What did, however, have significant influence was McCarty's notion of reasoning with precedents as *theory construction*. As he put it:

> Legal rules are not static, but dynamic. As they are applied to new situations, they are constantly modified to "fit" the new "facts". Thus the important process in legal reasoning is not theory application, but theory construction.

This idea was to prove influential in, for example, [Bench-Capon and Sartor, 2003] and [Chorley and Bench-Capon, 2005b], and was also the basis of [Prakken and Sartor, 1998], which used precedents to construct a theory comprising defeasible rules and priorities.

4.1.7 Heuristic Search

A rather different approach to argument generation was developed in BankXX. The system addressed the domain of bankruptcy (specifically the U.S. Bankruptcy Code, Chapter 13) [Rissland et al., 1996, 1997]. BankXX uses precedents (and other sources) to represent the domain knowledge as a highly interconnected network of building blocks, which is searched heuristically to gather *argument pieces*. The nodes in this network encompass a wide variety of ways of representing the domain knowledge, including cases as collections of facts, cases as dimensionally-analyzed fact situations, cases as bundles of citations, and cases as prototypical factual scripts, as well as legal theories represented in terms of domain dimensions. Thus cases are represented in several ways including, in its *Domain Factor Space*, "by a vector composed of the magnitudes of the case on each dimension that applies to it; non-applicable factors are encoded as NIL. This ... represents a case as a point in an n-dimensional space." Arguments are then formed by performing heuristic search over the network, using evaluation functions at the domain level, the argumentation piece level, and the overall argument level. The result is a highly sophisticated system, which provides a detailed analysis of the arguments available in a case. The approach is illustrated with a detailed case study of a particular case[11], and the system as a whole is subjected to a detailed evaluation in [Rissland et al., 1997].

The evaluation in [Rissland et al., 1997] is one of the most (if not the most) detailed examples of evaluation in AI and Law. It considers several different forms of the BankXX program, and the evaluation is conducted from several perspectives. A number of issues relating specifically to the evaluation of programs in the domain of law are noted.

BankXX has no obvious descendants in AI and Law research. Construction of cases by performing heuristic search was also carried out by AGATHA [Chorley and Bench-Capon, 2005a], but the search tree was over only a collection of cases represented as bundles of CATO-style

---

[11] In re Estus, 695 F.2d 311 (8th Cir. 1982)

factors, rather than the highly sophisticated network of knowledge used in BankXX.

4.1.8 Rule Based Representation of Precedents

Thus far we have seen how arguments can be generated from rule based representations using the proof trace of deductions from that rule base, and that arguments can be based on case based representations using the notion of similarity. The former had been primarily used for statute based reasoning and the latter for precedent based reasoning. The two were brought together in [Prakken and Sartor, 1998], which demonstrated a way of representing precedent cases and the decisions in these cases as a set of rules.

As was seen in Section 4.1.5, in CATO [Aleven, 1997], a case is associated with a set of factors, some pro-plaintiff and some pro-defendant, and an outcome. The pro-plaintiff factors offer reasons to find for the plaintiff and the pro-defendant factors offer reasons to find for the defendant. Now, if we have a decided case, C, containing pro-precedent factors P and pro-defendant factors D, then the conjunction of all factors in P will be the strongest[12] reason to decide C for the plaintiff and the conjunction of all factors in D the strongest reason to decide C for the defendant. The outcome in the case will show which of these two reasons is stronger. This means we have three rules:

r1 $P \rightarrow plaintiff$;

r2 $D \rightarrow defendant$;

r3 $C \rightarrow r2 \prec r1$ if the decision was for the plaintiff and $C \rightarrow r1 \prec r2$ if the decision was for the defendant.

---

[12]This assumes that the conjunction of two factors favouring the same side will always be stronger that the factors individually. This assumption is queried in [Prakken, 2005b], where an apparent counter example is given. However, such situations can be avoided by modelling the domain differently, using different factors for which the original "factors" are facts (e.g. [Horty and Bench-Capon, 2012], footnote 17). Arguably it is a necessary feature of factors as understood in [Aleven, 1997] they always favour a particular side, and this should hold whatever the context set by other factors [Bench-Capon, 2017].

This representation sees precedents as providing a one step argument from factors to outcome, which was the view taken in subsequent approaches such as [Bench-Capon, 1999] and the formalisations of precedential constraint stemming from [Horty, 2011]. In [Prakken and Sartor, 1998], however, Prakken and Sartor argue strongly that precedents should be seen in terms of multi-step arguments. Often the importance of a precedent will be with respect to a particular issue in the case [Branting, 1999]. Thus if we partition the factors according to the issues of the case, using, for example, the abstract factor hierarchy of [Aleven, 2003], we can get a finer grained representation of the argument. We now represent the case as $P_1 \cup D_1 \cup ... \cup P_n \cup D_n$, where P_i are the pro-plaintiff factors relating to issue i and D_i are the pro-defendant factors relating to issue i. We can now produce a set of three rules for each issue:

r4 $P_i \rightarrow I_i^P$, where I_i^P means that issue i is found for the plaintiff;

r5 $D_i \rightarrow I_i^D$; where I_i^D means that issue i is found for the defendant;

r6 $C \rightarrow r5 \prec r4$ if the issue was found for the plaintiff in C and $C \rightarrow r4 \prec r5$ if the issue was found for the defendant in case C.

We now write a set of three rules, using the issues in the antecedents to show how the issues determine the outcome. Suppose we have a case with three issues, of which two were found for the plaintiff and one for the defendant but the defendant won the case. This would give the rules:

r7 $I_1^P \wedge I_2^P \rightarrow \textit{Plaintiff}$

r8 $I_3^D \rightarrow \textit{Defendant}$

r9 $C \rightarrow r7 \prec r8$

Not only does this more faithfully reflect the reasoning in the case, but it has the practical advantage that an inference is not blocked by a distinction which is irrelevant because it pertains to a different issue. This two step reasoning, from factors to issues and then from issues to outcome was later used in IBP [Brüninghaus and Ashley, 2003] and Grabmair's VJAP [Grabmair, 2017]. More recently it has been argued that adopting this finer grained representation would improve the formal accounts of precedential constraint [Bench-Capon and Atkinson, 2021]. Even finer granularity would be possible, to give rise to three step arguments, but that will often associate too few factors with each sub-issue to be useful.

Using this representation we can generate arguments for both sides for a given issue, and also arguments based on precedents for which argument is the stronger.

4.2 Evaluation of Arguments

As we saw in Section 3.2, little had been done about the evaluation of arguments on the 1980s. In this decade, however, techniques for assessing competing arguments began to be developed.

4.2.1 Reason-Based Logic

In the 1990s, Hage developed Reason-based logic [Hage, 1993, 1996].[13] Hage presents Reason-based logic as an extension of first-order predicate logic in which reasons play a central role. Reasons are the result of the application of rules. Treating rules as individuals allows the expression of properties of rules. Whether a rule applies depends on the rule's conditions being satisfied, but also on possible other reasons for or against applying the rule. Consider, for instance, the rule that thieves are punishable:

punishable: thief(x) $\Rightarrow$ punishable(x)

Here 'punishable' before the colon is the rule's name. When John is a thief (expressed as thief(john)), the rule's applicability can follow:

Applicable(thief(john) $\Rightarrow$ punishable(john))

This gives a reason that the rule ought to be applied. If there are no reasons against the rule's application, this leads to the obligation to apply the rule. From this it will follow that John is punishable.

---

[13]Reason-based logic exists in a series of versions, some introduced in collaboration with Verheij (e.g. [Verheij, 1996]). The discussion here follows [van Eemeren and Verheij, 2018].

A characteristic aspect of Reason-based logic is that it models the weighing of reasons. In this system, there is no numerical mechanism for weighing; rather it can be explicitly represented that certain reasons for a conclusion outweigh the reasons against the conclusion. When there is no weighing information the conflict remains unresolved and no conclusion follows.

The formalization of Reason-based logic uses elements from classical logic and non-monotonic logic. Because of the emphasis on philosophical and legal considerations, the flavour of Reason-based logic is less that of formal logic, and comes closer to formally representing the actual ways of reasoning in the domain of law.

Reason-based logic has been applied, for instance, to a well-known distinction made by the legal theorist Dworkin [1978]: whereas legal rules seem to lead directly to their conclusion when they are applied, legal principles are not as direct, and merely give rise to a reason for their conclusion. Only a subsequent weighing of possibly competing reasons leads to a conclusion. Different models of the distinction between rules and principles in Reason-based logic have been proposed. Hage [1996] follows Dworkin and makes a strict formal distinction between rules and principles, whereas Verheij et al. [1998] show how the distinction can be softened by presenting a model in which rules and principles are the extremes of a spectrum.

4.2.2 Most Specific Argument

In Law, sometimes the following three principles are used to resolve conflicts between laws:

- *Lex superior*: prefer the law issued by the higher authority. Thus a national statute is preferred to a local by-law.

- *Lex specialis*: prefer the more specific law. Thus a law expressing an exception is preferred to the more general law.

- *Lex posterior*: Prefer the more recent law. Thus a new law overrules an existing law, and, in case law, the more recent decision is preferred.

Two of these, specificity and recency, had also been commonly used to resolve conflicts in Production Rule systems [Bench-Capon, 1990].

Inspired by [Poole, 1985], [Prakken, 1991] developed a formal theory based on preferring the most specific argument. In this paper, Prakken was mainly motivated by the need to handle exceptions to laws, which are indeed very common in law. The combination with the other two principles was addressed in [Prakken, 1993].

4.2.3 An Abstract Account of Argumentation

Later in the 90s, the world of computational argumentation was transformed by the introduction of Dung's notion of abstract argumentation frameworks [Dung, 1993b, 1995]. Dung's seminal idea was to represent a set of arguments and the attack relations between them, and then apply various semantics to identify acceptable sets of arguments. The key principle was that an argument is acceptable if and only if all its attackers are themselves attacked. Thus an argument may be defended by another argument which attacks its attacker, and these arguments may form an admissible set. To be *admissible*, a set must be conflict free (the members must not attack one another), and for all members of the set any attacker must be attacked by some member of the set. Two of the most important semantics are *preferred* semantics, which defines alternative sets of acceptable arguments as any *subset-maximal* admissible set, and *grounded* semantics, which defines a unique set of acceptable arguments as the least fixpoint of an operator that for any set of arguments returns the set of arguments defended by that set. There is always a single grounded extension, although it may be empty, but there may be multiple preferred extensions. In preferred semantics, therefore, sceptical acceptance, where an argument is in all preferred extensions, is distinguished from credulous acceptance, where an argument is in at least one preferred extension.

This abstract account of argumentation was taken up in AI and Law with papers by [Prakken and Sartor, 1996, 1997], [Kowalski and Toni, 1996] and [Jakobovits and Vermeir, 1999]. Prakken & Sartor defined their system for argument-based logic programming by defining the structure of arguments, an attack relation between arguments, and the use of priorities to determine which attacks result in defeats. By regarding the resulting defeat relation as the attack relation of abstract argumentation frameworks, this allows the use of abstract argumentation semantics for evaluating the arguments.[14]. This approach was later also applied in the ASPIC+ framework.

[Kowalski and Toni, 1996] developed a similar approach in the context of assumption-based argumentation. Instead of explicitly using priorities, they proposed to encode them in rule-exception structures.

[Jakobovits and Vermeir, 1999] defined various types of dialogue games to verify the acceptability of arguments in an abstract argumentation framework. They also studied dynamic games in which the argumentation framework can be extended during a dialogue, motivated by the observation that legal reasoning typically is an evolving process.

4.2.4 Burden of Proof

[Freeman and Farley, 1996] designed and implemented a model of legal argument in which arguments can be evaluated relative to a given level of proof. The language of their system divides rules into three epistemic categories: 'sufficient', 'evidential' and 'default', in decreasing order of priority. Arguments are structured as a variant of Toulmin's argument structures (see Section 2.2.1 above) and can be of various types. Firstly, besides modus ponens the system also allows modus tollens. Moreover, it allows certain types of nondeductive arguments, viz. abductive ($p \Rightarrow q$ and q imply p) and a contrario arguments ($p \Rightarrow q$ and $\neg p$ imply $\neg q$). Taken by themselves these inferences clearly are the well-known fallacies of 'affirming the consequent' and 'denying the antecedent' but Freeman & Farley deal with this by also defining how such arguments can be attacked.

The strength of arguments is measured in terms of the four values 'valid', 'strong', 'credible' and 'weak', in decreasing order of priority. The strength depends both on the type of rule and on the type of argument. For instance, modus tollens results in a valid argument when applied to sufficient rules, but is a weak argument when applied to de-

---

[14]Strictly speaking Prakken & Sartor defined their system as an application of [Dung, 1993a] instead of [Dung, 1995], but their system can easily be recast as generating Dung's abstract argumentation frameworks and applying grounded semantics to it

fault or evidential rules. Abduction and a contrario always result in just a weak argument. Finally, modus ponens yields a valid argument when applied to sufficient rules, a strong argument with default rules, and a credible argument with evidential rules. The strength of arguments is used to compare conflicting arguments, resulting in defeat relations among arguments, which in turn determine whether a move is allowed in a dispute.

Arguments can then be evaluated in terms of five different levels of proof, depending on which level is suitable in the given problem context:

- *scintilla of evidence* (find at least one defendable argument);

- *preponderance of the evidence* (find at least one defendable argument that outweighs the other side's rebutting arguments);

- *dialectical validity* (find at least one credible, defendable argument that defeats all of the other side's rebutting arguments);

- *beyond a reasonable doubt* (find at least one strong, defendable argument that defeats all of the other side's rebutting arguments);

- *beyond a doubt* (find at least one valid, defendable argument that defeats all of the other side's rebutting arguments).

Freeman & Farley motivate this approach by the observation that different legal problem solving contexts require different levels of proof. For instance, for the question whether a case can be brought before court, only a 'scintilla of evidence' is required, while for a decision in a case 'dialectical validity' is needed. Later, Tom Gordon incorporated these five levels of proof in his Carneades argumentation system [Gordon et al., 2007; Gordon and Walton, 2009] (see Section 5.1.4).

4.2.5 Social Values and Time Dependence

In their work on case-based argumentation in the law, Berman and Hafner emphasise the role of social values in the decision making of courts [Berman and Hafner, 1991, 1993; Hafner and Berman, 2002]. Such decision making is often purpose-oriented or teleological, in the sense that the purpose of promoting one social value may have to be balanced with the purpose of promoting another, competing value. Berman and Hafner write that legal precedents are 'embedded in a political context, where competing policies and values are balanced by the courts, and where legal doctrines evolve to accommodate new social and economic realities' [Hafner and Berman, 2002].

As an example of the balancing of social values, Hafner and Berman discuss cases about hunting wild animals. In one case, the plaintiff was a fisherman closing his large net, whereupon the defendant entered through the remaining opening and caught the fish inside (Young v Hitchens, 1844). Here there was a conflict between the competing social values of the pursuit of livelihood through productive work and economic competition. By deciding for the plaintiff or the defendant, a court can achieve the promotion of one value, but at the price of demoting the other. Here the court found for the defendant, but the judges' opinions show the careful balancing in the background. This case and the other wild animal cases have been extensively studied in Artificial Intelligence and Law, starting with [Bench-Capon, 2002b].

A specific theme addressed by Hafner and Berman is that the relevance of a case as an authoritative source to base new decisions on can evolve over time. The precedential value is not cast in stone, but develops over time influenced by societal changes. As their main example, they discuss a series of New York tort cases about car accidents. The issue was whether a driver should repair a passenger's damages. The series of cases are about what should be done when different jurisdictions are relevant, each with a different authoritative solution. For instance, when the driver and passenger are from New York, where the trip starts, and the accident happens in Ontario, Canada, should then the Ontario rule be followed—barring a law suit in such a case—or the New York rule where negligent driving could imply recovery of damages? Hafner and Berman discuss a series of cases that show the tension between a territory perspective, where the location of the accident (the *situs*) is leading, and a forum perspective, where the place of litigation determines the applicable law. Gradually, the cases shift from a strict territorial rule to a center-of-gravity rule, where the circumstances are weighed. Inspired by the work of Berman and Hafner, Verheij [2016b] developed a formalization of the example using techniques for the formal connection of qualitative and quantitative primitives (and the discussion here follows that paper). That formalization was used by Zheng et al. [2021] in an analysis of the hardness of case-based decisions and how that hardness changes over time.

4.3 Use of Arguments

4.3.1 Pleadings Game

In 1993 the idea of using nonmonotonic logics as a tool for formalising legal argument was already somewhat established. In [Gordon, 1993b,a] Tom Gordon added a new topic to the research agenda of the formalists in AI & Law: formalising the procedural context of legal argument. Gordon attempted to formalise a set of procedural norms for civil pleading by a combination of a nonmonotonic logic and a formal dialogue game for argumentation. The resulting Pleadings Game was not meant to formalise an existing legal procedure but to give a "normative model of pleading, founded on first principles", derived from Robert Alexy's [Alexy, 1978] discourse theory of legal argumentation.

The Pleadings Game had several sources of inspiration. Formally it was inspired by formal dialogue games for monotonic argumentation of e.g. [Mackenzie, 1979] and philosophically by the ideas of procedural justice and procedural rationality as expressed in e.g. [Alexy, 1978], [Rescher, 1977] and [Toulmin, 1958]. For example, Toulmin claimed that outside mathematics the validity of an argument does not depend on its syntactic form but on whether it can be defended in a rational dispute. The task for logicians is then to find procedural rules for rational dispute and they can find such rules by drawing analogies to legal procedures.

Besides a theoretical goal, Gordon also had the aim to lay the formal foundations for a new kind of advanced IT application for lawyers, namely, mediation systems, which support discussions about alternative theories by making sure that the rules of procedure are obeyed and by keeping track of the arguments exchanged and theories constructed.

The objective of the Pleadings Game is to support 'issue spotting', that is, to allow two human parties in a law suit to state the arguments and facts that they believe to be relevant, so that they can determine where they agree and where they disagree. The residual disagreements will go on to form the issues when the case is tried. The system plays two roles in this process: it acts as a referee to ensure that the proper procedure is followed, and records the facts and arguments that are presented and what points are disputed, so as to identify the issues that require resolution. The Pleadings Game has a built-in proof mechanism for an early argumentation-based nonmonotonic logic [Geffner and Pearl, 1992], which is applied to check the logical well-formedness of the arguments stated by the user, and to compute which of the stated arguments prevail, on the basis of the priority arguments also stated by the user and a built-in specificity checker. The Pleadings Game is truly dialogical since not only the content of the arguments is relevant but also the attitudes expressed towards the arguments and their premises.

Let us illustrate this with the following simplified dispute about whether a valid contract was concluded by the parties.

Plaintiff: I claim (1) we have a contract.
Defendant: I deny (1).
Plaintiff: We have a valid contract since (2) I made an offer and (3) you accepted it, so we have a contract.
Defendant: I concede (2) but I deny (3).
Plaintiff: (4) you said "I accept...", so you accepted my offer.
Defendant: I concede (4), but (5) my statement "I accept ..." was followed by terms that do not match the terms of your offer. This point takes priority (6) so I did not accept your offer.
Plaintiff: I concede the priority (6) but I deny (5).
Defendant: You required payment upon delivery (7) while I offered payment 30 days following delivery (8), so there is a mismatch.
Plaintiff: I concede (7) and the argument but I deny (8).

At this point, there is an argument for the conclusion that a contract was created using premises (2) and (4). The intermediate conclusion (3) of this argument that there was an acceptance is defeated by a counterargument using (7) and (8). So one outcome of the dispute is that no contract exists between the parties. However, in the Pleadings Game it also matters that the plaintiff has denied defendant's claim (8). This is a factual issue making the case hard, to be decided in court.

4.3.2 Other Dialogue Approaches

Other dialogue models of argumentation in AI and law have been proposed by Prakken and Sartor [1996, 1998], Hage et al. [1993], and Lodder [1999]. In Prakken and Sartor's approach (1996, 1998), dialogue models are presented as a kind of proof theory for their argumentation model (cf. Section 4.2.3). Prakken and Sartor interpret a proof as a dialogue between a proponent and opponent. An argument is justified when there is a winning strategy for the proponent of the argument. Hage et al. [1993], Lodder [1999] and Lodder and Herczog [1995] propose models of argumentation dialogues with the purpose of establishing the law in a concrete case. They are inspired by the idea of law as a pure procedure (though not endorsing it): when the law is purely procedural, there is no criterion for a good outcome of a legal procedure other than the procedure itself.

Some models emphasize that the rules of argumentative dialogue can themselves be the subject of debate. An actual example is a parliamentary discussion about the way in which legislation is to be discussed. In philosophy, Suber has taken the idea of self-amending games to its extreme by proposing the game of Nomic, in which the players can gradually change the rules.[15] Vreeswijk [1995] studied the game in a context of formal models of argumentative dialogues allowing self-amendments.

In an attempt to clarify how logic, defeasibility, dialogue and procedure are related, Prakken [1995] proposed to distinguish four layers of argumentation models. The first is the logical layer, which determines contradiction and support. The second layer is dialectical, which defines what counts as attack, counterargument, and also when an argument is defeated. The third layer is procedural and contains the rules constraining a dialogue, for instance, which moves parties can make, when parties can make a move, and when the dialogue is finished. The fourth and final layer is strategic. At this layer, one finds the strategies and heuristics used by a good, effective arguer.

Further dialog approaches from this period include [Bench-Capon et al., 2000; Bench-Capon and Staniford, 1995; Freeman and Farley, 1996; Bench-Capon, 1998].

---

[15]http://en.wikipedia.org/wiki/Nomic

4.3.3 Toulmin's Argument Model

Toulmin's argument model has been used in the context of information retrieval [Dick, 1991] and of the explanation of neural networks [Zeleznikow and Stranieri, 1995].

Dick [1991]'s starting point was that Boolean search could be enriched by the use of the conceptual structure underlying legal text. A proposal was made to analyze cases involving contract law using a frame-based representation of the elements of Toulmin's argument model. Case retrieval could then be achieved by matching frames.

Zeleznikow and Stranieri [1995] developed the Split-Up system in which knowledge-based modeling is combined with a neural network approach. The system addresses Australian family law, which by its discretionary nature cannot be fully represented in rule-based form. It is claimed that since neural network models can learn weights of relevant factors, they are well-suited for discretionary domains. However, in order to address the lack of explanations of decisions suggested by neural networks, a hybrid approach is developed in which the structure of Toulmin's argument model is used as an explanation format.

4.3.4 Argumentation and dialogue software

The theoretical developments on the modeling of argumentation and dialogue also led to various implemented software tools intended for support and guidance [Loui et al., 1997; Gordon and Karacapilidis, 1997; Verheij, 1999].

Room 5 [Loui et al., 1997] was intended as a testbed for public interactive legal argumentation. The user interface consisted of a web-based form that can be used to add reasons for and against claims in a public debate. The interface could list open cases, with also access to cases that are no longer argued. As an example, a local freedom of speech case was used. Argument structure is not visualized—as is more common—using boxes-and-arrows, but instead uses nested boxes ('encapsulated subargument frames'). Each box represents a claim, and a box in another box represents a reason relevant for a claim. Nested boxes to the left represent supporting reasons and to the right attacking. The representation format was developed to avoid the 'pointer spaghetti' that arises in a

3 - COMPUTATIONAL MODELS OF LEGAL ARGUMENT

boxes-and-arrows format. The project's goals were not technical but to develop a web community of arguers trained in the use of the dialogue format, thereby building a database of semi-structured arguments.

The Zeno project [Gordon and Karacapilidis, 1997] was also meant to support and mediate online discussion. Its representation and interaction format combines elements of Toulmin's argument model with Rittel's Issue-Based Information System (IBIS). The approach includes issues, alternatives, positions (either for or against) and constraints that allow for the expression of preferences. The information entered by users is represented in a tree-like structure. Motivation for the specific approach in the Zeno project included the conceptual and computational complexity of the formalisms for nonmonotonic reasoning of that time.

ArguMed [Verheij, 1999] was intended as an argument-assistance system supporting a user's reasoning, to be distinguished from an automated reasoning system replacing the reasoning of a user. Statements entered could be assumptions or issues, for which supporting and attacking reasons could be given. It could be debated whether a reasoning step made was appropriate ('step warrants'; inspired by Toulmin's warrants) but also whether an attack of a reasoning step was appropriate ('undercutter warrants'). The system evaluated which statements were justified or not given the state of the discussion, graphically visualized using a boxes-and-arrows format. ArguMed was intended as a realization and testbed for theoretical argumentation models, and as step towards being a practical aid.

5   2000s: Deepening of the knowledge-data gap

In the decade following the year 2000, computational argumentation is becoming a field in itself, with its own conference series (Computational Models of Argument, COMMA),[16] still with significant influence from the field of law. Argumentation schemes and diagrams take off, and reasoning with values is formally analyzed. Burdens of proof are further studied, and evidence and fact finding in the law receive more attention. In the general field of AI, the gap between knowledge-based and data-driven methods is deepening. In connection with argumentation, there

---

[16]http://comma.csc.liv.ac.uk

is some work on prediction methods (see in particular Section 5.1.3), but this is not like the big data approaches that are arising. Knowledge-based approaches are on decline, while ontologies are an attempt to make a bridge with data-driven approacges. In the rest of the AI world, machine learning is gradually taking over, although rather neglected in AI and Law. Argument mining starts for real.

5.1 Argument Generation

5.1.1 Argumentation Schemes

In 2003, argumentation schemes were introduced in AI & Law by two articles in AI & Law journal [Bex *et al.*, 2003; Verheij, 2003c]. Argumentation schemes are forms of argument that represent stereotypical patterns of human reasoning in a conditional form like rules. The idea of defining recurring patterns of reasoning through argumentation schemes originated with [Walton, 1996], who also studied it for legal and evidential reasoning ([Walton, 2002]). A well-known example of an argumentation scheme is the scheme for argument from expert opinion:

Argumentation scheme from expert opinion
Source e is an expert in domain d
e asserts that proposition a is known to be true (false)
a is within d
Therefore, a may plausibly be taken to be true (false)

In addition to a general rule or inference scheme, argumentation schemes also have associated critical questions that point to typical sources of doubt in an argument based on the scheme. For the scheme from expert opinion, the following six critical questions have been proposed ([Bex *et al.*, 2003]):

1. Expertise Question: How credible is e as an expert source?

2. Field Question: Is e an expert in d?

3. Opinion Question: What did e assert that implies a?

4. Trustworthiness Question: Is e personally reliable as a source?

5. Consistency Question: Is *a* consistent with what other experts assert?

6. Backup Evidence Question: Is *e*'s assertion based on evidence?

Answers to critical questions can lead to various types of counterarguments. For example, a negative answer to the "field question" would undercut an argument from expert opinion and a negative answer to the "consistency question" points to a possible rebutting counterargument with an opposite conclusion.

The idea of argumentation schemes is very closely related to Toulmin's notion of warrants (Section 2.2.1) and logical rules (Section 3.1.2). In fact, [Verheij, 2003c] argued that argumentation schemes should be used as the basis for a *logic of law* that focuses on the specific domain rules and contextual reasoning patterns in law. Both he and [Bex et al., 2003] formalised the argumentation schemes and their critical questions in logics for structured argumentation (namely [Verheij, 2003a] and [Prakken, 1993], respectively). Note that not all such schemes are schemes for reasoning with evidence and facts. For example, [Verheij, 2003c] also provides an example of a more legal scheme:

Person p has committed crime c
Crime c is punishable by n years of imprisonment
Therefore, person p can be punished with up to n years in prison

5.1.2 Stories and explanations

In addition to the rule-based and case-based approaches to (legal) argumentation, the 2000's also saw the rise of *story-based* argumentation, mainly in reasoning with evidence (see Section 5.3.3). The story-based approach to reasoning stems from legal psychology ([Pennington and Hastie, 1993; Wagenaar et al., 1993]), and focuses on stories about what happened in a legal case, that is, the facts in the case. These hypothetical stories, coherent sequences of events connected by (sometimes implicit) causal links of the form *c is a cause for e*, are used to explain the observed evidence in a case. When explaining some observed event e, we perform what is commonly called *causal–abductive reasoning* ([Josephson and Josephson, 1996]): If we have a general causal rule

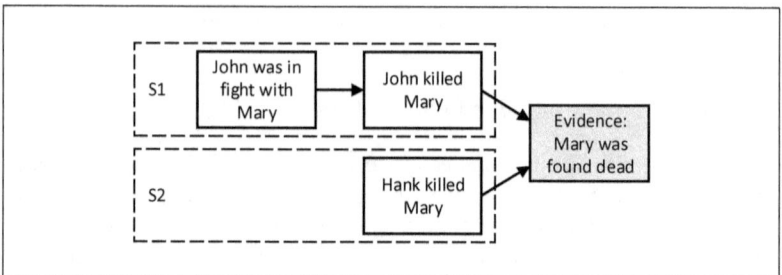

Figure 7: Different stories explaining the evidence, where the arrows indicate causal links.

$c \rightarrow e$ and some observed evidence e, we can infer cause c as a possible explanation of effect e. This cause can be a single event, but it can also be a sequence of events, a story. Taken by itself the abductive scheme is nothing but the fallacy of affirming the consequent. However, in a setting where alternative abductive explanations are generated and compared, it can still be rational to accept an explanation if no better other explanation is available.

Like argumentation, reasoning with stories is dialectical, in that different competing explanatory stories are compared. Consider the two stories in Figure 7, where two possible explanations for the observation that *Mary was found dead* are provided: one story where John killed Mary, and another one where Hank killed Mary. These two stories are alternative explanations for the evidence, and we have to choose between them by, for example, looking for new evidence that supports or attacks the different stories. This reasoning with, and about, stories was first formalised by [Keppens and Zeleznikow, 2003], and later in a series of articles by Bex, Prakken and Verheij ([Bex *et al.*, 2006, 2007b]), in which the *hybrid theory* of stories and arguments is proposed, where individual arguments based on evidence are used to support and attack the different hypothetical stories in the process of inference to the best explanation. This will be further discussed in Section 5.3.3 on evidence.

In addition to stories explaining the evidence as in Figure 7, it is also important that a story is *plausible*: irrespective of the evidence, does the story fit with our ideas about how things generally happen?

Plausibility plays a big part in our reasoning. For example, we would not seriously consider the scenario 'Aliens killed Mary' because this is highly implausible. Furthermore, elements which are implausible at first sight might warrant further investigation: for example, if John has no history of violent behaviour, it seems implausible that he would immediately kill Mary after getting into a (verbal) fight with her (i.e. the causal link between *John was in a fight with Mary* and *John killed Mary* is implausible). Furthermore, stories can contain gaps, missing elements that make them less plausible. One way to look for such gaps is to compare the story to story schemes ([Pennington and Hastie, 1993; Bex, 2009]) or scripts ([Schank and Abelson, 1977]), stereotypical patterns that serve as a scheme for particular stories. Take, for example, a general scenario scheme for intentional action: a motive leads to an action, which has certain consequences. In our example, S2 is less plausible than S1 because it does not include a motive for why Hank killed Mary.

Such reasoning about motives in stories was the subject of further work by Bex, Atkinson and Bench-Capon ([Bex *et al.*, 2009], who define an argumentation scheme for *abductive practical reasoning* based on the regular (non-abductive) scheme for practical reasoning (Section 5.2.2):

Argumentation scheme for abductive practical reasoning
The current circumstances C
are explained by the performance of action A
in the previous circumstances R
with motivation M

Possible critical questions for this scheme are, for example, *Are there alternative ways of explaining the current circumstances S?* or *Can the current explanation be induced by some other motivation?*. Following [Atkinson and Bench-Capon, 2007], the abductive scheme and its critical questions were formalised as action based alternating transition systems, providing a formal semantics for abductive reasoning about motives using stories.

5.1.3 Issue-Based Prediction

[Brüninghaus and Ashley, 2003] and [Ashley and Brüninghaus, 2009] proposed a system for Issue-Based Prediction (IBP) as a descendant of the HYPO and CATO systems (see Sections 3.1.1 and 4.1.5). It predicts outcomes of US trade secret misappropriation cases and provides explanations for its predictions in terms of an argumentation model that combines rule- and case-based reasoning. Cases are as in CATO represented as two sets of factors favouring, respectively, the plaintiff and the defendant. IBP's knowledge model combines a logical decision tree with lists of pro-plaintiff and pro-defendant factors for each of the five leaves of the tree, called the issues (e.g. *did the plaintiff maintain secrecy*, and *did the defendant obtain the secret by improper means?*), as shown in Figure 8. Issues are addressed with a prediction model that according to Ashley and Brüninghaus applies a kind of scientific evidential reasoning. Roughly, if all factors in the case favour the same side for that issue, then IBP predicts a win for that side on the issue (unless all these factors are 'weak'). Otherwise it retrieves precedents that contain all case factors on that issue. If all have the same outcome, then IBP predicts that outcome, otherwise it tests the hypothesis that the side that won the majority of precedents will win, by trying to explain away each precedent won by the other side; this attempt succeeds if the precedent contains a 'knock-out' factor that is not in the current case. IBP's notions of weak and knock-out factors are a refinement of the CATO factor model and are defined in terms of low, respectively, high predictive power for the side they favour. Finally, IBP's predictions on all the issues are combined in an overall prediction.

In an evaluation experiment IBP outperformed 11 other outcome predictors and achieved a high accuracy score of 92%. Although, strictly speaking, IBP does not reason about what to *decide* but about what to *predict* as a decision, [Ashley, 2019], quoting [Aleven, 2003], claims that predictive accuracy is a good (although not the only) measure of the reasonableness of a computational model of argument.

5.1.4 Carneades

In [Gordon et al., 2007; Gordon and Walton, 2009] Tom Gordon and co-authors proposed a new formal argumentation system, with various sources of inspiration. One was [Freeman and Farley, 1996]'s model of argument evaluation with five alternative levels of proof, and another was [Walton, 1996]'s dialogical theory of argumentation schemes. Like Gor-

3 - COMPUTATIONAL MODELS OF LEGAL ARGUMENT

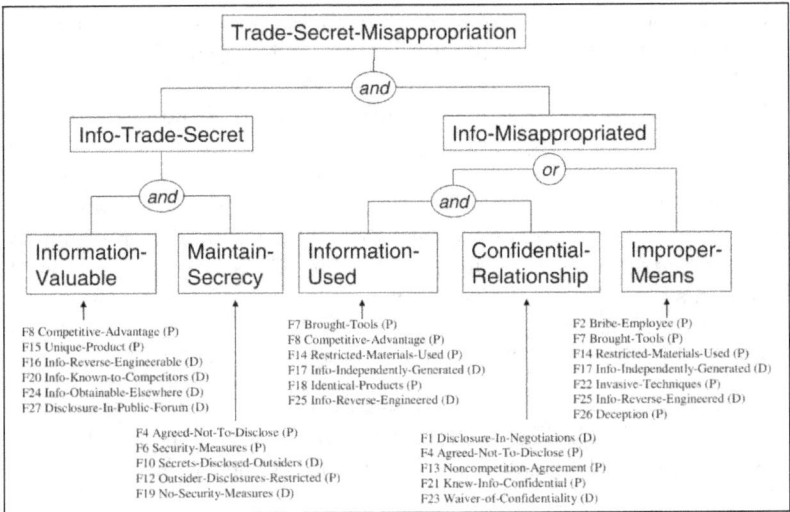

Figure 8: Hierarchy of Issues and Factors, taken from [Ashley and Brüninghaus, 2009]

don's earlier Pleadings Game (see Section 4.3.1) the Carneades system is meant to be used in a dialogical context, although unlike the Pleadings Game it does not explicitly generate dialogues but only records which statements have been accepted or rejected by a given audience. It then incorporates this information in its evaluation of arguments and statements.

Unlike, for instance, ASPIC+, Carneades does not evaluate arguments by generating Dung-style abstract argumentation frameworks. Instead, in Carneades each statement can be assigned its own standard of proof. The system takes not proof burdens but proof standards as the primary concept, and encodes proof burdens with particular assignments of proof standards. A Carneades argument has a set of premises P, a set of exceptions E and a conclusion c, which is either pro or con a statement. Carneades does not assume that premises and conclusions are connected by inference rules but it does allow that arguments instantiate argument schemes. Also, all arguments are elementary, that is, they contain a single inference step; they are combined in recursive definitions of *applicability* of an argument and *acceptability* of its conclusion. In essence, an *argument* is *applicable* if (1) all its premises are given as a fact or are else an acceptable conclusion of another argument and (2), none of its exceptions is given as a fact or is an acceptable conclusion of another argument. A *statement* is *acceptable* if it satisfies its proof standard. Facts are stated by an *audience*, which also provides numerical *weights* of each argument plus *thresholds* for argument weights and differences in argument weights. Three of Carneades' proof standards are then defined as follows:

Statement p satisfies:

- *preponderance of the evidence* iff there exists at least one applicable argument pro p for which the weight is greater than the weight of any applicable argument con p.
- *clear-and-convincing evidence* iff there is an applicable argument A pro p for which:
  * p satisfies *preponderance of the evidence* because of A; and
  * the weight for A exceeds the threshold α, and
  * the difference between the weight of A and the maximum weight of the applicable con arguments exceeds the threshold β.
- *beyond-reasonable-doubt* if and only if p satisfies *clear-and-convincing evidence* and the maximum weight of the applicable con arguments is less than the threshold γ.

Although Carneades was not set up with the aim to generate Dung-style abstract argumentation frameworks, [Van Gijzel and Prakken, 2012] translated Carneades into Dung's frameworks via the ASPIC+ framework of [Prakken, 2010], showing that Carneades induces a unique extension in all semantics. [Brewka and Gordon, 2010] give an alternative reconstruction of Carneades in terms of [Brewka and Woltran, 2010]'s abstract dialectical frameworks. In [Gordon, 2013] a web-based implementation of Carneades is described, including an argument visualisation tool. Some attempts have been made to connect Carnaedes to data by combining ontologies with argumentation [Gordon, 2011].

5.1.5 A Rule-Based Approach: Defeasible Logic

Governatori and others developed an approach to legal knowledge representation and reasoning in the context of Defeasible Logic [Nute, 1994]. This logic does not have an explicit notion of an argument but models recursive notions of defeasible and strict derivability in terms of the application of possibly conflicting prioritized rules. Governatori and colleagues paid much attention to various aspects of legal knowledge representation, such as deontic notions, time and change [Governatori et al., 2005, 2007; Rotolo and Governatori, 2009].

5.1.6 Argumentation Mining

The idea of argumentation mining first arose in AI & Law: [Grover et al., 2003; Hachey and Grover, 2005, 2006] developed machine learning models for the automatic detection of pieces of text that represent rhetorical roles in UK House of Lords judgements. Such rhetorical roles are the elements of arguments in legal texts, for example, facts, citations, but also more direct argumentative types of roles such as refutations against or argumentations for claims. The classification results (i.e. what rhetorical role does this sentence play?) were quite good for a 7-class problem, with F-scores around 55-60%.

Argumentation mining as a separate task (as opposed to a sub-task of summarization as with [Hachey and Grover, 2006]) was widely popularised by the seminal work of [Moens et al., 2007] and [Palau and Moens, 2009] (which was later extended and published as [Mochales and Moens, 2011]) . Where [Moens et al., 2007] detected just elements (i.e. premises, (sub)conclusions) of arguments, [Mochales and Moens, 2011] also detected the structure of arguments, that is, the inference relations between the premises and conclusions. Such detection of structures was generally more difficult than element detection: an accuracy of 60% was achieved when detecting argumentation, while an F1-score of 70% was achieved for recognizing premises and conclusions.

5.1.7 Machine Learning

During this decade machine learning approaches were rather neglected. There was little interest in neural networks as used in the 1990s (e.g. [Bench-Capon, 1993] and [Zeleznikow and Stranieri, 1995]) and the availability of large datasets and techniques for learning from them had not yet reached the stage where they could support later big data approaches [Villata et al., 2022]. Still the decade did produce two interesting examples of using machine learning for argumentation.

The first example concerns Argument Based CN2 (ABCN2), [Možina et al., 2005], is an extension of the well-known rule-learning algorithm CN2 of [Clark and Niblett, 1989]. CN2 is an inductive logic programming algorithm that produces a set of rules that can be used to classify instances in the domain. The central idea is to augment CN2 to accept, along with data, arguments explaining the classification of a small number of instances to improve both the efficiency of the learning process, and the quality of the rules learned. The arguments constrain the search space, and, it is claimed, induce rules that make more sense to domain experts. The process is iterative. After the first pass, the most frequently misclassified example is presented to the expert to give an argument as to why it should be classified in the correct way. The second pass then begins from a rule induced from the expert's argument. The process continues until there are no problematic examples.

The example study was the fictional welfare benefit data first used in [Bench-Capon, 1993] and later used in other projects and now publicly available [Steging et al., 2023]. This example took seven iterations. The rules induced were close to the ideal set, except that two thresholds were slightly low: 59 rather than 60 and 2900 instead of 3000, because there were no examples in the dataset to identify the thresholds precisely.

Further experiments tested robustness in the face of incorrect data. This is important since we cannot guarantee that all examples will have been correctly classified, especially in a domain like welfare benefits, where the error rate is notoriously high. Various noise levels were tested up to 40%. The results showed that ABCN2 outperformed the original CN2 at every level, with the gap widening as noise exceeded 10%.

A second example of the use of Machine Learning in the 2000s was [Ashley and Brüninghaus, 2009]. This paper describes the augmentation of IBP (See Section 5.1.3) with a program, SMILE, which will ascribe factors given natural language input. This means that there is no need to manually analyse the cases: together the programs can predict an outcome based on a natural language description of the facts. SMILE used a dedicated set of rules for each factor. There is no need to dwell on details here, since natural language techniques are now vastly better.

The performance using SMILE fell off drastically from that which had been achieved using IBP with manually ascribed factors outcome-prediction accuracy dropped from 92% to 64%. This suggests that the learning to ascribe factors is rather hard. A better comparison is with machine learning approaches such as [Medvedeva et al., 2019]. Such approaches also fall well short of 90% accuracy, typically achieving something in the 70-85% range. This is true for [Medvedeva et al., 2019], which was tested in the domain of the European Convention on Human Rights, although it fell to 52% for Article 10 with the best performance of 84% on Article 16. Average performance across all articles was 74%. It should be noted that these recent approaches do not learn to extract factors from case texts but instead immediately relate the natural-language case texts to outcomes.

Since the logical model can be constructed to a high degree of accuracy - and, importantly, can provide arguments to justify the prediction - some argue that the machine learning should be used for factor ascription and the outcome determined with a logical model (e.g. [Mumford et al., 2022]). Whether, however, sufficiently accurate performance in ascribing factors can be achieved has yet to be shown. An alternative approach in [Prakken and Ratsma, 2022] is to use a logical model to provide explanatory arguments for the decision reached by a machine learning program. The problem with this approach is that it will provide arguments to justify the 20% or so of incorrect decisions.

5.2 Evaluation of Arguments

5.2.1 Abstract Accounts of Argumentation

In the 1990's Dung's abstract account of argumentation was mainly used as the final stage of a three-stage model of argumentation: construction of arguments, identifying their conflict relations and resolving the conflicts with preference information [Prakken, 1995]. This results in a set of arguments with a defeat relation, to which any semantics of [Dung, 1995] can be applied. Around 2000 an alternative approach emerged, in which arguments are directly encoded in abstract argumentation frameworks [Bench-Capon, 2002a] and in which preference information that is needed to resolve conflicts is added to these abstract frameworks after they have been constructed [Amgoud and Cayrol, 2002; Bench-Capon, 2003]. Briefly, the idea is to say that if argument A attacks argument B and is not inferior to B, then A defeats B. The semantics of abstract argumentation frameworks is then applied with this defeat relation. This idea is explicit in [Amgoud and Cayrol, 2002] and is indirectly modelled by [Bench-Capon, 2003] by attaching a (legal, ethical or societal) value to each argument and evaluating the success of attacks in terms of an ordering on the set of values.

In a series of subsequent papers Atkinson, Bench-Capon and colleagues applied the latter approach to frameworks where the arguments instantiate practical-reasoning argument schemes and in which the critical questions of these schemes are pointers to attacking arguments [Atkinson *et al.*, 2005; Atkinson and Bench-Capon, 2005, 2007]. This approach is very attractive as long as arguments do not have an internal inferential structure, because of the simplicity and elegance of the theory of abstract argumentation frameworks. However, [Modgil and Prakken, 2013] argue that when arguments do have an internal inferential structure, an explicit account of the structure of arguments should be given, in order to apply the preference information to the points at which the arguments conflict. Thus the use of preference information to resolve conflicts comes before the generation of abstract argumentation frameworks. This approach is formalised in the ASPIC+ framework.

5.2.2 Values

[Bench-Capon, 2002a]'s addition of values to abstract argumentation frameworks was inspired by a preceding body of work of himself and others on the use of values in models of case-based reasoning [Bench-Capon and Sartor, 2001], [Bench-Capon, 2002b] [Prakken, 2002] [Sartor, 2002], which work was in turn inspired by [Berman and Hafner, 1993]. Criticising purely factor-based models of case-based reasoning, Berman

3 - COMPUTATIONAL MODELS OF LEGAL ARGUMENT

and Hafner argued that often a factor can be said to favour a decision by virtue of the purposes served or values promoted by taking that decision because of the factor. A choice in case of conflicting factors is then explained in terms of a preference ordering on the purposes, or values, promoted or demoted by the decisions suggested by the factors[17]. Cases can then be compared in terms of the values at stake rather than on the factors they contain.

The role of purpose and value is often illustrated with some well-known cases from Anglo-American property law on ownership of wild animals that are being chased. Here we follow the analysis of three of these cases given by [Bench-Capon, 2002b]. In *Pierson* plaintiff was hunting foxes for sport on open land when defendant killed the chased fox and carried it away. The court held for defendant. In *Keeble* a pond owner placed a duck decoy in his pond with the intention to sell the caught ducks for a living. Defendant used a gun to scare away the ducks, for no other reason than to damage plaintiff's business. Here the court held for plaintiff. Finally, in *Young* both plaintiff and defendant were fishermen fishing in the open sea. Just before plaintiff closed his net, defendant came in and caught the fishes with his own net.

Let us assume that the task is to argue for a decision in *Young* on the basis of *Pierson* and *Keeble*. If cases are only compared on the factors they contain, then no ruling precedent can be found. *Young* contains pro-plaintiff factors absent in *Pierson*, namely, that the plaintiff was pursuing his livelihood, so *Young* can be distinguished from *Pierson*. Moreover, *Young* lacks a pro-plaintiff factor of *Keeble*, namely, that the plaintiff was hunting on his own land, and contains a pro-defendant factor that is not in *Keeble*, namely, that the defendant was also hunting for a living. So *Young* can also be distinguished from *Keeble*.

However, Berman & Hafner convincingly argue that skilled lawyers do not confine themselves to factor-based comparisons, but often frame their arguments in terms of the values that are at stake.[18] [Bench-Capon, 2002b] applies this view to the above cases and assumes that three values are at stake in these cases, viz. economic benefit for society

---

[17]The need for values to resolve issues requiring choice is also found in [Perelman *et al.*, 1980] and Searle [Searle, 2003].

[18]Below we will use 'values' to cover also purposes, policies, interests etc.

($Eval$), legal certainty ($Cval$), and the protection of property ($Pval$). Then a key idea is to specify how case decisions advance values.

- Deciding for a side because that side was hunting for a living advances $Eval$.
- Deciding for a side because that side was hunting on his own land advances $Pval$.
- Deciding for a side because that side had caught the animal advances $Pval$.
- Deciding for a side because the other side had not caught the animal advances $Cval$.

We can then say that *Pierson* was decided for defendant to promote legal certainty and since no values are served by deciding for plaintiff: he was not hunting for a living so economic benefit would not be advanced, and he had not yet caught the fox and was hunting on open land, so there are no property rights to be protected. Further, we can say that *Keeble* was decided for plaintiff since the value of economic benefit and the protection of property are together more important than the value of certainty. Thus *Keeble* also reveals part of an ordering of the values. Finally, in this interpretation of *Pierson* and *Keeble*, *Young* should be decided for defendant: the value of economic benefit does not support plaintiff since defendant was also fishing for his living, the value of protecting property does not apply since plaintiff had not yet caught the fish and was not on his own land, so the only value at stake is certainty, which is served by finding for the defendant.

We now give a general argument-scheme account of the reasoning involved, which captures the essence of how the above-cited papers analyse these cases and which can be formalised in ASPIC+ along the lines of [Bench-Capon *et al.*, 2013]. The first idea is that the specification of how case decisions advance values can be used in the following argument scheme.

Argument scheme from case decisions promoting values
> Deciding *Current Pro* promotes set of values V_1
> Deciding *Current Con* promotes set of values V_2
> V_1 is preferred over V_2
> ――――――――――――――――――――
> Therefore (presumably), *Current* should be decided *Pro*.

Here *Pro* and *Con* are variables ranging over {*Plaintiff*,*Defendant*}. Another idea is that whether a set of values is preferred over another set of values, can be derived from a precedent (as in our example from *Keeble*).

Argument scheme from preference from precedent
Deciding *Precedent Pro* promotes set of values V_1
Deciding *Precedent Con* promotes set of values V_2
Precedent was decided *Pro*
——————————————————————
Therefore (presumably), V_1^+ is preferred over V_2^-

Here the notation V_1^+ denotes any superset of V_1 of values while V_2^- denotes any subset of V_2. This notation captures *a fortiori* reasoning in that if in a new case deciding *Pro* promotes at least V_1 and possibly more values, while deciding *Con* promotes at most V_2, then the new case is even stronger for *Pro* than the precedent.

If it is also given that a proper superset of values is always preferred over a proper subset, then the first scheme directly applies to *Young*, since deciding *Young* for the defendant promotes {*Pval, Eval*} while deciding *Young* for the plaintiff promotes {*Eval*}. However, imagine another new case in which deciding for the plaintiff promotes {*Pval, Eval*} or a superset thereof, while deciding for the defendant promotes {*Cval*}: then the second scheme is needed to infer the preference of the first value set over the second (for instance, from *Keeble*), after which the first scheme can be applied to conclude that the plaintiff should win.

[Bench-Capon and Sartor, 2001, 2003] employ a similar way to express that factor-decision rules promote values, and a similar way to derive rule preferences from the preference ordering on the sets of values they promote. But then they embed this in a method for constructing theories that explain a given set of cases, inspired by [McCarty, 1976]'s view of legal case-based reasoning as theory construction (see Section 4.1.6 above). Theory construction is modelled by Bench-Capon and Sartor as an adversarial process, where both sides take turns to modify the theory so that it explains the current case in the way they want. The process starts with a set of factor-value pairs and a set of cases represented in terms of factors and an outcome. Then the theory is constructed by creating rules plus rule priorities derived from value preferences. This continues until the theory can be applied to give an outcome for the case under consideration. At this point the onus moves to the other party, who must attempt to extend the theory to produce a better theory with an outcome for its favoured side, whereupon, it is again the turn of the original side. This process of extending and refining the theory continues until there is no possible extension of the theory which changes the outcome.

This approach was tested empirically in [Chorley and Bench-Capon, 2005b] and [Chorley and Bench-Capon, 2005a]. The first of these papers explored the use of CATE (CAse Theory Editor) in a series of experiments intended to explore a number of issues relating to the theories constructed using the operators of [Bench-Capon and Sartor, 2003], including how the theories should be constructed, how sets of values should be compared, and the representation of cases using structured values (which are akin to dimensions) as opposed to factors. In CATE, the construction of theories is done by the user, supported by the CATE toolset.

The second paper described AGATHA (Argument Agent for Theory Automation) which was designed to automate the theory construction process, by constructing the theory first as a search over the space of possible theories, and then as a two player dialogue game (which could be played with the AGATHA program playing both sides). A set of search operators and argument moves are defined in terms of the theory constructors and the resulting theories are evaluated according to their explanatory power and their simplicity. The search or game continues until it is not possible to produce a better theory. Several search methods were investigated: brute force and heuristic search using A* and adversarial search using α/β pruning. The results proved to be good, as reported in [Chorley and Bench-Capon, 2005a]:

> AGATHA produces better theories than hand constructed theories as reported in [Chorley and Bench-Capon, 2005b], and theories comparable in explanatory power to the best performing reported technique, IBP [Ashley and Brüninghaus, 2009]. Note also that AGATHA can be used even when there is no accepted structural model of the domain, whereas IBP relies on using the structure provided by the Restatement of Torts.

The attention for the role of value and purpose led to accounts of legal interpretation as a decision problem, namely, as a choice between alternative interpretations on the basis of the likely consequences of these interpretations in terms of promoting and demoting values.

Bench-Capon and Atkinson have studied legal practical reasoning in the context of [Bench-Capon, 2003]'s value-based abstract argumentation frameworks. As explained above in Section 5.2.1, such VAFs extend abstract argumentation frameworks by giving each argument a value that it promotes and by defining a total ordering on these values. Attacks are then resolved by comparing the relative preference of the values promoted by the conflicting arguments. In e.g. [Atkinson et al., 2005; Atkinson and Bench-Capon, 2005] the instantiation is studied of the arguments in VAFs with the following so-called argument scheme for practical reasoning:

In the current circumstances R
Action A should be performed
To bring about new circumstances S
Which will realise goal G
And promote value V

The scheme comes with a list of critical questions that can be used to critique each element of this scheme and to generate counterarguments to uses of the scheme. For example:

CQ1: Are the believed circumstances true?
CQ2: Assuming the circumstances, does the action have the stated consequences?
CQ7: Are there alternative ways of promoting the same value?
CQ9: Does doing the action have a side effect which demotes some other value?
CQ16: Is the value indeed a legitimate value?

This generates a VAF as follows. Instantiations of this scheme are arguments, while arguments for incompatible actions and 'bad' answers to critical questions are counterarguments to such arguments. Then each argument is assigned a value.

Atkinson and Bench-Capon have applied this approach both to legal interpretation and to legislative and policy debates. In [Atkinson et

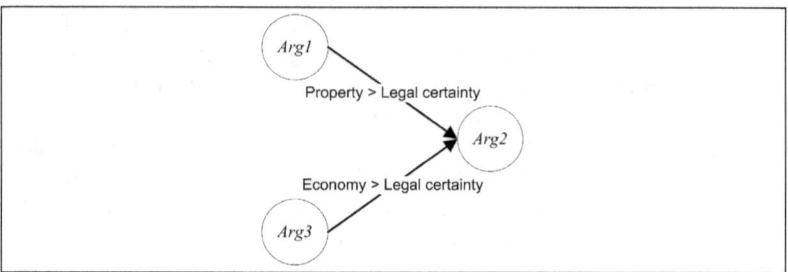

Figure 9: Dung-style AF for the wild animals case

al., 2005] they applied it to reasoning with precedents, representing the *Keeble* case as follows (note that they equate circumstances S and goal G):

Arg1:
Where plaintiff is hunting on his own land
find ownership established
as plaintiff's property is thus respected
which promotes the protection of property

Arg2:
Where plaintiff is hunting for a living
find ownership established
as plaintiff's activities are thus encouraged
Which promotes the economy

Arg3:
Where there is no possession
find ownership not established
as this will reduce litigation
which promotes legal certainty.

Here both Arg1 and Arg3 and Arg2 and Arg3 attack each other. To explain the decision in *Keeble*, the values of protection of property and the economy should be preferred to the value of legal certainty, so that Arg3 is defeated by either Arg1 and Arg2. The resulting abstract argumentation framework is displayed in Figure 9.

3 - COMPUTATIONAL MODELS OF LEGAL ARGUMENT

While this approach has its merits, it also has some limitations. First, it does not deal naturally with aggregation of values promoted by the same decision, unlike the above-discussed *Argument scheme from case decisions promoting values*. Second, different parts of the scheme model different kinds of inference steps. That action A will result in consequences S is (causal) epistemic reasoning, while the step to the value is evaluative and the conclusion that A should be performed is practical reasoning. Now a conflict on whether the action has a certain result is different from a conflict on whether the action should be performed. The latter indeed requires value comparisons but the former is a conflict of epistemic reasoning, to which value considerations do not apply.

An alternative approach is to formulate practical reasoning as a combination of various elementary argument schemes and to embed their use in a framework for argumentation that allows for the stepwise construction of arguments. This approach, briefly discussed above, was applied in the next decade, to be discussed in Section 6 below.

5.2.3 Burden of Proof

Above we saw that [Freeman and Farley, 1996] and [Gordon *et al.*, 2007; Gordon and Walton, 2009] incorporate standards of proof in their models of legal argument. However, [Prakken and Sartor, 2009] argue that standards of proof cannot be applied on their own but are relative to *burdens* of proof, and different phases of a legal proceeding can be about different proof burdens. Generally[19] a distinction is made between a burden to provide evidence on an issue during a proceeding (in common-law systems often called the burden of production) and a burden to prove that a claim is true or justified beyond a given standard of proof (in common-law systems often called the burden of persuasion). If the burden of production on an issue is not met, the issue is decided as a matter of law against the burdened party, while if it is met, the issue is decided in the final stage of the proceeding according to the burden of persuasion. In the law the burdens of production and persuasion are usually determined by the 'operative facts' for a legal claim, i.e. the facts that legally are ordinarily sufficient reasons for the claim. The

---

[19]Much of this Section is taken from [Prakken and Sartor, 2015b].

law often designates the operative facts with rule-exception structures. For instance, for manslaughter the operative facts are that there was a killing and that it was done with intent, while an exception is that it was done in self-defence. Therefore, at the start of a criminal proceeding, the prosecution has the burden to produce evidence on 'killing' and 'intent'; if this burden is fulfilled, the defence's burden to produce evidence for 'self-defence' is activated. For operative facts the burdens of production and persuasion usually go together so in our example the prosecution also has the burden of persuasion for 'killing' and 'intent'. However, for exceptions things are more complicated. In criminal proceedings usually the defence only has a burden of production for an exception while if fulfilled, the prosecution then has an active burden of persuasion against the exception. For instance, once the defence has produced evidence for 'self-defence', the prosecution has the burden of persuasion that there was no self-defence. By contrast, in civil cases often the burden of persuasion holds for an exception also: for instance, in Dutch and Italian law insanity at the time of accepting an offer is an exception to the rule that offer and acceptance create a binding contract, but if the evidence on insanity is balanced, the party claiming insanity will lose on that issue.

This account fits rather well with argumentation-based logics for defeasible reasoning. The idea is that a burden of persuasion for a claim is fulfilled if at the end of a proceeding the claim is sceptically acceptable according to the argumentation logic applied to the then available evidence [Prakken and Sartor, 2009]. However, there is a complication, namely, the just-mentioned possibility in civil cases that the burden of persuasion is distributed over the adversaries. The complication can best be explained in terms of abstract argumentation frameworks. Consider again the above contract example, and consider the following arguments:

P_1: The contract was concluded because there was an offer and acceptance (assuming there is no exception)

O_1: There is an exception since the offeree was insane when accepting the offer (evidence provided)

P_2: The offeree was not insane when accepting the offer, since (evidence provided)

3 - COMPUTATIONAL MODELS OF LEGAL ARGUMENT

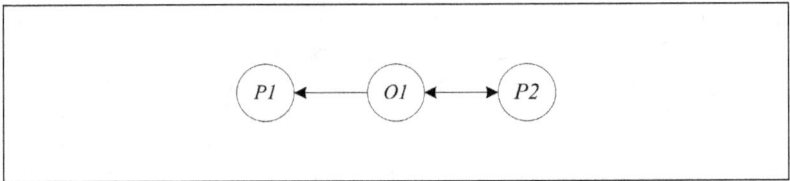

Figure 10: A Dung graph

It seems reasonable to say that argument O_1 strictly defeats P_1, since it refutes P_1's assumption that there is no exception. Assume, furthermore, that O_1 and P_2 are regarded as equally strong (according to any suitable notion of strength). Then it seems reasonable to say that both arguments defeat each other. The resulting Dung graph is displayed in Figure 10:

The grounded extension is empty, while two preferred extensions exist: one with P_1 and P_3 and one with O_1. So the plaintiff has no sceptically acceptable argument for his main claim. Yet according to the law the plaintiff wins, since the defendant has not fulfilled her burden of persuasion as regards her insanity: O_1 is also just defensible.

This is one challenge for a Dung-style approach. Another challenge is to account for the fact that different kinds of legal issues can have different standards of proof. For example, in common-law jurisdictions claims must in criminal cases be proven 'beyond reasonable doubt' while in civil cases usually proof 'on the balance of probabilities' suffices. Consider now again the killing-in-selfdefence example, and assume that the prosecutor has an argument P_1 that the accused killed, the accused has an argument O_1 that he killed in selfdefence, and the prosecutor has an argument P_2 against this argument, which is considerably stronger than its target but not strong enough to satisfy the 'beyond reasonable doubt' proof standard. In a Dungean account defeat is an all-or-nothing matter, so to obtain the legally correct outcome that the accused must be acquitted, in this case O_1 and P_2 must be said to defeat each other (resulting in Figure 10).

One approach to deal with these problems is to give up a Dungean approach. As we saw above in Section 4.2.4, this is what Tom Gordon did in his Carneades system. However, as noted above, his approach arguably conflates the distinction between proof standards and proof burdens. Prakken and Sartor have made various attempts to deal with these problems in a Dungean setting, e.g. in [Prakken and Sartor, 2011; Calegari *et al.*, 2021]. Their most recent attempt is [Prakken and Sartor, 2023], in which they use ASPIC+ as a metalevel formalism to specify decompositions of reasoning problems, where each subproblem can be solved by its own reasoning method, which can be of any kind. Then shifts in the burden of persuasion can be modelled in metalevel rules that explicitly indicate which propositions should be proven, and degrees of defeat can be modelled in evidential problem-solving modules that apply some numerical model of reasoning under uncertainty, such as Bayesian probability theory.

A concept closely related to that of burden of proof is the notion of **presumption**. Legal presumptions obligate a fact finder to draw a particular inference from a proved fact. Typical examples are a presumption that the one who possesses an object in good faith is the owner of the object, or a presumption that when a pedestrian or cyclist is injured in a collision with a car, the accident was the driver's fault. Some presumptions are rebuttable while others are irrebuttable.

The logical interpretation of (rebuttable) presumptions is less complicated than for burdens of proof but not completely trivial. [Prakken and Sartor, 2008] argue that the function of legal presumptions is not to *allocate* a proof burden but to *fulfil* it. More precisely, the interpret presumptions are default rules or default conditionals, which can be used in arguments for claims that have a proof burden attached to them.

Further logical issues concerning presumptions and burdens of proof are discussed in [Prakken and Sartor, 2007a, 2008, 2009].

5.2.4 Accrual

One recurring theme in the computational study of argumentation is that of accrual of arguments, or how several arguments for the same conclusion should be combined. This issue has been especially (although not exclusively) been studied in the context of AI & Law. The main issue is whether accrual should be modelled at the knowledge representation level, by combining different reasons for the same conclusion in antecedents of rules (in [Prakken, 2005b] called the *KR approach*), or whether it should be modelled at the logical level as a logical operation on arguments (in [Prakken, 2005b] called the *inference approach*). Early work on the inference approach was [Verheij, 1996]'s Cumula system and [Hage, 1996]'s Reason-Based Logic. [Prakken, 2005b] proposed three principles that any model of argument accrual should satisfy, namely:

- An accrual is sometimes weaker than its elements (since reasons can interact, as in 'both heat and rain are a reason not to go jogging, but the combination is so pleasant that it is a reason to go jogging');

- an accrual makes its elements inapplicable (for instance, if it is hot and rainy, then the individual fact that it is hot cannot be used any more in an argument);

- flawed reasons or arguments may not accrue (for instance, if the argument that it rains can be refuted, then the argument 'it rains, therefore I should not go jogging' cannot be accrued with the argument 'it is hot, therefore I should not go jogging'; only the latter argument should be considered).

[Prakken, 2005b] then proposed an inference-based model that satisfied these three principles in terms of a combination of Dung's theory of abstract of argumentation frameworks with [Pollock, 1995]'s theory of defeasible reasons. The key idea was to label conclusions of individual arguments and to have a defeasible accrual reason $\varphi^{l1}, \ldots, \varphi^{ln} \Rightarrow \varphi$ that can be applied to the conclusions of a set of arguments with the same conclusion when unlabelled.

5.3 Use of Arguments

5.3.1 Dialogue Games

Research on dialogue systems continued in this decade. Partly motivated by the earlier AI and Law work on dialogue systems, [Prakken, 2005a] proposed a general framework for specifying systems for two-party persuasion dialogue, and then instantiated it with some example protocols. The framework largely abstracts from the logical language, the logic and

Acts	Attacks	Surrenders
claim φ	why φ	concede φ
φ since S	why $\psi (\psi \in S)$	concede ψ $(\psi \in S)$
	φ' since S' (φ' since S' defeats φ since S)	concede φ
why φ	φ since S	retract φ
concede φ		
retract φ		

Table 1: An example communication language [Prakken, 2005a]

the communication language but the logic is assumed to be argument-based (in fact a preliminary version of the ASPIC+ framework) and to conform to grounded semantics.

A main motivation of the framework is to ensure focus of dialogues while yet allowing for freedom to move alternative replies and to postpone replies. This is achieved with two main features of the framework. Firstly, an explicit reply structure on the communication language is assumed, where each move either *attacks* or *surrenders to* its target. An example language of this format is displayed in Table 1. Secondly, winning is defined for each dialogue, whether terminated or not, and it is defined in terms of a notion of *dialogical status* of moves. The *dialogical status* of a move is recursively defined as follows, exploiting the tree structure of dialogues. A move is *in* if it is surrendered or else if all its attacking replies are *out*. (This implies that a move without replies is *in*). And a move is *out* if it has a reply that is *in*. Then a dialogue is (currently) won by the proponent if its initial move is *in* while it is (currently) won by the opponent otherwise.

Together, these two features of the framework allow for a notion of relevance that ensures focus while yet leaving the desired degree of freedom: a move is *relevant* just in case making its target *out* would make the speaker the current winner. Termination is defined as the situation that a player is to move but has no legal moves.

[Prakken et al., 2005] applied the framework to specify a protocol for dialogues about who has the burden of proof for a given claim. [Prakken, 2008a] extended an instance of the framework with a neutral third party in order to model so-called adjudicator dialogues, in which an adjudicator monitors whether the adversaries respect the protocol and in the end decides the dispute. The main feature of the model is a division into an argumentation phase, where the adversaries plea their case and the adjudicator has a largely mediating role, and a decision phase, where the adjudicator decides the dispute on the basis of the claims, arguments and evidence put forward in the argumentation phase. The model allows for explicit decisions on admissibility of evidence and burden of proof by the adjudicator in the argumentation phase. Adjudication is modelled as putting forward arguments, in particular undercutting and priority arguments, in the decision phase. [Prakken, 2008b] applied this model in a case study to a Dutch civil ownerships dispute.

When a dialogue protocol is fully specified in some formal language, then its metatheory can be investigated with the help of automated reasoning tools. [Brewka, 2001] specified his protocols in a dialect of the situation calculus and [Artikis et al., 2003] formalised variations of Brewka's protocols in the C^+ language of [Giunchiglia et al., 2004]. They then used implemented tools to verify various properties, such as the minimal length of dialogues that reach a certain state given a certain initial state. Another benefit of a logical formalisation of a dialogue protocol is that this supports the automatic execution of protocols. To this end, [Bodenstaff et al., 2006] formalised an instance of [Prakken, 2005a]'s framework in [Shanahan, 1999]'s version of the 'full' Event Calculus and then implemented it as a Prolog program. The implementation computes in any state of a dialogue the players' commitments, whether the moves made were legal, who is to move and what are the legal next moves. It can thus be used as a 'dialogue consultant' by a player, adjudicator or external observer.

Another strand of work was the dialogue protocols developed at the University of Liverpool. [Atkinson et al., 2006a] embedded [Atkinson et al., 2005; Atkinson and Bench-Capon, 2005]'s modelling of practical-reasoning argument schemes in a dialogue protocol in which the critical questions of the schemes drive the dialogue.

[Wardeh et al., 2009] used datamining for extraction association rules from case bases concerning the classification of routine claims for a hypothetical welfare benefit. They defined a dialogue game for refining the mined association rules through a dialogue with moves based on case-based reasoning systems such as CATO (see Section 4.1.5 above), including moves for citing, distinguishing, giving counter examples, and for pointing out unwanted consequences of a rule. The main idea is that during the course of the dialogue the rule is refined so that when the dialogue was complete, the winning rule is available to justify and explain an outcome. The authors also defined game strategies and tactics.

5.3.2 Dynamics of Case Law

It is a feature of case law that it evolves over time, with decisions being refined, and overruled in response to novel fact situations and social practices and values. It had received some attention in AI and Law (e.g. [Rissland and Friedman, 1995] and [Berman and Hafner, 1995]). In [Henderson and Bench-Capon, 2001] argumentation was used to address the topic. The paper was based on some remarks by Levi [Levi, 1948]:

> "Reasoning by example shows the decisive role which the common ideas of the society and the distinctions made by experts can have in shaping the law. The movement of common or expert concepts into the law may be followed. The concept is suggested in arguing difference or similarity in a brief ... In subsequent cases, the idea is given further definition and is tied to other ideas which have been accepted by courts. It is no longer the idea which was commonly held by society"

The particular domain was the notion of whether a person owes a duty of care in virtue of their occupation. The idea was to start from a "common sense" ontology of occupations. Then given a set of cases stating whether occupation owed a duty of care or not arguments could be constructed for new occupations. Arguments for could be found by finding the closest common ancestors covering both the current case and a pro-case and a contra-case. Arguments for and against could then be generated by pointing to the similarities and differences between the current and previous cases. The decision indicates which arguments are accepted and hence which rules are established, thus developing a specifically legal concept. As more cases are decided these rules are narrowed and broadened and exceptions identified. The implemented program illustrated this process of reinterpretation and modification. One result was that the order in which cases were presented was shown to matter: given a different sequence, different arguments will be available, and the final rule may be different also.

5.3.3 Evidence

Work in AI and Law on evidential reasoning started in 2003 with articles by [Keppens and Zeleznikow, 2003], [Bex et al., 2003] and [Prakken, 2004]. The latter two of these focus on (logical) evidential argumentation, where the knowledge base contains the evidence in a case and non-legal, common-sense rules are used to reason towards a conclusion. [Bex et al., 2003] discuss a number of argumentation schemes for evidential reasoning, such as the *scheme from appeal to expert opinion* (Section 5.1.1), the *scheme from appeal to general knowledge* and the *scheme from appeal to witness testimony*. Furthermore, they also discuss general schemes for inferences from perception or memory ([Pollock, 1995]), and their relation to, for example, the witness testimony scheme. Inspired by [Loui and Norman, 1995] (Section 4.1.2), [Bex and Prakken, 2004] show that the witness testimony scheme "if a witness testifies that P is the case then usually P is the case" can be unpacked into "if a witness testifies that they observed P then usually they believe that they observed P", "if a witness believes that they observed P then usually their senses gave evidence of P" and "if a witness' senses gave evidence of P then usually P is the case".

Where [Bex et al., 2003] focus mostly on arguments from evidence towards some conclusion, [Keppens and Zeleznikow, 2003] use scenarios (or stories, cf. Section 5.1.2) to explain the evidence. [Bex et al., 2006, 2007b, 2010] combined argument-based and story-based reasoning in one hybrid theory, where arguments based on evidence can be used to support or attack explanatory stories. Figure 11 shows how the two stories explaining Mary's death can be supported and attacked by arguments based on evidence. For story S1, we have two supporting arguments

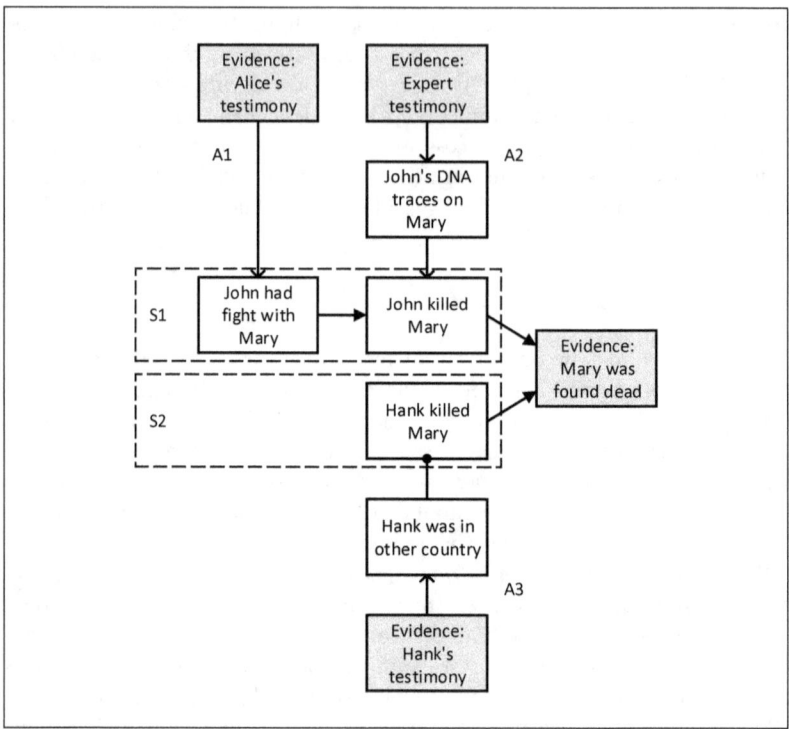

Figure 11: Arguments based on evidence supporting (A1, A2) and attacking (A3) the two stories about Mary's death.

based on a witness testimony and expert DNA evidence. Story S2 is attacked by the fact that Hank was in another country as testified by himself. The work on the hybrid theory was used as a basis for sensemaking using argument and scenario diagramming in the AVERs tool [van den Braak *et al.*, 2007; Bex *et al.*, 2007a] (see also Section 5.3.4).

5.3.4 Implemented argument structure and evaluation

Following the work in the 1990s on argumentation and dialogue software (Section 4.3.4), work continued on systems for various argumentation

3 - COMPUTATIONAL MODELS OF LEGAL ARGUMENT

tasks, often using formalized graphical representation formats.

Verheij [2003b] continued the work on the ArguMed system (see Section 4.3.4) by extending its expressiveness and developing a corresponding formalization of the logic of argumentation (DefLog, Verheij [2003a]). In the resulting system (ArguMed based on DefLog), the graphical elements in a diagram correspond to formal sentences: each box (representing a statement) corresponds to an elementary proposition in the logic, and each arrow to a conditional sentence. There are conditional representations of supporting and attacking reasons, and the conditional relations can themselves be supported and attacked (using nested conditional sentences). By this mechanism, undercutting defeaters attacking the connection between a reason and the conclusion it supports [Pollock, 1995] can be modelled as an attack on a conditional and warrants supporting the connection between a reason and its conclusion [Toulmin, 1958] as support of a conditional. The software computes which statements are justified and which defeated by evaluating the *prima facie* assumptions in the system (using a formal generalization of the stable semantics of abstract argumentation frameworks [Dung, 1995]). For instance, the example argument in the introduction (Figure 1) can be represented in ArguMed using the following formal sentences:

```
Mary is original owner (an elementary sentence)
John is the buyer
if(Mary is original owner, Mary is owner)
(a conditional sentence expressing support)
if(John is the buyer, John is owner)
if(John was not bona fide, x(if(John is the
                            buyer, John is owner)))
(a conditional sentence expressing attack, indicated using the
so-called dialectical negation x)
inc(John is owner, Mary is owner)
(abbreviating incompatibility, i.e. that the two staments attack each other)
if(John bought the bike for 20 euros, John was
                            not bona fide)
John bought the bike for 20 euros
if(John bought the bike for 25 euros,
```

```
         x(John bought the bike for 20 euros))
   John bought the bike for 25 euros
   if(John bought the bike for 25 euros, John was
                                          not bona fide)
```

Evaluating this set of sentences as *prima facie* assumptions in DefLog, gives the result visualized in Figure 1: all sentences listed above are justified, except for the sentences John bought the bike for 20 euros and if(John was not bona fide, x(if(John is the buyer, John is owner))), which are defeated sentences in the *prima facie* theory. The sentences Mary is owner and John was not bona fide are justified by derivation from other justfied sentences (using Modus Ponens) and sentences John is owner and John bought the bike for 20 euros are defeated by derivation from other justfied sentences.

The Carneades model of argument [Gordon *et al.*, 2007] continued on the formalization of argument structure and evaluation. The model design was aimed as the basis for software development. A characteristic feature of the model is that it included proof standards. For each premise, a burden of proof can be allocated to a proponent and opponent. See Section 5.1.4 for an extended discussion.

Ashley *et al.* [2007] study the use of argument diagramming in the context of intelligent tutoring systems. The system described, LARGO, is developed to train the legal reasoning skills of first year law students. The focus is on hypothetical legal reasoning on the basis of US Supreme Court cases. Students can propose a rule deciding a case, challenge such a rule, and continue the discussion by proposing analogies or distinctions (cf. also HYPO, see Section 3.1.1). The system includes domain-specific critical questions thereby allowing for system feedback on weaknesses in an argument move. The system's user interface provides a graphical representation of the argument structure.

The AVERs system [van den Braak *et al.*, 2007] was developed in order to support the process of making sense of the evidence collected in crime investigation. The approach is hybrid in the sense that it combines argumentation and scenario elements. Initially observed facts can be connected to hypothesized events, and combined into stories. The elements of stories can be supported by evidence using arguments. The AVERs system supports the visualization of hybrid diagrams of events and evidence connected by arguments. The system is connected to the work on a hybrid combination of arguments, stories and evidence discussed in Section 5.3.3 [Bex et al., 2007b; Bex, 2011].

Verheij [2007] discusses boxes-and-arrows diagramming in the context of argumentation support software, discussing both opportunities and limitations. Suggestions to go beyond boxes-and-arrows diagram include the need for more expressiveness than what boxes and arrows allow, the inclusion of argumentation schemes (see Section 5.1.1), refocusing on natural language, and simplified diagram structures that may be more helpful.

5.3.5 Web Based Dialogues

Enabling citizens to engage in dialogue with their governments is an important feature of a democracy. Whilst this had for some time been conducted by traditional means, such as writing letters, attending town hall debates and holding individual local 'surgeries', new web-based methods of interaction have been developed to exploit emerging digital technologies. A number of such tools have been developed that make use of computational models of argument. One key example is the Parmendies tool [Atkinson et al., 2006b], where the aim was to present to members of the public a policy proposal for their review and critique.

The policy was presented as an instantiation of the practical reasoning argumentation scheme with values [Atkinson et al., 2006a], discussed earlier in this section. Using the scheme enabled presentation of the current situation the policy scenario was arising in and what the policy proposed was meant to achieve in terms of facts, goals and values. This policy proposal was all presented to the user through a webpage. They were then given the opportunity to critique the policy in terms of relevant critical questions characteristic of the practical reasoning argumentation scheme. The critical questions were posed systematically through navigation to subsequent webpages, to tease apart the precise points of disagreements and motives for these that a user may wish to express about the policy. Different people might disagree with how the current policy situation was expressed, others might question whether the policy would achieve the intended ends, and yet others might oppose these ends because the do not subscribe to the values the ends promote.

The tool was thus intended to enable a form of web-based dialogue between citizens and policy makers. Parmenides later formed the basis for the development of a richer 'Structured Consultation Tool (SCT)' [Bench-Capon et al., 2015] produced as part of the IMPACT project[20].

5.3.6 Game Theory

In a dialogical setting issues of strategy and choice naturally arise but in a legal context they have not been much investigated. In [Riveret et al., 2007; Roth et al., 2007; Riveret et al., 2008] game theory was applied to the problem of determining optimal strategies in adjudication debates (see also [Sartor et al., 2009]). In such debates, a neutral third party (for example, a judge or a jury) decides at the end of the debate whether to accept the statements that the opposing parties have made during the debate, so the opposing parties must make estimates about how likely it is that the premises of their arguments will be accepted by the adjudicator. Moreover, they may have preferences over the outcome of a debate, so that optimal strategies are determined by two factors: the probability of acceptance of their arguments' premises by the adjudicator and the costs/benefits of such arguments. In [Riveret et al., 2007] the logical basis is Defeasible Logic [Antoniou et al., 2000]; in [Roth et al., 2007] it is a dynamic version of the argument game of [Prakken and Sartor, 1997]; and in [Riveret et al., 2008] an abstract argument game.

6 2010s: Computational Argumentation as a Field

By the 2010s, argumentation approaches have become influential in Artificial Intelligence and attempts are made at standardization. This decade saw a deepening of understanding of many themes and topics, such as machine learning and evidence. In the study of the latter, Bayesian networks and other probabilistic approaches receive more attention.

---

[20]Integrated Method for Policy Making Using Argument Modelling and Computer Assisted Text Analysis, in the European Framework 7 project (Grant Agreement No 247228) in the ICT for Governance and Policy Modeling theme (ICT-2009.7.3).

6.1 Argument Generation

6.1.1 Rule-Based Approaches

Governatori and colleagues continued their work on applying Defeasible Logic to legal reasoning [Governatori and Sartor, 2010; Governatori et al., 2013; Rotolo et al., 2015]. Satoh and colleagues developed an alternative rule-based approach close to logic programming called PROLEG [Satoh et al., 2012].

6.1.2 Argumentation Schemes

In the 2010s the idea of using argument schemes for case-based reasoning in the context of a general structured account of argumentation (see Section 5.2.2) was further developed. This was first done by [Bench-Capon and Prakken, 2010], who semiformally sketched how a collection of argument schemes involving value-based reasoning can be formalised in argumentation logics. They applied these schemes in a semiformal reconstruction of various US Supreme Court cases concerning the automobile exception to the constitutional protection against unreasonable searches and seizures. Next, [Prakken et al., 2015] (first online in 2013) gave a full formalisation in ASPIC+ of a set of argument schemes modelling CATO-style case-based reasoning. Building on this work, [Bench-Capon et al., 2013] added schemes for value-based reasoning, while [Atkinson et al., 2013] added schemes for reasoning with dimensions. In all this work the idea is that argument schemes can be formalised as defeasible rules in ASPIC+ (or a similar system) while critical questions are pointers to rebutting or undercutting counterarguments.

6.1.3 Machine Learning

Even though the focus in AI & Law was very much still on formal logical models, at least when it concerned argumentation in AI & Law, there were already a few authors that included machine learning in their work on argumentation.

[Ashley and Walker, 2013] use logical models of argument to represent legal rules and the reasoning from evidence to some (legal) conclusion. Thus, more high-level legal concepts are decomposed into facts (entities, events) that can be more readily mined from texts using machine learning methods. They further annotate such facts and confidence levels in a legal corpus about vaccine injury tort cases. Their ultimate aim is to develop a QA-system.

[Schraagen and Bex, 2018] also propose a QA-system based on argumentation to assist the Dutch Police in the assessment of crime reports submitted by civilians. Similar to [Ashley and Walker, 2013], they use a logical argumentation model to decompose legal concepts into facts, the latter of which can be gathered by extracting them from the initial user input using machine learning NLP ([Schraagen et al., 2017; Schraagen and Bex, 2019] provide explorations in this regard), or by asking the user relevant questions based on the conclusions that can be drawn from the argumentation model at any time. The aim of [Schraagen and Bex, 2018] was to learn policies for such question-asking using reinforcement learning. Further information about the ideas expressed in this paper are discussed in Section 6.1.3.

6.1.4 Case Models and Argument Validity

Verheij [2017a] develops the formal connections between arguments, rules and cases, building on the idea that legal argument has two typical kinds of backing, namely cases and rules. Cases are used as the formal semantics of rule-based arguments, in the sense that cases formally determine which rules and arguments hold ('are valid').

An argument from certain premises to certain conclusions can have one of three types of validity, given a so-called case model. An argument is coherent if there exists a case (in the given case model) in which the argument's premises and the conclusions both hold. A coherent argument is conclusive if in each case in which the premises hold also the conclusions hold. In a case model, cases also come with an ordering, for instance representing their exceptionality. A case lower in the ordering is more exceptional. Using the ordering relation, a third kind of validity can be defined, corresponding to the idea of defeasible reasoning: A coherent argument is presumptively valid if there is a case in which both the premises and conclusions hold, such that the case is at least as high in the ordering as each case in which the argument's premises hold.

3 - COMPUTATIONAL MODELS OF LEGAL ARGUMENT

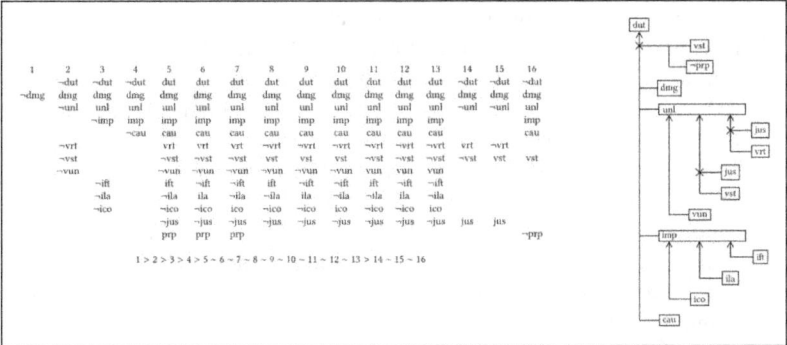

Figure 12: A case model (left) validating key arguments in Dutch tort law (right) [Verheij, 2017a]

The paper discusses formal analogy and distinction between cases (but not using the more fine-grained apparatus of factors and dimensions, as in HYPO; cf. Section 3.1.1). The approach also allows for the formalization of rebutting, undercutting and undermining attack.

A case model is presented that is the formal basis of a concrete, realistic legal domain, namely Dutch tort law about the duty to pay for damages (Figure 12). In that model, the least exceptional case (represented in the figure in the far left column, numbered 1) is a case in which there are no damages (¬**dmg**), so the question about a duty to pay damages does not arise. As a result, a presumptively valid conclusion is that there are no damages. Cases in which there are damages are more exceptional (the cases numbered 2 to 16 in the figure). The least exceptional cases in which there is a duty to pay for the damages (**dut**) correspond to the various combinations of conditions that determine such a duty (cases 5–13). An argument with such conditions as premises (e.g. **dmg** ∧ **unl** ∧ **imp** ∧ **cau** representing that there are damages by an unlawful act imputable to the actor causing the damages) presumptively justifies the conclusion that damages have to be paid. Cases with special circumstances, such as grounds of justification (**jus**, in cases 14 and 15), are more exceptional and do not imply a duty to pay. If such grounds are among the premises, there are so-called defeating circumstances attacking the argument to pay.

The case model approach suggests new ways to the formal integration of cases, rules and arguments, in which formalized cases provide a formal semantics for rule-based argument structure. Qualitative and quantitative representation results are given in [Verheij, 2016a]. Initial ideas of the case model approach to argument validity were developed in the context of evidence [Verheij, 2014, 2017b], further discussed in Section 6.4.3. The case model approach was also applied to the modeling of value-based argumentation [Verheij *et al.*, 2016], focusing on time dependence as in the work by Berman and Hafner on New York tort cases (see Section 4.2.5). See Section 6.2.1 for other work on value-based argumentation in the 2010s.

6.2 Evaluation of Arguments

6.2.1 Values

Values received attention in a variety of different works by different groups of authors during this decade.

A series of papers by Grabmair and colleagues [Grabmair and Ashley, 2010] [Grabmair and Ashley, 2011], plus Grabmair's PhD thesis [Grabmair, 2016], feature values in a formal account of legal reasoning. The formal models set out in this suite of work capture the notion of one factual situation being preferable over another by virtue of the situations' respective effects on values. In [Grabmair and Ashley, 2010] legal sources are modelled as sets of value judgments and legal methodologies as collections of argumentation schemes. This model is then put to use to enable hypothetical reasoning within decision making on legal cases. The work was then extended in [Grabmair and Ashley, 2011] to further enable fine-grained case comparison whereby intermediate legal concepts were captured to determine their impact on the applicable values. A full 'value judgment formalism' is set out in detail to capture the interaction between facts and values, yielding the ability to produce arguments comparing cases within the task of legal case-based reasoning. Grabmair's PhD thesis [Grabmair, 2016] adds to the formalism an experimental implementation of the value judgment formalism and demonstrates how this implementation is capable of arguing about, and predicting outcomes of, cases from the CATO trade secret misappropri-

3 - COMPUTATIONAL MODELS OF LEGAL ARGUMENT

ation dataset.

In a separate strand of work on values, [Sartor, 2010] studied how legal choices, and in particular legislative determinations, need to consider multiple rights and values, and can be assessed accordingly. Recognising that legal norms often prescribe the pursuit of conflicting goals, Sartor sets out a model of teleological reasoning through which legislative action is guided not only by constitutional 'action-norms', but also by constitutional 'goal-norms', that inform the legislator's teleological reasoning about which values should be advanced. The formal model provided is intended to capture the space of legislative and administrative actions by evaluating the teleological appropriateness of legislative choices.

With the increasing prevalence of multi-agent systems being deployed in real world scenarios, the work in [Bench-Capon and Modgil, 2017] advocates for such agents being equipped with the ability to reason about a system's norms, achieved by reasoning about the social and moral values that norms are designed to serve. A specific focus is placed on reasoning in circumstances where it can be argued that the rules should be broken and a decision should be made on whether compliance with the norms should be upheld and, if not, how best to violate the norms. To enable this reasoning to be undertaken, the practical reasoning argumentation scheme with values is used to generate arguments for and against actions such that agents can choose between actions based on their preferences over the values.

The key focus of the paper is on this argument-based account of practical reasoning, which can be used to consider when norms should be violated. The approach is illustrated using a road traffic example that characterises scenarios where the quandary on norm violation may occur. A second and related contribution of the paper is the consideration of what makes an ordering on values acceptable and how such an ordering might be determined.

A final key piece of work from the decade is [Bench-Capon and Atkinson, 2017], which looks at the interaction between dimensions and values. Building on well-established formalisations of factor-based reasoning [Horty, 2017] [Rigoni, 2018], it is shown how values can play several distinct roles in these accounts of legal reasoning, both by explaining preferences between factors and indicating the purposes of the law.

6.2.2 Accrual

Above in Section 5.2.4 early work was described on the accrual of reasons or arguments for the same conclusion. More recently, Gordon's proposed a new model of accrual in a new version of his Carneades system [Gordon and Walton, 2016; Gordon, 2018], with added expressiveness. Then [Prakken, 2019] proposed a new approach in the context of ASPIC+, motivated by some shortcomings in [Prakken, 2005b] and in Gordon's new proposals. [Prakken, 2019] shows that the new proposal satisfies the three proposals of accrual proposed by [Prakken, 2005b] while avoiding the shortcomings of the earlier work.

6.3 Precedential constraint

Early AI and Law work on case-based reasoning was primarily rhetorical in nature in that it was not about computing an 'outcome' or 'winner' of a dispute but instead about generating debates as they can take place between 'good' lawyers. Later work was more logic- and outcome-oriented [Loui *et al.*, 1993; Hage, 1993; Prakken and Sartor, 1998]. This later work inspired a line of research initiated by [Horty, 2011], which aims to formalise the common-law concept of *precedential constraint*, that is, to characterise the conditions under which a decision in a new case is forced or at least allowed by a body of precedents. This is a problem hardly addressed in the initial work on the HYPO and CATO systems. Initially Horty only studied factor-based reasoning but recently he has adapted his approach to dimensions [Horty, 2017, 2019, 2021].

In the factor-based models precedents are simply represented as in CATO, with two sets of factors, respectively, pro and con a boolean decision. Horty's simplest model of precedential constraint is the *result model*, which regards a decision in a new case as *forced* if the precedent cannot be distinguished in the new case, that is, if the new case contains at least all factors pro the decision that the precedent has, while it contains at most all factors con that decision in the precedent. Then a decision is *allowed* if the opposite decision is not forced.

Horty's *reason model* is somewhat more involved. First, it allows to say that in a precedent a subset of the pro decision factors was sufficient for the court to outweigh all the con-decision factors in the case. This subset is called the *rule* of the case. Next, Horty adapts the idea of [Prakken and Sartor, 1998] that a case decision expresses a preference for the pro-decision factors over the con-decision factors in the case. In the following definition, $pro(c)$ and $con(c)$ denote, respectively, the factors pro and con the decision s in case c, $\overline{s}$ is the opposite decision, while $ppro(c)$, which is a subset of $pro(c)$, is the rule of the case. Hence a case decision expresses a preference for any pro-decision set containing at least the pro-decision factors of the case over any con-decision set containing at most the con-decision factors of the case. As in [Prakken and Sartor, 1998], this allows *a fortiori* reasoning from a precedent adding pro-decision factors and/or deleting con-decision factors.

Let $(ppro(c) \cup con(c), pro(c), s)$ be a case, CB a case base and X and Y sets favouring s and $\overline{s}$, respectively. Then

1. $Y <_c X$ iff $Y \subseteq con(c)$ and $X \supseteq pro(c)$;
2. $Y <_{CB} X$ iff $Y <_c X$ for some $c \in CB$.

Next Horty defines a case base CB to be *inconsistent* if and only if there are factor sets X and Y such that $X <_{CB} Y$ and $Y <_{CB} X$. Then CB is *consistent* if and only if it is not inconsistent. Then Horty defines a decision to be *forced* according to the *reason model* if that is the only way to keep the case base consistent when updated with the new case. Horty proves that for consistent cases bases his result and reason model are equivalent on the assumption that $pro(c) = ppro(c)$ for all cases c.

Quite recently, [van Woerkom *et al.*, 2023] generalised the factor-based result model to deal with hierarchical relations between factors as in CATO's factor hierarchy (see Section 4.1.5), while [Canavotto and Horty, 2023] have done the same for the factor-based reason model. In both cases the definition of precedential constraint is made recursive to allow for reuse of precedents for intermediate decisions.

[Horty, 2019]'s *result model* for *dimension-based* case-based reasoning is quite simple. Cases are now represented as a set of value assignments to dimensions plus a boolean decision. More formally, a *dimension* is a set of values V with two partial orders $\leq_s$ and $\leq_{\overline{s}}$ on V such that $v \leq_s v'$ iff $v' \leq_{\overline{s}} v$. These orderings capture the extent to which different values of a dimension are better for one side and so worse for the other side. Note the difference with [Ashley, 1990]: while Ashley regarded a dimension as always favouring a particular side (although to different degrees), in Horty's approach all that can be said is whether one value favours a side more than another value.

A *value assignment* is a pair (d, v). The functional notation $v(d) = x$ denotes the value x of dimension d. Then a (dimension-based) *case* is a pair $c = (F, outcome(c))$ such that D is a set of dimensions, F is a set of value assignments to all dimensions in D and $outcome(c) \in \{s, \overline{s}\}$. Then a (dimension-based) *case base* is as before a set of cases, but now explicitly assumed to be relative to a set D of dimensions in that all cases assign values to a dimension d iff $d \in D$. Likewise, a (dimension-based) *fact situation* is now an assignment of values to all dimensions in D. As for notation, $v(d, c)$ denotes the value of dimension d in case or fact situation c.

In Horty's *dimension-based result model* of precedential constraint a decision in a fact situation is *forced* iff there exists a precedent c for that decision such that on each dimension the fact situation is at least as favourable for that decision as the precedent. This is formalised with the help of the following preference relation between sets of value assignments.

> Let F and F' be two fact situations with the same set of dimensions. Then $F \leq_s F'$ iff for all $(d, v) \in F$ and all $(d, v') \in F'$ it holds that $v(d) \leq_s v'(d)$.

Then, given a case base CB, deciding fact situation F for s is *forced* iff there exists a case $c = (F', s)$ in CB such that $F' \leq_s F$.

Defining a *dimension-based reason model* of precedential constraint is far more complicated. The main problem is how to define that a subset of the value assignments 'pro' a decision outweighs the value assignments 'con' the decision given that formally value assignments are not categorically pro or con a decision but only better or worse for a decision than other value assignments. In fact, Horty had to revise his initial proposal of [Horty, 2017, 2019] in [Horty, 2021], because of some counterintuitive outcomes of the initial proposal. [Rigoni, 2018] proposes an alternative dimension-based reason model. See [Prakken, 2021] for a formal analysis of these and some other factor- and reason-based models of precedential constraint.

6.4 Use of Arguments

6.4.1 Applications

Numerous example applications have been produced to evaluate evolving computational models of argument, including some scenarios posed by real world problems in an industrial setting. Four such characteristic examples are highlighted here.

The first is a feasibility study conducted in collaboration with a large law firm, and reported in [Al-Abdulkarim *et al.*, 2019], to build a practical system using the ANGELIC methodology [Al-Abdulkarim *et al.*, 2016b] described earlier in this section. In the application, a body of case law relevant for the business was captured as a so-called Abstract Dialectical Framework (ADF) using the ANGELIC methodology. ADFs [Brewka and Woltran, 2010] are a generalisation of Dungian abstract argumentation frameworks. The domain of case law was claims for noise-induced hearing loss against employers. The study involved identification of usable arguments that are key to guiding case handlers in assessing the strength of a negligent hearing loss claim and whether or not it had reasonable prospect of defence. The application of the methodology, and thus the use of ADFs, in this application scenario was shown to be very effective in modelling the domain and providing assistance to case handlers in identifying the arguments relevant for deciding the cases. Subsequently, this line of research was extended [Atkinson *et al.*, 2019] to investigate how the ANGELIC methodology could be used to capture reasoning about factors with magnitude, expanding the range of industrial scenarios that the methodology could be applied in.

A second exemplar real world application setting is given in [Contissa *et al.*, 2013], which uses argument maps to assess liability in the field of air traffic management. In this setting, a 'Legal Case' methodology is used to assist an interdisciplinary team to foresee and mitigate legal problems that may occur through the proposed use of automated technology. The methodology, as described in [Contissa *et al.*, 2013], covers steps to mapping and classifying possible automated technology failures, produce a set of hypotheses of liability link to the failures, and analyse legal rules and arguments supporting the attribution of liability for each of the hypotheses. Although not in full deployment, the argument maps produced are intended for presentation to stakeholders in the domain, including lawyers, to facilitate the cooperative design and assessment of new technologies for air traffic management.

The third and fourth examples of real-world applications both concern legal argumentation in the domain of law enforcement. [Bex *et al.*, 2016] looked at an AI system for citizen complaint intake about online trade fraud, for example, false web shops or malicious traders on eBay not delivering products to people. In the paper, the first ideas are provided for an argument-based recommender system that, given a complaint form, uses argumentation to determine whether a case is possibly fraud, and then only recommends filing an official report if it is. More information about the intake system can be found in Sections 6.1.3 and 7.3.1. [Testerink *et al.*, 2019] use the same argument-based system, but instead of recommending whether or not to file a report to citizens, it provides responses to messages from international police partners given what is in then police database and certain policy rules.

6.4.2 Policy

Policy making is a domain in which argumentation naturally occurs. Political disputes can turn on disagreements as to objective facts and subjective values, so computational models of argument are well suited to representing these different types of debate. In [Atkinson *et al.*, 2011] a demonstration was given as to how to construct a semantic model on the basis of responses received to a Green Paper, which is a government publication released as part of a consultation process that details specific issues, and then points out possible courses of action in terms of policy and legislation in order to receive feedback from interested parties. An example of the type of debate that has been modelled is an issue in UK Road Traffic policy. The starting situation is that the number of fatal road accidents is an obvious cause for concern on UK roads. There are already speed restrictions in place on various types of road, in the belief that excessive speed causes accidents. The policy issue to be considered is how to reduce road deaths. One option is to introduce speed cameras to discourage speeding. Another is to educate motorists on the dangers of speeding. In [Atkinson *et al.*, 2011] it is shown how a semantic model of this debate can be built from which different policy options can be considered for implementation, representing issues of importance to different stakeholders, such as road safety organisations, motoring lobby groups and civil liberties groups. From the semantic model, arguments for different policy options can be identified with the policy selected being depending up the preferred values being promoted by that option.

6.4.3 Arguments, stories and probabilities in evidential reasoning

Continuing from research done in the 2000s (Section 5.3.3), further work was done on evidential reasoning, enhancing the understanding on how various approaches to the rational handling of the evidence are connected. Three types of normative approaches for the handling of evidence can be distinguished: arguments, scenarios and probabilities [Kaptein et al., 2009; Dawid et al., 2011; Anderson et al., 2005; Verheij et al., 2016; Di Bello and Verheij, 2018]. Each approach can help systematize and regulate how to examine, analyse and weigh the evidence. Where an argument-based approach focuses on dialogue, support and attack, scenario-based approaches highlight explanatory sense-making in coherent, holistic accounts of what has happened, while probabilistic approaches enable the quantitative analysis of evidential strength by connecting to Bayesian modeling and statistics. In the 2010s, the three approaches were studied in various combinations.

One idea was to embed argument structure in Bayesian networks using 'legal idioms' [Fenton et al., 2013; Lagnado et al., 2013; Neil et al., 2019]. The approach aims to systemize the embedding of legally relevant argument structure (for instance about witness reliability and alibi testimony) in Bayesian networks. Inspired by and extending the work on so-called object-oriented Bayesian networks [Hepler et al., 2007], Fenton et al. [2013]; Lagnado et al. [2013] propose reusable graphical network structures aimed at the modeling of legal evidential arguments. Idioms are provided for the modeling of evidence accuracy, motive and opportunity, evidential dependencies, alibi evidence and explaining away (downplaying alternative explanations). In its attempt to provide a catalog of reusable argument structure, the idiom approach is similar to work on argumentation schemes [Walton et al., 2008; Walton, 1996], but now in the context of Bayesian network analysis.

This idiom approach was applied to the embedding of evidential stories in Bayesian networks by Vlek *et al.* [2014, 2016]. In this work, a design method is proposed aimed at alleviating three common difficulties in reasoning with evidence: (1) tunnel vision, (2) the problem of a good story pushing out a true story and (3) finding the relevant variables for a model of the case. The design method uses four idioms: the scenario idiom, representing the events in a story and how they depend on one another (for instance how a burglary developed); the subscenario idiom, allowing for the embedding of one story in another (for instance how the house was entered during a burglary); the variation idiom, used for the modeling of different versions of a story (for instance, entering after smashing a window or picking a lock); and the merged scenarios idiom, used for combining different stories in one Bayesian network model. During the design of a model, four steps are iterated: collecting the relevant scenarios; unfolding scenarios by considering for which story elements evidence can be added; merging the scenarios; and finally adding the evidence. The design method comes with a corresponding explanation format, which allows for the explanation of a Bayesian network model built with the method in terms of the scenarios modeled, the quality of those scenarios (interpreted probabilistically) and the evidential support that is available. The method is evaluated using case studies of real crime cases in the Netherlands.

Work continued on the hybrid integration of argument-based and story-based approaches (Sections 5.1.2 and 5.3.3): [Bex *et al.*, 2010; Bex, 2011] discuss the hybrid theory for stories and arguments about evidence in detail. [Bex and Verheij, 2013; Bench-Capon and Bex, 2015] connect evidential reasoning with stories and arguments to legal reasoning with arguments and cases. [Bex, 2015] further integrates reasoning with stories and arguments, allowing for both of them to be represented and evaluated as elements of a Dung-style argumentation framework [Dung, 1995]. This then allows for different types of reasoning with causality in argumentation frameworks.

There was also work on combining probabilities and arguments in various different ways. [Wieten *et al.*, 2018] discusses an approach for transforming arguments into so-called argument graphs, containing the same kind of causality information as the integrated framework by [Bex,

2015], to Bayesian network structures by using the specific causality information. [Wieten et al., 2019] takes a different approach: arguments are not transformed to Bayesian networks, but they are used in a dialogue to critically analyse Bayesian networks. The paper provides different argument schemes and a dialogue structure for Bayesian network analysis.

[Keppens, 2012] extracts argument diagrams from a Bayesian network so that the evidential reasoning can be scrutinised better by people who do not have in-depth knowledge of Bayesian networks. In a similar vein, [Timmer et al., 2015, 2017] investigates how an argumentation perspective can help in the interpretation of statistical dependency information as modeled by a Bayesian network. For this purpose, support graphs are proposed as an intermediate format. A support graph can disentangle the graphical properties of a Bayesian network and enhances the intuitive interpretation of statistical dependencies. By the use of support graphs, a succinct set of arguments can be generated, reducing superfluous elements.

The case model approach (Section 6.1.4) was originally conceived as a way to connect arguments, scenarios and probabilities in a single modeling approach. The informal ideas of combining the three approaches in one were presented in [Verheij, 2014]. Arguments were intended for addressing the adversarial setting of reasons for and against claims, scenarios for providing a globally coherent perspective, and probabilities for the modeling of gradual uncertainty. The combined approach aimed at keeping the strengths of each of the three separate approaches, while avoiding limitations. For instance, probabilistic approaches provide a well-known and widely useful account of rational evidence handling, they typically require more numerical, statistical information than is reasonably available (in particular Bayesian networks, which model probabilities for all possible combinations of all model variables). Balancing these, the case model approach is presented as 'with and without numbers' by providing an approach that is consistent with a probabilistic analysis but does not require full numerical information. Verheij [2017b] provide the further formal development of the approach. van Leeuwen and Verheij [2019] compares an analysis in terms of Bayesian networks with embedded scenarios and in terms of case models. Both kinds of analysis show how the gradual collection of the evidence has a stepwise influence on how strongly various hypotheses about what has happened are supported in comparison with one another.

Verheij *et al.* [2016] discuss different combinations of the modeling of evidential reasoning using arguments, scenarios and probabilities, explicating strengths and limitations of each. Prakken *et al.* [2020] introduce a special issue including various modeling approaches, separately and in hybrid combinations, all using the same real case as an example. Prakken [2020] provides an argumentative analysis, while Fenton *et al.* [2020] one in terms of Bayesian networks, and Dahlman [2020] applies Bayesian thinking. Koppen and Mackor [2020] use story analysis, and Bex [2020] uses the hybrid argumentative-narrative approach (Section 5.3.3, Figure 11), while Verheij [2020a] gives an analysis with and without probabilities using case models (Section 6.1.4).

6.4.4 Methodology

In [Al-Abdulkarim *et al.*, 2016b] a methodology for capturing case law (ANGELIC) was presented and it was shown how three domains could be modelled using the methodology to capture the factor-based reasoning within those domains and decide cases in accordance with the model. The three domains that were used in the evaluation of the methodology were: the CATO trade secrets cases, cases regarding warrantless search of automobiles, and cases concerning capture and possession of wild animals, which have been popular testbeds in the AI and law literature. The domains are all modelled as ADFs. Once defined for a domain, an ADF can easily be transformed into a logic program that, when instantiated with the facts of a case, can determine outcome for the case and the acceptable arguments leading to this decision. The programs reported in [Al-Abdulkarim *et al.*, 2016b] demonstrated a high degree of success in replicating the outcomes from the cases used in the experiments, yielding a success rate of over 96% accuracy.

The need to maintain the model in the face of evolving case law was recognised in [Al-Abdulkarim *et al.*, 2016a]. There it was shown how the highly modular nature of the ADF facilitated the addition, modification and re-odering of acceptance conditions, as well as the addition and removal of nodes.

6.4.5 Evolution of Case Law

Henderson and Bench-Capon [2019] returned to the topic earlier explored in Section 5.3.2. Again this was based on Levi [1948], and this time focused on changing rules rather than classes. The idea was that each side would present an argument, and the winning argument would form a rule to be applied to future cases.

A number of types of argument were identified:

- *stare decisis*: if a rule covering the current case exists, that rule can be cited as a reason for the decision;
- *class membership*: argues that the current case should be decided in virtue of membership of a particular class; this may broaden or narrow a class used used in an existing rule;
- *floodgate*: argues against a broadening or narrowing on the grounds that it is too big a step and would include or exclude too many cases;
- *exception*: points to a distinguishing feature of a case, and proposes that it should be an exception to the existing rule;
- *logical similarity*: combines two rules with the same outcome into a single rule.

The process is illustrated with three examples: a fictional example based on interpreting the phrase "expected to work", Levi's liability cases beginning with *Dixon v Bell*, and automobile exception to the US 4th Amendment cases involving luggage.

In [Bench-Capon and Henderson, 2019a] the process was made concrete as a set of dialogue moves. The dialogue was conducted over three plies: a proposal, a response and a rebuttal, followed by a judgement resolving the discussion. Each ply was associated with several moves expressing different kinds of argument. An example with sixteen cases is given in [Bench-Capon and Henderson, 2019b].

6.4.6 Statutory Interpretation

While most work on argumentation in AI and law concerns reasoning with legal cases, argumentation also has a role to play in the interpretation of statutes. Often the interpretation of a term in a statute is not clear: should it be given a literal interpretation, or interpreted according to its context and the purpose of the statute?

In [Sartor *et al.*, 2014] two jurisprudential sources are used to identify the kinds of arguments that can be used. MacCormick and Summers [2016] identify eleven types of arguments and Tarello [1980] identifies fourteen. Tarello's list complements MacCormick and Summers' list, since the latter focuses on the kinds of input on which the interpretive argument is based (ordinary language, technical language, statutory context, precedent, etc.) while the first focuses on the reasoning steps by which the interpretive argument is constituted. Where conflicting arguments of different types are available, criteria are needed: a list of such criteria is given in [Alexy and Dreier, 1991]. This jurisprudential work is used to provide a general logical structure for arguments based on interpretative canons, and for arguments about which should be followed in cases of conflict. This approach was formalised in defeasible deontic logic with canons taken as defeasible rules in [Rotolo *et al.*, 2015].

In [Walton *et al.*, 2016] the canons of MacCormick and Summers and Tarello were presented as argumentation schemes, some positive arguing for an interpretation and some negative arguing against an interpretation. Counterarguments can be generated using the three associated critical questions. It is illustrated with a case modelled using Carneades [Gordon, 2013].

7 Current developments: The breakthrough of AI in society

Currently, AI has become a wide-spread topic of discussion throughout society. Both its risks and its limitations are addressed in public debate and in research. Machine prediction of decisions is studied, both as a tool and as a risk. Notwithstanding widespread efforts, the aligning of learning and reasoning remains a research challenge. Large language models (in particular since ChatGPT's public release in November 2022) are used by virtually everyone. Ethical concerns are discussed, by philosophers, tech developers and AI researchers. Machine learning is now also 'good-old fashioned', a term before only applied to symbolic AI methods.

Attempts are made to arrive at new hybrid AI approaches connecting knowledge, reasoning, learning and language. Steps are made to use the argumentation approach as a model of such integration (cf. [Verheij, 2020b]).

7.1 Argument Generation

7.1.1 Argumentation Schemes

In [Atkinson and Bench-Capon, 2021] a summary is given of the impact of Walton's argumentation schemes on research in AI and Law research. Within that discussion it is shown how the systematisation of natural patterns of arguments can be done into schemes to enable arguments to be generated from these schemes. There are a number of ways in which schemes can be encoded. Logic programming can be used straighforwardly to represent the schemes as rules, or they could be captured as defeasible rules within a framework such as ASPIC+ [Prakken, 2010]. Hand coding is required to encode schemes as rules in this way, but other works have taken a more general approach by building inference engines to execute the schemes. One of the richest tools developed to meet this aim is the Carneades system [Gordon, 2013]. The system is an integrated set of tools for argument (re)construction, evaluation, mapping and interchange. Carneades provides a library of 106 schemes but with the ability to extend this list with the specification of additional schemes. Utilising these schemes, and their critical questions, allows for the generation of arguments and counter arguments. Carneades also allows for the evaluation of arguments using several different standards of proof required for acceptance of a given argument.

7.1.2 Rationales

Building on work in the 1990s on rationales and argument moves [Loui and Norman, 1995] (discussed in Section 4.1.2), Bench-Capon and Verheij [2022] show how methods of computational argumentation developed later can be applied to the unpacking of arguments. Concretely, examples of a compression rationale and of a resolution rationale were analyzed. The compression rationale example involves the unpacking of an intermediate step (as discussed in Section 4.1.2, Figure 5) and that of a resolution rationale the unpacking of an implicit conflict resolution involving a preference. Methods applied include structured argumentation approach (similar to ASPIC+ [Prakken, 2010]) and a sentence-based approach (DefLog [Verheij, 2003a]). It is concluded that modern approaches can make the unpackings explicit, but do not retain the formal connection between the unpacked and the unpacking arguments.

7.2 Evaluation of Arguments

Whilst teleological reasoning is a well established feature of AI and law research, as discussed within this chapter, new models for reasoning about values continue to be developed. In [Maranhão et al., 2021] an additive model of balancing values is set out whereby factors intensify or attenuate impacts on values and values are assigned degrees of relative importance. What results is an assessment of an action's impacts on single values, which are then aggregated to determine the action's total impact on the sets of values promoted and demoted. Comparing an action's impacts on the promoted and demoted values then yields a determination as to whether the action is either permitted or prohibited.

Supplementing the balancing model are formal definitions of change functions that induce shifts in the balance of values through addition or subtraction of factors, or by additions or subtractions of values in the model. These operations are intended to have some resemblance to argument moves, where new features of the legal case or moral considerations are brought into play to oppose previously justified conclusions.

7.3 Use of Arguments

7.3.1 Applications in law enforcement

In the 2020's, the ideas and prototypes of argument-based applications for law enforcement (cf. Sections 6.1.3. 6.4.1) are implemented and used at scale at the Dutch Police. The intake and analysis of citizen crime reports regarding online trade fraud are handled by an online recommender system that uses argumentation ([Odekerken et al., 2020, 2022]). Furthermore, [Odekerken and Bex, 2020] adapt the case-based reasoning model of precedential constraint (see Section 6.3) for the classification of possibly fraudulent webshops, extending the model of [Horty, 2011] to also deal with incomplete cases and inconsistent case bases [Odekerken et al., 2023a,b].

7.3.2 Machine learning and explanations

Following the earlier work of [Ashley and Brüninghaus, 2009; Ashley and Walker, 2013] and [Schraagen and Bex, 2018], further research was done into models that extract basic facts or factors from text using data driven NLP methods and then reason with these fact(ors) using logical models of argumentation and case-based reasoning. [Mumford et al., 2021] use a BERT model to extract factors from cases of the ECHR, after which they reason with these cases using logical models of argument (cf. Section 7.3.3).

[Prakken and Ratsma, 2022] provide an argumentation method with which the outcome of black box (machine learning) systems can be explained post-hoc. Based on the model of precedential constraint (6.3), they define a dialogue-based argumentation model in which the proponent cites the case that is to be explained (i.e. why did case c have outcome o?), and the opponent then tries to defeat this cited case by arguing, for example, that it has missing or additional factors that might influence the outcome o. The model was later extended by [Peters et al., 2023] so that it could reason with inconsistent case bases and also have the case bases be constructed by the black box model.

Steging et al. [2021a,b] address the issue that machine learning systems can draw the right conclusions for the wrong reasons, in the sense that high accuracies do not imply that the correct conditions are used in a trained model. Whereas in image classification, such a mismatch between reasons and conclusion may not be problematic, the use of unsound rationales is unwanted in legal applications where the justification of a decision is of central relevance.

Steging et al. [2021a] develops a human-in-the-loop approach to investigate and improve the rationale used by machine learning models. The method is hybrid in two ways. First it is an example of a hybrid intelligence approach [Akata et al., 2020] in which humans and machines augment each others' performance. In this case, the human knowledge that is available (although perhaps incomplete) can be used for improvement of the rationale used by a machine learning model. Second the approach is hybrid by combining different methods in AI, in particular by the use of both machine learning and knowledge representation methods. Steging *et al.* [2021b] applies explainable AI methods that detect the features used for decision making in a trained model. The paper shows that even with high accuracy and good relevant feature detection, the use of a correct rationale is not guaranteed.

In this research [Steging *et al.*, 2021a,b], synthetic data sets with a known structure are used. The data sets are generated using a given knowledge structure, so that a correct ground truth rationale is known beforehand and can be used for method evaluation. One data set concerns a fictional welfare benefit domain about eligibility of a person for a welfare benefit to cover the expenses for visiting their spouse in the hospital. The domain and data set were introduced by Bench-Capon [1993] in order to investigate whether a neural network can correctly learn a rule from data. The other data set models actual Dutch tort law based on the articles 6:162 and 6:163 of the Dutch civil code about when an unlawful act legally determines a duty to repair the damages caused (cf. also Figure 12). The data sets are publicly available [Steging *et al.*, 2023].

7.3.3 Methodology

The first full account of the ANGELIC methodology was set out in 2016 [Al-Abdulkarim *et al.*, 2016b] and since then it has been extended and applied to a range of legal domain scenarios, most recently the popular domain of the European Convention on Human Rights. In [Collenette *et al.*, 2023] it was shown how Article 6 of the convention, covering the right to a fair trial, could be modelled using the ANGELIC methodology. The model was then evaluated using forty cases heard in the European Court of Human Rights (ECHR) to determine whether the ANGELIC model could produce the same outcomes that the judges had in the original cases. A 97% success rate was reported with this exercise and, crucially, the program was able to give easily digestible explanations as to why it had arrived at its outcome of whether or not there had been a violation of Article 6 in each of the cases under consideration. The current version of the ANGELIC methodology is presented in [Atkinson and Bench-Capon, 2023].

Despite the success in terms of both accuracy and explainability, there were still parts of the process of constructing the domain model that rely on manual analysis, specifically the ascription of factors from cases to the model. To automate this task within the overall process, a model was proposed in [Mumford *et al.*, 2022] to use machine learning for the factor ascription task, such that once factors present in a case are ascribed, the outcome follows from reasoning over the domain model. The approach yields a hybrid AI model combining symbolic and data-driven approaches. The most recent line of work reported on a study involving the annotation of a corpus of Article 6 cases, yielding insight on the distribution of the factors relevant to the complaint of a potential violation of Article 6. The study produced an annotated data set for training models, using natural language processing techniques, to perform the factor ascription task in accordance with the ADF for Article 6. Encouraging results were reported from experiments on this task, providing impetus for further exploring the hybrid use of AI techniques for supporting automated reasoning about legal cases.

7.3.4 Statutory Interpretation

Statutory reasoning was discussed in Section 6.4.6: a different approach was proposed in [Araszkiewicz *et al.*, 2020]. This paper introduces the notion of reasoning protocol as a frame for a set of elements used by relevant agents to justify their claims. The model allows the representation of reasoning using not only factors, but also about the relevance of factors in deciding legal cases on the basis of statutory rules. After defining the various elements of the protocol, the paper investigates selected patterns of case-based judicial reasoning in the context of statutory interpretation as understood in continental legal culture.

A second paper by Araszkiewicz [2022] takes the work described in Section 6.4.6 as its starting point and extends it with a layer of case base reasoning reasoning with and about default preference relations between (classes of) interpretive canons. A set of factors supporting preference for linguistic canons over teleological canons and *vice versa* are identified. These are then used with rules extracted from precedent cases the manner of [Prakken and Sartor, 1998]. An argumentation scheme to represent the reasoning is provided.

Concluding remarks

This chapter has showcased the close interaction between research in the theory of computational argumentation and the field of AI and Law. By exploring the early days and historical developments decade by decade, a natural continuity of mutual inspiration between the fields has been presented.

The work described over the timeline has covered the generation, evaluation and use of arguments in AI and Law. In the earlier years, many of the approaches were of a semi-formal nature, then these were followed by the development of rule-based and case-based approaches that mirrored developments in the field of general AI. These approaches were then overtaken by significant advances in topics on computational argumentation, which developed into a field in its own right and brought forth a much more formal approach to modelling legal reasoning. The review of developments closed with coverage of how data-driven approaches to AI that have received significant attention in recent times are being brought to bear on tasks involving the modelling of arguments in legal settings.

In addition to the continued development of specific techniques for argument-based approaches to AI and law, more research is emerging demonstrating the integration of knowledge-based and data-driven approaches, with the aim of producing hybrid systems that seek to reap the benefits of these distinct approaches. With argumentation playing such a strong role in human-based legal reasoning, we can expect to see computational models of argument remaining of importance for driving forward research in AI and law, and leading to applications in the legal domain that contribute to important aims within the topic of explainable and trustworthy AI.

References

Z. Akata, D. Balliet, M. de Rijke, F. Dignum, V. Dignum, G. Eiben, A. Fokkens, D. Grossi, K. Hindriks, H. Hoos, H. Hung, C. Jonker, C. Monz, M. Neerincx, F. Oliehoek, H. Prakken, S. Schlobach, L. van der Gaag, F. van Harmelen, H. van Hoof, B. van Riemsdijk, A. van Wynsberghe, R. Verbrugge, B. Verheij, P. Vossen, and M. Welling. A research agenda for hybrid intelligence: Augmenting human intellect with collaborative, adaptive, responsible, and explainable artificial intelligence. *Computer*, 53(08):18–28, August 2020.

Latifa Al-Abdulkarim, Katie Atkinson, and Trevor Bench-Capon. Accommodating change. *Artificial Intelligence and Law*, 24:409–427, 2016.

Latifa Al-Abdulkarim, Katie Atkinson, and Trevor Bench-Capon. A methodology for designing systems to reason with legal cases using Abstract Dialectical Frameworks. *Artificial Intelligence and Law*, 24:1–49, 2016.

Latifa Al-Abdulkarim, Katie Atkinson, Trevor Bench-Capon, Stuart Whittle, Rob Williams, and Catriona Wolfenden. Noise induced hearing loss: Building an application using the ANGELIC methodology. *Argument and Computation*, 10(1):5–22, 2019.

Vincent Aleven and Kevin D Ashley. Doing things with factors. In *Proceedings of the 5th International Conference on Artificial Intelligence and Law*, pages 31–41. ACM Press, 1995.

Vincent Aleven and Kevin D Ashley. Evaluating a learning environment for case-based argumentation skills. In *Proceedings of the 6th International Conference on Artificial Intelligence and Law*, pages 170–179. ACM Press, 1997.

Vincent Aleven. *Teaching case-based argumentation through a model and examples*. Ph.D. thesis, University of Pittsburgh, 1997.

Vincent Aleven. Using background knowledge in case-based legal reasoning: a computational model and an intelligent learning environment. *Artificial Intelligence*, 150(1-2):183–237, 2003.

Robert Alexy and Ralf Dreier. Statutory interpretation in the Federal Republic of Germany. In D.N. MacCormick and R.S. Summers, editors, *Interpreting Statutes: A Comparative Study,*, pages 72–101. Dartmouth Aldershot/Brookfield USA/Hong Kong/Singapore/Sydney, 1991.

R. Alexy. *Theorie der juristischen Argumentation. Die Theorie des rationalen Diskurses als eine Theorie der juristischen Begründung.* Suhrkamp Verlag, Frankfurt am Main, 1978.

L. Amgoud and C. Cayrol. A model of reasoning based on the production of acceptable arguments. *Annals of Mathematics and Artificial Intelligence*, 34:197–215, 2002.

T. Anderson, D. Schum, and W. Twining. *Analysis of Evidence. 2nd Edition.* Cambridge University Press, Cambridge, 2005.

G. Antoniou, D. Billington, G. Governatori, and M. Maher. A flexible framework for defeasible logics. In *Proceedings of the 17th National Conference on Artificial Intelligence*, pages 405–410, 2000.

Michał Araszkiewicz, Tomasz Żurek, and Błażej Kuźniacki. Reasoning with and about factors in statutory interpretation. In *Proceedings of the 4th International Workshop on MIning and REasoning with Legal texts. CEUR Workshop Proceedings*, volume 2632, 2020.

M. Araszkiewicz. A hybrid model of argument concerning preferences between statutory interpretation canons. In E. Francesconi, G. Borges, and C. Sorge, editors, *Legal Knowledge and Information Systems. JURIX 2022: The Twenty-fifth Annual Conference*, pages 3–12. IOS Press, Amsterdam, 2022.

A. Artikis, M.J. Sergot, and J. Pitt. An executable specification of an argumentation protocol. In *Proceedings of the 9th International Conference on Artificial Intelligence and Law*, pages 1–11. ACM Press, 2003.

Kevin Ashley and Stefanie Brüninghaus. Automatically classifying case texts and predicting outcomes. *Artificial Intelligence and Law*, 17(2):125–165, 2009.

Kevin D Ashley and Vern R Walker. Toward constructing evidence-based legal arguments using legal decision documents and machine learning. In *Proceedings of the 14th International Conference on Artificial Intelligence and Law*, pages 176–180. ACM Press, 2013.

Kevin D Ashley, Niels Pinkwart, Collin Lynch, and Vincent Aleven. Learning by diagramming supreme court oral arguments. In *Proceedings of the 11th International Conference on Artificial Intelligence and Law*, pages 271–275. ACM Press, 2007.

Kevin D Ashley. Toward a computational theory of arguing with precedents. In *Proceedings of the 2nd International Conference on Artificial Intelligence and Law*, pages 93–102. ACM Press, 1989.

Kevin D Ashley. *Modeling legal arguments: Reasoning with cases and hypotheticals*. MIT press, Cambridge, Mass., 1990.

Kevin D Ashley. Reasoning with cases and hypotheticals in HYPO. *International Journal of Man-Machine Studies*, 34(6):753–796, 1991.

K.D. Ashley. A brief history of the changing roles of case prediction in AI and law. *Law in Context*, 36:93–112, 2019.

Katie Atkinson and Trevor Bench-Capon. Legal case-based reasoning as practical reasoning. *Artificial Intelligence and Law*, 13:93–131, 2005.

K. Atkinson and T. Bench-Capon. Practical reasoning as presumptive argumentation using Action based Alternating Transition Systems. *Artificial Intelligence*, 171:855–874, 2007.

Katie Atkinson and Trevor Bench-Capon. Argumentation schemes in AI and Law. *Argument and Computation*, 12(3):417–434, 2021.

Katie Atkinson and Trevor Bench-Capon. ANGELIC II: An improved methodology for representing legal domain knowledge. In *Proceedings of the Nineteenth International Conference on Artificial Intelligence and Law*, pages 12–21. ACM Press, 2023.

Katie Atkinson, Trevor Bench-Capon, and Peter McBurney. Arguing about cases as practical reasoning. In *Proceedings of the 10th International Conference on Artificial Intelligence and Law*, pages 35–44. ACM Press, 2005.

K. Atkinson, T. Bench-Capon, and P. McBurney. Computational representation of persuasive argument. *Synthese*, 152:157–206, 2006.

Katie Atkinson, Trevor Bench-Capon, and Peter McBurney. Parmenides: facilitating deliberation in democracies. *Artificial Intelligence and Law*, 14:261–275, 2006.

Katie Atkinson, Trevor Bench-Capon, D Cartwright, and Adam Wyner. Semantic models for policy deliberation. In *Proceedings of the 13th International Conference on Artificial Intelligence and Law*, pages 81–90. ACM Press, 2011.

K. Atkinson, T. Bench-Capon, H. Prakken, and A. Wyner. Argumentation schemes for reasoning about factors with dimensions. In K.D. Ashley, editor, *Legal Knowledge and Information Systems. JURIX 2013: The Twenty-sixth Annual Conference*, pages 39–48. IOS Press, Amsterdam, 2013.

K. Atkinson, T. Bench-Capon, T. Routen, A. Sánchez, S. Whittle, R. Williams, and C. Wolfenden. Realising ANGELIC designs using logiak. In M. Araszkiewicz and V. Rodríguez-Doncel, editors, *Legal Knowledge and Information Systems. JURIX 2019: The Thirty-second Annual Conference*, pages 151–156. IOS Press, 2019.

T Bench-Capon and K Atkinson. Dimensions and values for legal CBR. In A.Z. Wyner and G. Casini, editors, *Legal Knowledge and Information Systems. JURIX 2017: The Thirtieth Annual Conference*, pages 27–32, Amsterdam, 2017. IOS Press.

Trevor Bench-Capon and Katie Atkinson. Precedential constraint: The role of issues. In *Proceedings of the 18th International Conference on Artificial Intelligence and Law*, pages 12–21. ACM Press, 2021.

T. Bench-Capon and F.J. Bex. Cases and stories, dimensions and scripts. In A. Rotolo, editor, *Legal Knowledge and Information Systems. JURIX 2015: The Twenty-Eight Annual Conference*, pages 11–20. IOS Press, 2015.

T. Bench-Capon and T. Gordon. Implementing a theory of a legal domain. In E. Francesconi, G. Borges, and C. Sorge, editors, *Legal Knowledge and Information Systems. JURIX 2022: The Thirty-Fifth Annual Conference*, pages 13–22. IOS Press, Amsterdam, 2022.

T. Bench-Capon and J. Henderson. A dialogical model of case law dynamics. In M. Araszkiewicz and V. Rodriguez-Doncel, editors, *Legal Knowledge and Information Systems. JURIX 2019: The Thirty-Second Annual Conference*, pages 163–168. IOS Press, Amsterdam etc., 2019.

Trevor Bench-Capon and John Henderson. Modelling case law dynamics with dialogue moves. Technical report, Technical Report ULCS-19-003, University of Liverpool, 2019.

Trevor Bench-Capon and Sanjay Modgil. Norms and value based reasoning: justifying compliance and violation. *Artificial Intelligence and Law*, 25(1):29–64, 2017.

Trevor Bench-Capon and Henry Prakken. Using argument schemes for hypothetical reasoning in law. *Artificial Intelligence and Law*, 18:153–174, 2010.

T.J.M. Bench-Capon and E. L. Rissland. Back to the future: dimensions revisited. In B. Verheij, A. Lodder, R.P. Loui, and A. Muntjewerff, editors, *Legal Knowledge and Information Systems. JURIX 2001: The Fourteenth Annual Conference*, pages 41–52, Amsterdam etc, 2001. IOS Press.

Trevor Bench-Capon and Giovanni Sartor. Theory based explanation of case law domains. In *Proceedings of the 8th International Conference on Artificial Intelligence and Law*, pages 12–21. ACM Press, 2001.

Trevor Bench-Capon and Giovanni Sartor. A model of legal reasoning with cases incorporating theories and values. *Artificial Intelligence*, 150(1-2):97–143, 2003.

Trevor Bench-Capon and Marek Sergot. Towards a rule-based representation of open texture in law. In C Walter, editor, *Computer Power and Legal Language*, pages 39–61. Quorum Books: New York, 1988.

Trevor Bench-Capon and Geof Staniford. Plaid: proactive legal assistance. In *Proceedings of the 5th International Conference on Artificial Intelligence and Law*, pages 81–88. ACM Press, 1995.

T. Bench-Capon and B. Verheij. Unpacking arguments. In E. Francesconi, G. Borges, and C. Sorge, editors, *Legal Knowledge and Information Systems. JURIX 2022: The Twenty-fifth Annual Conference*, pages 145–150, Amsterdam, 2022. IOS Press.

Trevor Bench-Capon, Paul Dunne, and Paul Leng. Interacting with knowledge-based systems through dialogue games. In *Proceedings of the Eleventh International Conference on Expert Systems and Applications*, pages 123–140, 1991.

Trevor Bench-Capon, Tim Geldard, and Paul H. Leng. A method for the computational modelling of dialectical argument with dialogue games. *Artificial Intelligence and Law*, 8:233–254, 2000.

T.J.M. Bench-Capon, H. Prakken, and G. Sartor. Argumentation in legal reasoning. In I. Rahwan and G.R. Simari, editors, *Argumentation in Artificial Intelligence*, pages 363–382. Springer, Berlin, 2009.

Trevor Bench-Capon, Henry Prakken, Adam Wyner, and Katie Atkinson. Argument schemes for reasoning with legal cases using values. In *Proceedings of the 14th International Conference on Artificial Intelligence and Law*, pages 13–22. ACM Press, 2013.

Trevor Bench-Capon, Katie Atkinson, and Adam Wyner. Using argumentation to structure e-participation in policy making. *Trans. Large Scale Data Knowl. Centered Syst.*, 18:1–29, 2015.

Trevor Bench-Capon. *Knowledge Representation: An Approach to Artificial Intelligence*. Academic Press, 1990.

Trevor Bench-Capon. Neural networks and open texture. In *Proceedings of the 4th International Conference on Artificial Intelligence and Law*, pages 292–297. ACM Press, 1993.

T. Bench-Capon. Arguing with cases. In A. Oskamp, R.V. de Mulder, C. Van Noortwijk, C.A.F.M. Grütters, K. Ashley, and T. Gordon, editors, *Legal Knowledge-Based Systems. JURIX: The Tenth Conference*, pages 85–100. IOS Press, 1997.

3 - COMPUTATIONAL MODELS OF LEGAL ARGUMENT

Trevor Bench-Capon. Specification and implementation of Toulmin dialogue game. In J.C. Hage, T.J.M. Bench-Capon, A.W. Koers, C.N.J. de Vey Mestdagh, and C.A.F.M. Grütters, editors, *Legal Knowledge-Based Systems. JURIX: The Eleventh Conference*, pages 5–19, Nijmegen, 1998. Gerard Noodt Instituut.

Trevor Bench-Capon. Some observations on modelling case based reasoning with formal argument models. In *Proceedings of the 7th International Conference on Artificial Intelligence and Law*, pages 36–42. ACM Press, 1999.

T. Bench-Capon. Representation of case law as an argumentation framework. In T. Bench-Capon, A. Daskalopulu, and R. Winkels, editors, *Legal Knowledge and Information Systems. JURIX 2002: The Fifteenth Annual Conference*, pages 53–62, Amsterdam etc, 2002. IOS Press.

Trevor Bench-Capon. The missing link revisited: The role of teleology in representing legal argument. *Artificial Intelligence and Law*, 10:79, 2002.

T.J.M. Bench-Capon. Persuasion in practical argument using value-based argumentation frameworks. *Journal of Logic and Computation*, 13:429–448, 2003.

Trevor Bench-Capon. HYPO's legacy: introduction to the virtual special issue. *Artificial Intelligence and Law*, 25:205–250, 2017.

Trevor Bench-Capon. Before and after Dung: Argumentation in AI and Law. *Argument and Computation*, 11(1-2):221–238, 2020.

D. H. Berman and C. D. Hafner. Incorporating procedural context into a model of case-based legal reasoning. In *Proceedings of the Third International Conference on Artificial Intelligence and Law*, pages 12–20. ACM Press, 1991.

Donald H Berman and Carole D Hafner. Representing teleological structure in case-based legal reasoning: the missing link. In *Proceedings of the 4th International Conference on Artificial Intelligence and Law*, pages 50–59. ACM Press, 1993.

D. H. Berman and C. D. Hafner. Understanding precedents in a temporal context of evolving legal doctrine. In *Proceedings of the Fifth International Conference on Artificial Intelligence and Law*, pages 42–51. ACM Press, 1995.

F.J Bex and H Prakken. Reinterpreting arguments in dialogue: an application to evidential reasoning. In T.F. Gordon, editor, *Legal Knowledge and Information Systems: JURIX 2004, the Seventeenth Annual Conference*, pages 119–129. IOS Press, 2004.

Floris Bex and Bart Verheij. Legal stories and the process of proof. *Artificial Intelligence and Law*, 21:253–278, 2013.

Floris Bex, Henry Prakken, Chris Reed, and Douglas Walton. Towards a formal account of reasoning about evidence: argumentation schemes and generalisations. *Artificial Intelligence and Law*, 11:125, 2003.

F.J. Bex, H. Prakken, and B. Verheij. Anchored narratives in reasoning about evidence. In T. M. van Engers, editor, *Legal Knowledge and Information Systems: JURIX 2006: the Nineteenth Annual Conference*, volume 152, page 11. IOS Press, 2006.

F.J. Bex, S. Van den Braak, H. Van Oostendorp, H. Prakken, B. Verheij, and G. Vreeswijk. Sense-making software for crime investigation: how to combine stories and arguments? *Law, Probability and Risk*, 6(1-4):145–168, 2007.

Floris Bex, Henry Prakken, and Bart Verheij. Formalising argumentative story-based analysis of evidence. In *Proceedings of the 11th International Conference on Artificial Intelligence and Law*, pages 1–10. ACM Press, 2007.

Floris Bex, Trevor Bench-Capon, and Katie Atkinson. Did he jump or was he pushed? Abductive practical reasoning. *Artificial Intelligence and Law*, 17:79–99, 2009.

Floris Bex, Peter J Van Koppen, Henry Prakken, and Bart Verheij. A hybrid formal theory of arguments, stories and criminal evidence. *Artificial Intelligence and Law*, 18:123–152, 2010.

Floris Bex, Joeri Peters, and Bas Testerink. AI for online criminal complaints: From natural dialogues to structured scenarios. In *Artificial intelligence for justice workshop (ECAI 2016)*, 2016.

F.J. Bex. Analysing stories using schemes. In H. Kaptein, H. Prakken, and B. Verheij, editors, *Legal Evidence and Proof: Statistics, Stories, Logic*, pages 93–116. Ashgate Publishing, Farnham, 2009.

F Bex. *Arguments, Stories and Criminal Evidence: A Formal Hybrid Theory*. Springer, Berlin, 2011.

Floris Bex. An integrated theory of causal stories and evidential arguments. In *Proceedings of the 15th International Conference on Artificial Intelligence and Law*, pages 13–22. ACM Press, 2015.

F.J. Bex. The hybrid theory of stories and arguments applied to the Simonshaven case. *TopiCS in Cognitive Science*, 12(4):1152–1174, 2020.

L. Bodenstaff, H. Prakken, and G. Vreeswijk. On formalising dialogue systems for argumentation in the event calculus. In *Proceedings of the Eleventh International Workshop on Nonmonotonic Reasoning*, pages 374–382, Windermere, UK, 2006.

A. Bondarenko, P.M. Dung, R.A. Kowalski, and F. Toni. An abstract, argumentation-theoretic approach to default reasoning. *Artificial Intelligence*, 93:63–101, 1997.

L.K. Branting. *Reasoning with Rules and Precedents: A Computational Model of Legal Analysis*. Kluwer Academic Publishers, Dordrecht/Boston/London, 1999.

G. Brewka and T.F. Gordon. Carneades and Abstract Dialectical Frameworks: A reconstruction. In P. Baroni, F. Cerutti, M. Giacomin, and G.R. Simari, editors, *Computational Models of Argument. Proceedings of COMMA 2010*, pages 3–12. IOS Press, Amsterdam, 2010.

G. Brewka and S. Woltran. Abstract dialectical frameworks. In *Principles of Knowledge Representation and Reasoning: Proceedings of the Twelfth International Conference*, pages 102–111. AAAI Press, 2010.

G. Brewka. Dynamic argument systems: a formal model of argumentation processes based on situation calculus. *Journal of Logic and Computation*, 11:257–282, 2001.

Stephanie Brüninghaus and Kevin Ashley. Predicting outcomes of case based legal arguments. In *Proceedings of the 9th International Conference on Artificial Intelligence and Law*, pages 233–242. ACM Press, 2003.

Bruce G Buchanan and Edward H Shortliffe. *Rule based expert systems: the Mycin experiments of the Stanford heuristic programming project*. Addison-Wesley Longman Publishing Co., Inc., 1984.

Roberta Calegari, Régis Riveret, and Giovanni Sartor. The burden of persuasion in structured argumentation. In *Proceedings of the 18th International Conference on Artificial Intelligence and Law*, pages 180–184. ACM Press, 2021.

I. Canavotto and J. Horty. Reasoning with hierarchies of open-textured predicates. In *Proceedings of the 19th International Conference on Artificial Intelligence and Law*, pages 52–61. ACM Press, 2023.

Alison Chorley and Trevor Bench-Capon. Agatha: Using heuristic search to automate the construction of case law theories. *Artificial Intelligence and Law*, 13(1):9–51, 2005.

Alison Chorley and Trevor Bench-Capon. An empirical investigation of reasoning with legal cases through theory construction and application. *Artificial Intelligence and Law*, 13(3):323–371, 2005.

Peter Clark and Tim Niblett. The CN2 induction algorithm. *Machine Learning*, 3:261–283, 1989.

Joe Collenette, Katie Atkinson, and Trevor Bench-Capon. Explainable AI tools for legal reasoning about cases: A study on the European Court of Human Rights. *Artificial Intelligence*, page 103861, 2023.

G. Contissa, M. Laukyte, G. Sartor, and H. Schebesta. Assessing liability with argumentation maps: An application in aviation law. In

K.D. Ashley, editor, *Legal Knowledge and Information Systems. JURIX 2013: The Twenty-sixth Annual Conference*, pages 73–76, Amsterdam, 2013. IOS Press.

C. Dahlman. De-biasing legal fact-finders with bayesian thinking. *TopiCS in Cognitive Science*, 12(4):1115–1131, 2020.

A. P. Dawid, W. Twining, and M. Vasiliki, editors. *Evidence, Inference and Enquiry*. Oxford University Press, Oxford, 2011.

M. Di Bello and B. Verheij. Evidential reasoning. In G. Bongiovanni, G. Postema, A. Rotolo, G. Sartor, C. Valentini, and D. N. Walton, editors, *Handbook of Legal Reasoning and Argumentation*, pages 447–493. Springer, Dordrecht, 2018.

Judith P Dick. Representation of legal text for conceptual retrieval. In *Proceedings of the 3rd International Conference on Artificial Intelligence and Law*, pages 244–253. ACM Press, 1991.

Phan Minh Dung. An argumentation semantics for logic programming with explicit negation. In *Proceedings of the Tenth Logic Programming Conference*, pages 616–630, Cambridge, MA, 1993. MIT Press.

Phan Minh Dung. On the acceptability of arguments and its fundamental role in nonmonotonic reasoning and logic programming. In *Proceedings of the 13th International Joint Conference on Artificial Intelligence*, volume 93, pages 852–857, 1993.

Phan Minh Dung. On the acceptability of arguments and its fundamental role in nonmonotonic reasoning, logic programming and n-person games. *Artificial Intelligence*, 77(2):321–357, 1995.

R. Dworkin. *Taking rights seriously*. Duckworth, London, 1978.

N. E. Fenton, M. D. Neil, and D. A. Lagnado. A general structure for legal arguments about evidence using Bayesian Networks. *Cognitive Science*, 37:61–102, 2013.

N. E. Fenton, M. D. Neil, B. Yet, and D. A. Lagnado. Analyzing the Simonshaven case using Bayesian networks. *TopiCS in Cognitive Science*, 12(4):1092–1114, 2020.

Kathleen Freeman and Arthur M Farley. A model of argumentation and its application to legal reasoning. *Artificial Intelligence and Law*, 4:163–197, 1996.

D. M. Gabbay, C. J. Hogger, and J. A. Robinson, editors. *Handbook of Logic in Artificial Intelligence and Logic Programming. Volume 3. Nonmonotonic Reasoning and Uncertain Reasoning*. Clarendon Press, Oxford, 1994.

H. Geffner and J. Pearl. Conditional entailment: bridging two approaches to default reasoning. *Artificial Intelligence*, 53:209–244, 1992.

E. Giunchiglia, J. Lee, V. Lifschitz, N. McCain, and H. Turner. Nonmonotonic causal theories. *Artificial Intelligence*, 153:49–104, 2004.

Thomas F Gordon and Nikos Karacapilidis. The Zeno argumentation framework. In *Proceedings of the 6th International Conference on Artificial Intelligence and Law*, pages 10–18. ACM Press, 1997.

Thomas F Gordon and Douglas Walton. Legal reasoning with argumentation schemes. In *Proceedings of the 12th International Conference on Artificial Intelligence and Law*, pages 137–146. ACM Press, 2009.

T.F. Gordon and D.N. Walton. Formalizing balancing arguments. In P. Baroni, T.F. Gordon, T. Scheffler, and M. Stede, editors, *Computational Models of Argument. Proceedings of COMMA 2016*, pages 327–338. IOS Press, Amsterdam, 2016.

Thomas F Gordon, Henry Prakken, and Douglas Walton. The Carneades model of argument and burden of proof. *Artificial Intelligence*, 171(10-15):875–896, 2007.

Thomas F Gordon. The pleadings game. *Artificial Intelligence and Law*, 2(4):239–292, 1993.

Thomas F Gordon. The pleadings game: Formalizing procedural justice. In *Proceedings of the 4th International Conference on Artificial Intelligence and Law*, pages 10–19. ACM Press, 1993.

Thomas F Gordon. Combining rules and ontologies with Carneades. In S. Bragaglia et al., editor, *Proceedings of the 5th International RuleML2011@BRF Challenge, at the 5th International Web Rule Symposium*, volume 799, pages 103–110. CEUR Workshop Proceedings, 2011.

Thomas F Gordon. Introducing the Carneades web application. In *Proceedings of the 14th International Conference on Artificial Intelligence and Law*, pages 243–244, 2013.

T.F. Gordon. Defining argument weighing functions. *Journal of Applied Logics – IfCoLog Journal of Logics and their Application*, 5:747–773, 2018.

G. Governatori and G. Sartor. Burdens of proof in monological argumentation. In R.G.F. Winkels, editor, *Legal Knowledge and Information Systems. JURIX 2010: The Twenty-Third Annual Conference*, pages 37–46. IOS Press, Amsterdam etc., 2010.

G. Governatori, A. Rotolo, and G. Sartor. Temporalised normative positions in Defeasible Logic. In *Proceedings of the 10th International Conference on Artificial Intelligence and Law*, pages 25–34. ACM Press, 2005.

G. Governatori, A. Rotolo, R. Riveret, M. Palmirani, and G. Sartor. Variants of temporal defeasible logics for modelling norm modifications. In *Proceedings of 11th International Conference on Artificial Intelligence and Law*, pages 155–9. ACM Press, 2007.

G. Governatori, F. Olivieri, A. Rotolo, and S. Scannapieco. Computing strong and weak permissions in Defeasible Logic. *Journal of Philosophical Logic*, 42:799–829, 2013.

G. Governatori, B. Verheij, M. Araszkiewicz, T. Bench-Capon, E. Francesconi, and M. Grabmair. Thirty years of AI and Law: The first decade. *Artificial Intelligence and Law*, 30(4):481–519, 2022.

Matthias Grabmair and Kevin D Ashley. Argumentation with value judgments-an example of hypothetical reasoning. In R.G.F. Winkels, editor, *Legal Knowledge and Information Systems. JURIX 2010: The*

Twenty-Third Annual Conference, pages 67–76. IOS Press, Amsterdam, 2010.

M Grabmair and K Ashley. Facilitating case comparison using value judgments and intermediate legal concepts. In *Proceedings of the 13th International Conference on Artificial Intelligence and Law*, pages 161–170. ACM Press, 2011.

Matthias Grabmair. *Modeling Purposive Legal Argumentation and Case Outcome Prediction using Argument Schemes in the Value Judgment Formalism*. PhD thesis, University of Pittsburgh, 2016.

Matthias Grabmair. Predicting trade secret case outcomes using argument schemes and learned quantitative value effect tradeoffs. In *Proceedings of the 16th International Conference on Artificial Intelligence and Law*, pages 89–98. ACM Press, 2017.

Claire Grover, Ben Hachey, Ian Hughson, and Chris Korycinski. Automatic summarisation of legal documents. In *Proceedings of the 9th International Conference on Artificial Intelligence and Law*, pages 243–251. ACM Press, 2003.

Ben Hachey and Claire Grover. Automatic legal text summarisation: experiments with summary structuring. In *Proceedings of the 10th International Conference on Artificial Intelligence and Law*, pages 75–84. ACM Press, 2005.

Ben Hachey and Claire Grover. Extractive summarisation of legal texts. *Artificial Intelligence and Law*, 14:305–345, 2006.

Carole D Hafner and Donald H Berman. The role of context in case-based legal reasoning: teleological, temporal, and procedural. *Artificial Intelligence and Law*, 10(1-3):19–64, 2002.

J. C. Hage, R. Leenes, and A.R. Lodder. Hard cases: a procedural approach. *Artificial Intelligence and Law*, 2:113–167, 1993.

J. C. Hage. Monological reason-based logic: A low level integration of rule-based reasoning and case-based reasoning. In *Proceedings of the 4th International Conference on Artificial Intelligence and Law*, pages 30–39. ACM Press, 1993.

J. C. Hage. A theory of legal reasoning and a logic to match. *Artificial Intelligence and Law*, 4(2):199–273, 1996.

Charles Leonard Hamblin. *Fallacies*. Methuen, London, 1970.

John Henderson and Trevor Bench-Capon. Dynamic arguments in a case law domain. In *Proceedings of the 8th International Conference on Artificial Intelligence and Law*, pages 60–69. ACM Press, 2001.

John Henderson and Trevor Bench-Capon. Describing the development of case law. In *Proceedings of the 17th International Conference on Artificial Intelligence and Law*, pages 32–41. ACM Press, 2019.

A. B. Hepler, A. P. Dawid, and V. Leucari. Object-oriented graphical representations of complex patterns of evidence. *Law, Probability and Risk*, 6(1–4):275–293, 2007.

John F Horty and Trevor Bench-Capon. A factor-based definition of precedential constraint. *Artificial Intelligence and Law*, 20(2):181–214, 2012.

J. Horty. Rules and reasons in the theory of precedent. *Legal Theory*, 17:1–33, 2011.

J. Horty. Reasoning with dimensions and magnitudes. In *Proceedings of the 16th International Conference on Artificial Intelligence and Law*, pages 109–118. ACM Press, 2017.

J. Horty. Reasoning with dimensions and magnitudes. *Artificial Intelligence and Law*, 27:309–345, 2019.

J. Horty. Modifying the reason model. *Artificial Intelligence and Law*, 29:271–285, 2021.

Hadassa Jakobovits and Dirk Vermeir. Dialectic semantics for argumentation frameworks. In *Proceedings of the 7th International Conference on Artificial Intelligence and Law*, pages 53–62. ACM Press, 1999.

John R Josephson and Susan G Josephson. *Abductive inference: Computation, philosophy, technology*. Cambridge University Press, 1996.

H. Kaptein, H. Prakken, and B. Verheij, editors. *Legal Evidence and Proof: Statistics, Stories, Logic (Applied Legal Philosophy Series)*. Ashgate, Farnham, 2009.

Jeroen Keppens and John Zeleznikow. A model based reasoning approach for generating plausible crime scenarios from evidence. In *Proceedings of the 9th International Conference on Artificial Intelligence and Law*, pages 51–59. ACM Press, 2003.

Jeroen Keppens. Argument diagram extraction from evidential Bayesian networks. *Artificial Intelligence and Law*, 20:109–143, 2012.

Janet L Kolodner, Robert L Simpson, and Katia Sycara-Cyranski. *A process model of cased-based reasoning in problem solving*. School of Information and Computer Science, Georgia Institute of Technology, 1985.

P.J. van Koppen and A.R. Mackor. A scenario approach to the Simonshaven case. *TopiCS in Cognitive Science*, 12(4):1132–1151, 2020.

Robert A Kowalski and Francesca Toni. Abstract argumentation. *Artificial Intelligence and Law*, 4(3-4):275–296, 1996.

David A Lagnado, Norman Fenton, and Martin Neil. Legal idioms: a framework for evidential reasoning. *Argument and Computation*, 4(1):46–63, 2013.

Imre Lakatos. *Proofs and refutations*. Nelson London, 1963.

Edward H Levi. *An Introduction to Legal Reasoning*. University of Chicago Press, 1948.

Arno R Lodder and Aimée Herczog. DiaLaw: A dialogical framework for modeling legal reasoning. In *Proceedings of the 5th International Conference on Artificial Intelligence and Law*, pages 146–155. ACM Press, 1995.

A. R. Lodder. *DiaLaw: on legal justification and dialogical models of argumentation*. Kluwer Academic Publishers, Dordrecht, 1999.

Ronald Prescott Loui and Jeff Norman. Rationales and argument moves. *Artificial Intelligence and Law*, 3:159–189, 1995.

R.P. Loui, J. Norman, J. Olson, and A. Merrill. A design for reasoning with policies, precedents, and rationales. In *Proceedings of the 4th International Conference on Artificial Intelligence and Law*, pages 202–211. ACM Press, 1993.

Ronald P Loui, Jeff Norman, Joe Altepeter, Dan Pinkard, Dan Craven, Jessica Linsday, and Mark Foltz. Progress on Room 5: A testbed for public interactive semi-formal legal argumentation. In *Proceedings of the 6th International Conference on Artificial Intelligence and Law*, pages 207–214. ACM Press, 1997.

Leonard S Lutomski. The design of an attorney's statistical consultant. In *Proceedings of the 2nd International Conference on Artificial Intelligence and Law*, pages 224–233. ACM Press, 1989.

D Neil MacCormick and Robert S Summers. *Interpreting Statutes: A Comparative Study*. Routledge, 2016.

Jim D Mackenzie. Question-begging in non-cumulative systems. *Journal of philosophical logic*, pages 117–133, 1979.

Juliano Maranhão, Edelcio G de Souza, and Giovanni Sartor. A dynamic model for balancing values. In *Proceedings of the 18th International Conference on Artificial Intelligence and Law*, pages 89–98. ACM Press, 2021.

Catherine C Marshall. Representing the structure of a legal argument. In *Proceedings of the 2nd International Conference on Artificial Intelligence and Law*, pages 121–127. ACM Press, 1989.

L Thorne McCarty and NS Sridharan. The representation of an evolving system of legal concepts: II. phototypes and deformations. In *Proceedings of the 7th International Joint Conference on Artificial Intelligence*, pages 246–253. ACM Press, 1981.

L Thorne McCarty. Reflections on TAXMAN: An experiment in artificial intelligence and legal reasoning. *Harvard Law Review*, 90:837, 1976.

L Thorne McCarty. A language for legal discourse i. basic features. In *Proceedings of the 2nd International Conference on Artificial Intelligence and Law*, pages 180–189. ACM Press, 1989.

L Thorne McCarty. An implementation of Eisner v. Macomber. In *Proceedings of the 5th International Conference on Artificial Intelligence and Law*, pages 276–286. ACM Press, 1995.

Masha Medvedeva, Michel Vols, and Martijn Wieling. Using machine learning to predict decisions of the European Court of Human Rights. *Artificial Intelligence and Law*, pages 1–30, 2019.

Marvin Minsky. A framework for representing knowledge. In Winston. P, editor, *The Psychology of Computer Vision*, pages 211–277. McGraw Hill: New York, 1975.

Raquel Mochales and Marie-Francine Moens. Argumentation mining. *Artificial Intelligence and Law*, 19:1–22, 2011.

S. Modgil and H. Prakken. A general account of argumentation with preferences. *Artificial Intelligence*, 195:361–397, 2013.

Sanjay Modgil and Henry Prakken. The ASPIC+ framework for structured argumentation: a tutorial. *Argument and Computation*, 5(1):31–62, 2014.

Marie-Francine Moens, Erik Boiy, Raquel Mochales Palau, and Chris Reed. Automatic detection of arguments in legal texts. In *Proceedings of the 11th International Conference on Artificial Intelligence and Law*, pages 225–230. ACM Press, 2007.

Martin Možina, Jure Žabkar, Trevor Bench-Capon, and Ivan Bratko. Argument based machine learning applied to law. *Artificial Intelligence and Law*, 13:53–73, 2005.

Jack Mumford, Katie Atkinson, and Trevor Bench-Capon. Machine learning and legal argument. In *Proceedings of the 21st Workshop on Computational Models of Natural Argument. CEUR Workshop Proceedings*, volume 2937, pages 47–56, 2021.

J. Mumford, K. Atkinson, and T. Bench-Capon. Reasoning with legal cases: A hybrid ADF-ML approach. In E. Francesconi, G. Borges, and C. Sorge, editors, *Legal Knowledge and Information Systems. JURIX 2022: The Twenty-fifth Annual Conference*, pages 93–102. IOS Press, Amsterdam, 2022.

M. Neil, N. Fenton, D. Lagnado, and R. Gill. Modelling competing legal arguments using Bayesian model comparison and averaging. *Artificial Intelligence and Law*, 27:403–430, 2019.

D. Nute. Defeasible logic. In D. Gabbay, C.J. Hogger, and J.A. Robinson, editors, *Handbook of Logic in Artificial Intelligence and Logic Programming*, pages 253–395. Clarendon Press, Oxford, 1994.

D. Odekerken and F.J. Bex. Towards transparent human-in-the-loop classification of fraudulent web shops. *Legal Knowledge and Information Systems: JURIX 2020*, 334:239–242, 2020.

Daphne Odekerken, AnneMarie Borg, and Floris Bex. Efficient argument-based inquiry at the dutch police. *on Applications of AI to Forensics 2020 (AI2Forensics 2020)*, page 22, 2020.

Daphne Odekerken, Floris Bex, AnneMarie Borg, and Bas Testerink. Approximating stability for applied argument-based inquiry. *Intelligent Systems with Applications*, 16:200110, 2022.

D. Odekerken, F.J. Bex, and H. Prakken. Precedent-based reasoning with incomplete cases. In G. Sileno, J. Spanakis, and G. van Dijck, editors, *Legal Knowledge and Information Systems. JURIX 2023: The Thirty-Sixth Annual Conference*, pages 33–42. IOS Press, Amsterdam, 2023.

Daphne Odekerken, Floris Bex, and Henry Prakken. Justification, stability and relevance for case-based reasoning with incomplete focus cases. In *Proceedings if the 19th International Conference on Artificial Intelligence and Law*. ACM Press, 2023.

Raquel Mochales Palau and Marie-Francine Moens. Argumentation mining: the detection, classification and structure of arguments in text.

In *Proceedings of the 12th International Conference on Artificial Intelligence and Law*, pages 98–107. ACM Press, 2009.

Nancy Pennington and Reid Hastie. Reasoning in explanation-based decision making. *Cognition*, 49(1-2):123–163, 1993.

C. Perelman and L. Olbrechts-Tyteca. *The New Rhetoric. A Treatise on Argumentation.* University of Notre Dame Press, 1969.

Ch Perelman, Harold J Berman, Ch Perelman, and Harold J Berman. *Justice and Reason.* Springer, 1980.

Joeri GT Peters, Floris J Bex, and Henry Prakken. Model-and data-agnostic justifications with a fortiori case-based argumentation. In *Proceedings if the 19th International Conference on Artificial Intelligence and Law.* ACM Press, 2023.

J.L. Pollock. *Cognitive Carpentry. A Blueprint for How to Build a Person.* MIT Press, Cambridge, MA, 1995.

David Poole. On the comparison of theories: Preferring the most specific explanation. In *Proceedings of the 9th International Joint Conference on Artificial Intelligence*, volume 85, pages 144–147. ACM Press, 1985.

Henry Prakken and Rosa Ratsma. A top-level model of case-based argumentation for explanation: formalisation and experiments. *Argument and Computation*, 13:159–194, 2022.

H Prakken and G Sartor. A dialectical model of assessing conflicting arguments in legal reasoning. *Artificial Intelligence and Law*, 4:331–368, 1996.

H. Prakken and G. Sartor. Argument-based extended logic programming with defeasible priorities. *Journal of Applied Non-classical Logics*, 7:25–75, 1997.

Henry Prakken and Giovanni Sartor. Modelling reasoning with precedents in a formal dialogue game. *Artificial Intelligence and Law*, 6:231–287, 1998.

H. Prakken and G. Sartor. Formalising arguments about the burden of persuasion. In *Proceedings of the 11th International Conference on Artificial Intelligence and Law*, pages 97–106. ACM Press, 2007.

Henry Prakken and Giovanni Sartor. Formalising arguments about the burden of persuasion. In *Proceedings of the 11th International Conference on Artificial Intelligence and Law*, pages 97–106. ACM Press, 2007.

H. Prakken and G. Sartor. More on presumptions and burdens of proof. In E. Francesconi, G. Sartor, and D. Tiscornia, editors, *Legal Knowledge and Information Systems. JURIX 2008: The Twentyfirst Annual Conference*, pages 176–185. IOS Press, Amsterdam, 2008.

H. Prakken and G. Sartor. A logical analysis of burdens of proof. In H. Kaptein, H. Prakken, and B. Verheij, editors, *Legal Evidence and Proof: Statistics, Stories, Logic*, pages 223–253. Ashgate Publishing, Farnham, 2009.

H. Prakken and G. Sartor. On modelling burdens and standards of proof in structured argumentation. In K. Atkinson, editor, *Legal Knowledge and Information Systems. JURIX 2011: The Twenty-fourth Annual Conference*, pages 83–92. IOS Press, Amsterdam, 2011.

H. Prakken and G. Sartor. Law and logic: A review from an argumentation perspective. *Artificial Intelligence*, 227:214–245, 2015.

Henry Prakken and Giovanni Sartor. Law and logic: A review from an argumentation perspective. *Artificial Intelligence*, 227:214–245, 2015.

H. Prakken and G. Sartor. A formal framework for combining legal reasoning methods. In *Proceedings of the 19th International Conference on Artificial Intelligence and Law*, pages 22–236. ACM Press, 2023.

Henry Prakken, Chris Reed, and Douglas Walton. Dialogues about the burden of proof. In *Proceedings of the 10th International Conference on Artificial Intelligence and Law*, pages 115–124. ACM Press, 2005.

Henry Prakken, Adam Wyner, Trevor Bench-Capon, and Katie Atkinson. A formalization of argumentation schemes for legal case-based reasoning in ASPIC+. *Journal of Logic and Computation*, 25(5):1141–1166, 2015.

H. Prakken, F. J. Bex, and A. R. Mackor. Editors' review and introduction: Models of rational proof in criminal law. *TopiCS in Cognitive Science*, 12(4):1053–1067, 2020.

Henry Prakken. A tool in modelling disagreement in law: preferring the most specific argument. In *Proceedings of the 3rd International Conference on Artificial Intelligence and Law*, pages 165–174. ACM Press, 1991.

Henry Prakken. A logical framework for modelling legal argument. In *Proceedings of the 4th International Conference on Artificial Intelligence and Law*, pages 1–9. ACM Press, 1993.

Henry Prakken. From logic to dialectics in legal argument. In *Proceedings of the 5th International Conference on Artificial Intelligence and Law*, pages 165–174. ACM Press, 1995.

Henry Prakken. An exercise in formalising teleological case-based reasoning. *Artificial Intelligence and Law*, 10:113, 2002.

Henry Prakken. Analysing reasoning about evidence with formal models of argumentation. *Law, Probability and Risk*, 3(1):33–50, 2004.

H. Prakken. Coherence and flexibility in dialogue games for argumentation. *Journal of Logic and Computation*, 15:1009–1040, 2005.

Henry Prakken. A study of accrual of arguments, with applications to evidential reasoning. In *Proceedings of the 10th International Conference on Artificial Intelligence and Law*, pages 85–94. ACM Press, 2005.

H. Prakken. A formal model of adjudication dialogues. *Artificial Intelligence and Law*, 16:305–328, 2008.

H. Prakken. Formalising ordinary legal disputes: a case study. *Artificial Intelligence and Law*, 16:333–359, 2008.

H. Prakken. An abstract framework for argumentation with structured arguments. *Argument and Computation*, 1:93–124, 2010.

Henry Prakken. Modelling accrual of arguments in ASPIC+. In *Proceedings of the 17th International Conference on Artificial Intelligence and Law*, pages 103–112. ACM Press, 2019.

H. Prakken. An argumentation-based analysis of the Simonshaven case. *TopiCS in Cognitive Science*, 12(4):1068–1091, 2020.

H. Prakken. A formal analysis of some factor- and precedent-based accounts of precedential constraint. *Artificial Intelligence and Law*, 29:559–585, 2021.

C. Reed and T.J. Norman, editors. *Argumentation Machines. New Frontiers in Argument and Computation.* Kluwer Academic Publishers, Dordrecht, 2004.

Chris Reed and Glenn Rowe. Araucaria: Software for argument analysis, diagramming and representation. *International Journal on Artificial Intelligence Tools*, 13(04):961–979, 2004.

N. Rescher. *Dialectics: a Controversy-oriented Approach to the Theory of Knowledge.* State University of New York Press, Albany, N.Y., 1977.

A. Rigoni. Representing dimensions within the reason model of precedent. *Artificial Intelligence and Law*, 26:1–22, 2018.

Edwina L Rissland and Kevin D Ashley. A case-based system for trade secrets law. In *Proceedings of the 1st International Conference on Artificial Intelligence and Law*, pages 60–66. ACM Press, 1987.

E.L. Rissland and K.D. Ashley. A case-based system for trade secrets law. In *Proceedings of the First International Conference on Artificial Intelligence and Law*, pages 60–66. ACM Press, 1987.

Edwina L Rissland and Kevin D Ashley. A note on dimensions and factors. *Artificial Intelligence and law*, 10(1-3):65–77, 2002.

Edwina L Rissland and M Timur Friedman. Detecting change in legal concepts. In *Proceedings of the 5th International Conference on Artificial Intelligence and Law*, pages 127–136. ACM Press, 1995.

Edwina L Rissland, Eduardo M Valcarce, and Kevin D Ashley. Explaining and arguing with examples. In *Proceedings of the Fourth National Conference on Artificial Intelligence*, pages 288–294, 1984.

Edwina L Rissland, David B Skalak, and M Timur Friedman. Bankxx: Supporting legal arguments through heuristic retrieval. *Artificial Intelligence and Law*, 4(1):1–71, 1996.

Edwina L Rissland, David B Skalak, and M Timur Friedman. Evaluating a legal argument program: The bankxx experiments. *Artificial Intelligence and Law*, 5(1):1–74, 1997.

E. L. Rissland, K. D. Ashley, and R. P. Loui. AI and Law: A fruitful synergy. *Artificial Intelligence*, 150(1–2):1–15, 2003.

Edwina L Rissland. Examples in legal reasoning: Legal hypotheticals. In *Proceedings of the 8th International Joint Conference on Artificial Intelligence*, pages 90–93. ACM Press, 1983.

Edwina L Rissland. Dimension-based analysis of hypotheticals from supreme court oral argument. In *Proceedings of the 2nd International Conference on Artificial Intelligence and Law*, pages 111–120. ACM Press, 1989.

Regis Riveret, Antonino Rotolo, Giovanni Sartor, Roth Bram, and Henry Prakken. Success chances in argument games: a probabilistic approach to legal disputes. In A.R. Lodder and L. Mommers, editors, *Legal Knowledge and Information Systems. JURIX 2007: The Twentieth Annual Conference*, pages 99–108. IOS Press, Amsterdam, 2007.

R. Riveret, H. Prakken, A. Rotolo, and G. Sartor. Heuristics in argumentation: a game-theoretical investigation. In Ph. Besnard, S. Doutre, and A. Hunter, editors, *Computational Models of Argument. Proceedings of COMMA 2008*, pages 324–335. IOS Press, Amsterdam, 2008.

Bram Roth, Régis Riveret, Antonino Rotolo, and Guido Governatori. Strategic argumentation: a game theoretical investigation. In *Proceedings of the 11th International Conference on Artificial Intelligence and Law*, pages 81–90. ACM Press, 2007.

A. Rotolo and G. Governatori. Changing legal systems: Legal abrogations and annulments in Defeasible Logic. *Logic Journal of IGPL*, 18:157–94, 2009.

Antonino Rotolo, Guido Governatori, and Giovanni Sartor. Deontic defeasible reasoning in legal interpretation: two options for modelling interpretive arguments. In *Proceedings of the 15th International Conference on Artificial Intelligence and Law*, pages 99–108. ACM Press, 2015.

Giovanni Sartor, Michel Rudnianski, Antonino Rotolo, Régis Riveret, and Eunate Mayor. Why lawyers are nice (or nasty) a game-theoretical argumentation exercise. In *Proceedings of the 12th International Conference on Artificial Intelligence and Law*, pages 108–117. ACM Press, 2009.

G. Sartor, D.N. Walton, F. Macagno, and A. Rotolo. Argumentation schemes for statutory interpretation: a logical analysis. In R. Hoekstra, editor, *Legal Knowledge and Information Systems. JURIX 2014: The Twenty-seventh Annual Conference*, pages 11–20, Amsterdam, 2014. IOS Press.

Giovanni Sartor. Teleological arguments and theory-based dialectics. *Artificial Intelligence and Law*, 10(1-3):95–112, 2002.

Giovanni Sartor. Doing justice to rights and values: teleological reasoning and proportionality. *Artificial Intelligence and Law*, 18:175–215, 2010.

K. Satoh, T. Kogawa, N. Okada, K. Omori, S. Omura, and K. Tsuchiya. On generality of PROLEG knowledge representatio. In *Proceedings of the 6th International Workshop on Juris-informatics (JURISIN 2012)*, pages 115–128, Miyazaki, Japan, 2012.

Roger C Schank and Robert P Abelson. *Scripts, plans, goals, and understanding: An inquiry into human knowledge structures*. Psychology Press, 1977.

Marijn Schraagen and Floris Bex. Argumentation-driven information extraction for online crime reports. In *CKIM 2018 International Workshop on Legal Data Analysis and Mining (LeDAM 2018)*, 2018.

Marijn Schraagen and Floris Bex. Extraction of semantic relations in noisy user-generated law enforcement data. In *2019 IEEE 13th International Conference on Semantic Computing (ICSC)*, pages 79–86. IEEE, 2019.

MP Schraagen, MJS Brinkhuis, FJ Bex, et al. Evaluation of named entity recognition in dutch online criminal complaints. *Computational Linguistics in the Netherlands Journal*, 7:3–16, 2017.

John R Searle. *Rationality in action*. MIT press, 2003.

Marek Sergot, Fariba Sadri, Robert Kowalski, Frank Kriwaczek, Peter Hammond, and H Cory. The British Nationality Act as a logic program. *Communications of the ACM*, 29(5):370–386, 1986.

Murray Shanahan. The event calculus explained. In M.J. Wooldridge and M. Veloso, editors, *Artificial Intelligence Today. Recent Trends and Developments*, volume 1600 of *Lecture Notes in Computer Science*, pages 409–430. Springer, 1999.

David Skalak and Edwina Rissland. Arguments and cases: An inevitable intertwining. *Artificial Intelligence and Law*, 1(1):3–44, 1992.

C. Steging, S. Renooij, and B. Verheij. Discovering the rationale of decisions: Towards a method for aligning learning and reasoning. In *Proceedings of the 18th International Conference on Artificial Intelligence and Law*, pages 235–239. ACM Press, 2021.

C. Steging, S. Renooij, and B. Verheij. Rationale discovery and explainable AI. In E. Schweighofer, editor, *Legal Knowledge and Information Systems. JURIX 2021: The Thirty-fourth Annual Conference*, pages 225–234. IOS Press, Amsterdam, 2021.

Cor Steging, Silja Renooij, Bart Verheij, and Trevor Bench-Capon. Arguments, rules and cases in law: Resources for aligning learning and reasoning in structured domains. *Argument and Computation*, 14(2):1–8, 2023.

Giovanni Tarello. *L'interpretazione della legge*. Guiffrè, 1980.

Bas Testerink, Daphne Odekerken, and Floris Bex. AI-assisted message processing for the netherlands national police. In *ICAIL 2019 Workshop on AI and the Administrative State (AIAS 2019)*, 2019.

Sjoerd T Timmer, John-Jules Ch Meyer, Henry Prakken, Silja Renooij, and Bart Verheij. A structure-guided approach to capturing Bayesian reasoning about legal evidence in argumentation. In *Proceedings of the 15th International Conference on Artificial Intelligence and Law*, pages 109–118. ACM Press, 2015.

S. T. Timmer, J. J. Meyer, H. Prakken, S. Renooij, and B. Verheij. A two-phase method for extracting explanatory arguments from Bayesian Networks. *International Journal of Approximate Reasoning*, 80:475–494, 2017.

Stephen E Toulmin. *The Uses of Argument*. Cambridge University Press, 1958.

Susan W van den Braak, Gerard AW Vreeswijk, and Henry Prakken. AVERs: An argument visualization tool for representing stories about evidence. In *Proceedings of the 11th International Conference on Artificial Intelligence and Law*, pages 11–15. ACM Press, 2007.

F. H. van Eemeren and B. Verheij. Argumentation theory in formal and computational perspective. In P. Baroni, D. Gabbay, M. Giacomin, and L. van der Torre, editors, *Handbook of Formal Argumentation*, pages 3–73. College Publications, London, 2018.

F. H. van Eemeren, B. Garssen, E. C. W. Krabbe, A. F. Snoeck Henkemans, B. Verheij, and J. H. M. Wagemans. *Handbook of Argumentation Theory*. Springer, Berlin, 2014.

B. Van Gijzel and H. Prakken. Relating Carneades with abstract argumentation via the ASPIC+ framework for structured argumentation. *Argument and Computation*, 3:21–47, 2012.

L. van Leeuwen and B. Verheij. A comparison of two hybrid methods for analyzing evidential reasoning. In M. Araszkiewicz and V. Rodríguez-Doncel, editors, *Legal Knowledge and Information Systems: JURIX 2019: The Thirty-second Annual Conference*, pages 53–62. IOS Press, Amsterdam, 2019.

W. van Woerkom, D. Grossi, H. Prakken, and B. Verheij. Hierarchical precedential constraint. In *Proceedings of the 19th International Conference on Artificial Intelligence and Law*, pages 333–342. ACM Press, 2023.

B. Verheij, J. C. Hage, and H. J. van den Herik. An integrated view on rules and principles. *Artificial Intelligence and Law*, 6(1), 1998.

Bart Verheij, Floris Bex, Sjoerd T Timmer, Charlotte S Vlek, John-Jules Ch Meyer, Silja Renooij, and Henry Prakken. Arguments, scenarios and probabilities: connections between three normative frameworks for evidential reasoning. *Law, Probability and Risk*, 15(1):35–70, 2016.

B. Verheij. *Rules, reasons, arguments: formal studies of argumentation and defeat*. Doctoral dissertation University of Maastricht, 1996.

Bart Verheij. Automated argument assistance for lawyers. In *Proceedings of the 7th International Conference on Artificial Intelligence and Law*, pages 43–52, 1999.

B. Verheij. DefLog: on the logical interpretation of prima facie justified assumptions. *Journal of Logic and Computation*, 13(3):319–346, 2003.

Bart Verheij. Artificial argument assistants for defeasible argumentation. *Artificial intelligence*, 150(1-2):291–324, 2003.

Bart Verheij. Dialectical argumentation with argumentation schemes: An approach to legal logic. *Artificial Intelligence and Law*, 11(2):167–195, 2003.

B. Verheij. Argumentation support software: Boxes-and-arrows and beyond. *Law, Probability and Risk*, 6:187–208, 2007.

B. Verheij. To catch a thief with and without numbers: Arguments, scenarios and probabilities in evidential reasoning. *Law, Probability and Risk*, 13:307–325, 2014.

B. Verheij. Correct grounded reasoning with presumptive arguments. In L. Michael and A. Kakas, editors, *15th European Conference on Logics in Artificial Intelligence, JELIA 2016. Larnaca, Cyprus, November 9–11, 2016. Proceedings (LNAI 10021)*, pages 481–496. Springer, Berlin, 2016.

B. Verheij. Formalizing value-guided argumentation for ethical systems design. *Artificial Intelligence and Law*, 24(4):387–407, 2016.

Bart Verheij. Formalizing arguments, rules and cases. In *Proceedings of the 16th International Conference on Artificial Intelligence and Law*, pages 199–208. ACM Press, 2017.

Bart Verheij. Proof with and without probabilities: Correct evidential reasoning with presumptive arguments, coherent hypotheses and degrees of uncertainty. *Artificial Intelligence and Law*, 25:127–154, 2017.

B. Verheij. Analyzing the Simonshaven case with and without probabilities. *TopiCS in Cognitive Science*, 12(4):1175–1199, 2020.

B. Verheij. Artificial intelligence as law. Presidential address to the Seventeenth International Conference on Artificial Intelligence and Law. *Artificial Intelligence and Law*, 28(2):181–206, 2020.

Serena Villata, Michal Araszkiewicz, Kevin Ashley, Trevor Bench-Capon, L Karl Branting, Jack G Conrad, and Adam Wyner. Thirty years of artificial intelligence and law: the third decade. *Artificial Intelligence and Law*, 30(4):561–591, 2022.

C. S. Vlek, H. Prakken, S. Renooij, and B. Verheij. Building Bayesian Networks for legal evidence with narratives: a case study evaluation. *Artificial Intelligence and Law*, 22(4):375–421, 2014.

Charlotte S Vlek, Henry Prakken, Silja Renooij, and Bart Verheij. A method for explaining Bayesian networks for legal evidence with scenarios. *Artificial Intelligence and Law*, 24:285–324, 2016.

G. A. W. Vreeswijk. Formalizing Nomic. working on a theory of communication with modifiable rules of procedure. Technical report, CS 95-02, Vakgroep Informatica (FdAW), Rijksuniversiteit Limburg, Maastricht, 1995.

Willem A Wagenaar, Peter J Van Koppen, and Hans FM Crombag. *Anchored narratives: The psychology of criminal evidence.* St Martin's Press, 1993.

D. N. Walton, C. Reed, and F. Macagno. *Argumentation Schemes.* Cambridge University Press, Cambridge, 2008.

Douglas Walton, Giovanni Sartor, and Fabrizio Macagno. An argumentation framework for contested cases of statutory interpretation. *Artificial Intelligence and Law*, 24:51–91, 2016.

Douglas Walton. *Argumentation schemes for presumptive reasoning.* Lawrence Erlbaum Associates, 1996.

Douglas Walton. *Legal argumentation and evidence.* Penn State Press, 2002.

Maya Wardeh, Trevor Bench-Capon, and Frans Coenen. Padua: a protocol for argumentation dialogue using association rules. *Artificial Intelligence and Law*, 17:183–215, 2009.

R. Wieten, F.J Bex, H. Prakken, and S. Renooij. Exploiting causality in constructing Bayesian network graphs from legal arguments. In M. Palmirani, editor, *Legal Knowledge and Information Systems: JURIX 2018: The Thirty-First Annual Conference*, pages 151–160. IOS Press, 2018.

Remi Wieten, Floris Bex, Henry Prakken, and Silja Renooij. Supporting discussions about forensic bayesian networks using argumentation. In *Proceedings of the Seventeenth International Conference on Artificial Intelligence and Law*, pages 143–152. ACM Press, 2019.

Adam Wyner and Trevor Bench-Capon. Argument schemes for legal case-based reasoning. In A.R. Lodder and L. Mommers, editors, *Legal Knowledge and Information Systems. JURIX 2007: The Twentieth Annual Conference*, pages 139–149. IOS Press, Amsterdam, 2007.

Tangming Yuan, David Moore, and Alec Grierson. A conversational agent system as a test-bed to study the philosophical model "DC". In *Proceedings of the 3rd Workshop on Computational Models of Natural Argument (CMNA'03)*, 2003.

John Zeleznikow and Andrew Stranieri. The Split-Up system: integrating neural networks and rule-based reasoning in the legal domain. In *Proceedings of the 5th International Conference on Artificial Intelligence and Law*, pages 185–194. ACM Press, 1995.

H. Zheng, D. Grossi, and B. Verheij. Hardness of case-based decisions: A formal theory. In *Proceedings of the 18th International Conference on Artificial Intelligence and Law*, pages 149–158. ACM Press, 2021.

CHAPTER 4

APPLICATIONS OF ARGUMENTATION-BASED DIALOGUES

ANDREAS XYDIS
Lincoln Institute for Agri-Food Technology, University of Lincoln, United Kingdom
AXydis@lincoln.ac.uk

FEDERICO CASTAGNA
Department of Computer Science, Brunel University, United Kingdom
Federico.Castagna@brunel.ac.uk

ELIZABETH I SKLAR
Lincoln Institute for Agri-Food Technology, University of Lincoln, United Kingdom
ESklar@lincoln.ac.uk

SIMON PARSONS
Lincoln Centre for Autonomous Systems, University of Lincoln, United Kingdom
SParsons@lincoln.ac.uk

1 Introduction

Arguments are intrinsically intertwined with the notion of dialogue, since dialogue can be characterized as the interplay of arguments. As argued in an earlier Handbook chapter: " [...] *it is the idea of dialogue as an exchange between two or more individuals, an exchange which captures features of what would be informally called an "argument". That is, dialogue as the exchange of reasons [i.e., arguments] for or against some matter."*[Black et al., 2021]. Notice that, due to their potential expressivity (despite being subject to specific restrictions), dialogues have been advocated and often chosen as the standard communication protocol within the multi-agent system paradigm that views computation as predominately led by interaction [Luck et al., 2005]. Indeed, this new paradigm required the design of an appropriate means of communication between such intelligent agents [McBurney and Parsons, 2009], thus acknowledging the importance that dialogue exhibits in any kind of interplay, whether it occurs among humans, computational entities or both.

In this chapter we build on [Black et al., 2021], which covered some of the theoretical basis of argumentation-based dialogues by discussing applications of argumentation-based dialogues. We review and analyse a broad spectrum of proposed and existing implementations, ranging from fully-fledged software suites to rough sketches of architectures at an early stage of deployment. The key element in the distinction is that all this work is focused on the deployment of argumentation-based dialogue, rather than the development of new argumentation models.

Figure 1 illustrates our definition of argumentation-based dialogue systems and clarifies the scope of this chapter. We consider systems that have some kind of underlying *Knowledge Base (KB)*, some kind of *Argumentation Engine*, which builds arguments from the contents of the KB and/or computes the extensions of a set of arguments according to some semantics, and some kind of *User Interface (UI)*. A number of implemented argumentation systems have provided a UI whereby the user can interrogate the underlying argumentation engine and KB, but this functionality serves only to help the user gain understanding of the reasoning performed by the argumentation engine and the information it used for reasoning. While the user may be able to query the engine and KB, the user does not have the ability to *change* anything in the KB. This type of *one-way* system is illustrated in Figure 1a. The second type of system—the one which we focus on in this chapter—is specifically a *two-way* system, whereby the user engages in a *dialogue* with the system and therefore has the ability to *change* information in

4 - APPLICATIONS OF ARGUMENTATION-BASED DIALOGUES

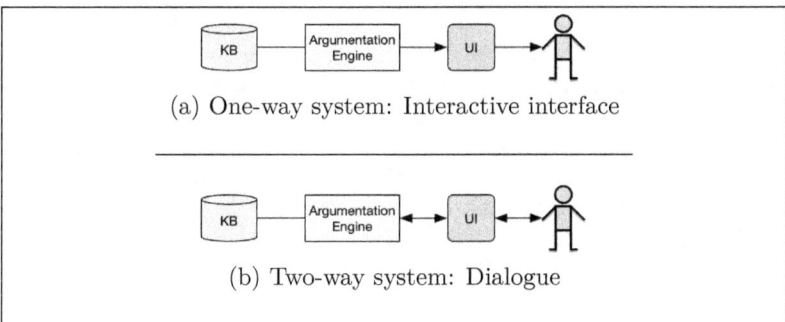

(a) One-way system: Interactive interface

(b) Two-way system: Dialogue

Figure 1: Scope of systems discussed here: "one-way" systems provide a user interface for interrogating the underlying argumentation engine and knowledge base (KB), but do not allow the user to change the knowledge base in the reasoning system; "two-way" systems support true *dialogue*, in the sense that the user is considered part of the system and can influence the knowledge and rules employed by the system. The direction of the arrows in the figure denote flow of information.

the system's KB and hence affect the arguments that the system can construct[1]. In some cases, the user also has the ability to change the behaviour of the argumentation system, by altering the rules used by the reasoner, through dialogue. This type of *two-way* system is illustrated in Figure 1b. Here, in this chapter, we mainly focus on the second type of system (but include a couple of examples of one way systems).

Another set of distinctions can be drawn between the groups of individuals that are engaged in the dialogue. A participant can be a human or an agent — we make no distinction between agents that are purely software, and agents that have a physical embodiment, such as a robot. As a result, we can imagine three kinds of dialogue: those that

---

[1] A number of argumentation-based dialogue systems make use of the notion of a "commitment store", which, in the context of Figure 1, would hold information presented by the user. Depending on the system, this commitment store might, from a theoretical perspective, be considered to be distinct from the knowledge base of the system. However, the contents of the commitment store can often be used by the argumentation system in the construction of arguments, and in such cases we would consider it to be a subset of the KB in a two-way system.

involve only humans, those that involve only agents, and those human-agent dialogues that involve both kinds of participant. We also consider that dialogues may only involve a single participant. For example, many of us are familiar with the kind of internal conversation that provides a mechanism for reflecting on some position or for ensuring that an argument will be convincing to an audience. We call such dialogues "human-self". Similarly we recognise "agent-self" dialogues, and point out that such dialogues are one way of computing argument acceptability [Caminada, 2015; Caminada *et al.*, 2014a; Castagna, 2023].

2 Theoretical Foundations

In this section we provide an overview of the fundamental notions underlying the remainder of the chapter. In particular, at first, we discuss argumentation frameworks and then formal models of dialogues. This then leads into Section 3 which sketches some of the components, such as argumentation solvers, that are used to construct applications of argumentation-based dialogues.

2.1 Argumentation frameworks

Dung's abstract argumentation theory [Dung, 1995] has a good claim to be the most known and used formalism in computational argumentation for AI at this point. The basic formalism from [Dung, 1995] has been extended in a number of different ways (e.g., see [Amgoud and Cayrol, 2002; Amgoud *et al.*, 2008; Bench-Capon, 2003]). Dung's paper [Dung, 1995] can be recognised as a supporting and analytical tool for non-monotonic reasoning [Bench-Capon and Dunne, 2007] due to the fact that arguments are represented as abstract entities, providing also important insights regarding the semantic acceptance of arguments. Specifically, Dung introduced a framework (known as an *abstract argument framework* — AAF), which is used to depict the attacking relationships (represented as directed edges) between arguments (nodes) in a graph G. A set of arguments S appearing in G is called *conflict-free* if and only if there are no arguments $A, B \in S$ such that A attacks B or B attacks A. Based on G more sets of arguments are defined with specific characteristics (called *extensions*) that determine which arguments are *acceptable* according to

4 - APPLICATIONS OF ARGUMENTATION-BASED DIALOGUES

different semantics such as *complete, preferred* and *grounded* (see [Dung, 1995] for details). Alternatives for computing extensions in AAFs have also been developed (e.g., the *labelling function* described in [Caminada, 2006]).

Although the analysis of arguments at the abstract level can provide many insights into the way that rational[2] arguers can and should behave, abstract argumentation is not always enough when tackling real argumentation problems. For example, using abstract argumentation we cannot examine how arguments are instantiated, how conclusions are inferred, what is the nature of the attacks between arguments or how arguments (and their supports or attacks) change over time. To support the justification process more naturally and examine an argument structurally, we need to access its internal parts. Thus, structured approaches to argumentation seem more appropriate to exploit.

Different frameworks for structured argumentation have been established, such as ABA [Bondarenko *et al.*, 1997; Dung *et al.*, 2009], $ASPIC^+$ framework [Modgil and Prakken, 2013], $DeLP$ [García and Simari, 2004] and *deductive argumentation* [Besnard and Hunter, 2001], for instantiating arguments with some internal structure (see [Baroni *et al.*, 2018] for details). These arguments are typically instantiated on the basis of a *knowledge base* (KB) — that contains certain and/or defeasible (i.e., uncertain) information (often called *premises*) represented in a logical language — and the application of *inference rules* to premises, leading to the *conclusion* of an argument — represented in the same logical language. Note, an argument may consist of multiple sub-arguments in which case *intermediate conclusions* are also part of an argument. Additionally, two types of inference rules may be defined: (1) *strict rules* whose inferences are certain to hold; and (2) *defeasible rules* whose inferences are, presumably, a consequence of the premises. Inference rules are explicit when more than one type are used, but may be implicit when only one type is employed. Templates that represent structures of common kinds of presumptive arguments used in everyday discourse as well as special contexts (e.g., legal argumentation), called *argument schemes*, may be perceived as a special type of inference rules since they connect the premises and the conclusion of an argument. The internal

---

[2] For a range of different instantiations of the concept "rational".

structure of arguments allows for different kinds of attacks: (1) attacking the premises of an argument (usually called *undermine*); (2) attacking the application of defeasible rules used to infer a conclusion (usually called *undercut*); and (3) attacking the conclusion–if this is inferred using a defeasible rule — of the argument (usually called *rebuttal*).

In real-life, however, human agents often use 'incomplete' arguments known as *enthymemes*. In the argumentation literature there have been works on frameworks which explore how an enthymeme can be constructed from the intended argument and how the intended argument can be reconstructed from an enthymeme. For example, in [Hunter, 2007] the author uses common knowledge (CK) between agents to show how real arguments can be encoded by the sender and decoded by a recipient. Specifically, an agent may omit premises of their intended argument that they assume to be part of CK between the agents. The recipient aims to understand these missing premises by referring to CK. In [Black and Hunter, 2012] (an extension of the work presented in [Hunter, 2007]), the authors propose a formal framework for constructing and reconstructing enthymemes based on relevance theory [Sperber and Wilson, 1986] which is grounded in two principles: *maximising cognitive effect* and *minimising cognitive effort*. There, two classes of enthymemes are defined: (1) the *implicit support enthymemes* which are enthymemes that do not include all the premises needed to entail the claim of the argument they are constructed from; and (2) the *implicit claim enthymemes* which are enthymemes missing some of the premises of the argument they are constructed from as well as the claim of that argument. In Section 6.1, we examine more works on enthymemes, which concentrate particularly on the handling of enthymemes in dialogues.

2.2 Formal models of dialogues

Both abstract and structured approaches to argumentation define binary attack (or defeat) relations between arguments where the claims of the winning (acceptable) arguments in the argument framework *AF* identify the non-monotonic inferences from the belief base instantiated in *AF*. These approaches, initially defined for single agent (monological) reasoning, can be generalised to dialogical models of distributed non-monotonic reasoning in which two or more agents exchange arguments and other locutions.

Walton and Krabbe's work [Walton and Krabbe, 1995] was one of the most influential regarding the typology of primary dialogue types. Each type of dialogue depends on the initial information that participants have, their individual goals and the objective of the dialogue (see Table 1). A number of these types of dialogue have been studied in detail by the argumentation community. For example, see [Prakken, 2006] for persuasion, [Fan and Toni, 2012] for information-seeking, [Black and Hunter, 2009] for inquiry, [Rahwan *et al.*, 2003] for negotiation and [McBurney *et al.*, 2007] for deliberation. Walton and Krabbe's list is not intended to be exhaustive, and not only is it possible to identify kinds of dialogue beyond those in Table 1^3, but also new dialogues can be formed by combining types of dialogues from Table 1.

Here, we briefly describe common constituents of such dialogue systems.

2.2.1 Constituents of dialogues

As described in [Black *et al.*, 2021], the most common constituents of argumentation-based dialogues found in the literature are *moves, dialogue history, protocol* and, possibly, *strategies*[4]. We review each of these here.

Moves. Although different variations of *moves* can be found in the literature, there are three basic pieces which constitute a move. These are the *sender* of the move[5], the *locution* of the move (which indicates the *type* of the move, i.e. what an agent is allowed to utter using this move), and the *content* of the move. Different locutions may be used in different

---

[3] For example [Cogan *et al.*, 2005] does this by considering different combinations of initial situation.

[4] Note that while we recognise that modelling the beliefs of other agents [Sklar *et al.*, 2004], and belief revision as a result of dialogue [Parsons and Sklar, 2005] are both important with relation to dialogues, we consider them to be out of scope for this chapter, not least because belief revision and argumentation are studied at length elsewhere [Falappa *et al.*, 2009].

[5] The sender is often considered to be either the *proponent* of an argument or the opponent, since most work on dialogue considers just two participants (assuming easy generalisation to many) and a more or less adversarial stance where one agent (proponent) is trying to have the other accept the argument that they are making.

Dialogue types	Initial situation	Individual goal	Dialogue goal
Persuasion	Conflict of opinions	Persuade other party	Resolve or clarify issue
Inquiry	Need for proof	Find and verify evidence	Prove/disprove hypothesis
Information-seeking	Need Information	Acquire or give information	Exchange information
Negotiation	Conflict of interests	Get what you most want	Reasonable settlement both can live with
Deliberation	Dilemma or practical choice	Co-ordinate goals and actions	Decide best available course of action

Table 1: Types of dialogues proposed in [Walton and Krabbe, 1995].

dialogue systems, but a common set include the following: *assert, accept, challenge, question, since* and *retract*. They allow for claims and their supporting arguments to be stated (or retracted), arguments to be requested and questions to be asked. The content of a move can be a formula ϕ built from some logical language, or even *null*.

Dialogue history. The *dialogue history* represents the moves made during the dialogue and it is usually formalised as a non-empty sequence of these moves. In most cases moves are indexed by the step of the dialogue, where d_k denotes the length k of the dialogue.

Protocol. The *protocol* of a dialogue determines the legal moves an agent can make during a dialogue. Although it is impractical to designate the moves permitted for every possible dialogue state, simpler rules can indicate what kind of move an agent is allowed to make. Such rules can be turn-taking rules, relationship between locutions (e.g., if an agent utters a *question* move, their interlocutor can reply only with a *since* move) or commitment rules stating conditions upon which moves can be made (e.g., if an agent has asserted ϕ and has not retracted it, then asserting $\neg\phi$ will produce an inconsistent *commitment store CS* for the agent).

Strategy. A *strategy* is a mechanism for deciding an agent's move. This can be determined by the agent's objective (e.g., preserving rationality principles or "winning" the dialogue). Formally, a strategy S_{Ag} of an agent Ag may be perceived as a function $S_{Ag} : D \times K_{Ag} \longrightarrow 2^M$, where D denotes the dialogue history, K_{Ag} the private knowledge base KB of Ag and M the set of moves. If there is no probability distribution over the possible dialogue moves of agents then a strategy is *deterministic*, whereas a strategy which returns only a single move is called *decisive*. A common practice for dialogue systems that employ strategies is for these to depend on the previous move made in the dialogue.

2.2.2 Persuasion dialogue systems

In persuasion dialogues, two or more participants try to resolve a conflict of opinion, each trying to persuade the other participant(s) to adopt their point of view. Many papers examine persuasion dialogues; and different features in combination with such dialogues are investigated, such as opponent modelling (e.g. [Hadjinikolis *et al.*, 2013; Oren and Norman, 2009]), planning (e.g. [Black *et al.*, 2017]), decision trees for strategising (e.g. [Hadoux and Hunter, 2017]), probabilities (e.g. [Hunter and Potyka, 2017]) and natural language processing (e.g. [Chalaguine and Hunter, 2020; Rosenfeld and Kraus, 2016]). We briefly introduce some important formal models of dialogue systems [Prakken, 2006; Prakken, 2009] that capture persuasion so that it is easier for the reader to understand the main aspects of practical applications concerning persuasion dialogues presented later.

Walton and Krabbe's paper [Walton and Krabbe, 1995] describes the "Permissive Persuasion Dialogues" (PPD) system (amongst other systems capturing different types of dialogues). In PPD, dialogues have no context, and include two participants (P and Op) which may declare assertions and concessions in an implicit preparation phase before the dialogue commences. Each one is considered the proponent of their own assertions and opponent of the other participant's initial assertions. The communication language includes challenges, (tree-structured) arguments, concessions, questions, resolution demands, and two types of retraction locutions for commitments. The logical language used is propositional logic, and the inference rules are deliberately incomplete to reflect the complexity of natural language. Participants commit to the premises of the arguments they move, but arguments may be incomplete, leaving room for further exploration (these may also be perceived as enthymemes, but a further discussion on dialogue systems that deal with enthymemes will follow later).

The protocol is guided by the participants' CSs and the content of the move made in their last turn. P starts the dialogue, and in the first turn, both P and Op either concede or challenge each other's initial assertions. Each turn, a participant is obliged to reply to all moves made in the other player's last turn, except concessions and retractions. Multiple replies are allowed, and alternative arguments for the same assertion can be made. Counterarguments are not permitted. The protocol is non-deterministic, multi-move and multi-reply, but postponement of replies is not allowed. Challenges, concessions, retractions and questions are always related to commitments. A participant cannot challenge or concede their own commitments. Inconsistent commitments can be resolved, and implications between commitments may require concessions or retractions. The outcome of the dialogue is determined by the participants' commitments, and the dialogue terminates after a predetermined number of turns.

In [Parsons et al., 2003], a dialogue system between two "players" is described. The dialogue concerns a single topic and each participant has their own KB which itself may be inconsistent. The communication language allows for participants to move claims, challenges, and concessions during the dialogue, but there is no explicit reply structure. Claims can pertain to individual propositions or sets of propositions, and the logic employed is non-monotonic. Arguments are classical proofs from consistent premises, and arguments attack other arguments by negating a premise of their target. Conflict between arguments is resolved using a preference relation on the premises. The system utilizes grounded semantics to decide acceptability of arguments. Arguments can be implicitly

4 - APPLICATIONS OF ARGUMENTATION-BASED DIALOGUES

moved as a claim ψ replying to a challenge of another claim ϕ, given that ψ is consistent and ψ contradicts ϕ. The system defines commitment rules, but (contrary to Walton and Krabbe's system [Walton and Krabbe, 1995] discussed above) these rules neither determine legal moves nor the outcome of the dialogue. A CS is only used as a supplementary KB to access an interlocutor's knowledge revealed during the dialogue.

An important aspect introduced in this system is the *assertion* and *acceptance* attitudes of players, which they must adhere to throughout the dialogue. These attitudes are defined in relation to the player's private KB, which does not change during the dialogue, and both players' CSs, which may change during the dialogue (see [Parsons et al., 2003] for details). Players' attitudes influence their moves during the dialogues. Although there are works (e.g. [Prakken, 2006; Walton and Krabbe, 1995]) defending the idea that a dialogue protocol should only enforce coherence of dialogues, [Parsons et al., 2003] argues that a dialogue protocol should refer to private KBs to ensure rationality and honesty of players. As a result, a formal definition of a protocol is given where the assertion and acceptance attitudes of players partly dictate the legal moves of the system, as well as termination of a dialogue. The winner of a dialogue is not defined, but the possible outcomes are defined in terms of the propositions claimed and conceded by participants. Finally, the protocol is unique-move, unique-reply and deterministic, with some exceptions (see [Parsons et al., 2003] for details).

In [Prakken, 2005], Prakken establishes a general dialogue framework, assuming two participants, whose purpose is to formally describe the components needed to formalise any kind of dialogue. Specifically, his initial dialogue framework is general enough to capture various kinds of dialogue from Walton and Krabbe's typology [Walton and Krabbe, 1995], whereas later he specifies locutions and rules which are used to model persuasion dialogues (called *liberal* dialogues). Prakken is non-committal on certain specifics, except for an explicit reply structure between moves. Moreover, he explores different protocols (of varying degrees of complexity) for regulating dialogues and the belief bases of participants do not influence the dialogues' protocol.

Prakken also defines the *dialogical status* of moves made in a dialogue so that: (a) different turn-taking and termination rules are examined as well as the relevance of moves; and (b) a correspondence is established between the dialogical status of the initial move of a dialogue (whose content is the topic of the dialogue) and the justified arguments in supporting the dialogue topic. An argument is presented as a tree whose nodes are elements of the logical language, the edges between them depict either strict or defeasible inference rules, the root of the tree is the conclusion of the argument and its leaves are the argument's premises, similar to $ASPIC^+$. The locutions introduced for modelling persuasion dialogues in [Prakken, 2005] are $claim, why, concede, retract$ and $argue$. The dialogical status of a move can be either IN or OUT. The author represents a dialogue as a tree where each node n is a move, and a child of n (if any) is a reply to n.

Additionally, the notion of *logical completion* of a dialogue is introduced, which means that if there is an argument A that can be constructed from the content of the locutions exchanged during the dialogue which defeats an argument B asserted by a participant, then A is moved against B for each occurrence of B in the dialogue (since an argument may be moved multiple times in a dialogue; and each time is represented as a different instance in the dialogue tree). Furthermore, the participant Ag making the first move m_0 in the dialogue is the *winner* of the dialogue if and only if m_0 is IN, otherwise its counterpart Ag' is the winner of the dialogue. Termination is defined as the situation where a player is supposed to move but has no legal moves. Alternative and postponed replies are allowed, and the instantiations of protocols described are multi-move and multi-reply. Finally, it is worth mentioning that the author produces some soundness and fairness results stating that, for a finite logically completed liberal dialogue, Ag wins the dialogue (under the grounded semantics) if and only if using the contents of the locutions exchanged during the dialogue, a justified argument in support of the topic of the dialogue (i.e., the content of the first move, m_0, made in the dialogue) can be constructed.

2.2.3 Inquiry dialogue systems

In inquiry dialogues, participants collaborate to answer some question(s) that they could not answer on their own. Comparatively little work has been done on inquiry dialogue protocols that employ argumentation.

4 - APPLICATIONS OF ARGUMENTATION-BASED DIALOGUES

We briefly describe two important papers related to this field. Later on in this chapter, we present practical applications concerning inquiry dialogues that implement characteristics discussed below.

As mentioned earlier, [Parsons et al., 2003] presents some protocols for different types of dialogues. Most of the characteristics discussed in this chapter, in Section 2.2.2, for the same paper [Parsons et al., 2003] also hold for inquiry dialogues, i.e. the number of participants and locutions allowed; the logic is non-monotonic; arguments are still classical proofs from consistent premises and not directly moved, but implicitly; and the commitment rules do not determine legal moves nor the outcome of the dialogue. However, since agents work together, they do not attack each other's arguments but only challenge them. Agents' assertion and acceptance attitudes influence the protocol of the dialogue in this case, too. Additionally, two different protocols are described, since the first one (assert) is so simple that it includes flaws which the authors try to address (e.g. a proof might not be allowed to be found even though it is available to the agents if they moved different sets of assertions). Although the second protocol (accept) presented deals with the issues of the first protocol, again the authors mention that there is room for improvement (e.g. in the second protocol, only one agent dictates assertions, which are also restricted to be connected to what is already uttered). Finally, although the first protocol presented is unique-move and unique-reply, the second one allows for multiple moves and replies.

In [Black and Hunter, 2009], a general inquiry dialogue system between two participants is introduced where a strategy for each agent for picking a unique move to make in each step of the dialogue is also developed. Specifically, the authors give a general definition of a dialogue which allows for other types to be considered within their framework, whereas later they provide protocols for two different types of inquiry dialogues: (1) an *argument inquiry dialogue*, which allows participants to share knowledge to jointly construct arguments; and (2) a *warrant inquiry dialogue*, which allows participants to share knowledge to jointly construct dialectical trees (i.e. a tree with an argument at each node in which a child node is a counterargument to its parent). In an argument inquiry dialogue, the agents exchange beliefs in order to jointly construct arguments for a particular claim, but the acceptability of the arguments constructed cannot be determined. However, in a warrant inquiry dialogue, the acceptability of a particular argument is examined by jointly constructing a dialectical tree that collects all the arguments that may be relevant to the acceptability of the argument in question.

The authors use $DeLP$ to represent not only beliefs and arguments, but also preferences over arguments to decide successful attacks. The dialectical status of an argument in a dialectical tree is either D, for defeated arguments, or U, for undefeated ones. The locutions allowed are: *open, assert* and *close*. For an argument inquiry dialogue, the topic of the dialogue is a defeasible rule, whereas for a warrant inquiry dialogue, its topic is a defeasible fact. The termination of a dialogue (in both cases) is defined as the consecutive appearance of two close moves. Essentially, this means that both participants must agree to the termination of the dialogue (as they alternate moves). Both agents have a CS, but there is also a shared *query store*, defined as the set of literals that could help construct an argument for the consequent of the topic of an argument inquiry dialogue (during the dialogue, participants try to provide arguments for the literals in the query store). The outcome of an argument inquiry dialogue is defined as the set of all arguments that can be constructed from the union of the CSs and whose claims are in the query store, whereas the outcome of a warrant inquiry dialogue is determined by the dialectical tree that is constructed from the union of the CSs: the topic of the dialogue is warranted if and only if the root of the dialectical tree is undefeated.

2.2.4 Information-seeking dialogue systems

In information-seeking dialogues, the goal of the dialogue is the exchange of information where a participant wants to acquire some information they are not aware of from their interlocutor who tries to fulfill that request. To the best of our knowledge there are not many works on dialogue systems designed specifically for information-seeking dialogues. Instead, general dialogue systems have been examined as to how they could be used to instantiate this type of dialogues. Below we present how an information-seeking framework has been considered in [Parsons et al., 2003].

Most of the characteristics discussed in this chapter, in Sections 2.2.2 and 2.2.3, for the paper [Parsons et al., 2003] also hold for information-seeking dialogues, i.e. the number of participants and locutions allowed; the logic is non-monotonic; arguments are still classical proofs from consistent premises and not directly moved, but implicitly; and the commitment rules do not determine legal moves nor the outcome of the dialogue. However, in this case, the dialogue starts with a question from participant A towards their interlocutor B regarding a proposition p. The dialogue is similar to an inquiry dialogue, where agents cannot attack each other's arguments, but only challenge them and provide support for them. Agents' assertion and acceptance attitudes influence the protocol of the dialogue in this case, too. Later on, authors discuss interesting properties that characterise their protocol, some of which are true for any assertion and acceptance attitudes of participants, whereas some others depend on these attitudes.

2.2.5 Negotiation dialogue systems

In negotiation dialogues, participants try to resolve a conflict of interest by reaching a deal that all the parties can live with. Although much work has been done in the development of theoretical negotiation dialogue frameworks, only a little work has been done on practical applications that consider negotiation dialogues which account for human agents. We briefly present two influential works in this topic, and later in Section 4.3.5 we examine related applications found in the literature.

In [Parsons et al., 1998] an argumentation-based framework for negotiation dialogues is proposed where the associated protocol that governs the dialogue is presented as a state machine. The locutions of this protocol are: *proposal* (used to open the dialogue suggesting a solution to the problem that the agents face, or offer a proposal at a different stage in the dialogue), *critique* (used to provoke an alternative proposal), *counter − proposal* (which is a proposal that is made as a response to a previous one), *accept* (used to accept a proposal showing that an agreement is reached and the dialogue terminates) and *withdraw* (used by a participant to leave the dialogue showing that no agreement is reached and the dialogue terminates). Note that agents may make counter-proposals without waiting for a response to a previous one, and the participants of the dialogue are assumed to be two although they can be more.

While describing their introduced locutions and protocol, the authors refer to the notion of *explanation* which they define as additional information explaining why a proposal, counter-proposal or critique was made. Essentially they present a pair $p = (\Gamma, \phi)$ consisting of an utterance ϕ and an explanation Γ as an argument where Γ is a set of formulae available to an agent. To construct arguments, the authors use classical first-order logic. Arguments are used as part of the content to the locutions described above. They also define rebut and undercut from an argument A towards an argument B as an attack from A to the conclusion and premise(s) of B, respectively, but also mention that attacks on stated inference rules can also take place without extending their discussion on this topic.

Notice that a reason for using first-order logic, is because the agents' architecture follows the BDI model (Belief-Desire-Intention), thus each of these components can be represented as a predicate in the communication language of the agents. Additionally, because of their BDI model, they define conflicts between agents as agents having opposite intentions, or an agent intending to change the mental state of their interlocutor. Finally, classes of acceptable arguments are also defined, so that the agent can determine how strongly it objects to a proposal as well as evaluate it internally before sending it as a proposal to the other agent.

In [Amgoud *et al.*, 2000], the authors provide another protocol for negotiation dialogues, which is based on abstract argumentation to instantiate arguments. Specifically, an argument is defined as a pair $A = (H, h)$ where h is a formula of a propositional language L, and H is a set of formulae of L such that H is consistent, $H \vdash h$ and H is minimal with respect to set inclusion. The KBs of the participants may be inconsistent and an *undercut* between arguments is defined as a case where the conclusion of one argument A contradicts one of the elements of the support of another argument B. A preference ordering between arguments is also taken into account to determine successful attacks and, thus, acceptable arguments.

Later, a protocol is given which assumes two participants, describes the legal moves of the dialogue and defines an argument dialogue as a

4 - APPLICATIONS OF ARGUMENTATION-BASED DIALOGUES

sequence of such moves. Argument dialogue trees are also defined, where each branch of this finite tree is an argument dialogue, and winning criteria are also discussed. Additionally, the authors assume that agents have a set of beliefs, desires and intentions (i.e., following a BDI system as the work examined above), but focus on the set of beliefs.

Finally, authors in [Amgoud et al., 2000] also discuss how they expand the logical language so that it includes implications, how CSs of agents are considered and how these are updated based on the moves the agents make during a dialogue. Particularly, the authors present the locutions they employ in their protocol, the rationality behind them, the available responses as well as the effect of these locutions in the CSs of participants.

2.2.6 Deliberation dialogue systems

In deliberation dialogues, participants need to jointly decide on an action or a course of action. Here, we briefly describe two important works related to this field. Later on in this chapter, applications with similar features are discussed.

In [McBurney et al., 2007] the authors develop the first formal framework for deliberation dialogues called Deliberation Dialogue Framework (DDF). After presenting the characteristics that differentiate deliberation dialogues from other types of dialogue, the authors present a formal model of deliberation dialogue which consists of eight stages. To define these stages the authors describe some necessary features (types) such as *actions, goals, constraints, perspectives, facts* and *evaluations*. Notice that later, using a sentential language, sentences moved during the dialogue are instances of these types. The stages characterising deliberation dialogues are: *Open* (i.e., where the dialogue starts with a question regarding what is to be done), *Inform* (i.e., where agents discuss desirable goals, constraints on actions, evaluation of proposals and facts relevant to evaluation), *Propose* (i.e., where agents suggest possible actions-options), *Consider* (i.e., where agents comment on proposals), *Revise* (i.e., where agents revise goals, constraints, perspectives and actions-options), *Recommend* (i.e., where agents suggest an option for action and either they accept it or reject it), *Confirm* (i.e., where agents confirm the acceptance of their choices), and *Close* (i.e., where agents close the dialogue). The authors also specify that the aforementioned stages may occur in any order and participants can visit them as often as they desire, as long as they obey to some rules that the authors describe in the paper (see [McBurney et al., 2007] for details).

Later on in [McBurney et al., 2007], the locutions that enable DDF are presented. These are:

> *open_dialogue, enter_dialogue, propose, assert, prefer, ask_justify, move, reject, retract, withdraw_dialogue*

Details on the contents of such locutions are also discussed, as well as the effects of these locutions in the CS of the participating agents (which are public, but only the participant's own utterances lead to additions into its CS). Of course, a protocol with rules on relationships between the locutions is also given (e.g., which locutions may be used as a response to *ask_justify*), but is considered relatively liberal. Additionally, the authors associate locutions to each stage given earlier. Finally, they evaluate DDF and the associated protocol by comparing it to human deliberation dialogues, considering their protocol from the perspective of the deliberation processes it implements and considering the outcomes, if any, that deliberation dialogues conducted under the DDF protocol achieve.

Another notable deliberation dialogue system is proposed in [Kok et al., 2011]. This work is based on [Prakken, 2005] described earlier in Section 2.2.2. Thus, the deliberation system proposed in [Kok et al., 2011] provides, similarly, an explicit reply structure, a turntaking function and a termination rule, dialogical statuses of moves similar to the ones discussed in Section 2.2.2 (ensuring coherent dialogues), different protocol rules (which can be added/discarded depending on the domain) and an anytime outcome which can also be used to decide the winner of the dialogue at that particular moment. Note, arguments here are formed using inference trees of strict and defeasible rules, grounded on the formalism of arguments in [Prakken, 2010].

The authors, however, had to make some modifications to the system presented in [Prakken, 2005] so that their system accommodates deliberation dialogues. Firstly, more than two participants are allowed in [Kok et al., 2011]. Additionally, notions such as relevance and protocol rules had to be revised accordingly, and multiple proposals (instead of just one claim) are discussed during the dialogue. Moreover, the dialogue outcome is no longer a direct result of the moves. Finally, a winning function is needed to select a single action from all actions that are proposed, or possible none if there is no acceptable option.

3 Implementation Building Blocks

As noted above, the previous section covered the theoretical basis of computational argumentation and the work on argumentation-based dialogues that was built on that theoretical basis. In this section we now look at some of the software tools that have been developed to support argumentation-based dialogue. Many of these have been developed as implementations of that theoretical basis — for example, given the description of formal dialogues, it is clear that an implementation could benefit from a tool that computed the extensions of a set of arguments — and the mainstream of work progressed assuming that the route to argumentation-based dialogue software systems would always be through implementing formal models. However, more recently, the successes of natural language processing[6] has led to a second branch of work on argumentation-based dialogues, that which is based on NLP-derived chatbots. We therefore take a quick look at some of the work in that direction.

3.1 Components of an argumentation-based dialogue system

In considering the tools, it is helpful to identify some common components of argumentation-based dialogue system architectures, as shown in Figure 2. Taken together, these components represent a super-set of the components implemented in the set of applications discussed in this chapter. The diagram is meant to be neither prescriptive nor exhaustive. As the remainder of the chapter unfolds, the reader may find it helpful to refer back to this figure to understand the types of components im-

---

[6]The confluence of argumentation and natural language processing (NLP) has been long in the making, with the Argument Mining workshop in its 11th year, as of the time of writing, and the workshop on Computational Models of Natural Argument about to have its 24th instantiation.

plemented in the systems discussed and how they are placed within an overall schema.

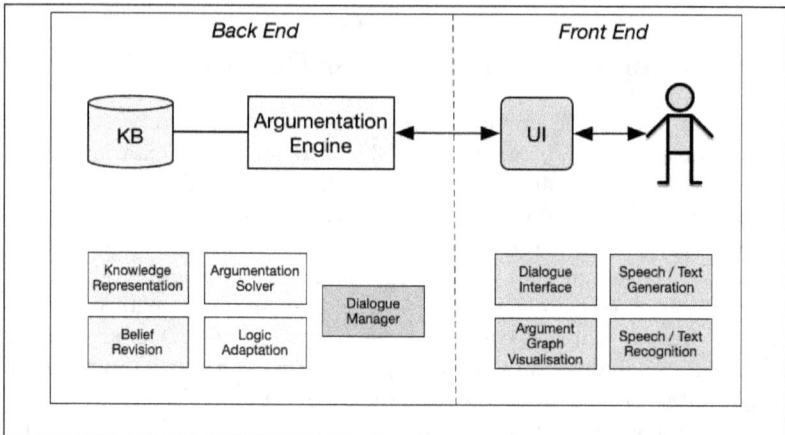

Figure 2: Architecture components. "Front End" and "Back End" separation is indicated.

Note that we separate the components into "Front End" and "Back End" elements. In programming, back-end is a commonly used term to describe the underlying infrastructure that drives the involved application. In the context of implementations of argumentation systems, reasoning engines, for example, fall into this category. Their task is to steer the decision-making processes, thus guiding the software towards its goal. We consider back end components in Section 3.2.

Computational argumentation is not only leveraged for back-end purposes. Argument graphs, for example, represent an informative way to display pieces of knowledge and the relations subsisting between those pieces. Such visual and interactive elements constitute (part of) the user interface, that is to say, the front-end component. For this reason, it is worth reviewing the existing argumentation-based dialogue applications that have both back and front-end argumentation-related components. We call these end-to-end argumentation components, and we consider these in Section 3.3.

3.2 Back-end argumentative implementations

One of the main purposes of computational argumentation is to enable the resolution of conflicting knowledge, thus allowing for a selection of the most appropriate (i.e. justified) pieces of information. *"A decision is a choice between competing beliefs about the world or between alternative courses of action. [...] Inference processes generate arguments for and against each candidate. Decision making then ranks and evaluates candidates based on the underlying arguments and selects one candidate as the final decision. Finally, the decision commits to a new belief about a situation, or an intention to act in a particular way."* [Fox et al., 2007]. Decision-making processes can be encoded as problems whose solutions are rendered by the computation and evaluation of AFs: an argumentation engine is essentially a reasoning tool driven by the same logic and process. The resulting acceptable entities provide a strong (logical) rationale for and against a given decision, while also leaving space for further deliberation [Dix et al., 2009]. Such an argumentative decision-making apparatus can be a useful addition to real-world software applications concerning defeasible reasoning, as advocated by the comprehensive study of Bryant and Krause [Bryant and Krause, 2008]. Without any claims of completeness, we now provide a brief overview of one of the most common types of component of reasoning engines leveraged by argumentation-based dialogue applications: the solvers.

A specialized piece of software that encodes and provides the solution to a particular computational problem is known as a 'solver'. Popular stages where a plethora of different solvers for abstract argumentation decision procedures are presented are the ICCMA (International Competition on Computational Models of Argumentation) events [Bistarelli et al., 2021a; Gaggl et al., 2020; Järvisalo et al., 2023; Lagniez et al., 2021; Thimm and Villata, 2017]. In this competition, various pieces of software are evaluated according to their capabilities of addressing computational argumentation-related reasoning challenges in connection with specific σ semantics: for example, the enumeration of σ-extensions in the AF and the credulous and sceptical membership of a particular argument to at least one (credulous) or each (sceptical) σ-extensions. Among these computational argumentation solvers, we can acknowledge *AFGCN v2* [Malmqvist, 2023] and *PYGLAF* [Alviano, 2021], both of which harness Python scripts to achieve the desired results. In particular, *AFGCN v2* leverages an approximation method based on the Graph Convolutional Network (GCN), whereas *PYGLAF* combines Circumscription [McCarthy, 1980] and SAT solvers.

Similarly, SAT encodings and solvers are employed by μ-*toksia* [Niskanen and Järvisalo, 2023] (either Glucose [Audemard and Simon, 2018] or Cryptominisat [Soos et al., 2009]), *FUDGE* [Thimm et al., 2023] (whose reduction-based method and sophisticated encodings ensure an optimized procedure over the benchmark) and *Crustabri* [Lagniez et al., 2023]. The latter stems from a rewriting of CoQuiAAS [Lagniez et al., 2015] developed using the Rust language. Other examples are *FARGO-LIMITED* [Thimm, 2023a], an approximate reasoning tool that relies on a variant of the standard DPLL search algorithm [Biere et al., 2009] and *HARPER++* [Thimm, 2023b], a solver whose operations hinge on the grounded semantics and its properties. It is also worth mentioning *ASPARTIX-V21* [Dvorák et al., 2021] and *ASPforABA* [Lehtonen et al., 2023], both of which make use of Answer-Set-Progrmamming (ASP) encodings for, respectively, abstract and structured (ABA [Dung et al., 2009]) computational argumentation tasks. Leveraging a different approach, *ConArg* [Bistarelli et al., 2021b] takes advantage of Constraint Programming (CP) techniques and heuristics (via a specific C++ toolkit) to provide its output. Finally, *A-Folio DPDB* [Fichte et al., 2021] resorts to an inventive solution by leveraging DPDB (i.e., a general framework designed to address counting tasks via dynamic programming and database management system [Fichte et al., 2022]) adapted for computational argumentation reasoning purposes. We conclude the list by mentioning *AGNN* [Craandijk and Bex, 2020], an Argumentation Graph Neural Network that learns how to predict the likelihood of an argument being credulously and sceptically accepted.

3.3 End-to-end argumentation implementations

The work in the previous section largely consisted of implementations of formal systems. Here we start by considering end-to-end systems that are based on formal models before turning to chatbots.

3.3.1 Panoptic Engines

Similarly to solvers, panoptic (or *all-encompassing*) engines are suites of different pieces of software that perform specific calculations concerning computational argumentation semantics. However, unlike standard solvers, those engines are designed to provide additional functionalities and customisation tools (e.g., knowledge base manipulation, domain selection, underlying logic adptation, graph visualization). Among these reasoning tools, we can include *ArguLab* [Podlaszewski *et al.*, 2011] which computes (and graphically visualizes) the extensions of an AF, engages in structured dialectical exchanges to prove the acceptability of the justified arguments and incorporates considerations of judgement aggregations [Caminada and Pigozzi, 2011] to handle the stance of groups of agents.

Other examples are *Prengine* [Hung, 2017] and *PyArg* [Borg and Odekerken, 2002], both implemented in Python. The latter is a comprehensive tool capable of executing different computational argumentation tasks, including AFs (either abstract or structured) generations, evaluations and visualization. *Prengine* is instead designed as a multi-purpose engine that handles Probabilistic Assumption-based Argumentation (PABA [Dung and Thang, 2010]) by translating Probabilistic Argumentation (PA) models into PABA, implementing inferences about arguments likelihood and computing their semantics. *NEXAS* [Dachselt *et al.*, 2022] harnesses Python (in particular, the pandas library[7]) to provide an interactive exploration of the solution space, statistical analysis and a correlation matrix for the acceptance of individual arguments for the selected semantics.

On the other hand, *Argue tuProlog* [Bryant *et al.*, 2006] leverages a reasoning core Java-based Prolog to specify whether a claim can be argumentative and evaluates the outcome by tracing an argument game envisaged to prove such a claim. Another Prolog-implemented engine is *CaSAPI* [Gaertner and Toni, 2007], whose features include supporting users' customisation regarding argument, semantics and domain selection within the ABA framework. Furthermore, we can also acknowledge *IACAS* [Vreeswijk, 1994] as being one of the oldest prototypes of an argumentation engine whose purpose concerns the evaluation of argu-

---
[7]https://pandas.pydata.org/

ments via two-party immediate response disputes. Finally, *ArgTrust v1.0* [Tang et al., 2012b] is an argumentation engine implemented in Java, whose underlying methodology [Tang et al., 2012a] reasons over data by assigning values to the arguments (and their relations) depending on how much the source is 'trusted'. A later version, *ArgTrust v2.0* [Sklar et al., 2016], was implemented in Python and MySQL and facilitated an interface for users to interrogate the underlying *AF*.

3.3.2 Chatbots

Finally, we look at chatbots. These are conversational software systems designed to mimic human discourse mostly to enable automated online guidance and support [Caldarini et al., 2022], thus allowing humans to interact with digital devices as if they were communicating with a real person [Oracle,]. These computer programs generate responses based on given inputs producing replies via text or speech format [Bala et al., 2017; Sojasingarayar, 2020] employing different architectures [Codecademy, 2022]. Indeed, the long history of such conversational agents stems from rule-based, scripted template chatbots (e.g. the famous ELIZA [Weizenbaum, 1966]), whose replies are predefined and returned according to a series of NLP-encoded rules. The field has advanced towards retrieval-based architectures (e.g. A.L.I.C.E. [Wallace, 2009]), in which responses are instead pulled from a corpus of sentences according to the received input, and most recently centres on generative models, for example the well-known ChatGPT[8]. The generative architecture, which grants an agent the ability to formulate its own original responses rather than relying on existing text, hinges on the recent Transformer technology [Vaswani et al., 2017] that revolutionized the entire field of chatbot research[9]. Interactive agents engineered upon such a Transformer-based structure convey impressive performances within open-domain conversations (although they are not immune from various shortcomings [OpenAI, 2024a]), while previous bot architectures could only aim at closed-domain

---

[8] https://chat.openai.com

[9] Arguably transformers have revolutionized the whole of NLP, as well as having found applications in related fields such as computer vision and genomics.

conversations[10]. While chatbots are not, in general, argumentation-based, we mention them here because, as we will see, there have been recent efforts to develop argumentation-based chatbots.

Note that chatbots can be considered end-to-end software implementations[11] where an underlying response architecture elaborates the replies to be sent into a specifically designed chatbox. Here, the user will be able to interact and dialogue with the bot in text or speech format. This ability is attractive from a user interface point of view and is one of the reasons that chatbots are an interesting element of an argumentation-based dialogue system.

4 Selected applications

Having briefly clarified the background notions underpinning the whole chapter, we are ready to dive into a discussion of existing applications of argumentation-based dialogue. We do this according to dimensions described in Section 4.1 as well as components of the dialogue and the employed argumentation framework. Overall, we did not draw a strict line, and we opted for a comprehensive review by including all the pertinent research we could find. Our survey has not been constrained by dates of publications (although, where feasible, we preferred the latest version of a particular line of work), domains or evaluation method: we were only concerned with the application of argumentation-based dialogues, whether fully-fledged developed or just sketched, whether their structures relate to back, front or end-to-end operations, irrespective of the dialogue protocol, software tool or dataset (if any) adopted.

We start, in Section 4.1 by presenting ways to structure the literature on argumentation-based dialogue systems — see Tables 2 and 3, justifying the analysis and discussing some aspects of it. Then we proceed to examine individual systems. First, in Section 4.2 we describe an application, implemented in the health sector by a team inclduing many of the authors, as a use-case study. We do so because it represents a complete

---

[10] A comprehensive survey of chatbot history can be found in the work of Adamopoulou and Moussiades [Adamopoulou and Moussiades, 2020], whereas a review of Transformer-based conversational models has been conducted in the study by Zhao et al. [Zhao et al., 2023].

[11] Hence their inclusion here.

system of a dialogical application, comprising of all the components introduced in Figure 2. Thus, it represents a good example of the desired pieces and features of a fully-developed (in the sense of both back and front-end) dialogical application in the field of argumentation.

Next, we broaden the discussion under two main headings. First, in Section 4.3 we discuss systems that are built on theoretical models like those introduced in Section 2. Then, in Section 4.4 we look at work on chatbots, reflecting the more recent work that has grown out of research in the natural language processing community. In both of these latter sections, we draw on the distinction between dialogue types (see Section 4.1) as a way of structuring the discussion.

4.1 Methodology

In this section, we describe ways of structuring the current literature on applications of argumentation-based dialogues that we use in the remainder of the chapter. Tables 2 and 3 identify six different dimensions that we use as a basis for comparison: **application domain**, **user interface**, **dialogue type**, **data sets**, **software tools** and **evaluation**. The references cited in this table are discussed in detail in Sections 4 and 5. Here we limit ourselves to a few, more obvious remarks: that health applications dominate (though this is perhaps skewed by our work on CONSULT, see below); that persuasion and inquiry dominate in terms of the Walton/Krabbe classification; that much work is evaluated formally, as one would expect from a literature with its roots in formal logic, but that an increasing number of papers include some kind of human-participant study; that there is no consensus on what software tools to use; that a large (and growing) number of systems make use of some form of chatbot, perhaps in response to the natural dialogic approach of argumentation; and few existing systems are data-driven despite the existence of a number of existing datasets.

The formal concepts of both abstract and structured argumentation frameworks, along with the notion of argumentative-based dialogue protocols, yield several software implementations that are reviewed in the following sections.

4.2 Consult: Argumentation-based dialogue in decision support

We start by discussing an argumentation-based dialogue system which was developed by the authors of this paper as part of the CONSULT decision support system.

Decision support systems (DSS) represent valuable tools that assist human users in making well-informed choices via the provision of pertinent recommendations. In the healthcare sector, such DSS prove to be especially useful for a number of reasons, including patient safety, cost containment and improved quality of documentation [Sutton *et al.*, 2020]. Indeed, there exists a long history of expert systems in the medical domain field [Saibene *et al.*, 2021] that can be traced back to MYCIN [Shortliffe, 1977]. *Clinical decision support systems (cDSS)* are mostly characterised by machine learning approaches, although the literature also comprises a number of cDSS driven by computational argumentation as the underlying reasoning procedure [Chapman *et al.*, 2019; Cyras *et al.*, 2018; Kökciyan *et al.*, 2021; Oliveira *et al.*, 2018]. Surely, as highlighted by Lindgren et al. [Lindgren *et al.*, 2020], this is a thriving area for argumentation since it can handle the conflict of knowledge occurring when multiple stakeholders are solicited with regard to specific medical cases.

In particular, CONSULT [Balatsoukas *et al.*, 2019; Essers *et al.*, 2018; Kökciyan *et al.*, 2019] is a data-driven cDSS that leverages an argumentation reasoning engine to help patients manage their conditions in collaboration with healthcare professionals[12]. The system receives multiple inputs (coming from different wearable wellness sensors, clinical guidelines, the patient's preferences and electronic health record), which then encodes and structures as arguments. The reasoning engine runs on the ASPARTIX solver [Egly *et al.*, 2008] and computes recommendations by instantiating textual explanation templates with acceptable (according to Dung's semantics) arguments [Kökciyan *et al.*, 2020]. The outcome of this operation is stored in an internal repository of the cDSS whose elements feed the EQRbot, the chatbot responsible for interacting with the patient [Castagna *et al.*, 2022; Castagna *et al.*, 2023].

---

[12]The overall (microservice) architecture of CONSULT can be found in [Chapman *et al.*, 2022].

Drawing from a previous dialogue framework [Sassoon et al., 2019], the EQRbot engages in an Explanation-Question-Response (EQR, hence the name of the chatbot) dialogue starting from the instantiations of the homonym argument scheme, embedding the initial CONSULT recommendation, and then proceeding by clarifying any additional follow-up user question. The implementation of the dialogue presented in [Castagna et al., 2023] is still limited and may be extended in the future by including the full spectrum of available locutions of an Explanation-Question-Response dialogue (originally sketched in [McBurney et al., 2021]) according to the formalization of [Castagna, 2022; Castagna et al., in press]. The advantages of such a protocol comprise the following. First, a simple design that avoids meta-level locutions to manage the dialectical interplay whilst conveniently embedding multiple dialogue types. Compared to other dialogues that require a Control Layer, the simplicity of the EQR design favours its implementation. Second, EQR exchanges of arguments result in interactions satisfying desirable properties of explanations (i.e. exhaustivity, selectivity, transfer of understanding and contextuality). Lastly, the information conveyed by a terminated EQR dialogue proves to be justified by a number of compelling reasons. Indeed, such an explanation produces sound and complete results with respect to Dung's AFs admissible semantics, thus allowing evaluation of the EQR dialogue moves using any proof theory, algorithmic procedures or methodologies semantically associated with computational argumentation.

The next two sub-sections contain more examples.

4.3 Applications based on formal models

This section reviews implementations of argumentation-based dialogues based on the theory summarized in Section 2. We start by covering software tools that can be used to implement dialogues and then move to look at individual applications. We structure this latter part of the current section using the Walton and Krabbe typology.

4.3.1 Tools for implementing a dialogue

The DGEP (Dialogue Game Execution Platform) [Bex et al., 2014] is a system capable of interpreting dialogue game specifications which are expressed in an amended version of the Dialogue Game Description Language (DGDL) [Wells and Reed, 2012] named DGDL+[13]. Based on these specifications, DGEP[14] generates dialogue templates [Bex and Reed, 2012], which are schematic representations of individual moves in a dialogue, along with their replies and connections to the underlying argument structure using the Argument Interchange Format (AIF) ontology [Chesnevar et al., 2006]. The AIF serves as an abstract core ontology for representing different theoretical and practical approaches to argumentation, acting as an interlingua between various argumentation approaches and enabling evaluation of arguments constructed in visualization packages using different argumentation theoretic semantics [Bex et al., 2013b]. The AIF also underlies the Argument Web [Bex et al., 2013a], a linked data Semantic Web structure containing numerous claims and arguments with different relations between them. Thanks to DGEP, dialogue histories with explicit reply structures can be formed by combining multiple templates, allowing existing argument structures in the Argument Web to be navigated and updated using dialogues.

DGDL+ includes several requirements as inbuilt predicates, such as CS checks and role checks, and it provides a general-purpose predicate for indicating arbitrary functions not defined in the protocol specification. DGEP processes DGDL+ specifications as if it were a compiler, converting them into an Abstract Syntax Tree (AST), and then further transforming it into a Python data structure representing the hierarchy of elements within the DGDL+ specification. The main goal of DGEP is to execute dialogue games specified in DGDL+, building AIF graphs and expanding the Argument Web. It develops the legal move list for each participant and handles the instantiation of rule effects and other parameters, like initial states, allowing agents to use the Argument Web as their KB. Every turn, DGEP generates all legal moves and delivers them to participants in the form of a 4-tuple, consisting of a moveID,

---

[13]The DGDL+ specification and example dialogue protocols can be found at https://www.arg.tech/index.php/research/dgdl/.

[14]DGEP is available at https://github.com/arg-tech.

opener (informal indication of the utterance), reply (formal structure of the move), and a fragment of AIF corresponding to the move. This fragment provides information about the structure of the move, allowing agents to create queries on AIF and extract relevant arguments. However, dialogue strategies for artificial agents are not fully implemented, so move selection is random. Finally, DGEP updates AIF structures during move execution based on theoretical accounts discussed in [Bex and Reed, 2012]. Note, a set of simple web service interfaces is provided which allows clients for both autonomous agents and human interfaces to connect and play instances of dialogue games.

DGEP is also the core of the modular architecture called Dialogue Utterance Generator[15] (DUG) introduced in [Snaith et al., 2020]. DUG finds propositional content to instantiate abstract move types, provided by DGEP, into concrete moves. Specifically, DUG uses content descriptors and associated content locators. Content descriptors describe how variables in the "reply" object should be populated, whereas a content locator provides content by querying a MySQL database. As a result, the reply is given to a participant of the dialogue as a concrete legal move they can make. If the result outputs multiple values, then a concrete move is created for each piece of content. AIFdb [Lawrence et al., 2012], argument mining [Lawrence and Reed, 2020], logical representations or other queriable sources may be considered for content instantiation.

4.3.2 Persuasion

Following the Walton and Krabbe typology and mirroring the previously introduced formal models of dialogues, this subsection discusses three systems for persuasion.

First we look at Polemicist [Lawrence et al., 2022]. Polemicist[16]. is a dialogical interface for exploring complex debates from the BBC Radio 4 programme The Moral Maze. It allows for the user to interact with software agents, who act as the participants in the original programme, and explore the topic as they wish, asking questions to delve into the areas they want. The agents' KBs are extracted from analysis of the

---

[15]DUG is available at https://github.com/arg-tech.

[16]Polemicist and other argumentation-related work can be found in http://www.johnlawrence.net/projects.php.

4 - APPLICATIONS OF ARGUMENTATION-BASED DIALOGUES

original episodes (represented in the AIF [Chesnevar et al., 2006]).

Polemicist translates navigation of the generated knowledge graph into a series of dialogical moves conducted according to the dialogue game for persuasion from [Prakken, 2006]. It uses a fixed protocol, defined in DGDL [Bex et al., 2014], where the user is the moderator of the debate, allowing agents to select topics and control the flow of the dialogue. Its interface contains two panels, where the one lists the participants with green and red highlighting showing their agreement or disagreement, respectively, with the most recent point made. The other depicts the history of the dialogue as well as a sub-panel which enables the user to either ask the opinion of a participant or question the reasons why the participant's opinion holds. Note that the dialogue history allows the user not only to view the dialogue but also return to previous points, and listen to the original audio associated with each text segment. Finally, it is worth mentioning that the Polemicist relies on pre-annotated material from AIFdb[17] [Lawrence et al., 2012] to provide the responses of the software agents in a dialogue.

Next we discuss DISCO [Booth et al., 2018]. DISCO, or, more correctly DIScussion COmputation[18] provides a web-based implementation of the Preferred [Caminada et al., 2014a] and Grounded Discussion Game [Caminada, 2015]. These two models [Caminada, 2015; Caminada et al., 2014a] are theoretical models of persuasion that build on the earlier work of [Modgil and Caminada, 2009]. DISCO is written in Javascript, and all computation is performed on the client side. The motivation of the authors is to implement these discussion games for the purpose of explanation. The user can choose to open an existing *AF* (in a JSON file format) or construct one manually by adding arguments and attacks to an initially empty canvas, so that they can play either the preferred or grounded discussion game. The user may choose to play as proponent or opponent and accordingly they take turns in moving arguments (where a move by the user can be typed into a text field, or can be selected by clicking on the relevant arguments). For both games, if the computer has the role of the proponent, it will win the game as it follows the associated winning strategy [Caminada, 2018]. Finally,

---

[17]http://www.aifdb.org/
[18]DISCO can be found at http://disco.cs.cf.ac.uk

DISCO provides additional features such as saving the *AF* (possibly as an image), allowing the user to ask for recommendations regarding the moves they can make and viewing the grounded labelling and associated min-max numbering for the Grounded Discussion Game.

Finally, we look at *Argument-Based Discussion using $ASPIC^-$* [Caminada and Uebis, 2020]. This is a variation of DISCO[19] where the construction of arguments is based on the $ASPIC^-$ framework, a variant of $ASPIC^+$ where the definition of attack is more suitable for interactive applications [Caminada *et al.*, 2014b]. Specifically, rule-based arguments are constructed from an underlying KB, stored in a text file, instead of abstract arguments. The demonstrator is written in Python3, it does not require any non-standard libraries, and has been tested to work under both Windows and Linux. Firstly, propositions and strict rules are specified in the file, followed by defeasible rules where those in the same block have the same strength and those in later blocks have a higher strength. Notice that defeasible rules have names for undercutting. To start the application, one starts from the command line adding as parameters $-wl$ (for weakest link principle) or $-ll$ (for last link principle) or $-do$ (for democratic order) or $-eo$ (for elitist order) [Modgil and Prakken, 2013]. After this step, a query to the inference engine is made regarding whether a statement is justified or not (i.e., if the statement is the conclusion of an argument in the grounded extension). After the engine's response, the user may ask for an explanation and start a discussion with the system. If the statement is justified, the system will assume the role of the proponent and the user the role of the opponent, otherwise the roles change correspondingly. At the moment, arguments played in the game are written in a nested, machine readable way and the target is for these to be given in natural language. The authors motivate the use of natural language by briefly suggesting the medical domain as an example for the implementation of their application.

4.3.3 Inquiry

Next we consider three argumentation-based dialogue systems that implement inquiry dialogues.

---

[19]ABDA can be found at https://github.com/Schirmi136/ABDA.

4 - APPLICATIONS OF ARGUMENTATION-BASED DIALOGUES

We start with [Yan et al., 2018], where the authors present a multi-agent framework designed to handle uncertain or inconsistent information in a distributed environment, and implemented in a clinical decision-support system (called DMSS-W) for diagnosing dementia. Specifically, the system involves a dialogue between a novice physician (PA) and a medical domain expert (DA), which is represented by the system, where the PA suggests a hypothetical diagnosis in a patient case. This is verified through the dialogue if sufficient patient information is present, otherwise the user is informed about the missing information and potential inconsistencies in the information as a way to support their medical education. Notice that pragmatic evaluation is left as part of future work and planned to take place in clinical practice. The framework builds upon the inquiry dialogue introduced by Black and Hunter in [Black and Hunter, 2009], allowing agents to collaborate in finding the best solutions and new knowledge while also addressing inconsistencies. The dialogue system consists of two participating agents, moves (which consist of the sender, the move type, the dialogue type, and the topic of the dialogue), a protocol that defines legal moves and other aspects of the dialogue system (such as its outcome), as well as a dialogue history (see [Yan et al., 2018] for details on the possible values of these components). It incorporates possibilistic logic [Dubois and Prade, 2004] to build an argumentation system which captures uncertain information and degrees of confidence in knowledge sources.

Practically, the system utilizes data from a platform called ACKTUS, a web-based tool for modeling medical knowledge into rules and claims in natural language. The multi-agent system (MAS) is developed using the Java Agent Development Framework (JADE). By using JADE, the authors can implement DA and PA as agents of MAS as well as define the components the agents have access to (e.g., their KBs and the arguments they can instantiate) and generate dialogues between them. Thus, JADE acts as the inference engine of DMSS-W. The dialogue between PA and DA leads to a diagnostic result, which are presented in the DMSS-W user interface. The domain experts model interaction objects (IOs) through ACKTUS and store them in the domain repository. Each IO contains scales with different values to determine the level of certainty. The user answers questions in the interface by clicking on scale values, which are used as state beliefs in reasoning. Rules are created based on premises, conclusions are derived from the IOs, and possibilistic values are assigned to these rules. As the reasoning process may lead to conflicting arguments due to uncertain and inconsistent data in the KBs, two different strategies are introduced so that the user may choose which one they will use to manage the conflict.

The second system we consider is that of [Nieves and Lindgren, 2015]. Here, a cooperative layer within a multi-agent system is presented, focusing on a scenario involving an older adult's needs and preferences for support in daily activities within a smart home environment. The agents in the system need to find optimal actions despite partial and inconsistent information, considering the changing needs and wishes of the older adult. The argumentation dialogues in this system are again inspired by [Black and Hunter, 2009] (although more participants are allowed to participate, and the components of a move are slightly different, resulting to changes in the CSs of agents too) and use default theories (extended logic programs) that can be mapped into Assumption-Based Argumentation (ABA) [Bondarenko et al., 1997; Fan and Toni, 2014] for dialogue inference. The intelligent infrastructure, called As-A-Pal, includes three instantiated agents: Environment Agent, Activity Agent, and Coach Agent. These agents, which possess rule-based KBs, collaborate to provide support to the older adult in conducting activities. Specifically, deliberation dialogues occur when they attempt to agree on actions to perform in certain situations, and agreement rules are used to reach a consensus. The system utilizes *Well-Founded Semantics (WFS)* [Dung, 1995] as a reasoning engine[20] to infer information from logic programs. Inquiry dialogues [Black and Hunter, 2009] are applied to validate the truth of "agreement atoms" or agreement rules. If an agreement atom holds true in a given state, it represents a particular belief's truth in the entire As-A-Pal system.

Finally, we consider the work of Bex and colleagues [Bex et al., 2016; Odekerken et al., 2020]. In [Bex et al., 2016], an initial sketch is given for an artificial agent handling the intake of internet trade fraud by combining natural language processing with symbolic techniques for reasoning about crime reports. The system serves two main types of

---

[20] Details can be found at https://github.com/esteban-g/wfsargengine.

4 - APPLICATIONS OF ARGUMENTATION-BASED DIALOGUES

users: complainants filing new criminal complaints and the police who want to analyse reports and build case files. Both interact with the system through the dialogue interface, which allows them to submit input and view the status of the dialogue, including open questions. A dialogue manager, based on the Dialogue Game Execution Platform (DGEP)[21] [Bex et al., 2014], specifies the dialogue protocol, such as turn-taking and legal moves, and keeps tracks of users' commitments. Multiple scenario reasoning agents can participate in a dialogue, using predefined fraud schemes from a library and crime report repository. These agents can match scenarios to typical fraud schemes, compare scenarios based on available evidence, and elicit further information from users. Determining the true scenario often requires additional evidence, turning the investigation into a process of inference to find the best explanation.

Later in [Odekerken et al., 2020] [22], the subsequent development of the intake agent is regarded as argument-based inquiry dialogue, once more inspired by [Black and Hunter, 2009]. $ASPIC^+$ is used to define an argumentation system where defeasible rules represent the laws and practices surrounding trade fraud are combined with the citizen's knowledge of the specific situation they observed, to build arguments for and against the main claim made by the citizen. Additionally, natural language processing techniques are used to extract automatically the initial observations from free-text user input [Schraagen and Bex, 2019], so that these observations can be combined with rules concerning trade fraud in the argumentation setting to build arguments for and against the claim "fraud". The notion of *Stability* is also discussed (from a theoretical point of view) which is used to decide whether the addition of more observations from the citizen in the future can change the acceptability status of the "fraud" claim. If not, the dialogue terminates; otherwise a question policy component finds the best question to ask given current observations. In other words, the stability component provides a termination criterion that prevents the agent from asking unnecessary questions. Notice that the stability component can also be

---

[21]See more details on DGEP in the next Section
[22]The Dutch Police's website, which implements the intake agent (in the Dutch language), can be found at https://aangifte.politie.nl/iaai-preintake.

perceived as a "tool" for dealing with enthymemes, since it essentially signifies whether or not an argument is complete. See [Odekerken et al., 2022] for further insights and applications of the intake (dialogue) agent, which also includes an empirical evaluation and provides a deeper comprehension on how it captures enthymemes that miss some of the necessary support to entail a conclusion.

4.3.4 Information-seeking

From the applications of information-seeking in the literature we pick that of [Panisson et al., 2022b] to examine in detail.[23] In this paper, an argumentation framework is described where agents are able to exchange shorter messages when engaging in dialogues by omitting information that is common knowledge. These messages are treated as enthymemes; and shared argumentation schemes are used, as well as common organisational knowledge, to build an enthymeme-based communication framework. Concerning the argumentation schemes, the "Argument from Position to Know" from [Walton, 1996] with associated critical questions are applied, but referring to organisational concepts. According to the authors, such argumentation schemes can be represented in structured argumentation, using defeasible inferences. Additionally, they use first-order logic to represent arguments, arguing that this is a reasonable choice given that most agent-oriented programming languages are based on logic programming. The authors also argue that instantiating arguments from argumentation schemes allows agents to use such arguments for both reasoning and communication processes, and so they use argument schemes to guide the decoding of enthymemes into the original sender's argument.

Note that all agents/participants are aware of other agents' roles in the organisation as well as the associated features/abilities related to them. Furthermore, the authors [Panisson et al., 2022b] show that their work addresses some of Grice's maxims, proving that agents can be brief in communication, without any loss in the content of the intended arguments. To implement this enthymeme-based communication[24] framework

---

[23]Note that the same implementation can be used to model inquiry dialogues.

[24]The implementation of the framework is available open source in https://github.com/AlisonPanisson/EBCF.

the JaCaMo Platform [Boissier *et al.*, 2013] was employed. Finally, to evaluate their framework, the authors use scenarios of argumentation-based dialogues that use different argumentation schemes and argumentation-based protocols from the literature. At first the information-seeking protocol specified in [Panisson and Bordini, 2017] was used, but later on the inquiry protocol specified in [Panisson *et al.*, 2021] was also employed. However, in both scenarios, the same locutions are leveraged, which are described within [Panisson *et al.*, 2022b] together with their effects on the CSs of the agents and the dialogue.

4.3.5 Negotiation

Representing negotiation, we have [Bentahar and Labban, 2011] and [Koit and Õim, 2015]. In [Bentahar and Labban, 2011], the authors propose a formal description and implementation of a negotiation protocol between autonomous agents using persuasive argumentation. In persuasive negotiation, an agent is trying to influence the behaviour of another agent using arguments supporting the proposed offers[25]. The logical language the authors use comprises of propositional Horn clauses, i.e. disjunction of literals with at most one positive literal which can be also written as implications. An argument is a a pair $A = (H, h)$ where h is a formula of a propositional language L, and H is a set of formulae of L such that H is consistent, $H \vdash h$ and H is minimal, similar to [Amgoud *et al.*, 2000]. The KBs of the participants however are assumed to be consistent but attacks between arguments are defined as in [Amgoud *et al.*, 2000], i.e. an argument A attacks another argument B on its premise(s). Additionally, a commitment store is used to track the arguments that have been publicly exchanged. Notice that agents can reason about trust and use trustworthiness to decide, in some cases, about the acceptance of arguments.

In regards to formalising their protocol, the authors use small computational *dialogue games*, i.e. a logical rule indicating that if Ag_1 performs action Act_1, and a formula of the logical language is satisfied, then Ag_2 will perform Action Act_2 afterwards. Five types of dialogue games are considered, namely: *entry, defence, challenge, justification and attack*,

---

[25] As a result, one might therefore consider this to be a form of persuasion dialogue as well.

where the entry game allows agents to open the dialogue, and the rest represent the *chaining games* which constitute the main negotiation process between the agents. The protocol terminates either by a final acceptance or by a refusal of the proposal discussed. The locutions and moves that the agents can use depend on the dialogue game played. Moreover, different properties of the protocol are proved, and discussion over the complexity efficiency of the protocol takes place.

Finally, the authors describe the implemented prototype of their system, where they use the $Jack^{TM}$ platform [AOS, 2005]. $Jack^{TM}$ is an agent-oriented language based on the Belief/Desire/Intention (BDI) model, and offering a framework for multi-agent system development. It supports Java and includes all components of Java offering specific extensions to implement agents' behaviours, including support for logical variables and cursors which is helpful when querying the state of an agent's beliefs. In addition, both the agents and their KBs are implemented using $Jack^{TM}$, where the agents communicate with $MessageEvents$ representing actions that an agent applies to a commitment or to its content. A dialogue game is implemented as a set of events and plans, where a plan describes a sequence of actions that an agent can perform when an event occurs. An agent Ag_1 starts a dialogue game by generating an event and by sending it to the addressee Ag_2, then Ag_2 executes the plan corresponding to the received event and answers by generating another event and by sending it back to Ag_1, and so on. Note, to start the entry game, an agent chooses a goal that it tries to achieve which is to persuade its interlocutor that a given propositional formula is true. This is why a BDI event is used as it models goal-directed behaviour in agents, rather than plan-directed behaviours.

In [Koit and Õim, 2015], the authors describe a computational model of agreement negotiation processes, which involves natural reasoning. The general type of interaction the authors deal with represents a kind of directive interaction where the goal of one participant, Ag_1, is to get another participant, Ag_2, to carry out a certain action D. One of the authors' aims is to investigate actual dialogues and this is why they selected to analyse, three sub-corpora of the Estonian Dialogue Corpus (EDiC) which although includes mainly information-seeking dialogues, typical sequences of dialogue acts (DAs) were found in human–human spoken

4 - APPLICATIONS OF ARGUMENTATION-BASED DIALOGUES

dialogues that form agreement negotiations and reflect reasoning of the participant who has to make a decision about an action. Their model is implemented in an experimental dialogue system as an application where a user participates in communication training sessions.

The application is implemented in Java supporting (text-based) interaction with a user in Estonian and employs only predefined set of sentences which they can select from a menu. The sentences are only classified semantically according to their possible functions and contributions in a dialogue (e.g., the sentences leveraged by Ag_1 to increase the usefulness of the action, the sentences harnessed by Ag_2 to indicate harmfulness of the action, etc.). These sentences are dealt as arguments, and private and public information are considered in each information state of a conversational agent. The private information of an agent Ag_1 contains: a model of their interlocutor Ag_2, a reasoning procedure[26] which Ag_1 is trying to trigger in Ag_2 to persuade them positively for the decision D, aspects of D under consideration, a set of DAs (including the proposal and statements for increasing or decreasing weights of different aspects of D for Ag_2), and a set of utterances for increasing or decreasing the weights (i.e., arguments for/against). Every utterance can be chosen only once by Ag_1 and so Ag_1 has to abandon its initial goal if there are no more arguments to move. The shared part of information contains a set of reasoning models, a set of tactics (such as enticing, persuading and threatening) and a dialogue history, i.e. the utterances together with participants' signs and DAs. Furthermore, update rules used for transitioning from an information state into another are also defined. However, notice that the usual aspects considered in this chapter, such as instantiating arguments via logical language or traditional protocol representation, do not take place in this work. Finally, an evaluation occurs where a group of volunteers used the application, and a user needs to accept to do D, but 65% of the dialogues did not have this result.

---

[26]The reasoning model of an agent in [Koit and Õim, 2015] is analogous to a BDI model, but more kinds of motivational inputs are considered for creating the intention of an action in an actor in order to understand the effects that these factors –namely *wish, needed, must*– will have on the reasoning process.

4.3.6 Deliberation

Here we discuss [Kampik and Gabbay, 2020], which primarily handles deliberation dialogues, though it can also support inquiry. [Kampik and Gabbay, 2020] presents the implementation of the DiArg argumentation-based dialogue engine. It focuses on automating sequential argumentation, i.e. the iterative resolution of sequences of *AF*s (mainly for deliberation but, as previously stated, also for inquiry dialogues). By resolution, the authors mean that extensions of an *AF* are determined where one is selected as the *AF*'s conclusion, either automatically or manually by a human user. Specifically, DiArg resolves abstract *AF*s. In DiArg dialogues, an *AF* sequence is created by expanding an initial *AF*, i.e. by adding new arguments and attack relations to it (and again resolved, and so forth). DiArg can also ensure that results derived from an *AF* sequence preserve *Reference independence* and *Cautious monotony* principles.

In software terms, DiArg is an open-source Java library[27], where the program code and its documentation allow for inspection of the underlying data structures and algorithms. DiArg also utilises Tweety [Thimm, 2014], that provides Java libraries to define and resolve different types of formal argumentation frameworks, to implement argumentation-based dialogue systems. A scenario of a digital assistant for stress management [Guerrero *et al.*, 2016] is described in the paper, where the assistant recommends stress-relieving activities (in the form of arguments) to a user who can then either accept the suggestion of the system and add it to their schedule, or reject the activity by attacking it with an argument. Finally, the authors discuss limitations of DiArg that relate to context support, integration with recommend systems approaches and interoperability enhancements in alignment with the AIF [Chesnevar *et al.*, 2006].

---

[27]The DiArg reasoner as well as an implemented dialogue example is available at https://github.com/Interactive-Intelligent-Systems/diarg

4.3.7 Other[28]

The work presented by Fazzinga et al. engineered a privacy-preserving dialogue system based on computational modes of arguments [Fazzinga *et al.*, 2022]. This architecture focuses on data protection and explainability to address the mistrust that current dialogue systems can raise in their users. By means of an *Argumentation Module*, it is possible to probe the rationales behind the dialogue system responses and understand the supporting and conflicting reasons underpinning them. A Covid-19 vaccination case study illustrates how such an architecture can fit a real-world scenario. The system has also been formally evaluated by proving specific formal properties (such as consistency, well-formedness, and termination).

Also in this category is the Multi-Agent Intentional Dialogue System (MAIDS) framework. (Arguably this could appear in several of the previous sections since it supports peruasion, information-seeking and inquiry dialogues.) This combines argumentation theories with other features to support complex dialogue [Engelmann *et al.*, 2023]. Several agents are instantiated and each provides unique expertise in the system. The *assistant* engages in argumentation-based reasoning (following the approach developed in [Panisson *et al.*, 2014]) the results of which are then translated into natural language and conveyed to the human user by the *communication expert(s)*. *Ontology expert(s)* handle various ontologies (e.g. OWL), whereas *domain* agents address the specificity of different domain applications.

4.4 Argumentation-based conversational agents

As previously stated, chatbots are interactive pieces of software with a specific history and recognizable features: a virtual chatbox (or log, especially for speech-to-speech agents) and a strategy to provide messages. Given their well-defined structure and characteristics, which further diversify according to the internal architecture and the operational domains, we choose to dedicate a separate section to examine the combination of

---

[28] As Dawkins notes in "The Selfish Gene" [Dawkins, 2016], any attempt at a taxonomy other than that based on evolutionary history will end up with a "miscellaneous" category. This is ours.

such conversational agents with argumentative dialogues. While chatbots grew out of work on natural language processing, they may handle and deliver their responses by leveraging the protocols and the formalism of argumentation-based dialogues. Harnessing the dialogue logic, the conversational agent can optimize its strategy and move only the arguments that prove to be necessary for achieving its final goal[29]. We discuss work on what we might call "argumentation-enabled chatbots" using the same structure based on dialogue-type that we used in Section 4.3. However, it is interesting to notice how all the reviewed works concern persuasion protocols or a mixture of dialogues that include persuasion.

4.4.1 Persuasion

As a first example, we could examine the work introduced by Hadoux et al. [Hadoux *et al.*, 2021], which expands upon previous studies from the same authors [Hadoux and Hunter, 2019; Hunter, 2018; Hunter *et al.*, 2019], and depicts an overall framework for modelling beliefs and concerns in a persuasion dialogue. An implementation of such a framework is then envisaged via an *automated persuasion system (APS)*, a software application aiming at convincing the interacting agent to accept some argument. Following the asymmetric persuasion dialogue protocol illustrated therein (i.e. unlike the system, the user is restricted in choosing replies among the provided options), the proposed chatbot proves to be capable of identifying, within its knowledge base embedded in an argument graph, the most appropriate argument to posit. Essentially, the APS performs a Monte Carlo Tree Search coupled with a reward function to maximize the addressing of concerns (paired with the arguments of the graph) and the user's beliefs.

Similarly, the bot presented in [Chalaguine and Hunter, 2020] aims at persuading the interlocutor via a free-text interaction where the user's inputs are matched (by vector rendering and cosine similarity) with the (crowdsourced) arguments of the graph representing the knowledge base. The chatbot trains a classifier to detect the most common concerns of the

---

[29]Some of the authors have recently written an extensive review of argumentation-based chatbots [Castagna *et al.*, 2024a]. We invite interested readers to refer to such a study for a detailed list of conversational agents employing computational argumentation beyond dialectical delivery.

persuadee and employs it to select counterarguments that will produce a result more compelling than a random choice. If no argument similarity is detected, then the conversational agent will resort to a default reply based on the user's concerns. Furthermore, the same authors presented an analogous architecture for a persuasion bot with the addition of a particular concern-argument graph [Chalaguine and Hunter, 2021]. By incorporating the knowledge base within such a small graph, it can be proved that no large amount of data is needed to generate effective persuasive dialogues. Interestingly, a preliminary analysis of the impact (appeal) of arguments addressing the users' concerns in a persuasion dialogue performed by a chatbot has also been conducted in a dedicated investigation [Chalaguine et al., 2019]. A different example of such a concern-based approach may be represented by Argumate, a chatbot designed to facilitate students' production of persuasive statements [Guo et al., 2022]. To provide appropriate suggestions, the bot retrieves its replies from an underlying argument graph, whose edges denote attack and support relations, via a concern identification method. Notice that the interactions between Argumate and the users occur both by typing and selecting predefined options.

A common trait amongst most of the above argumentation-based conversational agents is that, although the corpus from which they extract their replies is organized as an argument graph, there is no interest in any particular acceptability semantics [Dung, 1995]. That is to say, the knowledge base is organized and considered as a plain AF, where arguments and attacks are the only relevant features. In addition, most of these studies also account for a baseline chatbot which exploits a random strategy for selecting counterarguments from the available choices within the underlying knowledge base. The reason for this is to provide a means for comparing the developed bots which employ more fine-grained strategies for choosing their replies.

Another conversational agent that focuses on the delivery of persuasion dialogues is the chatbot designed by Andrews et al. [Andrews et al., 2008]. Implemented by harnessing the AIML markup language [Wallace, 2003], the bot comprises a planning component that searches over an argumentation model for the optimal dialectical path to pursue in order to persuade the user. The agent records the user's beliefs and updates this information whenever its interlocutor agrees/disagrees during the interaction. Such beliefs-revisions play an important role in the strategic planning of the chatbot.

Finally, one last chatbot (SPA), envisaged in the study of Rosenfeld and Kraus [Rosenfeld and Kraus, 2016], employs an *AF* as the basis of a reasoning procedure to perform persuasive interactions. In particular, it embeds its knowledge base into a Weighted Bipolar AF (WBAF) and computes the argument that maximizes the framework evaluation function according to the user input. The score returned by the valuation function represents the reasoner's ability to support that argument and defend it against potential attacks. The dialectical interaction with the user follows a strategical persuasion dialogue protocol (optimized via Monte Carlo Planning [Silver and Veness, 2010]) that might involve updating the argument frameworks of both the persuader and the persuadee.

4.4.2 Information-seeking and Inquiry

As noted above, all the chatbots that we cover have some element of persuasion. Here we consider those which have some non-persuasion element. First, we consider the conversational agent implemented by Sassoon et al. [Sassoon *et al.*, 2019], within the context of explanation for wellness consultation. This exploits deliberation and information seeking protocols, in addition to persuasion whilst exchanging instantiations of acceptable argument schemes with its interlocutor. The adoption of diversified dialogue protocols (i.e. persuasion, inquiry and information seeking) also characterises the chatbot-equipped robot proposed by Sklar and Azhar [Sklar and Azhar, 2015]. Retrieving the most appropriate argument constructed from its beliefs, an operation facilitated by the restricted options available to the user, the robot communicates with its human interlocutor in order to strategize about a treasure-hunting game and explain the rationale behind its decisions.

Finally, we consider the bot introduced in [Hauptmann *et al.*, 2024]. This German-language conversational agent, following the formalisation of [Hadoux and Hunter, 2019], makes use of an argument graph to encode its knowledge base from which it retrieves main stances and counterarguments to engage the users in discussions concerning the ethical challenges of AI implementations. The delivery strategy is somehow

4 - APPLICATIONS OF ARGUMENTATION-BASED DIALOGUES

ambiguous but seems to balance a mixture of persuasion and information-seeking, according to the specific stage of the conversation.

4.4.3 Evaluation of chatbots

The argumentation-based chatbots described above have typically been evaluated via specifically designed user studies. Since this differs from the way that much work on argumentation-based dialogue is assessed, we think it worth discussing in detail.

The SPA conversational agent introduced in [Rosenfeld and Kraus, 2016] outplayed the baseline chatbot (which harnessed a different heuristic strategy) when tested in its persuasion task, thus proving capable of delivering human-like conversations. Similarly outperforming the baseline agent is the bot presented by Chalaguine et al. [Chalaguine et al., 2019]. Indeed, the paper includes an experiment that shows how such a chatbot, by positing arguments that address the users' concerns, is more likely to positively change the users' attitude in comparison with another agent that does not employ such a strategy. An analogous interest in users' concerns is encompassed in a study implemented by the same authors [Chalaguine and Hunter, 2020]. The results (conjointly supported by the experiments in [Hadoux and Hunter, 2019] and confirmed by [Hadoux et al., 2021]) conclude that a strategic chatbot accounting for concerns is more likely to provide relevant and cogent arguments.

Moreover, it is also worth mentioning the evaluation outcome of the other two aforementioned persuasive agents presented [Andrews et al., 2008; Chalaguine and Hunter, 2021]. The former bot provides fluent conversations with its interlocutors performing generally better than a purely task-oriented system. The latter, instead, shows how an interactive chatbot yields more compelling information than a static webpage. Resorting to pre- and post-dialogue Likert-scale questionnaires is the preferred evaluation choice of the work presented in [Hauptmann et al., 2024]. The results record successful shifts of the opinions of 40-50% of the participants after engaging with the chatbot. Overall, the users acknowledged the quality of the arguments and the design of the conversational system. Lastly, the dialectical agent designed in [Sklar and Azhar, 2015], implemented and evaluated on a robot in [Azhar and Sklar, 2017], was further investigated in [Sklar and Azhar, 2018], where discussions conducted within the previous user study [Azhar and Sklar, 2017] were evaluated from the viewpoint of explanations provided. The results show how leveraging argumentation-based dialogue improves system performance and users' satisfaction, although no particular correlation was detected between these metrics and the possibility of receiving explanations.

5 Discussion

Just as the exchange of arguments influences our reasoning [Mercier and Sperber, 2011], so the engagement in dialogues considerably affects human lives in a plethora of different scenarios. Argumentation-based dialogues formalise inter-agent communication protocols and strategies, and their applications are likewise broad in scope and modalities. Whether chatbots, recommender systems, end-to-end software or just blueprints of future implementations, the literature reviewed highlights some common patterns that can be harnessed to underpin the following analysis.

Reading through our survey, it is clear that persuasion is the type of argumentation-based dialogue protocol that is most embedded in interactive software architecture, such as chatbots or cDSS (e.g., [Andrews et al., 2008; Booth et al., 2018; Caminada and Uebis, 2020; Chalaguine and Hunter, 2020; Chalaguine and Hunter, 2021; Hadoux et al., 2021; Hauptmann et al., 2024; Lawrence et al., 2022; Rosenfeld and Kraus, 2016; Sassoon et al., 2019; Sklar and Azhar, 2015]). This is rather natural since argumentation-based formalisms prove to be quite effective in providing compelling strategies and replies to induce belief change, as suggested by the results of several studies [Andrews et al., 2008; Chalaguine and Hunter, 2020; Chalaguine and Hunter, 2021; Chalaguine et al., 2019; Hadoux and Hunter, 2019; Hadoux et al., 2021; Hauptmann et al., 2024; Rosenfeld and Kraus, 2016]. Another trend that emerges from our survey is the connection between eXplainable AI, argumentation-based dialogues and their applications. Indeed, providing clarifications about the inner workings of black box algorithms seems to be a thriving area of application for dialectical protocols that involve argumentation[30] [Čyras et al., 2021;

---

[30]Doubtless this popularity is a result of the recent interest in eXplainable AI and its link with computational models of arguments [Čyras et al., 2021; McBurney et al., 2021;

4 - APPLICATIONS OF ARGUMENTATION-BASED DIALOGUES

Vassiliades *et al.*, 2021]. In particular, a frequent procedure to reveal the underpinning rationales of AI systems' decisions consists of retrieving acceptable information (from the pertinent knowledge base) according to specific argumentation semantics [Castagna *et al.*, 2023; Fazzinga *et al.*, 2022; Sassoon *et al.*, 2019].

Although it is persuasion that has been mostly considered in dialogical applications for argumentation, there are works that investigate the implementation of other types of dialogues, too. Inquiry is an example of a dialogue type that has been studied several times as a practical application [Bex *et al.*, 2016; Kampik and Gabbay, 2020; Nieves and Lindgren, 2015; Odekerken *et al.*, 2020; Panisson *et al.*, 2022a; Sklar and Azhar, 2015; Yan *et al.*, 2018]. The cooperative nature of inquiry allows agents to combine their knowledge to find the truth regarding the matter discussed, and this is why it has been found useful in applications concerning various domains, such as healthcare [Kampik and Gabbay, 2020; Nieves and Lindgren, 2015; Yan *et al.*, 2018], fraud investigation [Bex *et al.*, 2016; Odekerken *et al.*, 2020] and communication in organisations [Panisson *et al.*, 2022b] as well as human-robot teams [Sklar and Azhar, 2015]. As a side note, we observe that most of the applications we have found, are concerned with the healthcare domain[31] [Balatsoukas *et al.*, 2019; Essers *et al.*, 2018; Castagna *et al.*, 2023; Castagna *et al.*, 2022; Chalaguine and Hunter, 2021; Kampik and Gabbay, 2020; Kökciyan *et al.*, 2019; Nieves and Lindgren, 2015; Sassoon *et al.*, 2019; Yan *et al.*, 2018], thus stressing the importance that efficient communication tools (such as argumentation-based dialogues) assume within the medical context.

Information-seeking and deliberation are only appear to exist in applications that include more than one dialogue type, where persuasion or inquiry take precedence (e.g., [Sklar and Azhar, 2015] refers to persuasion, inquiry and information-seeking, [Panisson *et al.*, 2022b] refers to inquiry and information-seeking, [Sassoon *et al.*, 2019] refers to persuasion, information-seeking and deliberation, and [Kampik and Gabbay, 2020; Nieves and Lindgren, 2015] refer to inquiry and deliberation). This can be explained taking into account that: (1) information-seeking and

---

Sklar and Azhar, 2018; Vassiliades *et al.*, 2021].

[31]This holds even taking into account the biases we introduced by describing multiple aspects of the CONSULT system.

inquiry dialogues are similar types, with the difference being mainly that information-seeking dialogues should start with a question [Parsons et al., 2003]; (2) deliberation can be examined both in conflicting (persuasion) and cooperative (inquiry) scenarios between agents, with the difference being that it focuses on deciding about an action that agents should take rather than the validity of a topic of discussion. With regard to negotiation dialogical applications, [Bentahar and Labban, 2011] focused on the persuasive aspect of negotiations and the goal in [Koit and Õim, 2015] was primarily the study of human real-life communication rather than the application itself. The number of works and what they concentrate on demonstrates that practical implementations of such dialogues has been under-examined.

Most of the applications that concern dialogue types other than persuasion come with a simple User Interface [Booth et al., 2018; Caminada and Uebis, 2020; Kampik and Gabbay, 2020; Koit and Õim, 2015; Lawrence et al., 2022; Yan et al., 2018]. In most cases this is because the main focus is either developing or examining a theoretical argumentation framework for dialogues and/or investigating if it is feasible to implement it as an actual application [Booth et al., 2018; Caminada and Uebis, 2020; Kampik and Gabbay, 2020; Koit and Õim, 2015; Yan et al., 2018], or inspecting specific argumentation software tools [Lawrence et al., 2022].

Sections 3 and 4 discussed tools for building argumentation-based reasoners, for example DGEP (discussed in [Bex et al., 2014; ?; Odekerken et al., 2020; Snaith et al., 2020]), WFS (discussed in [Nieves and Lindgren, 2015]) and Tweety (discussed in [Kampik and Gabbay, 2020]), as well as tools used for instantiating agents, for example JADE (discussed in [Yan et al., 2018]), $Jack^{TM}$ (discussed in [Bentahar and Labban, 2011]) and JaCaMo (discussed in [Panisson et al., 2022b]). Concerning the latter, we have also encountered Dial4JaCa (leveraged by [Engelmann et al., 2023]), a communication interface integrating JaCaMo and Google Dialogflow[32]. These are more sophisticated software tools compared to the simple use of programming languages for application development purposes (e.g., [Booth et al., 2018; Caminada and Uebis, 2020; Koit and Õim, 2015]), and bring elements of agent theory into the implementations. Notice that only a small number of existing applications have attempted to use

---

[32]https://dialogflow.com/.

4 - APPLICATIONS OF ARGUMENTATION-BASED DIALOGUES

NLP (i.e., [Bex et al., 2016; Fazzinga et al., 2022; Odekerken et al., 2020]) or adopt a chatbot-like approach (i.e., [Engelmann et al., 2023]) in these dialogical applications.

It is worth observing that, within the surveyed literature, only a handful of argumentation-based dialogue implementations clearly harnessed panoptic engines or solvers as described in Sections 3.2, and 3.3. In particular, two of such research [Castagna et al., 2023; Castagna et al., 2022] incorporate the ASPARTIX solver ([Egly et al., 2008] an older version of the latest [Dvořák et al., 2021]), whereas a third study [Sklar and Azhar, 2015] structures its main argumentative module (ArgHRI) by embedding the results of ArgTrust v1.0 [Tang et al., 2012b]. Although this does not exclude dialogue systems that merge reasoning engine components with other elements in their overall architecture (which constitutes the majority of our findings), it is still surprising that we did not identify more dialectical applications employing argumentation engines, given the subsisting straightforward connection between the two.

One of the factors included in our analysis methodology is the use of data sets in dialogical applications. It is interesting to see that this component is not taken into account by all the implementations reviewed as we might have expected. Instead, applications such as the ones described in [Booth et al., 2018; Caminada and Uebis, 2020; Kampik and Gabbay, 2020] deal with arguments leaving out of the conversation the employment of specific domains. Regarding the data sets visited, [Bex et al., 2014; Lawrence et al., 2022; Snaith et al., 2020] use the AIFdb [Lawrence et al., 2012] database which deals with the storage and access of AIF argument structures [Chesnevar et al., 2006], whereas non-argumentative data sets were also found, such as ACKTUS (a web-based tool for modelling medical knowledge into rules and claims in natural language [Lindgren and Winnberg, 2012; Lindgren and Yan, 2015]), fraud scenarios from the scenario library and the repository of crime report from the National Service Centre E-Crime Dutch Police (discussed in [Bex et al., 2016; Odekerken et al., 2020]), and EdiC: the Estonian Dialogue Corpus which comprises of different kinds of human-human dialogues (discussed in [Koit and Õim, 2015]).

On the evaluation side, many of the works assessed use formal proofs for appraising their applications as it is common that they implement existing dialogue systems from the literature, or prove different properties for their systems, for example [Bentahar and Labban, 2011; Booth et al., 2018; Caminada and Uebis, 2020; Fazzinga et al., 2022; Kampik and Gabbay, 2020; Lawrence et al., 2022; Odekerken et al., 2020; Panisson et al., 2022b; Yan et al., 2018]. Note, even in papers where this is not explicitly stated, we assume that this occurs as the dialogue systems employed come with proven features. The use of formal proofs demonstrates the value of the results of theoretical dialogue systems investigated in the argumentation research. However, it is also important to assess the functionality of an application itself, especially when it involves interactive systems such as chatbots. Indeed, their primary goal is direct communication with the user, thus, the most suited evaluation should occur via tests with human participants, as, for example, is done in [Andrews et al., 2008; Chalaguine and Hunter, 2020; Chalaguine and Hunter, 2021; Hauptmann et al., 2024; Koit and Õim, 2015; Rosenfeld and Kraus, 2016].

Finally, many of the works reviewed either describe the dialogue protocol they follow (e.g., [Bentahar and Labban, 2011; Fazzinga et al., 2022; Nieves and Lindgren, 2015; Panisson et al., 2022a; Sklar and Azhar, 2015; Yan et al., 2018]), or this is implicit as the authors refer the dialogue system they leverage (e.g., [Booth et al., 2018; Caminada and Uebis, 2020; Koit and Õim, 2015; Lawrence et al., 2022]). The ones that do not refer to a protocol are concentrated on describing software tools (e.g., [Bex et al., 2014; Bex et al., 2016; Snaith et al., 2020]), or other theoretical properties of the dialogue discussed (e.g., [Kampik and Gabbay, 2020; Odekerken et al., 2020]). The characteristics of the moves as well as the dialogue history (also referred to as commitment store) are specified too in the works where the protocol of the dialogue is examined. The component of strategy, however, is not visited that often in applications that concern non-persuasion dialogue types. For example, [Booth et al., 2018] and [Caminada and Uebis, 2020] refer to winning strategies based on the dialogue games they implement, but both of these papers examine persuasion dialogues. One exception is [Yan et al., 2018], which examines inquiry dialogues, but provides strategies for avoiding endless dialogues, finishing a dialogue quickly and resolving conflicts. Finally, ,we note that structured argumentation is mainly employed in the applications reviewed (e.g., [Bentahar and Labban, 2011; Caminada and Uebis, 2020; Castagna et al., 2023; Castagna et al., 2022; Lawrence et al., 2022; Nieves and Lindgren, 2015; Odekerken et al., 2020; Panisson et al., 2022b; Sassoon et al., 2019; Yan et al., 2018]) in comparison to abstract argumentation (e.g., [Booth et al., 2018; Fazzinga et al., 2022; Kampik and Gabbay, 2020; Sklar and Azhar, 2015]).

6 Future Directions

This section focuses on two key emerging areas for future work in the application of argumentation-based dialogues. The first is the use of enthymemes (Section 6.1), to handle incomplete arguments. The second is the use of argumentation to resolve current issues with LLM implementations (Section 6.2).

6.1 Enthymemes

As mentioned earlier, enthymemes are arguments that lack a complete logical structure. This means that one may omit one or more premises or inference rules, or the claim of the argument they intend to get across to their discussant. This might be because they expect the recipient of the 'incomplete' argument to understand its missing elements based on information they share, or previous conversations they had. Nevertheless, it is not always certain that the recipient of an enthymeme E is able to reconstruct correctly the intended 'complete' argument A from which E was generated. There might be multiple ways to complete E, e.g. the recipient of E might assume that E is part of an intended 'complete' argument B and fill the gaps with parts of B instead of A. Consider for example the following dialogue [Xydis et al., 2020]:

Example 1.
1. **Bob:** *You can't afford to eat at a restaurant today.*
2. **Alice:** *Why not?*
3. **Bob:** *Because you owe money and if you owe money then you probably can't afford to eat at a restaurant.*
4. **Alice:** *I made a deal with my creditors.*
5. **Bob:** *So what?*

6. Alice: *So I don't need to pay the bills today.*
7. Bob: *Why is that relevant?*
8. Alice: *I thought that the reason you thought I owe money is because I have bills to pay today. Hence, I can't afford to eat at a restaurant today.*
9. Bob: *No! I meant that you owe money because you need to pay Kate back today. So, you can't afford to eat at a restaurant today.*

Bob first asserts a claim without any supporting premises (1). The reasons for believing the claim are not clear to Alice, so she asks for clarification (2), which Bob provides (3). Notice that, when combined, (1) and (3) form a complete argument, hence they can both be considered enthymemes for this complete argument. Alice then presents an enthymeme (4) for an argument that she believes counters the argument Bob is making. Note that the enthymeme Alice presents does not explicitly contradict anything that Bob has said, and so Bob asks for clarification (5) on what she is meant to infer from this enthymeme, which Alice provides (6). However, Alice's clarification still does not explicitly contradict anything Bob has said. Since Bob does not understand why Alice's enthymeme is relevant to what he said, he asks Alice to explain what she thought he meant (7). Alice explains the assumption she had made (8), which Bob then corrects (9).

This simple example illustrates the need for a dialogue system that allows human and/or computational agents to both 'backward extend' enthymemes (where missing premises are provided in 3 above) and 'forward extend' enthymemes (where missing inferences are given, as in 6), and to request such extensions (2 and 5). It also warrants the need for allowing agents to ask what another agent has assumed was intended by an enthymeme (7), to answer such a question (8), and to correct any erroneous assumptions (9).

Work on how enthymemes are handled during dialogues between human and/or computational agents is another area that is not heavily studied. Notable exceptions include the work of Black and Hunter [Black and Hunter, 2008], De Saint-Cyr [de Saint-Cyr, 2011], Hosseini [Hosseini, 2017], Xydis et al. [Xydis *et al.*, 2020; Xydis *et al.*, 2021; Xydis *et al.*, 2022], Odekerken et al. [Odekerken *et al.*, 2022] and Leiva et al. [Leiva *et al.*, 2023]. From these works, [Black and Hunter, 2008; Hosseini, 2017; Leiva *et al.*, 2023; Odekerken *et al.*, 2022] employ locutions that capture only the backward extension of enthymemes, [Xydis et al., 2021] makes use of locutions used to handle only the forward extension of enthymemes, and [Xydis et al., 2022] focuses on capturing the misunderstandings that may occur during the dialogue, whereas [Odekerken et al., 2022] does not specify locutions, but explores the notion of "queryable literals" which essentially enable dealing with backward extension. Only [de Saint-Cyr, 2011] and [Xydis et al., 2020] address both backward and forward extension of enthymemes, whereas [Xydis et al., 2020] additionally enables resolution of misunderstandings that arise due to use of enthymemes.

Note that Prakken's dialogue system for persuasion [Prakken, 2005] (described previously in Section 2.2.2) can also be perceived as a dialogue system which accounts for enthymemes since it includes locutions which support backward extension of enthymemes (e.g. *why* and *since*), as does the work discussed in [Modgil, 2017]. Both of these works, also, refer to how the outcome of the dialogue relates to the AF that is instantiated based on contents of the enthymemes moved during a dialogue, with the former providing soundness and completeness results and the latter making a conjecture that such results hold for their system. Likewise, the authors in [Xydis et al., 2021; Xydis et al., 2022] show soundness and completeness results for their respective systems. This is important as it confirms that there is no disadvantage to the use of enthymemes in dialogue and ensures that the dialogue can be played out such that an enthymeme moved in the dialogue is only justified in the case that its intended argument is justified by the contents of the moves made in the dialogue.

Not many practical applications on argumentation-based dialogues account for the use of enthymemes. We believe that more applications implementing argumentation-based dialogues that allow the handling of enthymemes should be developed. Although enthymemes' ubiquity poses a significant challenge when it comes to applying them in formal dialogues and verifying their acceptability status during these dialogues (e.g., in [Odekerken et al., 2022] it is explained how querying –or else requesting a backward extension for– all possible premises can be computationally challenging, however a sound approximation alternative is presented), humans are able to manage the use of enthymemes in their everyday life and assess them correctly during their communication (as displayed in Example 1). Therefore, if we are to develop computational dialogue systems and applications implementing them which reflect people's dialogical interactions and produce accurate results on the evaluation of their utterances, we need to formally incorporate enthymemes in sound and complete dialogues.

In [Sklar and Azhar, 2015], a persuasion dialogue protocol is presented where a participating agent Ag_1 commences a dialogue by *asserting* an argument $A = (S, c)$, where S is the set of premises of A and c is the conclusion of A. Then, their interlocutor Ag_2 can either *accept*, or *challenge*, or *attack* A. In case Ag_2 challenges A, it means that Ag_2 requests a supporting argument for either a premise $p \in S$ of A or the claim c of A. Ag_1 can fulfill the request of Ag_2 by *asserting* (i.e., providing) an argument B that either supports p (i.e., $B = (S', p)$) or c (i.e., $B = (S', c)$) depending on the request of Ag_2. If B supports p, it is easy to see that by combining arguments A and B, an argument $C = (S' \cup S, c)$ can be instantiated.

Although the authors in [Sklar and Azhar, 2015] assume that their arguments are complete, the locutions employed in the aforementioned scenario can be used to model backward extension, both requesting (with a *challenge* move) and providing it (with an *assert* move). Specifically, we can consider A and B as enthymemes of the intended complete argument C. In other words, [Sklar and Azhar, 2015] already captures an instance of enthymeme handling in argumentation-based dialogues. In [Xydis et al., 2020], the authors present a dialogue protocol which accounts for both backward and forward extension of enthymemes, as well as resolution of misunderstandings that may occur between the participants of a dialogue. We believe that by expanding the set of locutions in [Sklar and Azhar, 2015] and the persuasion dialogue protocol introduced in that work, it will be possible to additionally capture and deal with both forward extension of enthymemes and misunderstandings that may occur during the dialogue due to the use of enthymemes. The latter is of particular importance, as in case that a misunderstanding has already taken place, the participants can backtrack to that point of the dialogue, resolve the misunderstanding and still reach the "correct" conclusions/decisions based on the knowledge they have shared[33]. Modifying the locutions

---

[33] See [Xydis et al., 2022] for a system that focuses on dealing with misunderstand-

and the protocol introduced in [Sklar and Azhar, 2015] are two lines of research which the authors of this chapter are actively exploring as part of our ongoing work.

6.2 Improving the performance of large-language models

The recent significant increase in popularity of Artificial Intelligence is largely due to the surge of *Large Language Models (LLMs)* and their outstanding performance against multiple benchmarks. Essentially, a *language model (LM)* is primarily designed to predict tokens based on the likelihood of their occurrences given previous word sequences. Stemming from statistical learning methods and recurrent neural networks, it was eventually the Transformer architecture [Vaswani *et al.*, 2017] that consolidated the paradigm shift of 'pre-training' and 'fine-tuning' a language model on large datasets, ultimately leading to the development of LLMs [Zhao *et al.*, 2023]. Indeed, researchers discovered how scaling the internal structure or the training data size results in enhanced capabilities compared to smaller versions of the same model [Chowdhery *et al.*, 2022; Hoffmann *et al.*, 2022; Kaplan *et al.*, 2020]. For example, LLMs prove to outperform most of the previous standards and predecessors within the scope of information extraction [Li *et al.*, 2023], natural language inference, question answering, dialogue tasks [Qin *et al.*, 2023] and machine translation [Jiao *et al.*, 2023].

A noteworthy instance of this new technology is the well-known ChatGPT[34], which hinges on the GPT model family [Brown *et al.*, 2020; OpenAI, 2024a; OpenAI, 2024b], although many other LLMs are regarded as having similar performance levels [Anthropic, 2024; Google, 2024; Meta, 2024]. The trade-off for such impressive accomplishments consists of multiple shortcomings that likewise affect each large language model. Among these weaknesses, we highlight: hallucinations [Ji *et al.*, 2023], emergent abilities [Wei *et al.*, 2022], biased and toxic output (along with the challenging task of models-humans values alignment) [Brown

---

ings, and discusses soundness and completeness results in persuasion dialogues. Such results concern the acceptability of arguments and enthymemes moved in the dialogue and the argument framework instantiated by the contents of the moves, under some semantics σ

[34]https://chat.openai.com/

et al., 2020], lack of transparency in response generation, high cost of training and carbon footprint emissions. In addition, every LLM is limited in its knowledge of the world to its pre-training data, thus leaving a gap concerning up-to-date information that can only be covered by resorting to external tools or plugins (usually involving web search or retrieval-augmented-generation, RAG, capabilities [Gao *et al.*, 2023]). Furthermore, it has also been shown how models such as GPT-3 fall short of producing adequate and compelling arguments [Hinton and Wagemans, 2023]. The authors of such a study elaborate this conclusion after a thorough application of the Comprehensive Assessment Procedure for Natural Argumentation (CAPNA) protocol [Hinton, 2021]. GPT-3 is able to produce different argument types (thus identifying common human dialectical patterns), but it fails when it comes to providing their acceptability, mostly generating fallacious arguments. The entailed consequence is that the capability of arguing, intended as an exchange of reasoning between intelligent entities, should be learnt by AIs if their purpose aims for more than just acquiring and repeating information.

In the following, drawing from the insights outlined in [Castagna *et al.*, 2024a], we show how the employment of computational argumentation-based dialogical approaches may result in promising solutions for issues in current LLMs. Aside from an overall improvement in the quality of the posited arguments, LLMs can achieve different benefits from combining with computational argumentation [Castagna *et al.*, 2024b].

Transparency in response generation. Given the current 'black-box' nature of LLMs and the complexity of understanding their output generation (especially for laypeople), there is a present urge to provide clear explanations about what drives AIs' decisions. The goal of overcoming this lack of transparency is among the reasons that foster research within the thriving field of eXplaninable AI (XAI), where argumentative strategies are proposed as adequate forms of justifications [Čyras *et al.*, 2021; Vassiliades *et al.*, 2021]. These intuitions are backed by studies such as [Castagna, 2022; Castagna *et al.*, in press; McBurney *et al.*, 2021] where it is suggested that AI systems should adopt an argumentation-based approach to explanations consisting of dialogue protocols characterising the interactions between an explainer and an explainee. Embedded in LLMs, such a dialectical interplay would provide an informative post hoc method to deliver deliberated explanations to end-users while also ensuring detailed replies to follow-on queries. Contrary to the study of Turpin et al. [Turpin et al., 2024], we believe such a formal argumentative approach to be capable of producing and rationalising unbiased explanations by filtering, following Dung's semantics [Dung, 1995], the unacceptable ones.

Notice that even the renowned GPT-4 exhibits drawbacks when dealing with the *process consistency* of its explanations: it provides a plausible account of the rationale behind the generation of its output, but it often fails in representing a more general justification able to predict the outcome of the model given similar inputs [OpenAI, 2024a]. An argumentative dialogue (such as the Explanation-Question-Response, EQR, protocol [Castagna et al., 2023; Castagna et al., in press], previously mentioned in Section 4.2) designed for explanation purposes would allow solving the process-consistency issues by providing conversations where more information can be retrieved and thus eschewing the limited explanation length and language constraints deemed to be the leading causes of the problem [Bubeck et al., 2023].

Hallucination. Defined as *"generated content that is nonsensical or unfaithful to the provided source content"* [Ji et al., 2023], the phenomenon of hallucination in natural language generation can be divided into *intrinsic* and *extrinsic*. The former refers to generated output that contradicts the source upon which the LLM was trained. The second, instead, represents an output that cannot be verified. The employment of argumentative XAI dialogical methods can assist in probing the model replies, thus, potentially identifying and filtering out hallucinating contents, or granting, in the worst-case scenario, the retrieval of additional information over the produced content.

Emergent Abilities. The occurrence of these unpredictable phenomena consists of the unexpected appearance of specific competencies in large-scale models that do not manifest in smaller ones. Thus, it is not possible to anticipate the "emergence" of these abilities[35] (e.g. improved arithmetic, multi-task understanding, enhanced multilingual operations) by simply examining smaller-scale models [Wei *et al.*, 2022]. Leveraging argumentative XAI dialogical methods (e.g. the aforementioned EQR protocol [Castagna *et al.*, in press]) could indirectly help as a post hoc solution: although it cannot identify the reasons why emergent abilities originate, it could nonetheless provide explanations that would clarify their functioning. Notice that, although inexplicable, emergent abilities usually characterise useful competencies acquired by a model, in contrast with hallucinations that only refer to contradictory or made-up textual facts provided by the LLM as a reply to a user prompt.

7 Conclusion

This chapter set out to review applications of argumentation-based dialogue, and took a broad view as to what this meant. Viewing "dialogue" as meaning "an exchange of ideas and opinions"[36], we see it as covering any such exchange between two or more humans or agents (in any combination) or even the internal reasoning process of a single human or agent (though we do not focus on the latter). To make this chapter relatively self-contained, we briefly covered (Section 2) the elements of formal argumentation and dialogue games that we felt were required to understand the rest of the paper, before beginning the review proper by discussing (Section 3) components of argumentation-based dialogue systems such as solvers of various kinds and chatbots.

Section 4 then contains the main body of the review, looking at current work on applications of argumentation-based dialogue. It starts (see Section 4.1) by providing a description of the way that we went about the analysis of the systems that we found in the literature. Next (Section 4.2), we look at one specific application, indulging ourselves by taking this from our work on the CONSULT project, which we think nicely illustrates many of the features of a typical use of argumentation-

---

[35] Emergent abilities constitute a controversial topic and some studies even argue against their existence [Schaeffer *et al.*, 2023].

[36] Meaning 2b in Merriam Webster https://www.merriam-webster.com/dictionary/dialogue at the time of writing.

4 - Applications of Argumentation-based Dialogues

based dialogue. Following that, we look (Section 4.3) at a number of applications that are built on top of work on formal dialogue models, many of them fitting neatly into the typology introduced by Walton and Krabbe. As we argue, these are systems that fit the more traditional approach in the computational argumentation community. Then, finally (Section 4.4), we consider chatbots that are based around the use of argumentation. These we consider to be a more recent development, following the growth in ML-based chatbots.

Section 5 then discusses key themes that cut across all this work, and Section 6 digs into the detail of two areas of future work — enthymemes and the benefit of combining large language models with argumentation-based dialogues. These two areas are ones we find particularly exciting, and plan to pursue work in them ourselves.

Acknowledgements

This work was partially supported by the UK Engineering & Physical Sciences Research Council (EPSRC) under grant #EP/P010105/1, and by a grant from the University of Lincoln.

References

[Adamopoulou and Moussiades, 2020] Eleni Adamopoulou and Lefteris Moussiades. Chatbots: History, technology, and applications. *Machine Learning with applications*, 2:100006, 2020.

[Alviano, 2021] Mario Alviano. The PYGLAF argumentation reasoner (ICCMA2021), 2021. http://argumentationcompetition.org/2021/downloads/pyglaf.pdf.

[Amgoud and Cayrol, 2002] Leila Amgoud and Claudette Cayrol. A reasoning model based on the production of acceptable arguments. *Annals of Mathematics and Artificial Intelligence*, pages 197–215, 2002.

[Amgoud et al., 2000] Leila Amgoud, Simon Parsons, and Nicolas Maudet. Arguments, dialogue, and negotiation. In *Proceedings of the 14th European Conference on Artificial Intelligence*, pages 338–342, 2000.

[Amgoud et al., 2008] Leila Amgoud, Claudette Cayrol, Marie-Christine Lagasquie-Schiex, and Pierre Livet. On bipolarity in argumentation frameworks. *International Journal of Intelligent Systems*, pages 1062–1093, 2008.

[Andrews et al., 2008] Pierre Andrews, Suresh Manandhar, and Marco De Boni. Argumentative human computer dialogue for automated persuasion. In *Proceedings of the 9th SIGdial Workshop on Discourse and Dialogue*, pages 138–147, 2008.

[Anthropic, 2024] Anthropic. Claude 3.5. *Anthropic Blog*, 2024. https://www.anthropic.com/news/claude-3-5-sonnet (last accessed 24/07/2024).

[AOS, 2005] AOS. Agent manual, release 5.3. http://https://aosgrp.com.au/jack/, June 2005.

[Audemard and Simon, 2018] Gilles Audemard and Laurent Simon. On the glucose SAT solver. *International Journal on Artificial Intelligence Tools*, 27(01):1840001, 2018.

[Azhar and Sklar, 2017] M Q Azhar and Elizabeth I Sklar. A study measuring the impact of shared decision making in a human-robot team. *International Journal of Robotics Research (IJRR)*, 36:461–482, 2017.

[Bala et al., 2017] K Bala, Mukesh Kumar, Sayali Hulawale, and Sahil Pandita. Chat-bot for college management system using ai. *International Research Journal of Engineering and Technology*, 4(11):2030–2033, 2017.

[Balatsoukas et al., 2019] Panos Balatsoukas, Talya Porat, Isabel Sassoon, Kai Essers, Nadin Kokciyan, Martin Chapman, Archie Drake, Sanjay Modgil, Mark Ashworth, Elizabeth Sklar, Simon Parsons, and Vasa Curcin. User involvement in the design of a data-driven self-management decision support tool for stroke survivors. In *18th IEEE International Conference on Smart Technologies*, pages 1–6, 2019.

[Baroni et al., 2018] Pietro Baroni, Dov Gabbay, Massimilino Giacomin, and Leendert van der Torre, editors. *Handbook of Formal Argumentation*, volume 1. College Publications, London, England, 2018.

[Bench-Capon and Dunne, 2007] Trevor JM Bench-Capon and Paul E Dunne. Argumentation in artificial intelligence. *Artificial intelligence*, pages 619–641, 2007.

[Bench-Capon, 2003] Trevor JM Bench-Capon. Persuasion in practical argument using value-based argumentation frameworks. *Journal of Logic and Computation*, pages 429–448, 2003.

[Bentahar and Labban, 2011] Jamal Bentahar and Jihad Labban. An argumentation-driven model for flexible and efficient persuasive negotiation. *Group Decision and Negotiation*, 20:411–435, 2011.

[Besnard and Hunter, 2001] Philippe Besnard and Anthony Hunter. A logic-based theory of deductive arguments. *Artificial Intelligence*, 128(1-2):203–235, 2001.

[Bex and Reed, 2012] Floris Bex and Chris Reed. Dialogue templates for automatic argument processing. In *Proceedings of the Conference on Computational Models of Argument*, pages 366–377, 2012.

[Bex et al., 2013a] Floris Bex, John Lawrence, Mark Snaith, and Chris Reed. Implementing the argument web. *Communications of the ACM*, pages 66–73, 2013.

[Bex et al., 2013b] Floris Bex, Sanjay Modgil, Henry Prakken, and Chris Reed. On logical reifications of the argument interchange. *Journal of Logic and Computation*, 2013.

[Bex et al., 2014] Floris Bex, John Lawrence, and Chris Reed. Generalising argument dialogue with the dialogue game execution platform. In *Proceedings of the Conference on Computational Models of Argument*, pages 141–152, 2014.

[Bex et al., 2016] Floris Bex, Joeri Peters, and Bas Testerink. AI for online criminal complaints: From natural dialogues to structured scenarios. In *Workshop on Artificial intelligence for Justice (ECAI 2016)*, page 22, 2016.

[Biere et al., 2009] Armin Biere, Marijn Heule, and Hans van Maaren. *Handbook of Satisfiability*, volume 185. IOS press, 2009.

[Bistarelli et al., 2021a] Stefano Bistarelli, Lars Kotthoff, Francesco Santini, and Carlo Taticchi. Summary report for the Third International Competition on Computational Models of Argumentation. In *AI Magazine*, volume 42, pages 70–73, 2021.

[Bistarelli et al., 2021b] Stefano Bistarelli, Fabio Rossi, Francesco Santini, and Taticchi Carlo. CONARG: A constraint-programming solver for abstract argumentation problems, 2021. http://argumentationcompetition.org/2021/downloads/conarg.pdf.

[Black and Hunter, 2008] Elizabeth Black and Anthony Hunter. Using enthymemes in an inquiry dialogue system. In *Proceedings of the 7th International Conference on Autonomous Agents and Multi-agent Systems*, pages 437–444, 2008.

[Black and Hunter, 2009] Elizabeth Black and Anthony Hunter. An inquiry dialogue system. *Autonomous Agents and Multi-Agent Systems*, pages 173–209, 2009.

[Black and Hunter, 2012] Elizabeth Black and Anthony Hunter. A relevance-theoretic framework for constructing and deconstructing enthymemes. *Logic and Computation*, pages 55–78, 2012.

[Black et al., 2017] Elizabeth Black, Amanda J Coles, and Christopher Hampson. Planning for persuasion. In *Proceedings of the 16th International Conference on Autonomous Agents and Multiagent Systems*, pages 933–942, 2017.

[Black et al., 2021] Elizabeth Black, Nicolas Maudet, and Simon Parsons. Argumentation-based dialogue. In Dov Gabbay, Massimiliano Giacomin, Guillermo R. Simari, and Matthias Thimm, editors, *Handbook of Formal Argumentation*, volume 2, page 511. College Publications, 2021.

[Boissier et al., 2013] Olivier Boissier, Rafael H Bordini, Jomi F Hübner, Alessandro Ricci, and Andrea Santi. Multi-agent oriented programming with JaCaMo. *Science of Computer Programming*, pages 747–761, 2013.

[Bondarenko et al., 1997] Andrei Bondarenko, Phan Minh Dung, Robert A Kowalski, and Francesca Toni. An abstract, argumentation-theoretic approach to default reasoning. *Artificial Intelligence*, pages 63–101, 1997.

[Booth et al., 2018] Richard Booth, Martin Caminada, and Braden Marshall. Disco: A web-based implementation of discussion games for grounded and preferred semantics. In *Proceedings of the Conference on Computational Models of Argument*, pages 453–454. 2018.

[Borg and Odekerken, 2002] AnneMarie Borg and Daphne Odekerken. PyArg for solving and explaining argumentation in python: Demonstration. In *Proceedings of the Conference on Computational Models of Argument*, pages 349–350. 2002.

[Brown et al., 2020] Tom Brown, Benjamin Mann, Nick Ryder, Melanie Subbiah, Jared D Kaplan, Prafulla Dhariwal, Arvind Neelakantan, Pranav Shyam, Girish Sastry, Amanda Askell, et al. Language models are few-shot learners. *Advances in neural information processing systems*, 33:1877–1901, 2020.

[Bryant and Krause, 2008] Daniel Bryant and Paul Krause. A review of current defeasible reasoning implementations. *The Knowledge Engineering Review*, 23(3):227–260, 2008.

[Bryant et al., 2006] Daniel Bryant, Paul J Krause, and Gerard Vreeswijk. Argue tuProlog: A lightweight argumentation engine for agent applications. In *Proceedings of the Confernece on Computational Models of Argument*, pages 27–32. 2006.

[Bubeck et al., 2023] Sébastien Bubeck, Varun Chandrasekaran, Ronen Eldan, Johannes Gehrke, Eric Horvitz, Ece Kamar, Peter Lee, Yin Tat Lee, Yuanzhi Li, Scott Lundberg, Harsha Nori, Hamid Palangi, Marco Tulio Ribeiro, and Yi Zhang. Sparks of artificial general intelligence: Early experiments with GPT-4. *arXiv preprint arXiv:2303.12712*, 2023.

[Caldarini et al., 2022] Guendalina Caldarini, Sardar Jaf, and Kenneth McGarry. A literature survey of recent advances in chatbots. *Information*, 13(1):41, 2022.

[Caminada and Pigozzi, 2011] Martin Caminada and Gabriella Pigozzi. On judgment aggregation in abstract argumentation. *Autonomous Agents and Multi-Agent Systems*, 22:64–102, 2011.

[Caminada and Uebis, 2020] Martin Caminada and Sören Uebis. An implementation of argument-based discussion using ASPIC. In *Proceedings of the Conference on Computational Models of Argument*, pages 455–456. 2020.

[Caminada et al., 2014a] Martin Caminada, Wolfgang Dvořák, and Srdjan Vesic. Preferred semantics as socratic discussion. *Journal of Logic and Computation*, pages 1257–1292, 2014.

[Caminada et al., 2014b] Martin Caminada, Sanjay Modgil, and Nir Oren. Preferences and unrestricted rebut. In *Proceedings of the Conference on Computational Models of Argument*, pages 209–220, 2014.

[Caminada, 2006] Martin Caminada. On the issue of reinstatement in argumentation. In *European Workshop on Logics in Artificial Intelligence*, pages 111–123, 2006.

[Caminada, 2015] Martin Caminada. A discussion game for grounded semantics. In *Proceedings of the 3rd International Workshop on Theory and Applications of Formal Argumentation*, pages 59–73, 2015.

[Caminada, 2018] Martin Caminada. Argumentation semantics as formal discussion. In Pietro Baroni, Dov Gabbay, Massimiliano Giacomin, and Leendert van der Torre, editors, *Handbook of Formal Argumentation*, volume 1, pages 487—518. College Publications, 2018.

[Castagna et al., 2022] Federico Castagna, Simon Parsons, Isabel Sassoon, and Elizabeth I. Sklar. Providing explanations via the EQR argument scheme. In *Proceedings of the Conference on Computational Models of Argument*, 2022.

[Castagna et al., 2023] Federico Castagna, Alexandra Garton, Peter Mcburney, Simon Parsons, Isabel Sassoon, and Elizabeth I Sklar. EQRbot: a chatbot delivering EQR argument-based explanations. In *Frontiers in Artificial Intelligence*, volume 6. Frontiers, 2023.

[Castagna et al., 2024a] Federico Castagna, Nadin Kokciyan, Isabel Sassoon, Simon Parsons, and Elizabeth Sklar. Computational Argumentation-based Chatbots: a Survey. *Journal of Artificial Intelligence Research*, 80, 2024.

[Castagna et al., 2024b] Federico Castagna, Isabel Sassoon, and Simon Parsons. Can formal argumentative reasoning enhance LLM's performances? https://arxiv.org/abs/2405.13036, 2024.

[Castagna et al., in press] Federico Castagna, Peter McBurney, and Simon Parsons. Explanation–Question–Response dialogue: An argumentative tool for explainable AI. *Argument & Computation*, (in press).

[Castagna, 2022] Federico Castagna. Towards a fully-fledged formal protocol for the Explanation-Question-Response dialogue. In *Online Handbook of Argumentation for AI*, pages 17–21, 2022.

[Castagna, 2023] Federico Castagna. Dialectical argument game proof theories for classical logic. *Journal of Applied Logics*, 2631(3):279, 2023.

[Chalaguine and Hunter, 2020] Lisa Andreevna Chalaguine and Anthony Hunter. A persuasive chatbot using a crowd-sourced argument graph and concerns. In *Proceedings of the Conference on Computational Models of Argument*, pages 9–20, 2020.

[Chalaguine and Hunter, 2021] Lisa Andreevna Chalaguine and Anthony Hunter. Addressing popular concerns regarding COVID-19 vaccination with natural language argumentation dialogues. In *European Conference on Symbolic and Quantitative Approaches with Uncertainty*, pages 59–73, 2021.

[Chalaguine et al., 2019] Lisa Andreevna Chalaguine, Anthony Hunter, Henry Potts, and Fiona Hamilton. Impact of argument type and concerns in argumentation with a chatbot. In *Proceedings of the 31st IEEE International Conference on Tools with Artificial Intelligence*, pages 1557–1562, 2019.

[Chapman et al., 2019] Martin Chapman, Panagiotis Balatsoukas, Nadin Kökciyan, Kai Essers, Isabel Sassoon, Mark Ashworth, Vasa Curcin, Sanjay Modgil, Simon Parsons, and Elizabeth I Sklar. Computational argumentation-based clinical decision support. In *Proceedings of the 18th International Conference on Autonomous Agents and Multiagent Systems*, pages 2345–2347, 2019.

[Chapman et al., 2022] Martin Chapman, G Abigail, Isabel Sassoon, Nadin Kökciyan, Elizabeth I Sklar, Vasa Curcin, et al. Using microservices to design patient-facing research software. In *Proceedings of the 18th IEEE International Conference on e-Science*, pages 44–54, 2022.

[Chesnevar et al., 2006] Carlos Chesnevar, Sanjay Modgil, Iyad Rahwan, Chris Reed, Guillermo Simari, Matthew South, Gerard Vreeswijk, Steven Willmott, et al. Towards an argument interchange format. *The Knowledge Engineering Review*, pages 293–316, 2006.

[Chowdhery et al., 2022] Aakanksha Chowdhery, Sharan Narang, Jacob Devlin, Maarten Bosma, Gaurav Mishra, Adam Roberts, Paul Barham, Hyung Won Chung, Charles Sutton, Sebastian Gehrmann, et al. Palm: Scaling language modeling with pathways. *arXiv preprint arXiv:2204.02311*, 2022.

[Codecademy, 2022] Codecademy. What are chatbots? 2022. https://www.codecademy.com/article/what-are-chatbots (last accessed 11/10/2023).

[Cogan et al., 2005] Eva Cogan, Simon Parsons, and Peter McBurney. What kind of argument are we going to have today? In *Proceedings of the 4th International Conference on Autonomous Agents and Multiagent Systems*, pages 544–551, 2005.

[Craandijk and Bex, 2020] Dennis Craandijk and Floris Bex. Deep learning for abstract argumentation semantics. *arXiv preprint arXiv:2007.07629*, 2020.

[Cyras et al., 2018] Kristijonas Cyras, Brendan Delaney, Denys Prociuk, Francesca Toni, Martin Chapman, Jesús Domínguez, and Vasa Curcin. Argumentation for explainable reasoning with conflicting medical recommendations. In *Proceedings of the CEUR Workshop*, volume 2237. CEUR-WS, 2018.

[Čyras et al., 2021] Kristijonas Čyras, Antonio Rago, Emanuele Albini, Pietro Baroni, and Francesca Toni. Argumentative XAI: a survey. In *Proceedings of the Thirtieth International Joint Conference on Artificial Intelligence, Survey Track*, pages 4392–4399, 2021.

[Dachselt et al., 2022] Raimund Dachselt, Sarah Alice Gaggl, Markus Krötzsch, Julian Mendez, Dominik Rusovac, and Mei Yang. Nexas: A visual tool for navigating and exploring argumentation solution spaces. In *Proceedings of the Conference on Computational Models of Argument*, pages 116–127, 2022.

[Dawkins, 2016] Richard Dawkins. *The Selfish Gene*. Oxford University Press, 40th anniversary edition, 2016.

[de Saint-Cyr, 2011] Florence Dupin de Saint-Cyr. Handling enthymemes in time-limited persuasion dialogs. In *Proceedings of the International Conference on Scalable Uncertainty Management*, pages 149–162, 2011.

[Dix et al., 2009] Jürgen Dix, Simon Parsons, Henry Prakken, and Guillermo Ricardo Simari. Research challenges for argumentation. *Computer Science Research and Development*, 23(1):27–34, 2009.

[Dubois and Prade, 2004] Didier Dubois and Henri Prade. Possibilistic logic: a retrospective and prospective view. *Fuzzy sets and Systems*, pages 3–23, 2004.

[Dung and Thang, 2010] Phan Minh Dung and Phan Minh Thang. Towards (probabilistic) argumentation for jury-based dispute resolution. In *Proceedings of the Conference on Computational Models of Argument*, pages 171–182, 2010.

[Dung et al., 2009] Phan Minh Dung, Robert A Kowalski, and Francesca Toni. Assumption-based argumentation. In I. Rahwan and G. R. Simari, editors, *Argumentation in Artificial Intelligence*, pages 199–218. 2009.

[Dung, 1995] Phan Minh Dung. On the acceptability of arguments and its fundamental role in nonmonotonic reasoning, logic programming and n-person games. *Artificial Intelligence*, pages 321–357, 1995.

[Dvorák et al., 2021] Wolfgang Dvorák, Matthias König, Johannes P. Wallner, and Stefan Woltran. ASPARTIX-V21, 2021. http://argumentationcompetition.org/2021/downloads/aspartix-v21.pdf.

[Egly et al., 2008] Uwe Egly, Sarah Alice Gaggl, and Stefan Woltran. ASPARTIX: Implementing argumentation frameworks using answer-set programming. In *International Conference on Logic Programming*, pages 734–738. Springer,

2008.

[Engelmann et al., 2023] Débora C Engelmann, Alison R Panisson, Renata Vieira, Jomi Fred Hübner, Viviana Mascardi, and Rafael H Bordini. MAIDS — a framework for the development of multi-agent intentional dialogue systems. In *Proceedings of the 2023 International Conference on Autonomous Agents and Multiagent Systems*, pages 1209–1217, 2023.

[Essers et al., 2018] Kai Essers, Martin Chapman, Nadin Kokciyan, Isabel Sassoon, Talya Porat, Panagiotis Balatsoukas, Peter Young, Mark Ashworth, Vasa Curcin, Sanjay Modgil, et al. The CONSULT system. In *Proceedings of the 6th International Conference on Human-Agent Interaction*, pages 385–386, 2018.

[Falappa et al., 2009] Marcelo Alejandro Falappa, Gabriele Kern-Isberner, and Guillermo Ricardo Simari. Belief revision and argumentation theory. In I Rahwan and G. R. Simari, editors, *Argumentation in Artificial Intelligence*, pages 341–360. 2009.

[Fan and Toni, 2012] Xiuyi Fan and Francesca Toni. Agent strategies for ABA-based information-seeking and inquiry dialogues. In *Proceedings of the 20th European Conference on Artificial Intelligence*, pages 324–329, 2012.

[Fan and Toni, 2014] Xiuyi Fan and Francesca Toni. A general framework for sound assumption-based argumentation dialogues. *Artificial Intelligence*, pages 20–54, 2014.

[Fazzinga et al., 2022] Bettina Fazzinga, Andrea Galassi, and Paolo Torroni. A privacy-preserving dialogue system based on argumentation. *Intelligent Systems with Applications*, 16:200113, 2022.

[Fichte et al., 2021] Johannes K. Fichte, Markus Hecher, Piotr Gorczyca, and Ridhwan Dewoprabowo. A-FOLIO DPDB – system description for ICCMA 2021, 2021. http://argumentationcompetition.org/2021/downloads/a-folio-dpdb.pdf.

[Fichte et al., 2022] Johannes K Fichte, Markus Hecher, Patrick Thier, and Stefan Woltran. Exploiting database management systems and treewidth for counting. *Theory and Practice of Logic Programming*, 22(1):128–157, 2022.

[Fox et al., 2007] John Fox, David Glasspool, Dan Grecu, Sanjay Modgil, Matthew South, and Vivek Patkar. Argumentation-based inference and decision making–a medical perspective. *IEEE Intelligent Systems*, 22(6):34–41, 2007.

[Gaertner and Toni, 2007] Dorian Gaertner and Francesca Toni. CaSAPI: a system for credulous and sceptical argumentation. In *Proceedings of the Workshop on Nonmonotonic Reasoning*, pages 80–95, 2007.

[Gaggl et al., 2020] Sarah A Gaggl, Thomas Linsbichler, Marco Maratea, and

Stefan Woltran. Design and results of the second international competition on computational models of argumentation. *Artificial Intelligence*, 279:103193, 2020.

[Gao et al., 2023] Yunfan Gao, Yun Xiong, Xinyu Gao, Kangxiang Jia, Jinliu Pan, Yuxi Bi, Yi Dai, Jiawei Sun, and Haofen Wang. Retrieval-augmented generation for large language models: A survey. *arXiv preprint arXiv:2312.10997*, 2023.

[García and Simari, 2004] Alejandro J García and Guillermo R Simari. Defeasible logic programming: An argumentative approach. *Theory and Practice of Logic Programming*, 4(1-2):95–138, 2004.

[Google, 2024] Google. Gemini 1.5: Unlocking multimodal understanding across millions of tokens of context. https://arxiv.org/abs/2403.05530, 2024.

[Guerrero et al., 2016] Esteban Guerrero, Juan Carlos Nieves, and Helena Lindgren. An activity-centric argumentation framework for assistive technology aimed at improving health. *Argument & Computation*, pages 5–33, 2016.

[Guo et al., 2022] Kai Guo, Jian Wang, and Samuel Kai Wah Chu. Using chatbots to scaffold EFL students' argumentative writing. *Assessing Writing*, 54:100666, 2022.

[Hadjinikolis et al., 2013] Christos Hadjinikolis, Yiannis Siantos, Sanjay Modgil, Elizabeth Black, and Peter McBurney. Opponent modelling in persuasion dialogues. In *Proceedings of the 32nd International Joint Conference on Artificial Intelligence*, pages 164–170, 2013.

[Hadoux and Hunter, 2017] Emmanuel Hadoux and Anthony Hunter. Strategic sequences of arguments for persuasion using decision trees. In *Proceedings of the 31st AAAI Conference on Artificial Intelligence*, pages 1128–1134, 2017.

[Hadoux and Hunter, 2019] Emmanuel Hadoux and Anthony Hunter. Comfort or safety? gathering and using the concerns of a participant for better persuasion. *Argument & Computation*, 10(2):113–147, 2019.

[Hadoux et al., 2021] Emmanuel Hadoux, Anthony Hunter, and Sylwia Polberg. Strategic argumentation dialogues for persuasion: Framework and experiments based on modelling the beliefs and concerns of the persuadee. *arXiv preprint arXiv:2101.11870*, 2021.

[Hauptmann et al., 2024] Christian Hauptmann, Adrian Krenzer, Justin Völkel, and Frank Puppe. Argumentation effect of a chatbot for ethical discussions about autonomous AI scenarios. *Knowledge and Inf. Systems*, 66:3607–3637, 2024.

[Hinton and Wagemans, 2023] Martin Hinton and Jean HM Wagemans. How persuasive is AI-generated argumentation? an analysis of the quality of an argumentative text produced by the GPT-3 AI text generator. *Argument &*

Computation, 14(1):59–74, 2023.

[Hinton, 2021] Martin Hinton. CAPNA—the comprehensive assessment procedure for natural argumentation. In *Evaluating the Language of Argument*, volume 37 of *Argumentation Library*, pages 167–194. 2021.

[Hoffmann et al., 2022] Jordan Hoffmann, Sebastian Borgeaud, Arthur Mensch, Elena Buchatskaya, Trevor Cai, Eliza Rutherford, Diego de Las Casas, Lisa Anne Hendricks, Johannes Welbl, Aidan Clark, Tom Hennigan, Eric Noland, Katie Millican, George van den Driessche, Bogdan Damoc, Aurelia Guy, Simon Osindero, Karen Simonyan, Erich Elsen, Oriol Vinyals, Jack W. Rae, and Laurent Sifre. Training compute-optimal large language models. https://arxiv.org/pdf/2203.15556, 2022.

[Hosseini, 2017] Seyed Ali Hosseini. *Dialogues incorporating enthymemes and modelling of other agents' beliefs*. PhD thesis, King's College London, 2017.

[Hung, 2017] Nguyen Duy Hung. Inference procedures and engine for probabilistic argumentation. *International Journal of Approximate Reasoning*, 90:163–191, 2017.

[Hunter and Potyka, 2017] Anthony Hunter and Nico Potyka. Updating probabilistic epistemic states in persuasion dialogues. In *Proceedings of the 14th European Conference on Symbolic and Quantitative Approaches to Reasoning with Uncertainty*, pages 46–56, 2017.

[Hunter et al., 2019] Anthony Hunter, Lisa Chalaguine, Tomasz Czernuszenko, Emmanuel Hadoux, and Sylwia Polberg. Towards computational persuasion via natural language argumentation dialogues. In *Proceedings of the Joint German/Austrian Conference on Artificial Intelligence (Künstliche Intelligenz)*, pages 18–33. Springer, 2019.

[Hunter, 2007] Anthony Hunter. Real arguments are approximate arguments. In *Proceedings of the 22nd AAAI Conference on Artificial Intelligence*, pages 66–71, 2007.

[Hunter, 2018] Anthony Hunter. Towards a framework for computational persuasion with applications in behaviour change. *Argument & Computation*, 9(1):15–40, 2018.

[Järvisalo et al., 2023] Matti Järvisalo, Tuomo Lehtonen, and Andreas Niskanen. Design of ICCMA 2023, 5th International Competition on Computational Models of Argumentation: A preliminary report. In *Proceedings of the First International Workshop on Argumentation and Applications*, pages 4–10, 2023.

[Ji et al., 2023] Ziwei Ji, Nayeon Lee, Rita Frieske, Tiezheng Yu, Dan Su, Yan Xu, Etsuko Ishii, Ye Jin Bang, Andrea Madotto, and Pascale Fung. Survey of hallucination in natural language generation. *ACM Computing Surveys*, 55(12):248, 2023.

[Jiao et al., 2023] Wenxiang Jiao, Wenxuan Wang, JT Huang, Xing Wang, and ZP Tu. Is ChatGPT a good translator? yes with GPT-4 as the engine. https://arxiv.org/pdf/2301.08745, 2023.

[Kampik and Gabbay, 2020] Timotheus Kampik and Dov Gabbay. Towards DiArg: An argumentation-based dialogue reasoning engine. In *Proceedings of the Third International Workshop on Systems and Algorithms for Formal Argumentation*, pages 14–21, 2020.

[Kaplan et al., 2020] Jared Kaplan, Sam McCandlish, Tom Henighan, Tom B Brown, Benjamin Chess, Rewon Child, Scott Gray, Alec Radford, Jeffrey Wu, and Dario Amodei. Scaling laws for neural language models. https://arxiv.org/pdf/2001.08361, 2020.

[Koit and Õim, 2015] Mare Koit and Haldur Õim. A computational model of argumentation in agreement negotiation processes. *Argument & Computation*, 6(2):101–129, 2015.

[Kok et al., 2011] Eric M Kok, John-Jules Ch Meyer, Henry Prakken, and Gerard AW Vreeswijk. A formal argumentation framework for deliberation dialogues. In *Proceedings of the 7th International Workshop on Argumentation in Multi-Agent Systems*, pages 31–48, 2011.

[Kökciyan et al., 2019] Nadin Kökciyan, Martin Chapman, Panagiotis Balatsoukas, Isabel Sassoon, Kai Essers, Mark Ashworth, Vasa Curcin, Sanjay Modgil, Simon Parsons, and Elizabeth I Sklar. A collaborative decision support tool for managing chronic conditions. In *Proceedings of the 17th World Congress on Medical and Health Informatics*, pages 644–648, 2019.

[Kökciyan et al., 2020] Nadin Kökciyan, Simon Parsons, Isabel Sassoon, Elizabeth Sklar, and Sanjay Modgil. An argumentation-based approach to generate domain-specific explanations. In *Proceedings of the European Conference on Multi-Agent Systems*, pages 319–337. Springer, 2020.

[Kökciyan et al., 2021] Nadin Kökciyan, Isabel Sassoon, Elizabeth Sklar, Sanjay Modgil, and Simon Parsons. Applying metalevel argumentation frameworks to support medical decision making. *IEEE Intelligent Systems*, 36(2):64–71, 2021.

[Lagniez et al., 2015] Jean-Marie Lagniez, Emmanuel Lonca, and Jean-Guy Mailly. CoQuiAAS: A constraint-based quick abstract argumentation solver. In *Proceedings of the 27th IEEE International Conference on Tools with Artificial Intelligence*, pages 928–935, 2015.

[Lagniez et al., 2021] Jean-Marie Lagniez, Emmanuel Lonca, Jean-Guy Mailly, and Julien Rossit. Design and results of ICCMA 2021. https://arxiv.org/pdf/2109.08884, 2021.

[Lagniez et al., 2023] Jean-Marie Lagniez, Emmanuel Lonca, and Jean-Guy Mailly. Crustabri, the evolution of CoQuiAAS. In *Solver and Benchmark De-*

scriptions of ICCMA 2023 : 5th International Competition on Computational Models of Argumentation, page 20. 2023. http://hdl.handle.net/10138/565357.

[Lawrence and Reed, 2020] John Lawrence and Chris Reed. Argument mining: A survey. *Computational Linguistics*, 45(4):765–818, 2020.

[Lawrence et al., 2012] John Lawrence, Floris Bex, Chris Reed, and Mark Snaith. Aifdb: Infrastructure for the argument web. In *Proceedings of the Conference on Computational Models of Argument*, pages 515–516, 2012.

[Lawrence et al., 2022] John Lawrence, Jacky Visser, and Chris Reed. Polemicist: A dialogical interface for exploring complex debates. In *9th International Conference on Computational Models of Argument, COMMA 2022*, pages 365–366, 2022.

[Lehtonen et al., 2023] Tuomo Lehtonen, Johannes P Wallner, and Matti Järvisalo. ASPforABA–ASP-based algorithms for reasoning in ABA. In *Solver and Benchmark Descriptions of ICCMA 2023 : 5th International Competition on Computational Models of Argumentation*, page 18. 2023. http://hdl.handle.net/10138/565357.

[Leiva et al., 2023] Diego S Orbe Leiva, Sebastian Gottifredi, and Alejandro J García. Automatic knowledge generation for a persuasion dialogue system with enthymemes. *International Journal of Approximate Reasoning*, 160:108963, 2023.

[Li et al., 2023] Bo Li, Gexiang Fang, Yang Yang, Quansen Wang, Wei Ye, Wen Zhao, and Shikun Zhang. Evaluating ChatGPT's information extraction capabilities: An assessment of performance, explainability, calibration, and faithfulness. https://arxiv.org/abs/2304.11633, 2023.

[Lindgren and Winnberg, 2012] Helena Lindgren and Peter Winnberg. A model for interaction design of personalised knowledge systems in the health domain. In *Proceedings of the 3rd International Conference on Electronic Healthcare*, pages 235–242, 2012.

[Lindgren and Yan, 2015] Helena Lindgren and Chunli Yan. Acktus: A platform for developing personalized support systems in the health domain. In *Proceedings of the 5th International Conference on Digital Health 2015*, pages 135–142, 2015.

[Lindgren et al., 2020] Helena Lindgren, Timotheus Kampik, Esteban Guerrero Rosero, Madeleine Blusi, and Juan Carlos Nieves. Argumentation-based health information systems: A design methodology. *IEEE Intelligent Systems*, 36(2):72–80, 2020.

[Luck et al., 2005] Michael Luck, Peter McBurney, Onn Shehory, and Steve Willmott. Agent technology, Computing as interaction: A roadmap for agent based computing. AgentLink III, 2005.

[Malmqvist, 2023] Lars Malmqvist. AFGCN v2: A GCN-based approximate solver. In *Solver and Benchmark Descriptions of ICCMA 2023 : 5th International Competition on Computational Models of Argumentation*, page 8. 2023. http://hdl.handle.net/10138/565357.

[McBurney and Parsons, 2009] Peter McBurney and Simon Parsons. Dialogue games for agent argumentation. In I Rahwan and G. R. Simari, editors, *Argumentation in Artificial Intelligence*, pages 261–280. 2009.

[McBurney et al., 2007] Peter McBurney, David Hitchcock, and Simon Parsons. The eightfold way of deliberation dialogue. *International Journal of Intelligent Systems*, 22(1):95–132, 2007.

[McBurney et al., 2021] Peter McBurney, Simon Parsons, et al. Argument schemes and dialogue protocols: Doug Walton's legacy in Artificial Intelligence. *Journal of Applied Logics*, 8(1):263–286, 2021.

[McCarthy, 1980] John McCarthy. Circumscription—a form of non-monotonic reasoning. *Artificial intelligence*, 13(1-2):27–39, 1980.

[Mercier and Sperber, 2011] Hugo Mercier and Dan Sperber. Why do humans reason? Arguments for an argumentative theory. *Behavioral and Brain Sciences*, 34(2):57–74, 2011.

[Meta, 2024] Meta. Introducing Llama 3.1: Our most capable models to date. Meta Blog, 2024. https://ai.meta.com/research/publications/the-llama-3-herd-of-models/.

[Modgil and Caminada, 2009] Sanjay Modgil and Martin Caminada. Proof theories and algorithms for abstract argumentation frameworks. In I Rahwan and G. R. Simari, editors, *Argumentation in Artificial Intelligence*, pages 105–129. 2009.

[Modgil and Prakken, 2013] Sanjay Modgil and Henry Prakken. A general account of argumentation with preferences. *Artificial Intelligence*, 195:361–397, 2013.

[Modgil, 2017] Sanjay Modgil. Towards a general framework for dialogues that accommodate reasoning about preferences. In *Proceedings of the International Workshop on Theories and Applications of Formal Argumentation*, pages 175–191, 2017.

[Nieves and Lindgren, 2015] Juan Carlos Nieves and Helena Lindgren. Deliberative argumentation for service provision in smart environments. In *Proceedings of the 12th European Conference on Multi-Agent Systems*, pages 388–397, 2015.

[Niskanen and Järvisalo, 2023] Andreas Niskanen and Matti Järvisalo. µ-TOKSIA in ICCMA 2023. In *Solver and Benchmark Descriptions of ICCMA 2023 : 5th International Competition on Computational Models of Argumen-*

tation, page 31. 2023. http://hdl.handle.net/10138/56535.

[Odekerken *et al.*, 2020] Daphne Odekerken, AnneMarie Borg, and Floris Bex. Estimating stability for efficient argument-based inquiry. In *Proceedings of the Conference on Computational Models of Argument*, pages 307–318, 2020.

[Odekerken *et al.*, 2022] Daphne Odekerken, Floris Bex, AnneMarie Borg, and Bas Testerink. Approximating stability for applied argument-based inquiry. *Intelligent Systems with Applications*, 16:200110, 2022.

[Oliveira *et al.*, 2018] Tiago Oliveira, Jérémie Dauphin, Ken Satoh, Shusaku Tsumoto, and Paulo Novais. Argumentation with goals for clinical decision support in multimorbidity. In *Proceedings of the 17th International Conference on Autonomous Agents and MultiAgent Systems*, pages 2031–2033, 2018.

[OpenAI, 2024a] OpenAI. GPT-4 Technical Report. https://arxiv.org/abs/2303.08774, 2024.

[OpenAI, 2024b] OpenAI. Hello GPT-4o. https://openai.com/index/hello-gpt-4o/, 2024.

[Oracle,] Oracle. What is a chatbot? https://www.oracle.com/chatbots/what-is-a-chatbot/.

[Oren and Norman, 2009] Nir Oren and Timothy J Norman. Arguing using opponent models. In *Proceedings of the International Workshop on Argumentation in Multi-Agent Systems*, pages 160–174, 2009.

[Panisson and Bordini, 2017] Alison R Panisson and Rafael H Bordini. Argumentation schemes in multi-agent systems: A social perspective. In *Proceedings of the 5th International Workshop on Engineering Multi-Agent Systems*, pages 92–108, 2017.

[Panisson *et al.*, 2014] Alison R Panisson, Felipe Meneguzzi, Renata Vieira, and Rafael H Bordini. An approach for argumentation-based reasoning using defeasible logic in multi-agent programming languages. In *Proceedings of the 11th International Workshop on Argumentation in Multiagent Systems*, pages 1–15, 2014.

[Panisson *et al.*, 2021] Alison R Panisson, Peter McBurney, and Rafael H Bordini. A computational model of argumentation schemes for multi-agent systems. *Argument & Computation*, 12(3):357–395, 2021.

[Panisson *et al.*, 2022a] Alison R Panisson, Peter McBurney, and Rafael H Bordini. Towards an enthymeme-based communication framework. In *Proceedings of the 21st International Conference on Autonomous Agents and Multiagent Systems*, pages 1708–1710, 2022.

[Panisson *et al.*, 2022b] Alison R Panisson, Peter McBurney, and Rafael H Bordini. Towards an enthymeme-based communication framework in multi-agent systems. In *Proceedings of the International Conference on Principles* of *Knowledge Representation and Reasoning*, pages 267–277, 2022.

[Parsons and Sklar, 2005] S. Parsons and E. I. Sklar. How agents revise their beliefs after an argumentation-based dialogue. In *Proceedings of the 2nd International Workshop on Argumentation in Multiagent Systems*, 2005.

[Parsons et al., 1998] Simon Parsons, Carles Sierra, and Nick Jennings. Agents that reason and negotiate by arguing. *Journal of Logic and Computation*, 8(3):261–292, 1998.

[Parsons et al., 2003] Simon Parsons, Michael Wooldridge, and Leila Amgoud. Properties and complexity of some formal inter-agent dialogues. *Journal of Logic and Computation*, 13(3):347–376, 2003.

[Podlaszewski et al., 2011] Mikolaj Podlaszewski, Martin Caminada, and Gabriella Pigozzi. An implementation of basic argumentation components. In *Proceedinsg of the 10th International Conference on Autonomous Agents and Multiagent Systems-Volume*, pages 1307–1308, 2011.

[Prakken, 2005] Henry Prakken. Coherence and flexibility in dialogue games for argumentation. *Journal of Logic and Computation*, 15(6):1009–1040, 2005.

[Prakken, 2006] Henry Prakken. Formal systems for persuasion dialogue. *The Knowledge Engineering Review*, 21(2):163–188, 2006.

[Prakken, 2009] Henry Prakken. Models of persuasion dialogue. In I Rahwan and G. R. Simari, editors, *Argumentation in Artificial Intelligence*, pages 281–300. 2009.

[Prakken, 2010] Henry Prakken. An abstract framework for argumentation with structured arguments. *Argument & Computation*, 1(2):93–124, 2010.

[Qin et al., 2023] Chengwei Qin, Aston Zhang, Zhuosheng Zhang, Jiaao Chen, Michihiro Yasunaga, and Diyi Yang. Is ChatGPT a general-purpose natural language processing task solver? In *Proceedings of the Conference on Empirical Methods in Natural Language Processing*, pages 1339—1384, 2023.

[Rahwan et al., 2003] Iyad Rahwan, Sarvapali D Ramchurn, Nicholas R Jennings, Peter McBurney, Simon Parsons, and Liz Sonenberg. Argumentation-based negotiation. *The Knowledge Engineering Review*, 18(4):343–375, 2003.

[Rosenfeld and Kraus, 2016] Ariel Rosenfeld and Sarit Kraus. Strategical argumentative agent for human persuasion. In *Proceedings of the 22nd European Conference on Artificial Intelligence*, pages 320–328, 2016.

[Saibene et al., 2021] Aurora Saibene, Michela Assale, and Marta Giltri. Expert systems: Definitions, advantages and issues in medical field applications. *Expert Systems with Applications*, 177:114900, 2021.

[Sassoon et al., 2019] Isabel Sassoon, Nadin Kökciyan, Elizabeth Sklar, and Simon Parsons. Explainable argumentation for wellness consultation. In *Proceedings of the International Workshop on Explainable, Transparent Au-*

tonomous Agents and Multi-Agent Systems, pages 186–202, 2019.

[Schaeffer et al., 2023] Rylan Schaeffer, Brando Miranda, and Sanmi Koyejo. Are emergent abilities of large language models a mirage? In *Advances in Neural Information Processing 36*, 2023.

[Schraagen and Bex, 2019] Marijn Schraagen and Floris Bex. Extraction of semantic relations in noisy user-generated law enforcement data. In *Proceedings of the 13th IEEE International Conference on Semantic Computing*, pages 79–86, 2019.

[Shortliffe, 1977] Edward H Shortliffe. Mycin: A knowledge-based computer program applied to infectious diseases. In *Proceedings of the Annual Symposium on Computer Application in Medical Care*, pages 66–69. American Medical Informatics Association, 1977.

[Silver and Veness, 2010] David Silver and Joel Veness. Monte-Carlo planning in large POMDPs. In *Advances in Neural Information Processing Systems 23*, 2010.

[Sklar and Azhar, 2015] Elizabeth I Sklar and Mohammad Q Azhar. Argumentation-based dialogue games for shared control in human-robot systems. *Journal of Human-Robot Interaction*, 4(3):120–148, 2015.

[Sklar and Azhar, 2018] Elizabeth I Sklar and Mohammad Q Azhar. Explanation through argumentation. In *Proceedings of the 6th International Conference on Human-Agent Interaction*, pages 277–285, 2018.

[Sklar et al., 2004] E. I. Sklar, S. Parsons, and M. Davies. When is it okay to lie? A simple model of contradiction in agent-based dialogues. In *Proceedings of the First Workshop on Argumentation in Multiagent Systems*, 2004.

[Sklar et al., 2016] Elizabeth I Sklar, Simon Parsons, Zimi Li, Jordan Salvit, Senni Perumal, Holly Wall, and Jennifer Mangels. Evaluation of a trust-modulated argumentation-based interactive decision-making tool. *Journal of Autonomous Agents and Multi-Agent Systems*, 30(1):136–173, 2016.

[Snaith et al., 2020] Mark Snaith, John Lawrence, Alison Pease, and Chris Reed. A modular platform for argument and dialogue. In *Proceedings of the Conference on Computational Models of Argument*, pages 473–474, 2020.

[Sojasingarayar, 2020] Abonia Sojasingarayar. Seq2Seq AI chatbot with attention mechanism. Master's thesis, IA School Groupe GEMA, Boulogne-Billancourt, May 2020.

[Soos et al., 2009] Mate Soos, Karsten Nohl, and Claude Castelluccia. Extending SAT solvers to cryptographic problems. In *International Conference on Theory and Applications of Satisfiability Testing*, pages 244–257, 2009.

[Sperber and Wilson, 1986] Dan Sperber and Deirdre Wilson. *Relevance: Communication and cognition*. Harvard University Press, 1986.

[Sutton et al., 2020] Reed T Sutton, David Pincock, Daniel C Baumgart, Daniel C Sadowski, Richard N Fedorak, and Karen I Kroeker. An overview of clinical decision support systems: benefits, risks, and strategies for success. *npj Digital Medicine*, 3(17), 2020.

[Tang et al., 2012a] Yuqing Tang, Kai Cai, Peter McBurney, Elizabeth Sklar, and Simon Parsons. Using argumentation to reason about trust and belief. *Journal of Logic and Computation*, 22(5):979–1018, 2012.

[Tang et al., 2012b] Yuqing Tang, Elizabeth Sklar, and Simon Parsons. An argumentation engine: Argtrust. In *Ninth International Workshop on Argumentation in Multiagent Systems*, 2012.

[Thimm and Villata, 2017] Matthias Thimm and Serena Villata. The first international competition on computational models of argumentation: Results and analysis. In *Artificial Intelligence*, volume 252, pages 267–294. Elsevier, 2017.

[Thimm et al., 2023] Matthias Thimm, Federico Cerutti, and Mauro Vallati. Fudge v3. 2.8. *ICCMA 2023*, page 25, 2023. https://helda.helsinki.fi/server/api/core/bitstreams/2e6d8485-fa34-48e5-9e24-eefe0cc1572c/content.

[Thimm, 2014] Matthias Thimm. Tweety: A comprehensive collection of java libraries for logical aspects of artificial intelligence and knowledge representation. In *Fourteenth International Conference on the Principles of Knowledge Representation and Reasoning*, pages 528–537, 2014.

[Thimm, 2023a] Matthias Thimm. Fargo-limited v1. 1.1. *ICCMA 2023*, page 21, 2023. https://helda.helsinki.fi/server/api/core/bitstreams/2e6d8485-fa34-48e5-9e24-eefe0cc1572c/content.

[Thimm, 2023b] Matthias Thimm. Harper++. 1.1. *ICCMA 2023*, page 27, 2023. https://helda.helsinki.fi/server/api/core/bitstreams/2e6d8485-fa34-48e5-9e24-eefe0cc1572c/content.

[Turpin et al., 2024] Miles Turpin, Julian Michael, Ethan Perez, and Samuel Bowman. Language models don't always say what they think: unfaithful explanations in chain-of-thought prompting. *Advances in Neural Information Processing Systems*, 36, 2024.

[Vassiliades et al., 2021] Alexandros Vassiliades, Nick Bassiliades, and Theodore Patkos. Argumentation and explainable artificial intelligence: a survey. *The Knowledge Engineering Review*, 36, 2021.

[Vaswani et al., 2017] Ashish Vaswani, Noam Shazeer, Niki Parmar, Jakob Uszkoreit, Llion Jones, Aidan N Gomez, Lukasz Kaiser, and Illia Polosukhin. Attention is all you need. *Advances in Neural Information Processing Systems*, 30, 2017.

[Vreeswijk, 1994] Gerard Vreeswijk. IACAS: An interactive argumentation system. User manual, version 1.0. Technical Report 94-03, University of Limburg, Department of Computer Science, 1994.

[Wallace, 2003] Richard Wallace. *The elements of AIML style.* Alice AI Foundation, 2003.

[Wallace, 2009] Richard S Wallace. The anatomy of A.L.I.C.E. In R. Epstein, G. Roberts, and G. Beber, editors, *Parsing the Turing Test*, pages 181–210. Springer, 2009.

[Walton and Krabbe, 1995] D. Walton and E. C. W. Krabbe. *Commitment in Dialogue: Basic Concepts of Interpersonal Reasoning.* State University of New York Press, 1995.

[Walton, 1996] Douglas Walton. *Argumentation schemes for presumptive reasoning.* Routledge, 1996.

[Wei et al., 2022] Jason Wei, Yi Tay, Rishi Bommasani, Colin Raffel, Barret Zoph, Sebastian Borgeaud, Dani Yogatama, Maarten Bosma, Denny Zhou, Donald Metzler, et al. Emergent abilities of large language models. *arXiv preprint arXiv:2206.07682*, 2022.

[Weizenbaum, 1966] Joseph Weizenbaum. Eliza — a computer program for the study of natural language communication between man and machine. *Communications of the ACM*, 9(1):36–45, 1966.

[Wells and Reed, 2012] Simon Wells and Chris A Reed. A domain specific language for describing diverse systems of dialogue. *Journal of Applied Logic*, pages 309–329, 2012.

[Xydis et al., 2020] Andreas Xydis, Christopher Hampson, Sanjay Modgil, and Elizabeth Black. Enthymemes in dialogues. In *Proceedings of the Conference on Computational Models of Argument*, pages 395–402, 2020.

[Xydis et al., 2021] Andreas Xydis, Christopher Hampson, Sanjay Modgil, and Elizabeth Black. Towards a sound and complete dialogue system for handling enthymemes. In *International Conference on Logic and Argumentation*, pages 437–456, 2021.

[Xydis et al., 2022] Andreas Xydis, Christopher Hampson, Sanjay Modgil, and Elizabeth Black. A sound and complete dialogue system for handling misunderstandings. In *4th International Workshop on Systems and Algorithms for Formal Argumentation*, pages 19–32, 2022.

[Yan et al., 2018] Chunli Yan, Helena Lindgren, and Juan Carlos Nieves. A dialogue-based approach for dealing with uncertain and conflicting information in medical diagnosis. *Journal of Autonomous Agents and Multi-Agent Systems*, 32:861–885, 2018.

[Zhao et al., 2023] Wayne Xin Zhao, Kun Zhou, Junyi Li, Tianyi Tang, Xiaolei

Wang, Yupeng Hou, Yingqian Min, Beichen Zhang, Junjie Zhang, Zican Dong, et al. A survey of large language models. https://arxiv.org/pdf/2303.18223, 2023.

Application domain	
health	[Balatsoukas et al., 2019; Castagna et al., 2023; Castagna et al., 2022; Chalaguine and Hunter, 2021; Essers et al., 2018; Kampik and Gabbay, 2020; Kökciyan et al., 2019; Nieves and Lindgren, 2015; Sassoon et al., 2019; Yan et al., 2018]
other	[Andrews et al., 2008; Bentahar and Labban, 2011; Bex et al., 2016; Booth et al., 2018; Caminada and Uebis, 2020; Chalaguine and Hunter, 2020; Hadoux et al., 2021; Koit and Õim, 2015; Lawrence et al., 2022; Panisson et al., 2022b; Rosenfeld and Kraus, 2016; Sklar and Azhar, 2015]
Dialogue type	
inquiry	[Bex et al., 2016; Kampik and Gabbay, 2020; Nieves and Lindgren, 2015; Odekerken et al., 2020; Panisson et al., 2022b; Sklar and Azhar, 2015; Yan et al., 2018]
deliberation	[Kampik and Gabbay, 2020; Nieves and Lindgren, 2015; Sassoon et al., 2019]
persuasion	[Andrews et al., 2008; Booth et al., 2018; Caminada and Uebis, 2020; Chalaguine and Hunter, 2020; Chalaguine and Hunter, 2021; Hadoux et al., 2021; Hauptmann et al., 2024; Lawrence et al., 2022; Rosenfeld and Kraus, 2016; Sassoon et al., 2019; Sklar and Azhar, 2015]
information-seeking	[Hauptmann et al., 2024; Panisson et al., 2022b; Sassoon et al., 2019; Sklar and Azhar, 2015]
negotiation	[Bentahar and Labban, 2011; Koit and Õim, 2015]
Evaluation	
formal proofs	[Booth et al., 2018; Caminada and Uebis, 2020; Fazzinga et al., 2022; Kampik and Gabbay, 2020; Lawrence et al., 2022; Odekerken et al., 2020; Panisson et al., 2022b; Yan et al., 2018]
human participants	[Andrews et al., 2008; Chalaguine and Hunter, 2020; Chalaguine and Hunter, 2021; Koit and Õim, 2015; Rosenfeld and Kraus, 2016]

Table 2: Ways to structure the literature on argumentation-based dia-

4 - APPLICATIONS OF ARGUMENTATION-BASED DIALOGUES

Software tools	
ASPARTIX	[Castagna *et al.*, 2023; Castagna *et al.*, 2022]
ArgTrust v1.0	[Sklar and Azhar, 2015]
JADE	[Yan *et al.*, 2018]
DGEP	[Bex *et al.*, 2014; Bex *et al.*, 2016; Odekerken *et al.*, 2020; Snaith *et al.*, 2020]
WFS	[Nieves and Lindgren, 2015]
JaCaMo/Dial4JaCa	[Engelmann *et al.*, 2023; Panisson *et al.*, 2022a]
Tweety	[Kampik and Gabbay, 2020]
$Jack^{TM}$	[Bentahar and Labban, 2011]
programming languages only	[Booth *et al.*, 2018; Caminada and Uebis, 2020; Koit and Õim, 2015]
User Interface (UI)	
chatbot	[Andrews *et al.*, 2008; Balatsoukas *et al.*, 2019; Castagna *et al.*, 2023; Castagna *et al.*, 2022; Chalaguine and Hunter, 2020; Chalaguine and Hunter, 2021; Guo *et al.*, 2022; Hadoux *et al.*, 2021; Hauptmann *et al.*, 2024; Kökciyan *et al.*, 2019; Rosenfeld and Kraus, 2016; Sassoon *et al.*, 2019; Sklar and Azhar, 2015]
(other) use of NLP	[Bex *et al.*, 2016; Engelmann *et al.*, 2023; Fazzinga *et al.*, 2022; Odekerken *et al.*, 2020]
simple UI	[Bex *et al.*, 2016; Booth *et al.*, 2018; Caminada and Uebis, 2020; Kampik and Gabbay, 2020; Koit and Õim, 2015; Lawrence *et al.*, 2022; Odekerken *et al.*, 2020; Yan *et al.*, 2018]
Data sets	
ACKTUS	[Yan *et al.*, 2018]
National Service Centre E-Crime Dutch Police	[Bex *et al.*, 2016; Odekerken *et al.*, 2020]
AIFdb	[Bex *et al.*, 2014; Lawrence *et al.*, 2022; Snaith *et al.*, 2020]
EDiC	[Koit and Õim, 2015]

Table 3: Ways to structure the literature on argumentation-based dialogues: Software tools, user interface, and data sets.

CHAPTER 5

ARGUMENTATIVE AGENT-BASED MODELS

Louise Dupuis de Tarlé
Université Paris Dauphine-PSL
louise.dupuis@dauphine.eu

Matteo Michelini
Eindhoven University of Technology/Ruhr University Bochum
matteo.michelini@edu.ruhr-uni-bochum.de

Dunja Šešelja
Ruhr University Bochum/Eindhoven University of Technology
dunja.seselja@rub.de

Christian Strasser
Ruhr University Bochum
christian.strasser@rub.de

1 Introduction

Communication influences many social phenomena, from belief polarization to scientific inquiry and other forms of collective problem solving. Many of these phenomena are highly complex, so complex that they are difficult to study with analytic methods 'from the armchair'. Instead, researchers turned to computational methods, in particular computer simulations. Agent-based models (ABMs) are well suited for this task.

We are grateful to two anonymous reviewers for valuable feedback on an earlier draft of this entry. Many thanks also to Felix Kopecky and to Carlo Proietti for helpful comments.

ABMs are computer simulations used to study how and why local interactions among individual agents produce emergent phenomena at the level of the given community (Šešelja (2023)), that is, how micro-level processes lead to emergent macro-level phenomena. These models have been applied in various disciplines, from ecology to medicine, from social science to philosophy.[1] ABMs are particularly useful whenever –due to computational limitations– the collective behavior of the community cannot be analytically and linearly inferred from the individual behavior.

Communication can take many forms, from nonverbal signaling to stating opinions. The focus of this entry is on argumentative communication. Indeed, agents frequently go beyond merely making claims; they often bolster their assertions with supporting reasons, aiming to enhance persuasiveness and transparency. In cases of disagreement, merely stating an opposing view is typically less convincing than providing reasons to elucidate one's differing opinion or doubts. Beyond persuasion, argumentation serves numerous other goals: it can inform collective decision-making, facilitate problem solving in the context of scientific inquiry, and foster better understanding among discussants. To study emergent macrophenomena in the context of agents exchanging arguments, scholars have increasingly utilized ABMs. We will call these simulation methods *argumentative ABMs*, in short *AABMs*. AABMs tackle questions such as: Why do some cases of argumentative exchange result in consensus, whereas others result in polarization and radicalization? Moreover, if a community of discussants reaches consensus, which factors increase the chance that the consensus is warranted or that it tracks the truth? And what makes such a community likely to maintain the discussion among all its members, rather than to split into fragmented camps?

Recent interest in AABMs follows the tradition of ABMs focusing on simpler forms of communication, such as the exchange of opinions (classical models include DeGroot (1974); Friedkin and Johnsen (1990);

---

[1] Depending on the target domain of the simulation, agents can represent trees (in ecology), humans or institutions (in social science and philosophy), etc. Despite many similarities (e.g. being based on interacting, at least partially independent agents), ABMs are to be distinguished from multi-agent systems (see Carrera and Iglesias (2015)), studied mainly in computer science, whose purpose often lies in problem solving. In contrast, ABMs are aimed at explaining complex emergent phenomena.

Hegselmann and Krause (2002); Deffuant *et al.* (2002)) or of data (e.g., bandit models Zollman (2007, 2010)). AABMs have since been employed in several domains: from computational social science to artificial intelligence, from cognitive science to social epistemology. Each of these communities has developed different modeling frameworks, leaving the overall study of AABMs largely fragmented, with little communication across the domains. As a result, both the proposed frameworks and the obtained results have often remained disconnected, despite substantial similarities and synergies between them.

The aim of this chapter is to provide a systematic overview of AABMs and to bridge some of these disciplinary gaps. As we show, the underlying methodology employed by different modeling frameworks is often similar, allowing fruitful comparisons of both modeling assumptions and the obtained results. We will highlight the various roles argumentation plays in these models to address a variety of research questions.

The chapter is structured as follows. We start by presenting the key elements of AABMs in terms of a blueprint, explaining how social reasoning processes are represented in such models (Section 2). In Section 3 we provide the background on the study of opinion dynamics, central to the majority of AABMs. In Sections 4–6 we survey concrete models, grouping them according to basic modeling assumptions and the research questions they aim to address. We conclude by highlighting open questions and prospects for future research in the field (Section 7).

2 Argumentative ABMs: a Blueprint

When designing an AABM many design choices have to be made: How to model argumentative exchange? How to design social networks? How to model agents' cognitive abilities, including their knowledge representation and commitments? In the following section, we will provide a schematic overview on such design choices. In a nutshell, AABMs are round-based computer simulations in which a number of agents interact with their social and possibly physical environment and in this way engage in learning processes. To simulate learning processes, AABMs incorporate cognitive models for their agents, allowing for knowledge representation. The general procedure for AABMs is represented in the schematic Algorithm 1.

Algorithm 1 Simple Schematic Algorithm for AABMs

1: **procedure** ARGUMENTATIVEABM
2:    **initialize** social (and physical) environment
3:    **initialize** cognitive model of agents
4:    $t \leftarrow 0$ ▷ *init round counter*
5:    **repeat**
6:        $t \leftarrow t + 1$ ▷ *increment round counter*
7:        **for all agent** ∈ Agents **do** ▷ each turn agents act and reason
8:            interact with social (and physical) environment   ▷ *communicate, obtain data, etc.*
9:            update cognitive model   ▷ *knowledge representation and reasoning*
10:       **end for**
11:   **until** termination criterion triggered
12: **end procedure**

What is common to all AABMs is that the interaction with the social environment contains argumentative exchanges. The point at which the procedure ends —the termination criterion— can be a point at which the community has reached a stable state of polarization or consensus, a fixed time point (such as a certain round of the simulation), a point at which the community has fully explored their environment, etc. Models in the literature differ in how they represent argumentative exchange, in their cognitive models, etc. So, let us take a closer look at the various building blocks of AABMs.

B1. Modeling Arguments in AABMs

Real-life arguments have a lot of structure. Argumentation theorists such as Toulmin (1956) consider arguments to consist of a claim (what is argued for) and premises (the assumptions an argument is based on) that support the claim in terms of an underlying warrant (such as a logical inference rule or a default). Moreover, discourse is often organized around one or several issues (e.g. 'Should we go to the cinema?'; 'Who of

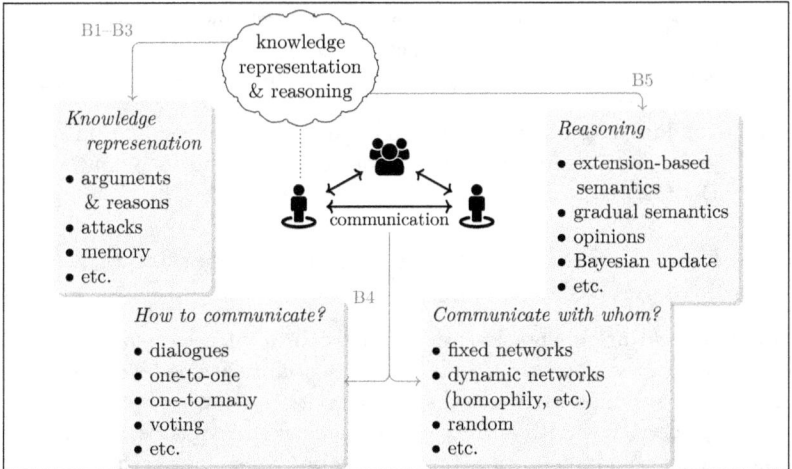

Figure 1: Blueprint for an argumentative ABM. In the green box are listed paradigmatic design decisions for a modeler for the building blocks B1–B5.

n candidates should we employ for the professorship?') and arguments are generated by discussants in favor or disfavor of the given issues.

For computational reasons and/or to keep the model streamlined, AABMs often abstract away from some or all of this structure. We can roughly distinguish the following approaches:

Abstract arguments. Some models follow Dung's abstract perspective by modeling arguments in terms of abstract units without internal structure (an approach pioneered by Gabbriellini and Torroni (2014)). Other models introduce (abstract) issues in whose favor or disfavor arguments are presented, but the latter are also often kept abstract (this approach is pioneered in Mäs and Flache (2013))

Formally structured arguments Other models incorporate some logical structure into arguments. For example, Kopecky (2022) models arguments as pairs ⟨Sup, Con⟩ where Sup is a set of propositional literals and Con is a literal. Introducing structure makes sense when aspects of argumentative exchange are analyzed in which structure matters, such as different types of argumentative attacks and different argumentative strategies in Kopecky (2022).

Natural language arguments Some models consider natural language arguments produced by machine learning techniques (e.g. Betz (2022)).

B2. Modeling Relations between and Properties of Arguments in AABMs

In real-world argumentation, we can observe a wide variety of potential relations between arguments. Two essential relations between arguments are argumentative attacks and support relations. Attacks can be further categorized into attacks of the conclusion (rebuttals), attacks of the warrant (undercuts) or attacks of the premises of an argument (undermining). In addition, arguments can often be compared in relation to their strength. AABMs typically simplify many of these aspects.

Schematically, we can model relations between arguments as follows. We equip a given set of arguments with relations and properties, resulting in *argumentation frameworks* (in short, AFs). The frameworks most frequently studied in formal argumentation are those introduced by Dung (1995), including only a binary attack relation. Subsequently, other relations such as support (bipolar argumentation Cayrol and Lagasquie-Schiex (2013)), preferences (Amgoud and Cayrol (1998)), explanatory relation (Šešelja and Straßer (2013)), and other relations were added. So, structures of the following type can be utilized in AABMs:

$$\mathsf{AF} = \langle \mathsf{Args}, R_1, \ldots, R_n, f_1, \ldots, f_m \rangle, \qquad (1)$$

where Args is a set of arguments (see B1); $R_1, \ldots, R_n$ are relations between arguments (typically binary) such as attack, support, preferences, etc., or relations of arguments to other entities such as values or information sources; and $f_1, \ldots, f_m$ are functions associating arguments with numerical values that indicate properties such as their strength. The information encoded in argumentation frameworks underlies the agents' decision as to what arguments to accept and/or reject (see B4 below). In concrete AABMs, this scheme may be simplified to Dung's classical frameworks of the type $\langle \mathsf{Args}, R \rangle$ where R is an attack relation. However, not all AABMs feature attack relations: we discuss ABMs without an attack relation in Section 4 and those with an attack relation in Section 5.

As mentioned above, many AABMs consider discursive situations in which agents discuss arguments in favor or against a given set of issues. In real-life debate, such issues can be read as "We should adopt a vegetarian diet" (Taillandier *et al.* (2021)) or "We should build a nuclear waste repository" (Stefanelli (2017)). We call AFs that model such situations *issue-based*. In these, issues are abstractly represented by a set Issues (where the "..." indicate possibly other structural entities, such as preferences; see Eq. (1)):

$$\mathsf{AF} = \langle \mathsf{Args}, \mathsf{Issues}, \ldots \rangle \qquad (2)$$

The agents then form opinions on these issues by means of arguments. One way to formally relate arguments to issues is by means of a function $w : \mathsf{Args} \times \mathsf{Issues} \to [-1, 1]$ that determines how much an argument supports (values closer to 1) or opposes (values closer to -1) a given issue. A very simple set-up is modeled in the AC model (Mäs and Flache (2013), see Section 4): only one issue is in place and each argument a in Args provides an argument strictly in favor (i.e., $w(a, \mathsf{i}) = 1$) or strictly disfavor of the given issue i (i.e., $w(a, \mathsf{i}) = -1$).

Example 1. *Consider the issue* i $=$ *"We should hold the workshop in a hybrid way." We have the following arguments:*

- a: *We should hold the workshop in a hybrid way because people will fly less and that is good for the environment.*

- b: *We should hold the workshop in a hybrid way since some scholars do not have a travel budget.*

- c: *We should not conduct the workshop in a hybrid way, since in-person communication allows for a more in-depth discussion.*

- d: *We should not conduct the workshop in a hybrid way, since video-conferencing tends to cause complications and delays.*

We may form an issue-based AF with $\mathsf{Args} = \{a, b, c, d\}$ *and let* $w(a, \mathsf{i}) = w(b, \mathsf{i}) = 1$ *and* $w(c, \mathsf{i}) = w(d, \mathsf{i}) = -1$.

Other approaches utilize, for instance, graded argumentation semantics for the purpose of determining argumentative support for issues. There, it is assumed that Issues is a subset of Args (e.g., specific arguments without outgoing attack relations, e.g. Dupuis de Tarlé et al. (2022); Butler et al. (2019)).

B3. Knowledge Representation: Cognitive Models of Agents

Agents in AABMs are modeled as epistemically autonomous and dynamic. This means that each agent has its own knowledge representational module which deals with incoming information and is used as the basis of agent communication. Depending on the application of an ABM, the knowledge of an agent may encompass many things, such as data points about its environment, knowledge about other agents (for instance, how trustworthy they are), etc. What is essential to AABMs is that agents utilize their knowledge for argumentative purposes. This means that each agent ag has its own *mental AF*, $\mathsf{AF}^t_{\mathsf{ag}}$, representing its unique mental model of the discursive situation at time point t.

The structural ingredients of mental AFs usually provide an incomplete representation of a central knowledge base given by an AF, which we call the *objective AF*. In particular, agents may not store all arguments available in the objective AF in their mental AF. The mental AFs are indexed by time points since ABMs are round-based: in each round, an agent may be exposed to new information (e.g., based on communication with other agents or interaction with its environment) and may therefore update its mental AF by incorporating new attacks and/or new relations. Each run of a simulation with n rounds and m agents therefore provides an epistemic trajectory described by the tuple

$$\left\langle \left\langle \mathsf{AF}^t_{\mathsf{ag}_i} : \langle \mathsf{Args}, \ldots \rangle \right\rangle_{i=1}^m \right\rangle_{t=1}^n$$

which is produced by an algorithm following the scheme in Algorithm 1. Given the terminology introduced in building blocks B1—B3, we can now refine it to Algorithm 2.

Algorithm 2 Schematic algorithm for argumentative ABMs (refined)

1: **procedure** ARGUMENTATIVEABM
2:    **initialize** global AF
3:    **initialize** mental AFs of agents: $\langle AF_{ag}^0 \rangle_{ag \in Agents}$
4:    $t \leftarrow 0$ ▷ *init round counter*
5:    **repeat**
6:       $t \leftarrow t + 1$ ▷ *increment round counter*
7:       **for all** $ag \in \{ag_1, \ldots, ag_n\}$ **do**
8:          ag interacts with physical environment ▷ *optional (e.g., obtain data)*
9:          ag argues   ▷ *argumentative communication (B4)*
10:         ag updates mental AF: $AF_{ag}^t \rightsquigarrow AF_{ag}^{t+1}$ ▷ *knowledge representation (B3)*
11:         ag forms argumentative commitments ▷ *reasoning (B5)*
12:       **end for**
13:    **until** termination criterion triggered
14: **end procedure**

B4. Argumentative Communication

When communicating, agents must make the following decisions: (a) with whom to communicate, (b) what and how to communicate, and (c) how to update the mental AF after the communication?

With whom to communicate? Some models implement one-to-one communication, where an agent exchanges one or more argument(s) with another agent (e.g. Mäs and Flache (2013); Banisch and Olbrich (2021)). Others allow for one-to-many communication, which is typical of deliberative spaces where a speaker addresses more than one discussant at the same time (e.g. Singer *et al.* (2019); Dupuis de Tarlé *et al.* (2024)). Some others combine these two and study their interactions (e.g. Butler *et al.* (2019, 2020)). These types of communication are based on specific protocols for the selection of communication partners. To this end, some models are based on fixed social networks (see Fig. 2 for an overview of frequently studied network types), which determine the social neighborhood of an agent (e.g. Liu *et al.* (2015); Fu and Zhang (2016); Baccini *et al.* (2023); Borg *et al.* (2019)). Others opt for proximity measures, which are used to obtain dynamic social networks, where proximity is often based on the similarity of argumentative commitments and / or opinions (e.g. Mäs and Flache (2013); Mäs *et al.* (2013); Mäs and Bischofberger (2015); Proietti and Chiarella (2023)). Finally, communication partners may be randomly produced (e.g., Banisch and Olbrich (2021); Banisch and Shamon (2023)).

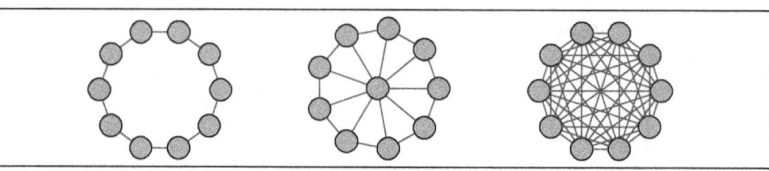

Figure 2: Three paradigmatic types of fixed social network structures with increasing density from left to right: The wheel, the star and the complete network.

What and how to communicate? In argumentative ABMs, agents typically share parts of their mental AF, in some they create new arguments. In this way, they introduce other agents to new arguments, or counter their stances in terms of attacks. In many ABMs, they communicate such parts of their AF as a one-shot communication, without engaging in a dialogue. Some ABMs, however, implement dialogue protocols in which agents engage in argumentative games (e.g., Dykstra *et al.* (2013); Gabbriellini and Torroni (2014)).

How to update? After the communication process, the receiving agent updates its mental AF to initialize a process of (argumentative) belief revision. For this, agents enhance their set of arguments and some relations such as support or attack.

Example 2 (Ex. 1 cont.). *Consider two agents, Woody and Zora with the initial mental AFs*

$$\mathsf{AF}^0_{\mathsf{Woody}} = \langle \{a,b,c\}, w \rangle \quad \textit{and} \quad \mathsf{AF}^0_{\mathsf{Zora}} = \langle \{b,c,d\}, w \rangle.$$

Woody contacts Zora to communicate the argument a to her. Subsequently, Zora updates her mental AF leading to

$$\mathsf{AF}^1_{\mathsf{Woody}} = \langle \{a,b,c\}, w \rangle \quad \text{and} \quad \mathsf{AF}^1_{\mathsf{Zora}} = \langle \{a,b,c,d\}, w \rangle.$$

Some ABMs include additional aspects of bounded rationality, such as *limited memory*, in which case agents may forget some parts of their mental AFs when updating in view of newly received information (e.g. Mäs and Flache (2013); Singer *et al.* (2019)). There are different ways of forgetting, such as random (forget a random piece of information in the mental AF), FIFO (first-in/first-out: forget the oldest piece of information), based on relevance or strength (where arguments are ranked according to relevance or strength: forget the argument ranked lowest), and so forth.

Example 3 (Ex. 1 cont.). *Suppose now our two agents have a (admittedly very limited) memory of 3 arguments. Assume now that the initialization mirrors the order in which Zora and Woody have received their arguments in the past:*

$$\mathsf{AF}^0_{\mathsf{Woody}} = \langle \langle a,b,c \rangle, w \rangle \quad \text{and} \quad \mathsf{AF}^0_{\mathsf{Zora}} = \langle \langle b,c,d \rangle, w \rangle.$$

In this case, Zora first learned d, then c and then b. When Woody contacts Zora to communicate the argument a to her, Zora will incorporate a into her mental AF, but –assuming FIFO– she forgets the 'oldest' argument d leading to:

$$\mathsf{AF}^1_{\mathsf{Woody}} = \langle \langle a,b,c \rangle, w \rangle \quad \text{and} \quad \mathsf{AF}^1_{\mathsf{Zora}} = \langle \langle a,b,c \rangle, w \rangle.$$

B5. Reasoning: Argumentative Commitments

Depending on how much structure the given AF provides, different rationales of forming argumentative commitments can be applied. There are two types of such commitments: (a) commitments to arguments and (b) commitments to claims. The latter commitments may represent beliefs, opinions, preferences for decisions, etc. The formation of argumentative commitments represents the reasoning process of agents. We highlight several ways in which it can be implemented in AABMs: (1) by means of extension-based semantics, (2) by means of gradual semantics and/or quantitative comparisons of the strength of arguments, and (3) we zoom into issue-based argumentation as an instance of (2).

Many ABMs consider AFs with attack relations to which (variants of) extension-based argumentation semantics (Dung (1995)) are applied to select arguments and form commitments of type (a) (see Fig. 3 for an illustration). Argumentation semantics provide criteria for selecting arguments from a given AF. Let us quickly recall the central definitions (we refer to Baroni *et al.* (2018) for an in-depth discussion). Two central criteria are, on the one hand, *conflict-freeness* expressing that a selected set of arguments (often called an *extension*) should not contain arguments that attack each other. Furthermore, it is often required that a selection of arguments $\mathcal{S}$ is *admissible*, that is, $\mathcal{S}$ is conflict-free, and for every argument that attacks an argument in $\mathcal{S}$, there should be an argument in $\mathcal{S}$ that attacks the attacker. This expresses the idea that the selected set of arguments is able to defend itself. *Preferred extensions* are maximally admissible (there is no strict superset that is admissible), *complete extensions* are admissible and contain every argument they defend, the *grounded extension* is the unique minimal complete set, and *stable extensions* are conflict-free and attack every argument that is not contained in them (one can show that they are always preferred).

Instead of modeling the acceptance of arguments as a discrete in-or-out(-or-undecided) status, other approaches track the persuasiveness or strength that agents attach to arguments in a gradual way. For example, gradual semantics (Amgoud and Ben-Naim (2018)) provides functions that return an acceptability score for each argument. Similarly, in issue-based AFs without attack relations, weight functions are often used to track the individual persuasiveness of arguments for agents in the simulation. This design decision is frequently motivated on the basis of Persuasive Argument Theory (Isenberg (1986)), according to which individuals form their opinions based on the persuasiveness of the pro and con arguments to which they are exposed.[2] In Bayesian frameworks, agents may have different beliefs about the weight of evidence and update differently (based on their Bayes ratio) when obtaining new

---

[2]The persuasiveness is itself based on an argument's validity (how acceptable is it to a given group of discussants) and novelty (to the group of discussants). In Masterton and Olsson (2014), the reader finds a contrastive discussion of Persuasive Argument Theory and Dung-style extensional argumentation semantics.

5 - Argumentative Agent-Based Models

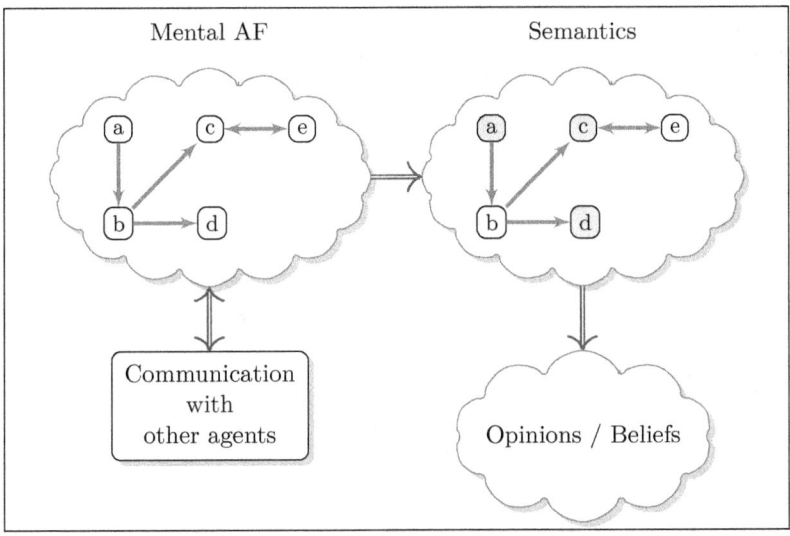

Figure 3: Diagram representing a reasoning process of agents based on deriving argumentative commitments from mental AFs by means of argumentation semantics. The selected set (in green on the right) is stable (therefore also preferred), but not grounded. The grounded extension is $\{a, d\}$.

evidence (see Section 6.3).

Based on the selection of arguments obtained in this way, agents form commitments to claims, that is they commit to the claims made by their argumentative commitments or the issues supported by them.

In the context of issue-based AFs, the commitment formation process is often more simplistic: an agent favors the issue which obtains the strongest support from the arguments in its mental AF. Typically, the strength of an issue is calculated in terms of a weighted sum. Where $\mathsf{AF} = \langle \mathsf{Args}, \mathsf{Issues}, w \rangle$ is the mental AF of a given agent, it considers the issue $i \in \mathsf{Issues}$ to have the strength:

$$\mathsf{Strength}(i) = \frac{\sum_{a \in \mathsf{Args}} w(a, i)}{|\mathsf{Args}|} \qquad (3)$$

building blocks	AC model (Section 4.1)	Taillandier et al. (2021) (Section 5.1.1)	Kopecky (2022) (Section 5.2)	argABM (Section 6.1)
B1	abstract reasons	abstract arguments	structued arguments	abstract arguments
B2	issue-based	values, issues	attacks, support	attacks between arguments in theory trees
B3	limited memory	limited memory	no memory, strategic	no limitations
B4	homophily	homophily	random	fixed networks
B5	quantitative semantics to determine opinion	extension-based semantics combined with quantitative strength calculation	claim-based commitments consistent with the available arguments	extension-based semantics

Table 1: Some implementations of the blueprint presented in Section 2.

The agent then selects/commits to an issue in $\text{argmax}_{\text{Issues}}(\text{Strength})$.

Example 4 (Ex. 3 cont.). *In round 0 of our run, Woody obtains the strength $2/3 = w(a) + w(b) + w(c)$ for the issue, while Zora has the strength $-2/3 = w(b) + w(c) + w(d)$. In round 1, after the communication took place, Zora evaluates the issue with $2/3 = w(a) + w(b) + w(c)$. In view of this, Zora now supports holding the conference hybrid, just like Woody.*

⋆ ⋆ ⋆

The above survey of building blocks highlights the variety of approaches available to develop AABMs. One's concrete modeling choices are usually determined by the underlying research question and the specific target phenomenon. In Table 1, we present some concrete implementations of our blueprint.

There are various platforms that help with ABM programming. We list three examples that offer tutorials and a smooth entry point for users with minimal programming background. The first is the GAMA platform (Taillandier *et al.* (2018)). It comes with its own modeling language (GAML) and offers extensions for argumentation. Another widely used platform is NetLogo (Tisue and Wilensky (2004)). Since it first appeared in 1999 it has been in steady development and, therefore, offers users a rich variety of resources, from introduction books and tutorials to online libraries of concrete models coded in NetLogo. Finally, some general-purpose languages offer specialized frameworks for agent-based modeling. We mention julia's `agents.jl` framework (Datseris *et al.* (2022)), known for its speed advantages over, for instance, NetLogo (Burbach *et al.* (2020)).

In what follows, we present specific ABMs, using the above terminology to classify them and highlight similarities and differences between them. To lay the groundwork for such an overview, we first look at the models of opinion dynamics, which provide the backdrop for AABMs.

3 Some Background on Opinion Dynamics

Opinion dynamics, a specific branch of agent-based modeling, explores how communication and other factors influence opinion formation (see Grabisch and Rusinowska (2020) for a survey). Social influence is a dominant force on people's opinions (Akers *et al.* (1995); Abelson (1964); Asch (1956)). In addition, social influence can have a significant and potentially disruptive impact on collectives. To study the complex interplay between individual (micro-level) and societal (macro-level) processes, scholars have turned to agent-based modeling (Latané (1996); Flache *et al.* (2017)). Paradigmatic phenomena studied in models of opinion dynamics are:

- *Consensus*: most individuals have (roughly) the same opinion.

- *Belief/Issue Polarization*: Individuals cluster in groups that have (roughly) the same opinions, and the opinions adopted by the groups oppose each other. A subphenomenon of particular interest is *bi-polarization*, a polarizing population around two opinion poles that increasingly radicalize.[3]

- *Social polarization/Fragmentation*: the population is divided into several subgroups that rarely interact with each other.

- *Dispersion*: individuals have a wide range of opinions.

In particular, the phenomenon of polarization has received significant research attention due to its relation to political radicalization, echo chambers in social networks, and other kinds of fragmentation.

Very commonly, ABMs of opinion dynamics represent opinions as numerical values, e.g., between 0 and 1 (DeGroot (1974); Friedkin and Johnsen (1990); Hegselmann and Krause (2002); Deffuant et al. (2002)).[4] Every round in the simulation agents meet, state their opinions, and then update their opinions on the basis of received information. In Fig. 4 we illustrate simulation runs in the well-known 'bounded-confidence model' (Hegselmann and Krause (2002); Deffuant et al. (2000)). These models often assume *homophily*: agents only communicate their opinions with those agents who are sufficiently close in opinion and then adopt an opinion by averaging between their communication partners and their own opinion. More precisely, homophily is modeled by a confidence interval. An agent's opinion is only influenced by agents that are within her confidence interval, that is, those whose opinion does not differ more than a given parameter ϵ.

Example 5 (Homophily-based communication). *Suppose we have three agents, Anne, Maeve and Wilma. They start in round 0 with opinions $opinion^0_{Anne} = .3$, $opinion^0_{Maeve} = .6$ and $opinion^0_{Wilma} = .9$. We work with a confidence interval of $\epsilon = .3$ and assume that agents average their opinion in each round with all agents in their neighborhood. This will lead to $opinion^1_{Anne} = .45$, $opinion^1_{Maeve} = .6$ and $opinion^1_{Wilma} =$*

---

[3]In the literature, the term 'polarization' describes a diverse array of phenomena. For instance, it can refer to a process where agents' beliefs, initially diverse, begin to coalesce and shift towards an extreme position collectively. In this entry, we adopt a definition that aligns with the predominant conceptual usage found in the reviewed literature.

[4]In some cases, opinions may be multi-dimensional, as in Lorenz (2003); Deffuant et al. (2000) that model them as a binary vector. See also Urbig and Malitz (2005) for a criticism of the one-dimensional approach to opinions.

5 - ARGUMENTATIVE AGENT-BASED MODELS

.75. Note that Anne will not listen to Wilma, since Wilma is outside of her confidence interval. Instead, she listens to Maeve and averages her own opinion with Maeve's, resulting in her new opinion of .45. The situation is different for Maeve, who listens to both Anne and Wilma, who averages their opinions with her own. As a result, her opinion remains unchanged. As can be easily seen, the more rounds pass by, the more the opinions will converge on Maeve's. In fact, convergence will speed up as soon as Anne and Wilma are within each other's confidence interval.

In Fig. 4 we have $\epsilon = 0.15, \epsilon = 0.2$ and $\epsilon = 0.4$. The first two settings lead to polarization, and the latter to consensus. The bottom row illustrates the model that includes a type of negative social influence, such that agents update away from other agents with whom their disagreement is large. This leads to an additional effect: the polarized groups *radicalize*.

Now, where do AABMs come in when studying opinion dynamics? In many real-life situations, such as political debates, social influence is also based on the exchange of reasons: it is argumentative. Argumentative exchange is not modeled in simple opinion-exchange models such as the bounded-confidence model. Thus, a key question is: Can we observe different dynamics once agents do not (only) exchange opinions, but also exchange arguments? For instance, can in this case the radicalization of polarized groups (Fig. 4, bottom row) be explained without recourse to negative influence – distancing from dissimilar others? For this purpose, the integration of argumentation into ABMs proved to be a promising modeling option.

In the following, we will discuss two types of argumentative ABMs of opinion dynamics. The first group, presented in Section 4, models debates in which participants exchange reasons pro and con on a given set of issues. These models are mainly based on issue-based argumentation frameworks of the form Eq. (2) and do not feature an attack relation. The second group of ABMs, presented in Section 5, is based on AFs that also feature attack relations (and that therefore are often in the tradition of Dung-style abstract argumentation).

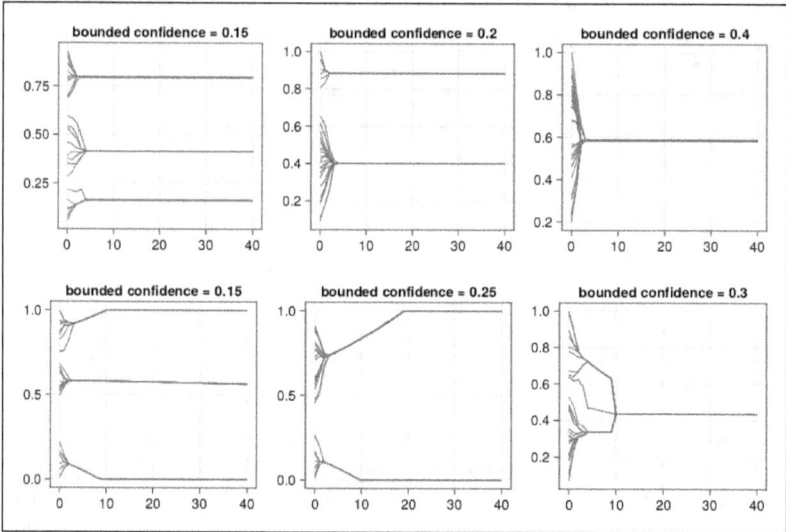

Figure 4: Typical runs in the bounded confidence model. Each line represents the dynamics of the opinion of a single agent in a population of 20, where the x-axis represents the number of turns and the y-axis the numerical opinion spectrum (from 0 to 1). Top row: without negative influence. Bottom row: with negative influence. The three columns present three confidence intervals, from .15 to .4. On the right we have runs with consensus, while the other columns present runs with polarization. As can be seen, negative influence leads to radicalization in polarized societies, while positive influence on its own does not account for it. The graphs are obtained by the implementation of the Hegselmann & Krause model in julia (the code can be obtained at https://juliadynamics.github.io/Agents.jl/v4.0/examples/hk/).

4 Enriching ABMs of Opinion Dynamics with Reasons

In the following, we describe how ABMs of opinion dynamics have been enhanced with argumentative devices such as the addition of explicitly modeled reasons or arguments (in this section) and of attack relations and sometimes dialogical deliberation (in the next section). We proceed as follows. In Section 4.1 we present the influential argument communication model of Mäs and Flache. Then, in Section 4.2 we show how the cognitive models underlying agent architectures have been made more realistic by including bias (Section 4.2.1) and aspects of bounded rationality, such as forgetting (Section 4.2.2). Finally, in 4.3 we consider applications to real-world debates (Section 4.3.1), natural language and machine learning (Section 4.3.2). We provide a summary in Section 4.4.

4.1 The Argument Communication Model of Opinion Dynamics

4.1.1 Mäs and Flache's Argumentative Model of Opinion Dynamics

Mäs and Flache (2013) were the first to explicitly model argument exchange in an ABM of opinion dynamics in order to explain the emergence of *bi-polarization*, a problem that had previously been considered unresolved (Bonacich and Lu (2012)). Mäs and Flache consider bi-polarization as a type of polarization in which the polarized groups radicalize (Esteban and Ray (1994), see the bottom row of Fig. 4). Three candidate mechanisms were available as potential explanantia: positive social influence (Abelson (1964)), homophily (Hegselmann and Krause (2002)), and negative social influence (Salzarulo (2006); Flache and Mäs (2008a)). However, neither positive social influence nor homophily could explain the distancing of opinions (that is, the radicalization of opinions in the group). In addition, empirical support for negative social influence has been mixed (Krizan and Baron (2007)).

To fill this gap, Mäs and Flache (2013) explain bi-polarization in situations in which like-minded agents exchange reasons (rather than exchanging mere opinions). In particular, they argue that if agents learn new reasons mainly from other agents with similar attitudes, this type of argumentative exchange can lead to the formation of two polarized groups whose opinions radicalize over time. This new explanation, which they call 'Argument Communication Theory of Bi-Polarization' (ACTB), is inspired by insights from Persuasive Argument Theory (Isenberg (1986); Burnstein and Vinokur (1977)) and related research on polarization and homophily (Sunstein (1999); McPherson et al. (2001)).

Mäs and Flache (2013)) test the explanatory power of ACTB by an AABM (which we call 'AC model' for Argument Communication model) and compare their simulated results with those obtained by an empirical experiment. The AC model is based on an issue-based AF $\langle \mathsf{Args}, \{\mathsf{i}\}, w \rangle$ with a single issue i (see Section 2). Each argument $a \in \mathsf{Args}$ either fully supports the issue (formally, $w(a, \mathsf{i}) = 1$), or its opposite (formally, $w(a, \mathsf{i}) = -1$). Agents have limited memory (governed by FIFO) and learn new arguments from interacting with agents that hold a similar opinion (i.e., the communication is homophily-based). The opinions of the agents are numerically represented as values in $[-1, 1]$ (where -1 is strictly disfavoring i while 1 favors i) and are calculated as in Eq. (3). In this way, the results of the AC model can be easily compared with those of classical models of opinion dynamics (Hegselmann and Krause (2002); Deffuant et al. (2002); Douven and Kelp (2011)).

While opinions remain numerically represented in the AC-model and many of the models discussed below, a key difference to non-argumentative models of opinion dynamics is that

(a) in the argumentative models opinions are functions of and explainable by reasons, and

(b) agents engage in reasoning processes (in terms of aggregating and weighing reasons) when forming opinions instead of merely conforming to their social environment.

Example 6. *We consider a situation with three agents: Ann, Bob, and Lucy, who are generally in favor of a given issue* i. *Each agent has a* memory for storing 5 reasons. Initially their mental AFs are:

$$\mathsf{AF}^0_{\mathsf{Ann}} = \langle\langle p_1, p_2, p_3, c_1, p_4\rangle, \mathsf{i}\rangle,$$
$$\mathsf{AF}^0_{\mathsf{Bob}} = \langle\langle p_1, p_5, p_6, c_2, c_3\rangle, \mathsf{i}\rangle \text{ and}$$
$$\mathsf{AF}^0_{\mathsf{Lucy}} = \langle\langle p_1, p_6, p_7, c_4, p_2\rangle, \mathsf{i}\rangle,$$

where p_i for $i = 1,\ldots,7$ are reasons pro i, while c_i for $i = 1,\ldots,4$ are reasons con. In view of Eq. (3), Ann's opinion on i is $\frac{1+1+1+1-1}{5} = 3/5$. An analogous calculation applies to Lucy, while Bob's opinion is given by $\frac{1+1+1-1-1}{5} = 1/5$.

Fig. 5 illustrates a sequence of argumentative exchanges between our three agents. When communicating, the sender picks a random argument from its mental AF and the receiver adds it as top element to their memory queue, while forgetting the oldest element. After 5 exchanges, we can see that Bob's and Lucy's mental AFs only contain pro-arguments leading to an opinion of $5/5 = 1$, fully in favor of i. As a consequence, our group of like-minded agents radicalized, without any negative influence.

As suggested by Example 6, if agents have a finite memory and they mainly interact with like-minded peers, they tend to radicalize. Indeed, the presence of homophily guarantees that agents are more likely to hear arguments that reinforce and radicalize their views than opposing ones. Consequently, the population tends to divide in two groups whose attitudes towards the issue at stake tend to shift to extreme positions. A typical simulation run is illustrated in Fig. 6.

The empirical experiment in Mäs and Flache (2013) further supports this theory: when participants exchange arguments, bi-polarization emerges, while it does not if participants only communicate opinions. In sum, the exchange of arguments can explain bi-polarization in the presence of homophily and limited memory, without recursing to negative influence (as is needed in traditional opinion dynamics models).

4.1.2 Subsequent Studies with the AC Model

The ACTB theory describes the evolution of opinions as a product of the exchange of arguments. As shown above and as will be further elaborated in the following, the AC model offers new explanations of specific

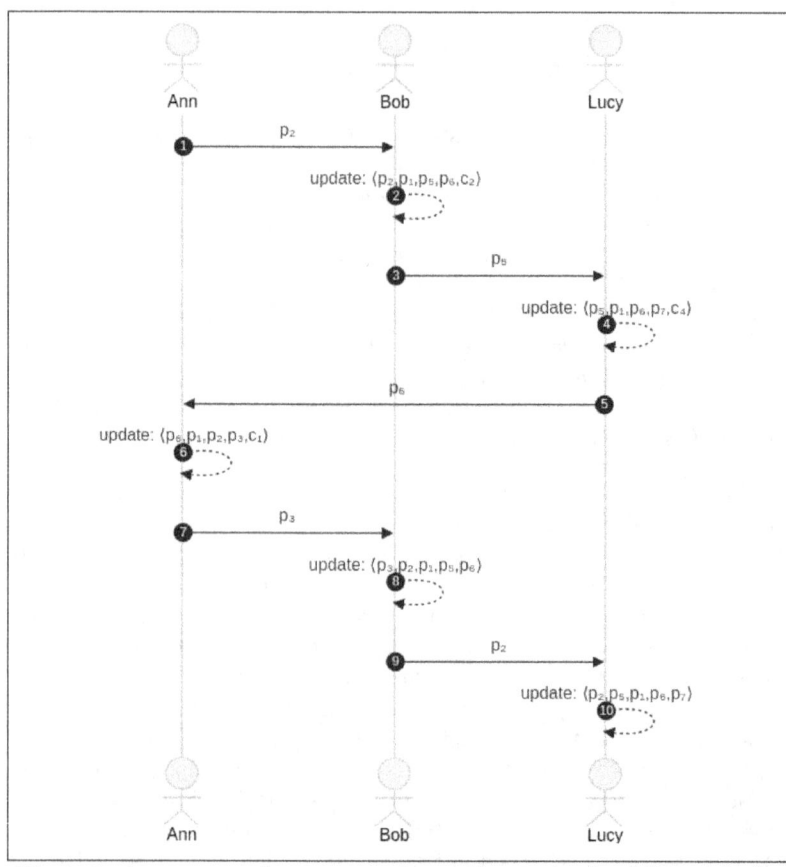

Figure 5: A sequence of argumentative exchanges in the AC-model. See Example 6. Odd numbers refer to the exchange of arguments, while even numbers refer to the updating of commitments.

5 - ARGUMENTATIVE AGENT-BASED MODELS

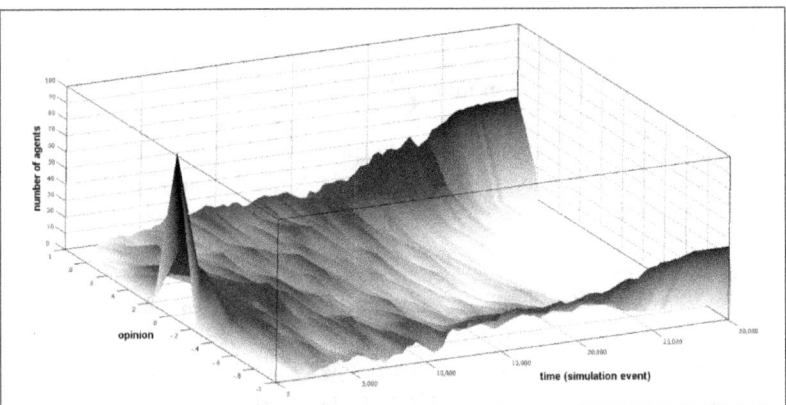

Figure 6: Bi-polarization in (Mäs and Flache, 2013, p. 7). The model is run with 100 agents, 30 pro and 30 con arguments, a memory of 10 and a relatively high homophily factor. Although at time point 0 all agents are of opinion 0 they radicalize as time goes by. At the end of the run (at 30.000 time steps) they are fully polarized at the extreme ends of the opinion spectrum.

types of opinion dynamics compared to traditional models (Friedkin and Johnsen (1990); Weisbuch *et al.* (2002); Hegselmann and Krause (2002); Flache and Mäs (2008b)). What does the ACTB theory imply for the broader understanding of social dynamics? The following works answer this question by applying the AC model to various issues.

Demographic faultlines. According to Lau and Murnighan (1998), social cohesion in a diverse group is threatened by the presence of demographic faultlines. The latter are hypothetical dividing lines that partition groups into homogeneous subgroups based on demographic attributes (such as age, sex, ethinicity). For example, a strong faultline is given in a team consisting of two highly educated African-American women and two Caucasian men with a low level of education. It is a strong faultline, since all three demographic dimensions (race, sex and education) split the team along the same line. Lau and Murnighan argue that diverse groups only polarize in the presence of strong demographic faultlines.

To test this theory, Mäs et al. model a team's opinion dynamics as the result of argument communication and homophily. In this case, homophily leads agents with similar opinions and demographic attributes to interact more often. As such, the present work of Mäs et al. complements the perspective provided by Flache and Mäs (2008b), which employed a negative influence model to explain the detrimental effect of faultlines on consensus formation.

In line with Lau and Murnighan (1998), Mäs *et al.* (2013) show that subgroup polarization can be caused by demographic faultlines only when faultlines are extraordinarily strong, and no 'criss-crossing actors' are present. Criss-crossing actors are agents that connect different demographic groups (Colson (1953)), and their presence ensures that even if polarization occurs in the short run, it lapses in the long run. Moreover, Mäs et al. refine the theory of Lau and Murnighan. First, they find that, although strong faultlines are often detrimental, they may, at times, bring about faster consensus. Secondly, they show that–in contrast with Lau and Murnighan's hypothesis–demographic faultlines may bring about polarization even in the absence of an initial correlation between demographic attributes and opinions. Finally, explicit modeling of the exchange of arguments provides a novel set of predictions. For example, Mäs et al.'s model suggests that the timing of contacts between different members of the group can either accelerate the process of polarization or slow it down to the point of stopping it.

Homophily. Mäs and Bischofberger (2015) discuss the effect of homophily on group bi-polarization. They demonstrate that strong homophily has a different impact on group dynamics depending on the mechanism that drives bi-polarization. In light of this, they warn sociologists against condemning homophily as the cause of polarization in online social networks. In particular, they compare the results of increasing the strength of homophily in the AC model and in the negative influence model (Salzarulo (2006); Flache and Mäs (2008a)). The negative influence model takes interacting agents to intensify opinion differences when opinions already diverge, and explains bi-polarization in virtue of this behavior (see also Fig. 4, bottom row).

Mäs and Bischofberger (2015) show that for high values of homophily the AC model predicts an increase in the likelihood of bi-polarization, while the negative influence model predicts the opposite. In the negative influence model, if like-minded agents only interact among themselves they have no reason to settle on more extreme opinions. Instead, in the AC model this brings opposite groups to accumulate more and more one-sided arguments, and, eventually to diverge. It is worth noting that Mäs and Bischofberger slightly modify the AC model by allowing agents who interact to modify the strength of each argument.[5]

Feliciani et al. (2021) offer a comparison of different ways in which homophily is represented in bi-polarization models of argument exchange. Mäs and Flache (2013); Mäs et al. (2013) and Mäs and Bischofberger (2015) implement homophily as biased partner selection following previous work by Carley (1991). This is different from the classical representation of homophily given by bounded confidence models (Hegselmann and Krause (2002); Weisbuch et al. (2002)), where the probability of interaction between any two agents depends only on those agents. The former type of homophily is adopted by models with implicit argument exchange (La Rocca et al. (2014); Feliciani et al. (2017)) and by Banisch and Olbrich (2021) and Banisch and Shamon (2023, 2022), who use a model obtained by slightly modifying the AC model (see below Section 4.2.1). Notably, Feliciani et al. found that when homophily affects the likelihood of interaction rather than its effectiveness, the agents' opinions become more extreme. This is because the likelihood of interaction has a greater impact on the evolution of opinions.

---

[5]The ACTB theory has also inspired models of argument communication that represent argument exchange implicitly (La Rocca et al. (2014); Pinasco et al. (2017); Feliciani et al. (2017)), and have been used to address different social phenomena. These models avoid representing arguments as distinct entities, but still describe agents' interactions as if the agents exchanged arguments. Since there is an abundance of AABMs of opinion dynamics, we do not consider these models in our entry. When comparing models with explicit to those with implicit representation of arguments, Feliciani et al. (2021) observe that in the former, bi-polarization tends to emerge slower and only in these models moderate consensus can stabilize.

4.2 Towards More Realistic Cognitive Models

4.2.1 Biased Processing and Bi-Polarization

Biased processing occurs when agents ignore arguments that undermine their current positions. Is biased processing of arguments sufficient to bring about bi-polarization? This question is addressed by different research teams (Liu *et al.* (2015); Fu and Zhang (2016); Proietti and Chiarella (2023); Banisch and Shamon (2023)) with different models of biased processing. Although different implementations produce different results, the models reviewed here come to the conclusion that bi-polarization *can* be explained by the combination of biased processing and argumentative exchange.

Ignoring new arguments. In the variant of the AC model presented in Liu *et al.* (2015), agents keep track of the number of pro- and con-arguments for a given issue. Biased agents ignore new arguments with fixed probability. The authors analytically show that for any social network, the opinions of all agents converge to extremes with probability one, and that agents will achieve an extreme opinion exponentially fast. Further simulations indicate that the degree of bi-polarization, measured as size parity in the extremes (Bramson *et al.* (2017)), decreases as either the network connectivity or the size of agents' memory increases.

Fu and Zhang (2016) replace the fixed social networks of Liu *et al.* (2015) by homophily-based dynamic networks. They report that homophily plays a greater role in the emergence of bi-polarization than biased processing. They also find that biased processing speeds up convergence when the bias is not too strong. Finally, in accordance with Liu *et al.* (2015), they observe that as network connectivity increases, the likelihood of bi-polarization decreases.

The GAAC Model. Banisch and Olbrich (2021) introduce a model we will dub 'global awareness argumentation communication' - GAAC model for short. It is a version of the AC model in which agents are always aware of all arguments but commit only to a selection of them. Hence, agents form opinions analogous to the AC model (recall Example 6), but instead of relying on the arguments that they are aware of, they rely only on those which they believe to be valid. Each agent starts out believing a subset of facts and committing to the corresponding arguments, and updates her beliefs when she interacts with another agent. When two agents interact, each agent convinces the other to change her mind with respect to one specific argument. For example, agent i may give agent j an argument for fact p_3, resulting in agent j adjusting her opinion accordingly.

Given this basic setup of the GAAC model, the authors extend it to study how communication affects agents' opinions on multiple issues. For this, the GAAC model is enriched with the presence of two non-contrastive issues i_1 and i_2, and take each argument to influence the opinion of an agent with respect to both issues.[6] So, for example, an argument for p_3 may favor i_1 while disfavoring i_2, or it may favor both of the issues. Hence, if agent j commits to argument p_3, this will affect her opinion on both i_1 and i_2.

When the two issues are only weakly correlated, an argument favoring i_1 is equally likely to favor or disfavor i_2; if two issues are strongly correlated, any argument favoring i_1 counts also as an argument favoring i_2 (positive correlation), or any argument disfavoring i_1 is also an argument against i_2 (negative correlation). For example, one may consider veganism and vegetarianism as strongly positively correlated issues. These correlations capture the presence of cognitive evaluative structures, inspired by the theory of expectancy value (Fishbein (1963)). In addition, the model assumes that agents interact only with those who hold similar opinions, where the authors offer four different ways to compute similarity.

The results of the model suggest that when two issues are strongly correlated, communication, in the long run, leads to a highly polarized state. On the contrary, when the two issues are only weakly correlated, communication leads to consensus. Because the two issues are only weakly correlated, an agent who believes in many arguments in favour of one specific issue is not completely unlikely to interact with an agent that mostly believes in arguments against the issue, because they may

---

[6]The version of the GAAC model produced by Banisch and Olbrich (2021) resembles the work by Urbig and Malitz (2005), who also study the effect of multidimensional opinions on a specific attitude.

have similar opinion concerning the second issue. This prevents the community from reaching a completely polarized state. Besides this finding, the introduction of a cognitive structure into the model provides a step toward representing real-world debates, as empirically informed studies of polarization indicate that alignment of attitudes across several issues reveals a polarized community (Baldassarri and Goldberg (2014)).

Enhancing GAAC with Biased Processing. Banisch and Shamon (2023) enhance GAAC with biased processing by allowing agents to ignore information they receive from others in the community. An argument shared with a biased agent, which is incoherent with the biased agent's opinion, is more likely to be ignored than an argument that coheres with their view. Following Shamon *et al.* (2019), this mechanism is empirically calibrated. Similarly to Liu *et al.* (2015), Banisch and Shamon (2023) also find that the combination of biased processing and argument exchange is sufficient to generate a process of bi-polarization. However, contrary to Liu *et al.* (2015) and Mäs and Flache (2013), they observe that any bi-polarized state is always transient. This means that while the community may split into two groups whose opinions have become increasingly different, these two groups will eventually converge and reach a consensus in the long run. This suggests that biased processing generates a bi-polarization process, but is never sufficiently strong to keep the two groups permanently separated. In accordance with Fu and Zhang (2016), Banisch and Shamon report that weakly biased processing speeds up the process of consensus, while more biased communities traverse long periods of bi-polarization.

4.2.2 Forgetting Mechanisms

Mäs and Flache (2013) assume that agents do not base their opinion on all the arguments they have encountered, but only on those they memorize. Therefore, when agents learn a new argument, they also forget one. Singer *et al.* (2019) discuss more refined mechanisms: weight-minded, coherence-minded and simple-minded forgetting.[7] Weight-minded agents forget the weakest argument upon learning a new one.

---

[7]Similar ideas are to be found in Mäs and Bischofberger (2015).

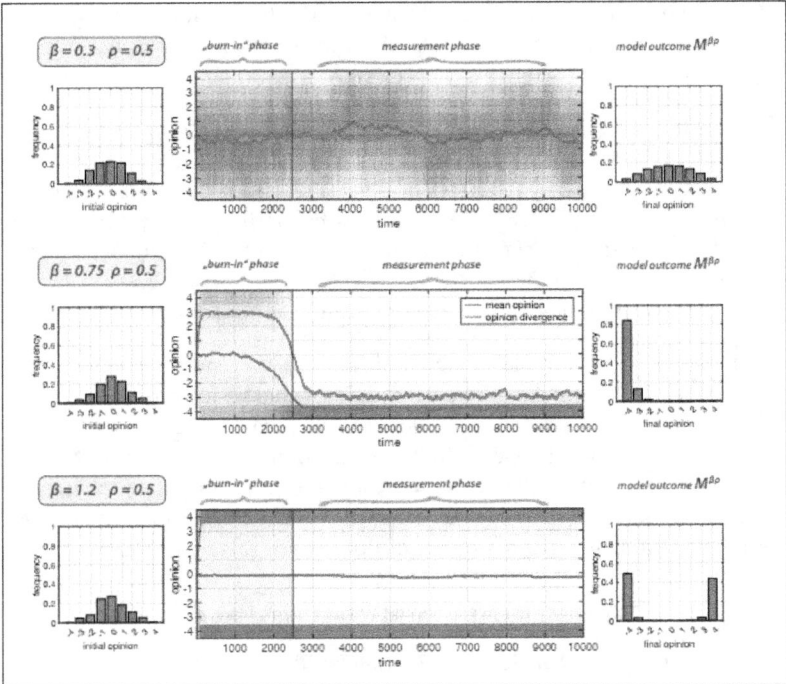

Figure 7: Bi-polarization in (Banisch and Shamon, 2022, p. 11). The model NACM is run with 500 agents, 8 arguments and for 10000 steps. The parameter ρ controls the probability of receiving a random communication from outside: in each turn an agent receives a random communication from outside with probability ρ, and a communicate with another agent with probability $1 - \rho$. The parameter β determines the degree of biased processing. For a low value of β (weak bias), agents' opinions remain normally distributed. For a medium value (moderate bias), agents all converge on the same opinion. For a high value of β (strong bias), bi-polarization emerges stably.

Coherence-minded agents forget argument with the least support for their current positions. Finally, simple-minded agents forget arguments at random. The authors explore under which conditions such mechanisms are sufficient to bring about bi-polarization in the absence of homophily.

Example 7. *Suppose that our model contains three reasons pro r_1^+, r_2^+, r_3^+ and three reasons con r_1^-, r_2^-, r_3^-, where $\mathsf{strength}(r_i^*) = i$ for $* \in \{+, -\}$ and $i \in \{1, 2, 3\}$. Moreover, suppose that our agent has a memory of size 4 containing $r_1^+, r_2^+, r_3^+, r_2^-$ and is communicated another reason r_3^-. If our agent is weight-minded, she will remove r_1^+ from her memory (since its strength is only 1) and then incorporate r_3^-, while if she is coherence-minded, she will remove r_2^- and then incorporate r_3^-.*

In Singer et al.'s model, weight-minded agents tend to converge onto consensus. In contrast, coherence-minded agents undergo a process of bi-polarization, and reach a stable state of polarization. Coherence is an epistemic value defended in the philosophical literature (Thagard and Verbeurgt (1998)), which is why Singer et al. argue that coherence-minded agents are rational. In view of this, polarization can be the result of argumentative reasoning of boundedly rational agents.

4.3 Real World Debates and Natural Language

4.3.1 Applying Models to Real World Debates

In this section, we will take a look at various developments that apply AABMs of opinion dynamics to real world debates.

External Sources of Information. Banisch and Shamon (2022) employ the GAAC model to validate its results against empirical data on opinions about renewable energy collected by Shamon *et al.* (2019). They slightly modify the version provided by Banisch and Olbrich (2021), by introducing in the debate the presence of an external information source (such as a newspaper). Agents can thus receive arguments not only from other agents but also from an external source.

The model also introduces bias in the way agents evaluate new arguments. The bias mechanism is the same as that used by Banisch and Shamon (2023) and relies on the notion of coherence (see above). The data in Shamon *et al.* (2019) concerns both the degree of biased processing that each individual exhibits, as well as the general opinion distribution for each of several different subtopics regarding renewable energy.

Banisch and Shamon (2022) show that their model can account for all empirically observed opinion distributions: a normal distribution of opinions, a uniform one, a one-sided one and a bipolar one. The likelihood of each of them depends on how likely communication from an external source is and on the degree to which agents are biased.

Online Product Reviews. Gabbriellini and Santini (2017) study online product reviews by an AABM. In many cases, V-shaped distributions of ratings can be observed in online product reviews, with the majority weight distributed among low and high ratings. The underlying dynamics are not well understood and simple opinion dynamics models seem to be ill-suited as explanatory devices due to the argumentative nature of reviews. After all, reviews often are argumentative in that they contain reasons that support an overall rating. Gabbriellini and Santini provide both an empirical and a simulation study. In their AABM, each agent represents a customer and a potential reviewer. Every turn, agents have the opportunity to post a review with a certain probability (calibrated relative to the empirical study). Reviews are lists of pro and con arguments. Gabbriellini and Santini study different reviewing strategies the agents follow, differing in how much the rating is influenced by the ratio of pro- and con-arguments it contains. The variant that allows for some, but a minority of, negative arguments in positive reviews is shown to be the most predictive of the outcome observed in the empirical study.

Social Judgment Theory and Argumentation. Stefanelli and Seidl (2014) (and in the follow-up paper Stefanelli and Seidl (2017)) present a model studying opinion dynamics by incorporating insights from Social Judgment Theory (Sherif and Hovland (1961)).[8] The latter

---

[8]Social Judgment Theory has been previously considered in the context of models of opinion dynamics, e.g. in Jager and Amblard (2005), although without an emphasis

explains attitudes and their relationship to self-identity. People do not just hold opinions, they more broadly categorize stances into those they find acceptable, rejectable, and those for which they are non-committal.

Stefanelli and Seidl build on an issue-based AF in which arguments weigh in on the given issue, and come in various types depending on which criteria they are based: e.g., some concern risks (opposed to the issue), some concern benefits (in favor of the issue), etc. Agents are receptive to these types of argument to different degrees. For instance, some may be more swayed by arguments based on risks, others may be more responsive to arguments based on benefits. The strength of an argument in favor of an issue is a function of how responsive an agent is to the given type of argument and how much the argument supports the issue (the valence of the argument).

The way in which agents influence each other's opinions is modeled in terms of social judgment theory. Agents come with latitudes of acceptance, rejection, and noncommitment for each argument type. The larger the latitude of rejection, the more ego-involved the individual is on the given topic, and the harder it is for them to be swayed by the opinion of others. When agents receive arguments from other agents, depending on their latitudes, they update the importance they ascribe to the given argument type. As a consequence, their overall opinions (calculated as the weighted average of the strength of their arguments) on the issues potentially shift.

Although a systematic study of how different latitude profiles generally influence the dynamics of opinion is still missing, the follow-up paper (Stefanelli and Seidl (2017)) offers a validation of an empirically calibrated version of their model with respect to a debate on the construction of a deep geological repository for nuclear waste in Switzerland. Through a questionnaire, they identify the acceptance, rejection, and noncommitment latitudes for 10 presented arguments, as well as the general strength of each argument for or against the given issue of building the repository.

---

on argumentative exchanges. In Section 5, we present models by Butler *et al.* (2019, 2020) that enhance Jager and Amblard (2005) with argumentative exchanges and voting.

4.3.2 Natural language

In Betz (2022), Betz presents an AABM in which both the production of arguments and the formation of opinions are powered by a large language model (in short, LLM; in this specific case GPT-2, Radford *et al.* (2019)).[9] Every round, agents exchange arguments for and against a given issue in the form of posts generated by the LLM. Agents have limited memory to keep track of previous posts. The memory is initialized with posts from a Kialo (`kialo.com`) debate on the legalization of drugs. When updating with new posts, others are forgotten. When creating new posts, the LLM uses a concatenation of the posts in the memory of an agent as a prompt, combined with the following (Betz, 2022, p. 14):

> *I more or less agree with what my peers are saying here. Regarding the legalization of drugs, I'd just add the following thought: ...*

The opinion of an agent ag at time t on the given issue is calculated on the basis of the conditional probability of producing a claim pro or con the given issue:

$$\mathsf{opinion}(\mathsf{memory}_{\mathsf{ag}}^t) = \frac{1 - P(\mathsf{con} \mid \mathsf{memory}_{\mathsf{ag}}^t)}{(1 - P(\mathsf{pro} \mid \mathsf{memory}_{\mathsf{ag}}^t)) + (1 - P(\mathsf{con} \mid \mathsf{memory}_{\mathsf{ag}}^t))}$$

Betz simulates societies of agents that each round update on all new posts (baseline scenario), or whose communication is based on homophily (agents that only listen to agents in a bounded confidence interval of their opinion), or who come with confirmation bias (they update only on posts that are close to their opinions). In addition, agents are equipped with various argumentative strategies that influence the way they post. There are passive agents that share posts from their memory but that do not create new posts, as well as (to varying degrees) creative agents (who create and share more or less surprising posts). As

---

[9] Research on ABMs using LLM is still in its infancy; the only other model we know of is Hamilton (2023). Since it does not feature an explicit representation of arguments, we omitted it in our discussion.

for passive societies, the model gives rise to similar dynamics as known from other opinion dynamics models, such as Mäs and Flache (2013); Hegselmann and Krause (2002) and Singer *et al.* (2019) (e.g., no polarization in the baseline scenario, strongest polarization for homophily, etc.). Interestingly, in societies with creative agents, clear differences can be observed (e.g., stronger clustering already in the baseline scenario but less clustering for homophily).

4.4 Summing Up

Mäs and Flache (2013) introduced argumentation into the study of opinion dynamics with the aim of explaining bi-polarization. Inspired by the success of the AC model, more models have been developed to study different aspects of bi-polarization in the context of argumentative communication, enriching the vast literature on polarization based on non-argumentative models of opinion dynamics (Table 2).

The models reviewed here aim to find minimal explanations for a several phenomena underlying opinion dynamics. Model development is often characterized by attempts to either subtract unnecessary parts from already existing models or to add new elements to explain new phenomena and to investigate new candidate mechanisms. This can be seen, for example, in the work of Banisch and Shamon (2023, 2022), who study the role of biased processing is studied. For this, the authors streamlined the argument learning process, proposed by Mäs and Flache (2013), and introduced a more elaborate model of biased processing, allowing them to show that biased processing is sufficient to produce bi-polarization.

While the search for minimal explanations of opinion dynamics motivated models with a simple representation of agents' mental AFs, argumentative exchange allows for additional modeling assumptions. In what follows, we will take a look at models that introduce more structure in terms of attack relations between arguments.

model framework	paper	communication	outside learning	memory forgetting	biased processing
AC model	Mäs and Flache (2013)	pairwise: homophily	no	limited recency	no
	Mäs et al. (2013)	pairwise: homophily + demographic	no	limited recency	no
	Mäs and Bischofberger (2015)	pairwise: homophily	no	limited recency	no
	Liu et al. (2015)	pairwise: network	no	limited recency	fixed probability
	Fu and Zhang (2016)	pairwise: homophily network	no	limited recency	fixed probability
GAAC model	Banisch and Olbrich (2021)	pairwise: homophily	no	no	no
	Banisch and Shamon (2023)	pairwise: random	no	no	coherence-based non-linear
	Banisch and Shamon (2022)	pairwise: random	yes	no	coherence-based non-linear
Other	Singer et al. (2019)	public announcement	yes	limited recency / strength / coherence	no
	Stefanelli and Seidl (2014, 2017)	pairwise: random	no	no	no

Table 2: Overview of the different features of the models in the tradition of the AC model reviewed in Section 4. Although all these models are rather similar in the structure of mental AFs, and in the evaluation of arguments, they employ a wide range of assumptions concerning the way agents interact. Notably, three large families of models can be obtained.

5 Enriching ABMs with Argumentative Attacks

In the previous section, we discussed several models that enhance opinion dynamic models through argumentative exchange. In these models, argumentative exchange comes down to presenting reasons pro and con a given issue. However, real-life argumentative exchange is richer. For example, we also use arguments to counter other arguments.

Example 8 (Example 1 cont.). *The discussion is on whether to hold the conference in a hybrid way. The argument b gave a reason based on the fact that some scholars do not have a traveling budget. Someone could reply:*

e : *All scholars attending are funded by an academic network (so b does not state a valid con-reason).*

Argument e is not a direct pro or con argument for the given issue, but it attacks a con-argument. (In this role, it becomes an indirect pro argument for the defended issue.) This kind of dynamics is not captured by issue-based AFs without attack relations.

In what follows, we present various ABMs of opinion dynamics based on AFs that feature argumentative attacks (often based on Dung's work on abstract argumentation, Dung (1995)).

5.1 Enhancing the AC model with Argumentative Attacks

In Section 4, we saw that enhancing the communication of agents with the exchange of reasons in the AC model leads to novel types of opinion dynamics, such as radicalization of group opinions on the basis of homophily and absence of negative influence. In this section, we stay close to the spirit of the AC model, but we go one step further by incorporating not just reasons that speak in favor/disfavor of some given issues but also explicitly represented argumentative attacks. An overview of ABMs of opinion dynamics can be found in Fig. 8.

5 - ARGUMENTATIVE AGENT-BASED MODELS

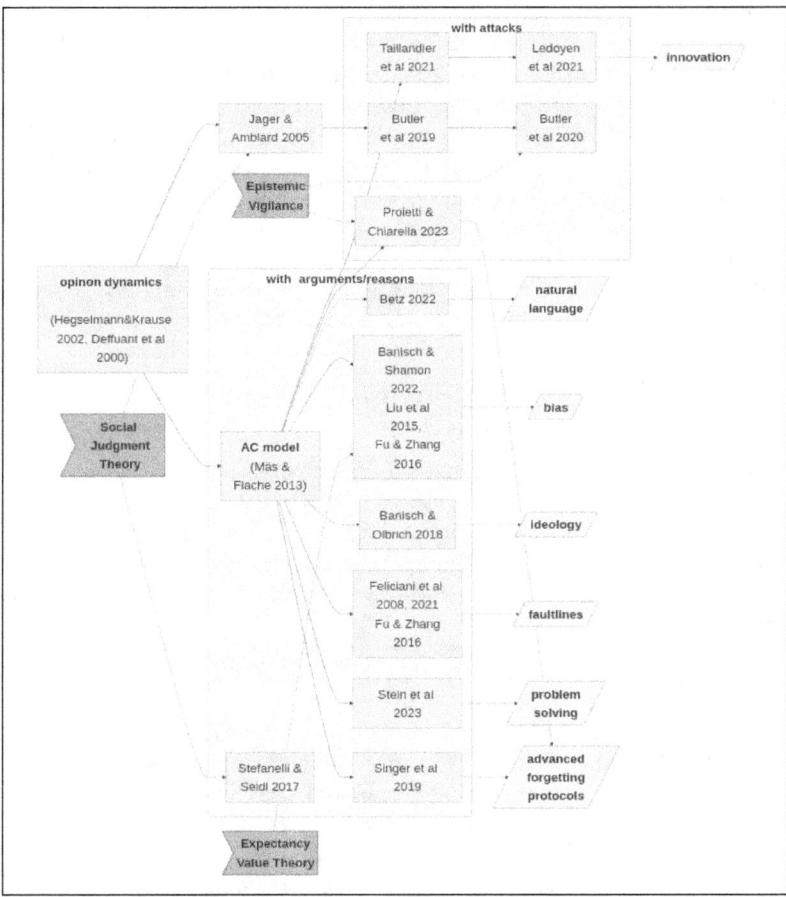

Figure 8: Evolutionary tree of AABMs in the tradition of opinion dynamics models such as Hegselmann and Krause (2002). In orange, social and psychological theories that directly influenced the design of the models. In yellow, characteristic design principles and applications.

5.1.1 Studying polarization with argumentative attacks

The model of Taillandier *et al.* (2021) investigates polarization under homophily for issue-based debates, similar to Mäs and Flache (2013). The authors generalize the AC model, inter alia, by introducing argumentative attacks and by giving more structure to their arguments. In addition to issues (in the paper called "options"), the model also includes values (in the paper called "criteria"). Arguments rely more or less on a given value (measured by a function **value**). For instance, an argument in favor of veganism may rely on an 'environmentalist' value, while an argument in favor of meat consumption may rely on a 'nutritional' value. Moreover, each value has more or less importance for a given agent (measured by a function **importance**), so that the subjective strength of an argument for an agent **ag** is calculated by:

$$\text{strength}(\text{ag}, a) = \sum_{v \in \text{Values}} \text{importance}(v, \text{ag}) \times \text{value}(v, a)$$

Example 9. *Suppose the presence of the values* health *and* environmental protection. *The model allows some agents to be more convinced of arguments that appeal to health (e.g. "Becoming vegetarian reduces the risk of cardiovascular disease.") than of arguments appealing to environmental values (e.g. "The meat industry is a key factor in methane emission."), and vice versa.*

An agent considers an attack successful if the attacking argument is at least as strong as the one attacked. Each agent chooses a preferred extension from their mental AFs for which the (normalized) sum of the strength of arguments favoring a given issue is maximal among all preferred extensions. The resulting normalized sum is a value between -1 and 1 representing the opinion of the agents, in analogy to Mäs and Flache (2013). As a consequence, homophily is implemented with the same underlying mechanisms as in the latter model. Also, similar to the AC model, agents have limited memory, exchange arguments when communicating, and update using FIFO.

The authors instantiate the model with data to simulate the diffusion of the vegetarian diet. They use arguments extracted during a previous project (Salliou *et al.* (2019)), and initialize the agents to match the distribution found in an opinion survey. They show that the system evolves towards an increase in the acceptance of vegetarian diet and extreme opinions. They also study the dynamics which result from the introduction of new arguments in the population.

5.1.2 Studying Innovation Diffusion with ABMs

Ledoyen *et al.* (2021) investigate innovation diffusion, that is, the process by which a *novel* practice or idea spreads through a population. Their formal model is similar to the one developed in Taillandier *et al.* (2021), except that the agents now argue about the adoption of innovations. In addition, *decision variables* are added to the model to supplement the opinion of agents and to allow for a fine-grained model of how agents take the decision of whether to adopt an innovation. Moreover, a notion of trust in sources of information is included. The opinion of the agents on the merits of an innovation is computed as an average of their *personal benefit*, and a perceived *social value*, which corresponds to an evaluation of the opinion of other agents. The model is applied to a real-life scenario: the adoption of new water meters by farmers in the South-West of France. They show that argument exchange increases the adoption of innovations over time.

5.1.3 Biased and vigilant agents

Proietti and Chiarella (2023) adapt the AC model by Mäs and Flache (2013) to simulate more realistic argumentative interactions. They present three modifications. First, they introduce attack relations to vary the strength of the arguments: agents measure the strength of an argument using gradual semantics. Unlike Taillandier *et al.* (2021), these attack relations are shared between all mental AFs. Proietti and Chiarella study scenarios with a single issue relative to which pro- and con-arguments are exchanged among the agents. They observe that, in scenarios in which pro-arguments are stronger than con-arguments,[10] the average opinion of the agents also supports the issue.

---

[10]This assumption is meant to represent a case in which the issue at stake is factually correct, and consequently, pro-argument are more convincing. The setting resembles models of truth tracking communities in which agents try to identify the true theory among a set of competing candidates (see Section 6).

In addition, Proietti and Chiarella extend the AC model to account for two forms of biased processing: preferential and vigilant updates. In the preferential update, an agent discards an argument if it does not support the agent's present opinion. In the *vigilant update*, which is inspired by the notion of *epistemic vigilance* from cognitive science (Sperber et al. (2010)), an agent includes an argument contrary to their current opinion only by also adding a number of other arguments in favor of their current opinion. This captures the core idea of epistemic vigilance: agents are usually suspicious of new arguments that would undermine their view, and may react by looking further for arguments that support their opinions. Their results indicate that only when such updates are employed consistently whenever a new argument is encountered, biased processing generates stable bi-polarization in the absence of homophily. This is in line with the results obtained by Liu *et al.* (2015) and Banisch and Shamon (2023). However, when agents use biased processing inconsistently, any polarized state is transient and eventually all the communities converge to a consensus.

Finally, similarly to Singer *et al.* (2019), Proietti and Chiarella (2023) also consider the possibility for an agent to forget the weakest argument instead of the oldest one. This introduces a form of biased forgetting. In the same way as Singer *et al.* (2019), Proietti and Chiarella find that this forgetting protocol is not sufficient to bring about bi-polarization in the absence of homophily. This is the case even under the assumption that agents consistently only communicate their strongest arguments.

5.2 Polarization and strategic argumentation

In Kopecky (2022), Kopecky investigates polarization effects in the absence of both negative influence and homophily. The underlying question is whether different types of argumentative strategies give rise to polarization or have the potential to depolarize a society of agents. Argumentative strategies concern the production of arguments in a dialogical situation, such as attacking the position of another agent or fortifying one's own position.

Kopecky's model is based on structured arguments of the form

$$\langle \ell_1, \ldots, \ell_n; \ell \rangle$$

	premises	conclusion
allocentric		
convert	opponent: accept	proponent: accept
undercut	opponent: accept	opponent: reject + proponent: accept
egocentric		
fortify	proponent: accept	proponent: accept
attack	proponent: accept	opponent: reject

Table 3: Argumentation strategies in Kopecky (2022). The proponent of a new argument makes sure that the premises and the conclusion have the indicated acceptance status for the proponent and/or the opponent of a dialogue.

where $\ell_1, \ldots, \ell_n, \ell$ are propositional literals, $\ell_1, \ldots, \ell_n$ are the premises of the argument, and ℓ its conclusion. We are dealing with argumentation frameworks of the form $\langle \mathsf{Arg}, \mathsf{Att}, \mathsf{Sup} \rangle$, with an attack and a support relation (also known as bipolar AFs, Cayrol and Lagasquie-Schiex (2013)).[11] An argument $\langle \ell_1, \ldots, \ell_n; \ell \rangle$ attacks $\langle \ell'_1, \ldots, \ell'_m; \ell' \rangle$ in case $\ell = \neg \ell'_i$ or $\neg \ell = \ell_i$ for some $i \in \{1, \ldots, m\}$, while it supports the latter in case $\ell = \ell_i$ for some $i \in \{1, \ldots, m\}$.[12] All arguments are considered sound by the agents but not necessarily valid. This means that if an agent learns about an argument whose premises she believes but whose conclusion she disbelieves, she has to revise her beliefs. This revision has to result in a belief state as close as possible to her previous belief state that is consistent with all known arguments.[13]

Each round, agents meet and exchange newly generated arguments according to an underlying argumentative strategy. The newly generated arguments must be consistent with the arguments already exchanged. Therefore, throughout the run, the space of possible belief states contin-

---

[11] The framework is based on the theory of dialectical structures in Betz (2009).

[12] Similar attack forms can be found in various frameworks of logical/structured argumentation, such as assumption-based argumentation (Dung *et al.* (2009)) or sequent-based argumentation (Arieli and Straßer (2015)). For an overview, see Arieli *et al.* (2021).

[13] In more technical terms, the belief state of each agent is given by a Boolean assignment to the logical atoms. The distance between two belief states is given by the Hamming distance.

ually shrinks. Agents utilize one of two classes of argumentative strategies: *allocentric* and *egocentric* strategies. In allocentric strategies, the premises of the generated arguments are acceptable to the opponent, while in egocentric strategies they are acceptable to the opponent. Table 3 details the four resulting strategies in more detail.

Example 10. *Suppose that we have four atoms $p_1, \ldots, p_4$ and the argument pool already contains $\langle \neg p_1; p_3 \rangle, \langle \neg p_1; \neg p_3 \rangle, \langle p_1; p_4 \rangle$. Note that this argument pool renders beliefs in $\neg p_1$ and in $\neg p_4$ impossible, since beliefs have to cohere with all arguments in the given pool. We consider two agents with the following belief states: Agent 1 believing $\{p_1, \neg p_2, p_3, p_4\}$ and Agent 2 believing $\{p_1, p_2, \neg p_3, p_4\}$. Suppose now that Agent 1 proposes an argument to Agent 2. The argument $\langle p_3; \neg p_2 \rangle$ would be an attack, as its premise is accepted by the proponent, while its conclusion is rejected by the opponent. It also fortifies since the conclusion is accepted by the proponent. The argument $\langle p_2; p_4 \rangle$ could be proposed according to the convert strategy since its premise is accepted by the opponent and its conclusion is accepted by the proponent. The argument $\langle p_2; p_3 \rangle$ follows the undercut strategy: its premise is accepted by the opponent, although the conclusion is not (while being accepted by the proponent).*[14]

By utilizing three polarization measures from Bramson *et al.* (2017), Kopecky shows that allocentric strategies have depolarizing effects, while egocentric strategies tend to polarize (with the aggressive stragies, undercut and attack, each having a stronger effect than their non-aggressive counterparts, convert and fortify).

5.3 Modelling dialogical argumentation in ABMs

In this section, we present two models in which dialogue protocols are utilized.

Strong and weak ties in social networks. Gabbriellini and Torroni (2014) present a general framework for AABMs in which agents engage in dialogical exchanges. Agents are connected in a social network, and,

---

[14]Undercutting resembles Caminada's *hang-yourself-arguments* (also called Socratic-style arguments, see Caminada (2008)).

5 - ARGUMENTATIVE AGENT-BASED MODELS

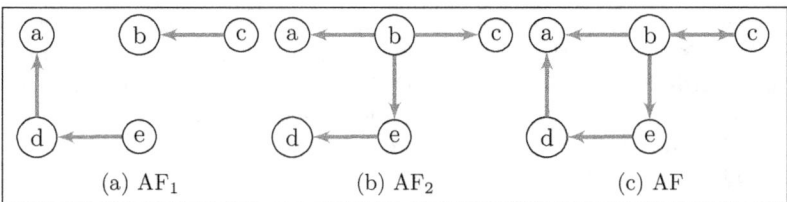

Figure 9: (Left) and (Center): the mental AFs of two agents in Gabbriellini (2014). (Right): The objective AF.

each round, an agent is assigned to a random neighboring agent. She then attacks an argument of her opponent according to her mental AF. Her opponent can either accept the attack and incorporate it in her framework in case the proponent is trusted, or otherwise reply with a counterattack.

The framework allows for different types of social networks and different underlying argumentation semantics. In the experiments of Gabbriellini and Torroni (2014), complete semantics is utilized. Each agent commits to a complete extension of its mental AF. The authors study social networks, which are characterized by strong and weak ties: locally dense subnets (so-called "caves") are sparsely connected by weak ties. The experiments show that polarization decreases with the presence of weak ties. Another question investigated in the paper is whether specific mental AFs are more successful in being adapted than others. For this, Gabbriellini and Torroni start off with a bipartition of their populations: one part is equipped with AF_1 (with unique complete extension $\{c, e, a\}$, see Fig. 9) and the other with AF_2 (with unique complete extension $\{b, d\}$). As it turns out, in the presence of weak ties, the population equipped with AF_1 is more successful in converting the other part of the population, although $\{b, d\}$ remains a complete extension of the objective, that is, the fully merged framework AF (see Fig. 9). As the authors indicate, more research is necessary to understand the underlying dynamics.

Building Reputation. In Dykstra *et al.* (2013), the authors present an ABM called DIAL to study the dynamics of network and opinion with a focus on social mechanisms to attribute reputation to agents. Two such mechanisms are at the core of their study: on the one hand, reputation is rewarded by holding opinions that conform with the ones that are dominant in the social environment of an agent and, on the other hand, reputation is rewarded from winning argumentative disputes with other agents.

Instead of opting for a one-dimensional opinion space (e.g. $[0, 1]$) to represent the agents' opinions, the authors opt for what they call 'topic spaces'. These are modeled as two-dimensional Euclidean spaces in which agents can move. The proximity of two agents in this space symbolizes their distance. Proportional to an agent's reputation, its communication has a certain social reach (the authors call it the 'loudness'). As agents take turns, they travel, initiate dialogues, or participate in dialogues. When initiating a dialogue, an agent announces a statement S that it supports. Furthermore, the agent announces a bet in terms of stating how many reputation points p it would pay if it were to lose the dialogue and how many reputation points r it wants to be rewarded with if it wins. Agents who are close enough in the topic space to hear the message (given its loudness, it only has a limited reach) can chime in the dialogue attacking S. The winner of the debate is decided by a majority vote of the neighboring agents of the debaters. After this, the bets are resolved by a transfer of reputation points from the loser to the winner. In addition to the game-theoretic role described here of the stakes p and r, they also serve an epistemic function. That is, the p/r-ratio expresses the degree of belief an agent has in the statement S.

Simulation runs in DIAL typically stabilize in two types of configurations. On the one hand, the authors observe societies with not much variance in the distribution of reputation and which are segregated into diverse but internally homogeneous clusters. On the other hand, societies may form 'authoritarian configurations' in which a minority of agents ('leaders') have most of the reputation and there is not much clustering. The main result is that the way in which the reputation is mainly gained by agents is essential in producing one of the two stable configurations. If winning dialogues rewards most of the reputation, mostly authoritarian societies emerge, while if reputation is mostly earned by conformity with the environment, segregated configurations are more likely to emerge.[15]

5.4 Social Judgment Theory

In Section 4.1.2 we discussed a model (Stefanelli and Seidl (2014)) that incorporates aspects of Social Judgment Theory for the study of opinion dynamics. We now discuss models on this topic that enhance argumentative communication with an explicit representation of attacks.

Modelling Social Judgment Theory with Abstract Argumentation. In Butler *et al.*; Butler *et al.* the opinions of the agents are contained in the interval $[-1, 1]$. Following Jager and Amblard (2005), each agent is equipped with a pair $(U, T) \in [0, 2]^2$ (with $T > U$), representing the latitudes of acceptance (U) and rejection (T). When Bob communicates his opinion o_{Bob} on the given issue to Alice, Alice calculates the distance to her opinion o_{Alice}. Bob's opinion falls within Alice's latitude of acceptance if it is within distance U_{Alice}, while it falls in her latitude of rejection if it is further away than T_{Alice}. In the former case Alice will move closer to Bob's opinion, in the latter case she will distance herself.

Example 11. *To make this more concrete, suppose that Alice has latitude of acceptance .5 and latitude of rejection 1. Her opinion is currently .2. Now Bob communicates his opinion of .3 to her. He falls within her latitude of acceptance, so she moves her opinion closer to Bob. Alternatively, if Bob's opinion would have been $-.9$, she would move her opinion further away from Bob's, since his opinion would be situated in her latitude of rejection.*

In addition to dyadic one-on-one interactions, Butler et al.'s model also features argumentative group deliberations in which agents have the opportunity to exchange arguments, including the opportunity to attack arguments forwarded by other agents. The model builds on an issued-based AF equipped with grounded semantics. Regularly, agents meet in groups to jointly deliberate on some given issue. The group size,

---

[15]The authors extend their model with fuzzy opinion formation and insights from Social Judgment Theory in a follow-up paper (Dykstra *et al.* (2015)).

composition, minimal and maximal length, as well as the regularity of the meetings, are decided by a central authority in the model (i.e., they are fixed parameters). The central authority issues the initial argument for the deliberation and agents take turns forwarding arguments defending or attacking arguments from previous rounds. In their turn, agents always advance arguments from their mental AF that are among the strongest in support of their respective opinions and that have not yet been included in the debate. This way, the participating agents build a collective AF in the process of deliberating.[16] The debate terminates if it reaches a maximal length (decided by the authority) or if the initial argument is part of or attacked by the grounded extension of the collective AF. Subsequently, agents update their opinions depending on whether the initial argument is in or attacked by the grounded extension, again respecting attraction and repulsion tendencies in accordance with social judgment theory.

One takeaway from the experiments conducted in Butler *et al.* (2019) is that opting for an intermediate frequency and size of deliberation phases increases consensus formation and the overall level of correctness of the populations opinion (as measured on the basis of the objective underlying AF).

Argumentation Strategies and Voting. The deliberation phase in Butler *et al.* (2019) is optionally enhanced with two features. First, similar to Kopecky (2022) the paper studies argumentation strategies. For this, the authors introduce two profiles, naive and strategic agents (called 'focused'). The latter are, for instance, able to anticipate attacks by other agents and therefore prospectively forward arguments in defense. Societies of focused agents, as it turns out, have higher variance in their opinions and tend to be more extreme, displaying less shifts of opinion. The second enhancement concerns the social deliberation phase. A majority voting mechanism is added, in which agents can vote whether they agree with the acceptance states of the initial arguments (that was determined by the grounded semantics). It can be observed

---

[16]This is similar to the dialogues in Gabbriellini and Torroni (2014) (see Section 5.3), but follows a different protocol and, like Dykstra *et al.* (2013), it also allows for groups larger than 2.

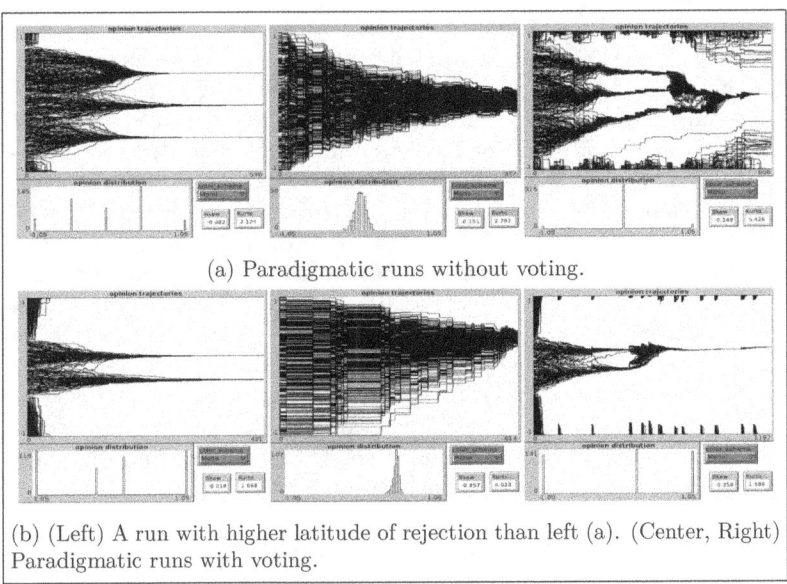

(a) Paradigmatic runs without voting.

(b) (Left) A run with higher latitude of rejection than left (a). (Center, Right) Paradigmatic runs with voting.

Figure 10: Some paradigmatic runs in Butler *et al.* (2019). The x-axis is the number of turns while the y-axis presents the numeric opinion spectrum between -1 and 1. Each line presents the opinion of an agent. Runs with only dyadic opinion exchanges resemble the dynamics of the simpler model in Jager and Amblard (2005). A higher latitude of rejection leads to more extremism. (Center) Only social deliberation. Runs typically converge, but voting can shift the convergence from the centrist position. (Right) Hybrid. Voting leads to accentuated extremist poles and less variance.

that two-third majority voting leads to fewer dynamics in opinion, while a simple majority rule tends to cause polarization. Fig. 10 illustrates some paradigmatic simulation runs.

Epistemic Vigilance. In Butler *et al.* (2020), the authors introduce another psychological property of agents: epistemic vigilance (Sperber *et al.* (2010)). The basic idea is that agents are cautious when updating based on information from other agents to reduce the risk of being misinformed. They take into account markers such as the competence, the interests and the honesty of their communication partners, instead of blindly trusting them.[17] In Butler *et al.* (2019), in dyadic exchanges, agents communicate opinions, whereas in Butler *et al.* (2020), one-to-one communication is based on argumentative one-shot exchanges. The proponent presents an argument that supports her opinion. The vigilant recipient then checks whether the received argument is sufficiently close to the opinion of the sender, if not, she is distrustful and does not update on it. Similarly, also in social debates, the notion of vigilance is implemented. In contrast to Butler *et al.* (2019), agents in Butler *et al.* (2020) have limited memory and voting is not considered. Interestingly, vigilant societies show a heightened tendency to extreme polarization and to do so quickly, an effect that can be amplified by social deliberation (without voting). Moreover, epistemic vigilance has a stronger tendency to produce stable subclustering.

5.5 Modelling Opinions with Gradual Semantics

The work of Dupuis de Tarlé *et al.* (2022) is also an opinion dynamic model where the opinions of the agents are derived from their mental AFs. The initial motivation of this model is the study of the behavior of a particular semantics: they apply a gradual semantics to a special argument, the "issue" of the debate. Gradual semantics (Amgoud *et al.* (2022)), unlike the extension-based semantics used in Taillandier *et al.* (2021), Gabbriellini and Torroni (2014) and Butler *et al.* (2019), are functions that return a *score* in terms of a real number for every argument. These scores represent the agents' opinions of each argument, similar to opinion dynamics models. The agents argue strategically in order to influence the general debate in the direction of their own opinion. The agents are not part of a social network, but they 'publish'

---

[17]One may wonder whether this does not lead to overall distrustful societies of vigilant agents. As argued in Sperber *et al.* (2010), it is exactly the fact that we assume other agents to be vigilant that we also direct a basic level of trust towards them. They compare the situation with a walk down a crowded street and the risk of collision. Since we assume agents to be careful, we trust them and do not hesitate to walk among them.

their arguments on a public gameboard, where all of the other agents have a certain probability of perceiving the published arguments. The communication method allows Dupuis de Tarlé *et al.* (2022) to model confirmation bias: the probability that an agent will learn a new argument is higher if this argument favors their own opinion of the issue.

The authors measure the convergence of the opinions of the agents. They find that, although formally communication between agents does not warrant convergence, in practice, the higher the probability of learning new arguments, the more the opinion of agents converges. They also present a supplemented version of their protocol where agents vote on the arguments they endorse, and they offer robustness analyzes.

6 Problem-solving and Truth-tracking

In this section, we turn to ABMs that study the impact of argumentative dynamics on collective problem-solving and decision-making. More specifically, these models examine the effect of social networks, homophily, and other factors on the performance of groups who face a certain problem, such as choosing the best of the available scientific theories, or making an optimal decision after collective deliberation.

6.1 Modeling network effects with abstract argumentation

ArgABM The model ArgABM(Borg *et al.* (2017a, 2018, 2019, 2017b)) is based on the idea that scientific interaction among proponents of rival scientific theories is largely argumentative. Scientists exchange not only bare data, but rather arguments —for instance in the form of scientific papers— for and against their theories. In light of these arguments, they determine the best theories. Borg and colleagues investigate how the following factors impact the efficiency of the community in reaching consensus about which is the best theory:

(a) social networks among scientists (see Fig. 2);

(b) types of information sharing (such as reliable and deceptive);

(c) ways of conducting inquiry (such as cautious or incautious decision-making about changing one's current theory); and

(d) ways of evaluating scientific theories (such as measuring the degree to which a theory is 'defensible', or the degree to which it is 'anomalous').

To model scientific inquiry, ArgABM employs a specific kind of epistemic landscape —an 'argumentative landscape'— representing rival theories (or research programs) in a given domain. Each theory is represented by a rooted tree, with nodes as abstract arguments and edges as a 'discovery relation', representing paths agents take to move from one argument to another. In addition to the discovery relation, arguments in one theory can attack arguments in another theory (see Figure 11). The representation of arguments is based on Dung's abstract argumentation, enriched with the discovery relation. At the start of a simulation, scientists are randomly positioned at the roots of the theory trees. Thereon, they (i) gradually explore the argumentative landscape and so discover arguments and attacks, (ii) exchange arguments and attacks within their social network and (iii) update their assessment of the given theories by utilizing argumentation semantics. This allows them to determine which theory they deem worthy of investigation.

The simulations show that (given some constraints) the more connected a scientific community is, the more likely it is to converge on the best theory. Moreover, some norms for theory evaluation lead to a significantly more successful inquiry than others (for details see Borg *et al.* (2019)). Finally, although communities consisting of deceptive agents perform worse than those consisting of reliable ones, increasing their connectivity increases their chance of converging on the best theory.

Modeling Scientific Inquiry with Gradual Semantics Dupuis de Tarlé *et al.* (2023) also present an ABM which explores network effects on the outcome of scientific inquiry. Like in Dupuis de Tarlé *et al.* (2022), the agent's opinions are real numbers derived from their mental AFs by applying a gradual semantics. Unlike in ArgABM, the agents do not explore a preexisting epistemic landscape of arguments, but rather, they create it by generating new arguments. The principle is inspired

5 - ARGUMENTATIVE AGENT-BASED MODELS

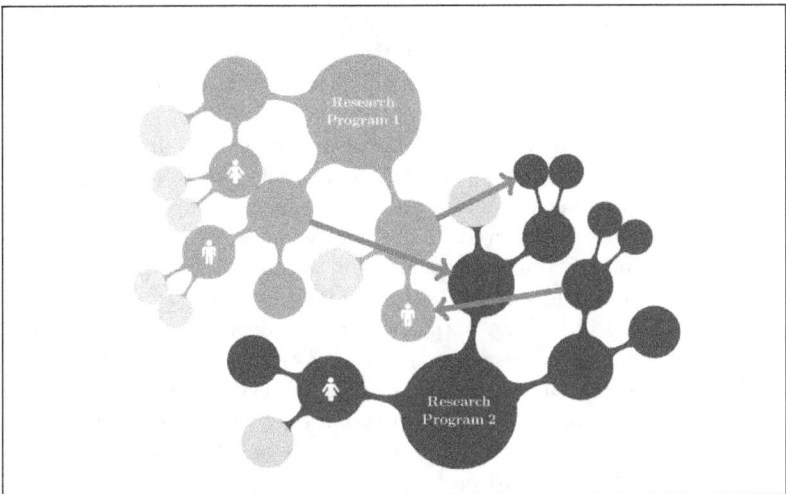

Figure 11: An example of an argumentative landscape consisting of 2 theories (or research programs). Darker shaded nodes represent arguments that have been investigated by agents and are thus visible to them; brighter shaded nodes stand for arguments that aren't visible to agents. The largest node in each theory is the root argument, from which agents start their exploration via the discovery relation, which connects arguments within one theory. Arrows represent attacks from an argument in one theory to an argument in another theory (Borg *et al.* (2019)).

by previous models of scientific inquiry such as Zollman (2007). In these models, Bayesian agents use sampled results from a probability distribution to estimate the value of the mean. In Dupuis de Tarlé *et al.* (2023) this principle is emulated by defining a 'truth value' for each argument that determines the probability for agents to attack or support this argument. By sampling results from a normal distribution centered on the truth value, agents generate new arguments and correct their assessment of the issue. Because the truth values are numbers between 0 and 1, the authors can directly compare the opinion of the agents to the truth value of the issue, measuring their *epistemic success*.

The agents are placed in a social network in which they share arguments. The authors investigate the influence of the degree of connectivity of the social network on two factors: the diversity of the agent's opinions and the epistemic success of the agents (how good they are at approximating the truth).

They observe that the relation between connectivity and opinion diversity is not monotonic: increasing connectivity up to a tipping point leads to more diversity in opinions, while past the tipping point a decrease in diversity is observed. This result is noteworthy because so far, few mechanisms of communication could account for both an increase and a decrease in the similarity of opinions (see, e.g., the discussion of Kopecky (2022) in Section 5.2). They also observe that the epistemic success of agents decreases when the diversity of their opinions increases. This is at odds with a well-known result of Bayesian models of scientific inquiry: the 'Zollman Effect' which states that less connectivity in the social network, by inducing more diversity, is beneficial for the agent's epistemic success (for hard problems).

6.2 Myside Bias

Myside bias is a cognitive phenomenon according to which humans are biased towards their own opinion. Different emphasis has been put on various aspects. On the one hand, following Mercier and Sperber (2017), it has been described as a bias that leads to a disproportionate production of arguments in support of one's own beliefs while neglecting the search for defeating arguments. On the other hand, for instance following Stanovich *et al.* (2013), the bias consists of evaluating evidence disproportionally favorable to ones own beliefs. AABMs have been produced following both perspectives.

Myside Bias as Production Bias. Since myside bias might have disruptive effects on scientific research, Dupuis de Tarlé *et al.* (2024) model it in the context of scientific debates. Mercier and Heintz (2014) conjecture that the presence of shared beliefs, intended as shared standards for the quality of the arguments, mitigates the disruptive power of the bias (construed as production bias following Mercier and Sperber). Dupuis de Tarle et al. test this conjecture using an AABM. They represent a scientific debate as a process of argument production that revolves around the central claims of two theories. In each turn, an agent observes the state of the debate and tries to produce a new argument. In addition, agents evaluate the state of the debate using argumentation semantics. This allows them to form a preference for one of the two theories.

Dupuis de Tarle et al. observe that unbiased communities typically converge on the strongest theory, since it is harder to attack and easier to defend. Biased agents, on the other hand, tend to always attack the opposing theory and are more likely to produce arguments against it. As a consequence, if the initial group supporting the weakest theory is large enough, it usually convinces the rest of the community that this is the strongest theory, for it is able to produce an overwhelming amount of low- and high-quality arguments in favor of it. This highlights the importance of the starting distribution of the agents' preferences, which is also discussed by Baccini *et al.* (2023). Although an equal initial distribution of biased agents is likely to result in the acceptance of the strongest theory, unequal distributions are not equally successful.

Dupuis de Tarle et al. observe two other features of biased communities. First, biased communities are more likely to reach a consensus than non-biased ones. Since biased agents only produce arguments supporting their positions, whenever a consensus is reached, it remains stable. This is not the case for non-biased communities. Secondly, they characterize the effect of shared beliefs, which are modeled as a collective filter that weeds out low-quality arguments. They find that shared beliefs mitigate the detrimental effect of bias, as hypothesized by Mercier and Heintz (2014), although not completely. In fact, strong shared epistemic standards are likely to prevent the majority of agents from making the weakest theory appear as the strongest one, by blocking a large number of low-quality arguments in favor of the weakest theory. However, even strict epistemic standards may not be enough for the group to identify the strongest theory, whenever the bias is very strong and the difference between the two theories is minimal.

Myside bias as an evaluation bias. Baccini *et al.* (2023) explore the effect of the combination of a production bias, following Mercier and Sperber (2017), and an evaluation bias, following Stanovich *et al.* (2013).

Similarly to Dupuis de Tarlé et al. (2024), they study how effective biased communities are in determining whether a certain hypothesis is true. In their framework, the evaluation bias is modeled by Baccini and Hartmann (2022), who provides an account for it based on the notion of diagnosticity (Hahn and Hornikx (2016)). The diagnosticity of a proposition A with respect to a proposition B corresponds to how much more the agent should increase its credibility in B after learning A. Accordingly, Baccini and Hartmann (2022) take unbiased agents to adapt their beliefs in accordance with the diagnosticity of the new proposition learnt, while agents affected by an evaluation bias would attribute lower diagnosticity to arguments that are against their present position, and higher diagnosticity to arguments in favour. Baccini et al. (2023) adopt this same mechanism for the evaluation bias, and consider agents who also have a production bias, insofar as they only share with other agents arguments that are in favor of their position.

Baccini et al. mainly observe two possible outcomes: bi-polarization and consensus. In particular, they find that, given that the bias is distributed equally among agents, correct consensus is more likely when the community is smaller and —in contrast to Dupuis de Tarlé et al. (2024)— when the strength of the bias is weaker. Moreover, they observe that biased agents are better at tracking truth than non-biased agents in scenarios in which the agents' beliefs at the beginning of the inquiry are already moderately accurate. If agents start with inaccurate priors, the presence of bias makes things drastically worse.

6.3 Bayesian Models of Inquiry and Opinion Dynamics

A Bayesian Model of Scientific Inquiry. NormAN (Normative Argument Exchange Across Networks) is a Bayesian model of scientific inquiry presented by Assaad et al. (2023).[18] Agents modeled as Bayesian reasoners gather and communicate evidence within a social network. Evidence is the basis of arguments for or against some given hypotheses. NormAN employs a Bayesian network accessible to all agents to rep-

---

[18]The NormAN model has been extended by Schöppl and Hahn (2024) to explore how self-censorship, i.e. a tendency to hide one's own evidence that could cause disagreement, affect a population's beliefs, and by Assaad et al. (2023) showing that rational agents who only share their best evidence with others may polarize.

resent the causal structure of the world and conditional probabilities. Disagreements arise in the model when agents obtain different evidence. Each agent's inquiry is governed by two parameters: the maximum number of pieces of evidence they may collect during a simulation, and their 'curiosity', which determines the probability that they gather evidence in a new round. When sharing evidence, agents may share a random piece of evidence, the most recently acquired piece of evidence or the most 'impactful' piece of evidence (which is the evidence that would make the greatest change relative to the agent's initial prior belief, upon update).

Assaad et al. use their model to demonstrate radicalization of opinions. In particular, they show how the mean group belief of Bayesian reasoners shifts to a more extreme position merely as a result of deliberation. Furthermore, they show that communication based on sharing only impactful arguments leads the community away from reaching a consensus, since it makes it less likely for agents to share all their gathered evidence.

Trust and Testimony. Another group of models that studies the truth-tracking ability of epistemic communities originates from the Laputa model of Olsson (2013). In this framework, agents aim to determine whether a certain proposition is true. They do so by investigating the value of this proposition on their own and by exchanging testimonies with other agents. Both testimony and personal inquiry are modeled as receiving a signal, which either supports or negates the proposition. The degree to which such a signal affects the belief of the agent that receives it depends on both her prior belief and how much she trusts the source. Whenever an agent receives a signal, she also adjusts her trust in the source of the signal. If the signal received is at odds with her prior belief, the trust in the source is decreased; if it is in line with her prior belief, the trust is increased.[19]

---

[19]One way to interpret agents' communications is that of an exchange of arguments (Olsson (2013, 2020); Pallavicini *et al.* (2021)) in line with Persuasive Argument Theory (Isenberg (1986)). Olsson (2013) argues that the exchanged signals can be interpreted as novel and sound arguments, which affect an agent's belief based on whether or not they support the issue at stake.

Olsson (2013) shows that regardless of the starting distribution of opinions, group discussion may lead to two possible outcomes. The first one, which they call 'polarization' is a case in which agents' opinions all move towards the same position about the issue at stake, ending up agreeing on whether or not the proposition is true or false. In this case, although the group reaches a consensus, the initial opinions of the agents are 'radicalized' insofar as they become more extreme than their initial opinions. The second one is a situation of 'divergence' in which the group splits, and part of the agents get to support the proposition at stake while the other part firmly rejects it.[20] Which of the two outcomes emerges depends on the initial degree of trust agents have in each other, and on their ability to obtain correct information. Pallavicini *et al.* (2021) further extends the study of polarization carried out by Olsson (2013), and argues that the rational agents of the Laputa model 'appear to polarize more than human agents' (p.34). In view of this, they hold that one may either have to concede that polarization is indeed rational or that Laputa agents may fail to capture some crucial aspects of individual rationality. In particular, they conjecture that Laputa agents may not be able to effectively learn from higher-order evidence (Whiting (2021)) when communicating with each other.

6.4 Social decision making and problem solving

Stein *et al.* (2023) use an AABM to study social decision making. Although diversity can improve decision making, it can also lead to polarization and in this way hinder good decision making. A diversifying factor is homophily in communication (see 3): If agents favor communication with agents they find agreeable, they form isolated subgroups. An isolated subgroup may miss out on information that is crucial for reasoning toward an optimal decision. However, temporary disagreements may prevent a society of agents to prematurely converge on suboptimal decisions and give them the necessary time to agree on the right decision. AABMs are useful methods to better understand such tradeoffs between deliberation time and decision quality. What is the right

---

[20]This outcome resembles closely 'bi-polarization' as discussed above (see Section 4), i.e., a process in which two opposing groups form and the distance between their opinions grows.

amount of homophilic communication that optimally contributes to the type of transient diversity that underlies optimal decision-making?

In order to investigate this question, Stein et al. model a hidden profile scenario (Lu *et al.* (2012)) with only partially shared information. The agents are divided into groups $g_1, \ldots, g_M$ and have the task of finding the optimal decision within a given set of options $o_1, \ldots, o_J$ (where $J > M$). Each group g_i initially favors the decision o_i. When communicating, agents exchange arguments that support a decision with a specific weight. The goodness of a decision is determined by the sum of the weights of those arguments that support it.[21] Communication follows the homophily principle and agents preferably communicate strong arguments in favor of their favored decision.

The obtained results show that increasing homophily increases the quality of the decision making and its effect is more visible the harder the problem (a problem is the harder, the closer the strength of the second best option is to the best). Also, the model clearly shows the expected trade-off: the higher homophily, the more time agents need to converge.

7 Outlook and Conclusion

Argumentation forms the basis of a variety of social reasoning processes: from ethical debates such as those surrounding vegetarianism (Taillandier *et al.* (2021)), to political debates such as those surrounding nuclear waste, to online product reviews (Gabbriellini and Santini (2017)), to scientific inquiry (Borg *et al.* (2017b); Assaad *et al.* (2023); Dupuis de Tarlé *et al.* (2024)). Although it can improve collective decision making, it is also susceptible to bias and can lead societies to radicalization. It is therefore important to understand socio-epistemic mechanisms that underlie such dynamic phenomena. How does the way agents argue (e.g. in a biased way or in a strategic way) influence emergent phenomena on the group level (such as polarization)? How do specific choices of one's communication partners (such as homophily-based) impact the opinion dynamics in an argumentative exchange? Such ques-

---

[21]In the terminology of Section 2, we are dealing with an issue-based AF of the type $\mathsf{AF} = \langle \mathsf{Args}, \mathsf{Decisions}, w \rangle$ (just that the issues are now decisions).

tions cannot be easily answered from the armchair and, for reasons of complexity, they can rarely be tackled by analytic methods. Researchers have therefore turned to simulation methods, to argumentative agent-based models, to what we have called in this entry AABMs.

Since Mäs and Flache (2013) included an explicit representation of the exchange of arguments in their opinion dynamics model, a large literature has emerged containing various combinations of agent-based modeling and argumentation. Many questions have been investigated for communities whose information flow is characterized by an argumentative exchange. In fact, focusing on argumentative communication produced novel insights even for phenomena that had previously been extensively studied by ABMs. Take, for example, investigations of opinion dynamics and research questions on the causes of (de-)polarization. What agents communicate in traditional models of opinion dynamics is opinions, often simplified to a real number between 0 and 1. Considering scenarios in which agents also give reasons, that is, in which they argue, opened the study of novel questions. For example, Mäs and Flache (2013) showed that bi-polarization (incl. radicalization) can occur absent negative influence simply due to the fact that agents exchange reasons rather than simple opinions and the fact that agents have limited memory (see Section 4.1.1). Similarly, Kopecky (2022) demonstrated that if agents argue strategically in an allocentric way (see Section 5.2), societies composed of such agents can depolarize. This suggests that combining argumentation and agent-based modeling is fruitful and that more is to come.

In this chapter, we have provided a systematic overview of the growing field of AABMs. Since the plurality of approaches may seem overwhelming at first, we also offered a structural blueprint in terms of several building blocks and typical modeling choices underlying AABMs (see Section 2). These building blocks span from the cognitive model of agents, to knowledge representation of argumentative information, to reasoning defeasibly with arguments, and to the (possibly dynamic) structure of the underlying social network in which arguments are exchanged. We have seen that most models address questions within the domain of opinion dynamics. They investigate the role of possible causal factors of polarization, such as bias, strategic argumentation, forgetting,

5 - Argumentative Agent-Based Models

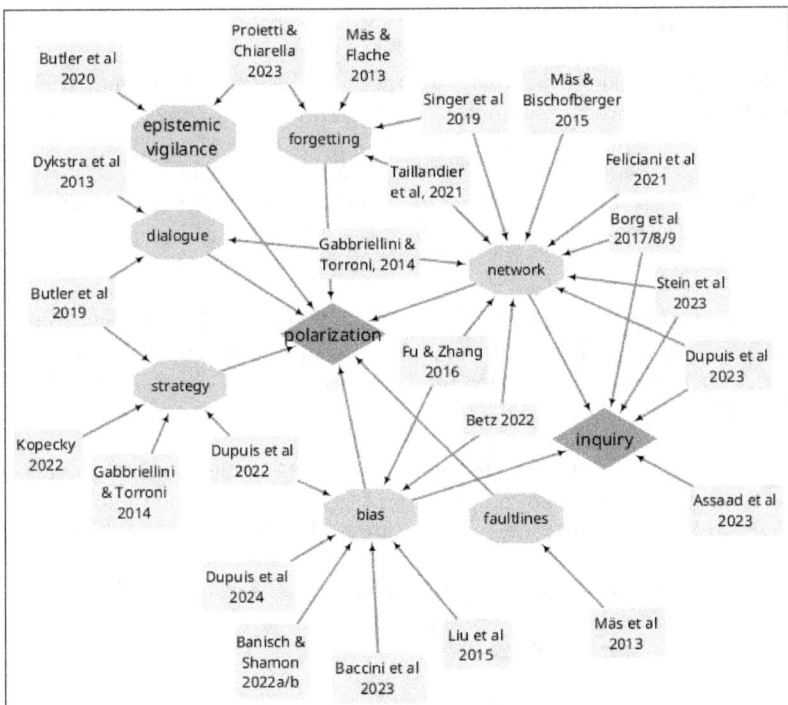

Figure 12: Overview on the topics 'polarization' and 'inquiry' (diamond-shaped) and their potential causal factors (in green, octagon-shaped) investigated by AABMs. The papers covered in this entry are represented by blue rectangular nodes.

etc. But other topics also are increasingly getting the attention of modelers, such as scientific inquiry, innovation diffusion, online review, and so forth. We conclude by giving an overview of the AABMs that focus on two central topics, polarization and inquiry in Fig. 12.

Acknowledgements The research for this paper is supported by the Deutsche Forschungsgemeinschaft (DFG, German Research Foundation) – project number 426833574.

References

Robert P Abelson. Mathematical models of the distribution of attitudes under controversy. *Contributions to mathematical psychology*, 1964.

Ronald L Akers, Marvin D Krohn, Lonn Lanza-Kaduce, and Marcia Radosevich. Social learning and deviant behavior: A specific test of a general theory. *Contemporary masters in criminology*, pages 187–214, 1995.

Leila Amgoud and Jonathan Ben-Naim. Weighted bipolar argumentation graphs: Axioms and semantics. In *Twenty-Seventh International Joint Conference on Artificial Intelligence-IJCAI 2018*, pages 5194–5198, 2018.

Leila Amgoud and Claudette Cayrol. On the acceptability of arguments in preference-based argumentation. In Gregory F. Cooper and Serafín Moral, editors, *UAI*, pages 1–7. Morgan Kaufmann, 1998.

Leila Amgoud, Dragan Doder, and Srdjan Vesic. Evaluation of argument strength in attack graphs: Foundations and semantics. *Artificial Intelligence*, 302:103607, 2022.

Ofer Arieli and Christian Straßer. Sequent-based logical argumentation. *Argument and Computation.*, 6(1):73–99, 2015.

Ofer Arieli, AnneMarie Borg, Jesse Heyninck, and Christian Straßer. Logic-based approaches to formal argumentation. *Journal of Applied Logics-IfCoLog Journal*, 8(6):1793–1898, 2021.

Solomon E Asch. Studies of independence and conformity: I. a minority of one against a unanimous majority. *Psychological monographs: General and applied*, 70(9):1, 1956.

Leon Assaad, Rafael Fuchs, Ammar Jalalimanesh, Kirsty Phillips, Leon Schoeppl, and Ulrike Hahn. A Bayesian agent-based framework for argument exchange across networks. *arXiv preprint arXiv:2311.09254*, 2023.

Edoardo Baccini and Stephan Hartmann. The myside bias in argument evaluation: A Bayesian model. In *Proceedings of the Annual Meeting of the Cognitive Science Society*, volume 44, 2022.

Edoardo Baccini, Zoé Christoff, Stephan Hartmann, and Rineke Verbrugge. The wisdom of the small crowd: Myside bias and group discussion. *Journal of Artificial Societies and Social Simulation*, 26(4), 2023.

Delia Baldassarri and Amir Goldberg. Neither ideologues nor agnostics: Alternative voters' belief system in an age of partisan politics. *American Journal of Sociology*, 120(1):45–95, 2014.

Sven Banisch and Eckehard Olbrich. An argument communication model of polarization and ideological alignment. *Journal of Artificial Societies and Social Simulation*, 24(1):nil, 2021.

Sven Banisch and Hawal Shamon. Validating argument-based opinion dynamics with survey experiments, 2022.

Sven Banisch and Hawal Shamon. Biased processing and opinion polarization: Experimental refinement of argument communicationtheory in the context of the energy debate. *Sociological Methods & Research*, 0(0):00491241231186658, 2023.

Pietro Baroni, Martin Caminada, and Massimiliano Giacomin. Abstract argumentation frameworks and their semantics. *Handbook of formal argumentation*, 1:157–234, 2018.

Gregor Betz. Evaluating dialectical structures. *Journal of Philosophical Logic*, 38:283–312, 2009.

Gregor Betz. Natural-language multi-agent simulations of argumentative opinion dynamics. *Journal of Artificial Societies and Social Simulation*, 25(1):nil, 2022.

Phillip Bonacich and Philip Lu. *Introduction to mathematical sociology*. Princeton University Press, 2012.

AnneMarie Borg, Daniel Frey, Dunja Šešelja, and Christian Straßer. *An Argumentative Agent-Based Model of Scientific Inquiry*, pages 507–510. Springer International Publishing, Cham, 2017.

AnneMarie Borg, Daniel Frey, Dunja Šešelja, and Christian Straßer. *Examining Network Effects in an Argumentative Agent-Based Model of Scientific Inquiry*, pages 391–406. Springer Berlin Heidelberg, Berlin, Heidelberg, 2017.

AnneMarie Borg, Daniel Frey, Dunja Šešelja, and Christian Straßer. Epistemic effects of scientific interaction: approaching the question with an argumentative agent-based model. *Historical Social Research*, 43(1):285–309, 2018.

AnneMarie Borg, Daniel Frey, Dunja Šešelja, and Christian Straßer. Theory-choice, transient diversity and the efficiency of scientific inquiry. *European Journal of Philosophy of Science*, (26), 2019.

Aaron Bramson, Patrick Grim, Daniel J Singer, William J Berger, Graham Sack, Steven Fisher, Carissa Flocken, and Bennett Holman. Understanding polarization: Meanings, measures, and model evaluation. *Philosophy of science*, 84(1):115–159, 2017.

Laura Burbach, Poornima Belavadi, Patrick Halbach, Lilian Kojan, Nils Plettenberg, Johannes Nakayama, Martina Ziefle, and André Calero Valdez. *Netlogo vs. Julia: Evaluating Different Options for the Simulation of Opinion Dynamics*, pages 3–19. Lecture Notes in Computer Science. Springer International Publishing, 2020.

Eugene Burnstein and Amiram Vinokur. Persuasive argumentation and social comparison as determinants of attitude polarization. *Journal of Experimental Social Psychology*, 13(4):315–332, 1977.

George Butler, Gabriella Pigozzi, and Juliette Rouchier. Mixing dyadic and deliberative opinion dynamics in an agent-based model of group decision-making. *Complexity*, 2019:1–31, Aug 2019.

George Butler, Gabriella Pigozzi, and Juliette Rouchier. *An Opinion Diffusion Model with Vigilant Agents and Deliberation*, pages 81–99.

Multi-Agent-Based Simulation XX. Springer International Publishing, 2020.

Martin WA Caminada. A formal account of socratic-style argumentation. *Journal of Applied Logic*, 6(1):109–132, 2008.

Kathleen Carley. A theory of group stability. *American sociological review*, pages 331–354, 1991.

Álvaro Carrera and Carlos A. Iglesias. A systematic review of argumentation techniques for multi-agent systems research. *Artificial Intelligence Review*, 44(4):509–535, 2015.

Claudette Cayrol and Marie-Christine Lagasquie-Schiex. Bipolarity in argumentation graphs: Towards a better understanding. *International Journal of Approximate Reasoning*, 54(7):876–899, 2013.

Elizabeth Colson. Social control and vengeance in plateau tonga society1. *Africa*, 23(3):199–212, 1953.

George Datseris, Ali R. Vahdati, and Timothy C. DuBois. Agents.jl: a performant and feature-full agent-based modeling software of minimal code complexity. *SIMULATION*, 0(0):003754972110688, January 2022.

Guillaume Deffuant, David Neau, Frederic Amblard, and Gérard Weisbuch. Mixing beliefs among interacting agents. *Advances in Complex Systems*, 3(01n04):87–98, 2000.

Guillaume Deffuant, Frédéric Amblard, Gérard Weisbuch, and Thierry Faure. How can extremism prevail? a study based on the relative agreement interaction model. *Journal of artificial societies and social simulation*, 5(4), 2002.

Morris H DeGroot. Reaching a consensus. *Journal of the American Statistical association*, 69(345):118–121, 1974.

Igor Douven and Christoph Kelp. Truth approximation, social epistemology, and opinion dynamics. *Erkenntnis (1975-)*, 75(2):271–283, 2011.

P.M. Dung, R.A. Kowalski, and F. Toni. Assumption-based argumentation. *Argumentation in Artificial Intelligence*, pages 199–218, 2009.

Phan Minh Dung. On the acceptability of arguments and its fundamental role in nonmonotonic reasoning, logic programming and n-person games. *Artificial intelligence*, 77(2):321–357, 1995.

Luise Dupuis de Tarlé, Elise Bonzon, and Nicolas Maudet. Multiagent dynamics of gradual argumentation semantics. In *Adaptive Agents and Multi-Agent Systems*, 2022.

Luise Dupuis de Tarlé, Gabriella Pigozzi, and Juliette Rouchier. Exploring opinion diversity and epistemix success with an argumentative model. In *Social Simulation Conference*, 2023.

Louise Dupuis de Tarlé, Matteo Michelini, Annemarie Borg, Gabriella Pigozzi, Juliette Rouchier, Dunja Šešelja, and Christian Strasser. An agent-based model of myside bias in scientific debates. *JASSS*, 2024.

Piter Dykstra, Corinna Elsenbroich, Wander Jager, Gerard Renardel de Lavalette, and Rineke Verbrugge. Put your money where your mouth is: Dial, a dialogical model for opinion dynamics. *Journal of Artificial Societies and Social Simulation*, 16(3):nil, 2013.

Piter Dykstra, Wander Jager, Corinna Elsenbroich, Rineke Verbrugge, and Gerard Renardel de Lavalette. An agent-based dialogical model with fuzzy attitudes. *Journal of Artificial Societies and Social Simulation*, 18(3):nil, 2015.

Joan-Maria Esteban and Debraj Ray. On the measurement of polarization. *Econometrica: Journal of the Econometric Society*, pages 819–851, 1994.

Thomas Feliciani, Andreas Flache, and Jochem Tolsma. How, when and where can spatial segregation induce opinion polarization? two competing models. *Journal of Artificial Societies and Social Simulation*, 20(2):6, 2017.

Thomas Feliciani, Andreas Flache, and Michael Mäs. Persuasion without polarization? modelling persuasive argument communication in teams with strong faultlines. *Computational and Mathematical Organization Theory*, 27:61–92, 2021.

Martin Fishbein. An investigation of the relationships between beliefs about an object and the attitude toward that object. *Human relations*, 16(3):233–239, 1963.

Andreas Flache and Michael Mäs. How to get the timing right? a computational model of how demographic faultlines undermine team performance and how the right timing of contacts can solve the problem. *Comput Math Organ Theory*, 14:23–51, 2008.

Andreas Flache and Michael Mäs. Why do faultlines matter? a computational model of how strong demographic faultlines undermine team cohesion. *Simulation Modelling Practice and Theory*, 16(2):175–191, 2008.

Andreas Flache, Michael Mäs, Thomas Feliciani, Edmund Chattoe-Brown, Guillaume Deffuant, Sylvie Huet, and Jan Lorenz. Models of social influence: Towards the next frontiers. *Journal of Artificial Societies and Social Simulation*, 20(4), 2017.

Noah E Friedkin and Eugene C Johnsen. Social influence and opinions. *Journal of mathematical sociology*, 15(3-4):193–206, 1990.

Guiyuan Fu and Weidong Zhang. Opinion formation and bi-polarization with biased assimilation and homophily. *Physica A: Statistical Mechanics and its Applications*, 444:700–712, 2016.

Simone Gabbriellini and Francesco Santini. From reviews to arguments and from arguments back to reviewers' behaviour. In *Agents and Artificial Intelligence: 8th International Conference, ICAART 2016, Rome, Italy, February 24-26, 2016, Revised Selected Papers 8*, pages 56–72. Springer, 2017.

Simone Gabbriellini and Paolo Torroni. *A New Framework for ABMs Based on Argumentative Reasoning*, pages 25–36. Advances in Intelligent Systems and Computing. Springer Berlin Heidelberg, 2014.

Simone Gabbriellini. The evolution of online forums as communication networks: an agent-based model. *Revue française de sociologie*, (4):805–826, 2014.

Michel Grabisch and Agnieszka Rusinowska. A survey on nonstrategic models of opinion dynamics. *Games*, 11(4):65, 2020.

Ulrike Hahn and Jos Hornikx. A normative framework for argument quality: argumentation schemes with a bayesian foundation. *Synthese*, 193(6):1833–1873, 2016.

Sil Hamilton. Blind judgement: Agent-based supreme court modelling with gpt. *arXiv preprint arXiv:2301.05327*, 2023.

Rainer Hegselmann and Ulrich Krause. Opinion dynamics and bounded confidence models, analysis, and simulation. *Journal of artificial societies and social simulation*, 5(3), 2002.

Daniel J Isenberg. Group polarization: A critical review and meta-analysis. *Journal of personality and social psychology*, 50(6):1141, 1986.

Wander Jager and Frederic Amblard. Uniformity, bipolarization and pluriformity captured as generic stylized behavior with an agent-based simulation model of attitude change. *Computational & Mathematical Organization Theory*, 10(4):295–303, 2005.

Felix Kopecky. Arguments as drivers of issue polarisation in debates among artificial agents. *Journal of Artificial Societies and Social Simulation*, 25(1), 2022.

Zlatan Krizan and Robert S Baron. Group polarization and choice-dilemmas: how important is self-categorization? *European Journal of Social Psychology*, 37(1):191–201, 2007.

Cristian E La Rocca, Lidia A Braunstein, and Federico Vazquez. The influence of persuasion in opinion formation and polarization. *Europhysics Letters*, 106(4):40004, 2014.

Bibb Latané. Dynamic social impact: The creation of culture by communication. *Journal of communication*, 46(4):13–25, 1996.

Dora C Lau and J Keith Murnighan. Demographic diversity and faultlines: The compositional dynamics of organizational groups. *Academy of management review*, 23(2):325–340, 1998.

Francois Ledoyen, Rallou Thomopoulos, Stephane Couture, Loic Sadou, and Patrick Taillandier. An agent-based model representing the exchanges of arguments to accurately simulate the process of innovation diffusion. In *2021 RIVF International Conference on Computing and Communication Technologies (RIVF)*, page nil, 8 2021.

Qipeng Liu, Jiuhua Zhao, and Xiaofan Wang. Multi-agent model of group polarisation with biased assimilation of arguments. *IET Control Theory & Applications*, 9(3):485–492, 2015.

Jan Lorenz. Multidimensional opinion dynamics when confidence changes. *Economic Complexity, Aix-en-Provence*, 2003.

Li Lu, Y Connie Yuan, and Poppy Lauretta McLeod. Twenty-five years of hidden profiles in group decision making: A meta-analysis. *Personality and Social Psychology Review*, 16(1):54–75, 2012.

Michael Mäs and Lukas Bischofberger. Will the personalization of online social networks foster opinion polarization? *Available at SSRN 2553436*, 2015.

Michael Mäs and Andreas Flache. Differentiation without distancing. explaining bi-polarization of opinions without negative influence. *PLoS ONE*, 8(11):e74516, 2013.

Michael Mäs, Andreas Flache, Károly Takács, and Karen A Jehn. In the short term we divide, in the long term we unite: Demographic crisscrossing and the effects of faultlines on subgroup polarization. *Organization science*, 24(3):716–736, 2013.

George Masterton and Erik J Olsson. Argumentation and belief updating in social networks: a Bayesian model. *Trends in belief revision and argumentation dynamics. Cambridge: College Publications*, 2014.

Miller McPherson, Lynn Smith-Lovin, and James M Cook. Birds of a feather: Homophily in social networks. *Annual review of sociology*, 27(1):415–444, 2001.

Hugo Mercier and Christophe Heintz. Scientists' argumentative reasoning. *Topoi*, 33:513–524, 2014.

Hugo Mercier and Dan Sperber. *The enigma of reason*. Harvard University Press, 2017.

Erik J Olsson. A Bayesian simulation model of group deliberation and polarization. *Bayesian argumentation: The practical side of probability*, pages 113–133, 2013.

Erik J Olsson. Why Bayesian agents polarize. In *The Epistemology of Group Disagreement*, pages 211–229. Routledge, 2020.

Josefine Pallavicini, Bjørn Hallsson, and Klemens Kappel. Polarization in groups of Bayesian agents. *Synthese*, 198:1–55, 2021.

Juan Pablo Pinasco, Viktoriya Semeshenko, and Pablo Balenzuela. Modeling opinion dynamics: Theoretical analysis and continuous approximation. *Chaos, Solitons & Fractals*, 98:210–215, 2017.

Carlo Proietti and Davide Chiarella. The role of argument strength and informational biases in polarization and bi-polarization effects. *Journal of Artificial Societies and Social Simulation*, 26(2), 2023.

Alec Radford, Jeffrey Wu, Rewon Child, David Luan, Dario Amodei, Ilya Sutskever, et al. Language models are unsupervised multitask learners. *OpenAI blog*, 1(8):9, 2019.

Nicolas Salliou, Patrick Taillandier, and Rallou Thomopoulos. Vitamin project (vegetarian transition argument modelling). *Dataset available at: https://doi. org/10.15454/HOBUZH*, 2019.

Laurent Salzarulo. A continuous opinion dynamics model based on the principle of meta-contrast. *Journal of Artificial Societies and Social Simulation*, 9(1), 2006.

Klee Schöppl and Ulrike Hahn. Exploring effects of self-censoring through agent-based simulation. In *Proceedings of the Annual Meeting of the Cognitive Science Society*, volume 46, 2024.

Dunja Šešelja and Christian Straßer. Abstract argumentation and explanation applied to scientific debates. *Synthese*, 190:2195–2217, 2013.

Dunja Šešelja. Agent-Based Modeling in the Philosophy of Science. In Edward N. Zalta and Uri Nodelman, editors, *The Stanford Encyclopedia of Philosophy*. Metaphysics Research Lab, Stanford University, Winter 2023 edition, 2023.

Hawal Shamon, Diana Schumann, Wolfgang Fischer, Stefan Vögele, Heidi U Heinrichs, and Wilhelm Kuckshinrichs. Changing attitudes and conflicting arguments: Reviewing stakeholder communication on electricity technologies in germany. *Energy research & social science*, 55:106–121, 2019.

Muzafer Sherif and Carl I Hovland. Social judgment: Assimilation and contrast effects in communication and attitude change. 1961.

Daniel J. Singer, Aaron Bramson, Patrick Grim, Bennett Holman, Jiin Jung, Karen Kovaka, Anika Ranginani, and William J. Berger. Rational social and political polarization. *Philosophical Studies: An International Journal for Philosophy in the Analytic Tradition*, 176(9):2243–2269, 2019.

Dan Sperber, Fabrice Clément, Christophe Heintz, Olivier Mascaro, Hugo Mercier, Gloria Origgi, and Deirdre Wilson. Epistemic vigilance. *Mind & language*, 25(4):359–393, 2010.

Keith E Stanovich, Richard F West, and Maggie E Toplak. Myside bias, rational thinking, and intelligence. *Current Directions in Psychological Science*, 22(4):259–264, 2013.

Annalisa Stefanelli and Roman Seidl. Moderate and polarized opinions. using empirical data for an agent-based simulation. In *Social Simulation Conference*, 2014.

Annalisa Stefanelli and Roman Seidl. Opinion communication on contested topics: How empirics and arguments can improve social simulation. *Journal of Artificial Societies and Social Simulation*, 20(4), 2017.

Annalisa Stefanelli. *Opinions on Contested Infrastructures Over Time: A Longitudinal, Empirically Based Simulation*, pages 393–400. Advances in Intelligent Systems and Computing. Springer International Publishing, 2017.

Jonas Stein, Vincenz Frey, and Andreas Flache. Temporary disagreements foster better solutions: How homophilic interactions in diverse teams can improve collective decision-making. In *Social Simulation Conference*, 2023.

Cass R Sunstein. The law of group polarization. *University of Chicago Law School, John M. Olin Law & Economics Working Paper*, (91), 1999.

Patrick Taillandier, Benoit Gaudou, Arnaud Grignard, Quang-Nghi Huynh, Nicolas Marilleau, Philippe Caillou, Damien Philippon, and Alexis Drogoul. Building, composing and experimenting complex spatial models with the gama platform. *GeoInformatica*, 23(2):299–322, 2018.

Patrick Taillandier, Nicolas Salliou, and Rallou Thomopoulos. Introducing the argumentation framework within agent-based models to better simulate agents' cognition in opinion dynamics: Application to vegetarian diet diffusion. *Journal of Artificial Societies and Social Simulation*, 24(2):6, 2021.

Paul Thagard and Karsten Verbeurgt. Coherence as constraint satisfaction. *Cognitive science*, 22(1):1–24, 1998.

Seth Tisue and Uri Wilensky. Netlogo: A simple environment for modeling complexity. In *International conference on complex systems*, volume 21, pages 16–21. Citeseer, 2004.

Diemo Urbig and Robin Malitz. Dynamics of structured attitudes and opinions. In *Third Conference of the European Social Simulation Association*, pages 5–8, 2005.

Gérard Weisbuch, Guillaume Deffuant, Frédéric Amblard, and Jean-Pierre Nadal. Meet, discuss, and segregate! *Complexity*, 7(3):55–63, 2002.

Daniel Whiting. Higher-Order Evidence. *Analysis*, 80(4):789–807, 01 2021.

Kevin J. S. Zollman. The communication structure of epistemic communities. *Philosophy of Science*, 74(5):574–587, 2007.

Kevin J. S. Zollman. The epistemic benefit of transient diversity. *Erkenntnis*, 72(1):17–35, 2010.

CHAPTER 6

ARGUMENTATION-BASED APPLICATIONS FOR DECISION-MAKING

SRDJAN VESIC
CRIL, CNRS - Univ. Artois, Lens, France
`vesic@cril.fr`

BRUNO YUN
*Univ Lyon, UCBL, CNRS, INSA Lyon, LIRIS, UMR5205, F-69622
Villeurbanne*
`bruno.yun@univ-lyon1.fr`

1 Introduction

In the realm of decision-making, the utilization of argumentation-based applications has witnessed a proliferation of diverse tools and technologies designed to enhance decision-making. From aiding in policy formulation to guiding strategic choices, these applications have addressed various fields. The abundance of these tools makes it impractical to compile an exhaustive list, as the landscape continues to evolve dynamically.

This chapter illuminates argumentation-based applications for decision-making by discussing a selection of relevant papers. The intention is not to provide an exhaustive inventory, but rather to offer readers an exploration of contributions to the field and guide them towards resources available in the field. Interested readers are encouraged to stay updated on the most recent advancements through the proceedings of the International Conference on Computational Models of Argument series (COMMA)[1], and the Argument and Computation journal (A&C)[2].

---

[1] https://comma.csc.liv.ac.uk/
[2] https://www.iospress.com/catalog/journals/argument-computation

Decision-making can be studied from many different perspectives, including psychological, cognitive, normative/descriptive, group decision making and decision theory, with a lot of literature available. In this chapter, we focus on argumentation-based decision-making with a perspective on the reasoning technology and mathematics behind them. Indeed, while decision theory is quantitative (mainly using probability with utility), argumentation captures reasoning with defeasible knowledge, which is a form of qualitative uncertainty. This means that argumentation can handle situations where information is incomplete, evolving, or contradictory. Argumentation frameworks allow decision-makers to work with partial knowledge and construct arguments based on the available evidence, even if it is incomplete, to reach provisional conclusions. They also provide a structured way to compare and evaluate conflicting arguments, weighing their strengths and weaknesses to reach rational conclusions. This is especially important in contexts like legal reasoning, where multiple perspectives and pieces of evidence need to be considered. Real-world environments are often dynamic, with new information emerging that can affect decisions. Argumentation allows for the continuous revision of decisions as new evidence becomes available, crucial for making informed decisions in rapidly changing situations. Furthermore, argumentation supports the process of explaining and justifying decisions, making it easier to communicate the reasoning behind decisions to stakeholders and enhancing trust and understanding. It also enables collaborative decision-making by providing a framework for multiple stakeholders to present, challenge, and refine their arguments, ensuring diverse perspectives are considered and leading to more robust and accepted decisions. Finally, argumentation allows for the explicit representation and reasoning about qualitative factors, such as ethical considerations, values, and preferences, which are difficult to quantify. By integrating argumentation with decision theory, decision-makers can achieve more comprehensive and well-reasoned outcomes.

One of the main motivations for writing this chapter is, on one hand, the large number of frameworks, and on the other hand, the fact that many of the frameworks and tools follow a fairly similar approach. Namely, they allow us to model the possible options, and to each option are attached the arguments in favor and against it, with possibly some preferences. This chapter is an effort to make a taxonomy of the existing approaches and classify them into sub-categories. In doing so, we are inspired by the categorization by Ouerdane *et al.* (2010), which we use and adapt in our chapter. Due to the large number of papers and tools, in this chapter, we focus on the papers in decision-making and omit (or only briefly mention) the papers about e.g. collaborative forecasting of what will happen (Irwin *et al.*, 2022).

We would like to finish this introduction section by pointing out an important issue we stumbled upon while preparing this chapter. Namely, despite the plethora of applications, there remains a remarkable scarcity of argument-based tools that have been introduced in scholarly publications and have stood the test of time, continuing to offer accessible source code or maintenance.

Also, there is a lack of fielded or commercial systems based on computational argumentation (with few exceptions, e.g., Al-Abdulkarim *et al.* (2019)). We hypothesize that this might be the case because of multiple reasons. First, many semantics used in argumentation are computationally expensive to use and hardly scalable and the machine learning heuristics only give very approximate answers with no guarantee of correctness. Second, while many theoretical results show the normative behaviour of argumentation semantics, there is still a lack of work about whether they are intuitive or understandable by human users (and in what form) (Vesic *et al.*, 2022). Lastly, computational argumentation manly assumes that the graph is provided or that the debate has some existing structure. However, in real life, the debates are often unstructured and provided in natural language and the argument-mining research is not yet mature enough to bridge the gap between symbolic argumentation and machine-learning argumentation.

We also note that recent tools tend to be developed as online applications, allowing argumentation tools to be accessible to a wide range of users, regardless of their technical expertise or location. Those observations prompt a necessary discussion about the longevity, sustainability, and maintenance of such software and tools, both in a broader context and specifically within the academic environment.

2 A survey of argumentation-based techniques for decision-making

Argumentation has proven its value in facilitating reasoning when faced with uncertainty and/or inconsistency, serving as a valuable addition to numerical techniques for assessing beliefs and providing a comprehensive framework that enables the comprehension of various competing approaches. Many works, including those of Parsons and Fox (1996) and Amgoud and Prade (2006), elaborate on how argumentation presents a diverse range of advantages for practical decision-making systems in the presence of uncertainty.

Additionally, multiple experiments have suggested that argumentation, along with its graph-based representations, can enhance people's ability to make more effective inferences. Hua and Kimbrough (1998) introduced a graphical representation tool known as Hypermedia-based Argumentation DSS (HADSS) and demonstrated its utility in assisting individuals with deductive reasoning. By employing graphical representations, the process of reasoning becomes more straightforward, as it involves navigating through graphs, making it readily comprehensible and accessible to humans. A more recent study by Vesic *et al.* (2022) shows that people respect the principles for gradual argumentation semantics more when they are shown a graphical representation of the corresponding dialogue. As a result, in this section, we review different types of argumentation frameworks as well as the argumentation techniques that can be leveraged on them to make decisions.

In the realm of decision-making with abstract/structured argumentation, a typical three-step process unfolds, encompassing (1) the acquisition of arguments, (2) the application of a semantics, usually based on Dung framework or ranking-based approaches, to gauge the acceptability of these arguments, and (3) a comparison of the available options predicated on the arguments that support them.

In the rest of the section, we study the decision systems based on abstract argumentation (sub-section 2.1), extended Dung frameworks (sub-section 2.2), and the systems based on structured argumentation (sub-section 2.3), such as ASPIC+, ABA, logic-based argumentation, and DeLP among others.

2.1 Abstract-argumentation

In his seminal paper, Dung [1995] introduced the abstract argumentation framework, composed of a set of abstract arguments and a binary attack relation between those arguments. In his approach, the arguments are atomic and "abstract", meaning that their content is unspecified. Dung's semantics (e.g., complete, grounded, preferred, etc.) are considered extension-based, as they yield sets of extensions, comprising arguments that can be collectively accepted. Many extension-based semantics have been defined and we refer the reader to the paper by Baroni *et al.* (2011) for an introduction.

In contrast, ranking-based semantics produce an ordered ranking of arguments. This ranking-centric approach revolves around the notion that arguments can be independently evaluated and ranked, leading to a more nuanced assessment compared to the conventional tripartite categorization of skeptical acceptance (in all extensions), credulous acceptance (in some extensions), and rejection (in none of the extensions) that characterizes individual argument evaluation using extension-based semantics.

Within the realm of ranking-based semantics, some methods directly establish rankings for individual arguments Amgoud and Ben-Naim (2013); Bonzon *et al.* (2016). Conversely, the majority of them, called *gradual semantics*, assign a score to each argument, often falling within the range of $[0,1]$ Amgoud *et al.* (2022); Yun and Vesic (2021). These scores naturally give rise to a ranking of the arguments. The value associated with an argument through ranking-based semantics is referred to as its "acceptability degree". This metric encapsulates the overall strength of an argument and is often derived recursively (until convergence) from the aggregation of its intrinsic strength and the strengths of its attackers (e.g., see Mossakowski and Neuhaus (2018)).

Compared to extension-based approaches, ranking-based semantics "flatten" the whole output, making it impossible to see which arguments are actually jointly acceptable or not. One might wonder what the relevance of ranking-based semantics in decision-making is. In the literature, the application of ranking-based semantics in sorting or refining options obtained through extension-based semantics has received substantial attention. In the work of Yun *et al.* (2018c), a comprehensive framework is proposed, comprising a *selection function* to determine sets of arguments that can be collectively accepted, a ranking-based semantics to score individual arguments, and a *lifting function* to aggregate these individual scores into scores for sets of arguments. This framework facilitates option comparison by assessing the corresponding sets of arguments. The work of Bonzon et al. (2018), which was developed at the same time, it presents three techniques to enhance extensions using ranking-based semantics. The first technique entails comparing the ranks of arguments to derive a score for the extension through an aggregation function. The second approach involves evaluating all pairs of extensions based on the number of arguments in one extension that are more acceptable than the arguments in another extension. The third approach, which is not detailed here, centers on disregarding attacks from weaker arguments to stronger ones, akin to the approaches in Amgoud and Cayrol (2002); Bench-Capon (2003).

In the study by Yun et al. (2018a), arguments for and against various packaging options are generated from knowledge bases expressed in Datalog+/-, Dung semantics are used to assess these arguments, and two approaches for selecting between the options are proposed: (1) leveraging preferences between packaging characteristics to rank the options based on Pareto, global, or local optimality, and (2) scoring the options using ranking-based semantics (Amgoud and Ben-Naim, 2013).

It is crucial to note that *preferences* are essential to all of the previous work, emerging at various stages. The approach to handling preferences between arguments developed by Amgoud and Cayrol (2002) involves removing attacks from less preferred to more preferred arguments. Preferences can also be employed differently, including preferences on arguments to compare sets of arguments, as explored in Yun et al. (2018c), or using preferences on underlying rules to assess arguments, typically through the means of a lifting function (Modgil and Prakken, 2014).

In the work of Amgoud and Prade (2009); Amgoud (2009), the authors construct and evaluate arguments for both beliefs and options, relying on some of the classical acceptability semantics (preferred and stable) introduced by Dung (1995). Subsequently, they compare pairs of options by applying decision principles that fall into three distinct categories: those considering only arguments in favor of or against a decision, those taking into account both types of arguments, and those involving an aggregation of these arguments into a meta-argument.

Müller and Hunter (2012) introduce the Argumentation Decision Framework for multi-criteria decision-making (MCDM) scenarios, to identify dominated decisions using a satisfaction function grounded in extensions. Additionally, they propose a novel method for decision generation employing a backward process.

In this sub-section, we have summarised some of the most well-known approaches for decision-making with Dung abstract argumentation frameworks. However, while Dung abstract argumentation framework elegantly captures the core of argumentation, it lacks real-world complexity. Extended Dung frameworks address this by incorporating additional structure (and semantics) that enable richer and more diverse reasoning. For example, extensions of Dung abstract argumentation framework like value-based or bipolar argumentation trade some theoretical simplicity for major gains in applied utility. As shown in the next sub-section, by leveraging these enriched models, it is possible to create decision-making systems that have more complex reasoning.

2.2 Extended Dung frameworks

An extended Dung framework is an abstract argumentation framework (with arguments and attacks) that also contains added features such as weights, values, supports, etc. Those added features can add expressivity in the decision-making process.

Amgoud and Cayrol (2002) introduced the notion of preference-based argumentation framework. The main idea is that some arguments might be stronger than others, which is captured by a binary preference relation between arguments. The authors propose to take these preferences into account by deleting an attack from argument a to argument b if b is strictly preferred to a. This framework can be used as the basis for a decision-making system. For instance, Amgoud et al. (2008) propose a system that takes as input different arguments, preferences and attacks between them, as well as the possible options (each option is supported by different arguments). The system returns as output the status of each option (skeptical, universal, argued, credulous, non-supported, or rejected) and the total order on the set of options based on the argument status. In this work, the status is defined on the basis of different inference mechanisms.

Value-based argumentation frameworks (VAFs), introduced by Bench-Capon (2003), is an abstract argumentation framework extended with three elements: a non-empty set of values, a function that associates arguments to values, and a preference relation on values. The general idea is that arguments can be compared based on an agent's preference on their underlying values. Based on VAFs, Atkinson *et al.* (2006b) describe how a Drama (Deliberative Reasoning with ArguMents about Actions) agent can be used for decision support. The process is composed of several steps. It obtains justifications for the available courses of action using the argument scheme of Atkinson and Bench-Capon (2007), generates counter arguments using the critical questions associated with that scheme and, once the factual issues have been resolved, uses a VAF to determine the actions compatible with various value orderings. The authors highlight the use of this agent in the medical domain with a decision example of a patient threatened by blood clotting. While not exactly within the domain of decision-making, there are notable uses of VAFs in related domains. For example, Bench-Capon *et al.* (2015); Wyner *et al.* (2012) show how VAFs can be implemented in tools to automatically and systematically critique policy proposals elicited from citizens or in persuasion in the legal domain Bench-Capon *et al.* (2005). In their original work, Bench-Capon (2003) also propose a strategic heuristics in VAFs which can be used to change the status of an argument by extending the VAF strategically. Atkinson and Bench-Capon (2007) study decision making (what is best for an agent to do in a given situation) using argument schemes and critical questions. They illustrate the system on several toy examples.

A bipolar argumentation framework is an abstract argumentation framework with an additional binary relation (apart from attacks) representing supports between arguments. In the human experiments conducted by Polberg and Hunter (2018), this additional support relation has been shown to be necessary to fully represent the complexity of human reasoning. It is not surprising that these frameworks have been extensively used for decision making and that regular techniques for Dung abstract framework were extended to it. For example, many works have adapted the acceptability of arguments to these bipolar frameworks Cayrol and Lagasquie-Schiex (2005); Potyka (2021). Similarly, ranking-based semantics have been developed for those frameworks. QuAD (Baroni et al., 2015) and Df-QuAD (Rago et al., 2016b,a) are semantics for bipolar argumentation frameworks that can act as automated decision support tools. They enable the quantification of the strength of alternative decision options, based on aggregation of the strengths of the arguments that support and attack them. An example of an application based on this framework is the tool by Karamlou et al. (2019b) which uses labelling algorithms Karamlou et al. (2019a) for deciding the winner in debates. We also mention the Attractor Java library (Potyka, 2022) implements several famous gradual semantics such as Df-QuAD, Euler-based Amgoud and Ben-Naim (2017), Quadratic Energy Potyka (2018), and MLP-based semantics Potyka (2020). In this setting, Himeur et al. (2021) propose a formalism to assess which agent has the most influence on a particular argument. The idea is that the impact of individual argument of an agent on a specific argument is assessed (similar to the work of Delobelle and Villata (2019) for Dung abstract frameworks) before they are aggregated. This is particularly important in the decision setting when agents have to decide which argument has to be advanced next. Gorur et al. (2023) propose the ArguCast tool, for humans to make predictions about specific future events. In his framework, natural language arguments are linked with supports and attacks and augmented with votes and forecasts (as numbers between 0 and 1). Irrational users are detected using a gradual semantics and their opinions are filtered out before obtaining the group forecasting predictions.

Please note that there exist many other extended Dung frameworks that we did not mention, such as the set of attacking argument (SETAF) frameworks (Nielsen and Parsons, 2007), the extended framework by Modgil (2009), or the weighted argumentation framework (Amgoud et al., 2022). Those frameworks are not mutually exclusive and can be combined, e.g., see the weighted bipolar SETAFs (Yun and Vesic, 2021).

2.3 Structured argumentation

Contrary to abstract argumentation, where arguments are left unspecified, structured argumentation systems formalize non-monotonic logical reasoning by emphasizing the dialectical interaction between arguments and counterarguments. Through the application of *structured argumentation frameworks*, one can discern the principles for constructing arguments, infer the relationships among arguments, compare conflicting arguments, and determine which arguments ultimately prevail in the face of competition. As a result, an argumentation process can be delineated into three distinct phases, as proposed by Prakken and Sartor (2002): the logical/structural layer, which deals with argument construction; the dialectical/relational layer, responsible for inferring argument relations; and the procedural layer, which addresses how arguments and their relations are employed to reach a decision. Note that Atkinson *et al.* (2017) distinguish 5 layers by further splitting the procedural layer into the dialogical (how to use arguments in dialogues), assessment (how to evaluate arguments), and rhetorical layer (how to persuade using argumentation).

The rest of this sub-section showcases examples of several structured argumentation frameworks and how they are used for decision-making: ASPIC+ (Modgil and Prakken, 2014; Prakken, 2010), Assumption-based Argumentation (ABA) (Toni, 2014), logic-based (sometimes called *deductive argumentation*) framework (Besnard and Hunter, 2008), DeLP García and Simari (2004), as well as custom-made frameworks.

In the Argumentation Service Platform with Integrated Components (ASPIC+) framework[3], basic arguments are atomic logical formulae and complex arguments are constructed using (defeasible or strict) rules and other arguments. It also possesses mechanisms to lift preferences from the logical elements (rules or formulae) to arguments. An online implementation of this framework, called TOAST (Snaith and Reed, 2012), was proposed but is currently not available.

There are many decision-making applications of ASPIC+ in the literature. For example, Croitoru *et al.* (2012) proposed the layout of a decision support system, in the context of the "EcoBioCap" project, which can aggregate the preferences and expertise of multiple project stakeholders through an ASPIC+ framework. Buche *et al.* (2013) implemented a first version of a decision system with a negotiation engine to

---

[3]The ASPIC+ is a development from the following project https://cordis.europa.eu/project/id/002307

help agents decide when a conflict occurs. Later on, Yun *et al.* (2018a) refined this structured argumentation system with the use of preferences and ranking-based semantics. They showcased the capabilities of this tool for decision-making in the context of the post-harvest environmental impact of fresh foods and presented a complete workflow that goes from the preference collection via polls to the ranking of solutions. There are also many applications of ASPIC+ in the medicine domain as it was initially funded by Cancer Research UK. For example, Fox *et al.* (2007) presented how Logic of Argumentation (LA) and ASPIC (the early version of ASPIC+) were used for decision-making in the medical domain. The web prototype tool is unfortunately not available anymore. We would also like to mention that ASPIC+ was also widely used to model legal decision making, e.g., Prakken *et al.* (2015) and Prakken (2020).

As outlined in Toni's tutorial (Toni (2014)), the assumption-based argumentation framework (ABA) Dung *et al.* (2009) provides a methodology for constructing arguments and establishing attacks within a deductive system using foundational elements: assumptions and their contraries. The central concept revolves around forming arguments grounded in a specific subset of assumptions. Each assumption is accompanied by a set of contraries, akin to "levers" that facilitate the initiation of debates centered around the given assumption. The attack relationship between two arguments in this framework is fundamentally rooted in the presence of opposing contraries. An extension of ABA, called ABA+, was introduced in the PhD thesis of Cyras (2017). ABA was used to create many decision-making frameworks and tools. For example, Fan and Toni (2013) show how to compute and explain decisions using ABA. Roughly speaking, they propose two frameworks. In the first one, they introduce strongly (resp. weakly) dominant ABA frameworks to compute strongly (resp. weakly) dominant decisions. In the second one, they show how to build an ABA framework when preferences over goals are introduced. We refer the reader to (Cyras *et al.*, 2017) for an overview of ABA, including its non-flat (without preferences) versions and its applications in non-monotonic and defeasible reasoning. A variety of argumentation tools from the Computational Logic and Argumentation group (CLArg) group at the Imperial College London is available at https://clarg.doc.ic.

ac.uk/clarg-design/website/content_software.html. One notable tool, in the context of this chapter, is Arg&Dec[4] which allows users to choose between several answers to an issue by making use of the QuAD (Baroni et al., 2015) and DF-QuAD (Rago et al., 2016b) semantics.

In the logic-based (or deductive) framework, introduced by Besnard and Hunter (2008), an argument consists of a couple: its support is a set of logical premises and its conclusion is a formula that is inferred from its support. An instantiation of this framework was proposed by Croitoru and Vesic (2013) for knowledge-based with existential rules and many generators were implemented using the Graal engine (Baget et al., 2015). For instance, Yun et al. (2018b) proposed a tool, called DAGGER, to generate those arguments (with binary attacks). A subsequent tool, called NAKED (Yun et al., 2019), was introduced to take into account n-ary attacks. Another example is the work of Kakas and Moraitis (2003) which presents a dynamic argumentation framework where arguments and their strength depend on the particular context that the agent finds himself, thus allowing the agent to adapt his decisions in a changing environment. Their arguments are subsets of an argumentation theory (in the background monotonic logic) that derives a particular literal.

Defeasible Logic Programming (DeLP) (García and Simari, 2004), is a formalism that combines results of logic programming and defeasible argumentation. Namely, a defeasible logic program is a pair consisting of a set of strict rules and facts and a set of defeasible rules (similar to ASPIC+). An argument from a program is a minimal non-contradictory (with the set of strict rules and facts) set of defeasible rules that derives a particular formula. Garcıa et al. (2000) introduce a multi-agent system for the stock market domain where several agents can deliberate, monitor the stock market, and perform actions. While the application is described thoroughly, there is no link to the source code nor the application in the paper. As another example on a system buliding on DeLP, consider the work of Williams and Hunter (2007). They extend the DeLP formalism to build ontology-based argumentation frameworks (OAFs), which consist of a set of defeasible rules and facts, where the facts are ground formulae in the A-Box of an ontology. They highlight the benefits of OAFs for decision-making in the medical domain (consistency, ease, transparency,

---

[4]http://www.arganddec.com/

6 - ARGUMENTATION-BASED APPLICATIONS FOR DECISION-MAKING

etc.) through a case study about early breast cancer decision-making by formalising 117 defeasible rules, 190 classes, and 31 properties from 57 medical papers.

In the rest of this section, we will describe some of the specialised structured argumentation frameworks, used for decision-making, that do not strictly make use of any of the usual aforementioned frameworks. The Carneades system is an online tool for generating and reasoning with argument graphs (using Dung semantics) introduced in Gordon *et al.* (2007); Gordon (2013). There are currently four versions of Carneades which are available at `https://github.com/carneades`. It is used in several applications, e.g., Gordon (2011), Gordon and Walton (2012), and Gordon and Walton (2016). (Tolchinsky *et al.*, 2006) developed an argumentation-based framework for deliberation and applied it to manage the human-organ transplant selection system between hospitals, called Carrel+. Butler *et al.* (2021, 2019) propose an agent-based decision-making model of opinion diffusion and voting where influence among individuals and deliberation in a group are mixed and inspired from social modeling. In their model, arguments are non-fallacious informational cues represented by a real number that stands for how much the argument respects or supports a principle. In his work, Hecham (2018); Hecham *et al.* (2020) proposes a graph-like structure, similar to structured argument graphs, called *statement graphs* for defeasible reasoning. The author proposed an online tool (DAMN) that implemented this framework for multi-agent decision-making, where each agent has its own knowledge base that can be combined with other agents to detect and visualize conflicts and potentially solve them using a defeasible semantic. Note that while the original tool has been taken down, another implementation is available at: `https://ico.iate.inra.fr/damn/`. Gorur *et al.* (2023) propose an argumentation framework, and the associated tool called ArguCast (available at `https://argucast.herokuapp.com/`), for humans to make predictions about specific future events. In his framework, natural language arguments are linked with supports and attacks and augmented with votes and forecasts (as numbers between 0 and 1). Irrational users are detected using a gradual semantics and their opinions are filtered out before obtaining the group forecasting predictions. In the work of Thomopoulos and Croitoru (2013); Croitoru and Vesic (2013), arguments are sequences of sets of facts such that each element in the sequence derives the next one using a logical rule. Let us also mention argumentation schemes, which provide an important insight into the structures of arguments and can be used in the decision-making context Atkinson and Bench-Capon (2007); Prakken *et al.* (2015); Tolchinsky *et al.* (2012). With these few illustrative examples, we aim to convey the notion that these frameworks, despite their divergence from conventional argumentation frameworks, pave the way for innovative approaches to decision-making across a wide spectrum of domains.

3 Types of decision-making systems

In this section, we offer an in-depth exploration of existing practical tools, classifying them according to the framework introduced by Ouerdane *et al.* (2010) while also shedding light on recent additions to the literature.

Specifically, we also categorize these applications based on their target user groups: experts (tools designed to aid expert decision-making or facilitate decision visualization), buyers (tools with recommendation capabilities), simple citizens (tools aimed at mediating public debates) as well as those that function autonomously on behalf of a user (tools for autonomous decision-making) or simulate the potential outcomes of various decisions.

In Tables 1, we summarise the papers and works including concrete software that apply argumentation in decision-making. We use the following abbreviations for the domain: G - general; AF - agrifood; M - medicine; P - policy. Regarding the aforementioned type of the application, we use the following abbreviations: **rs** - recommender system; **adm** - autonomous decision-making; **sed** - supporting expert decision; **ca** - acting as a collaborative assistant; **sim** - simulation; **viz** - visualization; **mfd** - mediator for public debate; and **o** - other.

3.1 Supporting expert decision

In this section, we survey the argumentation tools for supporting expert decisions in different domains such as agrifood, medicine, or law.

In the agrifood domain, we give the example of the ECOBIOCAP tool Yun *et al.* (2018a); Croitoru *et al.* (2012); Buche *et al.* (2013) which

6 - Argumentation-Based Applications for Decision-Making

Paper	Name	Dom.	Type	Availab.	Eval.	Use-case
Kakas and Moraitis (2003)	Argumentation Based Decision Making for Autonomous Agents	G	sim, adm	✗		
Kakas et al. (2019)	GORGIAS: Applying argumentation	G	ca, sed	✗		✓
Hecham et al. (2020)	DAMN: Defeasible Reasoning Tool for Multi-Agent Reasoning	G	adm, sed	✓		
Hua and Kimbrough (1998)	On hypermedia-based argumentation decision support systems	G	viz	✗	✓	
Butler et al. (2019)	Mixing dyadic and deliberative opinion dynamics in an agent-based model of group decision-making	G	sim	✉	✓	
Gorur et al. (2023)	ArguCast: A System for Online Multi-Forecasting with Gradual Argumentation	G	sed	✓		
Gordon (2013)	Introducing the Carneades web application	G	adm, sed	✓		✓
Toniolo et al. (2014)	Argumentation-based collaborative intelligence analysis in CISpaces	G	sed	✓	✓	✓
Buche et al. (2013)	Argumentation et négociation pour le choix d'emballages alimentaires biodégradables	AF	sed	✗		
Yun et al. (2018a)	Choice of environment-friendly food packagings through argumentation systems and preferences	AF	rs, sed	✗		✓
Glasspool et al. (2006)	Argumentation in decision support for medical care planning for patients and clinicians	M	sed	✗	✓	
Williams and Hunter (2007)	Harnessing ontologies for argument-based decision-making in breast cancer	M	sed	✗		✓
Tolchinsky et al. (2006)	Increasing human-organ transplant availability: Argumentation-based agent deliberation.	M	sed	✗		✓
Kökciyan et al. (2021)	Applying metalevel argumentation frameworks to support medical decision making	M	sed, rs	✗		✓
Grando et al. (2013)	Argumentation-logic for creating and explaining medical hypotheses	M	sed	✗	✓	
Chang et al. (2009)	Mixed-Initiative Argumentation: Group Decision Support in Medicine	M	sed	✗	✓	✓
Shankar et al. (2006)	Medical arguments in an automated health care system	M	sed	✗	✓	
Wyner et al. (2012)	A Model-Based Critique Tool for Policy Deliberation	P	mfd	✗		
Bench-Capon et al. (2015)	Using Argumentation to Structure E-Participation in Policy Making	P	mfd	✗		

Table 1: Argumentation-based tools for decision-making. The column Availab. uses the following symbols: ✗ not available, ✓ available, and ✉ available upon request. The column Eval. specifies whether the tool was evaluated.

was used to choose the most suitable packaging. Indeed, the challenge is complex as achieving an optimal balance between reducing environmental burdens (such as resource consumption and waste management) and maximizing the practical benefits (specifically, reducing food losses) when designing packaging for a specific application is crucial. Moreover, stakeholders from various sectors, including the food and packaging industries, health authorities, consumers, and waste management authorities, often have a range of considerations as conflicting arguments, related to safety, practicality, and perceptions of the packaging material. The tool allows the formalisation of the stakeholders' argument and their relations, as well as inferring the different possible decisions and corresponding outcomes. DAMN Hecham et al. (2020) is a similar tool (see Section 2.3) which has been used to formalise reasoning (using defeasible rules) and support expert decisions in the agrifood domain. While the former tool is no longer available, the latter is available for use but the source code can no longer be found. There are also many theoretical works using argumentation in agrifood that have not been implemented as tools, e.g. Thomopoulos and Croitoru (2013). While Thomopoulos et al. (2018) propose to combine argumentation and system dynamics simulation to enrich the deliberation process when considering various available options to agrifood chain stakeholders when considering the adoption of cereal-legume intercrops as an alternative to sole crops, their method was not implemented as an end-to-end tool but used available software solutions (like Aspartix[5] or the Anylogic platform[6]) instead. Lastly, we note that argumentation can also be used by experts to analyse answers from a survey. Kurtz and Thomopoulos (2021) used argumentation to analyze consumer priorities and perceptions of hazards in infant foods using answers from 1,750 French citizens.

In the medical domain, several argumentation-based tools have also been implemented. Williams and Hunter (2007) present the Ontology-based Argumentation Framework (OAF) that connects a logic-based argumentation method with description logic ontologies. They show that this framework can be used to model a breast cancer treatment decision-making. The case study they present showcases the five advantages of

---

[5]https://www.dbai.tuwien.ac.at/research/argumentation/aspartix/
[6]https://www.anylogic.fr/

6 - ARGUMENTATION-BASED APPLICATIONS FOR DECISION-MAKING

this framework. Unfortunately, the link to the breast cancer ontology provided in the paper is no longer available. In their paper, Chang et al. (2009) presented a web prototype tool for group decision support. They applied the tool with oncologists to discuss treatment therapy for cancer cases in the head to neck region. Multiple argumentation-based decision-making cycles are performed and the final decision is produced based on the accrual of all the arguments over all cycles. While the authors provide screenshots of the application, neither the source code nor the the application is currently available. Grando et al. (2013) noticed that the EIRA (Moss et al., 2009), an existing system that can detect additional anomalies from some anomaly details entered by an intensive care unit clinician, could not describe the rationales behind its predictions. As a result, they developed a new tool, called argueEIRA, which can provide argumentation-based justification system that formalizes and communicates to the clinicians the reasons why a patient response is anomalous. They provide multiple examples of justifications but we note that their evaluation only consisted of three clinicians. While the authors provide screenshots of the application, neither the source code nor the the application is currently available. In their work, Glasspool et al. (2006) introduced a software application called REACT (Risks, Events, Action and their Consequences over Time). This tool is designed to assist both clinicians and patients in medical planning. Its approach serves as a versatile support system, aiding clinicians and patients alike in visualising, customising, evaluating, and discussing care plans. Shankar et al. (2006) presented WOZ as a framework for explaining a clinical decision-support system. Originally a component of the EON[7] architecture (Musen et al., 1996), WOZ became integrated into the ATHENA decision-support system (Goldstein et al., 2000) as an explanation module. While there are some screen captures in the paper, the software itself and its source code are not accessible.

We would also like to mention some related theoretical work and surveys about supporting expert decision in the medical domain. Hunter and Williams (2010) developed a general argumentation-based framework for representing medical data, including the case with multiple outcome

---

[7]EON is a knowledge-based system architecture and a set of software components with which developers can build robust guideline-based decision-support systems

indicators (e.g. disease-free survival, or overall survival). The goal is to compare available treatments by grounding the decision on evidence from clinical trials, systematic reviews, meta-analyses, etc. The rules from the knowledge base are used to construct arguments that claim that one treatment is better than another according to some evidence. A pertinent question arises when considering such argumentation-based models: how do they fare in predictive performance compared to established machine learning methods? Longo and Hederman (2013) tackled this by comparing the performance of argumentation-based decision aiding models (e.g. preferred semantics) with well-established machine learning classifiers (decision tables, bayesian network, etc.). Surprisingly, their findings showcased that most of these models either performed equally or, in some cases, worse than the argumentation-based approach. Atkinson *et al.* (2006b); Atkinson and Bench-Capon (2007) introduce a general theoretical framework for decision-making based on argumentation. In particular, they study the application in supporting expert decision in the medical domain (the treatment of a patient). Their agent, Drama, synthesized information from various sources to generate arguments aiding the decision-making process. Gorogiannis *et al.* (2009) proposed a language for encoding and synthesizing knowledge from clinical trials, especially in modeling treatment efficacy. Their theoretical framework facilitated the construction of arguments derived from clinical trial data, as evidenced in a small case study concerning chemotherapy regimens for ovarian cancer. For a more expansive view of works applying argumentation in the medical domain, comprehensive surveys are available. Atkinson *et al.* (2017) and Fox *et al.* (2007) have conducted surveys that delve deeper into the utilization of argumentation in medical decision-making, providing additional insights and perspectives on this evolving field. With these few studies, we wanted to showcase the breadth of theoretical frameworks and empirical analyses exploring the application of argumentation in medical decision support systems.

In the law domain, the core concept of employing argumentation-based decision-making is to utilize both supportive and opposing arguments, creating a framework that facilitates the formulation of informed legal judgments. While there are no end-to-end tools (from natural language text to decision), many machine learning models have been created for specific tasks. For example, in Prakken and Ratsma (2022), the authors develop a formal top-level model for explaining the outputs of machine-learning algorithms (decision trees, SVC, Gaussian naive Bayes, logistic regression, etc.) used for decision-making in law[8]. A similar approach to explanation in the law domain, using argumentation for the decision-making, is the work of Branting *et al.* (2021).

Collenette *et al.* (2023) presents a web-based time-saving tool to support the legal community by predicting Article 6 case outcomes and determining admissibility at the European Court of Human Rights. Designed as Abstract Dialectical Frameworks (ADF), this tool prioritizes explainability and adaptability to legal changes. Achieving 97% accuracy, the Article 6 prediction tool received positive feedback from ECtHR lawyers, particularly for admissibility determinations. Mumford *et al.* (2022, 2023) presented and evaluated several machine learning models (one based on H-BERT and ADF and another based on H-BERT and ADM) for reasoning with legal cases. To classify cases (violation or no-violation) in an explainable manner, they propose to first ascribe factors using NLP and then reason over them using an argumentation model.

There are also some domain-agnostic argumentation-based tools to support expert decisions. Gorgias (Kakas *et al.*, 2019) is a general argument-based supporting expert decision tool which has been applied in several scenarios like medical support, network security, business computing, and cognitive personal assistants. There also exists a tool, Gorgias-B, which aims at simplifying the development of decision-making applications by generating underlying argumentation code from high-level requirements, which can then be fed into Gorgias. CISpaces (Collaborative Intelligence Spaces) (Toniolo *et al.*, 2014) is a decision support tool which supports intelligence analysis by integrating argumentation, provenance and crowd sourcing AI techniques for effective interpretation of evidence. Although it was devised to support intelligence analysts in the military, the approach is general enough that it can be applied in any domain. An evaluation of the tool was performed (Toniolo *et al.*, 2023)

---

[8]However, we note that the actual experiments were not performed with the data from the legal domain, but involve the datasets about employee churn, poisonous mushrooms and graduate admissions.

showing that it improves analysts' daily activities and help them create more robust and credible hypotheses to improve understanding of complex situations. The source code of CISpaces is available on Github[9]. The SAsSy demonstrator tool (Oren *et al.*, 2020) is an argumentation-based tool to support experts in their decisions by providing visual explanation of complex plans. The tool describes how arguments can be generated from domain rules, and what arguments are justified through dialogues. While not mentioned in the paper, the source code can be found on Bitbucket[10].

3.2 Mediator for public debate

A mediator in a public debate serves as a facilitator, responsible for guiding the discussion, ensuring fairness, and fostering communication between conflicting or diverse viewpoints.

Morge (2004) presents a computer-supported collaborative argumentation system for the public debate based on the Analytic Hierarchy Process (Saaty, 1980). He constructed a dialogue system of agents with reasoning abilities to support the group decision, where each user is assisted by an agent representing them in automated dialogues. While the author promises to implement (and evaluate) the system in the future, it is unknown whether an implementation exists. However, we recommend the interested reader to access the author's thesis Morge (2005) on dialectics multiagent system to support deliberation.

The "Risk Agora" (McBurney and Parsons, 2000, 2001; Rehg *et al.*, 2005) uses a model of dialectical argumentation and has been proposed as a system to support deliberations over the potential health and environmental risks of new chemicals and substances, and the appropriate regulation of these substances. While this system formally models debates in the risk domain, no implemented tool has been identified.

The PARMENIDES online tool (Atkinson *et al.*, 2006a; Atkinson, 2006) was designed to foster public participation and debate, specifically centered on the Government's justifications for proposed actions. The idea is to enable members of the public to submit their opinions about the Government's justification of a particular action. The authors provided

---

[9]https://github.com/CISpaces
[10]https://bitbucket.org/rkutlak/sassy

6 - Argumentation-Based Applications for Decision-Making

a prototype of the system on the topic ""Is Invasion of Iraq Justified?" but only the front page of the tool is currently working.

Wyner *et al.* (2012) present a novel approach in their argument-based mediator for policy proposals. Departing from the conventional model where institutions propose policies critiqued by users, they adopt a converse method. Here, users propose policies, triggering automated critiques by software agents. This innovative dialogue engages citizens, potentially deepening their comprehension of issues and constraints linked to proposed solutions. The authors also provided a web-based tool, which is, unfortunately, no longer available. Another related paper (Bench-Capon *et al.*, 2015) discusses the limitations of existing e-participation tools in the domain of computational argumentation. The authors introduce an argumentation system justifying policy proposals and a mechanism for auto-generating arguments. Through two prototype tools, they aim to present justifications and elicit user-proposed policy justifications.

These diverse studies highlight the evolving landscape of mediation within public discourse. From novel approaches in policy proposal critiques to tools fostering public participation, each contribution shows how the technology can enhance the dialogue and understanding in societal decision-making.

3.3 Acting as a collaborative assistant

In this section, we focus on argumentation-based tools that assist a human user by anticipating needs, performing the tasks they are well suited for, and leaving the remaining tasks to the human. Those tools have a strong focus on human-computer interaction, as they are based on a synergy between the human and the system.

Ferguson *et al.* (1996) implemented an interactive planning system for solving routing problems in transportation domains, where the interaction was implemented as a form of dialogue. We note that this work was part of the TRAINS project (1995), whose web page is still running [11]. Another system that was primarily aimed at supporting decision-making in urban planning is ZENO (Gordon and Karacapilidis, 1999), which is

---

[11] https://www.cs.rochester.edu/research/trains/

based on the model of argumentation developed by Toulmin and Rittel. This developed into the group decision support system developed by Karacapilidis and Papadias and called HERMES (Karacapilidis and Papadias, 1996, 1998, 2001), which was implemented in Java.

Kökciyan *et al.* (2018) developed an argumentation-based decision-support tool, called CONSULT, that helps patients to self-manage some chronic conditions. The system is personalized and can take into account the preferences of both the patient and the clinician. The arguments are first constructed and then analyzed, to resolve inconsistencies regarding different options for the treatment of the patient.

We already mentioned Gorgias (Kakas *et al.*, 2019) in the subsection about domain-agnostic argumentation-based tools that support expert decisions. We note that it can also be seen as a collaborative assistant.

We finish this subsection by mentioning the work by Cyras *et al.* (2019), who aim at bridging the gap between the black box optimization solvers and explainable systems. They define a new framework that uses argumentation to improve the interaction between an optimization solver and a user. Abstract argumentation is used as an intermediate layer to explain why a proposed schedule is good or not. The system is based on tractable explanations that support or attack the solutions. Namely, the authors show how to extract the explanations from the argumentation graphs and how to generate natural language explanations from the argumentative ones.

3.4 Recommendation system

A recommendation system is an algorithmic tool that suggests items or content based on user preferences and/or past interactions. In this setting, recommendation systems may utilize argumentation methods to offer logical recommendations or suggestions backed by a rationally justified approach. We now give some examples of work about such recommender systems. ArgueNet is a recommender system that classifies search results according to preference criteria specified by the user (e.g. "I always find newspapers written by X relevant"). The proposed approach operates on top of a traditional web search engine (usually based on traditional ranking algorithms) with a defeasible argumentation framework (Chesnevar and Maguitman, 2004). Unfortunately, the approach was not implemented as the authors noted that "a web question-answering system with deductive capabilities is still far from becoming a reality". We also refer the interested reader to the related book chapter "Recommender System Technologies based on Argumentation" (Chesñevar et al., 2007). Similarly, Briguez et al. (2012) proposed an argumentation-based music recommender system based on DeLP to model user preferences in terms of rules which are then used to create the recommendations. In the education domain, Rodríguez et al. (2015) proposed a hybrid recommendation method based on argumentation theory (DeLP) that combines content-based, collaborative and knowledge-based recommendation techniques to recommend learning objects fitted to the student's characteristics. In the medical domain, Kökciyan et al. (2021) proposed a system for patients to self-manage their condition and to adhere to agreed-upon treatment plans by giving recommendations and template-based explanations. As shown above, many other works have integrated argumentation techniques in recommender systems with a focus to improve the recommendation quality as its inference abilities can generate recommendations and its structured reasoning in a systematic manner (Bedi and Vashisth, 2011, 2014).

There was also a focus to emphasize explaining the recommendations in those systems. For example, Heras et al. (2009) proposed an argumentation-based social recommender system to provide justification of the recommendations to the user based on the preferences of their neighbors. Lamche et al. (2014) developed an interactive mobile shopping recommender system for fashion items. The system uses a multi-criteria decision-making method (MCDM) to assess which arguments are included in the explanations. The source code of the prototype is available on Github. Similar is the work of Rago et al. (2020) which also creates and evaluates a framework for interactive explanations for recommendations. By making use of the interaction between the system and the user, they provide better explanation but also feedback user's preferences to the system, allowing for better future recommendations. While there is still some debates on the usefulness of argumentation-based explanations in recommender systems, Naveed et al. (2018) performed some human experiments and shows that argumentation-based explanations increase the perceived explanation quality, information sufficiency, and overall satisfaction with the system.

3.5 Other useful visualisation tools for decision-making

There are many tools for debate and argument visualisation that are available online. In most of the tools presented in this section, arguments are represented within abstract argumentation frameworks. Each argument is accompanied by annotations, which may include natural language descriptions, pictures, and other types of information. Additionally, various binary relations, such as attack and support relations, are defined between the arguments. However, the level of automatic processing in those platforms is low or non-existent, as their main purpose is to help visualise the arguments, and not to act as a decision support system. In Table 2, we summarise a (non-exhaustive) list of tools, their link, and whether they are available (accessed on November 2024).

Kialo is a platform for rational debate where users can join or start discussions on various topics, such as politics, philosophy, science, ethics, gender, religion and more. Each debate is represented with a tree where each node is a textual argument. For each argument, users can see its history (how it was edited), votes on it (representing its perceived impact), its pro and con arguments, and similar arguments. Kialo was made for users to map discussions, see complex issues easily, learn from their insights and opinions. An education version of Kialo, called Kialo Edu, is also available for educators to host classroom debates and assess students' critical reasoning. The graphs from Kialo can easily be exported and have been used extensively by the argumentation community Boschi et al. (2022); Agarwal et al. (2022). In June 2024, Kialo has more than 19,000 debates for a total of more than 724,000 arguments.

IDebate's Debatabase is a repository of over 700 debates, mostly written by experienced debaters, on various topics, such as politics, economics, religion, culture, science and society. The website aims to help users learn about different perspectives and arguments on important issues, and to make informed decisions. Each debate is usually composed of a statement title (e.g. "This House would force feed sufferers of Anorexia Nervosa"), a short description, points for or against the statement, and a bibliography. While the website is still ongoing, the debates cannot be easily exported and no new debates are being produced (all of them

Name	Link	Ongoing
Kialo	https://www.kialo.com/	yes
IDebate	https://idebate.net/resources/debatabase	yes
DebateGraph	http://debategraph.org	yes
DebateHub	https://debatehub.net/	yes
Litemap	https://kmi.open.ac.uk/technologies/name/litemap	yes
Argunet	http://www.argunet.org/	no
Republique-Numérique	https://www.republique-numerique.fr	no
GrandDebat	https://granddebat.fr/	no
Futureu	https://futureu.europa.eu/	no
Arguman.org	https://github.com/arguman/arguman.org	no

Table 2: List of visualisation tools

have been created in 2022).

Debategraph is a platform that allows users to create, explore, contribute to, and share free networks of thought on various topics, especially complex public policy issues. Well-known contributors of maps are CNN, the White House, and the Foreign Office. To explore a map, users are presented with a star shape-directed graph with the center node being the current node. Each node has an associated description while the meaning of elements is conveyed through the shape and colors of the arrows/nodes.

DebateHub (Quinto *et al.*, 2021) is a tool for collective ideation (forming ideas or concepts) and deliberation. It allows users to set up, participate in, and vote on online challenges and debates on various topics. It aims to support democratic decision-making and collective knowledge creation. Namely, users can propose new debates, discuss and argue for or against the ideas by proposing debate issues, reduce and select the most promising debate issues based on the analysis of the arguments, and vote for the best issues to pursue further. DebateHub's community does not appear to be as active as Kialo's one with most debates having less than 10 debate issues and less than 10 members.

LiteMap (Zargayouna and Amara, 2006; Okada and De Liddo, 2018) is a web-based tool that allows online debates to be navigated and comprehended across diverse forums and platforms. It provides annotation and markup capabilities accessible via any web browser, allowing easy extraction and analysis of snippets of text from online conversations, and fostering deeper reflection and discussion.

Argunet (Schneider *et al.*, 2007) is a suite of visualisation tools that allows users to create their own argument maps (using Argunet Editor) and embed them as Javascript widgets into any webpage (using Argunet Browser), allowing user to interact with the arguments. Note that Argunet has been discontinued and no new versions are currently planned.

Republique-Numérique was a pivotal platform for French citizens, allowing engagement with the "République numérique" bill from September 26th to October 18th, 2015. This legislation encompassed three core objectives: fostering openness in data and knowledge, ensuring internet users' privacy, and enhancing citizens' digital accessibility. Through this platform, French citizens actively participated by proposing new ideas, casting votes (agree, neutral, disagree) on others' propositions, and presenting arguments supporting or challenging these proposals. Contributions underwent a ranking and aggregation process, followed by government responses. Crucially, citizens had the opportunity to witness how their input shaped the bill's evolution. Although the website no longer accepts new submissions, the valuable data remains accessible online, encapsulating a significant moment of citizen engagement with legislative processes. For more details on this action, we refer the interested reader to the PhD thesis by Gruson-Daniel (2018).

Le Grand Débat is a public debate launched by French President Emmanuel Macron on January 15, 2019, in response to the Yellow Vest movement. The debate was organized around the following themes: ecological transition, taxation, democracy and citizenship, and organization of the state and public services. The debate was open to all French citizens and was conducted through 10,000 local meetings with an average of 70 participants per debate and nearly 2 million contributions on the

6 - ARGUMENTATION-BASED APPLICATIONS FOR DECISION-MAKING

dedicated website. The National Commission for Public Debate (CNDP) co-organized the debate and provided kits and presentations to help debate organizers lead conversations. The Grand Débat was intended to be a consultative tool to help the government understand the wishes of the French people regarding the policies that affect them. The website is currently no longer accessible.

The Conference on the Future of Europe was a multilingual platform that allowed European citizens to debate on Europe's challenges and priorities. The European Parliament, the Council, and the European Commission have committed to listen to the feedback of citizens and to act according to the recommendations, within their sphere of competences. The Conference concluded its work in May 2022 with the submission of 49 proposals to the European institutions[12]. The final report is accessible on their website.

Arguman.org is an argument analysis and mapping platform largely built by developers in Turkey. It is similar to Kialo but allows for a more fined-grained modelization with relations such as "but", "because", and "however". Other users can also comment, report fallacies and suggest changes. The website implementing the software is down as of 2023 but the source code is available on Github.

In this section, we gave you a glance at the landscape of online debate tools. These tools, from Kialo's structured debates to platforms like Debategraph and DebateHub fostering collective ideation, exemplify diverse approaches to enhancing discourse, yet their functionalities and focus vary significantly. Each platform caters to specific needs, providing unique avenues for engagement and insight, but a common challenge persists in achieving a higher level of automated processing within these debate frameworks.

---

[12]https://commission.europa.eu/strategy-and-policy/priorities-2019-2024/new-push-european-democracy/conference-future-europe_en

4 Limitations, challenges, and future directions

4.1 Limitations and challenges

Argumentation-based decision-making tools have been developed over several decades in various fields, offering a structured approach to weighing evidence and reaching informed decisions. However, there are several short-term and long-term limitations, which we mention in this section.

Subjectivity. The argumentation tools rely on the quality of the input arguments. Human bias and subjectivity can influence the creation and selection of arguments, especially in cases where they are gathered and assessed by non-experts. Namely, users often introduce personal beliefs or perspectives that are not always objective. We have seen that when a large number of stakeholders are involved in the decision-making process, mediators or moderators are often needed to lead the debate, refine arguments, or delete malicious ones.

Incomplete information. The effectiveness of argumentation tools depends on the availability and accuracy of information - incomplete or inaccurate data can compromise the reliability of arguments, and the overall ability of the tool to adequately take into account the evolving circumstances of unforeseen variables.

Formalization level. Very structured systems can take into account and model all the details and, consequently, provide a precise treatment of information. However, asking the users to comply with those standards is often unrealistic. This is why many systems use simplified versions of argumentation schemes, which, in turn, might lead to oversimplification and the danger of missing some sophisticated differences between available options.

Scarcity. The unfortunate scarcity of accessible tools raises questions about the continuity and accessibility of these innovative systems. As technology continues to intersect with public engagement, the need for sustained development and accessibility of these tools becomes crucial, ensuring that the promise of enriched dialogue and informed decision-making remains within public reach.

Ethical considerations. The use of argumentation involves biases and emotions and raises ethical concerns, particularly regarding the potential reinforcement of existing biases or the unintended consequences of

6 - ARGUMENTATION-BASED APPLICATIONS FOR DECISION-MAKING

automated decision-making. Ensuring ethical use and avoiding discrimination requires ongoing vigilance and careful modeling of the systems by the authors.

4.2 Future directions

We now comment on some future directions.

Traditional decision-making methods often necessitate users to formalize their data beforehand, demanding a high level of expertise. In contrast, visualization tools rely on natural language, offering a more accessible entry point for users. We anticipate that this void will be bridged by advancements in neuro-symbolic AI. While existing transformer-based machine learning models (Vaswani *et al.*, 2017) already excel in argument mining and inferring relationships between arguments, they tend to be tailored to specific domains (Mayer *et al.*, 2021; Goffredo *et al.*, 2022). However, with the rapid evolution of large language models (GPT-4, Claude AI, etc.), we foresee that their enhanced generalization capabilities will enable the development of user-friendly agnostic argumentation-based decision-making tools.

Ensuring the explainability of argument-based decision-making systems is paramount in gaining user trust and ensuring transparency (Cyras *et al.*, 2021; Anaissy *et al.*, 2024; Kampik *et al.*, 2024). Argumentation has the intrinsic power of being understandable; however, the creators of the systems and platforms must provide a conscious and well-defined effort to keep the interface intuitive and user-friendly. Moreover, while argumentation is often described as intrinsically explainable, it is not always the case for non-experts. We argue that more empirical evaluations are needed to assess the real explanatory power of such explanations. We now see more interest focusing on the explainability component of decision-making tools (Rago *et al.*, 2021).

Another issue linked with this is the fact that, even if the system looks understandable and explainable to its creators, it might not be understandable by everybody, given the individual differences between the users. User studies must be done to asses to which extent the system is aligned with human reasoning, and whether its result is similar to that of a (rational) human reasoner. Another related question is should the system be able to describe human reasoning (including factors such as emotions or biases) that deviate from the normative approach.

Acknowledgements

We would like to thank the reviewers for their useful feedback.

Srdjan Vesic benefited from the support of the project ANR AG-GREEY (ANR-22-CE23-0005) and the AI Chair project Responsible AI (ANR-19-CHIA-0008) of the French National Research Agency (ANR).

References

Vibhor Agarwal, Sagar Joglekar, Anthony P. Young, and Nishanth Sastry. Graphnli: A graph-based natural language inference model for polarity prediction in online debates. In Frédérique Laforest, Raphaël Troncy, Elena Simperl, Deepak Agarwal, Aristides Gionis, Ivan Herman, and Lionel Médini, editors, *WWW '22: The ACM Web Conference 2022*, pages 2729–2737. ACM, 2022.

Latifa Al-Abdulkarim, Katie Atkinson, Trevor J. M. Bench-Capon, Stuart Whittle, Rob Williams, and Catriona Wolfenden. Noise induced hearing loss: Building an application using the ANGELIC methodology. *Argument Comput.*, 10(1):5–22, 2019.

Leila Amgoud and Jonathan Ben-Naim. Ranking-Based Semantics for Argumentation Frameworks. In *Scalable Uncertainty Management - 7th International Conference, SUM 2013*, pages 134–147, 2013.

Leila Amgoud and Jonathan Ben-Naim. Evaluation of Arguments in Weighted Bipolar Graphs. In *Symbolic and Quantitative Approaches to Reasoning with Uncertainty - 14th European Conference, ECSQARU 2017*, pages 25–35, 2017.

Leila Amgoud and Claudette Cayrol. Inferring from inconsistency in preference-based argumentation frameworks. *J. Autom. Reason.*, 29(2):125–169, 2002.

Leila Amgoud and Henri Prade. Explaining qualitative decision under uncertainty by argumentation. In *Proceedings, The Twenty-First*

National Conference on Artificial Intelligence and the Eighteenth Innovative Applications of Artificial Intelligence Conference, July 16-20, 2006, Boston, Massachusetts, USA, pages 219–224. AAAI Press, 2006.

Leila Amgoud and Henri Prade. Using arguments for making and explaining decisions. *Artif. Intell.*, 173(3-4):413–436, 2009.

Leila Amgoud, Yannis Dimopoulos, and Pavlos Moraitis. Making decisions through preference-based argumentation. In Gerhard Brewka and Jérôme Lang, editors, *Principles of Knowledge Representation and Reasoning: Proceedings of the Eleventh International Conference, KR 2008*, pages 113–123. AAAI Press, 2008.

Leila Amgoud, Dragan Doder, and Srdjan Vesic. Evaluation of argument strength in attack graphs: Foundations and semantics. *Artificial Intelligence*, 302:103607, January 2022.

Leila Amgoud. Argumentation for decision making. In *Argumentation in artificial intelligence*, pages 301–320. Springer, 2009.

Caren Al Anaissy, Jérôme Delobelle, Srdjan Vesic, and Bruno Yun. Impact measures for gradual argumentation semantics, 2024.

Katie Atkinson and Trevor J. M. Bench-Capon. Practical reasoning as presumptive argumentation using action based alternating transition systems. *Artif. Intell.*, 171(10-15):855–874, 2007.

Katie Atkinson, Trevor J. M. Bench-Capon, and Peter McBurney. PARMENIDES: facilitating deliberation in democracies. *Artif. Intell. Law*, 14(4):261–275, 2006.

Katie Atkinson, Trevor J. M. Bench-Capon, and Sanjay Modgil. Argumentation for decision support. In Stéphane Bressan, Josef Küng, and Roland R. Wagner, editors, *Database and Expert Systems Applications, 17th International Conference, DEXA 2006*, volume 4080 of *Lecture Notes in Computer Science*, pages 822–831. Springer, 2006.

Katie Atkinson, Pietro Baroni, Massimiliano Giacomin, Anthony Hunter, Henry Prakken, Chris Reed, Guillermo Ricardo Simari, Matthias Thimm, and Serena Villata. Towards artificial argumentation. *AI Mag.*, 38(3):25–36, 2017.

Katie Atkinson. Value-based argumentation for democratic decision support. In Paul E. Dunne and Trevor J. M. Bench-Capon, editors, *Computational Models of Argument: Proceedings of COMMA 2006*, volume 144 of *Frontiers in Artificial Intelligence and Applications*, pages 47–58. IOS Press, 2006.

Jean-François Baget, Michel Leclère, Marie-Laure Mugnier, Swan Rocher, and Clément Sipieter. Graal: A Toolkit for Query Answering with Existential Rules. In *Rule Technologies: Foundations, Tools, and Applications - 9th International Symposium, RuleML 2015*, pages 328–344, 2015.

Pietro Baroni, Martin Caminada, and Massimiliano Giacomin. An introduction to argumentation semantics. *Knowledge Eng. Review*, 26(4):365–410, 2011.

Pietro Baroni, Marco Romano, Francesca Toni, Marco Aurisicchio, and Giorgio Bertanza. Automatic evaluation of design alternatives with quantitative argumentation. *Argument & Computation*, 6(1):24–49, 2015.

P Bedi and P Vashisth. Interest based recommendations with argumentation. *Journal of Artificial Intelligence, ANSI*, pages 119–142, 2011.

Punam Bedi and Pooja Bhatt Vashisth. Empowering recommender systems using trust and argumentation. *Inf. Sci.*, 279:569–586, 2014.

Trevor J. M. Bench-Capon, Katie Atkinson, and Alison Chorley. Persuasion and value in legal argument. *J. Log. Comput.*, 15(6):1075–1097, 2005.

Trevor J. M. Bench-Capon, Katie Atkinson, and Adam Z. Wyner. Using argumentation to structure e-participation in policy making. *Trans. Large Scale Data Knowl. Centered Syst.*, 18:1–29, 2015.

Trevor J. M. Bench-Capon. Persuasion in practical argument using value-based argumentation frameworks. *J. Log. Comput.*, 13(3):429–448, 2003.

6 - ARGUMENTATION-BASED APPLICATIONS FOR DECISION-MAKING

Philippe Besnard and Anthony Hunter. *Elements of Argumentation.* MIT Press, 2008.

Elise Bonzon, Jérôme Delobelle, Sébastien Konieczny, and Nicolas Maudet. A Comparative Study of Ranking-Based Semantics for Abstract Argumentation. In *Proceedings of the Thirtieth AAAI Conference on Artificial Intelligence*, pages 914–920, 2016.

Elise Bonzon, Jérôme Delobelle, Sébastien Konieczny, and Nicolas Maudet. Combining Extension-Based Semantics and Ranking-Based Semantics for Abstract Argumentation. In *Principles of Knowledge Representation and Reasoning: Proceedings of the Sixteenth International Conference*, pages 118–127, 2018.

Gioia Boschi, Anthony P. Young, Sagar Joglekar, Chiara Cammarota, and Nishanth Sastry. Who has the last word? understanding how to sample online discussions. In Frédérique Laforest, Raphaël Troncy, Elena Simperl, Deepak Agarwal, Aristides Gionis, Ivan Herman, and Lionel Médini, editors, *Companion of The Web Conference 2022, April 25 - 29, 2022*, page 390. ACM, 2022.

Luther Karl Branting, Craig Pfeifer, Bradford Brown, Lisa Ferro, John S. Aberdeen, Brandy Weiss, Mark Pfaff, and Bill Liao. Scalable and explainable legal prediction. *Artif. Intell. Law*, 29(2):213–238, 2021.

Cristian E. Briguez, Maximiliano Celmo Budán, Cristhian A. D. Deagustini, Ana Gabriela Maguitman, Marcela Capobianco, and Guillermo Ricardo Simari. Towards an argument-based music recommender system. In Bart Verheij, Stefan Szeider, and Stefan Woltran, editors, *Computational Models of Argument - Proceedings of COMMA 2012*, volume 245 of *Frontiers in Artificial Intelligence and Applications*, pages 83–90. IOS Press, 2012.

Patrice Buche, Madalina Croitoru, Jérôme Fortin, Patricio Mosse, Nouredine Tamani, and Rallou Thomopoulos. Argumentation et négociation pour le choix d'emballages alimentaires biodégradables. *Rev. d'Intelligence Artif.*, 27(4-5):515–537, 2013.

George H. Butler, Gabriella Pigozzi, and Juliette Rouchier. Mixing dyadic and deliberative opinion dynamics in an agent-based model of group decision-making. *Complex.*, 2019:3758159:1–3758159:31, 2019.

George H. Butler, Juliette Rouchier, and Gabriella Pigozzi. Corrigendum to "Mixing dyadic and deliberative opinion dynamics in an agent-based model of group decision-making". *Complex.*, 2021:5832319:1–5832319:1, 2021.

Claudette Cayrol and Marie-Christine Lagasquie-Schiex. On the Acceptability of Arguments in Bipolar Argumentation Frameworks. In *Symbolic and Quantitative Approaches to Reasoning with Uncertainty, 8th European Conference, ECSQARU 2005*, pages 378–389, 2005.

Chee Fon Chang, Andrew Miller, and Aditya Ghose. Mixed-initiative argumentation: Group decision support in medicine. In Patty Kostkova, editor, *Electronic Healthcare - Second International ICST Conference, eHealth 2009*, volume 27 of *Lecture Notes of the Institute for Computer Sciences, Social Informatics and Telecommunications Engineering*, pages 43–50. Springer, 2009.

Carlos Iván Chesnevar and Ana G Maguitman. Arguenet: An argument-based recommender system for solving web search queries. In *2004 2nd International IEEE Conference on'Intelligent Systems'. Proceedings (IEEE Cat. No. 04EX791)*, volume 1, pages 282–287. IEEE, 2004.

Carlos Iván Chesñevar, Ana Gabriela Maguitman, and Guillermo Ricardo Simari. Recommender system technologies based on argumentation 1. In Ilias Maglogiannis, Kostas Karpouzis, Manolis Wallace, and John Soldatos, editors, *Emerging Artificial Intelligence Applications in Computer Engineering - Real Word AI Systems with Applications in eHealth, HCI, Information Retrieval and Pervasive Technologies*, volume 160 of *Frontiers in Artificial Intelligence and Applications*, pages 50–73. IOS Press, 2007.

Joe Collenette, Katie Atkinson, and Trevor J. M. Bench-Capon. Explainable AI tools for legal reasoning about cases: A study on the european court of human rights. *Artif. Intell.*, 317:103861, 2023.

6 - ARGUMENTATION-BASED APPLICATIONS FOR DECISION-MAKING

Madalina Croitoru and Srdjan Vesic. What Can Argumentation Do for Inconsistent Ontology Query Answering? In *Scalable Uncertainty Management - 7th International Conference, SUM 2013*, pages 15–29, 2013.

Madalina Croitoru, Jérôme Fortin, and Nir Oren. Arguing with preferences in EcoBioCap. In Bart Verheij, Stefan Szeider, and Stefan Woltran, editors, *Computational Models of Argument - Proceedings of COMMA 2012, Vienna, Austria, September 10-12, 2012*, volume 245 of *Frontiers in Artificial Intelligence and Applications*, pages 51–58. IOS Press, 2012.

Kristijonas Cyras, Xiuyi Fan, Claudia Schulz, and Francesca Toni. Assumption-based argumentation: Disputes, explanations, preferences. *FLAP*, 4(8), 2017.

Kristijonas Cyras, Dimitrios Letsios, Ruth Misener, and Francesca Toni. Argumentation for explainable scheduling. In *The Thirty-Third AAAI Conference on Artificial Intelligence, AAAI 2019, The Thirty-First Innovative Applications of Artificial Intelligence Conference, IAAI 2019, The Ninth AAAI Symposium on Educational Advances in Artificial Intelligence, EAAI 2019, Honolulu, Hawaii, USA, January 27 - February 1, 2019*, pages 2752–2759. AAAI Press, 2019.

Kristijonas Cyras, Antonio Rago, Emanuele Albini, Pietro Baroni, and Francesca Toni. Argumentative XAI: A survey. In Zhi-Hua Zhou, editor, *Proceedings of the Thirtieth International Joint Conference on Artificial Intelligence, IJCAI 2021, Virtual Event / Montreal, Canada, 19-27 August 2021*, pages 4392–4399. ijcai.org, 2021.

Kristijonas Cyras. *ABA+: assumption-based argumentation with preferences*. PhD thesis, Imperial College London, UK, 2017.

Jérôme Delobelle and Serena Villata. Interpretability of gradual semantics in abstract argumentation. In Gabriele Kern-Isberner and Zoran Ognjanovic, editors, *Symbolic and Quantitative Approaches to Reasoning with Uncertainty, 15th European Conference, ECSQARU 2019, Belgrade, Serbia, September 18-20, 2019, Proceedings*, volume

11726 of *Lecture Notes in Computer Science*, pages 27–38. Springer, 2019.

Phan Minh Dung, Robert A. Kowalski, and Francesca Toni. Assumption-based argumentation. In Guillermo Ricardo Simari and Iyad Rahwan, editors, *Argumentation in Artificial Intelligence*, pages 199–218. Springer, 2009.

Phan Minh Dung. On the Acceptability of Arguments and its Fundamental Role in Nonmonotonic Reasoning, Logic Programming and n-Person Games. *Artif. Intell.*, 77(2):321–358, 1995.

Xiuyi Fan and Francesca Toni. Decision making with assumption-based argumentation. In Elizabeth Black, Sanjay Modgil, and Nir Oren, editors, *Theory and Applications of Formal Argumentation - Second International Workshop, TAFA 2013, Beijing, China, August 3-5, 2013, Revised Selected papers*, volume 8306 of *Lecture Notes in Computer Science*, pages 127–142. Springer, 2013.

George Ferguson, James F. Allen, and Bradford W. Miller. TRAINS-95: towards a mixed-initiative planning assistant. In Brian Drabble, editor, *Proceedings of the Third International Conference on Artificial Intelligence Planning Systems, Edinburgh, Scotland, May 29-31, 1996*, pages 70–77. AAAI, 1996.

John Fox, David Glasspool, Dan Grecu, Sanjay Modgil, Matthew South, and Vivek Patkar. Argumentation-based inference and decision making–a medical perspective. *IEEE Intell. Syst.*, 22(6):34–41, 2007.

Alejandro Javier García and Guillermo Ricardo Simari. Defeasible Logic Programming: An Argumentative Approach. *TPLP*, 4(1-2):95–138, 2004.

Alejandro J Garcıa, Devender Gollapally, Paul Tarau, and Guillermo R Simari. Deliberative stock market agents using jinni and defeasible logic programming. In *Proceedings of esaw'00 engineering societies in the agents' world, workshop of ecai 2000*, 2000.

David Glasspool, John Fox, Ayelet Oettinger, and James Smith-Spark. Argumentation in decision support for medical care planning for patients and clinicians. In *Argumentation for Consumers of Healthcare, Papers from the 2006 AAAI Spring Symposium, Technical Report SS-06-01, Stanford, California, USA, March 27-29, 2006*, pages 58–63. AAAI, 2006.

Pierpaolo Goffredo, Shohreh Haddadan, Vorakit Vorakitphan, Elena Cabrio, and Serena Villata. Fallacious argument classification in political debates. In Luc De Raedt, editor, *Proceedings of the Thirty-First International Joint Conference on Artificial Intelligence, IJCAI 2022, Vienna, Austria, 23-29 July 2022*, pages 4143–4149. ijcai.org, 2022.

Mary K. Goldstein, Brian B. Hoffman, Robert W. Coleman, Mark A. Musen, Samson W. Tu, Aneel A. Advani, Ravi D. Shankar, and Martin J. O'Connor. Implementing clinical practice guidelines while taking account of changing evidence: ATHENA dss, an easily modifiable decision-support system for managing hypertension in primary care. In *AMIA 2000, American Medical Informatics Association Annual Symposium, Los Angeles, CA, USA, November 4-8, 2000*. AMIA, 2000.

Thomas F. Gordon and Nikos I. Karacapilidis. The zeno argumentation framework. *Künstliche Intell.*, 13(3):20–29, 1999.

Thomas F. Gordon and Douglas Walton. A carneades reconstruction of popov v hayashi. *Artif. Intell. Law*, 20(1):37–56, 2012.

Thomas F. Gordon and Douglas Walton. Formalizing balancing arguments. In Pietro Baroni, Thomas F. Gordon, Tatjana Scheffler, and Manfred Stede, editors, *Computational Models of Argument - Proceedings of COMMA 2016, Potsdam, Germany, 12-16 September, 2016*, volume 287 of *Frontiers in Artificial Intelligence and Applications*, pages 327–338. IOS Press, 2016.

Thomas F. Gordon, Henry Prakken, and Douglas Walton. The carneades model of argument and burden of proof. *Artif. Intell.*, 171(10-15):875–896, 2007.

Thomas F. Gordon. Analyzing open source license compatibility issues with carneades. In Kevin D. Ashley and Tom M. van Engers, editors, *The 13th International Conference on Artificial Intelligence and Law, Proceedings of the Conference, June 6-10, 2011, Pittsburgh, PA, USA*, pages 51–55. ACM, 2011.

Thomas F. Gordon. Introducing the carneades web application. In Enrico Francesconi and Bart Verheij, editors, *International Conference on Artificial Intelligence and Law, ICAIL '13, Rome, Italy, June 10-14, 2013*, pages 243–244. ACM, 2013.

Nikos Gorogiannis, Anthony Hunter, Vivek Patkar, and Matthew Williams. Argumentation about treatment efficacy. In David Riaño, Annette ten Teije, Silvia Miksch, and Mor Peleg, editors, *Knowledge Representation for Health-Care: Data, Processes and Guidelines, AIME 2009 Workshop KR4HC 2009, Verona, Italy, July 19, 2009, Revised Selected and Invited Papers*, volume 5943 of *Lecture Notes in Computer Science*, pages 169–179. Springer, 2009.

Deniz Gorur, Antonio Rago, and Francesca Toni. Argucast: A system for online multi-forecasting with gradual argumentation. In Oana Cocarascu, Sylvie Doutre, Jean-Guy Mailly, and Antonio Rago, editors, *Proceedings of the First International Workshop on Argumentation and Applications co-located with 20th International Conference on Principles of Knowledge Representation and Reasoning (KR 2023), Rhodes, Greece, September 2-8, 2023*, volume 3472 of *CEUR Workshop Proceedings*, pages 40–51. CEUR-WS.org, 2023.

María Adela Grando, Laura Moss, Derek H. Sleeman, and John Kinsella. Argumentation-logic for creating and explaining medical hypotheses. *Artif. Intell. Medicine*, 58(1):1–13, 2013.

Célya Gruson-Daniel. *Numérique et régime français des savoirs en-action: l'open en sciences. Le cas de la consultation République numérique (2015). (The french regime of knowledge and its dynamics: open in sciences and digital technologies in debate. The case study of the french bill for a "digital republic" (2015))*. PhD thesis, Paris Descartes University, France, 2018.

Abdelraouf Hecham, Madalina Croitoru, and Pierre Bisquert. DAMN: defeasible reasoning tool for multi-agent reasoning. In *The Thirty-Fourth AAAI Conference on Artificial Intelligence, AAAI 2020, The Thirty-Second Innovative Applications of Artificial Intelligence Conference, IAAI 2020, The Tenth AAAI Symposium on Educational Advances in Artificial Intelligence, EAAI 2020, New York, NY, USA, February 7-12, 2020*, pages 13612–13613. AAAI Press, 2020.

Abdelraouf Hecham. *Defeasible reasoning for existential rules. (Raisonnement defaisable dans les règles existentielles)*. PhD thesis, Université de Montpellier, 2018.

Stella Heras, Martí Navarro, Vicente J. Botti, and Vicente Julián. Applying dialogue games to manage recommendation in social networks. In Peter McBurney, Iyad Rahwan, Simon Parsons, and Nicolas Maudet, editors, *Argumentation in Multi-Agent Systems, 6th International Workshop, ArgMAS 2009, Budapest, Hungary, May 12, 2009. Revised Selected and Invited Papers*, volume 6057 of *Lecture Notes in Computer Science*, pages 256–272. Springer, 2009.

Areski Himeur, Bruno Yun, Pierre Bisquert, and Madalina Croitoru. Assessing the impact of agents in weighted bipolar argumentation frameworks. In Max Bramer and Richard Ellis, editors, *Artificial Intelligence XXXVIII - 41st SGAI International Conference on Artificial Intelligence, AI 2021, Cambridge, UK, December 14-16, 2021, Proceedings*, volume 13101 of *Lecture Notes in Computer Science*, pages 75–88. Springer, 2021.

Gary H. Hua and Steven O. Kimbrough. On hypermedia-based argumentation decision support systems. *Decis. Support Syst.*, 22(3):259–275, 1998.

Anthony Hunter and Matthew Williams. Argumentation for aggregating clinical evidence. In *22nd IEEE International Conference on Tools with Artificial Intelligence, ICTAI 2010, Arras, France, 27-29 October 2010 - Volume 1*, pages 361–368. IEEE Computer Society, 2010.

Benjamin Irwin, Antonio Rago, and Francesca Toni. Forecasting argumentation frameworks. In Gabriele Kern-Isberner, Gerhard Lakemeyer, and Thomas Meyer, editors, *Proceedings of the 19th International Conference on Principles of Knowledge Representation and Reasoning, KR 2022, Haifa, Israel. July 31 - August 5, 2022*, 2022.

Antonis C. Kakas and Pavlos Moraitis. Argumentation based decision making for autonomous agents. In *The Second International Joint Conference on Autonomous Agents & Multiagent Systems, AAMAS 2003, July 14-18, 2003, Melbourne, Victoria, Australia, Proceedings*, pages 883–890. ACM, 2003.

Antonis C. Kakas, Pavlos Moraitis, and Nikolaos I. Spanoudakis. *GORGIAS*: Applying argumentation. *Argument Comput.*, 10(1):55–81, 2019.

Timotheus Kampik, Nico Potyka, Xiang Yin, Kristijonas Cyras, and Francesca Toni. Contribution functions for quantitative bipolar argumentation graphs: A principle-based analysis. *CoRR*, abs/2401.08879, 2024.

Nikos I. Karacapilidis and Dimitris Papadias. A group decision and negotiation support system for argumentation based reasoning. In Grigoris Antoniou, Aditya Ghose, and Miroslaw Truszczynski, editors, *Learning and Reasoning with Complex Representations, PRICAI'96 Workshops on Reasoning with Incomplete and Changing Information and on Inducing Complex Representations, Cairns, Australia, August 26-30, 1996, Selected Papers*, volume 1359 of *Lecture Notes in Computer Science*, pages 188–205. Springer, 1996.

Nikos I. Karacapilidis and Dimitris Papadias. Hermes: Supporting argumentative discourse in multi-agent decision making. In Jack Mostow and Chuck Rich, editors, *Proceedings of the Fifteenth National Conference on Artificial Intelligence and Tenth Innovative Applications of Artificial Intelligence Conference, AAAI 98, IAAI 98, July 26-30, 1998, Madison, Wisconsin, USA*, pages 827–832. AAAI Press / The MIT Press, 1998.

Nikos I. Karacapilidis and Dimitris Papadias. Computer supported argumentation and collaborative decision making: the HERMES system. *Inf. Syst.*, 26(4):259–277, 2001.

Amin Karamlou, Kristijonas Cyras, and Francesca Toni. Complexity results and algorithms for bipolar argumentation. In Edith Elkind, Manuela Veloso, Noa Agmon, and Matthew E. Taylor, editors, *Proceedings of the 18th International Conference on Autonomous Agents and MultiAgent Systems, AAMAS '19, Montreal, QC, Canada, May 13-17, 2019*, pages 1713–1721. International Foundation for Autonomous Agents and Multiagent Systems, 2019.

Amin Karamlou, Kristijonas Cyras, and Francesca Toni. Deciding the winner of a debate using bipolar argumentation. In Edith Elkind, Manuela Veloso, Noa Agmon, and Matthew E. Taylor, editors, *Proceedings of the 18th International Conference on Autonomous Agents and MultiAgent Systems, AAMAS '19, Montreal, QC, Canada, May 13-17, 2019*, pages 2366–2368. International Foundation for Autonomous Agents and Multiagent Systems, 2019.

Nadin Kökciyan, Isabel Sassoon, Anthony P. Young, Martin Chapman, Talya Porat, Mark Ashworth, Vasa Curcin, Sanjay Modgil, Simon Parsons, and Elizabeth Sklar. Towards an argumentation system for supporting patients in self-managing their chronic conditions. In *The Workshops of the The Thirty-Second AAAI Conference on Artificial Intelligence, New Orleans, Louisiana, USA, February 2-7, 2018*, volume WS-18 of *AAAI Technical Report*, pages 455–462. AAAI Press, 2018.

Nadin Kökciyan, Isabel Sassoon, Elizabeth Sklar, Sanjay Modgil, and Simon Parsons. Applying metalevel argumentation frameworks to support medical decision making. *IEEE Intell. Syst.*, 36(2):64–71, 2021.

Amélie Kurtz and Rallou Thomopoulos. Safety vs. sustainability concerns of infant food users: French results and european perspectives. *Sustainability*, 13(18):10074, 2021.

Béatrice Lamche, Ugur Adıgüzel, and Wolfgang Wörndl. Interactive explanations in mobile shopping recommender systems. In *Joint Workshop on Interfaces and Human Decision Making in Recommender Systems*, volume 14, 2014.

Luca Longo and Lucy Hederman. Argumentation theory for decision support in health-care: A comparison with machine learning. In Kazuyuki Imamura, Shiro Usui, Tomoaki Shirao, Takuji Kasamatsu, Lars Schwabe, and Ning Zhong, editors, *Brain and Health Informatics - International Conference, BHI 2013, Maebashi, Japan, October 29-31, 2013. Proceedings*, volume 8211 of *Lecture Notes in Computer Science*, pages 168–180. Springer, 2013.

Tobias Mayer, Elena Cabrio, and Serena Villata. Extraction d'arguments basée sur les transformateurs pour des applications dans le domaine de la santé (transformer-based argument mining for healthcare applications). In Pascal Denis, Natalia Grabar, Amel Fraisse, Rémi Cardon, Bernard Jacquemin, Eric Kergosien, and Antonio Balvet, editors, *Actes de la 28e Conférence sur le Traitement Automatique des Langues Naturelles. Volume 1 : conférence principale, TALN 2021, Lille, France, June 28 - July 2, 2021*, pages 265–267. ATALA, 2021.

Peter McBurney and Simon Parsons. Risk agoras: Dialectical argumentation for scientific reasoning. In Craig Boutilier and Moisés Goldszmidt, editors, *UAI '00: Proceedings of the 16th Conference in Uncertainty in Artificial Intelligence, Stanford University, Stanford, California, USA, June 30 - July 3, 2000*, pages 371–379. Morgan Kaufmann, 2000.

Peter McBurney and Simon Parsons. Representing epistemic uncertainty by means of dialectical argumentation. *Ann. Math. Artif. Intell.*, 32(1-4):125–169, 2001.

Sanjay Modgil and Henry Prakken. The ASPIC+ framework for structured argumentation: a tutorial. *Argument & Computation*, 5(1):31–62, 2014.

Sanjay Modgil. Reasoning about preferences in argumentation frameworks. *Artif. Intell.*, 173(9-10):901–934, 2009.

Maxime Morge. Computer supported collaborative argumentation. In *Proceedings of 4th workshop on Computational Models of Natural Argument (CNMA04)*, pages 1–4, 2004.

Maxime Morge. *Syst'eme dialectique multi-agents pour l'aide 'a la concertation. (Dialectics multiagent system to support deliberation)*. PhD thesis, École nationale supérieure des mines de Saint-Étienne, France, 2005.

Laura Moss, Derek H. Sleeman, Malcolm Booth, Malcolm Daniel, Lyndsay Donaldson, Charlotte J. Gilhooly, Martin Hughes, Malcolm Sim, and John Kinsella. Explaining anomalous responses to treatment in the intensive care unit. In Carlo Combi, Yuval Shahar, and Ameen Abu-Hanna, editors, *Artificial Intelligence in Medicine, 12th Conference on Artificial Intelligence in Medicine, AIME 2009, Verona, Italy, July 18-22, 2009. Proceedings*, volume 5651 of *Lecture Notes in Computer Science*, pages 250–254, 2009.

Till Mossakowski and Fabian Neuhaus. Modular Semantics and Characteristics for Bipolar Weighted Argumentation Graphs. *CoRR*, abs/1807.06685, 2018.

Jann Müller and Anthony Hunter. An argumentation-based approach for decision making. In *IEEE 24th International Conference on Tools with Artificial Intelligence, ICTAI 2012, Athens, Greece, November 7-9, 2012*, pages 564–571. IEEE Computer Society, 2012.

Jack Mumford, Katie Atkinson, and Trevor J. M. Bench-Capon. Reasoning with legal cases: A hybrid ADF-ML approach. In Enrico Francesconi, Georg Borges, and Christoph Sorge, editors, *Legal Knowledge and Information Systems - JURIX 2022: The Thirty-fifth Annual Conference, Saarbrücken, Germany, 14-16 December 2022*, volume 362 of *Frontiers in Artificial Intelligence and Applications*, pages 93–102. IOS Press, 2022.

Jack Mumford, Katie Atkinson, and Trevor J. M. Bench-Capon. Combining a legal knowledge model with machine learning for reasoning with legal cases. In Matthias Grabmair, Francisco Andrade, and Paulo Novais, editors, *Proceedings of the Nineteenth International Conference on Artificial Intelligence and Law, ICAIL 2023, Braga, Portugal, June 19-23, 2023*, pages 167–176. ACM, 2023.

Mark A. Musen, Samson W. Tu, Amar K. Das, and Yuval Shahar. Synthesis of research: EON: A component-based approach to automation of protocol-directed therapy. *J. Am. Medical Informatics Assoc.*, 3(6):367–388, 1996.

Sidra Naveed, Tim Donkers, and Jürgen Ziegler. Argumentation-based explanations in recommender systems: Conceptual framework and empirical results. In Tanja Mitrovic, Jie Zhang, Li Chen, and David Chin, editors, *Adjunct Publication of the 26th Conference on User Modeling, Adaptation and Personalization, UMAP 2018, Singapore, July 08-11, 2018*, pages 293–298. ACM, 2018.

Søren Holbech Nielsen and Simon Parsons. A Generalization of Dung's Abstract Framework for Argumentation: Arguing with Sets of Attacking Arguments. In Nicolas Maudet, Simon Parsons, and Iyad Rahwan, editors, *Argumentation in Multi-Agent Systems*, pages 54–73. Springer Berlin Heidelberg, 2007.

Alexandra Okada and Anna De Liddo. Litemap outreach and usage by brazilian universities. *Brazilian Universities*, 2018.

Nir Oren, Kees van Deemter, and Wamberto Weber Vasconcelos. Argument-based plan explanation. In Mauro Vallati and Diane E. Kitchin, editors, *Knowledge Engineering Tools and Techniques for AI Planning*, pages 173–188. Springer, 2020.

Wassila Ouerdane, Nicolas Maudet, and Alexis Tsoukias. Argumentation theory and decision aiding. *Trends in multiple criteria decision analysis*, pages 177–208, 2010.

Simon Parsons and John Fox. Argumentation and decision making: A position paper. In Dov M. Gabbay and Hans Jürgen Ohlbach, editors, *Practical Reasoning, International Conference on Formal and Applied Practical Reasoning, FAPR '96, Bonn, Germany, June 3-7, 1996, Proceedings*, volume 1085 of *Lecture Notes in Computer Science*, pages 705–709. Springer, 1996.

Sylwia Polberg and Anthony Hunter. Empirical evaluation of abstract argumentation: Supporting the need for bipolar and probabilistic approaches. *Int. J. Approx. Reason.*, 93:487–543, 2018.

Nico Potyka. Continuous dynamical systems for weighted bipolar argumentation. In Michael Thielscher, Francesca Toni, and Frank Wolter, editors, *Principles of Knowledge Representation and Reasoning: Proceedings of the Sixteenth International Conference, KR 2018, Tempe, Arizona, 30 October - 2 November 2018*, pages 148–157. AAAI Press, 2018.

Nico Potyka. Interpreting neural networks as gradual argumentation frameworks (including proof appendix). *CoRR*, abs/2012.05738, 2020.

Nico Potyka. Generalizing complete semantics to bipolar argumentation frameworks. In Jirina Vejnarová and Nic Wilson, editors, *Symbolic and Quantitative Approaches to Reasoning with Uncertainty - 16th European Conference, ECSQARU 2021, Prague, Czech Republic, September 21-24, 2021, Proceedings*, volume 12897 of *Lecture Notes in Computer Science*, pages 130–143. Springer, 2021.

Nico Potyka. Attractor - A java library for gradual bipolar argumentation. In Francesca Toni, Sylwia Polberg, Richard Booth, Martin Caminada, and Hiroyuki Kido, editors, *Computational Models of Argument - Proceedings of COMMA 2022, Cardiff, Wales, UK, 14-16 September 2022*, volume 353 of *Frontiers in Artificial Intelligence and Applications*, pages 369–370. IOS Press, 2022.

Henry Prakken and Rosa Ratsma. A top-level model of case-based argumentation for explanation: Formalisation and experiments. *Argument Comput.*, 13(2):159–194, 2022.

Henry Prakken and Giovanni Sartor. The role of logic in computational models of legal argument: A critical survey. In Antonis C. Kakas and Fariba Sadri, editors, *Computational Logic: Logic Programming and Beyond, Essays in Honour of Robert A. Kowalski, Part II*, volume 2408 of *Lecture Notes in Computer Science*, pages 342–381. Springer, 2002.

Henry Prakken, Adam Z. Wyner, Trevor J. M. Bench-Capon, and Katie Atkinson. A formalization of argumentation schemes for legal case-based reasoning in ASPIC+. *J. Log. Comput.*, 25(5):1141–1166, 2015.

Henry Prakken. An abstract framework for argumentation with structured arguments. *Argument Comput.*, 1(2):93–124, 2010.

Henry Prakken. An argumentation-based analysis of the simonshaven case. *Top. Cogn. Sci.*, 12(4):1068–1091, 2020.

Ivana Quinto, Luca Iandoli, and Anna De Liddo. Designing online collaboration for the individual and social good: A collective argumentation approach. In *AMCIS*, 2021.

Antonio Rago, Kristijonas Cyras, and Francesca Toni. Adapting the df-quad algorithm to bipolar argumentation. In Matthias Thimm, Federico Cerutti, Hannes Strass, and Mauro Vallati, editors, *Proceedings of the First International Workshop on Systems and Algorithms for Formal Argumentation (SAFA) co-located with the 6th International Conference on Computational Models of Argument (COMMA 2016), Potsdam, Germany, September 13, 2016*, volume 1672 of *CEUR Workshop Proceedings*, pages 34–39. CEUR-WS.org, 2016.

Antonio Rago, Francesca Toni, Marco Aurisicchio, and Pietro Baroni. Discontinuity-free decision support with quantitative argumentation debates. In Chitta Baral, James P. Delgrande, and Frank Wolter, editors, *Principles of Knowledge Representation and Reasoning: Proceedings of the Fifteenth International Conference, KR 2016, Cape Town, South Africa, April 25-29, 2016*, pages 63–73. AAAI Press, 2016.

Antonio Rago, Oana Cocarascu, Christos Bechlivanidis, and Francesca Toni. Argumentation as a framework for interactive explanations for recommendations. In Diego Calvanese, Esra Erdem, and Michael Thielscher, editors, *Proceedings of the 17th International Conference on Principles of Knowledge Representation and Reasoning, KR 2020, Rhodes, Greece, September 12-18, 2020*, pages 805–815, 2020.

Antonio Rago, Oana Cocarascu, Christos Bechlivanidis, David A. Lagnado, and Francesca Toni. Argumentative explanations for interactive recommendations. *Artif. Intell.*, 296:103506, 2021.

William Rehg, Peter McBurney, and Simon Parsons. Computer decision-support systems for public argumentation: assessing deliberative legitimacy. *AI Soc.*, 19(3):203–228, 2005.

Paula Rodríguez, Stella Heras, Javier Palanca Cámara, Néstor D. Duque, and Vicente Julián. Argumentation-based hybrid recommender system for recommending learning objects. In Michael Rovatsos, George A. Vouros, and Vicente Julián, editors, *Multi-Agent Systems and Agreement Technologies - 13th European Conference, EUMAS 2015, and Third International Conference, AT 2015, Athens, Greece, December 17-18, 2015, Revised Selected Papers*, volume 9571 of *Lecture Notes in Computer Science*, pages 234–248. Springer, 2015.

Thomas Saaty. The analytic hierarchy process (ahp) for decision making. In *Kobe, Japan*, volume 1, page 69, 1980.

David C Schneider, Christian Voigt, and Gregor Betz. Argunet-a software tool for collaborative argumentation analysis and research. In *7th Workshop on Computational Models of Natural Argument (CMNA VII)*, 2007.

Ravi D. Shankar, Samson W. Tu, and Mark A. Musen. Medical arguments in an automated health care system. In *Argumentation for Consumers of Healthcare, Papers from the 2006 AAAI Spring Symposium, Technical Report SS-06-01, Stanford, California, USA, March 27-29, 2006*, pages 96–104. AAAI, 2006.

Mark Snaith and Chris Reed. TOAST: online aspic[+] implementation. In Bart Verheij, Stefan Szeider, and Stefan Woltran, editors, *Computational Models of Argument - Proceedings of COMMA 2012, Vienna, Austria, September 10-12, 2012*, volume 245 of *Frontiers in Artificial Intelligence and Applications*, pages 509–510. IOS Press, 2012.

Rallou Thomopoulos and Madalina Croitoru. Aide à la décision en ingénierie inverse. une approche s'appuyant sur l'argumentation. *Rev. d'Intelligence Artif.*, 27(4-5):493–513, 2013.

Rallou Thomopoulos, Bernard Moulin, and Laurent Bedoussac. Supporting decision for environment-friendly practices in the agri-food sector: When argumentation and system dynamics simulation complete each other. *Int. J. Agric. Environ. Inf. Syst.*, 9(3):1–21, 2018.

Pancho Tolchinsky, Ulises Cortés, Sanjay Modgil, Francisco Caballero, and Antonio López-Navidad. Increasing human-organ transplant availability: Argumentation-based agent deliberation. *IEEE Intell. Syst.*, 21(6):30–37, 2006.

Pancho Tolchinsky, Sanjay Modgil, Katie Atkinson, Peter McBurney, and Ulises Cortés. Deliberation dialogues for reasoning about safety critical actions. *Auton. Agents Multi Agent Syst.*, 25(2):209–259, 2012.

Francesca Toni. A tutorial on assumption-based argumentation. *Argument & Computation*, 5(1):89–117, 2014.

Alice Toniolo, Timothy Dropps, Wentao Robin Ouyang, John A. Allen, Timothy J. Norman, Nir Oren, Mani B. Srivastava, and Paul Sullivan. Argumentation-based collaborative intelligence analysis in cispaces. In Simon Parsons, Nir Oren, Chris Reed, and Federico Cerutti, editors, *Computational Models of Argument - Proceedings of COMMA 2014, Atholl Palace Hotel, Scottish Highlands, UK, September 9-12, 2014*, volume 266 of *Frontiers in Artificial Intelligence and Applications*, pages 481–482. IOS Press, 2014.

Alice Toniolo, Federico Cerutti, Timothy J. Norman, Nir Oren, John A. Allen, Mani Srivastava, and Paul Sullivan. Human-machine collaboration in intelligence analysis: An expert evaluation. *Intell. Syst. Appl.*, 17:200151, 2023.

Ashish Vaswani, Noam Shazeer, Niki Parmar, Jakob Uszkoreit, Llion Jones, Aidan N Gomez, Łukasz Kaiser, and Illia Polosukhin. Attention is all you need. *Advances in neural information processing systems*, 30, 2017.

Srdjan Vesic, Bruno Yun, and Predrag Teovanovic. Graphical representation enhances human compliance with principles for graded argumentation semantics. In Piotr Faliszewski, Viviana Mascardi, Catherine Pelachaud, and Matthew E. Taylor, editors, *21st International Conference on Autonomous Agents and Multiagent Systems, AAMAS 2022, Auckland, New Zealand, May 9-13, 2022*, pages 1319–1327. International Foundation for Autonomous Agents and Multiagent Systems (IFAAMAS), 2022.

Matthew Williams and Anthony Hunter. Harnessing ontologies for argument-based decision-making in breast cancer. In *19th IEEE International Conference on Tools with Artificial Intelligence (ICTAI 2007), October 29-31, 2007, Patras, Greece, Volume 2*, pages 254–261. IEEE Computer Society, 2007.

Adam Z. Wyner, Maya Wardeh, Trevor J. M. Bench-Capon, and Katie Atkinson. A model-based critique tool for policy deliberation. In Burkhard Schäfer, editor, *Legal Knowledge and Information Systems - JURIX 2012: The Twenty-Fifth Annual Conference, University of Amsterdam, The Netherlands, 17-19 December 2012*, volume 250 of *Frontiers in Artificial Intelligence and Applications*, pages 167–176. IOS Press, 2012.

Bruno Yun and Srdjan Vesic. Gradual semantics for weighted bipolar setafs. In Jirina Vejnarová and Nic Wilson, editors, *Symbolic and Quantitative Approaches to Reasoning with Uncertainty - 16th European Conference, ECSQARU 2021, Prague, Czech Republic, September 21-24, 2021, Proceedings*, volume 12897 of *Lecture Notes in Computer Science*, pages 201–214. Springer, 2021.

Bruno Yun, Pierre Bisquert, Patrice Buche, Madalina Croitoru, Valérie Guillard, and Rallou Thomopoulos. Choice of environment-friendly food packagings through argumentation systems and preferences. *Ecological Informatics*, 48:24–36, November 2018.

Bruno Yun, Madalina Croitoru, Srdjan Vesic, and Pierre Bisquert. DAGGER: Datalog+/- Argumentation Graph GEneRator. In *Proceedings of the 17th International Conference on Autonomous Agents and MultiAgent Systems, AAMAS 2018, Stockholm, Sweden, July 10-15, 2018*, pages 1841–1843, 2018.

Bruno Yun, Srdjan Vesic, Madalina Croitoru, and Pierre Bisquert. Viewpoints using ranking-based argumentation semantics. In *Proceedings of the 7th International Conference on Computational Models of Argument, COMMA 2018, 11th - 14th September, 2018, Warsaw, Poland.*, 2018.

Bruno Yun, Madalina Croitoru, and Srdjan Vesic. NAKED: N-Ary graphs from Knowledge bases Expressed in Datalog+/-. In *Proceedings of the 18th International Conference on Autonomous Agents and MultiAgent Systems, AAMAS 2019, Montreal, Canada, May 13-17, 2019*, 2019.

Haïfa Zargayouna and Nejla Amara. Litemap: An ontology mapping approach for mobile agents' context-awareness. In *OTM Confederated International Conferences" On the Move to Meaningful Internet Systems"*, pages 1934–1943. Springer, 2006.

Part II

Cross-field Connections

CHAPTER 7

SYNTACTIC AND SEMANTIC CONNECTIONS BETWEEN LOGIC PROGRAMMING AND ARGUMENTATION SYSTEMS

SAMY SÁ
Universidade Federal do Ceará, Brazil
samy@ufc.br

WOLFGANG DVOŘÁK
TU Wien, Institute of Logic and Computation, Austria
dvorak@dbai.tuwien.ac.at

MARTIN CAMINADA
Cardiff University, United Kingdom
CaminadaM@cardiff.ac.uk

1 Introduction

In the current book chapter, we will examine the connections between logic programming and a series of formalisms for non-monotonic reasoning which we gather under the umbrella of *argumentation systems*. As one would expect, the first connections are historical: the investigations leading to the proposal of Abstract Argumentation Frameworks (AFs) [Dung, 1995b], often taken as the seminal work in computational argumentation theory, happened amidst the development of semantics for logic programming with negation as failure [Dung, 1991; Kakas *et al.*, 1994; Dung, 1995a]. Just alike, several core argumentation systems find their roots in the literature of logic programming, including the approaches of Defeasible Logic Programming (DeLP) [Simari and Loui, 1992; García and Simari, 2004; García and Simari, 2018], Assumption-Based Argumentation (ABA) [Bondarenko et al., 1997; Dung et al., 2009; Čyras et al., 2018] and Abstract Dialectical Frameworks (ADFs) [Brewka and Woltran, 2010; Brewka et al., 2013]. Immediately, it was argued that some of these systems could model the semantic entailment relations from logic programming using translations.

One well-known example can be found in [Dung, 1995b], where the author showed how to translate a normal logic program (NLP) into an AF. Based on this translation, Dung proved that the *stable models* (resp. the *well-founded model*) of an NLP correspond to the *stable extensions* (resp. the *grounded extension*) of its corresponding AF. These results led to several studies concerning connections between them [Dung, 1995b; Nieves et al., 2008; Caminada and Gabbay, 2009; Egly et al., 2010; Toni and Sergot, 2011; Caminada et al., 2015b; Sá and Alcântara, 2021a; Caminada et al., 2022] . Notably, in [Caminada and Gabbay, 2009], it was proved that the *three-valued stable models* of a NLP correspond to the *complete extensions* of its corresponding AF. Later, [Caminada et al., 2015b] investigated whether the same results would hold for the particular cases of the three-valued models and complete extensions semantics, showing there are some exceptions to the expected correspondences.

Just like in these works, here we will primarily be concerned with semantics, investigating whether each argumentation system we examine is able (or not) to model the entailment of logic programming (and vice-versa). On that matter, we will provide a detailed overview of the connections between logic programming with AFs, ABA, ADFs and Argumentation Frameworks with Sets of Attacking Arguments (SETAFs) [Nielsen and Parsons, 2006; Bikakis et al., 2021]. Other notable systems will be mentioned in connection to logic programs only briefly, as required of each case.

Besides motivation and semantics, some connections naturally arise in the implementation of argumentation systems using logic programming solvers. More specifically, given the non-monotonic nature and the fact that typical reasoning tasks in argumentation are NP/coNP-hard, answer-set programming [Gelfond and Lifschitz, 1991] is an obvious choice for the implementation of argumentation systems. This connection has been explored and developed in multiple works, such as [Nieves *et al.*, 2008; Wakaki and Nitta, 2008; Egly *et al.*, 2008; Egly *et al.*, 2010; Dvořák *et al.*, 2011; Dvořák *et al.*, 2015; Ellmauthaler and Strass, 2014; Dvořák and Wallner, 2020; Sakama and Rienstra, 2017; Cyras and Toni, 2016; Lehtonen *et al.*, 2017; Lehtonen *et al.*, 2021a].

The current chapter is structured as follows. First, in Section 2, we introduce the necessary concepts and notation we require to discuss logic programs. Then, in Section 3, we examine how Abstract Argumentation can be used to model the entailment of Logic Programming and vice versa. The next couple of sections follow the same approach to compare logic programming to other notable argumentation systems: in Section 4, we examine how Assumption-Based Argumentation can be used to model the entailment of Logic Programming (and vice-versa) and in Section 5 we do the same for Abstract Dialectical Frameworks and for Frameworks with Sets of Attacking Arguments. In Section 6, we discuss possibilities for the implementation of argumentation systems based on Answer-Set Programming (ASP) solvers. We round off with a discussion in Section 7, where we mention a few other systems worth notice and their connection to logic programming.

2 Logic Programs: Syntax and Semantics

We start with formally introducing the notion of a logic program. For our current purposes, we restrict ourselves to normal logic programs, which are logic programs without strong negation where the head of each rule consists of a single atom.

Definition 1. *A logic programming rule (or simply a rule, for short) is an expression*

$$x \leftarrow y_1, \ldots, y_n, \text{not } z_1, \ldots, \text{not } z_m \quad (n \geq 0, m \geq 0)$$

where x, each y_i ($1 \leq i \leq n$) and each z_j ($1 \leq j \leq m$) is an atom, and **not** *represents negation as failure (NAF). A logic program (or simply a program) P consists of a finite set of rules.*

Intuitively, a rule r such as $x \leftarrow y_1, \ldots, y_n, \text{not } z_1, \ldots, \text{not } z_m$ expresses there is a proof for x if each of $y_1, \ldots, y_n$ can be proven while each of $z_1, \ldots, z_n$ cannot. Moving forward, we may refer to x as the *head* or *consequent* of the rule (writing $head(r) = x$) and to $y_1, \ldots, y_n$, not $z_1, \ldots,$ not z_m as its *body* (writing $body(r) = \{y_1, \ldots, y_n,$ not $z_1, \ldots,$ not $z_m\}$). Moreover, we say that $body^+(r) = \{y_1, \ldots, y_n\}$ is the strong part of the body and that $body^-(r) = \{$not $z_1, \ldots,$ not $z_m\}$ is the weak part of the body. Each expression not w, where w is an atom, is called a *NAF literal*. Then, a rule is NAF-free iff it does not contain NAF literals (i.e., iff $m = 0$). Similarly, a program is NAF-free iff all of its rules are NAF-free. Finally, the Herbrand Base of a logic program P (written as HB_P) is the set of all atoms in P.

In the following, we recall the definitions of logic programming semantics found in [Przymusinski, 1990; Caminada *et al.*, 2015b; Caminada and Schulz, 2017], but in slightly different fashion for the sake of uniformity. We will comment on what is different as we advance through the concepts.

Definition 2. *A* 3-valued interpretation *of a logic program P with respect to a set of atoms Atms $\supseteq HB_P$ is a triple $I = (T, F, U)$ [1] such that T, F and U are pairwise disjoint and $T \cup F \cup U = Atms$.*

Intuitively, a *3-valued interpretation* (or simply *an interpretation*) of P w.r.t. $Atms \supseteq HB_P$ evaluates the atoms in *Atms* according to the truth values *true*, *false* and *undecided*. Then, given an interpretation $I = (T, F, U)$, the atoms in T are said to be *true*, the atoms in F are said to be *false* and those in U are said to be *undecided*. Further, each interpretation I can be characterized as a function $I : Atms \to \{true, false, undecided\}$.

Given an interpretation, a program may be transformed into a corresponding NAF-free program through the notion of *reduct*:

Definition 3. *The* reduct *of a logic program P w.r.t. an interpretation $I = (T, F, U)$, written as P^I, is obtained by replacing in all rules of P the occurrences of each NAF literal* not x *by* **t** *if $x \in F$, by* **f** *if $x \in T$, and by* **u** *otherwise.*

---

[1] Traditionally, program interpretations are presented as a pair (T, F), where the result of $U = HB_P \setminus (T \cup F)$ is left implicit. Making U explicit helps with uniformity in our text.

7 - Logic Programming and Argumentation

In the context of a program reduct, **t**, **f** and **u** are auxiliary terms interpreted as positive literals (atoms) not in HB_P. As such, the reduct of any program P is necessarily a NAF-free program. Semantically, each interpretation $I = (T, F, U)$ of P w.r.t. HB_P is extended into a corresponding interpretation $\check{I} = (\check{T}, \check{F}, \check{U})$ of P w.r.t. $HB_P \cup \{\mathbf{t}, \mathbf{f}, \mathbf{u}\}$ such that $\check{T} = T \cup \{\mathbf{t}\}$, $\check{F} = F \cup \{\mathbf{f}\}$, and $\check{U} = U \cup \{\mathbf{u}\}$.

When restricted to NAF-free programs, the notion of a 3-valued model easily follows:

Definition 4. *Given a NAF-free program P, an interpretation $I = (T, F, U)$ of P w.r.t. Atms $\supseteq HB_P$ is a 3-valued model (or simply a model) of P if, for each rule $x \leftarrow y_1, \ldots, y_n$ in P, it holds that*

$$I(x) \geq \mathtt{min}(\{\check{I}(y_i) \mid i \in \{1, \ldots, n\}\})$$

following the truth order true $>$ undecided $>$ false.

Intuitively, in a model of P, the head of each rule is at least as true as the least true literal in its body.

When P is a NAF-free logic program (possibly containing **t**, **f** or **u**), the existence of a unique *minimal* 3-valued model $\Phi(P) = (T, F, U)$ with minimal T and maximal F (w.r.t. $\subseteq$) among all 3-valued models of P is ensured [Przymusinski, 1990]. Then, since any program reduct P^I is NAF-free, it necessarily has a unique minimal 3-valued model $\Phi(P^I)$. This leads to the core semantics we will discuss for logic programs in this work:

Definition 5. *([Przymusinski, 1990]) Let* Mod *be a model of P, then*

Mod *is a 3-valued stable model of P iff* $\Phi(P^{\mathtt{Mod}}) = \mathtt{Mod}$.

We now recall various logic programming semantics which are based on 3-valued interpretations. The presentation below was originally provided in [Caminada et al., 2015b; Caminada et al., 2022], where correspondences to other equivalent concepts for the same semantics were discussed. It is heavily based on Przymusinski's three-valued stable semantics [Przymusinski, 1990] and tailored to ease the comparison to argumentation semantics[2].

---

[2] A similar characterization of these semantics (except well-founded) as special cases of the three-valued stable semantics (there called P-stable models) can be found in [Saccà and Zaniolo, 1997]

Definition 6. *([Caminada et al., 2015b; Caminada et al., 2022])* Let P be a logic program and $\text{Mod} = (T, F, U)$ be a 3-valued interpretation of P w.r.t. HB_P. We say that Mod is:

1. *the* well-founded model *of P iff Mod is the 3-valued stable model where T is $\subseteq$-minimal among all 3-valued stable models of P*

2. *a* regular model *of P iff Mod is a 3-valued stable model where T is $\subseteq$-maximal among all 3-valued stable models of P*

3. *a* 2-valued stable model *of P iff Mod is a 3-valued stable model where $T \cup F = HB_P$*

4. *an* L-stable model *of P iff Mod is a 3-valued stable model where U is $\subseteq$-minimal among all 3-valued stable models of P*

5. *a* pre-ideal model *of P iff Mod is a 3-valued stable model where $T \subseteq T_{re}$ for each regular model $\text{Mod}_{re} = (T_{re}, F_{re}, U_{re})$ of P*

6. *the* ideal model *of P iff Mod is the pre-ideal model where T is $\subseteq$-maximal among all pre-ideal models of P*

7. *a* pre-eager model *of P iff Mod is a 3-valued stable model where $T \subseteq T_{ls}$ for each L-stable model $\text{Mod}_{ls} = (T_{ls}, F_{ls}, U_{ls})$ of P*

8. *the* eager model *of P iff Mod is the pre-eager model where T is $\subseteq$-maximal among all pre-eager models of P*

Each logic program has one or more 3-valued stable models, a unique well-founded model, one or more regular models, zero or more 2-valued stable models, one or more L-stable models, one or more pre-ideal models, a unique ideal model, one or more pre-eager models and a unique eager model.

3 On the Connection between Abstract Argumentation and Logic Programming

In the current section, we will show how abstract argumentation and logic programming are related. Each system has its own syntax and a particular variety of semantics, which relate to the evaluation of a particular set of sentences. For the logic programs we are interested in, the sentences are propositional atoms, whereas in argumentation frameworks, the sentences are called arguments. To proceed, we must consider back and forth translations: one will show how an argumentation framework can be encoded as a logic program (which we will call **AA2LP**), the other will show how a logic program can be encoded by an argumentation framework (which we will call **LP2AA**). In each case, we will show whether some semantics for the destiny system captures some semantics from the origin system. The discussion we conduct in this section summarizes results obtained in [Dung, 1995b; Caminada and Gabbay, 2009; Caminada *et al.*, 2015b; Sá and Alcântara, 2021a; Caminada *et al.*, 2022].

3.1 Abstract Argumentation: Syntax and Semantics

Intuitively, an argumentation framework portrays a set of arguments and the conflicts among them. The conflicts are modelled as a binary relation over the arguments, leading to the instantiation of an argument graph. For current purposes, we restrict ourselves to finite argumentation frameworks.

Definition 7. *([Dung, 1995b])* An argumentation framework *is a pair* $AF = (Ar, att)$ *where Ar is a finite set of arguments and $att \subseteq Ar \times Ar$.*

Argumentation semantics are commonly presented in the form of *argument extensions* [Dung, 1995b] or *argument labellings*, which, as explained in [Caminada, 2006; Caminada and Gabbay, 2009; Baroni *et al.*, 2018], coincide with their respective extension-based variants. The core semantics for argumentation frameworks is commonly considered to be the *complete semantics*, especially because a variety of other argumentation semantics can be obtained as particular cases of that semantics.

Definition 8. *Let $AF = (Ar, att)$ be an argumentation framework. An* argument labelling *is a function* $\text{ArgLab} : Ar \to \{\text{in}, \text{out}, \text{undec}\}$.

Given an argument labelling ArgLab, we write $\text{val}(\text{ArgLab})$ to refer to the set of arguments labelled as $\text{val} \in \{\text{in}, \text{out}, \text{undec}\}$ in ArgLab. For convenience, we may as well refer to ArgLab as the 3-tuple $(\text{in}(\text{ArgLab}),$

out(ArgLab), undec(ArgLab)). We invite the reader to notice how argument labellings are inherently similar to the interpretations found in logic programming (see Definition 2).

Definition 9. *Let* $AF = (Ar, att)$ *be an argumentation framework. An argument labelling* ArgLab *is called a* complete argument labelling *iff for each* $A \in Ar$ *it holds that:*

- *if* ArgLab(A) = in *then for every* $B \in Ar$ *that attacks* A *it holds that* ArgLab(B) = out

- *if* ArgLab(A) = out *then there exists some* $B \in Ar$ *that attacks* A *such that* ArgLab(B) = in

- *if* ArgLab(A) = undec *then (i) not every* $B \in Ar$ *that attacks* A *has* ArgLab(B) = out *and (ii) no* $B \in Ar$ *that attacks* A *has* ArgLab(B) = in

It was shown in [Caminada and Gabbay, 2009] that the complete semantics for abstract argumentation corresponds to the 3-valued stable model semantics of logic programming. The result is obtained through suitable translations from abstract argumentation frameworks to logic programs (and back) and mapping the complete labellings of the framework to the 3-valued stable models of the corresponding program (and vice-versa). From there, the same translations can be used to obtain a series of additional results [Caminada et al., 2015b; Caminada et al., 2022] regarding the correspondence of different argumentation and logic programming semantics.

Before we introduce the translations and properly show how those results are obtained, we must define the other semantics we require for argumentation frameworks. On that matter, we invite the reader to observe the similarities between Definition 6 and Definition 10.

Definition 10. *Let* ArgLab *be an argument labelling of argumentation framework* $AF = (Ar, att)$. ArgLab *is called:*

- *the* grounded argument labelling *iff* ArgLab *is a complete argument labelling where* in(ArgLab) *is* $\subseteq$-*minimal among all complete argument labellings of* AF

7 - LOGIC PROGRAMMING AND ARGUMENTATION

- *a* preferred argument labelling *iff* ArgLab *is a complete argument labelling where* in(ArgLab) *is* $\subseteq$-*maximal among all complete argument labellings of AF*

- *a* stable argument labelling *iff* ArgLab *is a complete argument labelling where* undec(ArgLab) $= \emptyset$

- *a* semi-stable argument labelling *iff* ArgLab *is a complete argument labelling where* undec(ArgLab) *is* $\subseteq$-*minimal among all complete argument labellings of AF*

- *a* pre-ideal argument labelling[3] *iff* ArgLab *is a complete argument labelling where* in(ArgLab) $\subseteq$ in(ArgLab$_{pr}$) *for each preferred argument labelling* ArgLab$_{pr}$ *of AF*

- *the* ideal argument labelling *iff* ArgLab *is a pre-ideal argument labelling where* in(ArgLab) *is* $\subseteq$-*maximal among all pre-ideal argument labellings of AF*

- *a* pre-eager argument labelling *iff* ArgLab *is a complete argument labelling where* in(ArgLab) $\subseteq$ in(ArgLab$_{sem}$) *for each semi-stable argument labelling* ArgLab$_{sem}$ *of AF*

- *the* eager argument labelling *iff* ArgLab *is a pre-eager argument labelling where* in(ArgLab) *is* $\subseteq$-*maximal among all pre-eager argument labellings of AF*

Each argumentation framework has one or more complete labellings, a unique grounded labelling, one or more preferred labellings, zero or more stable labellings, one or more semi-stable labellings[4], one or more pre-ideal labellings, a unique ideal labelling, one or more pre-eager labellings and a unique eager labelling.

---

[3]A similar concept is also present in [Dung *et al.*, 2007] where the ideal extension is defined in terms of *ideal sets*. We opted for a slightly narrower concept based on the complete semantics (instead of the admissible semantics) following [Caminada *et al.*, 2022]. This is merely a matter of choice, since the ideal extension of an argumentation framework is necessarily complete [Dung *et al.*, 2007].

[4]This is because we only consider finite argumentation frameworks. An infinite argumentation framework can have zero or more semi-stable labellings [Caminada and Verheij, 2010; Weydert, 2011; Baumann and Spanring, 2015; Baumann, 2018].

3.2 From Abstract Argumentation to Logic Programming

Now we will turn our attention to how argumentation semantics can be modelled by logic programming semantics via a suitable translation.

Intuitively, given an argumentation framework, an argument can be *accepted* only if all of its attackers are *rejected* (i.e., *not accepted*). That comprehension leads to a straightforward translation from argumentation frameworks to logic programs which can be found in [Wu et al., 2009; Caminada et al., 2015b]:

Definition 11. *([Wu et al., 2009]) Let $AF = (Ar, att)$ be an argumentation framework, the logic program associated to AF is*

$$\mathtt{AA2LP}(AF) = P_{AF} = \{A \leftarrow \mathtt{not}\ B_1, \ldots, \mathtt{not}\ B_m \mid A \in Ar\ and$$
$$\{B_i \mid (B_i, A) \in att\} = \{B_1, \ldots, B_m\}\}.$$

The program P_{AF} lists one rule for each argument a in AF, expressing there is a proof for a (semantically, that it is *true*) if each of its attackers cannot be proven (semantically, if they are *false*). A special case is an argument (say A) that has no attackers, which translates to a rule "$A \leftarrow$" with an empty body. Because there is only one rule for each argument, the exclusive set of conditions allows the rules to be read as "if and only if".

Example 1. *Let $AF = (Ar, att)$ be the argumentation framework with*

$$Ar = \{A_1, A_2, A_3, A_4, A_5, A_6, A_7, A_8, A_9\}\ and$$
$$att = \{(A_1, A_1), (A_1, A_4), (A_1, A_6), (A_2, A_3), (A_2, A_4), (A_2, A_7),$$
$$(A_3, A_2), (A_3, A_5), (A_3, A_8), (A_4, A_1), (A_4, A_4), (A_4, A_6),$$
$$(A_5, A_5), A_5, A_9)\},$$

as depicted below:

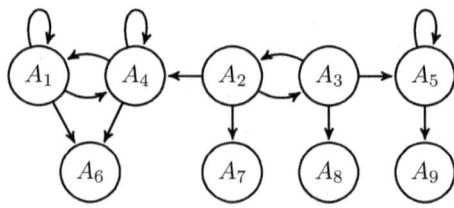

Its associated logic program P_{AF} is[5]:

$r_1: \quad A_1 \leftarrow \text{not } A_1, \text{not } A_4 \qquad r_2: \quad A_2 \leftarrow \text{not } A_3$
$r_3: \quad A_3 \leftarrow \text{not } A_2 \qquad\qquad\qquad r_4: \quad A_4 \leftarrow \text{not } A_1, \text{not } A_2, \text{not } A_4$
$r_5: \quad A_5 \leftarrow \text{not } A_3, \text{not } A_5 \qquad r_6: \quad A_6 \leftarrow \text{not } A_1, \text{not } A_4$
$r_7: \quad A_7 \leftarrow \text{not } A_2 \qquad\qquad\qquad r_8: \quad A_8 \leftarrow \text{not } A_3$
$r_9: \quad A_9 \leftarrow \text{not } A_5$

Notice how the rules in P_{AF} immediately describe the conditions for each argument in AF to be accepted based on the attack relation. For instance, the rule $r_2 : A_2 \leftarrow \text{not } A_3$ expresses that A_2 should be proven (i.e. accepted) if and only if A_3 is not. Indeed, A_3 is the only attacker of A_2 in AF.

Programs such as P_{AF} pertain to the class of *AF-Programs* ([Caminada et al., 2015b]), which includes all logic programs corresponding to the description of an argumentation framework.[6]

Definition 12. *(AF-Program [Caminada et al., 2015b])* A logic program P is an AF-Program *if for each $c \in HB_P$ there is at most one rule with conclusion c.*

Albeit simple, the translation function **AA2LP** was shown to preserve the semantics of any input argumentation frameworks [Caminada et al., 2015b; Caminada et al., 2022] for all the *complete semantics* and its particular cases listed in Definition 10. In fact, **AA2LP** preserves the complete labelling semantics from the input argumentation framework without change[7]:

---

[5] Please notice that we use the names of the arguments as atoms in the associated logic program. Doing so brings no prejudice to our original definitions on logic programs. Instead of using the names of arguments, we could use atoms such as $a_1, a_2, ..., a_4$ or a, b, c, d in order to build an alike program. In that setting, each atom would represent one of the arguments.

[6] Please notice that although each logic program that is the result of translating an argumentation framework is an AF-Program, it is *not* the case that every AF-Program can be the result of translating an argumentation framework. A counter example would be an AF-Program with a strong literal in the body of one of its rules.

[7] The original results require a translation between argumentation labellings and program models, but only because they defined program interpretations as a pair

Theorem 1. *([Caminada and Gabbay, 2009]) Let $AF = (Ar, att)$ be an argumentation framework and* `ArgLab` $= ($`in`, `out`, `undec`$)$ *be a complete labelling of AF. Then* `ArgLab` *is a 3-valued stable model of P_{AF}.*

This result rules in favor of logic programming subsuming abstract argumentation, since it portrays the *coincidence* of semantics, not merely a way to map labellings to models. A range of results similar to that of Theorem 1, immediately follows.

Corollary 2. *Let $AF = (Ar, att)$ be an argumentation framework,* `ArgLab` $= ($`in`, `out`, `undec`$)$ *be a complete argument labelling of AF and* `AA2LP`$(AF) = P_{\text{AF}}$. *Then:*

- *If* `ArgLab` *is grounded, then* `ArgLab` *is the well-founded model of P_{AF}.*
- *If* `ArgLab` *is preferred, then* `ArgLab` *is a regular model of P_{AF}.*
- *If* `ArgLab` *is stable, then* `ArgLab` *is a 2-valued stable model of P_{AF}.*
- *If* `ArgLab` *is semi-stable, then* `ArgLab` *is a L-stable model of P_{AF}.*
- *If* `ArgLab` *is pre-ideal, then* `ArgLab` *is a pre-ideal model of P_{AF}.*
- *If* `ArgLab` *is ideal, then* `ArgLab` *is the ideal model of P_{AF}.*
- *If* `ArgLab` *is pre-eager, then* `ArgLab` *is a pre-eager model of P_{AF}.*
- *If* `ArgLab` *is eager, then* `ArgLab` *is the eager model of P_{AF}.*

3.3 From Logic Programming to Abstract Argumentation

Moving from normal logic programming to abstract argumentation requires more steps and intricate machinery.

We start with describing how the rules of a Logic Program can be used to construct arguments. For this, we revisit the approach of [Caminada *et al.*, 2015b] with a slight change to include *default arguments*, which were introduced in [Sá and Alcântara, 2021a].

(T, F) (leaving implicit the set of undecided atoms) instead of a tuple (T, F, U). Given an argumentation labelling (`in`, `out`, `undec`), their translation involved only omitting `undec` to provide (`in`, `out`) as the resulting interpretation.

7 - LOGIC PROGRAMMING AND ARGUMENTATION

Definition 13. *Let P be a logic program, we define the arguments and default arguments induced by P as follows:*

- *If c is an atom in HB_P and there is at least one $r \in P$ for which $head(r) = c$, then* **not** *c is a default argument (say D_c) with*
  - $\texttt{Conc}(D_c) = \textbf{not } c,$
  - $\texttt{Rules}(D_c) = \emptyset,$
  - $\texttt{Vul}(D_c) = \{c\},$ *and*
  - $\texttt{Sub}(D_c) = \{D_c\}.$

- *If $c \leftarrow \textbf{not } b_1, \ldots, \textbf{not } b_m$ is a rule in P, then it is also an argument (say A) with*
  - $\texttt{Conc}(A) = c,$
  - $\texttt{Rules}(A) = \{c \leftarrow \textbf{not } b_1, \ldots, \textbf{not } b_m\},$
  - $\texttt{Vul}(A) = \{b_1, \ldots, b_m\},$ *and*
  - $\texttt{Sub}(A) = \{A\}.$

- *If $c \leftarrow a_1, \ldots, a_n, \textbf{not } b_1, \ldots, \textbf{not } b_m$ is a rule in P and for each a_i ($1 \leq i \leq n$) there exists an argument A_i with $\texttt{Conc}(A_i) = a_i$ and such that $c \leftarrow a_1, \ldots, a_n, \textbf{not } b_1, \ldots, \textbf{not } b_m$ is not contained in $\texttt{Rules}(A_i)$, then $c \leftarrow (A_1), \ldots, (A_n), \textbf{not } b_1, \ldots, \textbf{not } b_m$ is an argument (say A) with*
  - $\texttt{Conc}(A) = c,$
  - $\texttt{Rules}(A) = \texttt{Rules}(A_1) \cup \ldots \cup \texttt{Rules}(A_n) \cup \{c \leftarrow a_1, \ldots, a_n, \textbf{not } b_1, \ldots, \textbf{not } b_m\}$
  - $\texttt{Vul}(A) = \texttt{Vul}(A_1) \cup \ldots \cup \texttt{Vul}(A_n) \cup \{b_1, \ldots, b_m\},$ *and*
  - $\texttt{Sub}(A) = \{A\} \cup \texttt{Sub}(A_1) \cup \ldots \cup \texttt{Sub}(A_n).$

In essence, an argument can be seen as a tree-like structure of rules (the only difference with a real tree is that a rule can occur at more than one place in the argument) corresponding to a possible proof for some atom in the language of the program. Following that idea, default arguments concern possible proofs that can be drawn from the program using no rules. Default arguments model the fact that NAF-literals are true by default in every semantics of a logic program, hence their name.

In the above, if A is an argument, $\texttt{Conc}(A)$ is referred to as the *conclusion* of A, $\texttt{Rules}(A)$ is referred to as the *rules* of A, $\texttt{Vul}(A)$ is referred to as the *vulnerabilities* of A and $\texttt{Sub}(A)$ is referred to as the *subarguments* of A.

The next step in constructing the argumentation framework is to determine the attack relation: an argument attacks another iff its conclusion is among the vulnerabilities of the attacked argument. With that we mind, we can propose:

Definition 14. *Let P be a logic program. The argumentation framework associated with P is*

$$\texttt{LP2AA}(P) = AF_P = (Ar_P, att_P)$$

where Ar_P is the set of arguments from P (Definition 13) and

$$att_P = \{(A, B) \mid \texttt{Conc}(A) \in \texttt{Vul}(B)\}$$

Concerning AF_P, please notice that:

- Each argument that is not a default argument attacks one and only one default argument.
- Each default argument attacks zero arguments[8].

Therefore, in AF_P, one can differentiate whether two arguments have the same conclusion or not [Sá and Alcântara, 2021a]: it suffices to check if they attack the same default argument, which in turn are the only arguments in AF_P that do not attack any arguments. This is an advantage over the original definition of [Caminada *et al.*, 2015b].

We can now apply argumentation semantics to the resulting argumentation framework and, based on the resulting argument labelling(s), obtain their associated conclusion labelling(s) using the approach of [Caminada *et al.*, 2015b; Caminada *et al.*, 2022].

---

[8]This is because the conclusion of a default argument is a NAF-literal, which are never among the vulnerabilities of arguments.

7 - LOGIC PROGRAMMING AND ARGUMENTATION

Definition 15. *([Caminada et al., 2015b; Caminada et al., 2022]) Let P be a logic program, a conclusion labelling of P is a function* ConcLab : $HB_P \to \{\text{in}, \text{out}, \text{undec}\}$. *Then, given an argument labelling* ArgLab *of* AF_P, *the conclusion labelling associated to* ArgLab *is*

$$\text{ConcLab}(c) = max(\{\text{ArgLab}(A) \mid \text{Conc}(A) = c\} \cup \{\text{out}\})$$

where in $>$ undec $>$ out.

We say that a conclusion labelling is complete *iff it is the associated conclusion labelling of a complete argument labelling. Moving forward, we will refer to a function* ArgLab2ConcLab *such that, for any complete argument labelling* ArgLab *of* AF_P, ArgLab2ConcLab(ArgLab) *provides the conclusion labelling associated with* ArgLab. *Further, we will refer to* ConcLab2ArgLab *as the inverse function of* ArgLab2ConcLab[9].

Fundamentally, conclusion labellings and program interpretations are the same, the only difference being the names of the truth values in use[10]. It was shown in [Wu et al., 2009; Caminada et al., 2015b] that complete conclusion labellings coincide with 3-valued stable models[11].

Theorem 3. *([Wu et al., 2009; Caminada et al., 2015b]) Let P be a logic program and $AF_P = (Ar_P, att_P)$ be its associated argumentation framework. It holds that:*

1. *if* Mod *is a 3-valued stable model of P then* Mod *is a complete conclusion labelling of P*

2. *if* ConcLab *is a complete conclusion labelling of P then* ConcLab *is a 3-valued stable model of P*

---

[9]It has been shown in [Caminada et al., 2015b] that when restricted to complete argument labellings and complete conclusion labellings, ArgLab2ConcLab and ConcLab2ArgLab are both bijective and each others inverse.

[10]In previous works, such as in [Wu et al., 2009; Caminada et al., 2015b], the authors considered special functions ConcLab2Mod and Mod2ConcLab to convert between conclusion labellings and logic programming models, but they are not required here due to our choice of notation for program interpretations.

[11]The results reported were obtained using the translation from [Wu et al., 2009; Caminada et al., 2015b]. The only difference is that we add default arguments to the set of arguments obtained from a program in Definition 13. It was shown in [Sá and Alcântara, 2021a] that the introduction of the extra arguments preserves all the results from previous works.

As before, we obtained a coincidence result. However, we highlight that conclusion labellings are not argument semantics *per se*, since it is not arguments they evaluate. Fortunately, the result holds just as well in regard to complete argument labellings:

Theorem 4. *([Wu et al., 2009; Caminada et al., 2015b]) Let P be a logic program and $AF_P = (Ar_P, att_P)$ be its associated argumentation framework. Then* ArgLab *is a complete argument labelling of AF_P if and only if* ArgLab2ConcLab(ArgLab) *is a 3-valued stable model of P.*

Differently from the previous result, this one does not involve an immediate *coincidence*, only a way to map corresponding models and labellings. This difference is rather significant[12]: given a program P,

- if one retrieves the complete *conclusion* labellings of AF_P having minimal/maximal **in/out/undec**, they *will coincide* with the 3-valued stable models having minimal/maximal *true/false/undec* (due to Theorem 3);

- but if one retrieves the complete *argument* labellings of AF_P having minimal/maximal **in/out/undec**, they *may or may not correspond* to the 3-valued stable models having minimal/maximal *true/false/undec* [Caminada et al., 2015b].

The following example[13] illustrates a scenario where minimizing undecidness at the argument level is not the same as minimizing undecidedness at the conclusion level.

Example 2. *Consider the logic program P with rules*

$$r_1 : \quad c \leftarrow \text{not } c$$
$$r_2 : \quad a \leftarrow \text{not } b$$
$$r_3 : \quad b \leftarrow \text{not } a$$
$$r_4 : \quad c \leftarrow \text{not } c, \text{not } a$$
$$r_5 : \quad g \leftarrow \text{not } g, \text{not } b$$

---

[12]These results are discussed at length in [Caminada et al., 2015b] and complemented in [Caminada et al., 2022].

[13]The program example and its corresponding AF are adapted from [Caminada et al., 2015b; Caminada et al., 2022] to include default arguments. One of the original arguments was also removed for compactness.

One can build the following arguments from P:

- $A_1 = r_1$, with $\text{Conc}(A_1) = c$ and $\text{Vul}(A_1) = \{c\}$
- $A_2 = r_2$, with $\text{Conc}(A_2) = a$ and $\text{Vul}(A_2) = \{b\}$
- $A_3 = r_3$, with $\text{Conc}(A_3) = b$ and $\text{Vul}(A_3) = \{a\}$
- $A_4 = r_4$, with $\text{Conc}(A_4) = c$ and $\text{Vul}(A_4) = \{c, a\}$
- $A_5 = r_5$, with $\text{Conc}(A_5) = g$ and $\text{Vul}(A_5) = \{g, b\}$
- $A_6 = \text{not } c$, with $\text{Conc}(A_6) = \text{not } c$ and $\text{Vul}(A_6) = \{c\}$
- $A_7 = \text{not } a$, with $\text{Conc}(A_7) = \text{not } a$ and $\text{Vul}(A_7) = \{a\}$
- $A_8 = \text{not } b$, with $\text{Conc}(A_8) = \text{not } b$ and $\text{Vul}(A_8) = \{b\}$
- $A_9 = \text{not } g$, with $\text{Conc}(A_9) = \text{not } g$ and $\text{Vul}(A_9) = \{g\}$

The argumentation framework AF_P associated with P correspond to the one we discussed in Example 1, as one can observe in Figure 1.

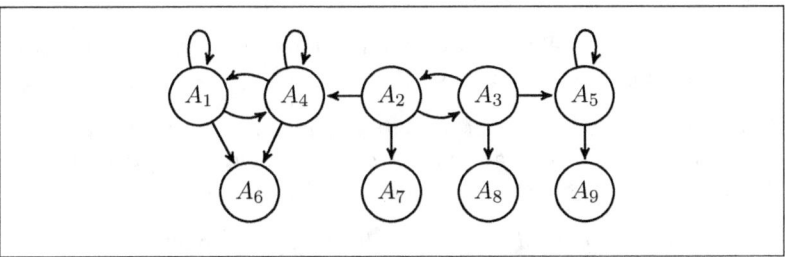

Figure 1: The argumentation framework AF_P associated with P.

The complete argument labellings of AF_P are

- $\text{ArgLab}_1 = (\emptyset, \emptyset, \{A_1, A_2, A_3, A_4, A_5, A_6, A_7, A_8, A_9\})$
- $\text{ArgLab}_2 = (\{A_2, A_8\}, \{A_3, A_4, A_7\}, \{A_1, A_5, A_6, A_9\})$
- $\text{ArgLab}_3 = (\{A_3, A_7, A_9\}, \{A_2, A_5, A_8\}, \{A_1, A_4, A_6\})$

The associated complete conclusion labellings are

- $\text{ConcLab}_1 = (\emptyset, \emptyset, \{a, b, c, g\})$,
- $\text{ConcLab}_2 = (\{a\}, \{b\}, \{c, g\})$, *and*
- $\text{ConcLab}_3 = (\{b\}, \{a, g\}, \{c\})$.

ArgLab_2 *and* ArgLab_3 *are semi-stable argument labellings, that is, complete argument labellings where* undec *is* $\subseteq$-*minimal. Hence, the associated conclusion labellings* ConcLab_2 *and* ConcLab_3 *are semi-stable conclusion labellings. But since* $\text{undec}(\text{ConcLab}_2) \supset \text{undec}(\text{ConcLab}_3)$, *we find that* ConcLab_2 *is not an L-stable model of* P. *So here we have an example of a logic program where the semi-stable and L-stable conclusion labellings do not correspond.*

Further, because ArgLab_2 *and* ArgLab_3 *are both semi-stable argument labellings,* ArgLab_1 *is the only pre-eager argument labelling and also the eager argument labelling of* AF_P. *On the other hand, since only* ConcLab_3 *is an L-stable model of* P, *both* ConcLab_1 *and* ConcLab_3 *are pre-eager models and* ConcLab_3 *is the eager model of* P. *Therefore, the pre-eager and the eager argument semantics may fail to capture the pre-eager and eager semantics for logic programs.*

Overall, the results found in [Caminada *et al.*, 2015b; Caminada *et al.*, 2022] can be summarized as follows:

Theorem 5. *([Caminada* et al., *2015b; Caminada* et al., *2022]) Let* P *be a logic program,* AF_P *be its associated argumentation framework and* ArgLab *be a complete argument labelling of* AF_P. *Then:*

1. ArgLab *is grounded if and only if* $\text{ArgLab2ConcLab}(\text{ArgLab})$ *is well-founded*

2. ArgLab *is preferred if and only if* $\text{ArgLab2ConcLab}(\text{ArgLab})$ *is regular*

3. ArgLab *is stable if and only if* $\text{ArgLab2ConcLab}(\text{ArgLab})$ *is stable*

4. ArgLab *is pre-ideal if and only if* $\text{ArgLab2ConcLab}(\text{ArgLab})$ *is pre-ideal*

5. `ArgLab` *is ideal if and only if* `ArgLab2ConcLab(ArgLab)` *is ideal*

6. *If* `ArgLab` *is semi-stable,* `ArgLab2ConcLab(ArgLab)` *may not be L-stable (and vice-versa).*

7. *If* `ArgLab` *is pre-eager,* `ArgLab2ConcLab(ArgLab)` *may not be pre-eager (and vice-versa).*

8. *If* `ArgLab` *is eager,* `ArgLab2ConcLab(ArgLab)` *may not be eager (and vice-versa).*

The results gathered so far allow us to observe the fundamental difference between logic programming and (instantiated) argumentation. Logic programming, in essence, does maximization and minimization at the *conclusion* level. That is, it (conceptually) takes all complete argument labellings, converts these to conclusion labellings, and then selects the maximal/minimal among these. Instantiated argumentation, on the other hand, does maximization and minimization at the *argument* level. That is, it (conceptually) takes all complete argument labellings, selects the maximal/minimal among these, and then converts these to conclusion labellings. So whereas logic programming does the maximization/minimization *after* converting argument labellings to conclusion labellings, instantiated argumentation does the maximization/minimization *before* converting argument labellings to conclusion labellings.[14]

3.4 The Conundrum of Minimizing Undecided Arguments vs Undecided Conclusions

The non-correspondence result concerning the semi-stable argument semantics and L-stable logic programming semantics (from [Caminada *et al.*, 2015b]) motivated further investigation in attempts to understand their differences and whether it is possible to devise an argumentation

---

[14]It has to be mentioned that formalisms such as ASPIC+ [Modgil and Prakken, 2013; Modgil and Prakken, 2014; Modgil and Prakken, 2018], ABA [Bondarenko *et al.*, 1997; Toni, 2014; Čyras *et al.*, 2018] and logic-based argumentation [Gorogiannis and Hunter, 2011; Besnard and Hunter, 2014] have been stated in terms of extensions instead of in terms of labellings. However, as extensions and labellings coincide [Caminada, 2006; Caminada and Gabbay, 2009; Baroni *et al.*, 2018] they could be viewed in terms of labellings as well.

semantics that captures the L-stable semantics. Among these efforts, Sá and Alcântara observed that the difference is always related to some arguments whose attackees coincide: in their redundancy, one or more of those arguments would become irrelevant to the evaluation of those arguments they mutually attack [Sá and Alcântara, 2021a]. They also identified that sink[15] arguments played a major role in pinpointing the culprit arguments. As we considered the instantiation of default arguments in Definition 13, these will become the sinks of the argumentation framework associated with an input program. Further, because Definition 13 ensures there is exactly one default argument for each atom that can be proven in a program, it allows for the proposal of new argumentation semantics that maximize/minimize the labels of default arguments. This led to the proposal of the L-stable argumentation semantics.

Definition 16. *Let* `ArgLab` *be an argument labelling of* AF. *Given*

$$\text{SINKS}_{AF} = \{A \in Ar \mid \forall B \in Ar, (A,B) \notin att\},$$

we say that `ArgLab` *is an* L-stable argument labelling *iff* `ArgLab` *is a complete argument labelling where* `undec(ArgLab)` $\cap$ SINKS_{AF} *is* $\subseteq$-*minimal among the complete argument labellings of* AF.[16]

The L-stable argument labellings share similar properties to those of the semi-stable argument labellings [Sá and Alcântara, 2021a]:

- Every AF has at least one L-stable labelling.
- If AF has at least one stable labelling, the L-stable, semi-stable and stable argument labellings of AF coincide.

Example 3. *Looking back at Example 2, we had that both*

$$\texttt{ArgLab}_2 = (\{A_2, A_8\}, \{A_3, A_4, A_7\}, \{A_1, A_5, A_6, A_9\}) \ and$$

$$\texttt{ArgLab}_3 = (\{A_3, A_7, A_9\}, \{A_2, A_5, A_8\}, \{A_1, A_4, A_6\})$$

---

[15]In graph theory terminology, a sink is a node from which no edges originate.
[16]The definition we provide differs from the original one in [Sá and Alcântara, 2021a], but they proved that the property we use in our definition is exclusively satisfied by L-stable argument labellings.

are semi-stable argument labellings, but only ArgLab2ConcLab(ArgLab$_3$) is an L-stable model of P. Given that SINKS$_{AF_P}$ = $\{A_6, A_7, A_8, A_9\}$, we have that only ArgLab$_3$ is an L-stable argument labelling. Hence, the L-stable argument labellings of AF$_P$ correspond to the L-stable models of P for this example.

It was proved in [Sá and Alcântara, 2021a] that the L-stable argument semantics indeed captures the L-stable program semantics.

Theorem 6. *([Sá and Alcântara, 2021a]) Let P be a logic program and* ArgLab *be a complete argument labelling of its associated argumentation framework AF$_P$. Then* ArgLab *is an L-stable argument labelling of AF$_P$ if and only if* ArgLab2ConcLab(ArgLab) *is an L-stable model of P.*

Furthermore, we can devise new semantics based on the L-stable labellings in a similar spirit to the pre-eager and eager semantics:

Definition 17. *Let* ArgLab *be an argument labelling of AF.* ArgLab *is called:*

- *a* L-pre-eager argument labelling *iff* ArgLab *is a complete argument labelling where* in(ArgLab) $\subseteq$ in(ArgLab$_{lst}$) *for each L-stable argument labelling* ArgLab$_{lst}$ *of AF*

- *the* L-eager argument labelling *iff* ArgLab *is a pre-eager argument labelling where* in(ArgLab) *is maximal (w.r.t. $\subseteq$) among all pre-eager argument labellings of AF*

Now, given that the L-stable argument semantics captures the L-stable program semantics, we obtain as a corollary of Theorem 6 that

Corollary 7. *Let P be a logic program* ArgLab *be a complete argument labelling of its associated argumentation framework AF$_P$. Then:*

1. ArgLab *is an L-pre-eager argument labelling of AF$_P$ if and only if* ArgLab2ConcLab(ArgLab) *is a pre-eager model of P.*

2. ArgLab *is the L-eager argument labelling of AF$_P$ if and only if* ArgLab2ConcLab(ArgLab) *is the eager model of P.*

These results ensure the existence of argument semantics able to model every logic programming semantics for which [Caminada et al., 2015b; Caminada et al., 2022] could not find correspondence results. However, they bring new questions: what logic programming semantics could model the L-stable, L-pre-eager and L-eager argument semantics?

From the discussion of translation AA2LP in Section 3.2, we can ensure that such logic programming semantics can definitely be obtained, however it is possible they have not been defined in the logic programming literature and will seem counter-intuitive. The reason is that we have one-to-one correspondence between arguments in a given argumentation framework AF and program rules in AA2LP(AF). This means that only a subset of the program rules corresponds to default arguments, therefore minimizing undecided default arguments in AF corresponds to minimizing undecided conclusions for only a subset of the atoms (or rules) in the program.

Defining such semantics anew could prove to be a complicated task, since the concept of sink nodes (or sink atoms) is not as obvious in the context of logic programs. Fortunately, given an argumentation framework AF and its corresponding program AA2LP(AF), it is rather simple to retrieve what atoms in AA2LP(AF) correspond to the sinks in AF: since all arguments that are not sinks by definition attack one or more arguments, the sinks correspond to those atoms in AA2LP(AF) for which their respective NAF-literals do not occur in the rules of AA2LP(AF).[17] [18]

Definition 18. *Given AF, let $P = $ AA2LP(AF) *and* $\text{Mod} = (T, F, U)$ *be a 3-valued interpretation of P w.r.t. HB_P. Further, given*

$$\text{SINKS}_P = \{c \in HB_P \mid \forall r \in P, \text{not } c \notin body(r)\},$$

we say that Mod *is*

---

[17]As an abuse of notation, we allow ourselves to reuse the function name SINKS both to retrieve from an AF its sink arguments and to retrieve from a program its atoms corresponding to sink arguments of a corresponding AF. The subscript text should be enough to indicate what is the case.

[18]As an example, notice how not A_6, not A_7, not A_8 and not A_9 do not occur in the rules of P_{AF} in Example 1.

- an L*-stable model *of P iff* Mod *is a 3-valued stable model where* $U \cap \text{SINKS}_P$ *is minimal (w.r.t.* $\subseteq$*) among all 3-valued stable models of P.*

- an L-pre-eager model *of P iff* Mod *is a 3-valued stable model where* $T \subseteq T_{l^*}$ *for each* L*-stable model* $\text{Mod}_{l^*} = (T_{l^*}, F_{l^*}, U_{l^*})$ *of P*

- the L-eager model *of P iff* Mod *is the pre-eager model where T is* $\subseteq$*-maximal among all pre-eager models of P*

Now, since the 3-valued stable models of AA2LP(AF) are precisely the complete argument labellings of AF (Theorem 1), we obtain the following results, which complement Corollary 2:

Corollary 8. *Let* $AF = (Ar, att)$ *be an argumentation framework,* ArgLab = (in, out, undec) *be a complete argument labelling of AF, and* $\text{AA2LP}(AF) = P_{\text{AF}}$. *Then:*

- *If* ArgLab *is L-stable, then* ArgLab *is an L*-stable model of* P_{AF}.

- *If* ArgLab *is L-pre-eager, then* ArgLab *is a L-pre-eager model of* P_{AF}.

- *If* ArgLab *is L-eager, then* ArgLab *is the L-eager model of* P_{AF}.

At this point, once again, we must seek different argumentation semantics capable of expressing the new program semantics. This game may stabilize around a few semantics based on minimal undecided literals and arguments, but it might just as well go on indefinitely. We should also mind how the translation functions have a fundamental role in the results obtained.[19] One can say that AA2LP is a definitive translation, given the coincidental results it provides starting from argumentation semantics, but since LP2AA can only provide correspondence results, it leaves open the possibility that it may somehow be improved for the sake of capturing logic programming semantics.

---

[19]For instance, a different translation including singleton arguments for *undefined* atoms in a program (those that are not in the heads of any rules) was proposed in [Cramer and Saldanha, 2020]. Their translation also includes default arguments for negative literals (just like ours), causing the undefined atoms to be labelled as undec by the grounded AF semantics, whereas in our approach, they will be labelled out. As an effect, they find that the grounded AF semantics captures the *weak completion semantics* [Hölldobler and Kencana Ramli, 2009] for logic programs.

4 On the Connection between Assumption-Based Argumentation and Logic Programming

Assumption-Based Argumentation (ABA) [Bondarenko et al., 1997; Dung et al., 2009] is a rule-based argumentation formalism where some special sentences called *assumptions*, which are true by default (that is, unless their contrary can be proved), have a central role in semantics. Just like in abstract argumentation, theories in ABA are proposed as *frameworks* whose semantics are primarily retrieved in terms of *extensions* and *labellings*, but, at the same time, ABA frameworks share similar syntax to logic programs, allowing their semantics to be understood in terms of *interpretations* and *models* [Sá and Alcântara, 2019; Sá and Alcântara, 2021b]. The similarities may be illustrated with matching examples even before we formally introduce ABA frameworks.

Example 4. *Take into consideration the logic program P^{20} as follows:*

$$P : \begin{array}{ll} a \leftarrow \text{not } b & c \leftarrow \text{not } c \\ b \leftarrow \text{not } a & d \leftarrow b, c \end{array}$$

Following [Bondarenko et al., 1997], P would be translated into the ABA framework $ABA(P) = \mathcal{F}^{21}$ below.

$$\mathcal{F} : \begin{array}{llll} a \leftarrow \beta & c \leftarrow \gamma & \overline{\alpha} = a & \overline{\beta} = b \\ b \leftarrow \alpha & d \leftarrow b, c & \overline{\gamma} = c & \overline{\delta} = d \end{array}$$

Notice how the rules in $\mathcal{F}$ (depicted in the first two columns) mirror the rules of P, while the operator $^-$ captures the semantics of **not***. The language of $\mathcal{F}$ is a bit different from the language of P, but not by much. It consists of eight sentences, namely $a, b, c, d, \alpha, \beta, \gamma, \delta$, where $\alpha, \beta, \gamma, \delta$ respectively model* **not** a, **not** b, **not** c, **not** d *as native elements in $\mathcal{F}$'s language. The sentences $\alpha, \beta, \gamma, \delta$ are called* assumptions *and each one has a unique contrary in our example: a is the contrary of α*

---

[20] This program is extracted from [Caminada and Schulz, 2017].

[21] At this time, to avoid the necessity of formal concepts, only the core syntactic elements of $\mathcal{F}$ are shown.

(which corresponds to **not** a), b is the contrary of β (which corresponds to **not** b), and so on.

Moving on, if one follows [Caminada and Schulz, 2017] to translate $\mathcal{F}$ into a logic program, one would obtain P as a result. However, this would also be the program obtained if the input is $\mathcal{F}'$ below, which corresponds to $\mathcal{F}$ except that δ is not in its language.

$$\mathcal{F}': \begin{array}{cccc} a \leftarrow \beta & c \leftarrow \gamma & \overline{\alpha} = a & \overline{\beta} = b \\ b \leftarrow \alpha & d \leftarrow b, c & \overline{\gamma} = c & \end{array}$$

While the difference between $\mathcal{F}, \mathcal{F}'$ may seem small, the semantics of these frameworks are significantly different just because $\mathcal{F}$ has one assumption more than $\mathcal{F}'$. And yet, the approach of [Caminada and Schulz, 2017] translates both $\mathcal{F}$ and $\mathcal{F}'$ to P.

The ABA frameworks above illustrate the main challenge when translating from ABA to logic programming: the semantics of logic programs evaluate all^{22} sentences in the language of the program (the atoms) while the mainstream semantics of ABA evaluate only *some* sentences in the language of each framework (the assumptions). Fortunately, the different results can only be observed for semantics that minimize undecided sentences over ABA frameworks for which some non-assumptions are not the contrary of any assumptions [Caminada and Schulz, 2017]. Even then the correspondence for those results can be restored by an operation called *semantic projection* [Sá and Alcântara, 2021b].

4.1 Assumption-Based Argumentation: Syntax and Semantics

We briefly restate the core concepts of ABA frameworks [Bondarenko et al., 1997; Dung et al., 2007; Dung et al., 2009] before we can proceed to discuss the translations and technical results regarding connections between Logic Programming and ABA.

---

[22]If one accounts for NAF-literals as sentences in the language of a program, the semantics would evaluate necessarily *half* the sentences, but traditionally the language of the program is defined in terms of its Herbrand Base. In either case, ABA frameworks are flexible regarding how many sentences in the language of a framework are assumptions.

Definition 19. *([Dung et al., 2009]) An ABA framework is a tuple* $\langle \mathcal{L}, \mathcal{R}, \mathcal{A}, ^- \rangle$ *where:*

- $\langle \mathcal{L}, \mathcal{R} \rangle$ *is a deductive system where* $\mathcal{L}$ *is a logical language and* $\mathcal{R}$ *is a set of inference rules on that language*

- $\mathcal{A} \subseteq \mathcal{L}$ *is a (non-empty) set, whose elements are referred to as assumptions*

- $^-$ *is a total mapping from* $\mathcal{A}$ *into* $\mathcal{L} \setminus \mathcal{A}$,[23] *where* $\overline{\alpha}$ *is called the contrary of* α.

For current purposes, we restrict ourselves to ABA frameworks that are *flat* [Bondarenko et al., 1997], meaning that no assumption appears in the head of an inference rule. Furthermore, we follow [Dung et al., 2009] in that each assumption has a unique contrary. This choice makes it easier to define some of the concepts we need.

Definition 20. *([Dung et al., 2009]) Given a deductive system* $\langle \mathcal{L}, \mathcal{R} \rangle$, *and a set of assumptions* $\mathcal{A} \subseteq \mathcal{L}$, *an* argument *for* $c \in \mathcal{L}$ *(the* conclusion *or* claim*) supported by* $S \subseteq \mathcal{A}$ *is a tree with nodes labelled by formulas in* $\mathcal{L}$ *or by the special symbol* $\top$ *such that:*

- *the root is labelled* c

- *for every node* N

    - *if* N *is a leaf then* N *is labelled either by an assumption or by* $\top$

    - *if* N *is not a leaf and* b *is the label of* N, *then there exists an inference rule* $b \leftarrow b_1, \ldots, b_m$ $(m \geq 0)$ *and either* $m = 0$ *and the child of* N *is labelled by* $\top$, *or* $m > 0$ *and* N *has* m *children, labelled by* $b_1, \ldots, b_m$ *respectively*

- S *is the set of all assumptions labelling the leaves.*

---

[23]In the ABA literature it is common that the contrary relation $^-$ allows an assumption to be the contrary of another assumption or even for an assumption to have multiple contraries. In each case, ABA frameworks can be rewritten into an equivalent ABA framework where each assumption has a single contrary that is not an assumption [Caminada and Schulz, 2017].

7 - LOGIC PROGRAMMING AND ARGUMENTATION

We say that a set of assumptions $\mathcal{A}sms \subseteq \mathcal{A}$ enables the construction of an argument A if A is supported by a subset of $\mathcal{A}sms$. A set of assumptions $\mathcal{A}sms_1$ is said to *attack* an assumption α iff $\mathcal{A}sms_1$ enables the construction of an argument for $\overline{\alpha}$. A set of assumptions $\mathcal{A}sms_1$ is said to attack a set of assumptions $\mathcal{A}sms_2$ iff $\mathcal{A}sms_1$ attacks some assumption $\alpha \in \mathcal{A}sms_2$.

The next step is to describe the various ABA semantics, which can be conveyed in the forms of *assumption extensions* [Bondarenko et al., 1997], *assumption labellings* [Schultz and Toni, 2014; Schulz and Toni, 2017] or *interpretations and models* [Sá and Alcântara, 2019; Sá and Alcântara, 2021b] with corresponding translations between them. For the sake of uniformity in our presentation, we opt to prioritize the discussion of ABA semantics in the form of *assumption labellings*.

Definition 21. *([Schultz and Toni, 2014; Schulz and Toni, 2017]) An assumption labelling of $\mathcal{F}$ is a total function $\mathcal{L} : \mathcal{A} \to \{\text{in}, \text{out}, \text{undec}\}$.*

The same conventions we applied before to interpretations and argument labellings apply to assumption labellings, since we can perceive all of those concepts as the same function applied over different domains.

Definition 22. *([Schultz and Toni, 2014; Schulz and Toni, 2017]) An assumption labelling $\mathcal{L} = (\text{in}(\mathcal{L}), \text{out}(\mathcal{L}), \text{undec}(\mathcal{L}))$ of $\mathcal{F}$ is complete iff for each $\alpha \in \mathcal{A}$ it holds that:*

- *if $\alpha \in \text{in}(\mathcal{L})$, then each $\mathcal{S} \subseteq \mathcal{A}$ attacking α has some $\beta \in \mathcal{S}$ such that $\beta \in \text{out}(\mathcal{L})$;*

- *if $\alpha \in \text{out}(\mathcal{L})$, then there exists some $\mathcal{S} \subseteq \mathcal{A}$ attacking α such that $\mathcal{S} \subseteq \text{in}(\mathcal{L})$;*

- *if $\alpha \in \text{undec}(\mathcal{L})$, then (i) there is at least one $\mathcal{S} \subseteq \mathcal{A}$ attacking α such that $\mathcal{S} \cap \text{out}(\mathcal{L}) = \emptyset$ and (ii) each $\mathcal{S} \subseteq \mathcal{A}$ attacking α is such that $\mathcal{S} \setminus \text{in}(\mathcal{L}) \neq \emptyset$.*

Notice how Definition 22 closely resembles the definition of complete argument labellings (Definition 9).

Example 5. *The ABAF $\mathcal{F}$ (Example 4) has three complete assumption labellings: $\mathcal{L}_1 = (\{\ \}, \{\ \}, \{\alpha, \beta, \gamma, \delta\})$, $\mathcal{L}_2 = (\{\beta, \delta\}, \{\alpha\}, \{\gamma\})$, and*

$\mathcal{L}_3 = (\{\alpha\}, \{\beta\}, \{\gamma, \delta\})$. Just alike, $\mathcal{F}'$ (Example 4) has three complete assumption labellings: $\mathcal{L}'_1 = (\{\ \}, \{\ \}, \{\alpha, \beta, \gamma\})$, $\mathcal{L}'_2 = (\{\beta\}, \{\alpha\}, \{\gamma\})$, and $\mathcal{L}'_3 = (\{\alpha\}, \{\beta\}, \{\gamma\})$.

The other labelling-based ABA semantics are defined as usual, as particular cases of the complete semantics:

Definition 23. Let $\mathcal{L}$ be an assumption labelling of ABA framework $\mathcal{F} = \langle \mathcal{L}, \mathcal{R}, \mathcal{A}, {}^- \rangle$. $\mathcal{L}$ is called[24]:

- *the* grounded assumption labelling *iff $\mathcal{L}$ is a complete assumption labelling where* $\text{in}(\mathcal{L})$ *is $\subseteq$-minimal among all complete assumption labellings of $\mathcal{F}$*

- *a* preferred assumption labelling *iff $\mathcal{L}$ is a complete assumption labelling where* $\text{in}(\mathcal{L})$ *is $\subseteq$-maximal among all complete assumption labellings of $\mathcal{F}$*

- *a* stable assumption labelling *iff $\mathcal{L}$ is a complete assumption labelling where* $\text{undec}(\mathcal{L}) = \emptyset$

- *a* semi-stable assumption labelling *iff $\mathcal{L}$ is a complete assumption labelling where* $\text{undec}(\mathcal{L})$ *is $\subseteq$-minimal among all complete assumption labellings of $\mathcal{F}$*

- *a* pre-ideal assumption labelling *iff $\mathcal{L}$ is a complete assumption labelling where* $\text{in}(\mathcal{L}) \subseteq \text{in}(\mathcal{L}_{pr})$ *for each preferred assumption labelling of $\mathcal{F}$*

- *the* ideal assumption labelling *iff $\mathcal{L}$ is the complete assumption labelling where* $\text{in}(\mathcal{L})$ *is $\subseteq$-maximal among all pre-ideal assumption labellings of $\mathcal{F}$*

- *a* pre-eager assumption labelling *iff $\mathcal{L}$ is a complete assumption labelling where* $\text{in}(\mathcal{L}) \subseteq \text{in}(\mathcal{L}_{sem})$ *for each semi-stable assumption labelling of $\mathcal{F}$*

---

[24]The semi-stable and the eager semantics for ABA were originally defined in [Caminada et al., 2015a] as extensions. We adapted the presentation of these semantics to favour uniformity with our previous definitions.

- *the eager assumption labelling iff $\mathcal{L}$ is the complete assumption labelling where $\text{in}(\mathcal{L})$ is $\subseteq$-maximal among all pre-eager assumption labellings of $\mathcal{F}$*

Example 6. *Among the complete assumption labellings of $\mathcal{F}$ (see Example 4 and Example 5), one can observe that: $\mathcal{L}_1$ is the grounded assumption labelling; $\mathcal{L}_2, \mathcal{L}_3$ are preferred assumption labellings; there are no stable assumption labellings; $\mathcal{L}_2$ is the only semi-stable assumption labelling; $\mathcal{L}_1$ is the only pre-ideal assumption labelling and therefore it is also ideal; $\mathcal{L}_1, \mathcal{L}_2$ are pre-eager and so $\mathcal{L}_2$ is the eager assumption labelling.*

4.2 From Logic Programming to Assumption-Based Argumentation

First, we will consider how logic programming semantics can be captured by ABA using a suitable translation. Intuitively, assumptions are sentences in the language of an ABA framework that are true unless their contrary can be proved. This describes precisely the behaviour of NAF-literals in the setting of logic programs. This intuition explains the translation proposed by [Bondarenko et al., 1997], which we introduce below.

Definition 24. *([Bondarenko et al., 1997]) Let P be a program. The ABA framework associated to P is*

$$\text{LP2ABA}(P) = \mathcal{F}_P = \langle \mathcal{L}_P, \mathcal{R}_P, \mathcal{A}_P, \overline{} \rangle$$

where[25]:

- $\mathcal{L}_P = HB_P \cup \{\text{not } a \mid a \in HB_P\}$;

- $\mathcal{R}_P = P$;

- $\mathcal{A}_P = \{\text{not } a \mid a \in HB_P\}$;

- $\overline{\text{not } a} = a$ *for each* $\text{not } a \in \mathcal{A}$.

---

[25]Remember we treat programs as sets of rules and that HB_P is the set of all atoms appearing in P.

We draw special attention to the fact that the NAF-literals of P become assumptions while the contrary relation $^-$ implements the NAF operator **not**. Further, the set of rules $\mathcal{R}_P$ coincides with P, therefore the framework keeps the rules from the original program unchanged.

Example 7. *Take the LP P from Example 4, then* $\mathsf{LP2ABA}(P) = \mathcal{F}_P = \langle \mathcal{L}_P, \mathcal{R}_P, \mathcal{A}_P, ^- \rangle$ *with* $\mathcal{L}_P = \{a, b, c, d, \text{not } a, \text{not } b, \text{not } c, \text{not } d\}$, $\mathcal{R}_P = P$, $\mathcal{A}_P = \{\text{not } a, \text{not } b, \text{not } c, \text{not } d\}$ *and* $^-$ *such that* $\overline{\text{not } a} = a$, $\overline{\text{not } b} = b$, $\overline{\text{not } c} = c$ *and* $\overline{\text{not } d} = d$. *We adopt the same visual queue used in the introduction of this section for easier reference:*

$$\mathcal{F}_P : \begin{array}{llll} a \leftarrow \text{not } b & c \leftarrow \text{not } c & \overline{\text{not } a} = a & \overline{\text{not } b} = b \\ b \leftarrow \text{not } a & d \leftarrow b, c & \overline{\text{not } c} = c & \overline{\text{not } d} = d \end{array}$$

Notice how $\mathcal{F}$ from Example 4 is the same as $\mathcal{F}_P$, only the assumptions were renamed as $\alpha, \beta, \gamma, \delta$.

ABA frameworks such as $\mathcal{F}_P$ are called *assumption spanning* in [Caminada and Schulz, 2017].

Definition 25. *([Caminada and Schulz, 2017]) Let $\mathcal{F} = \langle \mathcal{L}, \mathcal{R}, \mathcal{A}, ^- \rangle$ be an ABA framework. We say that $\mathcal{F}$ is assumption-spanning iff for each $x \in \mathcal{L} \setminus \mathcal{A}$ there exists some $\chi \in \mathcal{A}$ such that $\overline{\chi} = x$.*

The translation function $\mathsf{LP2ABA}$ was shown to model the semantics of any input logic programs for the *3-valued stable semantics* and all its particular cases listed in Definition 6 [Bondarenko et al., 1997; Caminada and Schulz, 2017] using an auxiliary translation between assumption labellings and program interpretations:

Definition 26. *Let $\mathcal{L} = (\text{in}, \text{out}, \text{undec})$ be an assumption labelling of $\mathcal{F}_P$. The program interpretation of P corresponding to $\mathcal{L}$ is $\mathsf{L2I}(\mathcal{L}) = (T, F, U)$ with $T = \{\overline{\alpha} \mid \alpha \in \text{out}(\mathcal{L})\}$, $F = \{\overline{\alpha} \mid \alpha \in \text{in}(\mathcal{L})\}$, and $U = HB_P \setminus (T \cup F)$.*

We are now ready to list the relevant results from [Bondarenko et al., 1997; Caminada and Schulz, 2017]:

Theorem 9. *Let P be a logic program. Then $\mathcal{L}$ is a complete assumption labelling of $\mathcal{F}_P$ if and only if $\mathsf{L2I}(\mathcal{L})$ is a 3-valued stable model of P.*

While this result is not coincidental, the function L2I becomes bijective when used to map assumption labellings of $\mathcal{F}_P$ to corresponding program interpretations of P.[26] In this setting, L2I^{-1} is also a function, therefore the result in Theorem 9 is equivalent to a coincidence result. The same reasoning holds for the results below.

Theorem 10. *Let P be a logic program and $\mathcal{L}$ be a complete assumption labelling of $\mathcal{F}_P$. Then:*[27]

- *If $\mathcal{L}$ is grounded, then* L2I$(\mathcal{L})$ *is the well-founded model of P.*
- *If $\mathcal{L}$ is preferred, then* L2I$(\mathcal{L})$ *is a regular model of P.*
- *If $\mathcal{L}$ is stable, then* L2I$(\mathcal{L})$ *is a stable model of P.*
- *If $\mathcal{L}$ is semi-stable, then* L2I$(\mathcal{L})$ *is a L-stable model of P.*
- *If $\mathcal{L}$ is pre-ideal, then* L2I$(\mathcal{L})$ *is a pre-ideal model of P.*
- *If $\mathcal{L}$ is ideal, then* L2I$(\mathcal{L})$ *is the ideal model of P.*
- *If $\mathcal{L}$ is pre-eager, then* L2I$(\mathcal{L})$ *is a pre-eager model of P.*
- *If $\mathcal{L}$ is eager, then* L2I$(\mathcal{L})$ *is the eager model of P.*

The results gathered ensure that flat ABA frameworks capture normal logic programs and their semantics, leaving it open whether logic programming also captures ABA frameworks and their semantics. We will delve into this question next.

4.3 From Assumption-Based Argumentation to Logic Programming

Moving from assumption-based argumentation to logic programming is trickier because logic programs have only one set of sentences (the atoms) and a matching number of corresponding NAF-literals, but assumptions and non-assumptions may appear in any proportion in an

---

[26]This is ensured by the fact that the contrary relation ^-_P obtained in $\mathcal{F}_P$ is necessarily bijective.

[27]The results concerning semi-stable, pre-ideal, pre-eager and eager ABA semantics are new, but their proofs may be dismissed since they follow from Theorem 9.

ABA framework. If we ignore this obstacle for a moment and only consider the semantics of assumptions, there are two straightforward ways to represent assumptions from an ABA framework in a corresponding logic program:

1. they can be mapped to the NAF-literals of the resulting program [Caminada and Schulz, 2017] or

2. they can be mapped to special atoms defined[28] by the negation of their contraries [Sá and Alcântara, 2021b].

The first option results from reversing the translation LP2ABA of [Bondarenko et al., 1997]. This approach works perfectly for the class of *assumption-spanning* ABA frameworks [Caminada and Schulz, 2017], even so that all the results of Theorem 9 and of Theorem 10 are mirrored for them. This much ensures that

Theorem 11. Normal logic programs *are equivalent*[29] *to* assumption-spanning ABA frameworks.

When it comes to ABA frameworks in general, this approach is enough to ensure that logic programming captures most of ABA semantics, but it finds exceptions in the ABA semantics that minimize undecided assumptions (such as semi-stable, pre-eager and eager). The discrepancy arises because this translation is only injective for assumption-spanning ABA frameworks, not in general.

The second option [Sá and Alcântara, 2021b] matches the language of the input ABA framework (all sentences) to the language of the output logic program (all atoms). By doing so, the models of the corresponding program will evaluate all sentences from the input ABA framework, both assumptions and non-assumptions. This intuition led to the proposal of a model theory with interpretation and model semantics for ABA in [Sá and Alcântara, 2019]. Then, in order to retrieve assumption labellings,

---

[28]If a program has only one rule for a given atom c such as $r : c \leftarrow body(r)$, that rule is understood as the *definition* of c in P and it may be read as "c if and only iff $body(r)$".

[29]Here, by saying two argumentation systems are *equivalent*, we mean that the sets of problems that can be expressed and solved in both systems coincide.

it is necessary to restrict the language of the models in the output program to the set of assumptions from the input ABA framework [Sá and Alcântara, 2021b]. This is achieved using an operation called *projection* in the logic programming literature. The results of [Sá and Alcântara, 2021b] ensure that

Theorem 12. *([Sá and Alcântara, 2021b])* Flat ABA frameworks *are equivalent to* normal logic programs *with* projection.

In the following, we will introduce the available translations and the results ensured by each one.

4.3.1 Mapping Assumptions to NAF-literals

We proceed to describe the approach and results obtained by [Caminada and Schulz, 2017].

Given an ABA framework $\mathcal{F} = \langle \mathcal{L}, \mathcal{R}, \mathcal{A}, ^- \rangle$ any translation from ABA to logic programming should be primarily based on $\mathcal{R}$. As such, the idea is to translate each rule in $\mathcal{R}$ to an associated Logic Programming rule. Further, in $\mathcal{F}$, if α is an assumption whose contrary is the sentence a, we will find that α can be labelled in as long as all arguments for a have at least one assumption in their support that is out. Intuitively, this means that α can be proved based on a labelling as long as a cannot. Therefore, not must be used to implement the contrary relation $^-$. In this first translation, each occurrence of α is replaced with "not a" or, using the contrary relation, "not $\overline{\alpha}$".

Definition 27. *([Caminada and Schulz, 2017])* Let $\mathcal{F} = \langle \mathcal{L}, \mathcal{R}, \mathcal{A}, ^- \rangle$ be *an ABA framework, the program corresponding to $\mathcal{F}$ is*

$$\text{ABA2LP}_{\text{CS17}}(\mathcal{F}) = P_{\mathcal{F}} = \{a \leftarrow b_1, \ldots, b_n, \text{not } \overline{\gamma_1}, \ldots, \text{not } \overline{\gamma_m} \mid$$
$$a \leftarrow b_1, \ldots, b_n, \gamma_1, \ldots, \gamma_m \in \mathcal{R}\}.$$

Example 8. *Remember $\mathcal{F}, \mathcal{F}'$ and P from Example 4. Both ABA frameworks present $a \leftarrow \beta$ as one of its inference rules. In each case, this rule will be translated to $a \leftarrow$ not b, given that $\overline{\beta} = b$. As we previously mentioned, we will find that $\text{ABA2LP}_{\text{CS17}}(\mathcal{F}) = \text{ABA2LP}_{\text{CS17}}(\mathcal{F}') = P$.*

As before, we must retrieve assumption labellings from the models of the resulting program, which is done by a specialized function:

Definition 28. *([Caminada and Schulz, 2017]) Let $I = (T, F, U)$ be a model of $P_\mathcal{F}$, the assumption labelling of $\mathcal{F}$ corresponding to I is*

$$\text{I2L}(I) = (\{\alpha \in \mathcal{A} \mid \overline{\alpha} \in F\}, \{\alpha \in \mathcal{A} \mid \overline{\alpha} \in T\}, \{\alpha \in \mathcal{A} \mid \overline{\alpha} \in U\}).$$

Notice how $\mathcal{A}$ is required for the computation of the function I2L. In the next section we will build over the relevance of this parameter to motivate a different translation from ABA to LP. For now, we opt to leave this parameter implicit here (as in [Caminada and Schulz, 2017]), since it can be retrieved from $\mathcal{F}$.

Theorem 13. *[Caminada and Schulz, 2017] Let $\mathcal{F}$ be an ABA framework. Then I is a 3-valued stable model of $P_\mathcal{F}$ if and only if $\text{I2L}(I)$ is a complete assumption labelling of $\mathcal{F}$.*

Differently from before, I2L cannot be inverted in this setting.

Example 9. *Take $\mathcal{F}'$ from Example 4, we find that $\text{ABA2LP}_{\text{CS17}}(\mathcal{F}') = P$. Now we have that $I = (\{a\}, \{b,d\}, \{c\})$ is a 3-valued stable model of P and that $\text{I2L}(I) = (\{\beta\}, \{\alpha\}, \{\gamma\})$ is a complete assumption labelling of $\mathcal{F}'$. However, $\text{L2I}(\text{I2L}(I)) = (\{a\}, \{b\}, \{c,d\})$, which is not the same as I. The difference follows from the fact that there is no $\delta \in \mathcal{A}'$ such that $\overline{\delta} = d$, so L2I leaves d undecided.*

Therefore, the result in Theorem 13 ensures a one-to-one correspondence, but it is not as strong as a coincidence result. This situation is quite similar to the one we found in Section 3.3, when trying to model logic programming semantics using abstract argumentation frameworks. In fact, the correspondence and exception results we will find in this scenario mimic those we found then.

On that matter, we find a similar discrepancy when trying to minimize undecided atoms in program models versus undecided assumptions in matching assumption labellings.

Example 10. *Once again, retrieve $\mathcal{F}, \mathcal{F}'$ from Example 4 and remember that $\text{ABA2LP}_{\text{CS17}}(\mathcal{F}) = \text{ABA2LP}_{\text{CS17}}(\mathcal{F}') = P$. Continuing from Examples 5 and 6, we have that $\mathcal{F}$ has a single semi-stable assumption labelling $\mathcal{L}_2 = (\{\beta, \delta\}, \{\alpha\}, \{\gamma\})$, whereas $\mathcal{F}'$ has two: $\mathcal{L}'_2 = (\{\beta\}, \{\alpha\}, \{\gamma\})$ and $\mathcal{L}'_3 = (\{\alpha\}, \{\beta\}, \{\gamma\})$. As expected, since $\text{LP2ABA}(P) = \mathcal{F}$, the semantics of P mirrors that of $\mathcal{F}$: P has a single L-stable model, namely*

$I_2 = (\{a\}, \{b,d\}, \{c\})$. But the L-stable models of $\mathtt{ABA2LP}(\mathcal{F}') = P$ do not correspond one-to-one to the semi-stable assumption labellings of $\mathcal{F}'$. Further, P has two pre-eager models, namely $I_1 = (\{\ \}, \{\ \}, \{a,b,c,d\})$ and I_2, so I_2 is the eager model of P, whereas the only pre-eager (and therefore eager) assumption labelling of $\mathcal{F}'$ is $\mathcal{L}'_1 = (\{\ \}, \{\ \}, \{\alpha, \beta, \gamma\})$. This means that the pre-eager and eager semantics of P do not correspond to the pre-eager and eager assumption labellings of $\mathcal{F}'$.

Overall, the results found in [Caminada and Schulz, 2017] can be summarized (and extended[30]) as follows:

Theorem 14. *Let $\mathcal{F}$ be an ABA framework, $\mathtt{ABA2LP}_{\mathrm{CS17}}(\mathcal{F}) = P_\mathcal{F}$ be its associated logic program and I be a 3-valued stable model of $P_\mathcal{F}$. Then:*

1. *I is well-founded if and only if $\mathtt{I2L}(I)$ is grounded*

2. *I is regular if and only if $\mathtt{I2L}(I)$ is preferred*

3. *I is stable if and only if $\mathtt{I2L}(I)$ is stable*

4. *I is pre-ideal if and only if $\mathtt{I2L}(I)$ is pre-ideal*

5. *I is ideal if and only if $\mathtt{I2L}(I)$ is ideal*

6. *I is L-stable, $\mathtt{I2L}(I)$ may not be semi-stable (and vice-versa)*

7. *I is pre-eager, $\mathtt{I2L}(I)$ may not be pre-eager (and vice-versa)*

8. *I is eager, $\mathtt{I2L}(I)$ may not be eager (and vice-versa)*

We highlight that for *assumption-spanning* ABA frameworks, I2L becomes bijective and the missing correspondence results are restored.

Theorem 15. *Let $\mathcal{F}$ be an ABA framework, $\mathtt{ABA2LP}_{\mathrm{CS17}}(\mathcal{F}) = P_\mathcal{F}$ be its associated logic program and I be a 3-valued stable model of $P_\mathcal{F}$. If $\mathcal{F}$ is assumption-spanning, then:*

1. *I is L-stable if and only if $\mathtt{I2L}(I)$ is semi-stable*

---

[30]The pre-ideal, pre-eager and eager ABA semantics are not mentioned in [Caminada and Schulz, 2017]. The result for pre-ideal follows as corollary of their proof about preferred semantics. The results for pre-eager and eager are justified in Example 10.

2. I is pre-eager if and only if $\texttt{I2L}(I)$ is pre-eager

3. I is eager if and only if $\texttt{I2L}(I)$ is eager

The results gathered ensure the conclusion in Theorem 11.

4.3.2 ABA as Logic Programming with Projection

The translation of [Caminada and Schulz, 2017] has a couple of drawbacks: (i) it is not injective over ABA frameworks in general and (ii) if the input ABA framework has more non-assumptions than assumptions, the output program may have more NAF-literals than the number of assumptions in the input. Further, the function $\texttt{I2L}$ requires knowledge of what the set of assumptions $\mathcal{A}$ is in the input ABA framework. Ideally, if we want logic programs to model ABA frameworks, the step where one retrieves assumption labellings from models of its corresponding program should *not* depend on knowledge of the input. This led [Sá and Alcântara, 2021b] to propose a different translation, mapping assumptions to positive literals instead of NAF-literals, thus avoiding those issues.

Given $\mathcal{F} = \langle \mathcal{L}, \mathcal{R}, \mathcal{A}, ^- \rangle$, the translation of [Sá and Alcântara, 2021b] produces a program P where:

1. $HB_P = \mathcal{L}$, which means that the assumptions of $\mathcal{F}$ now correspond to a subset of HB_P. This allows assumptions to be modelled in P regardless of the proportion between $\mathcal{A}$ and $\mathcal{L}$.

2. $P \supset \mathcal{R}$, so the inference rules from $\mathcal{F}$ are kept unaltered in the resulting program. Since **not** is alien to the native syntax of ABA, we can ensure that **not** does not appear in $\mathcal{R}$. A complimentary set of program rules using **not** is added to implement the contrary relation $^-$, ensuring that the translation is injective over ABA frameworks in general and that the set of assumptions from $\mathcal{F}$ can be retrieved syntactically from the complimentary rules.

Definition 29. *([Sá and Alcântara, 2021b]) Let $\mathcal{F} = \langle \mathcal{L}, \mathcal{R}, \mathcal{A}, ^- \rangle$ be an ABA framework, the program corresponding to $\mathcal{F}$ is*

$$\texttt{ABA2LP}_{\mathsf{SA21}}(\mathcal{F}) = P'_{\mathcal{F}} = \mathcal{R} \cup \{\alpha \leftarrow \textbf{not } \overline{\alpha} \mid \alpha \in \mathcal{A}\}.$$

Example 11. *Recover $\mathcal{F}, \mathcal{F}'$ from Example 4. We have that*
$$\text{ABA2LP}_{\text{SA21}}(\mathcal{F}) = P'_\mathcal{F}$$
$$= \mathcal{R} \cup \{\alpha \leftarrow \text{not } a,\ \beta \leftarrow \text{not } b,\ \gamma \leftarrow \text{not } c,\ \delta \leftarrow \text{not } d\}$$
and $\text{ABA2LP}_{\text{SA21}}(\mathcal{F}') = P'_{\mathcal{F}'} = P'_\mathcal{F} \setminus \{\delta \leftarrow \text{not } d\}$.

Given an $\mathcal{F}$ and its corresponding $P'_\mathcal{F}$, $\mathcal{A}$ can be retrieved from $P'_\mathcal{F}$: each $\alpha \in \mathcal{A}$ appears as the head of a single rule r_α in $P'_\mathcal{F}$ for which $body^-(r_\alpha) \neq \emptyset$, a condition only met by rules in $P'_\mathcal{F} \setminus \mathcal{R}$.

The results based on $\text{ABA2LP}_{\text{SA21}}$ include mappings regarding assumption labellings as well as *model semantics* for ABA frameworks, which were introduced in [Sá and Alcântara, 2019]. The general idea of their work consists of treating ABA frameworks just like logic programs, which includes the computation of semantics through a division operator and steps similar to what we introduced in Section 2. As such, interpretations and models evaluate all sentences in $\mathcal{L}$, not only the assumptions.

For the sake of uniformity and because ABA model semantics are not mainstream, we will only discuss in depth the results involving assumption labellings. Concerning ABA model semantics, to discuss it briefly, [Sá and Alcântara, 2021b] proved that the 3-valued stable models of $\text{ABA2LP}_{\text{SA21}}(\mathcal{F})$ *coincide* with the complete models of $\mathcal{F}$, ensuring that:

Theorem 16. *Flat-ABA frameworks under model semantics are equivalent to ABA-programs*[31].

The discussion of model semantics for ABA in [Sá and Alcântara, 2019] included comparative results to the assumption labellings of [Schulz and Toni, 2017]. The mapping from ABA models to corresponding ABA assumption labellings is performed by an operation they called tuple projection (or simply *projection*), which can be applied to interpretations, models and all labellings alike.

Definition 30. *Let S be a set and $T = (S_1, S_2, \ldots, S_k)$ be a tuple of sets. The projection of S on T is*
$$\sigma_S(T) = (S_1 \cap S, S_2 \cap S, \ldots, S_k \cap S).$$

---

[31]The fragment of ABA-programs describes all the programs that can be obtained as output of $\text{ABA2LP}_{\text{SA21}}$. See [Sá and Alcântara, 2021b] for their definition and properties.

Definition 31. *([Sá and Alcântara, 2021b]) Let $I = (T, F, U)$ be a model of $P'_\mathcal{F}$, the assumption labelling of $\mathcal{F} = \langle \mathcal{L}, \mathcal{R}, \mathcal{A}, ^-\rangle$ corresponding to I is obtained by projecting $\mathcal{A}$ on I,[32] i.e., $\text{I2L}_{\text{SA21}}(I) = \sigma_\mathcal{A}(I)$.*

We are now ready to list the main results from [Sá and Alcântara, 2021b]:

Theorem 17. *([Sá and Alcântara, 2021b]) Let $\mathcal{F}$ be an ABA framework. Then I is a 3-valued stable model of $P_\mathcal{F}$ if and only if $\text{I2L}_{\text{SA21}}(I) = \sigma_\mathcal{A}(I)$ is a complete assumption labelling of $\mathcal{F}$.*

Example 12. *Recover $\mathcal{F}, \mathcal{F}'$ from Example 4, for which*

$$\text{ABA2LP}_{\text{SA21}}(\mathcal{F}) = P'_\mathcal{F}$$
$$= \mathcal{R} \cup \{\alpha \leftarrow \text{not } a, \ \beta \leftarrow \text{not } b, \ \gamma \leftarrow \text{not } c, \ \delta \leftarrow \text{not } d\}$$

and $\text{ABA2LP}_{\text{SA21}}(\mathcal{F}) = P'_{\mathcal{F}'} = P'_\mathcal{F} \setminus \{\delta \leftarrow \text{not } d\}$ (Example 11).

- *The 3-valued stable models of $P'_\mathcal{F}$ are*

$$I_1 = (\{\}, \{\}, \{a, b, c, d, \alpha, \beta, \gamma, \delta\})$$
$$I_2 = (\{a, \beta, \delta\}, \{b, d, \alpha\}, \{c, \gamma\})$$
$$I_3 = (\{b, \alpha\}, \{a, \beta\}, \{c, d, \gamma, \delta\})$$

Now we can obtain $\text{I2L}_{\text{SA21}}(I_1) = \sigma_\mathcal{A}(I_1) = (\{\ \}, \{\ \}, \{\alpha, \beta, \gamma, \delta\})$, $\text{I2L}_{\text{SA21}}(I_2) = \sigma_\mathcal{A}(I_2) = (\{\beta, \delta\}, \{\alpha\}, \{\gamma\})$, and $\text{I2L}_{\text{SA21}}(I_3) = \sigma_\mathcal{A}(I_3) = (\{\alpha\}, \{\beta\}, \{\gamma, \delta\})$, which correspond precisely to the complete assumption labellings of $\mathcal{F}$, as seen in Example 5.

- *The 3-valued stable models of $P'_{\mathcal{F}'}$ are*

$$I'_1 = (\{\}, \{\}, \{a, b, c, d, \alpha, \beta, \gamma\})$$
$$I'_2 = (\{a, \beta\}, \{b, d, \alpha\}, \{c, \gamma\})$$
$$I'_3 = (\{b, \alpha\}, \{a, \beta\}, \{c, d, \gamma\})$$

---

[32] Remember that $\mathcal{A}$ can be retrieved directly from $P'_\mathcal{F}$ as $\mathcal{A} = \{head(r) \in P'_\mathcal{F} \mid body^-(r) \neq \emptyset\}$.

Now, we can obtain

$$\text{I2L}_{\text{SA21}}(I'_1) = \sigma_A(I'_1) = (\{\ \},\{\ \},\{\alpha,\beta,\gamma\})$$
$$\text{I2L}_{\text{SA21}}(I'_2) = \sigma_A(I'_2) = (\{\beta\},\{\alpha\},\{\gamma\})$$
$$\text{I2L}_{\text{SA21}}(I'_3) = \sigma_A(I'_3) = (\{\alpha\},\{\beta\},\{\gamma\})$$

which correspond precisely to the complete assumption labellings of $\mathcal{F}'$, as seen in Example 5.

When restricted to complete assumption labellings and 3-valued stable models of corresponding pair of $\mathcal{F}$ and $P'_\mathcal{F}$, I2L_{SA21} becomes bijective and admits an inverse $\text{I2L}^{-1}_{\text{SA21}}$. For this reason, similarly to what we observed in the discussion of Theorem 9 in Section 4.2, the result in Theorem 17 is equivalent to a coincidence result. The same reasoning holds for the results below.

Theorem 18. *Let $\mathcal{F}$ be an ABA framework, $\text{ABA2LP}_{\text{SA21}}(\mathcal{F}) = P'_\mathcal{F}$ be its associated logic program and I be a 3-valued stable model of $P'_\mathcal{F}$. Then:*

1. *I is well-founded if and only if $\text{I2L}(I)$ is grounded*

2. *I is regular if and only if $\text{I2L}(I)$ is preferred*

3. *I is stable if and only if $\text{I2L}(I)$ is stable*

4. *I is pre-ideal if and only if $\text{I2L}(I)$ is pre-ideal*

5. *I is ideal if and only if $\text{I2L}(I)$ is ideal*

6. *I is L-stable if and only if $\text{I2L}(I)$ is semi-stable*

7. *I is pre-eager if and only if $\text{I2L}(I)$ is pre-eager*

8. *I is eager if and only if $\text{I2L}(I)$ is eager*

The results gathered ensure that flat ABA framework and their semantics are captured by ABA-programs with the projection of assumptions. Combined with Theorem 16 ([Sá and Alcântara, 2021b]) and the results in Section 4.2, we obtain Theorem 11.

5 Equivalence for enhanced frameworks

In this section, we will briefly discuss some notable argumentation systems of which the connection to logic programming have been studied, namely Abstract Dialectical Frameworks (ADF) [Brewka and Woltran, 2010; Brewka et al., 2018] and Argumentation Frameworks with Sets of Attacking Arguments (SETAF) [Nielsen and Parsons, 2006; Bikakis et al., 2021]. We consider these extensions of Dung's AFs to be representative of the diverse enhanced frameworks proposed in the literature and have chosen them because of their relations to each other and to logic programming have been well studied. Other systems, for which not as much research has been conducted, will be discussed in the next section.

Differently from what we did in the previous sections, we will not fully introduce the definitions for systems we discuss here. Instead, we will focus only on their syntax to introduce appropriate translations and list the results concerning the preservation of semantics in each case.

5.1 Abstract Dialectical Frameworks

Abstract Dialectical Frameworks (ADFs) [Brewka and Woltran, 2010; Brewka et al., 2013] were proposed to treat arguments (called *statements* there) as abstract and atomic entities. The connections between ADFs and logic programming start from the original definitions in [Brewka and Woltran, 2010], where the authors adopted the standard terminology and the notion of *reduct* from logic programming (Definition 3) to obtain model semantics for ADFs. Their connection to logic programming was studied in depth by [Strass, 2013] and [Alcântara et al., 2019].

Similarly to Dung's AFs, an ADF can be perceived as a directed graph of which the nodes represent statements which can get *accepted* or not. But differently from Dung's AFs, where the edges represent conflicts between arguments, here the links represent the more general notion of *dependencies*: the status (accepted/not accepted) of a node s depends only on the status of its parents ($par(s)$), i.e., the nodes with a direct link to s. For simplicity, we will restrict ourselves to finite ADFs:

Definition 32. *([Brewka and Woltran, 2010]) An abstract dialectical framework is a tuple $D = (S, L, C)$ where*

- S is a finite set of statements (positions, nodes);

- $L \subseteq S \times S$ is a set of links, such that, for each each $s \in S$, we have $par(s) = \{t \in S \mid (t,s) \in L\}$;

- $C = \{C_s \mid s \in S\}$ is a set of total functions $C_s : 2^{par(s)} \to \{\mathbf{t}, \mathbf{f}\}$, one for each $s \in S$, where C_s is the acceptance condition *of s*.

The function C_s is intended to determine the acceptance status of a statement s, which only depends on the status of its parent nodes $par(s)$. Intuitively, s will be accepted if there exists $R \subseteq par(s)$ for which $C_s(R) = \mathbf{t}$, which means that every statement in R is accepted while each statement in $par(s) - R$ is not accepted.

The semantics of ADFs are primarily given by interpretations and models over the set of sentences. The definitions introduced by [Brewka and Woltran, 2010; Brewka et al., 2013] to obtain the *complete models* of an ADF [Brewka et al., 2013] closely resemble the ones we presented in Section 2 to obtain the 3-valued stable models of a program. As such, the complete models of an ADF D are obtained as the least fixed points of a reduction operator Γ_D [Brewka et al., 2013] which was adapted from logic programming to work with ADFs and was shown to always have a least model [Brewka and Woltran, 2010]. We opt not to introduce this operator here for the sake of brevity and simplicity in our presentation of ADFs.

Definition 33. *Let $D = (S, L, C^\varphi)$ be an ADF and v be a 3-valued interpretation over S.*[33] *Then v is a* complete model *of D iff $v = \Gamma_D(v)$.*

As usual, some of the mainstream semantics for an ADF are obtained as special cases of the complete semantics. Most of the semantics below were originally proposed in [Brewka et al., 2013], except for the L-stable ADF semantics, which was proposed by [Alcântara et al., 2019]. The text of this definition is adapted to our needs, to make its presentation uniform with our previous definitions of semantics.

---

[33]Similarly to interpretations of logic programs, given an ADF $D = (S, L, C)$, a 3-valued interpretation (or simply interpretation) over S is a mapping $v : S \to \{\mathbf{t}, \mathbf{f}, \mathbf{u}\}$ that assigns one of the truth values true ($\mathbf{t}$), false ($\mathbf{f}$) or unknown ($\mathbf{u}$), to each statement in S.

Definition 34. *Let* $D = (S, L, C)$ *be an ADF, and* v *a model of* D. *Then*

- v *is the* grounded model *of* D *iff* v *is the complete model of* D *for which* $\mathbf{t}(v) = \{s \in S \mid v(s) = \mathbf{t}\}$ *is* $\subseteq$-*minimal among complete models of* D.

- v *is a* preferred model *of* D *iff* v *is a complete model of* D *for which* $\mathbf{t}(v) = \{s \in S \mid v(s) = \mathbf{t}\}$ *is* $\subseteq$-*maximal among complete models of* D.

- v *is a* stable model *of* D *iff* v *is a 2-valued complete model of* D.

- v *is a* L-stable model *of* D *iff* v *is a complete model of* D *for which* $\mathbf{u}(v) = \{s \in S \mid v(s) = \mathbf{u}\}$ *is* $\subseteq$-*minimal among complete models of* D.

The above ADF semantics have been shown to capture corresponding logic programming semantics (resp. 3-valued stable models, well-founded, regular, stable and L-stable) in *Attacking Abstract Dialectical Frameworks* (ADF$^+$s) [Alcântara et al., 2019], a fragment of ADFs in which the unique relation involving statements is the attack relation.[34]

Definition 35. *An Attacking Abstract Dialectical Framework (ADF$^+$), is an ADF* (S, L, C) *such that every* $(r, s) \in L$ *is an* attacking link *[Brewka and Woltran, 2010], i.e., there is no* $R \subseteq par(s)$ *for which* $C_s(R) = \mathbf{f}$ *and* $C_s(R \cup \{r\}) = \mathbf{t}$. *This means that for every* $s \in S$ *and every* $M \subseteq par(s)$, *if* $C_s(M) = \mathbf{t}$, *then for every* $M' \subseteq M$, *we have* $C_s(M') = \mathbf{t}$.

[Alcântara et al., 2019] introduced a translation **LP2ADF**$^+$ inspired by the work of [Caminada et al., 2015b], on the basis of which they proved coincidental correspondence results between normal logic programs and ADF$^+$ for all ADF semantics in Definition 34. Here, we will resort to

---

[34]The class of ADF$^+$, sometimes also referred to as support-free ADFs [Dvořák et al., 2023b], is also a subclass of Bipolar Abstract Dialectical Frameworks [Brewka and Woltran, 2010].

7 - LOGIC PROGRAMMING AND ARGUMENTATION

Definition 13, which is used (indirectly) as part of the translation of [Alcântara et al., 2019].[35]

Definition 36. *Let P be a program, for each $a \in HB_P$, let*

$$\text{Sup}_P(a) = \{\text{Vul}(a) \mid A \text{ is an argument with } \text{Conc}(A) = a\}.$$

Then, the ADF^+ associated to P is

$$\text{LP2ADF}^+(P) = D_P = (HB_P, L_P, C_P)$$

where:

- $L_P = \{(b, a) \mid b \in B \text{ for some } B \in \text{Sup}_P(a)\}$;

- *For each $a \in HB_P$, $C_a = \Big\{ B' \subseteq par(a) \setminus B \;\Big|\; B \in \text{Sup}_P(a) \Big\}$.*

The intuition for C_a in the definition above is that if an interpretation of D_P accepts all $b \in B = \text{Vul}(A)$ for each argument A from P where $\text{Conc}(A) = a$, then in order to be a model of D_P, it must not accept a. Hence, the acceptance condition C_a for each $a \in HB_P$ requires that for each set of vulnerabilities $\text{Vul}(A) \in \text{Sup}_P(A)$, at least one $b \in par(a)$ is not accepted .

Based on the translation LP2ADF$^+$, [Alcântara et al., 2019] proved that the 3-valued stable model semantics for normal logic programs is equivalent to the complete model semantics of ADF for the class of ADF$^+$.

Theorem 19. *([Alcântara et al., 2019]) Let P be a program and D_P be its corresponding ADF^+. Then v is a 3-valued stable model of P iff v is a complete model of D_P.*

Until the study of [Alcântara and Sá, 2019; Alcântara et al., 2019], it was unclear if any ADF semantics could capture the 3-valued semantics

---

[35]Definition 25 of [Alcântara et al., 2019] speaks of *substatements*, which correspond to proofs in P and also to the arguments of our Definition 13. The *support* of a substatement A in their work is defined exactly as $\text{Vul}(A)$ from Definition 13. Finally, the notion of support is extended to each atom a as the set of supports from substatements with $\text{Conc}(A) = a$, which is the criterion we use here.

for normal logic programs. Theorem 20 ensures that the translation from logic programs to ADF in Definition 36 guarantees the equivalence between any semantics based on 3-valued stable models (at the logic program side) with any semantics based on complete models (at the ADF side). We highlight that, according to Theorem 19, the models of P and $\texttt{LP2ADF}^+(P)$ *coincide*, so the following results are immediate:

Corollary 20. *Let P be a program and $D_P = (A, L, C)$ be its corresponding ADF^+. We have*

- *v is a well-founded model of P iff v is a grounded model of D_P.*

- *v is a regular model of P iff v is a preferred model of D_P.*

- *v is a stable model of P iff v is a stable model of D_P.*

- *v is an L-stable model of P iff v is an L-stable model of D_P.*

Results such as the above can be extended to appropriate definitions of pre-ideal, ideal, pre-eager and eager semantics for ADF. Further, because $\texttt{LP2ADF}^+$ is injective and the models of P and D_P coincide for all programs P, $\texttt{LP2ADF}^+$ must be bijective and, therefore, it admits inversion. These results lead to the conclusion that

Theorem 21. *Normal logic programs* are equivalent to *attacking abstract dialectical frameworks.*

Given that ADF^+ is a fragment of ADFs in general, this means that the more general family of ADFs semantically subsume normal logic programs. On the other hand, previously, [Strass, 2013] showed a direct translation from ADFs to normal logic programs for which the program $\texttt{ADF}^+\texttt{2LP}(D)$, corresponding to an ADF D, would model the semantics of Definition 33 and Definition 34 (except for L-stable, which had not been defined) following the same correspondences we listed in Theorem 19 and Corollary 20. This means that normal logic programs semantically subsume ADFs (in general). Combining the results, we can gather that the two systems, NLPs and ADFs (in general), are inherently equivalent.

5.2 Argumentation Frameworks with Sets of Attacking Arguments

A *framework with sets of attacking arguments* (SETAF) [Nielsen and Parsons, 2006; Bikakis et al., 2021] is an extension of Dung's AFs (in the context of finite AFs) to allow joint attacks on arguments. Intuitively, the need for joint attacks arises from situations where an argument may not be enough to defeat another on its own, but two or more arguments, together, might suffice.

Definition 37. *([Nielsen and Parsons, 2006]) A Framework with Sets of Attacking Arguments (SETAF for short) is a pair $\mathfrak{A} = (Ar, att)$, in which Ar is a finite set of arguments and $att \subseteq (2^{Ar} - \{\emptyset\}) \times Ar$.*

The attack relation att is such that if $(\mathcal{B}, a) \in att$, there is no $\mathcal{B}' \subset \mathcal{B}$ such that $(\mathcal{B}', a) \in att$, i.e., $\mathcal{B}$ is a minimal set (w.r.t. $\subseteq$) attacking a. We write $att(a) = \{\mathcal{B} \subseteq Ar \mid (\mathcal{B}, a) \in att\}$ to retrieve the attackers of a.

In AFs, only individual arguments can attack arguments. In SETAFs, the novelty is that sets of two or more arguments can also attack arguments. This means that SETAFs (Ar, att) with $|\mathcal{B}| = 1$ for each $(\mathcal{B}, a) \in att$ amount to (standard Dung) AFs.

The semantics for SETAFs are generalisations of the corresponding semantics for AFs [Nielsen and Parsons, 2006] and can be defined equivalently in terms of extensions or labellings [Flouris and Bikakis, 2019; Caminada et al., 2024]. For our convenience, we will adhere to the presentation of labelling-based semantics as proposed in [Flouris and Bikakis, 2019].

Definition 38. *Let $\mathfrak{A} = (Ar, att)$ be a SETAF. A labelling is a function $\mathcal{L} : Ar \to \{\text{in}, \text{out}, \text{undec}\}$. A labelling is* complete *iff for each $a \in Ar$,*

- *If $\mathcal{L}(a) = \text{in}$, then for each $\mathcal{B} \in att(a)$, there is $b \in \mathcal{B}$ s.t. $\mathcal{L}(b) = \text{out}$*

- *If $\mathcal{L}(a) = \text{out}$, then there is a $\mathcal{B} \in att(a)$ s.t. $\mathcal{L}(b) = \text{in}$ for all $b \in \mathcal{B}$*

- *If $\mathcal{L}(a) = \text{undec}$, then there is a $\mathcal{B} \in att(a)$ s.t. $\mathcal{L}(b) \neq \text{out}$ for each $b \in \mathcal{B}$, and for each $\mathcal{B} \in att(a)$, it holds $\mathcal{L}(b) \neq \text{in}$ for some $b \in \mathcal{B}$.*

As usual, we may use as shorthand $\text{in}(\mathcal{L}) = \{a \in Ar \mid \mathcal{L}(a) = \text{in}\}$, $\text{out}(\mathcal{L}) = \{a \in Ar \mid \mathcal{L}(a) = \text{out}\}$, $\text{undec}(\mathcal{L}) = \{a \in Ar \mid \mathcal{L}(a) = \text{undec}\}$. Also as before, a labelling defines a partition of the set of arguments, so $\mathcal{L}$ can be written as a triple $(\text{in}(\mathcal{L}), \text{out}(\mathcal{L}), \text{undec}(\mathcal{L}))$. Intuitively, an argument labelled in is explicitly accepted; an argument labelled out is explicitly rejected; and one labelled undec is left undecided, i.e., it is neither accepted nor rejected. We can now describe the remaining SETAF semantics studied in [Alcântara et al., 2023]:

Definition 39. *([Flouris and Bikakis, 2019]) Let* $\mathfrak{A} = (Ar, att)$ *be a SETAF. A complete labelling* $\mathcal{L}$ *is called*

- grounded *iff* $\text{in}(\mathcal{L})$ *is* $\subseteq$-*minimal among all complete labellings of* $\mathfrak{A}$

- preferred *iff* $\text{in}(\mathcal{L})$ *is* $\subseteq$-*maximal among all complete labellings of* $\mathfrak{A}$

- stable *iff* $\text{undec}(\mathcal{L}) = \emptyset$.

- semi-stable *iff* $\text{undec}(\mathcal{L})$ *is* $\subseteq$-*minimal among all complete labellings of* $\mathfrak{A}$.

Let us consider the following example:

Example 13. *Consider the SETAF* $\mathfrak{A} = (Ar, att)$ *below:*

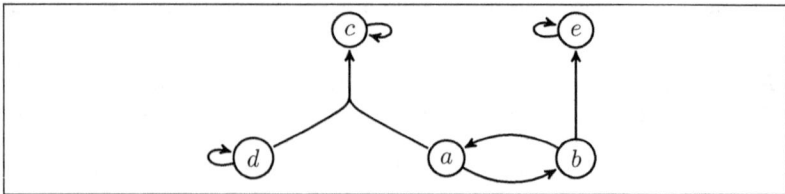

Figure 2: A SETAF $\mathfrak{A}$. Joint attacks are drawn as arrows with two or more origin nodes as, for instance, $\{d, a\}$ jointly attack argument c.

Concerning the semantics of $\mathfrak{A}$, *we have*

- *Complete labellings:* $\mathcal{L}_1 = (\emptyset, \emptyset, \{a, b, c, d, e\})$, $\mathcal{L}_2 = (\{a\}, \{b\}, \{c, d, e\})$ *and* $\mathcal{L}_3 = (\{b\}, \{a, e\}, \{c, d\})$;

- *Grounded labellings:* $\mathcal{L}_1 = (\emptyset, \emptyset, \{a, b, c, d, e\})$;

- *Preferred labellings:* $\mathcal{L}_2 = (\{a\}, \{b\}, \{c, d, e\})$ *and* $\mathcal{L}_3 = (\{b\}, \{a, e\}, \{c, d\})$;

- *Stable labellings:* none;

- *Semi-stable labellings:* $\mathcal{L}_3 = (\{b\}, \{a, e\}, \{c, d\})$.

The semantics of SETAF were studied in connection with Dung's AFs in [Flouris and Bikakis, 2019], with ADFs in [Polberg, 2016], [Dvořák et al., 2023b] and [Alcântara and Sá, 2021] (ADF^+'s[36]), with ABA and CAF in [König et al., 2022] and in connection with logic programming in [König et al., 2022] and [Alcântara et al., 2023]. For an overview of SETAFs and their properties, we refer to [Bikakis et al., 2021; Caminada et al., 2024]. In what follows, we will focus on the works of [Alcântara et al., 2023][37], which concerns the relation between SETAF and logic programming. The authors of both works devised the same translation functions between SETAF and LP. We will start with their translation from NLP to SETAF, which was shown in [Alcântara et al., 2023] to guarantee the equivalence between various kinds of NLPs models and SETAFs labellings, including the complete labellings, well-founded models and grounded labellings, regular models and preferred labellings, stable models and stable labellings, L-stable models and semi-stable labellings. Following [Alcântara et al., 2023], this translation is built upon the translation from NLP to AF of [Caminada et al., 2015b].

Definition 40. *Let P be a program and $\mathfrak{S}_P$ be the set of all non-default arguments constructed from P following Definition 13, the SETAF corresponding to P is* $\mathtt{LP2SETAF}(P) = \mathfrak{A}_P = (Ar_P, att_P)$ *with*

- $Ar_P = \{\mathtt{Conc}(s) \mid s \in \mathfrak{S}_P\}$

---

[36] Before that, [Alcântara et al., 2019] showed the translation from SETAF to ADF proposed by [Polberg, 2016] necessarily returns an ADF^+.

[37] The translations appearing in [Alcântara et al., 2023] were also proposed by [König et al., 2022], but they did not explore the connections between semantics of SETAF and logic programs beyond that. In contrast, [Alcântara et al., 2023] also offer translations between models and labellings and prove numerous semantic correspondence results.

- $att_P = \{(\mathcal{B}, a) \mid \mathcal{B} \text{ is a } \subseteq\text{-minimal set s.t.}$
  $\text{for each } A \in \mathfrak{S}_P \text{ there is some } b \in \mathcal{B} \cap \mathtt{Vul}(A)\}.$

A counterpart translation from SETAFs to NLPs is also considered:

Definition 41. *Let $\mathfrak{A} = (Ar, att)$ be a SETAF, the logic program corresponding to $\mathfrak{A}$ is $\mathtt{SETAF2LP}(\mathfrak{A}) = P_\mathfrak{A}$ with*

$$P_\mathfrak{A} = \{a \leftarrow \mathtt{not}\ b_1, \ldots \mathtt{not}\ b_n \mid a \in Ar \text{ and } \{b_1, \ldots, b_n\} \in \mathcal{V}_a\}$$

*where, $\mathcal{V}_{a \in Ar} = \{V \subseteq Ar \mid V \text{ is a } \subseteq\text{-minimal set s.t.}$
$\text{for each } \mathcal{B} \in att(a) \text{ there is some } b \in \mathcal{B} \cap V\}.$*

Example 14. *Recall the SETAF $\mathfrak{A}$ of Example 13 (depicted in Fig 2). The program corresponding to $\mathfrak{A}$ is $\mathtt{SETAF2LP}(\mathfrak{A}) = P_\mathfrak{A}$ comprising the rules:*

$$\begin{array}{ll} d \leftarrow \mathtt{not}\ d & c \leftarrow \mathtt{not}\ c, \mathtt{not}\ d \\ a \leftarrow \mathtt{not}\ b & b \leftarrow \mathtt{not}\ a \\ c \leftarrow \mathtt{not}\ c, \mathtt{not}\ a & e \leftarrow \mathtt{not}\ e, \mathtt{not}\ b \end{array}$$

Further, $\mathtt{LP2SETAF}(\mathtt{SETAF2LP}(\mathfrak{A})) = \mathfrak{A}$. As proven by [Alcântara et al., 2023], this holds for every SETAF $\mathfrak{A}$. This also suffices to show that

Theorem 22. *([Alcântara et al., 2023]) Let $\mathfrak{A}$ be a SETAF. Then $\mathcal{L}$ is a complete argument labelling of $\mathfrak{A}$ if and only if $\mathcal{L}$ is 3-valued stable model of $P_\mathfrak{A}$.*

Theorem 22 portrays a *coincidence* between the labellings of $\mathfrak{A}$ and the models of its corresponding program $P_\mathfrak{A}$. As before, other results immediately follow:

Corollary 23. *Let $\mathfrak{A}$ be a SETAF and $\mathcal{L} = (\mathtt{in}, \mathtt{out}, \mathtt{undec})$ be a complete argument labelling of $\mathfrak{A}$. Then:*

- *$\mathcal{L}$ is grounded if and only if $\mathcal{L}$ is the well-founded model of $P_\mathfrak{A}$.*
- *$\mathcal{L}$ is preferred if and only if $\mathcal{L}$ is a regular model of $P_\mathfrak{A}$.*
- *$\mathcal{L}$ is stable if and only if $\mathcal{L}$ is a stable model of $P_\mathfrak{A}$.*
- *$\mathcal{L}$ is semi-stable if and only if $\mathcal{L}$ is a L-stable model of $P_\mathfrak{A}$.*

On the other hand, we may obtain $\texttt{SETAF2LP}(\texttt{LP2SETAF}(P)) \neq P$ for some cases of a program P. This will only be observed if there is some $c \in HB_P$ for which there is no rule $r \in P$ with $head(r) = c$. In that case, there will be no argument for c in $\mathfrak{S}_P$, therefore c will not be in Ar_P of $\texttt{LP2SETAF}(P) = (Ar_P, att_P)$. Fortunately, this problem is easy to solve: if $\{r \in P \mid head(r) = c\} = \emptyset$, we observe that $I(c) = \mathbf{f}$ in every model I of P, so the extra atoms can be ignored [Alcântara et al., 2023]. In what follows, allow us to use the shorthand

$$HB'_P = \Big\{c \in HB_P \;\Big|\; \{r \in P \mid head(r) = c\} \neq \emptyset\Big\}.$$

Theorem 24. *([Alcântara et al., 2023]) Let P be a program. Then $I = (T, F, U)$ is 3-valued stable model of P if and only if $I' = (T, F \cap HB'_P, U)$ is a complete argument labelling of $\mathfrak{A}_P$.*

While the result in Theorem 24 is not coincidental, the models of P and labellings of $\mathfrak{A}_P$ are one-to-one related because $HB_P \setminus HB'_P$ can be retrieved from P. Results concerning the particular cases of the complete and 3-valued stable models immediately follow:

Corollary 25. *Let P be a program, $I = (T, U, F)$ be a 3-valued stable model of P and $I' = (T, F \cap HB'_P, U)$. Then:*

- *$I = (T, F, U)$ is well-founded if and only if I' is the grounded labelling of $\mathfrak{A}_P$.*

- *$I = (T, F, U)$ is regular if and only if I' is a preferred labelling of $\mathfrak{A}_P$.*

- *$I = (T, F, U)$ is stable if and only if I' is a stable labelling of $\mathfrak{A}_P$.*

- *$I = (T, F, U)$ is L-stable if and only if I' is a L-stable labelling of $\mathfrak{A}_P$.*

The results in Corollary 25 and Corollary 23 could be extended to appropriate definitions of pre-ideal, ideal, pre-eager and eager semantics for SETAF. Together, the results we presented ensure that

Theorem 26. *Normal logic programs are equivalent to frameworks with sets of attacking arguments.*[38]

---

[38]The only exceptions to this equivalence result are programs whose syntax include

6 Implementing Argumentation with Answer-Set Programming

In this section we discuss how logic programming can be used to implement solvers for argumentation formalisms. That is, we consider reduction-based approaches based on answer-set programming. Answer-set programming is based on the stable model semantics of logic programming, but, compared to normal logic programs considered so far, also allows for variables to denote collections of rules, for disjunction in rule heads, and several other extensions that deal with, for instance, constraints, aggregates, and minimization/maximization. Due to the availability of efficient solvers, answer-set programming is nowadays a successful declarative programming approach for NP-hard problems. We will distinguish two kinds of ASP-implementation:

1. In what one may call compiler-style implementations one uses a program that given a specific semantics transforms an argumentation framework into an equivalent ground, i.e., without any variables, logic program. That is, each argumentation framework has to be compiled into a logic program which can be solved by ASP-solvers in order to compute the extensions or labellings. This is the same schema as in typical SAT-based approaches.

2. In what we call query-based implementations or (in the context of argumentation) ASPARTIX-style implementations one encodes the argumentation framework as an input database, i.e., as facts of the LP, which is independent of the semantics and reasoning task. This input database can then be combined with fixed encodings of semantics and reasoning tasks in order to solve specific reasoning tasks.

Compiler-style implementations date back to [Nieves et al., 2008] and can also be found in a more recent paper by Sakama and Rienstra (2017). The query-based approach has been used for Dung style abstract argumentation by [Wakaki and Nitta, 2008] and the ASPARTIX system [Egly

---

irrelevant atoms, which are those $c \in HB_P \setminus HB'_P$. However, all occurrences of **not** c and each $r \in P$ for which $c \in body(r)$ can be removed from P without prejudice to the semantics of the remaining atoms.

et al., 2010], and similar approaches have also been used for richer abstract argumentation formalisms [Dvořák *et al.*, 2015; Ellmauthaler and Strass, 2014; Dvořák *et al.*, 2018] and structured argumentation [Lehtonen *et al.*, 2021b; Lehtonen *et al.*, 2021a]. Toni and Sergot (2011) provide a survey on the earlier works on ASP for abstract argumentation while later surveys on implementations techniques for argumentation by Charwat *et al.* (2015) and Cerutti *et al.* (2018) focus on ASPARTIX-style implementations.

In the remainder of this section we start with a brief introduction to answer-set programming, then discuss compiler style implementations for abstract argumentation frameworks, and finally discuss ASPARTIX-style approaches for different kinds of argumentation formalisms.

6.1 Answer-Set Programming

We give an overview of the syntax and semantics of disjunctive logic programs under the answer-set semantics [Gelfond and Lifschitz, 1991], generalizing our definitions from Section 2. We fix a countable set $\mathcal{U}$ of *(domain) elements*, also called *constants*. An *atom* is an expression $p(t_1, \ldots, t_n)$, where p is a *predicate* of arity $n \geq 0$ and each t_i is either a variable or an element from $\mathcal{U}$. An atom is *ground* if it is free of variables. $B_\mathcal{U}$ denotes the set of all ground atoms over $\mathcal{U}$. A *(disjunctive) rule* r is of the form

$$x_1 \vee \cdots \vee x_k \leftarrow y_1, \ldots, y_n, \text{not } z_1, \ldots, \text{not } z_m \qquad (1)$$

with $k \geq 0$, $n \geq 0$, $m \geq 0$, and in each rule at least one of k, n, m is non zero. $x_i, \ldots, x_k, y_1, \ldots, y_n, z_1, \ldots, z_m$ are atoms, and "not " stands for *negation as failure*. The *head* of r is the set $head(r) = \{x_1, \ldots, x_k\}$ and the *body* of r is $body(r) = \{y_1, \ldots, y_n, \text{not } z_1, \ldots, \text{not } z_m\}$. Furthermore, $body^+(r)r = \{b_1, \ldots, b_k\}$ and $body^-(r) = \{b_{k+1}, \ldots, b_m\}$. A rule r is *normal* if $k \leq 1$ and a *constraint* if $k = 0$. A rule r is *safe* if each variable in r occurs in $body^+(r)$. A rule r is *ground* if no variable occurs in r. A *fact* is a ground rule without disjunction and empty body. An *(input) database* is a set of facts. A program is a finite set of disjunctive rules. If each rule in a program is normal (resp. ground), we call the program normal (resp. ground).

For any program π, let the *Herbrand Literal Base* U_π be the set of all constants appearing in π (if no constant appears in π, we add an arbitrary constant to U_π). Moreover, $Gr(\pi)$ is the set of rules obtained by applying, to each rule $r \in \pi$, all possible substitutions σ from the variables in r to elements of U_π. We call $Ground(\pi, U_P)$ the grounding of π, and write $Gr(\pi)$ as a shorthand for $Ground(\pi, U_P)$. The semantics of a (non-ground) program π is defined via its grounding $Gr(\pi)$.

An *interpretation* $I \subseteq B_\mathcal{U}$ *satisfies* a ground rule r iff $head(r) \cap I \neq \emptyset$ whenever $body^+(r) \subseteq I$ and $body^-(r) \cap I = \emptyset$. I satisfies a ground program π, if each $r \in \pi$ is satisfied by I. A non-ground rule r (resp., a program π) is satisfied by an interpretation I iff I satisfies all groundings of r (resp., $Gr(\pi)$). $I \subseteq B_\mathcal{U}$ is an *answer-set* of π iff it is a subset-minimal set satisfying the *Gelfond-Lifschitz reduct* $\pi^I = \{head(r) \leftarrow body^+(r) \mid I \cap body^-(r) = \emptyset, r \in Gr(\pi)\}$. For a program π, we denote the set of its answer-sets by AS(π).

Modern ASP-solvers offer additional language features. Among them we make use of the *conditional literal* [Gebser et al., 2015]. In the head of a disjunctive rule literals may have conditions, e.g. consider the head of rule "$\mathbf{p}(X) : \mathbf{q}(X) \leftarrow$". Intuitively, this represents a head of disjunctions of atoms $\mathbf{p}(a)$ where also $\mathbf{q}(a)$ is true. As well rules might have conditions in their body, e.g. consider the body of rule "$\leftarrow \mathbf{p}(X) : \mathbf{q}(X)$", which intuitively represents a conjunction of atoms $\mathbf{p}(a)$ where also $\mathbf{q}(a)$ is true. Notice, that when using conditions in the head of a rule we have a disjunction of atoms while when using conditions in the body of a rule we have a conjunction of atoms.

6.2 Compiler-Style ASP Encodings

In this section we follow [Sakama and Rienstra, 2017] in order to illustrate the compiler style approach towards abstract argumentation. This approach is based on the labelling characterisation of complete semantics and thus works with three predicates $\mathbf{in}(x)$, $\mathbf{out}(x)$, $\mathbf{undec}(x)$ representing that an argument has the corresponding label.

Example 15. *As our running example for this section we will use the argumentation framework* $AF_{run} = (\{a, b, c\}, \{(a,b), (b,a), (b,c), (c,c)\})$ *as depicted in Figure 3. We have the three admissible sets $\emptyset$, $\{a\}$, $\{b\}$,*

7 - Logic Programming and Argumentation

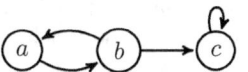

Figure 3: Illustration of our running example AF_{run} for Section 6.2.

with $\{a\}$, $\{b\}$ being the preferred extensions and $\{b\}$ being the only stable extension.

We start with the basic encoding of the **in** and **out** labels that applies to all semantics. That is, given an argumentation framework $AF = (Ar, att)$ we define $\pi_{basic}(AF)$ as follows:

$\pi_{basic}(AF) = \{\mathbf{out}(x) \leftarrow \mathbf{in}(y) \mid (y,x) \in att\} \cup$
$\{\mathbf{in}(x) \leftarrow \mathbf{out}(y_1), \ldots, \mathbf{out}(y_k) \mid x \in Ar, x^- = \{y_1, \ldots, y_n\}\} \cup$
$\{\leftarrow \mathbf{in}(x), \mathbf{not}\ \mathbf{out}(y) \mid (y,x) \in att\} \cup$
$\{\leftarrow \mathbf{out}(x), \mathbf{not}\ \mathbf{in}(y_1), \ldots, \mathbf{not}\ \mathbf{in}(y_k) \mid x \in Ar, x^- = \{y_1, \ldots, y_n\}\}$

$\pi_{basic}(AF)$ ensures the basic properties of labellings that if an argument is labelled in the all its neighbours are labelled out and that if an argument is labelled out it has an attacker that is labelled in. Notice that, so far, there is no restriction on the number of different labels an argument can have and no requirement to be labelled at all.

Example 16. *When considering our running example AF_{run} and apply π_{basic} we obtain the following ground logic program $\pi_{basic}(AF_{run})$:*

$\mathbf{out}(b) \leftarrow \mathbf{in}(a).$ $\qquad \leftarrow \mathbf{in}(b), \mathbf{not}\ \mathbf{out}(a).$
$\mathbf{out}(a) \leftarrow \mathbf{in}(b).$ $\qquad \leftarrow \mathbf{in}(a), \mathbf{not}\ \mathbf{out}(b).$
$\mathbf{out}(c) \leftarrow \mathbf{in}(b).$ $\qquad \leftarrow \mathbf{in}(c), \mathbf{not}\ \mathbf{out}(b).$
$\mathbf{out}(c) \leftarrow \mathbf{in}(c).$ $\qquad \leftarrow \mathbf{in}(c), \mathbf{not}\ \mathbf{out}(c).$
$\mathbf{in}(a) \leftarrow \mathbf{out}(b).$ $\qquad \leftarrow \mathbf{out}(a), \mathbf{not}\ \mathbf{in}(b).$
$\mathbf{in}(b) \leftarrow \mathbf{out}(a).$ $\qquad \leftarrow \mathbf{out}(b), \mathbf{not}\ \mathbf{in}(a).$
$\mathbf{in}(c) \leftarrow \mathbf{out}(b), \mathbf{out}(c).$ $\qquad \leftarrow \mathbf{out}(c), \mathbf{not}\ \mathbf{in}(b), \mathbf{not}\ \mathbf{in}(c).$

Stable Semantics. Let us now consider stable semantics. For stable semantics we only need in and out labels. In the following we extend the basic encoding by two constraints: (i) each argument must be labelled in or out and (ii) no argument can be labelled both in and out.

$$\pi_{st}(AF) = \{\mathbf{in}(x) \vee \mathbf{out}(x) \leftarrow \mid x \in Ar\} \cup$$
$$\{\leftarrow \mathbf{in}(x), \mathbf{out}(x) \mid x \in Ar\}$$

Example 17. *When considering our running example AF_{run} we obtain the following ground logic program $\pi_{st}(AF_{run})$:*

$\mathbf{in}(a) \vee \mathbf{out}(a) \leftarrow .$ $\qquad \leftarrow \mathbf{in}(a), \mathbf{out}(a).$
$\mathbf{in}(b) \vee \mathbf{out}(b) \leftarrow .$ $\qquad \leftarrow \mathbf{in}(b), \mathbf{out}(b).$
$\mathbf{in}(c) \vee \mathbf{out}(c) \leftarrow .$ $\qquad \leftarrow \mathbf{in}(c), \mathbf{out}(c).$

Theorem 27. *([Sakama and Rienstra, 2017]) For every argumentation framework $AF = (Ar, att)$, the stable labellings of AF are in one-to-one correspondence with the stable models of the logic program $\pi_{basic}(AF) \cup \pi_{st}(AF)$.*

The attentive reader may have noticed that $\pi_{basic}(AF) \cup \pi_{stb}(AF)$ uses disjunction in some rule heads and is thus not a normal logic program. However, often normal logic programs are preferable over disjunctive ones (e.g., due to limitations of solvers or computational advantages) and thus one might ask whether we can avoid disjunctive rules here. Indeed the addition for stable semantics can be reformulated as follows in order to obtain normal logic programs.

$$\pi_{stb'}(AF) = \{\mathbf{in}(x) \leftarrow \mathbf{not}\ \mathbf{out}(x) \mid x \in Ar\} \cup$$
$$\{\mathbf{out}(x) \leftarrow \mathbf{not}\ \mathbf{in}(x) \mid x \in Ar\}$$

Example 18. *When considering our running example AF_{run} we obtain the following ground logic program $\pi_{st'}(AF_{run})$:*

$\mathbf{in}(a) \leftarrow \mathbf{not}\ \mathbf{out}(a).$ $\qquad \mathbf{out}(a) \leftarrow \mathbf{not}\ \mathbf{in}(a).$
$\mathbf{in}(b) \leftarrow \mathbf{not}\ \mathbf{out}(b).$ $\qquad \mathbf{out}(b) \leftarrow \mathbf{not}\ \mathbf{in}(b).$
$\mathbf{in}(c) \leftarrow \mathbf{not}\ \mathbf{out}(c).$ $\qquad \mathbf{out}(c) \leftarrow \mathbf{not}\ \mathbf{in}(c).$

Theorem 28. *([Sakama and Rienstra, 2017]) For every argumentation framework $AF = (Ar, att)$, the stable labellings of AF are in one-to-one correspondence with the stable models of the normal logic program $\pi_{basic}(AF) \cup \pi_{stb'}(AF)$.*

Complete Semantics. Now we turn our attention to complete semantics. Here we have to deal with all three labels. Again we have two types of constraints: (i) each argument must be labelled **in**, **out**, or **undec**. (ii) no argument can be labelled with two of the labels.

$$\begin{aligned}\pi_{co}(AF) =& \{\mathbf{in}(x) \vee \mathbf{out}(x) \vee \mathbf{undec}(x) \leftarrow | \ x \in Ar\} \cup \\ & \{\leftarrow \mathbf{in}(x), \mathbf{out}(x) \mid x \in Ar\} \cup \\ & \{\leftarrow \mathbf{in}(x), \mathbf{undec}(x) \mid x \in Ar\} \cup \\ & \{\leftarrow \mathbf{out}(x), \mathbf{undec}(x) \mid x \in Ar\}\end{aligned}$$

Example 19. *When considering our running example AF_{run} we obtain the following ground logic program $\pi_{st}(AF_{run})$:*

$\mathbf{in}(a) \vee \mathbf{out}(a) \vee \mathbf{undec}(a) \leftarrow .$ $\leftarrow \mathbf{in}(a), \mathbf{undec}(a).$
$\mathbf{in}(b) \vee \mathbf{out}(b) \vee \mathbf{undec}(b) \leftarrow .$ $\leftarrow \mathbf{in}(b), \mathbf{undec}(b).$
$\mathbf{in}(c) \vee \mathbf{out}(c) \vee \mathbf{undec}(c) \leftarrow .$ $\leftarrow \mathbf{in}(c), \mathbf{undec}(c).$
$\leftarrow \mathbf{in}(a), \mathbf{out}(a).$ $\leftarrow \mathbf{out}(a), \mathbf{undec}(a).$
$\leftarrow \mathbf{in}(b), \mathbf{out}(b).$ $\leftarrow \mathbf{out}(b), \mathbf{undec}(b).$
$\leftarrow \mathbf{in}(c), \mathbf{out}(c).$ $\leftarrow \mathbf{out}(c), \mathbf{undec}(c).$

Theorem 29. *([Sakama and Rienstra, 2017]) For every argumentation framework $AF = (Ar, att)$, the complete labellings of AF are in one-to-one correspondence with the stable models of the logic program $\pi_{basic}(AF) \cup \pi_{co}(AF)$.*

Again we can modify $\pi_{com}(AF)$ in order to obtain a normal logic program.

$$\pi_{co'}(AF) = \{\mathbf{in}(x) \leftarrow \mathbf{not}\ \mathbf{out}(x), \mathbf{not}\ \mathbf{undec}(x) \mid x \in Ar\} \cup$$
$$\{\mathbf{out}(x) \leftarrow \mathbf{not}\ \mathbf{in}(x), \mathbf{not}\ \mathbf{undec}(x) \mid x \in Ar\} \cup$$
$$\{\mathbf{undec}(x) \leftarrow \mathbf{not}\ \mathbf{in}(x), \mathbf{not}\ \mathbf{out}(x) \mid x \in Ar\}$$

Example 20. *When considering our running example AF_{run} we obtain the following ground logic program $\pi_{co'}(AF_{run})$:*

$$\mathbf{in}(a) \leftarrow \mathbf{not}\ \mathbf{out}(a), \mathbf{not}\ \mathbf{undec}(a).$$
$$\mathbf{in}(b) \leftarrow \mathbf{not}\ \mathbf{out}(b), \mathbf{not}\ \mathbf{undec}(b).$$
$$\mathbf{in}(c) \leftarrow \mathbf{not}\ \mathbf{out}(c), \mathbf{not}\ \mathbf{undec}(c).$$
$$\mathbf{out}(a) \leftarrow \mathbf{not}\ \mathbf{in}(a), \mathbf{not}\ \mathbf{undec}(a).$$
$$\mathbf{out}(b) \leftarrow \mathbf{not}\ \mathbf{in}(b), \mathbf{not}\ \mathbf{undec}(b).$$

$$\mathbf{out}(c) \leftarrow \mathbf{not}\ \mathbf{in}(c), \mathbf{not}\ \mathbf{undec}(c).$$
$$\mathbf{undec}(a) \leftarrow \mathbf{not}\ \mathbf{in}(a), \mathbf{not}\ \mathbf{out}(a).$$
$$\mathbf{undec}(b) \leftarrow \mathbf{not}\ \mathbf{in}(b), \mathbf{not}\ \mathbf{out}(b).$$
$$\mathbf{undec}(c) \leftarrow \mathbf{not}\ \mathbf{in}(c), \mathbf{not}\ \mathbf{out}(c).$$

Theorem 30. *([Sakama and Rienstra, 2017]) For every argumentation framework $AF = (Ar, att)$, the complete labellings of AF are in one-to-one correspondence with the stable models of the normal logic program $\pi_{basic}(AF) \cup \pi_{co'}(AF)$.*

Preferred Semantics. Finally, for preferred semantics we introduce three new predicates $\mathbf{IN}(x), \mathbf{OUT}(x), \mathbf{UNDEC}(x)$ that will correspond to the actual labels of the preferred labelling. The main idea is that in the program we allow that an argument can satisfy both $\mathbf{in}(x)$ and $\mathbf{out}(x)$ and then map the resulting stable model to an argument labelling as follows. An argument a is: (i) labelled in if $\mathbf{in}(a)$ holds and $\mathbf{out}(x)$ does not hold; (ii) labelled out if $\mathbf{out}(a)$ holds and $\mathbf{in}(x)$ does not hold, (iii) labelled undec if $\mathbf{in}(a)$ and $\mathbf{out}(x)$ hold. We use the new predicates $\mathbf{IN}(x), \mathbf{OUT}(x), \mathbf{UNDEC}(x)$ to compute the argument labels.

$$\pi_{pr}(AF) = \{\mathbf{in}(x) \vee \mathbf{out}(x) \leftarrow | \; x \in Ar\} \cup$$
$$\{\mathbf{IN}(x) \leftarrow \mathbf{in}(x), \mathbf{not} \; \mathbf{out}(x) \mid x \in Ar\} \cup$$
$$\{\mathbf{OUT}(x) \leftarrow \mathbf{not} \; \mathbf{in}(x), \mathbf{out}(x) \mid x \in Ar\} \cup$$
$$\{\mathbf{UNDEC}(x) \leftarrow \mathbf{in}(x), \mathbf{out}(x) \mid x \in Ar\}$$

Example 21. *When considering our running example AF_{run} we obtain the following ground logic program $\pi_{pr}(AF_{run})$:*

$\mathbf{in}(a) \vee \mathbf{out}(a) \leftarrow .$	$\mathbf{OUT}(a) \leftarrow \mathbf{not} \; \mathbf{in}(a), \mathbf{out}(a).$
$\mathbf{in}(b) \vee \mathbf{out}(b) \leftarrow .$	$\mathbf{OUT}(b) \leftarrow \mathbf{not} \; \mathbf{in}(b), \mathbf{out}(b).$
$\mathbf{in}(c) \vee \mathbf{out}(c) \leftarrow .$	$\mathbf{OUT}(c) \leftarrow \mathbf{not} \; \mathbf{in}(c), \mathbf{out}(c).$
$\mathbf{IN}(a) \leftarrow \mathbf{in}(a), \mathbf{not} \; \mathbf{out}(a).$	$\mathbf{UNDEC}(a) \leftarrow \mathbf{in}(a), \mathbf{out}(a).$
$\mathbf{IN}(b) \leftarrow \mathbf{in}(b), \mathbf{not} \; \mathbf{out}(b).$	$\mathbf{UNDEC}(b) \leftarrow \mathbf{in}(b), \mathbf{out}(b).$
$\mathbf{IN}(c) \leftarrow \mathbf{in}(c), \mathbf{not} \; \mathbf{out}(c).$	$\mathbf{UNDEC}(c) \leftarrow \mathbf{in}(c), \mathbf{out}(c).$

Theorem 31. *([Sakama and Rienstra, 2017]) For every argumentation framework $AF = (Ar, att)$, the preferred labellings of AF are in one-to-one correspondence with the stable models of the logic program $\pi_{basic}(AF) \cup \pi_{pr}(AF)$.*

Again we have a disjunctive rule in our logic program. This time, in contrast with the previous encodings, we cannot replace these rules with normal rules (this is due to complexity results for preferred semantics [Sakama and Rienstra, 2017] [39]).

6.3 ASPARTIX-style ASP-Encodings

We will now discuss ASPARTIX-style implementations of argumentation formalisms. That is, we follow a query based approach where we

---

[39]This based on the fact that complexity of reasoning with preferred semantics is located on the second level of the polynomial hierarchy while reasoning with admissible, complete and stable semantics is located on the first level of the polynomial hierarchy [Dvořák and Dunne, 2018].

(a) have queries (that do not depend on the actual framework) encoding semantics and reasoning tasks, which are combined with (b) an input database that encodes the actual argumentation framework. First, we define an input format that encodes argumentation frameworks of the considered argumentation formalism as input database, an ASP program consisting solely of facts. This input format is independent of the actual semantics and reasoning task one aims to solve. Then for each semantics of interest one provides an ASP encoding, an ASP query consisting of non-ground rules that, when combined with an input database, results answer-sets that are in one-to-one correspondence with extensions (or labellings) of the argumentation framework represented by the input database. Moreover, in order to implement specific reasoning tasks one can add modules encoding theses reasoning tasks. In order to solve a reasoning task (for a given framework under a given semantics) one then combines the input encoding of the framework, the encoding of the semantics, and the encoding of the reasoning task and runs a state of the art ASP-solver on that to obtain the corresponding answer-sets. This answer-sets can then be easily interpreted in order to answer the reasoning task. This standard workflow is illustrated in Figure 4. What we described so far is the standard workflow when using one-shot solving via a single ASP encoding to solve reasoning problems. However, in particular in systems optimized towards performance, also multi-shot methods and incremental approaches have been exploited for argumentation systems [Dvořák et al., 2020b; Lehtonen et al., 2021a]. That is, instead of solving a reasoning task with a single call of an ASP-solver such methods might use several calls to an ASP-solver in order to solve a single argumentation reasoning task. However, as these approaches are often tailored to specific reasoning tasks one loses a bit of the flexibility and modularity of the standard workflow.

6.3.1 Abstract Argumentation

Here we start with encoding a given argumentation framework $AF = (Ar, att)$ as facts [Egly et al., 2010]. We use a unary predicate **arg**(x) to encode the arguments and a binary predicate **att**(x, y) to encode the attacks.

7 - Logic Programming and Argumentation

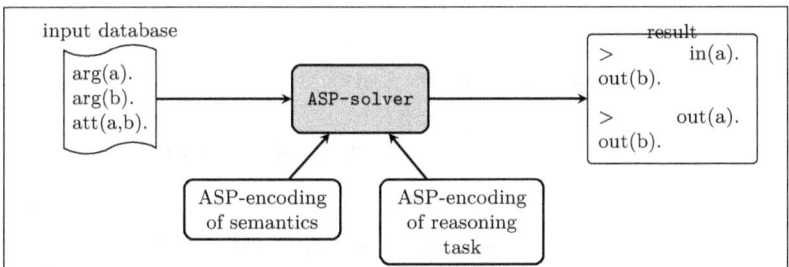

Figure 4: Basic workflow of ASPARTIX like implementations

$$\pi_{input}(AF) = \{\mathbf{arg}(x) \mid x \in Ar\}\} \cup$$
$$\{\mathbf{att}(x,y) \mid (x,y) \in att\}$$

Notice that $\pi_{input}(AF)$ only consists of facts and will act as input database for our queries. That is, $\pi_{input}(AF)$ is the only part that actually depends on the given argumentation framework.

Example 22. *When considering our running example AF_{run} we obtain the following input database $\pi_{input}(AF_{run})$:*

$$\begin{array}{ll} \mathbf{arg}(a). & \mathbf{att}(a,b). \\ \mathbf{arg}(b). & \mathbf{att}(b,a). \\ \mathbf{arg}(c). & \mathbf{att}(b,c). \\ & \mathbf{att}(c,c). \end{array}$$

In the following we introduce encodings of the central argumentation semantics. We start with conflict-free sets. This encoding will also act as the basis for most of the other semantics. We use a predicate **in** to encode that an argument is in the set and a predicate **out** to encode that an argument is not in the set. We first encode that each argument a is either in the set (**in**(a)) or it is not in the set (**out**(a)). That is, we simply generate subsets of the arguments. We then add a constraint to ensure that we do not select two arguments that appear in the same attack.

$$\pi_{cf} = \{\mathbf{in}(X) \leftarrow \text{not } \mathbf{out}(X).$$
$$\mathbf{out}(X) \leftarrow \text{not } \mathbf{in}(X).$$
$$\leftarrow \mathbf{in}(X), \mathbf{in}(Y), \mathbf{att}(X,Y).\}$$

If we now compute the answer-sets of $\pi_{AF}(AF) \cup \pi_{cf}$ we obtain the conflict-free sets AF from the answer-sets by, for each answer-set, forming a set with the arguments a that satisfy $\mathbf{in}(a)$.

Example 23. *The three answer-sets of $\pi_{input}(AF_{run}) \cup \pi_{cf}$ are (we neglect the input predicates $\mathbf{arg}$, $\mathbf{att}$):*

$$\{\mathbf{out}(a), \mathbf{out}(b), \mathbf{out}(c)\}$$
$$\{\mathbf{in}(a), \mathbf{out}(b), \mathbf{out}(c)\}$$
$$\{\mathbf{out}(a), \mathbf{in}(b), \mathbf{out}(c)\}$$

These answer-sets correspond to the three conflict-free sets $\emptyset$, $\{a\}$, $\{b\}$ of AF_{run}.

Proposition 32. *([Egly et al., 2010]) For every argumentation framework $AF = (Ar, att)$, the conflict-free sets of AF are in one-to-one correspondence with the stable models of the program $\pi_{input}(AF) \cup \pi_{cf}$.*

Now starting from the characterisation of conflict-free sets we can encode more evolved semantics by adding the additional constraints these semantics have for their extensions. For admissible semantics we define a unary predicate **defeated** that encodes that an argument is attacked by the selected extensions, i.e., it includes the arguments that would be labeled **out** in a labelling based characterisation. We then add a constraint stating that there is no argument that attacks the extensions but is not attacked by the extension which ensures that the extension defends all its arguments.

$$\pi_{ad} = \pi_{cf} \cup \{\mathbf{defeated}(X) \leftarrow \mathbf{in}(Y), \mathbf{att}(Y,X).$$
$$\leftarrow \mathbf{in}(X), \mathbf{att}(Y,X), \text{not } \mathbf{defeated}(Y).\}$$

7 - Logic Programming and Argumentation

Example 24. *The three answer-sets of $\pi_{input}(AF_{run}) \cup \pi_{ad}$ are (we neglect the input predicates* **arg** *and* **att***):*

$$\{\mathbf{out}(a), \mathbf{out}(b), \mathbf{out}(c)\}$$
$$\{\mathbf{in}(a), \mathbf{out}(b), \mathbf{out}(c), \mathbf{defeated}(b)\}$$
$$\{\mathbf{out}(a), \mathbf{in}(b), \mathbf{out}(c), \mathbf{defeated}(a), \mathbf{defeated}(c)\}$$

These answer-sets correspond to the three admissible sets $\emptyset$, $\{a\}$, $\{b\}$ of AF_{run}.

Again we obtain a one-to-one correspondence between admissible sets and the stable models of $\pi_{input}(AF) \cup \pi_{adm}$.

Proposition 33. *([Egly et al., 2010]) For every argumentation framework $AF = (Ar, att)$, the admissible sets of AF are in one-to-one correspondence with the stable models of the program $\pi_{input}(AF) \cup \pi_{ad}$.*

Next we extend the encoding of admissible semantics by (a) a predicate **undefended** that contains all arguments that are not defended by the extension and (b) a constraint that states that there is not argument outside the extension the is defended.

$$\pi_{com} = \pi_{adm} \cup \{\mathbf{undefended}(X) \leftarrow \mathbf{att}(Y, X), \mathrm{not}\ \mathbf{defeated}(Y).$$
$$\leftarrow \mathbf{out}(X), \mathrm{not}\ \mathbf{undefended}(X).\}$$

Example 25. *The three answer-sets of $\pi_{input}(AF_{run}) \cup \pi_{com}$ are (we neglect the input predicates* **arg** *and* **att***):*

$\{\mathbf{out}(a), \mathbf{out}(b), \mathbf{out}(c), \mathbf{undefended}(a), \mathbf{undefended}(b), \mathbf{undefended}(c)\}$
$\{\mathbf{in}(a), \mathbf{out}(b), \mathbf{out}(c), \mathbf{defeated}(b), \mathbf{undefended}(b), \mathbf{undefended}(c)\}$
$\{\mathbf{out}(a), \mathbf{in}(b), \mathbf{out}(c), \mathbf{defeated}(a), \mathbf{defeated}(c), \mathbf{undefended}(a),$
  $\mathbf{undefended}(c)\}$

These answer-sets correspond to the three complete extensions $\emptyset$, $\{a\}$, $\{b\}$ of AF_{run}.

Proposition 34. *([Egly et al., 2010]) For every argumentation framework $AF = (Ar, att)$, the complete extensions of AF are in one-to-one correspondence with the stable models of the program $\pi_{input}(AF) \cup \pi_{com}$.*

Along the same lines we one can extend the conflict-free encoding for stable semantics by again defining a predicate **defeated** and adding a constraint that rules out arguments that are neither in the extension nor attacked.

$$\pi_{st} = \pi_{cf} \cup \{\mathbf{defeated}(X) \leftarrow \mathbf{in}(Y), \mathbf{att}(Y,X).$$
$$\leftarrow \mathbf{out}(X), \mathtt{not}\ \mathbf{defeated}(X).\}$$

Example 26. *The only answer-set of $\pi_{input}(AF_{run}) \cup \pi_{st}$ is (we neglect the input predicates* **arg** *and* **att***):*

$$\{\mathbf{out}(a), \mathbf{in}(b), \mathbf{out}(c), \mathbf{defeated}(a), \mathbf{defeated}(c)\}$$

The unique answer-set correspond to the only stable extension $\{b\}$ of AF_{run}.

As before, this encoding explicitly encodes all the requirements of stable extensions as rules of the LP. That is, it does not use the close connection between stable semantics for AFs and stable model semantics for LPs. However, there is also a more direct approach using conditional literals [Dvořák et al., 2020b], that follows the correspondence result from Section 3.2.

$$\pi_{st'}(AF) = \{\mathbf{in}(Y) \leftarrow \mathbf{arg}(Y), \mathtt{not}\ \mathbf{in}(X) : \mathbf{att}(X,Y).\}$$

This encoding typically results in smaller grounding of the program in the ASP-solving process and has shown better performance on benchmark instances. Of course, both encodings provide the one-to-one correspondence between stable extensions and answer-sets.

Example 27. *The only answer-set of $\pi_{input}(AF_{run}) \cup \pi_{st'}$ is (we neglect the input predicates* **arg** *and* **att***) is $\{\mathbf{in}(b)\}$. This answer-set corresponds to the only stable extension $\{b\}$ of AF_{run}.*

Proposition 35. *([Egly et al., 2010; Dvořák et al., 2020b]) For every argumentation framework $AF = (Ar, att)$, we have that the stable extensions of AF are in one-to-one correspondence with the stable models of the programs $\pi_{input}(AF) \cup \pi_{st}$ and $\pi_{input}(AF) \cup \pi_{st'}$ respectively.*

Finally, let us consider preferred semantics. Due to complexity results, we know that we need disjunctive rules for preferred semantics. As preferred extensions are $\subseteq$-maximal admissible set we start from the encoding for admissible semantics (or alternatively from the encoding complete semantics) and can then apply ASP-techniques for $\subseteq$-maximization. Here we follow [Gaggl et al., 2015] and present an encoding based on saturation that uses conditional disjunction in the rule heads. The use of conditional disjunction allows for a compact encoding compared to earlier versions,

$\pi_{pr} = \pi_{adm} \cup$
{ **notTrivial** $\leftarrow$ **out**(X).
witness$(X) : $ **out**$(X) \leftarrow$ **notTrivial**.
spoil $\vee$ **witness**$(Z) : $ **att**$(Z,Y) \leftarrow$ **witness**$(X),$ **att**(Y,X).
spoil $\leftarrow$ **witness**$(X),$ **witness**$(Y),$ **att**(X,Y).
spoil $\leftarrow$ **in**$(X),$ **witness**$(Y),$ **att**(X,Y).
witness$(X) \leftarrow$ **spoil**, **arg**(X).
$\leftarrow$ **not spoil**, **notTrivial**.}

The admissible part generates admissible extensions and the additional rules deal with the $\subseteq$-maximality. The encoding first tests whether we are in the trivial case where all arguments are in the extension and thus the extension is clearly preferred. If not, we aim to construct a larger extension by adding some of the arguments which are not in the admissible set (at least one). These new arguments are stored in the predicate **witness**. We then have another rule that adds further arguments if those are required to defend the arguments already included as witness. If that is not possible, we obtain the constant **spoil** which indicates that the set we constructed is not admissible. Moreover we have two rules that check whether the constructed set is conflict-free and if not yield the constant **spoil**. The two final rules are then due to the saturation technique. First, whenever we obtain spoil we have to include all arguments as witnesses. This ensures that we only get one answer-set for each preferred extension and also makes sure that we rule out such potential answer-sets when we find a model for the same admissible set that does not lead **spoil**. Now assume we find a model M that satisfies all rules but the last one and does not include spoil. Of course, the last rule ensures that M is not an answer-set, but it also rules out all answer-sets M' that coincide on the admissible part but yield **spoil** as M is always a smaller model than M' on the reduct of M'.

Example 28. *The two answer-sets of $\pi_{input}(AF_{run}) \cup \pi_{pr}$ are (we neglect the input predicates **arg** and **att**):*

$\{\mathbf{in}(a), \mathbf{out}(b), \mathbf{out}(c), \mathbf{defeated}(b), \mathbf{spoil}, \mathbf{notTrivial},$
$\qquad\qquad\qquad\qquad \mathbf{witness}(a), \mathbf{witness}(b), \mathbf{witness}(c)\}$

$\{\mathbf{out}(a), \mathbf{in}(b), \mathbf{out}(c), \mathbf{defeated}(a), \mathbf{defeated}(c), \mathbf{spoil}, \mathbf{notTrivial},$
$\qquad\qquad\qquad\qquad \mathbf{witness}(a), \mathbf{witness}(b), \mathbf{witness}(c)\}$

These two answer-sets correspond to the preferred extensions $\{a\}$, $\{b\}$ of AF_{run}.

*Now consider the answer-set $S = \{\mathbf{out}(a), \mathbf{out}(b), \mathbf{out}(c)\}$ for the admissible encoding (which corresponds to the empty set). If we extend S by the atoms $\mathbf{witness}(a)$ and $\mathbf{notTrivial}$, i.e., we consider $S' = \{\mathbf{out}(a), \mathbf{out}(b), \mathbf{out}(c), \mathbf{witness}(a), \mathbf{notTrivial}\}$, we satisfy all but the last rule of π_{pr}. Moreover, the last rule is clearly violated as **spoil** is not included in S'. That is, S' is itself not an answer-set of the program but also excludes all other models extending S and containing **spoil** from being an answer-set, as this would be violating the minimal model property on the reduct.*

Proposition 36. *([Gaggl et al., 2015]) For every argumentation framework $AF = (Ar, att)$, we have that the preferred extensions of AF are in one-to-one correspondence with the stable models of the program $\pi_{input}(AF) \cup \pi_{pr}$.*

An alternative approach for encodings of preferred semantics [Dvořák et al., 2020b] is to use the encoding for admissible semantics and exploit clingo domain heuristics [Gebser et al., 2013] to perform the ⊆-maximization on the **in** predicate.

In this section we presented prototypical ASP encodings for the selected semantics. Indeed, there are ASP encodings for most of the argumentation semantics and those are integrated in the ASPARTIX system [Dvořák et al., 2020a]. For the interested reader we next provide the relevant pointers to the literature. Encodings for semi-stable and stage semantics are discussed in [Egly et al., 2010; Gaggl et al., 2015], ideal semantics are discussed in [Faber and Woltran, 2009; Dvořák et al., 2020b], encodings for cf2 semantics are presented in [Osorio et al., 2010; Gaggl and Woltran, 2013], for encodings of resolution-based grounded semantics see [Dvořák et al., 2011], and strongly admissible semantics and their minimization are discussed in [Dvořák and Wallner, 2020].

ASP queries to decide acceptance problems. Given the encoding of an argumentation semantics we can use an ASP-solver to compute an extension or even all extensions of an argumentation framework. However, sometimes we are even more interested in the acceptance status of an argument. Classic acceptance problems are the credulous/skeptical acceptance of an argument a (or a set of arguments $S = \{a_1, \ldots a_n\}$). That is, deciding whether an argument a (or a set S) is contained in one of the extensions, in all of the extensions respectively. We can simply address these problems by adding an additional rule to the encoding. For credulous acceptance we require that $\mathbf{in}(a)$ holds while for skeptical acceptance we are looking for a counter example by requiring that $\mathbf{in}(a)$ does not hold.

$$\pi_{cred}(a) = \{\leftarrow \mathbf{not\ in}(a)\}$$
$$\pi_{cred}(S) = \{\leftarrow \mathbf{not\ in}(a) \mid a \in S\}$$
$$\pi_{skept}(a) = \{\leftarrow \mathbf{in}(a)\}$$
$$\pi_{skept}(S) = \{\leftarrow \mathbf{in}(a_1), \ldots, \mathbf{in}(a_n) \mid S = \{a_1, \ldots a_n\}\}$$

If we now solve $\pi_{input}(AF) \cup \pi_\sigma \cup \pi_{cred}(a)$ the solver either provides an answer-set that corresponds to a σ-extension that contains the argument a or returns that the program is unsatisfiable. In the later case we know that no such σ-extension exists and thus can answer the query negatively. Similarly, if we now solve $\pi_{input}(AF) \cup \pi_\sigma \cup \pi_{skept}(a)$ the solver either provides an answer-set that corresponds to a σ-extension that does not include the argument a, i.e., argument a is not skeptically accepted, or returns that the program is unsatisfiable. In the later case we know that no σ-extension that acts as counter example exists and thus can answer the query positively.

ASP for generalizations of abstract argumentation frameworks.
ASP encodings have been provided for a number of generalizations of AFs. Some of the extensions are directly based on the encoding for AFs. For instance when dealing with preferences or support the encodings [Egly et al., 2010] construct a new attack relation that corresponds to standard AF and then use the encodings of semantics for AFs. Similarly, for argumentation frameworks with recursive attacks (AFRAs) the encoding of the ASPARTIX system[40] constructs an equivalent AF and exploits the existing encodings. However, for other generalizations careful adaptations or even new encodings are required. Popular examples are the encodings for extended argumentation frameworks (EAF) [Dvořák et al., 2015], SETAFs [Dvořák et al., 2018] and Abstract Dialectical Frameworks (ADF) [Ellmauthaler and Wallner, 2012; Ellmauthaler and Strass, 2014]. While a full discussion of these encodings is beyond the scope of this chapter we want to give a first impression how these generalizations can be approached via ASP by illustrating the input encodings of the formalisms in Figure 5.

6.3.2 Assumption-Based Argumentation

In this section we discuss the ASPARTIX-style approach towards assumption-based argumentation by Lehtonen et al. (2021b, 2021a). That is, we present an approach that encodes assumption-based argumentation frameworks (ABAFs) as input database and then provide fixed encodings of semantics that do not depend on the actual framework.

In the following let $\mathcal{F} = (\mathcal{L}, \mathcal{R}, \mathcal{A}, \overline{})$ be an ABAF such that $\mathcal{R} = \{r_1, r_2, ...r_k\}$. We use the following set of facts $\pi_{ABA}(\mathcal{F})$ to represent the ABAF $\mathcal{F}$ for our further investigations:

$$\pi_{ABA}(\mathcal{F}) = \{\textbf{assumption}(a) \mid a \in \mathcal{A}\} \cup$$
$$\{\textbf{head}(i, h) \mid r_i \in \mathcal{R}, h \in head(r_i)\} \cup$$
$$\{\textbf{body}(i, b) \mid r_i \in \mathcal{R}, b \in body(r_i)\} \cup$$
$$\{\textbf{contrary}(a, \overline{a}) \mid a \in \mathcal{A}\}.$$

---

[40] https://www.dbai.tuwien.ac.at/research/argumentation/aspartix/afra.html

7 - LOGIC PROGRAMMING AND ARGUMENTATION

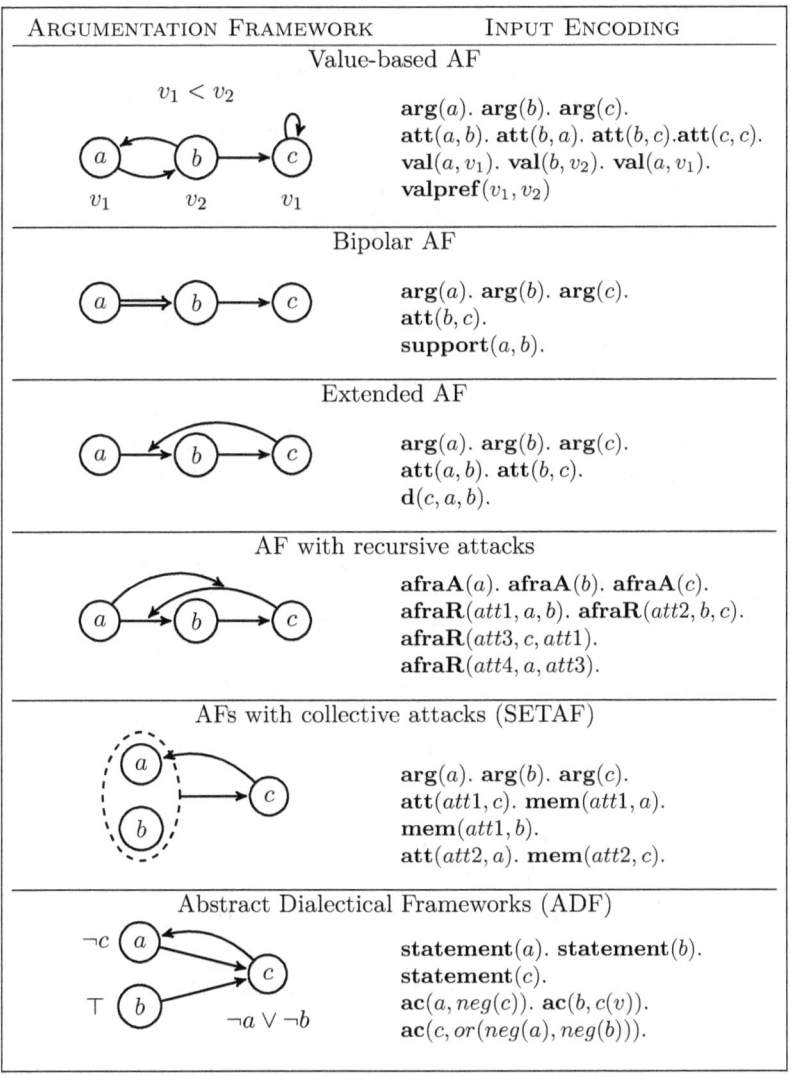

Figure 5: Input encodings for different generalizations of AFs.

Notice that $\pi_{ABA}(\mathcal{F})$ introduces a unique identifiers for each rules and then uses the binary predicate **head** to encode the head of a rule and a predicate **body** to encode the body atoms of the rule. Moreover, we can easily extend the input encoding to also deal with ABAFs that have a preference relation $\leq$ (commonly referred to as ABA$^+$ frameworks). That is we introduce a binary predicate **preferred** and add a fact **preferred**(a, b) to the input database whenever $a \leq b$, i.e., we obtain $\pi_{ABA^+}(\mathcal{F}, \leq) = \pi_{ABAF}(\mathcal{F}) \cup \{\textbf{preferred}(x, y) \mid y \leq x\}$. However, in this chapter we will focus on encodings of ABAFs without preferences and moreover restrict ourselves to flat ABAFs. The interested reader is referred to [Lehtonen et al., 2021b; Lehtonen et al., 2021a] for ASP encodings of ABA$^+$.

We start with encoding conflict-free assumption sets. Again this encoding will be the basis for the encodings of the other semantics. To this end we use two unary predicates **in** and **out** to encode that an assumption is in respectively outside the considered assumption set. We then introduce the unary predicate **supported** which encodes that a statement can be derived from the selected assumptions. In order to derive **supported** we use a unary predicate **triggered** to encode all rules whose body is satisfied by the selected assumptions and the already derived statements. Moreover, we have the unary **defeated** predicate that computes the assumptions which are attacked by the selected assumption set and is based on **supported** and the contrary function. Finally, to ensure conflict-freeness, we have a constraint that states that none of the selected assumptions is defeated.

$\pi_{cf} = \{\textbf{in}(X) \leftarrow \textbf{assumption}(X), \textbf{not } \textbf{out}(X).$
$\quad \textbf{out}(X) \leftarrow \textbf{assumption}(X), \textbf{not } \textbf{in}(X).$
$\quad \textbf{supported}(X) \leftarrow \textbf{assumption}(X), \textbf{in}(X).$
$\quad \textbf{supported}(X) \leftarrow \textbf{head}(R, X), \textbf{triggered}(R).$
$\quad \textbf{triggered}(R) \leftarrow \textbf{head}(R, \_), \textbf{supported}(X) : \textbf{body}(R, X).$
$\quad \textbf{defeated}(X) \leftarrow \textbf{supported}(Y), \textbf{contrary}(X, Y).$
$\quad \leftarrow \textbf{in}(X), \textbf{defeated}(X).\}$

The conflict-free sets of the ABAF correspond to the answer-set of the LP $\pi_{ABA}(D) \cup \pi_{cf}$, i.e., given an answer-set we can compute the corresponding assumption set by inspecting the **in** predicate. However, for ABAFs we are not only interested in the accepted assumptions but also in the statements that can be derived from them. These statements are accessible via the **supported** predicate, i.e., for each answer-set a statement s is a consequence of the selected assumptions if and only if **supported**(s) holds in that answer-set.

Proposition 37. *([Lehtonen et al., 2021b]) For every ABAF $\mathcal{F} = (\mathcal{L}, \mathcal{R}, \mathcal{A}, \overline{})$, the conflict-free assumption sets of $\mathcal{F}$ are in one-to-one correspondence with the answer-sets of $\pi_{ABA}(\mathcal{F}) \cup \pi_{cf}$. A statement s can be derived from a conflict-free assumption set iff* **supported**(s) *is in the corresponding answer-set.*

Next we consider admissible semantics. The idea here is to consider assumptions that are not defeated by the set of selected assumptions and test whether they attack the selected assumptions, which of course would violate admissibility. To this end we introduce the predicate **derivedFromUndef** that contains all undefeated assumptions and all statements that can be derived from them. Furthermore, we again use a predicate **triggeredByUndef** which collects the rules we use to derive these statements. Finally, with the predicate **attackedByUndef** we derive the assumptions attacked by the undefeated assumptions and add a constraint that none of the assumptions in our set is attacked by the undefeated assumptions.

$\pi_{ad} = \pi_{cf} \cup \{$
derivedFromUndef$(X) \leftarrow$ **assumption**$(X),$ **not defeated**$(X).$
derivedFromUndef$(X) \leftarrow$ **head**$(R, X),$ **triggeredByUndef**$(R).$
triggeredByUndef$(R) \leftarrow$ **head**$(R, \_),$
             derivedFromUndef$(X) : \text{body}(R, X).$
attackedByUndef$(X) \leftarrow$ **contrary**$(X, Y),$ **derivedFromUndef**$(Y).$
$\leftarrow$ **in**$(X),$ **attackedByUndef**$(X).\}$

Proposition 38. *([Lehtonen et al., 2021b])* *For every ABAF $\mathcal{F} = (\mathcal{L}, \mathcal{R}, \mathcal{A}, \overline{})$, the admissible assumption sets of $\mathcal{F}$ are in one-to-one correspondence with the answer-sets of $\pi_{ABA}(\mathcal{F}) \cup \pi_{ad}$. A statement s can be derived from an admissible assumption set iff $\mathbf{supported}(s)$ is in the corresponding answer-set.*

Now the step from admissible to complete semantics is rather easy. We just add a constraint that there are no assumptions that are neither in the selected set nor attacked by the undefeated assumptions.

$$\pi_{co} = \pi_{adm} \cup \{\leftarrow \mathbf{out}(X), \mathbf{not}\ \mathbf{attackedByUndef}(X).\}$$

Proposition 39. *([Lehtonen et al., 2021b])* *For every ABAF $\mathcal{F} = (\mathcal{L}, \mathcal{R}, \mathcal{A}, \overline{})$, the complete assumption sets of $\mathcal{F}$ are in one-to-one correspondence with the answer-sets of $\pi_{ABA}(\mathcal{F}) \cup \pi_{co}$. A statement s can be derived from a complete assumption set iff $\mathbf{supported}(s)$ is in the corresponding answer-set.*

Given the encoding for admissible and complete semantics one can use standard techniques for $\subseteq$-maximization to deal with preferred semantics. For enumeration Lehtonen *et al.* [2021b] propose to use preferential optimization statements while for skeptical reasoning algorithms iterative calls to the ASP-solver are used [Lehtonen *et al.*, 2021a].

Now for stable semantics, we only rely on the conflict-free encoding and add a constraint that each assumption that is not selected must be defeated by the selected assumptions.

$$\pi_{st} = \pi_{cf} \cup \{\leftarrow \mathbf{out}(X), \mathbf{not}\ \mathbf{defeated}(X).\}$$

Proposition 40. *([Lehtonen et al., 2021b])* *For every ABAF $\mathcal{F} = (\mathcal{L}, \mathcal{R}, \mathcal{A}, \overline{})$, the stable assumption sets of $\mathcal{F}$ are in one-to-one correspondence with the answer-sets of $\pi_{ABA}(\mathcal{F}) \cup \pi_{st}$. A statement s can be derived from a stable assumption set iff $\mathbf{supported}(s)$ is in the corresponding answer-set.*

We next provide an alternative encoding of stable semantics following the correspondence result between ABA stable semantics and stable model semantics of LPs in Section 4.3.2 (cf. Theorem 18). That is, we restate ABA rules as LP rules using conditional literals and then add rules stating that an assumption is supported if its contrary is not. In order to be compatible with the other encodings we then introduce rules that fills the **in** predicate with the supported assumptions.

$$\pi_{st'} = \{\mathbf{supported}(X) \leftarrow \mathbf{head}(R, X), \mathbf{supported}(Y) : \mathbf{body}(R, Y).$$
$$\mathbf{supported}(X) \leftarrow \mathbf{contrary}(X, Z), \mathbf{not}\ \mathbf{supported}(Z).$$
$$\mathbf{in}(X) \leftarrow \mathbf{assumption}(X), \mathbf{supported}(X).\}$$

ASP queries to decide acceptance problems. Given the characterisations above we can use ASP-solver to compute an assumption set or even all assumption sets of an argumentation framework (under a given semantics). However, sometimes we are even more interested in the acceptance status of an assumption or a statement. Classic acceptance problems are the credulous/skeptical acceptance of a statement s, i.e., deciding whether there is an assumption set that implies s or deciding whether s is implied by all assumption sets (under a given semantics). Again we can address these problems by adding the corresponding rule to the encoding. For credulous acceptance we require that **supported**(s) holds while for skeptical acceptance we are looking for a counter example by requiring that **supported**(s) does not hold.

$$\pi_{cred}(s) = \{\leftarrow \mathbf{not}\ \mathbf{supported}(s).\}$$
$$\pi_{skept}(s) = \{\leftarrow \mathbf{supported}(s).\}$$

If we now solve $\pi_{ABA}(\mathcal{F}) \cup \pi_\sigma \cup \pi_{cred}(s)$ the solver either provides an answer-set that corresponds to a σ-assumption set that implies the statement s or returns that the program is unsatisfiable. In the later case we know that there is no such σ-assumption set and thus can answer the query negatively. Similarly, if we now solve $\pi_{ABA}(\mathcal{F}) \cup \pi_\sigma \cup \pi_{skept}(s)$ the solver either provides an answer-set that corresponds to a σ-assumption set that does not imply the statement s, i.e., statement s is not skeptically accepted, or returns that the program is unsatisfiable. In the later case we know that no σ-assumption set that acts as counter-example exists and thus can answer the query positively.

7 Discussion

In this chapter we have discussed several argumentation systems (Dung's AFs, ABA, ADFs, SETAFs) focusing on their connections to logic programming. In each case, whenever it applies, we highlighted similarities concerning syntax, terminology and the representation and computation of semantic models, implementation in ASP as well as translations from theories in each system to and from normal logic programs. The highest interest when comparing logical systems usually concerns their relative expressive power, so we focused on the relation between the various systems based on semantics: for each argumentation system, we considered whether they can model inference from logic programming and vice-versa. The results gathered ensure that:

1. normal logic programs semantically subsume Dung's AFs (Section 3)[41]

2. normal logic programs are semantically subsumed by flat-ABA, but they become equivalent if the programs are equipped with semantic projection (Section 4)[42]

3. normal logic programs are semantically subsumed by ADFs[43], but they remain equivalent to the fragment of ADF^+s (Section 5.1)[44]

4. normal logic programs are equivalent to SETAF, with their semantics coinciding for all programs rid of irrelevant atoms (Section 5.2)[45]

Other systems worth mentioning regarding their relation to logic programming are Defeasible Logic Programming (DeLP) [Simari and

---

[41]The results were gathered primarily from [Caminada et al., 2015b; Sá and Alcântara, 2021a; Caminada et al., 2022].

[42]The results were gathered primarily from [Bondarenko et al., 1997; Caminada and Schulz, 2017; Sá and Alcântara, 2019; Sá and Alcântara, 2021b].

[43]If ADFs are equipped with three-valued acceptance conditions, which were proposed in [Alcântara and Sá, 2019].

[44]The results were gathered primarily from [Strass, 2013; Alcântara et al., 2019].

[45]The results were gathered primarily from [König et al., 2022; Alcântara et al., 2023].

Loui, 1992; García and Simari, 2004] and Claim-Augmented AFs (CAFs) [Dvořák and Woltran, 2020; König et al., 2022].

DeLP [Simari and Loui, 1992] is a rule-based argumentation system introduced in [García and Simari, 2004] as a formalism that combines results of logic programming and defeasible argumentation. Overall, the presentation of concepts in DeLP follow conventions from Logic Programming, but one important difference is that their semantics is inherently based on the possible *derivations* of claims, which intuitively amounts to the arguments constructed from a logic program following Definition 13. Further, the syntax of DeLP includes strong negation and the possibility of a priority relation between rules, which is extended to a set of preferences over arguments. These features pose as obstacles to a direct comparison between DeLP and normal logic programs or even AFs.

Intuitively, Claim-Augmented AFs (CAFs) [Dvořák and Woltran, 2020] are the same as AFs, but each argument is accompanied by a *claim* (or conclusion, as we consider in Definition 13). CAFs admit the argument extension semantics for AFs (including complete, grounded, preferred, etc.) and also claim-based variants of extensions where the claims of arguments in a σ-extension S of a CAF are extracted into corresponding σ-claim-extensions. The conversion between argument-based and claim-based extensions is identical to that observed in [Caminada et al., 2015b] to convert the argument labellings of LP2AA(P) (see Section 3.3) into *conclusion labellings*, which can be compared to the models of the program P. In fact, whatever P is, LP2AA(P) can be immediately perceived as a CAF, since each argument A built using Definition 13 has a single atom (a claim) assigned as Conc(A). From this connection, a series of results concerning the relationship between CAFs and normal logic programs are implied from the results of [Caminada et al., 2015b] and have been considered in [Dvořák et al., 2023a; König et al., 2022]: it means CAFs are subsumed by logic programs in the same way AFs are; but then translating from LPs to CAFs, we will find some differences between the L-stable program semantics and the CAF semi-stable claim-based semantics. Based on this observation, [Dvořák et al., 2023c] introduces alternative versions of semantics for CAFs based on the additional information about the claims of the argument. Prominently it provides a variant of semi-stable semantics that naturally maps to L-stable program semantics. Computational properties of these semantics have then been investigated in [Dvořák et al., 2023a]. [König et al., 2022] studied the connection between CAFs and logic programs, but they only prove correspondence for the stable semantics.[46]

It is also worth mentioning the investigation of [Caminada et al., 2015a] about the relationship between AFs and ABA. Their translations between AF and ABA could be combined with the translations from [Caminada et al., 2015b] between NLP and AF (see Section 3) to obtain results about the relationship between ABA and LP. We conjecture that the potential results obtained through the combined translations would prove themselves redundant to the results found in [Caminada and Schulz, 2017] and [Sá and Alcântara, 2021b] (Section 4).

We have also provided a detailed overview of the implementation of argumentation systems and the retrieval of their semantics using answer-set programs (ASP), an expressive class of logic programs based on the stable model semantics. Our discussion in Section 6 includes mapping theories of systems such as AFs and ABA to corresponding answer set programs, allowing the computation of their semantics using ASP solvers. On that matter, several authors contributed to this line of research. Nieves et al. [2008] provided a compiler-style approach to implement abstract argumentation in ASP, where the logic program is computed from the considered argumentation framework. On the other hand, Wakaki and Nitta [2008] and Egly et al. [2008] provided the first query-based implementations for abstract argumentation, where the argumentation framework is provided as input database. While Wakaki and Nitta followed the labelling-based characterisations of the semantics, Egly et al. followed the extension-based characterisation in their encodings. For a comparison of these early works the interested reader is referred to [Toni and Sergot, 2011]. The ASPARTIX system[47] of [Egly et al., 2008] was later extended to deal with several generalizations of abstract argumentation and new semantics that were intro-

---

[46][König et al., 2022] discusses back-and-forth translations between multiple systems, including CAFs, NLPs, ABA frameworks, ADFs and SETAFs, but the focus is primarily on the syntax of theories in each system.

[47]https://www.dbai.tuwien.ac.at/research/argumentation/aspartix/

duced in the literature (see, e.g., [Egly et al., 2010; Dvořák et al., 2015; Ellmauthaler and Strass, 2014; Dvořák and Wallner, 2020]). Alternative encodings in a compiler-style approach were introduced by Sakama and Rienstra [2017]. Moreover, ASP techniques have also been applied to structured argumentation formalisms, prominently assumption-based argumentation. The first approaches first build a corresponding abstract argumentation framework and then use ASP Encodings on these frameworks to deal with the argumentation semantics [Cyras and Toni, 2016; Lehtonen et al., 2017]. However, more recent systems directly approach the assumption-based argumentation semantics without constructing abstract frameworks [Lehtonen et al., 2021b; Lehtonen et al., 2021a].

References

[Alcântara and Sá, 2019] João Alcântara and Samy Sá. On three-valued acceptance conditions of abstract dialectical frameworks. *Electronic Notes in Theoretical Compututer Science*, 344(C):3–23, aug 2019.

[Alcântara and Sá, 2021] João Alcântara and Samy Sá. Equivalence results between SETAF and attacking abstract dialectical frameworks. In Leila Amgoud and Richard Booth, editors, *Proceedings of the 19th International Workshop on Nonmonotonic Reasoning (NMR 2021)*, pages 139–148, 2021.

[Alcântara et al., 2019] João Alcântara, Samy Sá, and Juan Acosta-Guadarrama. On the equivalence between abstract dialectical frameworks and logic programs. *Theory and Practice of Logic Programming*, 19(5-6):941–956, 2019.

[Alcântara et al., 2023] João Alcântara, Renan Cordeiro, and Samy Sá. On the equivalence between logic programming and SETAF. *(submitted, under review)*, 2023.

[Baroni et al., 2018] Pietro Baroni, Martin Caminada, and Massimiliano Giacomin. Abstract argumentation frameworks and their semantics. In *Handbook of Formal Argumentation*, volume 1. College Publications, 2018.

[Baumann and Spanring, 2015] Ringo Baumann and Christof Spanring. Infinite argumentation frameworks — on the existence and uniqueness of extensions. In Th. Eiter, H. Strass, M. Truszczyński, and S. Woltran, editors, *Advances in Knowledge Representation, Logic Programming, and Abstract Argumentation — Essays Dedicated to Gerhard Brewka on the Occasion of His 60th Birthday*, page 281–295. Springer, 2015.

[Baumann, 2018] Ringo Baumann. On the nature of argumentation semantics: Existence and uniqueness, expressibility, and replaceability. In *Handbook of Formal Argumentation*, volume 1. College Publications, 2018.

[Besnard and Hunter, 2014] Philippe Besnard and Anthony Hunter. Constructing argument graphs with deductive arguments: A tutorial. *Argument & Computation*, 5:5–30, 2014. Special Issue: Tutorials on Structured Argumentation.

[Bikakis et al., 2021] Antonis Bikakis, Andrea Cohen, Wolfgang Dvořák, Giorgos Flouris, and Simon Parsons. Joint attacks and accrual in argumentation frameworks. In D. Gabbay, M. Giacomin, G. Simari, and M. Thimm, editors, *Handbook of Formal Argumentation*, volume 2, chapter 2. College Publications, 2021.

[Bondarenko et al., 1997] Andrei Bondarenko, Phan Minh Dung, Bob Kowalski, and Francesca Toni. An abstract, argumentation-theoretic approach to default reasoning. *Artificial Intelligence*, 93:63–101, 1997.

[Brewka and Woltran, 2010] Gerhard Brewka and Stefan Woltran. Abstract dialectical frameworks. In *Twelfth International Conference on the Principles of Knowledge Representation and Reasoning*, pages 102–111. AAAI Press, 2010.

[Brewka et al., 2013] Gerhard Brewka, Stefan Ellmauthaler, Hannes Strass, Johannes Wallner, and Stefan Woltran. Abstract dialectical frameworks revisited. In *Proceedings of the Twenty-Third international joint conference on Artificial Intelligence*, pages 803–809. AAAI Press, 2013.

[Brewka et al., 2018] Gerhard Brewka, Stefan Ellmauthaler, Hannes Strass, Johannes P. Wallner, and Stefan Woltran. Abstract dialectical frameworks. In Pietro Baroni, Dov Gabbay, Massimiliano Giacomin, and Leendert van der Torre, editors, *Handbook of Formal Argumentation*, volume 1, chapter 5, pages 237–285. College Publications, 2018.

[Caminada and Gabbay, 2009] Martin Caminada and Dov Gabbay. A logical account of formal argumentation. *Studia Logica*, 93(2-3):109–145, 2009. Special issue: new ideas in argumentation theory.

[Caminada and Schulz, 2017] Martin Caminada and Claudia Schulz. On the equivalence between assumption-based argumentation and logic programming. *Journal of Artificial Intelligence Research*, 60:779–825, 2017.

[Caminada and Verheij, 2010] Martin Caminada and Bart Verheij. On the existence of semi-stable extensions. In G. Danoy, M. Seredynski, R. Booth, B. Gateau, I. Jars, and D. Khadraoui, editors, *Proceedings of the 22nd Benelux Conference on Artificial Intelligence*, 2010.

[Caminada et al., 2015a] Martin Caminada, Samy Sá, João Alcântara, and Wolfgang Dvořák. On the difference between assumption-based argumentation and abstract argumentation. *IfCoLog Journal of Logics and their Applications*, 2:15–34, 2015.

[Caminada et al., 2015b] Martin Caminada, Samy Sá, João Alcântara, and Wolfgang Dvořák. On the equivalence between logic programming semantics and argumentation semantics. *International Journal of Approximate Reasoning*, 58:87–111, 2015.

[Caminada et al., 2022] Martin Caminada, Sri Harikrishnan, and Samy Sá. Comparing logic programming and formal argumentation; the case of ideal and eager semantics. *Argument & Computation*, 13:91–120, 2022.

[Caminada et al., 2024] Martin Caminada, Matthias König, Anna Rapberger, and Markus Ulbricht. Attack semantics and collective attacks revisited. *Argument & Computation*, 2024. in print.

[Caminada, 2006] Martin Caminada. On the issue of reinstatement in argumentation. In M. Fischer, W. van der Hoek, B. Konev, and A. Lisitsa, editors, *Logics in Artificial Intelligence; 10th European Conference, JELIA 2006*, pages 111–123. Springer, 2006. LNAI 4160.

[Cerutti et al., 2018] Federico Cerutti, Sarah Gaggl, Matthias Thimm, and Johannes Wallner. Foundations of implementations for formal argumentation. In P. Baroni, D. Gabbay, M. Giacomin, and L. van der Torre, editors, *Handbook of Formal Argumentation*, chapter 15, pages 688–767. College Publications, 2018. also appears in IfCoLog Journal of Logics and their Applications 4(8):2623–2706.

[Charwat et al., 2015] Günther Charwat, Wolfgang Dvořák, Sarah Alice Gaggl, Johannes Peter Wallner, and Stefan Woltran. Methods for solving reasoning problems in abstract argumentation - A survey. *Artificial Intelligence*, 220:28–63, 2015.

[Cramer and Saldanha, 2020] Marcos Cramer and Emmanuelle-Anna Dietz Saldanha. Logic programming, argumentation and human reasoning. In Mehdi Dastani, Huimin Dong, and Leon van der Torre, editors, *Logic and Argumentation*, pages 58–79, Cham, 2020. Springer International Publishing.

[Cyras and Toni, 2016] Kristijonas Cyras and Francesca Toni. ABA+: assumption-based argumentation with preferences. In Ch. Baral, J. Delgrande, and F. Wolter, editors, *Principles of Knowledge Representation and Reasoning: Proceedings of the Fifteenth International Conference, KR 2016, Cape Town, South Africa, April 25-29, 2016*, pages 553–556. AAAI Press, 2016.

[Čyras et al., 2018] Kristijonas Čyras, Xiuyi Fan, Claudia Schulz, and Francesca Toni. Assumption-based argumentation: Disputes, explanations, preferences. In *Handbook of Formal Argumentation*, volume 1. College Publications, 2018.

[Dung et al., 2007] Phan Minh Dung, Paolo Mancarella, and Francesca Toni. Computing ideal sceptical argumentation. *Artificial Intelligence*, 171(10-15):642–674, 2007.

[Dung et al., 2009] Phan Minh Dung, Bob Kowalski, and Francesca Toni. Assumption-based argumentation. In G. Simari and I. Rahwan, editors, *Argumentation in Artificial Intelligence*, pages 199–218. Springer US, 2009.

[Dung, 1991] Phan Minh Dung. Negations as hypotheses: An abductive foundation for logic programming. In Koichi Furukawa, editor, *Logic Programming, Proceedings of the Eigth International Conference, Paris, France, June 24-28, 1991*, pages 3–17. MIT Press, 1991.

[Dung, 1995a] Phan Minh Dung. An argumentation-theoretic foundation for logic programming. *The Journal of Logic Programming*, 22(2):151–177, 1995.

[Dung, 1995b] Phan Minh Dung. On the acceptability of arguments and its fundamental role in nonmonotonic reasoning, logic programming and n-person games. *Artificial Intelligence*, 77:321–357, 1995.

[Dvořák and Dunne, 2018] Wolfgang Dvořák and Paul Dunne. Computational problems in formal argumentation and their complexity. In P. Baroni, D. Gabbay, M. Giacomin, and L. van der Torre, editors, *Handbook of Formal Argumentation*, chapter 14, pages 631–687. College Publications, 2018. also appears in IfCoLog Journal of Logics and their Applications 4(8):2557–2622.

[Dvořák and Wallner, 2020] Wolfgang Dvořák and Johannes Wallner. Computing strongly admissible sets. In H. Prakken, S. Bistarelli, F. Santini, and C. Taticchi, editors, *Computational Models of Argument - Proceedings of COMMA 2020, Perugia, Italy, September 4-11, 2020*, volume 326 of *Frontiers in Artificial Intelligence and Applications*, pages 179–190. IOS Press, 2020.

[Dvořák et al., 2011] Wolfgang Dvořák, Sarah Gaggl, Johannes Wallner, and Stefan Woltran. Making use of advances in answer-set programming for abstract argumentation systems. In H. Tompits, S. Abreu, J. Oetsch, J. Pührer, D. Seipel, M. Umeda, and A. Wolf, editors, *Applications of Declarative Programming and Knowledge Management - 19th International Conference, INAP 2011, and 25th Workshop on Logic Programming, WLP 2011, Vienna, Austria, September 28-30, 2011, Revised Selected Papers*, volume 7773 of *Lecture Notes in Computer Science*, pages 114–133. Springer, 2011.

[Dvořák et al., 2015] Wolfgang Dvořák, Sarah Gaggl, Thomas Linsbichler, and Johannes Wallner. Reduction-based approaches to implement Modgil's extended argumentation frameworks. In Th. Eiter, H. Strass, M. Truszczynski, and S. Woltran, editors, *Advances in Knowledge Representation, Logic Programming, and Abstract Argumentation - Essays Dedicated to Gerhard Brewka on the Occasion of His 60th Birthday*, volume 9060 of *Lecture Notes in Computer Science*, pages 249–264. Springer, 2015.

[Dvořák et al., 2018] Wolfgang Dvořák, Alexander Greßler, and Stefan Woltran. Evaluating SETAFs via Answer-Set Programming. In *Proceedings of the Second International Workshop on Systems and Algorithms for Formal Argumentation*, pages 10–21. CEUR-WS.org, 2018.

[Dvořák et al., 2020a] Wolfgang Dvořák, Sarah Gaggl, Anna Rapberger, Johannes Wallner, and Stefan Woltran. The ASPARTIX system suite. In H. Prakken, S. Bistarelli, F. Santini, and C. Taticchi, editors, *Computational Models of Argument - Proceedings of COMMA 2020, Perugia, Italy, September 4-11, 2020*, volume 326 of *Frontiers in Artificial Intelligence and Applications*, pages 461–462. IOS Press, 2020.

[Dvořák et al., 2020b] Wolfgang Dvořák, Anna Rapberger, Johannes Wallner, and Stefan Woltran. ASPARTIX-V19 - an answer-set programming based system for abstract argumentation. In A. Herzig and J. Kontinen, editors, *Foundations of Information and Knowledge Systems - 11th International Symposium, FoIKS 2020, Dortmund, Germany, February 17-21, 2020, Proceedings*, volume 12012 of *Lecture Notes in Computer Science*, pages 79–89. Springer, 2020.

[Dvořák et al., 2023a] Wolfgang Dvořák, Alexander Greßler, Anna Rapberger, and Stefan Woltran. The complexity landscape of claim-augmented argumentation frameworks. *Artif. Intell.*, 317:103873, 2023.

[Dvořák et al., 2023b] Wolfgang Dvořák, Atefeh Keshavarzi Zafarghandi, and Stefan Woltran. Expressiveness of setafs and support-free adfs under 3-valued semantics. *J. Appl. Non Class. Logics*, 33(3-4):298–327, 2023.

[Dvořák et al., 2023c] Wolfgang Dvořák, Anna Rapberger, and Stefan Woltran. A claim-centric perspective on abstract argumentation semantics: Claim-defeat, principles, and expressiveness. *Artif. Intell.*, 324:104011, 2023.

[Dvořák and Woltran, 2020] Wolfgang Dvořák and Stefan Woltran. Complexity of abstract argumentation under a claim-centric view. *Artificial Intelligence*, 285:103290, 2020.

[Egly et al., 2008] Uwe Egly, Sarah Alice Gaggl, and Stefan Woltran. ASPARTIX: implementing argumentation frameworks using answer-set programming. In M. de la Banda and E. Pontelli, editors, *Logic Programming, 24th International Conference, ICLP 2008, Udine, Italy, December 9-13 2008, Proceedings*, volume 5366 of *Lecture Notes in Computer Science*, pages 734–738. Springer, 2008.

[Egly et al., 2010] Uwe Egly, Sarah Gaggl, and Stefan Woltran. Answer-set programming encodings for argumentation frameworks. *Argument & Computation*, 1(2):147–177, 2010.

[Ellmauthaler and Strass, 2014] Stefan Ellmauthaler and Hannes Strass. The

DIAMOND system for computing with abstract dialectical frameworks. In S. Parsons, N. Oren, Ch. Reed, and F. Cerutti, editors, *Computational Models of Argument - Proceedings of COMMA 2014, Atholl Palace Hotel, Scottish Highlands, UK, September 9-12, 2014*, volume 266 of *Frontiers in Artificial Intelligence and Applications*, pages 233–240. IOS Press, 2014.

[Ellmauthaler and Wallner, 2012] Stefan Ellmauthaler and Johannes Wallner. Evaluating abstract dialectical frameworks with ASP. In B. Verheij, S. Szeider, and S. Woltran, editors, *Computational Models of Argument - Proceedings of COMMA 2012, Vienna, Austria, September 10-12, 2012*, volume 245 of *Frontiers in Artificial Intelligence and Applications*, pages 505–506. IOS Press, 2012.

[Faber and Woltran, 2009] Wolfgang Faber and Stefan Woltran. Manifold answer-set programs for meta-reasoning. In E. Erdem, F. Lin, and T. Schaub, editors, *Logic Programming and Nonmonotonic Reasoning, 10th International Conference, LPNMR 2009, Potsdam, Germany, September 14-18, 2009. Proceedings*, volume 5753 of *Lecture Notes in Computer Science*, pages 115–128. Springer, 2009.

[Flouris and Bikakis, 2019] Giorgos Flouris and Antonis Bikakis. A comprehensive study of argumentation frameworks with sets of attacking arguments. *International Journal of Approximate Reasoning*, 109:55–86, 2019.

[Gaggl and Woltran, 2013] Sarah Alice Gaggl and Stefan Woltran. The cf2 argumentation semantics revisited. *Journal of Logic and Computation*, 23(5):925–949, 2013.

[Gaggl et al., 2015] Sarah Gaggl, Norbert Manthey, Alessandro Ronca, Johannes Wallner, and Stefan Woltran. Improved answer-set programming encodings for abstract argumentation. *Theory and Practice of Logic Programming*, 15(4-5):434–448, 2015.

[García and Simari, 2004] Alejandro García and Guillermo Simari. Defeasible logic programming: an argumentative approach. *Theory and Practice of Logic Programming*, 4(1):95–138, 2004.

[García and Simari, 2018] Alejandro Javier García and Guillermo Ricardo Simari. Argumentation based on logic programming. In *Handbook of formal argumentation*, pages 409–436. College Publications, 2018.

[Gebser et al., 2013] Martin Gebser, Benjamin Kaufmann, Javier Romero, Ramón Otero, Torsten Schaub, and Philipp Wanko. Domain-specific heuristics in answer set programming. In M. desJardins and M. Littman, editors, *Proceedings of the Twenty-Seventh AAAI Conference on Artificial Intelligence, July 14-18, 2013, Bellevue, Washington, USA*, pages 350–356. AAAI Press, 2013.

[Gebser et al., 2015] Martin Gebser, Roland Kaminski, Benjamin Kaufmann,

7 - LOGIC PROGRAMMING AND ARGUMENTATION

Marius Lindauer, Max Ostrowski, Javier Romero, Torsten Schaub, and Sven Thiele. *Potassco User Guide*. University of Potsdam, second edition edition, 2015.

[Gelfond and Lifschitz, 1991] Michael Gelfond and Vladimir Lifschitz. Classical negation in logic programs and disjunctive databases. *New Generation Computing*, 9(3/4):365–386, 1991.

[Gorogiannis and Hunter, 2011] Nikos Gorogiannis and Anthony Hunter. Instantiating abstract argumentation with classical logic arguments: Postulates and properties. *Artificial Intelligence*, 175(9-10):1479–1497, 2011.

[Hölldobler and Kencana Ramli, 2009] Steffen Hölldobler and Carroline Dewi Puspa Kencana Ramli. Logic programs under three-valued łukasiewicz semantics. In *Logic Programming: 25th International Conference, ICLP 2009, Pasadena, CA, USA, July 14-17, 2009. Proceedings 25*, pages 464–478. Springer, 2009.

[Kakas et al., 1994] A. C. Kakas, P. Mancarella, and Phan Minh Dung. The acceptability semantics for logic programs. In *Proceedings of the Eleventh International Conference on Logic Programming*, page 504–519, Cambridge, MA, USA, 1994. MIT Press.

[König et al., 2022] Matthias König, Anna Rapberger, and Markus Ulbricht. Just a matter of perspective. In F. Toni, S. Polberg, R. Booth, M. Caminada, and H. Kido, editors, *Computational Models of Argument - Proceedings of COMMA 2022, Cardiff, Wales, UK, 14-16 September 2022*, volume 353 of *Frontiers in Artificial Intelligence and Applications*, pages 212–223. IOS Press, 2022.

[Lehtonen et al., 2017] Tuomo Lehtonen, Johannes Wallner, and Matti Järvisalo. From structured to abstract argumentation: Assumption-based acceptance via AF reasoning. In A. Antonucci, L. Cholvy, and O. Papini, editors, *Symbolic and Quantitative Approaches to Reasoning with Uncertainty - 14th European Conference, ECSQARU 2017, Lugano, Switzerland, July 10-14, 2017, Proceedings*, volume 10369 of *Lecture Notes in Computer Science*, pages 57–68. Springer, 2017.

[Lehtonen et al., 2021a] Tuomo Lehtonen, Johannes Wallner, and Matti Järvisalo. Harnessing incremental answer set solving for reasoning in assumption-based argumentation. *Theory and Practice of Logic Programming*, 21(6):717–734, 2021.

[Lehtonen et al., 2021b] Tuomo Lehtonen, Johannes Wallner, and Matti Järvisalo. Declarative algorithms and complexity results for assumption-based argumentation. *Journal of Artificial Intelligence Research*, 71:265–318, 2021.

[Modgil and Prakken, 2013] Sanjay Modgil and Henry Prakken. A general account of argumentation with preferences. *Artificial Intellligence*, 195:361–397, 2013.

[Modgil and Prakken, 2014] Sanjay Modgil and Henry Prakken. The ASPIC+ framework for structured argumentation: a tutorial. *Argument & Computation*, 5:31–62, 2014. Special Issue: Tutorials on Structured Argumentation.

[Modgil and Prakken, 2018] Sanjay Modgil and Henry Prakken. Abstract rule-based argumentation. In *Handbook of Formal Argumentation*, volume 1. College Publications, 2018.

[Nielsen and Parsons, 2006] Søren Holbech Nielsen and Simon Parsons. A generalization of Dung's abstract framework for argumentation: Arguing with sets of attacking arguments. In *International Workshop on Argumentation in Multi-Agent Systems*, pages 54–73. Springer, 2006.

[Nieves et al., 2008] Juan Carlos Nieves, Mauricio Osorio, and Ulises Cortés. Preferred extensions as stable models. *Theory and Practice of Logic Programming*, 8(4):527–543, 2008.

[Osorio et al., 2010] Mauricio Osorio, Juan Carlos Nieves, and Ignasi Gómez-Sebastià. CF2-extensions as answer-set models. In P. Baroni, F. Cerutti, M. Giacomin, and G. Simari, editors, *Computational Models of Argument: Proceedings of COMMA 2010, Desenzano del Garda, Italy, September 8-10, 2010*, volume 216 of *Frontiers in Artificial Intelligence and Applications*, pages 391–402. IOS Press, 2010.

[Polberg, 2016] Sylwia Polberg. Understanding the abstract dialectical framework. In *European Conference on Logics in Artificial Intelligence*, pages 430–446. Springer, 2016.

[Przymusinski, 1990] Teodor Przymusinski. The well-founded semantics coincides with the three-valued stable semantics. *Fundamenta Informaticae*, 13(4):445–463, 1990.

[Sá and Alcântara, 2019] Samy Sá and João Alcântara. Interpretations and models for assumption-based argumentation. In *Proceedings of the 34th ACM/SIGAPP Symposium on Applied Computing*, pages 1139–1146, 2019.

[Sá and Alcântara, 2021a] Samy Sá and João Alcântara. An abstract argumentation and logic programming comparison based on 5-valued labellings. In *Symbolic and Quantitative Approaches to Reasoning with Uncertainty: 16th European Conference, ECSQARU 2021, Prague, Czech Republic, September 21–24, 2021, Proceedings 16*, pages 159–172. Springer, 2021.

[Sá and Alcântara, 2021b] Samy Sá and João Alcântara. Assumption-based argumentation is logic programming with projection. In *Symbolic and Quantitative Approaches to Reasoning with Uncertainty: 16th European Conference, ECSQARU 2021, Prague, Czech Republic, September 21–24, 2021, Proceedings 16*, pages 173–186. Springer, 2021.

[Saccà and Zaniolo, 1997] Domenico Saccà and Carlo Zaniolo. Deterministic and non-deterministic stable models. *Journal of Logic and Computation*, 7(5):555–579, 1997.

[Sakama and Rienstra, 2017] Chiaki Sakama and Tjitze Rienstra. Representing argumentation frameworks in answer set programming. *Fundamenta Informaticae*, 155(3):261–292, 2017.

[Schultz and Toni, 2014] Claudia Schultz and Francesca Toni. Complete assumption labellings. In *Proceedings of COMMA 2014*, pages 405–412, 2014.

[Schulz and Toni, 2017] Claudia Schulz and Francesca Toni. Labellings for assumption-based and abstract argumentation. *International Journal of Approximate Reasoning*, 84:110–149, 2017.

[Simari and Loui, 1992] Guillermo Simari and Ronald Loui. A mathematical treatment of defeasible reasoning and its implementation. *Artificial Intelligence*, 53:125–157, 1992.

[Strass, 2013] Hannes Strass. Approximating operators and semantics for abstract dialectical frameworks. *Artificial Intelligence*, 205:39–70, 2013.

[Toni and Sergot, 2011] Francesca Toni and Marek Sergot. Argumentation and answer set programming. In Marcello Balduccini and Tran Cao Son, editors, *Logic Programming, Knowledge Representation, and Nonmonotonic Reasoning: Essays Dedicated to Michael Gelfond on the Occasion of His 65th Birthday*, pages 164–180. Springer, 2011.

[Toni, 2014] Francesca Toni. A tutorial on assumption-based argumentation. *Argument & Computation*, 5:89–117, 2014. Special Issue: Tutorials on Structured Argumentation.

[Wakaki and Nitta, 2008] Toshiko Wakaki and Katsumi Nitta. Computing argumentation semantics in answer set programming. In H. Hattori, T. Kawamura, T. Idé, M. Yokoo, and Y. Murakami, editors, *New Frontiers in Artificial Intelligence, JSAI 2008 Conference and Workshops, Asahikawa, Japan, June 11-13, 2008, Revised Selected Papers*, volume 5447 of *Lecture Notes in Computer Science*, pages 254–269. Springer, 2008.

[Weydert, 2011] Emil Weydert. Semi-stable extensions for infinite frameworks. In P. de Causmaecker, J. Maervoet, T. Messelis, K. Verbeeck, and T. Vermeulen, editors, *Proceedings of the 23rd Benelux Conference on Artificial Intelligence (BNAIC 2011)*, pages 336–343, 2011.

[Wu et al., 2009] Yining Wu, Martin Caminada, and Dov Gabbay. Complete extensions in argumentation coincide with 3-valued stable models in logic programming. *Studia Logica*, 93(1-2):383–403, 2009. Special issue: new ideas in argumentation theory.

CHAPTER 8

THE SEMANTICAL STRUCTURE OF CONDITIONALS, AND ITS RELATION TO FORMAL ARGUMENTATION

JESSE HEYNINCK
Open Universiteit, the Netherlands
jesse.heyninck@ou.nl

GABRIELE KERN-ISBERNER
TU Dortmund, Germany
gabriele.kern-isberner@cs.tu-dortmund.de

TJITZE RIENSTRA
Maastricht University, the Netherlands
t.rienstra@maastrichtuniversity.nl

KENNETH SKIBA
FernUniversität in Hagen, Germany
kenneth.skiba@fernuni-hagen.de

MATTHIAS THIMM
FernUniversität in Hagen, Germany
matthias.thimm@fernuni-hagen.de

1 Introduction

The study of conditionals has a long tradition in philosophy. This has resulted, among others, in a thorough and expansive formal study of conditionals, with a particular focus on semantical foundations of conditionals (see [Nute, 1984, 1980] for an overview, and Section 3 for a summary). The basic idea underlying these semantics is that a conditional $(\psi|\phi)$ is accepted if ψ is true in a subset of ϕ-worlds. The details of this selection depend on the specific semantics. For example, nonmonotonic conditionals of the form "if ϕ then typically ψ" define this selection in terms of the most plausible worlds [Kraus et al., 1990a].

The relation between argumentative formalisms and conditionals has been on the agenda of the computational argumentation community since its inception. On an intuitive level, an argument can be accepted only if it is sufficiently defended or supported, i.e., it seems that argumentation allows for a conditional interpretation. However, the investigation of connections between argumentation and conditionals on a more formal or semantical level provide a more nuanced perspective. It is well-known that argumentation and nonmonotonic resp. default logics are closely connected: In [Dung, 1995] it is shown that Reiter's default logic can be implemented by abstract argumentation frameworks, a most basic form of computational model of argumentation to which many existing approaches to formal argumentation refer. On the other hand, it is clear that argumentation allows for nonmonotonic, defeasible reasoning, and in [Rienstra et al., 2015a] computational models of argumentation are assessed by formal properties that have been adapted from nonmonotonic logics. Furthermore, answer set programming [Gelfond and Leone, 2002] as one of the most successful nonmonotonic logics has often been used to implement argumentation [Egly et al., 2010; Cerutti et al., 2017]. Nevertheless, argumentation and nonmonotonic reasoning are perceived as two different fields which do not subsume each other, and indeed, often attempts to transform reasoning systems from one side into systems of the other side have been revealing gaps that could not be closed (cf., e.g., [Thimm and Kern-Isberner, 2008b; Kern-Isberner and Simari, 2011; Heyninck, 2019]). While one might argue that this is due to the seemingly richer, dialectical structure of argumentation, in the end the evaluation of arguments often boils down to comparing arguments with their attackers, and comparing degrees of belief is a basic operation in qualitative nonmonotonic reasoning. In this chaper, we give a thorough introduction to the semantics of conditionals and provide an overview of work done on the comparison or incorporation of conditional semantics and argumentation.

Outline of this Chapter In Section 2, we introduce the necessary preliminaries on propositional logic (Section 2.1), Kleene's Three-valued logic (Section 2.2) and abstract dialectical frameworks (Section 2.3). We provide an overview of the semantics of conditionals in Section 3, looking at semantics using selection functions (Section 3.1), systems of spheres (Section 3.2) and preferential models (3.3), while also pointing out connections with belief dynamics (Section 3.4). In Section 4, we look at work that investigates syntactic similarities between conditionals and argumentative formalisms. In Section 5, we overview approaches integrating conditional semantics in argumentative formalisms. In Section 5.3, we look at connections that have been made between structured accounts of argumentation and conditonal logics. In Section 6, we summarize further works that thematise conditionals and argumentation. A summary and outlook is provided in Section 7.

2 Preliminaries

In this section, we introduce the necessary preliminaries on propositional logic (Section 2.1), Kleene's Three-valued logic (Section 2.2) and abstract dialectical frameworks (Section 2.3).

2.1 Propositional Logic

For a (finite) set At of atoms let $\mathcal{L}(\mathsf{At})$ be the corresponding propositional language constructed using the usual connectives $\wedge$ (*and*), $\vee$ (*or*), $\neg$ (*negation*) and $\rightarrow$ (*material implication*). A (classical) *interpretation* (also called *possible world*) ω for a propositional language $\mathcal{L}(\mathsf{At})$ is a function $\omega : \mathsf{At} \rightarrow \{\mathsf{T}, \mathsf{F}\}$. Let $\Omega(\mathsf{At})$ denote the set of all interpretations for At. We simply write Ω if the set of atoms is implicitly given. An interpretation ω *satisfies* (or is a *model* of) an atom $a \in \mathsf{At}$, denoted by

$\omega \models a$, if and only if $\omega(a) = \mathsf{T}$. The satisfaction relation $\models$ is extended to formulas as usual. As an abbreviation we sometimes identify an interpretation ω with its *complete conjunction*, i.e., if $a_1, \ldots, a_n \in \mathsf{At}$ are those atoms that are assigned T by ω and $a_{n+1}, \ldots, a_m \in \mathsf{At}$ are those propositions that are assigned F by ω we identify ω by $a_1 \ldots a_n \overline{a_{n+1}} \ldots \overline{a_m}$ (or any permutation of this). For example, the interpretation ω_1 on $\{a, b, c\}$ with $\omega(a) = \omega(c) = \mathsf{T}$ and $\omega(b) = \mathsf{F}$ is abbreviated by $a\bar{b}c$. For $\Phi \subseteq \mathcal{L}(\mathsf{At})$ we also define $\omega \models \Phi$ if and only if $\omega \models \phi$ for every $\phi \in \Phi$. We define the set of models $\mathsf{Mod}(X) = \{\omega \in \Omega(\mathsf{At}) \mid \omega \models X\}$ for every formula or set of formulas X. A formula or set of formulas X_1 *entails* another formula or set of formulas X_2, denoted by $X_1 \vdash X_2$, if $\mathsf{Mod}(X_1) \subseteq \mathsf{Mod}(X_2)$.

2.2 Kleene's Three-Valued Logic

Due to the three-valued nature of ADFs, we will need a three-valued logic to use as a basic logic underlying revision. Due to its high expressivity, we use Kleene's three-valued logic. A 3-valued interpretation for a set of atoms At is a function $v : \mathsf{At} \to \{\mathsf{T}, \bot, u\}$, which assigns to each atom in At either the value T (true, accepted), $\bot$ (false, rejected), or u (unknown). The set of all three-valued interpretations for a set of atoms At is denoted by $\mathcal{V}(\mathsf{At})$. We sometimes denote an interpretation $v \in \mathcal{V}(\{x_1, \ldots, x_n\})$ by $\dagger_1 \ldots \dagger_n$ with $v(x_i) = \dagger_i$ and $\dagger_i \in \{\mathsf{T}, \bot, u\}$, e.g., TT denotes $v(a) = v(b) = \mathsf{T}$ for $\mathsf{At} = \{a, b\}$. A 3-valued interpretation v can be extended to arbitrary propositional formulas $\phi \in \mathcal{L}(\mathsf{At})$ via the truth tables in Table 1. We furthermore extend the language with a second, *weak negation* $\sim$, which is evaluated to true if the negated formula is false or undecided (i.e. there is no positive information for the negated formula). Thus, $\sim\phi$ means that no explicit information for ϕ being true ($v(\phi) \neq \mathsf{T}$) is given, whereas $\neg\phi$ means that ϕ is false ($v(\phi) = \bot$). The truth table for $\sim$ can also be found in Table 1.[1]

It will prove convenient to define the connective $\odot$ which stipulates

---

[1] In the terminology of [Urquhart, 2001], the negation $\sim$ corresponds to Bochvar's *external negation* [Bochvar and Bergmann, 1981] and $\neg$ corresponds to Kleene's negation in his three-valued logic. $\sim$ is also referred to as Kleene's *weak negation* [Varzi and Warglien, 2003], since the conditions for $\sim\phi$ being satisfied are weaker than those for $\neg\phi$ being satisfied (i.e. $\{\neg\phi\} \models_\mathsf{K} \sim\phi$).

	$\neg$	$\sim$	$\odot$
$\top$	$\bot$	$\bot$	$\bot$
u	u	$\top$	$\top$
$\bot$	$\top$	$\top$	$\bot$

$\wedge$	$\top$	u	$\bot$
$\top$	$\top$	u	$\bot$
u	u	u	$\bot$
$\bot$	$\bot$	$\bot$	$\bot$

$\vee$	$\top$	u	$\bot$
$\top$	$\top$	$\top$	$\top$
u	$\top$	u	u
$\bot$	$\top$	u	$\bot$

Table 1: Truth tables for connectives in Kleene's K

a formula is undecided. We define $\odot \phi = \sim(\neg \phi \vee \phi)$. We define $\mathcal{L}^K(\mathsf{At})$ as the language based on At, the unary connectives $\langle \neg, \sim, \odot \rangle$ and the binary connectives $\langle \wedge, \vee, \rightarrow \rangle$.

The following facts about $\sim$, which show some similarities between $\sim$ and classical negation, will prove useful below:

Fact 1. *For any $\phi \in \mathcal{L}^K(\mathsf{At})$ and any $v \in \mathcal{V}(\mathsf{At})$: (1) $v(\sim \phi) \neq u$, and (2) $v(\sim \sim \phi) = \top$ iff $v(\phi) = \top$.*

We can show that $\odot$ expresses the undecidedness of any formula $\phi \in \mathcal{L}^K$:

Fact 2. *For any $\phi \in \mathcal{L}^K(\mathsf{At})$, $v(\odot \phi) = \top$ iff $v(\phi) = u$.*

We define the set of three-valued interpretations that satisfy a formula $\phi \in \mathcal{L}^K(\mathsf{At})$ as $\mathcal{V}(\phi) = \{v \in \mathcal{V}(\mathsf{At}) \mid v(\phi) = \top\}$. A formula X_1 K-*entails* another formula X_2, denoted $X_1 \models_K X_2$, if $\mathcal{V}(X_1) \subseteq \mathcal{V}(X_2)$. $X_1 \equiv_K X_2$ iff $X_1 \models_K X_2$ and $X_2 \models_K X_1$.

Given an interpretation $v \in \mathcal{V}(\mathsf{At})$, we define:

$$\mathsf{form}(v) = \bigwedge_{v(a)=\top} a \wedge \bigwedge_{v(a)=\bot} \neg a \wedge \bigwedge_{v(a)=u} \odot a$$

Clearly, $\mathsf{form}(v)$ expresses exactly the beliefs expressed by a three-valued interpretation:

Fact 3. *For any $v \in \mathcal{V}(\mathsf{At})$ and any $a \in \mathsf{At}$: (1) $\mathsf{form}(v) \models_K a$ iff $v(a) = \top$; (2) $\mathsf{form}(v) \models_K \neg a$ iff $v(a) = \bot$; (3) $\mathsf{form}(v) \models_K \odot a$ iff $v(a) = u$.*

2.3 Abstract Dialectical Frameworks

We briefly recall some technical details on ADFs following loosely the notation from [Brewka et al., 2013]. An ADF D is a tuple $D = (\mathsf{At}, L, C)$ where At is a finite set of atoms, $L \subseteq \mathsf{At} \times \mathsf{At}$ is a set of links, and $C = \{C_s\}_{s \in \mathsf{At}}$ is a set of total functions (also called acceptance functions) $C_s : 2^{par_D(\mathsf{At})} \to \{\top, \bot\}$ for each $s \in \mathsf{At}$ with $par_D(s) = \{s' \in \mathsf{At} \mid (s', s) \in L\}$. An acceptance function C_s defines the cases when the statement s can be accepted (truth value $\top$), depending on the acceptance status of its parents in D. By abuse of notation, we will often identify an acceptance function C_s by its equivalent *acceptance condition* which models the acceptable cases as a propositional formula. In more detail, C_s expresses the conditions that are to be accepted for s to be accepted. $\mathfrak{D}(\mathsf{At})$ denotes the set of all ADFs $D = (\mathsf{At}, L, C)$.

Example 1. *We consider the following ADF $D_1 = (\{a, b, c\}, L, C)$ with $L = \{(a, b), (b, a), (a, c), (b, c)\}$ and $C_a = \neg b$, $C_b = \neg a$ and $C_c = \neg a \vee \neg b$. Informally, the acceptance conditions can be read as "a is accepted if b is not accepted", "b is accepted if a is not accepted" and "c is accepted if a is not accepted or b is not accepted".*

An ADF $D = (\mathsf{At}, L, C)$ is interpreted through 3-valued interpretations $\mathcal{V}(\mathsf{At})$ (see Section 2.2). Recall that $\Omega(\mathsf{At})$ consists of all the two-valued interpretations (i.e. interpretations such that for every $s \in \mathsf{At}$, $v(s) \in \{\top, \bot\}$). We define the information order $\leq_i$ over $\{\top, \bot, u\}$ by making u the minimal element: $u <_i \top$ and $u <_i \bot$ and this order is lifted pointwise as follows (given two valuations v, w over At): $v \leq_i w$ iff $v(s) \leq_i w(s)$ for every $s \in \mathsf{At}$. The set of two-valued interpretations extending a valuation v is defined as $[v]^2 = \{w \in \Omega \mid v \leq_i w\}$. Given a set of valuations V, we denote with $\sqcap_i V$ the valuation defined by $\sqcap_i V(s) = v(s)$ if for every $v' \in V$, $v(s) = v'(s)$ and $\sqcap_i V(s) = u$ otherwise. $\Gamma_D : \mathcal{V}(\mathsf{At}) \mapsto \mathcal{V}(\mathsf{At})$ is defined as $\Gamma_D(v)(s) = \sqcap_i [v]^2(C_s)$. Intuitively, $\Gamma_D(v)$ assigns to an atom s the consensus of the truth values assigned by all completions of v to C_s.

For the definition of the stable model semantics, we need to define the reduct D^v of D given v, defined as: $D^v = (\mathsf{At}^v, L^v, C^v)$ with: (1) $L^v = L \cap (\mathsf{At}^v \times \mathsf{At}^v)$, and (2) $C^v = \{C_s[\{\phi \mid v(\phi) = \bot\}/\bot] \mid s \in \mathsf{At}^v\}$, where $C_s[\phi/\psi]$ is the formula obtained by substituting every occurrence

8 - CONDITIONALS AND ARGUMENTATION

of ϕ in C_s by ψ.

Definition 2.1. *Let $D = (\mathsf{At}, L, C)$ be an ADF with $v : \mathsf{At} \to \{\top, \bot, u\}$ an interpretation:*

- *v is a 2-valued model iff $v \in \Omega$ and $v(s) = v(C_s)$ for every $s \in \mathsf{At}$.*
- *v is admissible for D iff $v \leq_i \Gamma_D(v)$.*
- *v is complete for D iff $v = \Gamma_D(v)$.*
- *v is preferred for D iff v is $\leq_i$-maximally complete.*
- *v is grounded for D iff v is $\leq_i$-minimally complete.*
- *v is stable iff v is a model of D and $\{s \in \mathsf{At} \mid v(s) = \top\} = \{s \in \mathsf{At} \mid w(s) = \top\}$ where w is the grounded interpretation of D^{v}[2].*

With 2val(D), admissible(D), complete(D), prf(D), grounded(D), respectively stable(D) we denote the sets of two-valued, admissible, complete, preferred, grounded, respectively stable interpretations of D.

We finally define inference relations for ADFs:

Definition 2.2. *Given Sem $\in \{\text{prf}, \text{grounded}, \text{2val}, \text{stable}\}$, an ADF $D = (\mathsf{At}, L, C)$ and $\phi \in \mathcal{L}^K(\mathsf{At})$ we define: $D \mathrel{\vert\kern-0.3em\sim}^{\cap}_{\text{Sem}} \phi$ iff $v(\phi) = \top$ for all $v \in \text{Sem}(D)$.*

Example 2 (Example 1 continued). *The ADF of Example 1 has three complete models v_1, v_2, v_3 with:*

$$\begin{array}{lll} v_1(a) = \top & v_1(b) = \bot & v_1(c) = \top \\ v_2(a) = \bot & v_2(b) = \top & v_2(c) = \top \\ v_3(a) = u & v_3(b) = u & v_3(c) = u \end{array}$$

v_3 is the grounded interpretation whereas v_1 and v_2 are both preferred, two-valued and stable models.

---

[2][Brewka et al., 2013] has show the grounded interpretation is uniquely defined for any ADF.

Restricting ADFs to certain sub-classes based on the syntactic form of the acceptance conditions leads to representation of existing argumentative formalisms. One such formalism are the well known *Abstract argumentation frameworks* [Dung, 1995] where the only argumentative relation formalised is the one of attacks between arguments. In that case, acceptance conditions C_a are restricted to conjunctions of negations $\neg b_1 \wedge \ldots \wedge \neg b_n$, intuitively representing the attacks on the argument. For completeness, we include also the traditional definition of an argumentation framework:

Definition 2.3. *An abstract argumentation framework (AF) is a directed graph $AF = (A, R)$ where A is a finite set of* arguments *and R is an* attack relation $R \subseteq A \times A$.

For an AF $AF = (A, R)$, an argument a is said to *attack* an argument b if $(a, b) \in R$. We say that, an argument a is *defended by a set* $E \subseteq A$ if every argument $b \in A$ that attacks a is attacked by some $c \in E$. For $a \in A$ we define

$$a^-_{AF} = \{b \mid (b, a) \in R\} \quad \text{and} \quad a^+_{AF} = \{b \mid (a, b) \in R\}.$$

In other words, a^-_{AF} is the set of attackers of a and a^+_{AF} is the set of arguments attacked by a. For a set of arguments $E \subseteq A$ we extend these definitions to E^+_{AF} and E^-_{AF} via $E^+_{AF} = \bigcup_{a \in E} a^+_{AF}$ and $E^-_{AF} = \bigcup_{a \in E} a^-_{AF}$, respectively. If the AF is clear in the context, we will omit the index.

An argumentation framework AF can be represented as the ADF $D_{AF} = (A, C)$ where $C_a = \bigwedge_{b \in a^-} \neg b$ for every $a \in A$. In that case, all of the traditional *extension-based semantics* [Dung, 1995] coincide with the ADF-semantics. It is interesting to notice furthermore that two-valued models and stable models coincide in the case of ADFs based on AFs. We will also call the sets of arguments labelled T according to a certain kind of labelling as the respective extension. For example, if the grounded labelling assigns T to the arguments a and c, we say that $\{a, c\}$ is the grounded extension.

We also mention here the notions of conflict-freeness and admissibility:

Definition 2.4. *Given $AF = (A, R)$, a set $E \subseteq A$ is*

8 - CONDITIONALS AND ARGUMENTATION

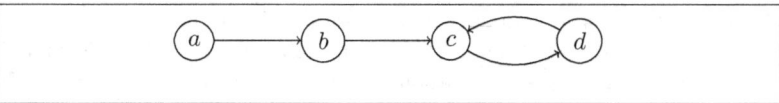

Figure 1: Abstract argumentation framework AF from Example 3.

- conflict-free *iff $\forall a, b \in E$, $(a,b) \notin R$;*
- admissible *iff it is conflict-free and it defends its elements.*

We use $cf(AF)$ and $ad(AF)$ for denoting the sets of conflict-free and admissible sets of an argumentation framework F, respectively.

Example 3. *Let $AF = (\{a,b,c,d\}, \{(a,b), (b,c), (c,d), (d,c)\})$ be an AF depicted as a directed graph in Figure 1. The sets $\{a,c\}$ and $\{a,d\}$ are the complete, preferred and stable extensions. While $\{a\}$ is the grounded extension.*

3 Semantics of conditionals

The study of the semantics of conditionals is concerned with statements of the form "if ϕ then ψ" as they are used in natural language. Several conditional logics have been developed with the aim of providing a semantics for conditional statements, and to study their properties. The aim of this section is to provide an introduction to the topic and an overview of the main approaches.

Of all the distinctions we can make among the types of conditionals that we use in everyday language, the most crucial distinction is that of *indicative* and *subjunctive* conditionals. While indicative conditionals make statements about what holds in the actual world, subjunctive conditionals make statements about hypothetical situations. The following example, due to [Adams, 1970], illustrates the difference.

1. If Oswald didn't kill Kennedy, then someone else did.

2. If Oswald hadn't killed Kennedy, then someone else would have.

Conditional (1) is an indicative conditional. It refers to the actual world where Oswald either did or did not kill Kennedy. It states that, in case

Oswald did not kill Kennedy, someone else killed him. Conditional (2) is a subjunctive conditional. It presumes that Oswald did in fact kill Kennedy, and makes a claim about the hypothetical situation in which Oswald did not kill Kennedy. Subjunctive conditionals are also referred to as *counterfactuals*. Clearly, despite the similarities between (1) and (2), they make two very different claims. While it is quite reasonable believe that (1) is true, the truth of (2) is more contentious.

Indicative conditionals may, as a first approximation, be interpreted as material implications in propositional logic. According to this interpretation, the conditional 'if ϕ then ψ" ($\phi \to \psi$) is true unless ϕ is true and ψ is false. While this definition provides an adequate interpretation for conditionals as they are used in mathematical proofs, it is not satisfactory for indicative conditionals as they are used in natural language. This is due to a number of unintuitive consequences of the definition, such as that $\phi \to \psi$ is implied by ψ and by $\neg\phi$, that $\neg(\phi \to \psi)$ implies ϕ, and that for any ϕ and ψ, either $\phi \to \psi$ or $\psi \to \phi$ is true. Truth-functionality represents another issue. The material implication is truth-functional, since the truth value of $\phi \to \psi$ is a function of the truth values of ϕ and of ψ. Conditionals as they are used in natural language are not truth-functional. To see why, consider the sentences "John is happy" and "Mary is happy". Knowing the truth values of these two sentences does not imply that we know the truth value of the conditional "If Mary is happy then John is happy".

The study of the semantics of conditionals is driven by the need for more sophisticated accounts of the relationship between premises and conclusions of conditional statements. In the following subsections, we look at the two main semantical accounts of conditionals that have been developed since the sixties of the last century: semantics using selection functions (Section 3.1) and semantics using systems of spheres (Section 3.2). Thereafter, we look at the main semantical account of non-monotonic conditionals, namely the preferential models (3.3) and survey connections with belief dynamics (Section 3.4).

3.1 Selection Functions

The *selection function* approach, due to [Stalnaker, 1968] and further developed by [Lewis, 1973b], represents one of the central ideas in the study of the semantics of conditionals. This approach is based on the idea that a conditional $\phi > \psi$ is true whenever ψ is true in the possible world where ϕ is true and which differs minimally from the actual world. The semantics is defined in terms of a *selection function* that represents a criterion to select such a possible world for any given antecedent ϕ and actual world w. Let L be a propositional language that, in addition to the usual propositional connectives, is closed under the conditional operator $>$. We will present the simplified formalisation of Stalnaker's semantics due to Nute [Nute, 1984]. A *Stalnaker model* is a quadruple $(I, R, s, [\cdot])$ where I is a set of possible worlds; $R \subseteq I \times I$ a binary reflexive accessibility relation; s a selection function; and $[\cdot]$ assigns to each sentence $\phi \in L$ a subset $[\phi]$ of I. The selection function s is a partial function that, if defined, assigns to a sentence ϕ and world $w \in I$ a world $s(\phi, w) \in I$. A selection function must satisfy the following conditions, which intuitively ensure that $s(\phi, w)$ can indeed be regarded as the world where ϕ is true that differ minimally from w.

1. $s(\phi, w) \in [\phi]$,

2. $(i, s(\phi, w)) \in R$,

3. If $s(\phi, w)$ is undefined then for all $w' \in I$ s.t. $(w, w') \in R$, $w' \notin [\phi]$,

4. If $w \in [\phi]$ then $s(\phi, w) = w$,

5. If $s(\phi, w) \in [\psi]$ and $s(\psi, w) \in [\phi]$, then $s(\phi, w) = s(\psi, w)$

6. $w \in [\phi > \psi]$ iff $s(\phi, w) \in [\psi]$ or $s(\phi, w)$ is undefined.

Example 4. *As an example, consider the model* $I = \{bf, b\overline{f}, \overline{b}f, \overline{b}\overline{f}\}$ *with* $R = I \times I$ *and s partially defined by* $s(b, b\overline{f}) = b\overline{f}$ *and* $s(b, w) = bf$ *for every* $w \in I \setminus \{b\overline{f}\}$. *Then we see that* $[b > f] = \{bf, \overline{b}f, \overline{b}\overline{f}\}$ *and thus $b > f$ is not true in $b\overline{f}$ but true in all other worlds.*

Given a Stalnaker model $(I, R, s, [\cdot])$, the conditional $\phi > \psi$ is true in world w whenever ψ is true in world $s(\phi, w)$ (more formally: $s(\phi, w) \in [\psi]$). The resulting logic, which Stalnaker refers to as **C2**, consists of those formulas that are true in every world of every model. The following set of properties provides an axiomatization of **C2**. More precisely, the logic **C2** coincides with the smallest set of formulas that is closed under the following two inference rules.

(RCEC) If $\phi \leftrightarrow \psi$ then $(\chi > \phi) \leftrightarrow (\chi > \psi)$.
(RCK)   If $(\phi_1 \wedge \cdots \wedge \phi_n) \to \psi$ then
$$((\chi > \phi_1) \wedge \cdots \wedge (\chi > \phi_n)) \to (\chi \to \psi) \text{ (for } n \geq 0).$$

and contains all instances of the axioms:

(ID) $\phi > \phi$
(MP) $(\phi > \psi) \to (\phi \to \psi)$
(MOD) $(\neg \phi > \phi) \to (\psi > \phi)$
(CSO) $((\phi > \psi) \wedge (\psi > \phi)) \to ((\phi > \chi) \leftrightarrow (\psi > \chi))$
(CV) $((\phi > \psi) \wedge \neg(\phi > \neg\chi)) \to ((\phi \wedge \chi) > \psi)$
(CEM) $(\phi > \psi) \vee (\phi > \neg\psi)$

The main point of contention in Stalnaker's account concerns the **CEM** (Conditional Excluded Middle) axiom. This axiom states that, for every ϕ and ψ, either $\phi > \psi$ holds or $\phi > \neg\psi$ holds. Lewis offers the following counterexample to **CEM** [Lewis, 1973a]: Let A stand for "Bizet and Verdi are compatriots", F for "Bizet and Verdi are French", and I for "Bizet and Verdi are Italian". According to Lewis, we may well accept the conditional $A > F \vee I$ but reject both $A > F$ and $A > I$. If we accept $A > F \vee I$, however, then **CEM** forces us to accept either $A > F$ or $A > I$. **CEM** is closely related to what Nute refers to as the *Uniqueness Assumption* in Stalnaker's semantics [Nute, 1984]: for every antecedent ϕ and world w, there is exactly *one* world where ϕ is true and which differs minimally from w. Dropping **CEM** amounts to letting the selection function be a function that maps every antecedent ϕ and world w to a *set* of possible worlds. This option was pursued by [Lewis, 1973a]. He formalises a logic similar to Stalnaker's except that a conditional $\phi > \psi$ is taken to be true in world w if ψ is true in *all worlds* where ϕ is true and which differ minimally from w. The resulting logic, which Lewis calls **VC**, is axiomatised by the same set of inference rules and axioms as that of **C2** outlined above, except that CEM is replaced with CS (Conjunctive Sufficiency):

(CS) $(\phi \wedge \psi) \to (\phi > \psi)$

While this logic drops the Uniqueness Assumption, it still relies on the questionable *Limit Assumption*: for every antecedent ϕ and world

w, there is *at least one* world where ϕ is true and which differs minimally from w. Lewis' *system of spheres* semantics, described in the next section, does not rely on the limit assumption.

3.2 Systems of Spheres

Recall that, according to the selection function account, $\phi > \psi$ is true in world w whenever ψ is true in all worlds where ϕ is true and which differ minimally from w. Recall that the Limit Assumption requires that, for any antecedent ϕ and world w, there is at least one world where ϕ is true that differs minimally from w. Lewis points out that this assumption disagrees with situations where worlds get closer and closer to the actual world without end. This may happen if we consider antecedents such as "I am over 7 feet tall", where for any possible world where I am $7 + \epsilon$ feet tall, there is an even closer possible world where I am $7 + \epsilon/2$ tall [Lewis, 1973a]. Lewis' *system of spheres* semantics provides an alternative semantics for conditionals that does not rely on the Limit Assumption, yet is characterised by the same axioms as the logic **VC** described above [Lewis, 1973a]. It is based on the idea that the conditional $\phi > \psi$ is true in world w whenever some world where both ϕ and ψ are true is closer then every world where ϕ and $\neg\psi$ are true. The formalisation of this idea requires a relative notion of closeness between worlds. A *sphere* around a world w is a set S that contains w and all worlds that are closer to w than every world not in S. A *system of spheres model* is a triple $(I, \$, [\cdot])$ where I and $[\cdot]$ are defined as before, and \$ maps each $w \in I$ to a nested set $\$_w$ of spheres around w. We can compare worlds according to their closeness to a world w as follows: if there is a sphere $S \in \$_w$ such that $w' \in S$ and $w'' \notin S$ then w' is more similar to w than w''. Given the model $(I, \$, [\cdot])$, the conditional $\phi > \psi$ is true in world w whenever either $\bigcup \$_w \cap [\phi]$ is empty, or there is an $S \in \$_w$ such that $S \cap [\phi]$ is not empty and $S \cap [\phi] \subseteq [\psi]$.

Example 5. *Consider the system of spheres (partially) defined by:* $\$_w = \{w\}, \{w, bf, \overline{b}\,\overline{f}, \overline{b}f\}, I$. *Then we see that in every world besides $b\overline{f}$, $b > f$ is true. In more detail, consider e.g. the world $\overline{b}\,\overline{f}$. As there is a sphere $\{bf, \overline{b}f, \overline{b}f\} \in \$_{\overline{b}\,\overline{f}}$ s.t. $\{bf, \overline{b}\,\overline{f}, \overline{b}f\} \cap [b] \subseteq [f]$. More informally, the b-world closest to $\overline{b}\,\overline{f}$ is also an f-world, and thus, in $\overline{b}\,\overline{f}$, the conditional*

$b > f$ is true.

3.3 Preferential Model Semantics

The preferential model semantics of [Kraus et al., 1990b] and [Lehmann and Magidor, 1992b] represents yet another approach to reasoning with conditionals. The main purpose of their approach, however, is to provide a semantics for non-monotonic consequence relations. A non-monotonic consequence relation is a relation $\mathrel|\!\sim$ between propositions having as its main characteristic that it violates, unlike the classical $\vdash$, the Monotony property:

(**Monotony**) If $\phi \mathrel|\!\sim \psi$ then $\phi \wedge \chi \mathrel|\!\sim \psi$.

Monotony means that we never retract conclusions when further information becomes known. However, in common sense reasoning, we often do so. We may, for instance, conclude that birds fly ($bird \mathrel|\!\sim flies$) but retract this conclusion if we learn that the bird in question is a penguin ($bird \wedge penguin \mathrel|\!\not\sim flies$). The preferential model approach represents one of the most influential approaches to the general problem of non-monotonic reasoning.

The connection between non-monotonic consequence relations and conditionals lies in the fact that we can regard a consequence relation $\mathrel|\!\sim$ as a "flat" (i.e., not allowing nested conditionals) conditional logic: $\phi > \psi$ if and only if $\phi \mathrel|\!\sim \psi$. Furthermore, as we will see, several properties considered in the preferential model approach correspond to properties that are discussed in the context of conditional logics. Monotony is the first example. [Nute, 1984] calls it *Strengthening Antecedents* and dismisses it as invalid for any logic of subjunctive conditionals.

We will now provide an overview of the approach. We start with the model theory, which provides a semantics for non-monotonic inference relations. These models consist of a preference relation $\prec$ over states, where each state is labelled with a set of possible worlds. The preference relation can be thought of as an agent's belief about the relative degree of normality of states: if $s \prec s'$ then state s is more normal than state s'. The agent is willing to conclude ψ from ϕ if all most preferred states that satisfy ϕ also satisfy ψ. There are four classes of such models, each putting additional restrictions on the preference relation or state

8 - CONDITIONALS AND ARGUMENTATION

labelling. *Cumulative models* form the most general class:

Definition 3.1. *A cumulative model over a set V of valuations is a triple $W = (S, \prec, l)$, where S is a set containing elements called states, $\prec$ is a binary relation over S, l is a function mapping every state $s \in S$ to a non-empty set $l(s) \subseteq V$, $(S, \prec, l)$ satisfies the smoothness condition defined below. For every formula $\phi \in$ lang we define $\widehat{\phi}$ by $\widehat{\phi} = \{s \in S \mid \forall v \in l(s), v \models \phi\}$. A state s is said to be $\prec$-minimal in a set $X \subseteq S$ iff $s \in X$ and there is no $s' \in X$ such that $s' \prec s$. Furthermore, W is called finite iff S is finite.*

The smoothness condition is related to the Limit Assumption discussed in Section 3.1. It ensures that, for every formula ϕ, it is possible to determine the preferred states in ϕ.

Definition 3.2. *A triple $(S, \prec, l)$ satisfies the smoothness condition iff for all $\phi \in$ lang and $s \in \widehat{\phi}$, either s is $\prec$-minimal in $\widehat{\phi}$, or there is some $s' \in \widehat{\phi}$ such that s' is $\prec$-minimal in $\widehat{\phi}$ and $s' \prec s$.*

The following definition defines three restricted classes of cumulative models. Ordered and preferential models were defined by [Kraus *et al.*, 1990b]. Ranked models were defined by [Lehmann and Magidor, 1992b].

Definition 3.3. *A cumulative model $W = (S, \prec, l)$ is:*

- ordered *if $\prec$ is a strict partial order.*

- preferential *if it is ordered and for all $s \in S$, $l(s)$ is a singleton.*

- ranked *if it is preferential and there exists a mapping $R : S \to \mathbb{N}$ such that $s \prec s'$ iff $R(s) < R(s')$.*

Definition 3.4. *A triple $W = (S, \prec, l)$ determines a consequence relation (denoted by $\mathrel|\!\sim_W$) by the following rule:*

$$\phi \mathrel|\!\sim_W \psi \text{ iff for all } s \prec\text{-minimal in } \widehat{\phi} \text{ we have } \forall v \in l(s), v \models \psi.$$

We now move on to the axiomatisation of the four classes of models just defined. Consider the following set of properties.

(Reflexivity)	$\phi \mid\!\sim \phi$
(Left Logical Equivalence)	If $\phi \equiv \psi$ and $\phi \mid\!\sim \chi$ then $\psi \mid\!\sim \chi$
(Right Weakening)	If $\phi \mid\!\sim \psi$ and $\psi \models \chi$ then $\phi \mid\!\sim \chi$
(Cut)	If $\phi \mid\!\sim \psi$ and $\phi \wedge \psi \mid\!\sim \chi$ then $\phi \mid\!\sim \chi$
(Cautious Monotony)	If $\phi \mid\!\sim \psi$ and $\phi \mid\!\sim \chi$ then $\phi \wedge \psi \mid\!\sim \chi$
(Loop)	If $\phi_0 \mid\!\sim \phi_1, \phi_1 \mid\!\sim \phi_2, \ldots,$
	$\phi_{k-1} \mid\!\sim \phi_k, \phi_k \mid\!\sim \phi_0$ then $\phi_0 \mid\!\sim \phi_k$
(Or)	If $\phi \mid\!\sim \chi$ and $\psi \mid\!\sim \chi$ then $\phi \vee \psi \mid\!\sim \chi$
(Rational Monotony)	If $\phi \not\mid\!\sim \neg\psi$ and $\phi \mid\!\sim \chi$ then $\phi \wedge \psi \mid\!\sim \chi$

Let us point out that Reflexivity corresponds to the ID axiom of C2 and that Rational Monotony corresponds to the CV axiom. The set of axioms Reflexivity, Right Weakening, Left Logical Equivalence, Cut, Cautious Monotony, and Or have become known as *system P* [Kraus *et al.*, 1990b] and is considered as kind of a gold standard for nonmonotonic inference relations. The correspondence between these axioms and the four classes of models is established by the following Theorem 3.6. The axiomatisation of cumulative, ordered and preferential models is due to [Kraus *et al.*, 1990b]. The axiomatisation of ranked models is due to [Lehmann and Magidor, 1992b].

Definition 3.5. *A consequence relation $\mid\!\sim$ is said to be:*

- cumulative *iff it satisfies Reflexivity, Right Weakening, Left Logical Equivalence, Cut and Cautious Monotony.*

- loop-cumulative *iff it is cumulative and satisfies Loop.*

- preferential *iff it is loop-cumulative and satisfies Or.*

- rational *iff it is preferential and satisfies Rational Monotony.*

Example 6. *As an example of a cumulative model, consider S consisting of all possible worlds over the signature $\{p, b, f\}$ and $\prec$ ordered as follows:*

$$\begin{array}{c}\overline{p}b f \\ \overline{p}\overline{b} f \\ \overline{p}\overline{b}\,\overline{f}\end{array} \rightrightarrows \begin{array}{c}\overline{p}b\overline{f} \\ pb\overline{f}\end{array} \rightrightarrows \begin{array}{c}p\overline{b}\,\overline{f} \\ pb f\end{array} \rightrightarrows p\overline{b}f$$

$$\xrightarrow{\prec}$$

Then we see that e.g. $pb \mathrel|\!\sim_W \overline{f}$ as the verifying world pbf is $\prec$-preferred to the only falsifying world $pb\overline{f}$, i.e., $pbf \prec pb\overline{f}$.

An example of a rational model is given by the following order:

$$\overline{p}bf, \quad \overline{p}\overline{b}f, \quad \overline{p}\overline{b}\,\overline{f} \quad \prec \quad pb\overline{f}, \quad \overline{p}b\overline{f} \quad \prec \quad pbf, \quad p\overline{b}\,\overline{f}, \quad p\overline{b}f$$

We see here again that $pb \mathrel|\!\sim_W \overline{f}$

Theorem 3.6. *Let $\mathrel|\!\sim\, \subseteq \mathcal{L} \times \mathcal{L}$. It holds that $\mathrel|\!\sim$ is cumulative (resp. loop-cumulative, preferential, rational) iff $\mathrel|\!\sim$ is defined by a cumulative (resp. cumulative-ordered, preferential, ranked) model. Furthermore, if $\mathcal{L}$ is logically finite (i.e., contains a finite number of atoms) and $\mathrel|\!\sim$ is cumulative (resp. loop-cumulative, preferential, rational) then $\mathrel|\!\sim$ is defined by a finite cumulative (resp. cumulative-ordered, preferential, ranked) model.*

We have seen that the KLM-framework offers a formal model of the semantics of defeasible conditionals. The framework, however, does not give an account of how to construct a cumulative model for a given conditional knowledge base (typically, many different cumulative, or even preferential or ranked models are possible). In more detail, given a set of conditionals Δ of the form $(\phi|\psi)$ (where $\phi, \psi \in \mathcal{L}$), we are interested in determining a unique cumulative model W s.t. for every $(\phi|\psi) \in \Delta$, $\phi \mathrel|\!\sim_W \psi$, i.e. W accepts every conditional. Several approaches for constructing such a model, sometimes called *inductive inference operators*, have been studied in the literature [Lehmann and Magidor, 1992b; Goldszmidt and Pearl, 1996b; Kern-Isberner et al., 2020]. Probably the best-known and most-studied (even though not necessarily the best-behaved) approach is known as *rational closure* [Lehmann and Magidor, 1992b] or system Z [Goldszmidt and Pearl, 1996b].

We focus on system Z defined as follows. A conditional $(\psi|\phi)$ is tolerated by a finite set of conditionals Δ if there is a possible world ω with $(\psi|\phi)(\omega) = 1$ and $(\psi'|\phi')(\omega) \neq 0$ for all $(\psi'|\phi') \in \Delta$, i.e. ω verifies $(\psi|\phi)$ and does not falsify any (other) conditional in Δ. The Z-partitioning $(\Delta_0, \ldots, \Delta_n)$ of Δ is defined as:

- $\Delta_0 = \{\delta \in \Delta \mid \Delta \text{ tolerates } \delta\}$;

- $\Delta_1, \ldots, \Delta_n$ is the Z-partitioning of $\Delta \setminus \Delta_0$.

For $\delta \in \Delta$ we define: $Z_\Delta(\delta) = i$ iff $\delta \in \Delta_i$ and $(\Delta_0, \ldots, \Delta_n)$ is the Z-partioning of Δ. Finally, the ranking function κ_Δ^Z is defined via: $\kappa_\Delta^Z(\omega) = \max\{Z(\delta) \mid \delta(\omega) = 0, \delta \in \Delta\} + 1$, with $\max \emptyset = -1$. Notice that this ranking correspond to a cumulative model, which we denote by $W^Z(\Delta)$.

We now illustrate ranked models in general and system Z in particular with the well-known "Tweety the penguin"-example.

Example 7. Let $\Delta = \{(f|b), (b|p), (\neg f|p)\}$. This conditional belief base has the following Z-partitioning: $\Delta_0 = \{(f|b)\}$ and $\Delta_1 = \{(b|p), (\neg f|p)\}$. This gives rise to the following κ_Δ^Z-ordering over the worlds based on the signature $\{b, f, p\}$:

ω	κ_Δ^Z	ω	κ_Δ^Z	ω	κ_Δ^Z	ω	κ_Δ^Z
pbf	2	$pb\bar{f}$	1	$p\bar{b}f$	2	$p\bar{b}\bar{f}$	2
$\bar{p}bf$	0	$\bar{p}b\bar{f}$	1	$\bar{p}\bar{b}f$	0	$\bar{p}\bar{b}\bar{f}$	0

As an example of a (non-)inference, observe that e.g. $\top \mathrel{|\!\sim}_{W^Z(\Delta)} \neg p$ and $p \wedge f \mathrel{|\!\not\sim}_{W^Z(\Delta)} b$.

3.4 Belief Revision and the Ramsey Test

Another important area in knowledge representation is that of *belief change*, which is concerned with supplying a formal model of the change of a belief base. In the context of this chapter, this is particularly interesting as there exist strong relationships between belief change and conditional reasoning, as we will explain below.

3.4.1 Belief Revision

We now recall the AGM-approach to belief revision [Alchourrón et al., 1985] as reformulated for propositional logic by [Katsuno and Mendelzon, 1991]. The following postulates for revision operators $\star : \mathcal{L} \times \mathcal{L} \to \mathcal{L}$ are formulated:

(R1) $\phi \star \psi \vdash \psi$
(R2)   If $\phi \wedge \psi$ is satisfiable, then $\phi \star \psi \equiv \psi \wedge \phi$
(R3)   If ψ is satisfiable, then so is $\phi \star \psi$
(R4)   If $\phi_1 \equiv \phi_2$ and $\psi_1 \equiv \psi_2$, $\phi_1 \star \psi_1 \equiv \phi_2 \star \psi_2$
(R5) $(\phi \star \psi) \wedge \mu \vdash \phi \star (\psi \wedge \mu)$
(R6)   If $(\phi \star \psi) \wedge \mu$ is satisfiable, then $\phi \star (\psi \wedge \mu) \vdash (\phi \star \psi) \wedge \mu$

An important result is the semantical characterisation of such a belief revision operator. For such a characterisation, a function $f : \mathcal{L}(\mathsf{At}) \to \wp(\Omega(\mathsf{At}) \times \Omega(\mathsf{At}))$ that assigns to each propositional formula $\phi \in \mathcal{L}$ a total preorder $\preceq_\phi$ over $\Omega(\mathsf{At})$ is used. The revision of a formula ϕ by a formula ψ is then defined as the formula which has as models exactly the $\preceq_\phi$-minimal models that satisfy ψ.

Definition 3.7 ([Katsuno and Mendelzon, 1991]). *Given a formula* $\phi \in \mathcal{L}(\mathsf{At})$, *a function* $f : \mathcal{L}(\mathsf{At}) \to \wp(\Omega(\mathsf{At}) \times \Omega(\mathsf{At}))$ *assigning preorders* $\preceq_\phi$ *over* $\Omega(\mathsf{At})$ *to every formula* $\phi \in \mathcal{L}(\mathsf{At})$ *is faithful iff:*

1. *For every* $\phi \in \mathcal{L}(\mathsf{At})$, *if* $\omega, \omega' \in \mathsf{Mod}(\phi)$ *then* $\omega \not\prec_\phi \omega'$,

2. *For every* $\phi \in \mathcal{L}(\mathsf{At})$, *if* $\omega \in \mathsf{Mod}(\phi)$ *and* $\omega' \notin \mathsf{Mod}(\phi)$ *then* $\omega \prec_\phi \omega'$,

3. *For every* $\phi, \phi' \in \mathcal{L}(\mathsf{At})$, *if* $\phi \equiv \phi'$ *then* $\preceq_\phi = \preceq_{\phi'}$.

In [Katsuno and Mendelzon, 1991] the following representation theorem for an AGM revision operator $\star$ was shown:

Theorem 3.8 ([Katsuno and Mendelzon, 1991]). *An operator* $\star : \mathcal{L}(\mathsf{At}) \times \mathcal{L}(\mathsf{At}) \to \mathcal{L}(\mathsf{At})$ *satisfies R1–R6 iff there exists a faithful mapping* $f^\star : \mathcal{L}(\mathsf{At}) \to \wp(\Omega(\mathsf{At}) \times \Omega(\mathsf{At}))$ *that maps each formula* $\phi \in \mathcal{L}(\mathsf{At})$ *to a total preorder s.t.:*

$$\mathsf{Mod}(\phi \star \psi) = \min_{f^\star(\phi)}(\mathsf{Mod}(\psi)) \qquad (1)$$

3.4.2 The Ramsey Test

Close relationships between belief revision and conditional logics were noticed by means of the *Ramsey test* [Ramsey, 1931], which says that a conditional $(\psi|\phi)$ is valid if ψ is believed after revision with the antecedent ϕ. The Ramsey test also gave rise to impossibility results on the compatibility of belief revision and conditional reasoning [Gärdenfors, 1986]. However, when [Katsuno and Mendelzon, 1991] showed that total preorders underlie AGM-belief revision in a fundamental and inevitable way, it was at once also established that belief revision, conditional logic, and nonmonotonic inference were shown to be fully compatible. They can thus be seen as two different sides of a single topic or mode of reasoning [Gärdenfors, 1990; Makinson, 1993], at least when restricted to propositional beliefs. Indeed, when moving to other kinds of belief revision (e. g. [Hansson, 1999; Delgrande and Peppas, 2015]), weaker kinds of conditionals [Hawthorne, 2007; Makinson, 2011] or other forms of nonmonotonic inference, these interrelations tend to break down or are not investigated. For example, for revision in Horn-theories, [Delgrande and Peppas, 2015] has shown that rational revision operators cannot be straightforwardly represented in terms of total preorders, thus severing the link between belief revision and nonmonotonic inference. It was shown that for revision operators in Horn theories satisfying additional postulates, semantics in terms of total preorders are sound and complete, but no investigations in corresponding non-monotonic inference relations have been made.

4 Syntactic similarities between Conditionals and Argumentative Formalisms

In this section, we survey work that explores syntactic similarities between formalisms in formal argumentation and conditionals, such as [Heyninck et al., 2020, 2021, 2019]. We explain where similarities have been identified and point to relevant differences.

The reason for looking at the syntactic similarities between abstract dialectical frameworks and conditional logic is the following. Syntactically, both frameworks focus on pairs of objects such as (ϕ, ψ). In conditional logic, these pairs are interpreted as conditionals with the informal meaning "if ϕ is true then, usually, ψ is true as well" and written as $(\psi|\phi)$. In abstract dialectical frameworks, these pairs are interpreted as acceptance conditions, and interpreted as "if ϕ is accepted then ψ is accepted as well". The resemblance of these informal interpretations is striking, but both approaches use fundamentally different semantics to formalise these interpretations. In several papers [Heyninck et al., 2020, 2021, 2019] these syntactical similarities formed the basis of a comparison between abstract dialectical frameworks and conditional logics. In more detail, they asked the question of whether, and how we can interpret abstract dialectical frameworks in terms of conditional logic so that acceptance in the argumentative system is defined by a nonmonotonic inference relation based on conditionals. The main insights are that there is a gap between argumentation and conditional semantics when applying several intuitive translations, but that there exists a class of translations that preserve the semantics for the 2-valued model semantics of **ADFs** (and for other semantics under certain conditions on the **ADFs**). Furthermore, none of the translations studied are adequate for the grounded semantics and for the preferred and stable semantics in general. In the rest of this section, we provide more details on these results.

The following summarizes the investigations by Heyninck et al on syntactic similarities between **ADFs** and conditionals [Heyninck et al., 2020]. Where S is a set of atoms and $\mathfrak{D}_S$ is the set of all **ADFs** defined on the basis of S (i.e. all **ADFs** $D = (S, L, C)$), and $(\mathcal{L}(S)|\mathcal{L}(S))$ is the set of all condtionals over the propositional language generated by S, we investigate mappings $\mathfrak{T} : \mathfrak{D}_S \to \wp((\mathcal{L}(S)|\mathcal{L}(S)))$ (for arbitrary S).

There is a whole family of translations from **ADFs** to conditional logics which are *prima facie* apt to express the links between nodes s and their acceptance conditions C_s:

- $\Theta_1(D) = \{(s|C_s) \mid s \in S\}$

- $\Theta_2(D) = \{(C_s|s) \mid s \in S\}$

- $\Theta_3(D) = \Theta_1(D) \cup \Theta_2(D)$

- $\Theta_4(D) = \Theta_1(D) \cup \{(\neg s|\neg C_s) \mid s \in S\}$

- $\Theta_5(D) = \{((C_s \equiv s)|\top) \mid s \in S\}$

- $\Theta_6(D) = \Theta_2(D) \cup \{(\neg C_s|\neg s) \mid s \in S\}$.

- $\Theta_7(D) = \{(\neg s|\neg C_s) \mid s \in S\} \cup \{(\neg C_s|\neg s) \mid s \in S\}$.

Notice that all of these translations are based on the idea that there is a strong connection between the acceptance of an acceptance condition C_s and the acceptance of the corresponding node s. Indeed, as [Brewka et al., 2013] puts it: "each node s has an associated acceptance condition C_s specifying the exact conditions under which s is accepted". However, in this formulation, it is not specified (1) when a formula is true according to a three-valued interpretation (i.e. is $a \vee \neg a$ true according to an interpretation v with $v(a) = u$? Different three-valued logics give different answers to this question), (2) what to accept when there are conflicts between different acceptance conditions (e.g. if $C_a = \neg b$ and $C_b = \neg a$) and (3) under which conditions we are justified in rejecting a node. Therefore, we systematically investigate different forms of conditionals based on the common idea that "the influence a node may have on another node is entirely specified through the acceptance condition" [Brewka et al., 2013].

We now explain in more detail every translation. Θ_1 formalizes the intuition that whenever the condition of a node s is believed, normally, s should be believed as well. Likewise, Θ_2 formalizes the idea that if a node is believed, its condition should be believed as well. Θ_3 combines the two aforementioned intuitions. Θ_4 is a slight variation on this idea, combining Θ_1 with the constraint that whenever the negation of a condition of a node is believed, the negation of the node itself should be believed as well. Θ_5 postulates that a node should be equivalent to its condition. Θ_6, formalizes the following intuition: if s is believed, C_s has to be believed, and if $\neg s$ is believed, $\neg C_s$ has to be believed as well. Finally, Θ_7 is a formalization of the idea that whenever the negation of a node, respectively the negation of the condition of a node is believed, the negation of the condition of the node, respectively the negation of the node should be believed. Note that Θ_1 has already been investigated to some small extent in [Kern-Isberner and Thimm, 2018]. These translations were investigated with respect to their adequacy in full detail in [Heyninck et al., 2020]. In more detail, the following notion of *adequacy* was used there:

Definition 4.1. *Let S be a set of atoms and $\mathfrak{T} : \mathfrak{D}_S \to \wp((\mathcal{L}(S)|\mathcal{L}(S)))$ be a translation from ADFs to conditional knowledge bases. We furthermore define $W \models \Delta$ iff $\phi \mid\!\sim_W \psi$ for every $(\psi|\phi) \in \Delta$. $\mathfrak{T}$ is:*

- OCF-adequate with respect to **Sem** *if: for every* $D = (S, L, C)$ *there is some ranked model* W *s.t. (1)* $W \models \mathfrak{T}(D)$ *and (2) for every* $s \in S$, $D \mathrel{|\!\!\sim}_{\mathsf{Sem}}^{\cap} s$ *iff* $\top \mathrel{|\!\!\sim}_W s$.[3]

- Z-adequate with respect to **Sem** *if: for every* $D = (S, L, C)$ *and every* $s \in S$ *it holds that:* $D \mathrel{|\!\!\sim}_{\mathsf{Sem}}^{\cap} s$ *iff* $\mathfrak{T}(D) \mathrel{|\!\!\sim}^Z s$.

Intuitively, a translation is OCF-adequate if the beliefs sanctioned by some ranking that is a model of the translation correspond to the consequences of the translated **ADF** D under some semantics **Sem**. The general picture that emerges in [Heyninck et al., 2020] is that:

- the translations Θ_1 and Θ_2 are *not* OCF-adequate or Z-adequate under any **ADF**-semantics,

- the translations $\Theta_3, \ldots, \Theta_7$ are OCF-adequate and Z-adequate under the two-valued model semantics, and

- the translations $\Theta_3, \ldots, \Theta_7$ are *not* OCF-adequate and Z-adequate under any other **ADF**-semantics.

We refer to [Heyninck et al., 2020] for full formal details, but illustrate this here with a few simple examples.

Example 8 (Z-Inadequacy of Θ_1 w.r.t. 2mod). *We consider the following **ADF*** $D_1 = (\{a, b, c\}, L, C)$ *. Notice that*

$$\Theta_1(D_1) = \{(b|\neg a), (a|\neg b), (c|\neg a \vee \neg b)\}$$

which has the following Z-ranking:

ω	κ_Δ^z	ω	κ_Δ^z	ω	κ_Δ^z	ω	κ_Δ^z
abc	0	$ab\overline{c}$	0	$a\overline{b}c$	0	$a\overline{b}\overline{c}$	1
$\overline{a}bc$	0	$\overline{a}b\overline{c}$	1	$\overline{a}\overline{b}c$	1	$\overline{a}\overline{b}\overline{c}$	1

We therefore see that $\Theta_1(D_1) \mathrel{\not|\!\!\sim}^Z c$ *even though* $D \mathrel{|\!\!\sim}_{\mathsf{2mod}}^{\cap} c$ *and thus* Θ_1 *is not Z-adequate with respect to the* **2mod***-semantics.*

---

[3]The term OCF-adequate comes from *ordinal conditional functions*, a particularly useful implementation of ranked models due to Spoh [Spohn, 1988].

Example 9 (Z-Inadequacy of Θ_2 w.r.t. 2mod). *We consider the following ADF $D_2 = (\{a, b, c\}, L, C)$ where:*

$$C_a = \neg b \qquad C_b = \neg a \qquad C_c = a \vee b$$

D_2 *has three complete models v_1, v_2, v_3 with: $v_1(a) = v_2(b) = v_1(c) = v_2(c) = \top$, $v_1(b) = v_2(a) = \bot$ and $v_3(a) = v_3(b) = v_3(c) = u$. Only v_1 and v_2 are 2-valued.*

Moving to $\Theta_2(D) = \{(\neg a|b), (\neg b|a), (a \vee b|c)\}$, we see that

$$(\kappa^Z_{\Theta_2(D)})^{-1}(0) = \{a\bar{b}c, a\bar{b}\bar{c}, \bar{a}bc, \bar{a}b\bar{c}, \bar{a}\bar{b}\bar{c}\}.$$

This means that $\Theta_2(D_2) \not\hspace{-2pt}\mid\hspace{-4pt}\sim^Z c$ even though $D \mid\hspace{-4pt}\sim^{\cap}_{\text{2mod}} c$, i.e. Θ_2 is not Z-adequate with respect to the 2mod-semantics.

Example 10. *If we look at $D_2 = (\{a, b, c\}, L, C)$ from the previous example again, we see that*

$$\Theta_3(D_2) = \{(a|\neg b), (b|\neg a), (c|a \vee b), (\neg a|b), (\neg b|a), (a \vee b|c)\}.$$

We see that $(\kappa^Z_{\Theta_3(D)})^{-1}(0) = \{a\bar{b}c, \bar{a}bc\}$. This illustrates OCF-adequacy and Z-adequacy of $\Theta_3(D_2)$. For the other translations $\Theta_4, \ldots, \Theta_7$, a similar result holds.

This example also lets us illustrate the Z-inadequacy of these translations for the complete and grounded semantics. Indeed, as there is one complete model of D_2 v_3 with $v_3(a) = v_3(b) = v_3(c) = u$, we see that $D \not\hspace{-2pt}\mid\hspace{-4pt}\sim_{\text{grounded}} a \vee b$ whereas $\Theta_2(D_2) \mid\hspace{-4pt}\sim^Z a \vee b$.

Likewise, for preferred semantics, all translations prove inadequate:

Example 11. *We consider the following ADF $D_3 = (\{a, b, c\}, L, C)$ where: $C_a = \neg b$; $C_b = \neg a$; $C_c = \neg b \wedge \neg c$; This ADF has the following unique 2-valued models: $v(a) = v(c) = \bot$ and $v(b) = \top$. If we consider e.g. $\Theta_3(D_3) = \{(a|\neg b), (b|\neg a), (c|\neg b \wedge \neg c), (\neg b|a), (\neg a|b), (\neg b \wedge \neg c|c)\}$, we see that $\kappa^Z_{\Theta_3(D)})^{-1}(0) = \{\bar{a}b\bar{c}\}$ which means $\Theta_3(D_3) \mid\hspace{-4pt}\sim^Z b$. However, D_3 has two preferred intepretations: one corresponds to the $\kappa^Z_{\Theta_3(D)}$-minimal world ($v(a) = v(c) = \bot$ and $v(b) = \top$), and a second preferred model is v' with $v'(a) = \top$, $v'(b) = \bot$ and $v'(c) = u$. Thus, $D \mid\hspace{-4pt}\sim^{\cap}_{\text{preferred}} b$.*

5 Conditional semantics in argumentation

In this section, we survey work that applies ideas from conditional logic in formal argumentation.

5.1 Abstract argumentation

In the following, we discuss the works [Skiba et al., 2021] and [Skiba and Thimm, 2022] which apply conditional logic semantics in abstract argumentation frameworks.

5.1.1 Non-Classical Semantics for Abstract Argumentation

Classical interpretations of propositional logic (and other classical logics) provide a simple interpretation for the elements in the signature of the logic: an interpretation (or possible world) ω either evaluates an atom $a \in \mathsf{At}$ to T or F. Similarly, classical interpretations of abstract argumentation frameworks (*extensions*) provide the same view on the acceptance status of arguments: either an argument a is contained in an extension E or it is not.

In conditional logics, interpretations provide more structure and are usually based on some form of rankings of classical interpretations wrt. their *plausibility*, such as with *ordinal conditional functions* (or *ranking functions*). In the following, we consider ranking functions for abstract argumentation, i.e., functions that assign a degree of plausibility to extensions. Such a ranking between sets of arguments allows us to reason in a more fine-grained manner than with extension-based semantics. Where in classical extension-based semantics, we can either say that a particular set of arguments is an extension or not, in the ranking-based approach, we can compare two sets (which are not necessarily extensions for a given semantics) on the basis of how close they are to being acceptable.

In order to approach the topic in a general manner, we first consider extension-ranking semantics by [Skiba et al., 2021].

Definition 5.1 (Extension-ranking semantics). *Let $AF = (A, R)$ be an AF. An extension ranking on AF is a preorder[4] $\sqsupseteq$ over the power set*

---
[4]A preorder is a (binary) relation that is *reflexive and transitive*.

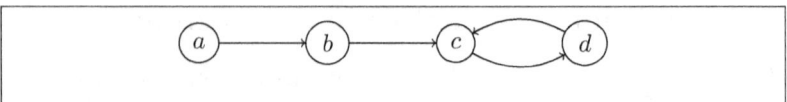

Figure 2: *AF* from Example 12.

of arguments 2^A. An extension-ranking semantics τ is a function that maps each *AF* to an extension ranking $\sqsupseteq^\tau_{AF}$ on *AF*.

Note that extension rankings are not necessarily total. For an AF $AF = (A, R)$, an extension-ranking semantics τ, an extension ranking $\sqsupseteq^\tau_{AF}$, $E, E' \subseteq A$, and for $E \sqsupseteq^\tau_{AF} E'$ we say that E is *at least as plausible as* E' with respect to τ in AF. We introduce the usual abbreviations:

- E is *strictly more plausible than* E', denoted $E \sqsupset^\tau_{AF} E'$, if $E \sqsupseteq^\tau_{AF} E'$ but not $E' \sqsupseteq^\tau_{AF} E$;

- E and E' are *equally plausible*, denoted $E \equiv^\tau_{AF} E'$, if $E \sqsupseteq^\tau_{AF} E'$ and $E' \sqsupseteq^\tau_{AF} E$;

- E and E' are *incomparable*, denoted $E \asymp^\tau_{AF} E'$, if neither $E \sqsupseteq^\tau_{AF} E'$ nor $E \sqsupseteq^\tau_{AF} E'$.

To motivate the need of extension rankings further consider the following example.

Example 12. *Lets recall the AF from Example 3 with*

$$AF = (\{a, b, c, d\}, \{(a, b), (b, c), (c, d), (d, c)\})$$

and depicted in Figure 2. To compare the two sets $\{b\}$ and $\{c, d\}$ extension-based semantics such as admissible semantics do not provide a suitable solution. Both these sets are not admissible extensions, however $\{c, d\}$ is not even conflict-free, while $\{b\}$ is conflict-free. Therefore we argue that $\{b\}$ is a "better" set than $\{c, d\}$, since conflict-freeness is an undisputed property in the area of abstract argumentation. Extension-ranking semantics provides a suitable approach to rank $\{b\}$ and $\{c, d\}$.

Extension-based semantics provide a naive way of defining extension-ranking semantics. A set of arguments E is "better" than another set E' if the first set satisfies an extension-based semantics and the second set does not.

Definition 5.2 (Least-discriminating extension-ranking semantics). *Let $AF = (A, R)$ be an AF. Given an extension-based semantics σ, we define the* least-discriminating extension-ranking semantics wrt. σ, *denoted LD^σ by:*

- $E \sqsupset_F^{LD^\sigma} E'$ *if $E \in \sigma(F)$ and $E' \notin \sigma(F)$;*
- *and $E \equiv_F^{LD^\sigma} E'$, if $E, E' \in \sigma(AF)$ or $E, E' \notin \sigma(F)$.*

Example 13. *Continuing Example 12. Consider the two sets $\{a, c\}$ and $\{c, d\}$. $\{a, c\}$ is an admissible set, while $\{c, d\}$ is not even conflict-free, by using LD^{ad} we have $\{a, c\} \sqsupset_{AF}^{LD^{ad}} \{c, d\}$. So, the least-discriminating extension-ranking semantics is behaves in line with the binary classification of extension-based semantics. A set is either accepted or not wrt. an extension-based semantics σ i.e. a set is either part of the upper level if that set satisfies semantics σ or on the lower level if the set does not satisfy σ.*

5.1.2 Ordinal Conditional Functions for Abstract Argumentation

[Skiba and Thimm, 2022] introduced ranking functions for abstract argumentation frameworks. These ranking functions are a starting point for fully capturing the ideas and concepts of conditional logics in abstract argumentation. A ranking function $\kappa(I, O)$ for an $AF = (A, R)$ is used to compute a numerical plausibility value for a set of arguments $I \subseteq A$ to be considered *in* under the assumption that the set of arguments $O \subseteq A$ is considered *out*. Unlike ranking functions for conditional logics, ranking functions for argumentation frameworks need two parameters to compute a numerical plausibility value, since AFs do not have a notion of negation. E.g. in Example 12, the pair $(\{a, c\}, \{b, d\})$ can be seen as an analogue to the world $a\bar{b}c\bar{d}$.

Definition 5.3. *Let $AF = (A, R)$ be an AF. A ranking function for AF is a function $\kappa : 2^A \to \mathbb{N} \cup \{\infty\}$ with $\kappa^{-1}(0) \neq \emptyset$. For sets $I, O \subseteq A$ we abbreviate*

$$\kappa(I, O) = \min\{\kappa(S) | I \subseteq S, S \cap O = \emptyset\}$$
$$\kappa(I, O) = \infty \text{ if } I \cap O \neq \emptyset$$

Example 14. Let $AF_2 = (\{a,b,c\}, \{(a,b),(b,c)\})$ be an AF and consider an exemplary ranking function κ. Since $\{a,c\}$ is the preferred extension we argue that $\{a,c\}$ should receive a plausibility value of 0, because no set is more plausible than a preferred extension, i.e. $\kappa(\{a,c\},\emptyset) = \kappa(\{a,c\},\{b\}) = 0$. Then the two admissible sets $\{a\}$ and $\emptyset$ should receive a plausibility value of 1, because these two sets are atleast admissible even-though they are not preferred, i.e. $\kappa(\{a\},\{b,c\}) = \kappa(\{a\},\{b\}) = \kappa(\{a\},\{c\}) = \kappa(\{a\},\emptyset) = 1$ and $\kappa(\emptyset,S) = 1$ for every $S \subseteq \{a,b,c\}$. The two conflict-free sets $\{b\}$ and $\{c\}$ receive a plausibility value of 2, i.e. $\kappa(\{b\},\{a,c\}) = \kappa(\{b\},\{a\}) = \kappa(\{b\},\{c\}) = \kappa(\{b\},\emptyset) = 2$ and $\kappa(\{c\},\{a,b\}) = \kappa(\{c\},\{a\}) = \kappa(\{c\},\{b\}) = \kappa(\{c\},\emptyset) = 2$. Since the two sets $\{a,b\}$ and $\{b,c\}$ each entail only one conflict each, they should receive a better plausibility value than $\{a,b,c\}$, i.e. $\kappa(\{a,b\},\{c\}) = \kappa(\{a,b\},\emptyset) = \kappa(\{b,c\},\{a\}) = \kappa(\{b,c\},\emptyset) = 3$ and $\kappa(\{a,b,c\},\emptyset) = 4$.

Conditional logics semantics follows one single principle for conditional acceptance ("a conditional is accepted if its verification is more plausible than its violation"). On the other hand in abstract argumentation and in particular for admissible-based reasoning two guiding principles can be found:

Conflict-freeness: An argument should not be accepted if one of its attackers is accepted.

Reinstatement: An argument should be accepted if all its attackers are not accepted.

Conflict-freeness describes that a set should not contain two arguments that attack each other. So conflicting sets should be less plausible than conflict-free sets. *Reinstatement* describes that if there is no reason to reject an argument, then that argument should not be rejected. So a set which defends itself against all possible attackers is at least as plausible as a set that does not defend itself. The implementation of these two principles for ranking functions κ is:

Definition 5.4. Let $AF = (A, R)$ be an AF, $a, b \in A$, and κ a ranking function.

- κ accepts an attack (a,b) with $a \neq b$ if $\kappa(\{a\},\{b\}) < \kappa(\{a,b\},\emptyset)$.

8 - CONDITIONALS AND ARGUMENTATION

i	$\kappa^{-1}(i)$
3	$(\{a,b\},\emptyset)$
2	$(\emptyset,\{a\}), (\emptyset,\{b\}), (\emptyset,\{a,b\})$
1	$(\{b\},\{a\}), (\{b\},\emptyset)$
0	$(\emptyset,\emptyset), (\{a\},\emptyset), (\{a\},\{b\})$

Table 2: Example ranking function for Example 15. Note κ is only partially defined.

- κ possibly reinstates *an argument a if* $\kappa(S \cup \{a\}, a^-) \leq \kappa(S, \{a\} \cup a^-)$ *for all* $S \subseteq A$ *with* $S \cap (a^- \cup a^+) = \emptyset$.

In other words, for an attack (a, b) to be accepted by a ranking function, it is more plausible for a to be *in* and b to be *out* than for both a and b to be *in* at the same time. An argument a is possibly reinstated by a ranking function if all attackers of a are *out*, then a being *in* should be at least as plausible as a being *out*.

If a ranking function satisfies the two principles for all arguments and all attacks of an AF, then that ranking function satisfies that AF.

Definition 5.5. *A ranking function κ satisfies an AF $AF = (A, R)$ if it accepts all attacks in R and possibly reinstates all arguments in A.*

Example 15. *Consider $AF_3 = (\{a,b\}, \{(a,b)\})$. So the following statements have to hold for a ranking function κ to satisfy AF_3:*

1. $\kappa(\{a\}, \{b\}) < \kappa(\{a,b\}, \emptyset)$

2. $\kappa(\{a\}, \emptyset) \leq \kappa(\emptyset, \{a\})$

3. $\kappa(\{b\}, \{a\}) \leq \kappa(\emptyset, \{a,b\})$

Table 2 depicts a ranking function that satisfies AF_3. The two admissible sets are $\emptyset$ and $\{a\}$, these two sets are also on the lowest level, meaning that these sets are the most plausible sets. If we compare the two not admissible sets $\{b\}$ and $\{a,b\}$ we see that $\{b\}$ is ranked higher than $\{a,b\}$. This behaviour is intuitive, since while both these sets are not admissible and $\{b\}$ is at least conflict-free.

Note that if an AF contains any self-attacking argument a, then there can be no ranking function that satisfies that AF. This is because in order to accept the attack (a,a), it must hold that $\kappa(\{a\},\{a\}) < \kappa(\{a\},\emptyset)$, which is impossible since $\kappa(\{a\},\{a\}) = \infty$.

5.1.3 System Z Ranking function for Abstract Argumentation Frameworks

Next, we discuss a ranking function for AFs inspired by system Z. Recall that the basic idea of system Z is that a conditional $(\phi|\psi)$ is *tolerated* by a set of conditionals if it is confirmed by a possible world ω and no other conditional is refuted. When investigating an attack (a,b) in an argumentation framework, it can be concluded that if a is part of an extension E, then b should not be part of the same extension. Thus attacks between two arguments within an argumentation framework represent a conditional relation between those two arguments, i.e. for an attack (a,b) we can formulate: "if a is acceptable then b should not be acceptable". Therefore, we interpret the attack relation of an argumentation framework as a set of conditionals and to model the idea of system Z it has to hold that in order to *tolerate* an attack (conditional) we have to find a set of arguments (interpretation), which verifies that attack while not violating any other attack.

Definition 5.6. *Let $AF = (A, R)$ be an AF.*

- *A set $S \subseteq A$ verifies an attack (a,b) iff $a \in S$ and $b \notin S$.*

- *A set $S \subseteq A$ violates an attack (a,b) iff $a \in S$ and $b \in S$.*

- *A set $S \subseteq A$ satisfies an attack (a,b) iff it does not violate it.*

Intuitively, a set of arguments satisfies an attack if this set does not contain both the attacker and the target of the attack. For AF $AF_2 = (\{a,b,c\}, \{(a,b),(b,c)\})$ from Example 14, we can observe that the set $S_1 = \{a\}$ verifies the attack (a,b) and does not violate the attack (b,c), while the set $S_2 = \{a,b\}$ verifies the attack (b,c), however S_2 violates attack (a,b).

Verifying an attack is not enough to capture the full picture of reasoning in abstract argumentation since only *conflict-freeness* is captured.

8 - CONDITIONALS AND ARGUMENTATION

To capture *reinstatement* as well the notion of defence has to be modelled with the toleration notion. For this purpose we add an another condition for an attack to be tolerated by a set of arguments, the so-called *attack admissibility*. To satisfy attack admissibility of an argument it has to hold that if all the attackers of the argument are *out*, then the argument itself should be *in*.

Definition 5.7. *Let $AF = (A, R)$ be an argumentation framework.*

- *A set $S \subseteq A$ verifies attack admissibility of $a \in A$ iff $a \in S$ and $b \notin S$ for all $b \in a^-$.*

- *A set $S \subseteq A$ violates attack admissibility of $a \in A$ iff $a \notin S$ and $b \notin S$ for all $b \in a^-$.*

- *A set $S \subseteq A$ satisfies attack admissibility of $a \in A$ iff it does not violate it.*

Recall $AF_2 = (\{a, b, c\}, \{(a, b), (b, c)\})$, then set $S_3 = \{a, c\}$ verifies attack admissibility of argument c, because the only attacker of c, b is not part of S_3 and one of the attackers of b is contained in S_3. For set $\{b\}$ we have the case where argument a is not part of $\{b\}$, however $\{b\}$ also does not contain any attacker of a, hence attack admissibility of a is violated.

By combining these two definitions, we define when an attack can be tolerated.

Definition 5.8. *Let $AF = (A, R)$ be an argumentation framework. A set $P \subseteq R$ tolerates an attack (a, b) iff there is a set $S \subseteq A$ that*

1. *verifies (a, b),*

2. *satisfies each attack in P, and*

3. *satisfies attack admissibility of each $c \in A$*

To tolerate an attack, we have to find a set of arguments S that is conflict-free and every argument not in S has to be attacked. Recall $AF_2 = (\{a, b, c\}, \{(a, b), (b, c)\})$, then attack (b, c) is not tolerated by $\{(a, b), (b, c)\}$. For (b, c) to be verified for any set S, it must hold that

$b \in S$. Then, to not violate (a,b) a is not allowed to be contained in S. However, then we have the problem that S does not contain any attackers of a, meaning that attack admissibility of a is violated.

With the help of the notion of toleration, we define a ranking function κ^Z inspired by system Z for AFs.

Definition 5.9. *Let $AF = (A, R)$ be an argumentation framework. Then the Z-attack-Partitioning $(R_0, \ldots, R_n)$ with $R_0 \cup \ldots \cup R_n \subseteq R$ is defined as*

- $R_0 = \{r \in R \mid R \text{ tolerates } r\}$

- $(R_1, \ldots, R_n)$ *is the Z-attack-Partitioning of* $R \setminus R_0$

For $r \in R$ define $Z_R(r) = i$ if $r \in R_i$ and

$$\kappa^Z(S, X) = \max\{Z(r) \mid S \text{ violates } r\} + 1$$

where $X \subseteq A$ is any set s.t. $S \cap X = \emptyset$.

Attacks in R_0 are tolerated with respect to the complete set of attacks R of an $AF = (A, R)$, while attacks in R_1 are tolerated only after removing the attacks in R_0. Using this partitioning of attacks we can rank the sets of arguments based on their plausibility with respect to the attacks. If a set violates an attack on level 0, while a different set violates an attack on level 1, then the first set is more plausible than the second one. The higher an attack is ranked, the worse its violation is. Thus, the partitioning of attacks can be interpreted as a split based on the impact of each attack in the AF, with attacks on lower ranks being considered better. It is therefore more important to satisfy a single high ranked attack than to satisfy several low ranked attacks.

Example 16. *Consider AF from Example 12. Then to tolerate attack (b, c) argument b has to be verified, however then attack admissibility of a is violated, hence $(b, c) \notin R_0$. The remaining attacks are tolerated by R, so the Z-attack-Partitioning of R is (R_0, R_1) with*

$$R_0 = \{(a,b), (c,d), (d,c)\}$$
$$R_1 = \{(b,c)\}$$

i	$\kappa^{-1}(i)$
2	$(\{b,c\},X),(\{a,b,c\},X),(\{b,c,d\},X),(\{a,b,c,d\},X)$
1	$(\{a,b\},X),(\{c,d\},X),(\{a,b,d\},X),(\{a,c,d\},X)$
0	$(\emptyset,X),(\{a\},X),(\{b\},X),(\{c\},X),(\{d\},X),(\{a,c\},X),(\{b,d\},X),(\{a,d\},X)$

Table 3: κ^Z, where for every pair (I,X) $X \subseteq A$ is any set s.t. $I \cap X = \emptyset$.

Consider sets $\{a,c,d\}$ and $\{b,c\}$, then $\{a,c,d\}$ violates $(a,b),(c,d)$ and (d,c) while $\{b,c\}$ violates (b,c). Since $(b,c) \in R_1$ it holds that

$$\kappa_{AF}^Z(\{a,c,d\},\emptyset) < \kappa^Z(\{b,c\},\emptyset).$$

Table 3 depicts κ_{AF}^Z for AF from Example 12.

5.1.4 Extension-ranking Semantics based on System Z

The ranking functions for AFs can be seen as a special instance of extension-ranking semantics. These functions allow us to rank sets of arguments based on their plausibility. So we can define an extension-ranking semantics based on the system Z ranking function for AFs by stating that a set of arguments E is at least as plausible as another set E' if κ^Z returns a lower value for E than for E' with respect to the remaining arguments.

Definition 5.10. Let $AF = (A,R)$ be an AF and $E, E' \subseteq A$. Define the system Z extension-ranking semantics $\sqsupseteq_{AF}^{\kappa^Z}$ via

$$E \sqsupseteq_{AF}^{\kappa^Z} E' \text{ iff } \kappa^Z(E, A \setminus E) \leq \kappa^Z(E', A \setminus E').$$

In other words, E is at least as plausible as E', if E being *in*, while all other arguments not in E are considered *out* is more plausible than E' being considered *in* while all arguments not in E' are considered *out*.

Example 17. *Consider again AF from Example 12. Then Table 4 depicts the ranking corresponding to $\sqsupseteq_{AF}^{\kappa^Z}$. All conflict-free sets are part of the most plausible sets, while sets with conflicts are ranked lower. The number of conflicts is not as important as in the approaches of [Skiba et al., 2021]. In their approaches, $\{b,c\}$ is always ranked strictly better than $\{b,c,d\}$. While for κ^Z these two sets are ranked equally.*

$$\emptyset \equiv_{AF}^{\kappa^Z} \{a\} \equiv_{AF}^{\kappa^Z} \{b\} \equiv_{AF}^{\kappa^Z} \{c\} \equiv_{AF}^{\kappa^Z} \{d\} \equiv_{AF}^{\kappa^Z} \{a,c\} \equiv_{AF}^{\kappa^Z} \{b,d\} \equiv_{AF}^{\kappa^Z} \{a,d\}$$
$$\sqsupset_{AF}^{\kappa^Z} \{a,b\} \equiv_{AF}^{\kappa^Z} \{c,d\} \equiv_{AF}^{\kappa^Z} \{a,b,d\} \equiv_{AF}^{\kappa^Z} \{a,c,d\}$$
$$\sqsupset_{AF}^{\kappa^Z} \{b,c\} \equiv_{AF}^{\kappa^Z} \{a,b,c\} \equiv_{AF}^{\kappa^Z} \{b,c,d\} \equiv_{AF}^{\kappa^Z} \{a,b,c,d\}$$

Table 4: Extension-ranking for AF based on $\sqsupset_{AF}^{\kappa^Z}$.

For further discussions about the system Z extension-ranking semantics we refer to [Skiba and Thimm, 2022].

In this subsection, we have seen that the ideas and concepts of ranking functions can be applied to abstract argumentation frameworks. Sets of arguments can be seen as interpretations and an attack between two arguments can be seen as a conditional i.e. for an attack (a,b) we say that if a is accepted, then b is not accepted. The results of this investigation are functions allowing us to compare sets of arguments based on their plausibility, which is in line with recent work on extension-ranking semantics [Skiba et al., 2021].

5.2 Dynamic Conditionals for Abstract Dialectical Frameworks

Partially based on the differences between the semantics of **ADFs** and conditionals (Section 4), [Heyninck et al., 2023] defined conditional inference relations for **ADFs**. They took inspiration from the propositional setting, where there exist strong connections between *conditional inference* and *belief revision* as explained in Section 3.4.

For simplicity, we explain here the main ideas for two-valued ADF-semantics (in particular, the two-valued model and stable semantics). We refer to full details, and analogous results for three-valued semantics (e.g. complete, grounded and preferred) to [Heyninck et al., 2023].

5.2.1 Revising ADFs

Informally, [Heyninck et al., 2021] study the revision of argumentative contexts, which are represented by an **ADF** D, by new information, represented as logical formula ϕ, resulting in a revised argumentative context $D \star \phi$.

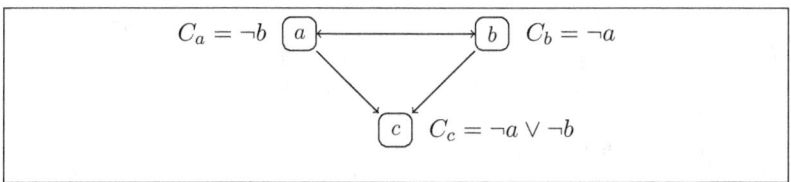

Figure 3: Argumentative representation of Example 18.

We concentrate on revising ADFs by formulas, resulting in a new ADF, i.e. revision operators $\star : \mathfrak{D}(\mathsf{At}) \times \mathcal{L}^K(\mathsf{At}) \to \mathfrak{D}(\mathsf{At})$. Revisions is always relative to a chosen semantics, and when this semantics is two-valued (e. g. two-valued models or stable models), we will restrict attention to revision by formulas in propositional logic in view of the two-valued nature of the mentioned semantics.

As an example of when this kind of revision can be useful, consider the following:

Example 18. *Consider making travel plans while being based in Germany. There are three candidate destinations: Addis Aba (Ethiopia), Boston (USA), and Cochem (Germany). There is not enough time to make two intercontinental travels, but when making at most one intercontinental travel, you will have enough money and time for an additional holiday in Germany. When you would make two intercontinental travels, no time for traveling to Cochem would be left.*

Argumentation can be used to make an informed decision in this scenario: there are three arguments a, b and c for the three respective destinations. a and b attack each other, whereas $\{a,b\}$ attack c. We have represented this as an ADF *consisting of three arguments a, b and c with their respective* acceptance conditions C_a, C_b and C_c. *This results in the* ADF $D_1 = (\{a,b,c\}, L, C)$ *with* $L = \{(a,b),(b,a),(a,c),(b,c)\}$ *and* $C_a = \neg b$, $C_b = \neg a$ *and* $C_c = \neg a \vee \neg b$. D_1 *is represent graphically in Figure 3. Informally, the acceptance conditions can be read as "a is accepted if b is not accepted", "b is accepted if a is not accepted" and "c is accepted if a is not accepted or b is not accepted".*

The argumentative formalisation does not tell us, however, how we should adapt our beliefs in view of changing information. For example, suppose that a highly infectious disease breaks out in Cochem. In that case, argumentative semantics do not give information about what can be expected, unless we change the ADF in view of this information and recalculate the semantic interpretations for this new ADF. However, it might be useful to have an indication of what can be expected in the face of dynamic information. For example, is it reasonable to expect we can still make an intercontinental travel when we do not travel to Cochem (i.e. $\neg c \mathrel{\smash{\mathop{\vert\!\sim}\limits_{a}^{\vee}}} b$)? The derivation of such statements about what can be expected requires the investigation of belief revision and the resulting dynamic conditionals in the setting of formal argumentation.

To give a formal account of such revision scenarios, We adapt the AGM-postulates for propositional revision to the setting of revision-operators $\star : \mathfrak{D}(\mathsf{At}) \times \mathcal{L}(\mathsf{At}) \to \mathfrak{D}(\mathsf{At})$ of ADFs by propositional formulas as follows:

Definition 5.11. *An operator $\star$ is a bivalent ADF revision operator (in short, $\mathsf{ADF}^2_\star$-operator) for an ADF $D = (\mathsf{At}, L, C)$ and a semantics Sem s.t. $\mathsf{Sem}(D) \subseteq \Omega(D)$[5] iff $\star$ satisfies:*

($\mathsf{ADF}^2_\star 1$) $D \star \psi \mathrel{\smash{\mathop{\vert\!\sim}\limits_{\mathsf{Sem}}^{\cap}}} \psi$
($\mathsf{ADF}^2_\star 2$)   If $\mathsf{Sem}(D) \cap \mathsf{Mod}(\psi) \neq \emptyset$ then
           $\mathsf{Sem}(D \star \psi) = \mathsf{Sem}(D) \cap \mathsf{Mod}(\psi)$
($\mathsf{ADF}^2_\star 3$)   If ψ is satisfiable, then $\mathsf{Sem}(D \star \psi) \neq \emptyset$
($\mathsf{ADF}^2_\star 4$)   If $\mathsf{Sem}(D) = \mathsf{Sem}(D')$ and $\psi_1 \equiv \psi_2$ then
           $\mathsf{Sem}(D \star \psi_1) = \mathsf{Sem}(D' \star \psi_2)$
($\mathsf{ADF}^2_\star 5$) $\mathsf{Sem}(D \star \psi) \cap \mathsf{Mod}(\mu) \subseteq \mathsf{Sem}(D \star (\psi \wedge \mu))$
($\mathsf{ADF}^2_\star 6$)   If $\mathsf{Sem}(D \star \psi) \cap \mathsf{Mod}(\mu) \neq \emptyset$, then
           $\mathsf{Sem}(D \star \psi) \cap \mathsf{Mod}(\mu) \supseteq \mathsf{Sem}(D \star (\psi \wedge \mu))$

Remark 5.12. *Equivalent formulations of ($\mathsf{ADF}^2_\star 5$) and ($\mathsf{ADF}^2_\star 6$) (that might be more intuitive to some readers) are:*

($\mathsf{ADF}^2_\star 5$) $D \star \psi \mathrel{\smash{\mathop{\vert\!\sim}\limits_{\mathsf{2mod}}^{\cap}}} \mu \to \bigvee \mathsf{Sem}(D \star (\psi \wedge \mu))$[6]
($\mathsf{ADF}^2_\star 6$)   If $\mathsf{Sem}(D \star \psi) \cap \mathsf{Mod}(\mu) \neq \emptyset$, then
           $D \star (\psi \wedge \mu) \mathrel{\smash{\mathop{\vert\!\sim}\limits_{\mathsf{2mod}}^{\cap}}} (\bigvee \mathsf{Sem}(D \star \psi) \wedge \mu)$

---

[5]The postulates ($\mathsf{ADF}^2_\star 1$)-($\mathsf{ADF}^2_\star 6$) can easily be generalised to a three-valued semantics by substituting $\mathsf{Sem}(D)$ by $\bigcup_{v \in \mathsf{Sem}(D)} [v]^2$. Since we define three-valued revisions below and for reasons of simplicity, we chose to restrict ourselves here to two-valued semantics.

These postulates are explained as follows. $\mathsf{ADF}_\star^2 1$ requires that any revision is successful, i.e. the formula that induces the revision should follow from the revised ADF. The second postulate $\mathsf{ADF}_\star^2 2$ requires that if some of the Sem-interpretations of the original ADF satisfy the formula inducing the revision, the revised ADF should have as Sem-interpretations exactly the Sem-interpretations of the original ADF that satisfy the formula inducing the revision. The third postulate states that revising by a consistent formula results in a Sem-consistent ADF, i.e. an ADF that admits Sem-interpretations. $\mathsf{ADF}_\star^2 4$ requires syntax independence: revising ADFs with the same Sem-interpretations by equivalent formulas results in Sem-equivalent revised ADFs. Finally, $\mathsf{ADF}_\star^2 5$ and $\mathsf{ADF}_\star^2 6$ are direct adaptations of the *super-* and *sub-expansion postulates*. They require, in the non-trivial case where $D \star \psi \not\models_{\mathsf{Sem}}^{\cap} \neg \mu$ (i.e. there is at least one Sem-interpretation of $D \star \psi$ that entails μ, or, in other words, $D \star \psi$ is consistent, under Sem, with μ), that the Sem-interpretations of $D \star (\psi \wedge \mu)$ are exactly the Sem-interpretations of $D \star \psi$ that satisfy μ.

As is usual in work on belief revision, a semantic characterisation in terms of plausibility-orders over interpretations is given. In more detail, we can semantically characterise revision of an ADF D with a formula ϕ in terms of total preorders over two-valued interpretations, in analogue to propositional revision. In order to do so, we consider mappings of the type $\mathfrak{D}(\mathsf{At}) \to \wp(\Omega(\mathsf{At}) \times \Omega(\mathsf{At}))$, i.e. functions mapping every ADF D to a total preorder $\preceq_D$ over possible worlds. We first modify Definition 3.7 of an assignment of preorders to be faithful w.r.t. an ADF D and a semantics Sem:

Definition 5.13. *Given a semantics Sem s.t. $\mathsf{Sem}(D) \subseteq \Omega(\mathsf{At})$ for every $D \in \mathsf{At}$, a function $f : D \mapsto \preceq_D$ assigning[7] a total preorder $\preceq_D$ over $\Omega(\mathsf{At})$ to every ADF $D \in \mathfrak{D}(\mathsf{At})$ is faithful w.r.t. the semantics Sem iff:*

1. *For every $D \in \mathfrak{D}(\mathsf{At})$, if $\omega, \omega' \in \mathsf{Sem}(D)$, then $\omega \preceq_D \omega'$;*

2. *For every $D \in \mathfrak{D}(\mathsf{At})$, if $\omega \in \mathsf{Sem}(D)$ and $\omega' \notin \mathsf{Sem}(D)$, then $\omega \prec_D \omega'$;*

3. *For every $D, D' \in \mathfrak{D}(\mathsf{At})$, if $\mathsf{Sem}(D) = \mathsf{Sem}(D')$ then $\preceq_D = \preceq_{D'}$.*

---

[7]Recall that $\Omega(\mathsf{At})$ is the set of all (two-valued) interpretations for S.

The intuition behind a faithful preorder for D (w.r.t. a two-valued semantics Sem) is that the beliefs justified on the basis of an ADF D can be represented as the formulas entailed by all interpretations in $\mathsf{Sem}(D)$ (which is in complete accordance with taking as beliefs all ϕ s.t. $D \mathrel{\mid\kern-0.4ex\sim}^{\cap}_{\mathsf{Sem}} \phi$). A faithful preorder then represents the relative plausibility of formulas (or equivalently, possible worlds) given the ADF D. Therefore, the interpretations sanctioned by D are on the lowermost level, and other interpretations are ranked according to their plausibility by $\preceq_D$.

Example 19. *We illustrate the above definitions by looking at the Dalal-revision operator [Dalal, 1988], adapted here to our setting. We first define the symmetric distance function between two possible worlds $\omega, \omega' \in \Omega(\mathsf{At})$ as: $\omega \triangle \omega' = |s \in \mathsf{At} \mid \omega(s) \neq \omega'(s)|$. We can then define $\preceq_D^{\triangle}$ over $\Omega(\mathsf{At})$ by setting*

$$\kappa_{\mathsf{d1}}(\omega) = \min\{\omega' \triangle \omega \mid \omega' \in 2\mathsf{mod}(D)\}$$

for any $\omega \in \Omega(\mathsf{At})$ and letting $\omega_1 \preceq_D^{\triangle} \omega_2$ iff $\kappa_{\mathsf{d1}}(\omega_1) \leq \kappa_{\mathsf{d1}}(\omega_2)$.
For the ADF of Example 1, we then obtain the following ranking:

ω	κ_{d1}	ω	κ_{d1}	ω	κ_{d1}	ω	κ_{d1}
abc	1	$ab\overline{c}$	2	$a\overline{b}c$	0	$a\overline{b}\overline{c}$	1
$\overline{a}bc$	0	$\overline{a}b\overline{c}$	1	$\overline{a}\overline{b}c$	1	$\overline{a}\overline{b}\overline{c}$	2

We can now semantically characterise revision of an ADF D (under the two-valued semantics Sem) by a formula $\psi \in \mathcal{L}(\mathsf{At})$ as the ADF $D \star \psi$ s.t. :

$$\mathsf{Sem}(D \star \psi) = \min_{\preceq_D}(\mathsf{Mod}(\psi)) \qquad (2)$$

Example 20. *Looking again at Example 19, we can use Equation 2 to obtain a revision operator $\star_{\mathsf{d1}}$, which we illustrate by revising D with $\neg c$ based on the preorder κ_{d1} which has as two-valued models: $2\mathsf{mod}(D \star_{\mathsf{d1}} \neg c) = \{a\overline{b}\overline{c}, \overline{a}\overline{b}\overline{c}\}$.*
As we will see below, this revision satisfies all $\mathsf{ADF}^2_\star$-postulates.

Notice firstly that strictly speaking the revision above does not determine a unique ADF. However, it does determine a unique ADF *up to semantical equivalence*. Indeed, in view of Postulate $\mathsf{ADF}^2_\star 4$, we are justified in thus restricting our attention, since the result of the revision of two ADFs D_1 and D_2 with the same Sem-interpretations will result in two ADFs $D_1 \star \phi$ and $D_2 \star \phi$ with the same Sem-interpretations. Secondly, notice that the revision operator defined above is a purely semantical characterisation of revision of ADFs, i.e. the revision of an ADF D by a formula ψ is identified with a set of models. Below we will describe one strategy for obtaining a specific ADF on the basis of the set of two-valued models of an ADF.

In [Heyninck et al., 2023], it is shown that the semantic characterisation outlined above is sound and complete:

Corollary 5.14. *Given a finite set of atoms* At, *an operator* $\star : \mathfrak{D}(\mathsf{At}) \times \mathcal{L}(\mathsf{At}) \to \mathcal{L}(\mathsf{At})$ *is an* $\mathsf{ADF}^2_\star$-*operator for two-valued model semantics* 2mod *iff there exists a function* $f : \mathfrak{D}(\mathsf{At}) \to \wp(\Omega(\mathsf{At}) \times \Omega(\mathsf{At}))$ *that is faithful w.r.t.* 2mod *s.t.:*

$$2\mathsf{mod}(D \star \psi) = \min_{\preceq_D}(\mathsf{Mod}(\psi))$$

We now move to revision under the stable semantics, where the semantic characterisation is more complicated. In more detail, not every set of two-valued interpretations is realisable under the stable semantics, which means that there might not exist an ADF that has exactly this set of two-valued interpretations as stable models. Indeed, the problem of realisability has been studied in depth by [Pührer, 2020]. To characterise revision under stable semantics, we need to ensure realisability of the corresponding faithful mappings. The basic idea is that every "layer" is a $\leq_T$-antichain. This ensures that every $\preceq_D$-minimal set of two-valued interpretations is realisable under the stable semantics [Pührer, 2020]. For example, it is shown there that a set of two-valued interpretations is realisable under the stable semantics if and only if it forms an anti-chain under $\leq_t$, i.e. every two interpretations in the set are $\leq_t$-incomparable. The need for an additional requirement on faithful orderings is shown by the following example

Example 21. *Consider the ADF D from Example 1 and consider $\preceq$ defined as:*

$$\overline{a}bc, a\overline{b}c \prec abc, \overline{a}b\overline{c}, \overline{a}\overline{b}c, a\overline{b}\overline{c} \prec \ldots$$

Notice that $\preceq$ is faithful w.r.t. stable. If we revise by $ab \vee \neg c$ by selecting the $\preceq$-minimal models satisfying $ab \vee \neg c$, we obtain $\mathsf{stable}(D \star (ab \vee \neg c)) = \{abc, \overline{a}b\overline{c}, a\overline{b}\overline{c}\}$. However, there exists no ADF $(D \star (ab \vee \neg c)) \in \mathfrak{D}(\{a,b,c\})$ with $\{abc, \overline{a}b\overline{c}, a\overline{b}\overline{c}\}$ as stable models, since $\overline{a}b\overline{c} <_\top abc$ contradicts $\mathsf{stable}(D \star (ab \vee \neg c))$ forming an $<_\top$-antichain (which we know in view of the results of [Pührer, 2020]).

This problematic behaviour can be avoided by requiring additionally that every layer of a faithful mapping is an $\leq_\top$-antichain:

Definition 5.15. *Given a semantics Sem s.t. $\mathsf{Sem}(D) \subseteq \Omega(\mathsf{At})$ for every $D \in \mathfrak{D}(\mathsf{At})$, a function $f : D \mapsto \preceq_D$ assigning a total preorder $\preceq_D$ over $\Omega(\mathsf{At})$ to every ADF $D \in \mathfrak{D}(\mathsf{At})$ is a $\top$-modular faithful assignment w.r.t. the semantics Sem iff:*

1. *if $\omega_1 \preceq_D \omega_2$ and $\omega_2 \preceq_D \omega_1$ then $\omega_1 \not<_\top \omega_2$ and $\omega_2 \not<_\top \omega_1$;*

2. *For every $D \in \mathfrak{D}(\mathsf{At})$, if $\omega, \omega' \in \mathsf{Sem}(D)$ then $\omega' \preceq_D \omega$;*

3. *for every $D \in \mathfrak{D}(\mathsf{At})$, if $\omega \in \mathsf{Sem}(D)$ and $\omega' \notin \mathsf{Sem}(D)$ then $\omega \prec_D \omega'$;*

4. *for every $D, D' \in \mathfrak{D}(\mathsf{At})$, if $\mathsf{Sem}(D) = \mathsf{Sem}(D')$ then $\preceq_D = \preceq_{\mathsf{Sem}(D')}$ for any ADF $D' = (\mathsf{At}, L', C')$.*

Thus, the above definition extends faithful mappings with the requirement that every layer is $\leq_t$-modular.

Example 22. *Consider again the preorder $\preceq$ from Example 21. We can turn this into a $\top$-modular faithful mapping $\preceq'$ as follows (among many other possibilities):*

$$\overline{a}bc, a\overline{b}c \prec' abc \prec' \overline{a}b\overline{c}, \overline{a}bc, a\overline{b}\overline{c} \prec' \dots$$

Revising D by $ab \vee \neg c$ now results in $\mathsf{stable}(D \star (ab \vee \neg c)) = \{abc\}$. By the results of [Pührer, 2020], $\{abc\}$ is realisable under stable semantics. This illustrates the usefulness of $\top$-modular faithful mappings, as now any selection is ensured to be realisable under stable semantics. This is further ilustrated by the following propositions.

Theorem 5.16. *An operator* $\star : \mathfrak{D}(\mathsf{At}) \times \mathcal{L}(\mathsf{At}) \to \mathcal{L}(\mathsf{At})$ *is a revision operator* $\star$ *for stable semantics iff there exists a function* $f : \mathfrak{D}(\mathsf{At}) \to \wp(\Omega(\mathsf{At}) \times \Omega(\mathsf{At}))$ *that is* $\top$-*modular faithful w.r.t.* stable *s.t.:*

$$\mathsf{stable}(D \star \psi) = \min_{\preceq_D}(\mathsf{Mod}(\psi)) \tag{3}$$

5.2.2 Dynamic Conditionals

On the basis of a revision operator, one can define conditional inference based on the *Ramsey-test*. In more detail, we can now stipulate that the conditional $(\psi|\phi)$ follows from the ADF D (relative to a revision operator $\star$ for some semantics Sem), in symbols $D \mid\!\sim_\star^{\mathsf{Sem}} (\psi|\phi)$, iff ψ is in all Sem-models of $D \star \phi$. More informally, the conditional 'if ϕ then usually ψ' is true in the argumentative context D if and only if ψ is true according to all argumentative positions that can be rationally taken in the argumentative context resulting from D revised by ϕ.

We first notice that, given a $\mathsf{ADF}_\star^2$-operator $\star$ and an ADF D, where $\star$ is based on the total preorder $f^\star(D) =\preceq_D$, see Theorem 5.14, a dynamical conditional consequence relation $D \mid\!\sim_\star^{\mathsf{Sem}}$ can be equivalently represented as conditional inference relation induced by the total preorder $\preceq_D$ over Ω. Given a $\mathsf{ADF}_\star^2$-operator satisfying $(\mathsf{ADF}_\star^2 1)$-$(\mathsf{ADF}_\star^2 6)$, we denote by $f^\star(D)$ the total preorder over Ω induced by $\star$ and D as in Theorem 5.14.[8]

Proposition 5.17. *Given a semantics* $\mathsf{Sem} \in \{\mathit{2val}, \mathsf{stable}\}$, *an ADF* D *and a* $\mathsf{ADF}_\star^2$-*operator* $\star$ *satisfying* $(\mathsf{ADF}_\star^2 1)$-$(\mathsf{ADF}_\star^2 6)$, $D \mid\!\sim_\star^{\mathsf{Sem}} (\psi|\phi)$ *iff* $\phi \mid\!\sim_{f^\star(D)} \psi$.

Thus, conditional inference based on ADFs w.r.t. two-valued semantics is a special case of preferential inference. This stands in contrast with dynamic conditionals based on three-valued semantics, for which an extension to a three-valued logic (such as Kleene's logic) is necessary. For more details, we refer to [Heyninck et al., 2023].

Example 23. *Continuing with Example 20, we see that* $D \mid\!\sim_{\star_{\mathsf{d1}}}^{\mathsf{Sem}} (a|\neg b)$ *in view of* $\mathsf{2mod}(D \star_{\mathsf{d1}} \neg c) = \{a\overline{b}\overline{c}, \overline{a}b\overline{c}\}$, *i.e. if Cochem is not a viable*

---

[8]Notice that the semantics Sem relative to which a $\mathsf{ADF}_\star^2$-operator is defined are implicitly taken into account in $f^\star(D)$, in the sense that the realisability of this semantics will be taken into account in the additional conditions on the total preorder.

travel option anymore, we will still go to Addis Aba if we don't go to Boston.

5.3 Structured Argumentation

In structured argumentation, arguments are not considered as abstract, atomic entities but are kind of rules, consisting of premises and conclusions, making the flow of reasoning more transparent. Since we focus on conditionals in argumentation here, we recall prominent approaches which make use of defeasible rules which are particularly well aligned to conditionals.

5.3.1 Defeasible Logic Programming and Ranking Functions

*Defeasible Logic Programming (*DeLP*)* [García and Simari, 2004] combines logic programming with defeasible argumentation. DeLP works in a highly dialectical way, allowing series of attacks and counterattacks to finally mark those statements as *warranted* for which all attackers could be invalidated. Attacks in DeLP are identified via logical contradictions, but the defeat relation needs a preference relation that originally was based on a notion of specificity. The paper [Kern-Isberner and Simari, 2011] makes use of ranking functions [Spohn, 1988] and more specific information from System Z [Pearl, 1990] to define preference (and hence defeat) between arguments. To this aim, the authors introduce the notions of examples and counterexamples of arguments via possible worlds which are evaluated on the base of ranking functions. The basic idea here is that arguments are as convincing and successful as their most plausible examples, and arguments with more plausible examples should prevail. We recall the basics of this approach from [Kern-Isberner and Simari, 2011], where the strict parts of defeasible logic programs are restricted to be facts.

Let $\mathcal{L}$ be a finitely generated propositional language with atoms $a, b, c, \ldots$, and with formulas $A, B, C, \ldots$, and let Ω denote the set of possible worlds over $\mathcal{L}$. A *defeasible logic program (de.l.p.)* $\mathcal{P} = (\Phi, \Delta)$ consists of a set Φ of facts[9] and a set Δ of defeasible rules which are written as conditionals $\delta = (L|B_1 \ldots B_n)$ with literals $L, B_1, \ldots, B_n$. In

---
[9]Note that in general, the strict part of a *de.l.p.* also may contain strict rules.

accordance with the notions in logic programming, we call L the *head* of the conditional ($L = head(\delta)$) and $B_1 \ldots B_n$ its *body*. Notice that the syntax of rules in DeLP is a special case of that of conditional logics, as the heads consist of single literals and the bodies consist of conjunctions of literals, whereas in conditional logic, any propositional formula is allowed in both the antecedent and the consequent. A literal L can be *defeasibly derived* from $\Delta' \subseteq \Delta$, $\Delta' \mid\sim L$, iff there exists a finite sequence $L_1, \ldots, L_n = L$ of ground literals, such that each L_i is either a fact in Π or there exists a rule in $\Pi \cup \Delta'$ with head L_i and body $\{B_1, \ldots, B_m\}$, and every literal B_j in the body is such that $B_j \in \{L_k\}_{k<i}$. $\Phi \cup \Delta'$ is called *contradictory* iff there is a literal L such that both L and $\overline{L}$ have defeasible derivations from $\Phi \cup \Delta'$. For any *de.l.p.* $\mathcal{P}$, we will presuppose that Φ is non-contradictory.

Given a *de.l.p.* $\mathcal{P} = (\Phi, \Delta)$ and a literal L, $\mathcal{A}$ is an argument for L, denoted $\langle \mathcal{A}, L \rangle$, if $\mathcal{A}$ is a minimal set of defeasible rules in Δ such that there exists a defeasible derivation of L from $\Phi \cup \mathcal{A}$, and $\Phi \cup \mathcal{A}$ is non-contradictory.

An argument $\langle \mathcal{B}, Q \rangle$ is a sub-argument of $\langle \mathcal{A}, L \rangle$ if $\mathcal{B}$ is subset of $\mathcal{A}$. Argument $\langle \mathcal{A}_1, L_1 \rangle$ *attacks*, or *counterargues* another $\langle \mathcal{A}_2, L_2 \rangle$ at a literal L if there exists a sub-argument of $\langle \mathcal{A}_2, L_2 \rangle$, $\langle \mathcal{A}, L \rangle$, i.e., $\mathcal{A} \subseteq \mathcal{A}_2$, such that there exists a literal L' verifying both $\Phi \cup \{L, L_1\} \mid\sim L'$ and $\Phi \cup \{L, L_1\} \mid\sim \overline{L'}$. Note that an argument $\langle \emptyset, L \rangle$ with $L \in \Pi$ can not be attacked since all arguments have to be consistent with Φ. Finally, another crucial notion involving consistency in DeLP is the notion of concordance. A set of arguments $\mathcal{A}_i, 1 \leq i \leq m$, of a defeasible logic program (Φ, Δ) is called *concordant* iff $\Phi \cup \bigcup_{i=1}^{n} \mathcal{A}_i$ is non-contradictory.

Example 24. *We consider the propositional variables b bird, p penguin, c chicken, s is_scared, f flies, w has_wings, and the set of conditionals:* $\Delta = \{\delta_1 = (b|c), \delta_2 = (b|p), \delta_3 = (f|b), \delta_4 = (\overline{f}|p), \delta_5 = (\overline{f}|c), \delta_6 = (f|cs), \delta_7 = (w|b)\}$. *For a de.l.p., this set of conditionals can be instantiated with various facts. For example, consider the defeasible logic program* $\mathcal{P}_1 = (\{cs\}, \Delta)$. *Then the following arguments can be built* supporting f resp. $\overline{f}$:

$$\langle \mathcal{A}_1, f\rangle, \quad \mathcal{A}_1 = \{(b|c),(f|b)\};$$
$$\langle \mathcal{A}_2, \overline{f}\rangle, \quad \mathcal{A}_2 = \{(\overline{f}|c)\};$$
$$\langle \mathcal{A}_3, f\rangle, \quad \mathcal{A}_3 = \{(f|cs)\}.$$

Clearly, $\langle \mathcal{A}_2, \overline{f}\rangle$ attacks $\langle \mathcal{A}_1, f\rangle$, and $\langle \mathcal{A}_3, f\rangle$ attacks $\langle \mathcal{A}_2, \overline{f}\rangle$. Note that $\{\langle \mathcal{A}_1, f\rangle, \langle \mathcal{A}_3, f\rangle\}$ is concordant, while $\{\langle \mathcal{A}_1, f\rangle, \langle \mathcal{A}_2, \overline{f}\rangle, \langle \mathcal{A}_3, f\rangle\}$ is not.

As usual in argumentation theory, an attacked argument may not be lost, but can be found to be stronger than its attacker(s). DELP makes use of a preference relation to compare arguments, and in the end, the crucial question in DELP is whether an argument is *warranted*. For the moment, we leave the exact instantiation of the preference relation open because the procedure to ensure warranty in DELP is the same for any suitable preference relation.

If $\langle \mathcal{A}_1, L_1\rangle$ and $\langle \mathcal{A}_2, L_2\rangle$ are two arguments $\langle \mathcal{A}_1, L_1\rangle$ is a *proper defeater* for $\langle \mathcal{A}_2, L_2\rangle$ at literal L iff there exists a sub-argument of $\langle \mathcal{A}_2, L_2\rangle$, $\langle \mathcal{A}, L\rangle$ such that $\langle \mathcal{A}_1, L_1\rangle$ counterargues $\langle \mathcal{A}_2, L_2\rangle$ at L and $\langle \mathcal{A}_1, L_1\rangle$ is strictly preferred over $\langle \mathcal{A}, L\rangle$. Alternatively, $\langle \mathcal{A}_1, L_1\rangle$ is a *blocking defeater* for $\langle \mathcal{A}_2, L_2\rangle$ at literal L iff there exists a sub-argument of $\langle \mathcal{A}_2, L_2\rangle$, $\langle \mathcal{A}, L\rangle$ such that $\langle \mathcal{A}_1, L_1\rangle$ counterargues $\langle \mathcal{A}_2, L_2\rangle$ at L and neither $\langle \mathcal{A}_1, L_1\rangle$ is strictly preferred over $\langle \mathcal{A}, L\rangle$ nor is $\langle \mathcal{A}, L\rangle$ preferred over $\langle \mathcal{A}_1, L_1\rangle$. If $\langle \mathcal{A}_1, L_1\rangle$ is either a proper or a blocking defeater of $\langle \mathcal{A}_2, L_2\rangle$, it is said to be a *defeater* of the latter.

In the warranty procedure, arguments $\mathcal{A}$ are evaluated in so-called dialectical trees where the root of such a tree is the argument to be evaluated. The paths of the tree consist of (finite) acceptable argumentation lines $[\mathcal{A} = \langle \mathcal{A}_0, L_0\rangle, \langle \mathcal{A}_1, L_1\rangle, \langle \mathcal{A}_2, L_2\rangle, \cdots]$ where each node is a defeater of its parent node, and acceptability of the argument lines is specified by further constraints. Here, it is presupposed that both the sets of *supporting* arguments $[\langle \mathcal{A}_0, L_0\rangle, \langle \mathcal{A}_2, L_2\rangle, \langle \mathcal{A}_4, L_4\rangle, \cdots]$ and *interfering* arguments $[\langle \mathcal{A}_1, L_1\rangle, \langle \mathcal{A}_3, L_3\rangle, \langle \mathcal{A}_5, L_5\rangle, \cdots]$ are concordant. Finally, the nodes are marked U (undefeated) or D (defeated), where a node is marked U iff every child is marked D; in particular, leaves are marked U.

As a novelty, in [Kern-Isberner and Simari, 2011], preference between arguments in DELP was given an example-based semantics.

Definition 5.18 (Examples, counterexamples). *Let* $\mathcal{P} = (\Phi, \Delta)$ *be a defeasible logic program. Let* $\omega \in \Omega$ *be a possible world, and let* $\langle \mathcal{A}, L \rangle$ *be an argument in* $\mathcal{P}$.

ω *is an* example *for* $\langle \mathcal{A}, L \rangle$ *iff* ω *satisfies all facts,* $\omega \models \Phi$, *and* ω *verifies all rules in* $\mathcal{A}$. ω *is a* counterexample *to* $\langle \mathcal{A}, L \rangle$ *iff* $\omega \models \Phi$ *and there is at least one rule in* $\mathcal{A}$ *that is falsified by* ω. ω *is a* supported counterexample *to* $\langle \mathcal{A}, L \rangle$ *iff* ω *is a counterexample to* $\langle \mathcal{A}, L \rangle$ *and there is an argument* $\langle \mathcal{A}', L' \rangle$ *such that* ω *is an example of* $\langle \mathcal{A}', L' \rangle$.

The set of examples of an argument $\langle \mathcal{A}, L \rangle$ *is denoted by* $\langle \mathcal{A}, L \rangle^+$, *the set of counterexamples by* $\langle \mathcal{A}, L \rangle^-$.

From the definition, it is immediately clear that $\langle \mathcal{A}, L \rangle^+ = Mod(\Phi \wedge \bigwedge_{\delta \in \mathcal{A}} head(\delta))$, and $\langle \mathcal{A}, L \rangle^- = Mod(\Phi \wedge \bigvee_{\delta \in \mathcal{A}} \overline{head(\delta)})$. By the definition of arguments, it is ensured that every argument has examples.

Example 25. *For the arguments* $\langle \mathcal{A}_1, f \rangle, \langle \mathcal{A}_2, \overline{f} \rangle, \langle \mathcal{A}_3, f \rangle$ *stated in example 24, examples and counterexamples are given as follows:*

$$\langle \mathcal{A}_1, f \rangle^+ = Mod(csbf) \quad \langle \mathcal{A}_1, f \rangle^- = Mod(cs(\overline{b} \vee \overline{f}))$$
$$\langle \mathcal{A}_2, \overline{f} \rangle^+ = Mod(cs\overline{f}) \quad \langle \mathcal{A}_2, \overline{f} \rangle^- = Mod(csf)$$
$$\langle \mathcal{A}_3, f \rangle^+ = Mod(csf) \quad \langle \mathcal{A}_3, f \rangle^- = Mod(cs\overline{f})$$

Hence, $\omega_1 = csb\overline{p}fw$ *is an example of* $\langle \mathcal{A}_1, f \rangle$ *and* $\langle \mathcal{A}_3, f \rangle$ *and a counterexample to* $\langle \mathcal{A}_2, \overline{f} \rangle$. *Reciprocally,* $\omega_2 = csb\overline{p}\overline{f}w$ *is an example of* $\langle \mathcal{A}_2, \overline{f} \rangle$, *and a counterexample to* $\langle \mathcal{A}_1, f \rangle$ *and* $\langle \mathcal{A}_3, f \rangle$.

Attacks can be characterized in terms of examples, as the next proposition shows.

Proposition 5.19. *Let* $\langle \mathcal{A}_1, L_1 \rangle, \langle \mathcal{A}_2, L_2 \rangle$ *be two arguments. If* $\langle \mathcal{A}_1, L_1 \rangle$ *attacks* $\langle \mathcal{A}_2, L_2 \rangle$, *then all examples of* $\langle \mathcal{A}_1, L_1 \rangle$ *are (supported) counterexamples to* $\langle \mathcal{A}_2, L_2 \rangle$, *i.e.* $\langle \mathcal{A}_1, L_1 \rangle^+ \subseteq \langle \mathcal{A}_2, L_2 \rangle^-$. *Conversely, if all examples of* $\langle \mathcal{A}_1, L_1 \rangle$ *are counterexamples to* $\langle \mathcal{A}_2, L_2 \rangle$, *then there is a sub-argument of* $\langle \mathcal{A}_1, L_1 \rangle$ *that attacks* $\langle \mathcal{A}_2, L_2 \rangle$.

Moreover, examples are also helpful to check the crucial notion of concordance in argumentation lines.

Proposition 5.20. *A set of arguments* $\langle \mathcal{A}_i, L_i \rangle, 1 \leq i \leq m$, *is concordant iff they have common examples, i.e. iff* $\bigcap_{1 \leq i \leq m} \langle \mathcal{A}_i, L_i \rangle^+ \neq \emptyset$.

By bringing ranking functions now into the play, plausibility degrees of arguments can be defined. Arguments are assumed to be as plausible as their most plausible examples, and the plausibilities of their counterexamples represent the degree to which they can be challenged. This makes comparisons between arguments easy.

Definition 5.21 (κ-values of arguments, κ-preference). *Let κ be an ordinal conditional function on Ω, let $\langle \mathcal{A}, L \rangle$ be an argument. Then $\kappa^+(\langle \mathcal{A}, L \rangle) = \min\{\kappa(\omega) \mid \omega \in \langle \mathcal{A}, L \rangle^+\}$, and $\kappa^-(\langle \mathcal{A}, L \rangle) = \min\{\kappa(\omega) \mid \omega \in \langle \mathcal{A}, L \rangle^-\}$.*

Let $\langle \mathcal{A}_1, L_1 \rangle$, $\langle \mathcal{A}_2, L_2 \rangle$ be two arguments. Then $\langle \mathcal{A}_1, L_1 \rangle \succeq^\kappa \langle \mathcal{A}_2, L_2 \rangle$ iff $\kappa^+(\langle \mathcal{A}_1, L_1 \rangle) \leq \kappa^+(\langle \mathcal{A}_2, L_2 \rangle)$.

From the remarks above, it is immediately clear that $\kappa^+(\langle \mathcal{A}, L \rangle) = \kappa(\Phi \wedge \bigwedge_{\delta \in \mathcal{A}} head(\delta))$ and $\kappa^-(\langle \mathcal{A}, L \rangle) = \kappa(\Phi \wedge \bigvee_{\delta \in \mathcal{A}} \overline{head(\delta)})$.

κ-preference yields a declarative criterion for warrant:

Proposition 5.22. *Let $\langle \mathcal{A}, L \rangle$ be an argument. If*

$$\kappa^+(\langle \mathcal{A}, L \rangle) < \kappa^-(\langle \mathcal{A}, L \rangle)$$

then $\langle \mathcal{A}, L \rangle$ is undefeated and hence warranted.

Of course, when we use a ranking function κ to assess the plausibility of arguments built over a de.l.p. $\mathcal{P}$, we expect κ to be a model of Δ. To find such a proper model, we may make use of the distinguished system Z approach [Goldszmidt and Pearl, 1996a] as a particularly well-behaved ranking model.

Example 26. *We apply system Z to Δ from $\mathcal{P}_1$ in Example 24. Here, the tolerance partitioning used by system Z is $\Delta_0 = \{\delta_3, \delta_7\}, \Delta_1 = \{\delta_1, \delta_2, \delta_4, \delta_5\}, \Delta_2 = \{\delta_6\}$. We compute the κ_z-values of the arguments in Example 25 as follows:*

$$\begin{array}{llll}
\kappa_z^+(\langle \mathcal{A}_1, f \rangle) = \kappa_z(csbf) = 2 & \kappa_z^-(\langle \mathcal{A}_1, f \rangle) = \kappa_z(cs(\overline{b} \vee \overline{f})) = 2 \\
\kappa_z^+(\langle \mathcal{A}_2, \overline{f} \rangle) = \kappa_z(cs\overline{f}) = 3 & \kappa_z^-(\langle \mathcal{A}_2, \overline{f} \rangle) = \kappa_z(csf) & = 2 \\
\kappa_z^+(\langle \mathcal{A}_3, f \rangle) = \kappa_z(csf) = 2 & \kappa_z^-(\langle \mathcal{A}_3, f \rangle) = \kappa_z(cs\overline{f}) & = 3
\end{array}$$

From Proposition 5.22, we may conclude immediately that $\langle \mathcal{A}_3, f \rangle$ is a warrant for the literal f.

Let us now consider the defeasible logic program $\mathcal{P}_2 = (\{p\}, \Delta)$. *In system Z, we have* $\kappa_z(pw) = \kappa_z(p\overline{w}) = 1$, *so, the status of the query* w *can not be determined by system Z. This means that it cannot be proved in system Z if penguins have wings. This effect has become known as the drowning effect (see, e.g. [Goldszmidt and Pearl, 1996a]).*

This problem can be solved in our argumentation framework: The only argument that can be built to connect p *and* w *is* $\langle\{(b|p),(w|b)\}, w\rangle$, *which is not attacked at all, so, in particular, is undefeated. Hence,* w *can be warranted. Note, however, that Proposition 5.22 would not be helpful here because* $\kappa_z^+(\langle\{(b|p),(w|b)\}, w\rangle) = \kappa_z(pbw) = 1 = \kappa_z(p(\overline{b} \vee \overline{w}) = \kappa_z^-(\langle\{(b|p),(w|b)\}, w\rangle)$.

We thus see here that insights from conditional logics can be made useful for argumentation (e.g. by supplying a preference relation as above), and that argumentation can help improve existing conditional logics (e.g. by helping in avoiding the drowning effect as demonstrated above).

Moreover, in [Kern-Isberner and Simari, 2011], the authors also proposed another preference relation between arguments which is based on system Z by, a bit more simply, comparing the Z-values of the conditionals contained in the involved arguments. This allows for a declarative criterion for ensuring warranty that just considers the (positive) examples of an argument. For further details, we refer to [Kern-Isberner and Simari, 2011].

5.3.2 Pollock's Defeasible Reasoning and Ranking Functions

Pollock developed a theory of defeasible reasoning [Pollock, 1995] where arguments consist of a set of premises and a conclusion which are connected by an inference rule, or reason-schema, respectively. In [Spohn, 2002], Spohn briefly discussed the basic ideas of Pollock's work and elaborated on possible connections to his own framework of ranking functions.

The core of Pollock's theory is a large set of defeasible inference rules which can be seen as specific proposals for a constructive theory of defeasible reasoning. Arguments have strengths and can be defeated, and Pollock proposed a formal theory of how defeats and strengths interact in an integrated graph with the aim of arriving at warranted beliefs.

Doxastic states are seen as huge networks of inferences and justifications, and all reasoning starts with perceptions.

Spohn appreciated the constructive and dynamic (regarding the flow of reasoning) nature of Pollock's theory, but criticizes it to be basically static because no new information (which are restricted to perceptions in Pollock's theory) can be taken into account in a way that makes the flow of change transparent. For each new perception, the whole reasoning machinery has to start again.

According to Spohn, Pollock's theory overlaps with ranking theory insofar as both approaches deal with justified and warranted belief. However, while ranking theory describes declaratively[10] how such beliefs behave, Pollock's defeasible reasoning implements how such beliefs emerge in many procedural ways. According to [Pollock and Cruz, 1999], all norms of rationality have to be procedural. This seems to be the most crucial difference between both approaches.

The problem with theories of defeasible reasoning like Pollock's approach where inference relies on intuitive procedural rules is that there is no independent assessment of the quality of their products, i.e., warranted beliefs. Spohn called it "normative defectiveness".

Defeasible logic programming (DeLP), as described in Section 5.3.1, is also mainly procedural in elaborating warranted beliefs, but it relies on basic logic by exploring contradictions and uses declarative notions like a (more or less) abstract preference relation to determine defeats between arguments. However, while it uses logic programming as kind of a base logic, the semantics of warranted beliefs in DeLP cannot be fully captured by answer set semantics [Thimm and Kern-Isberner, 2008a]. Nevertheless, DeLP appears to be a good compromise between procedural vs. declarative (or computational vs. regulative, as Spohn termed it in [Spohn, 2002]) approaches.

5.3.3 Structured Argumentation Based on Axiomatic Conditional Logic

In the paper [Besnard et al., 2013], the authors extend the deduction-based approach to argumentation from [Besnard and Hunter, 2001] by

---

[10]Spohn called ranking theory a *regulative theory*.

introducing an additional conditional connective ⇒ (giving rise to a logical language $\mathcal{L}_c$) and the novel concept of contrariety between arguments (formulas of $\mathcal{L}_c$). Conditional rules in $\mathcal{L}_c$ specified by ⇒ are meant to be hypotheses to be used for tentative reasoning, but which can be attacked by contrary rules in an argumentative process.

For implementing conditional reasoning, Besnard et al. make use of the conditional logic MP [Chellas, 1975] which is defined beyond Boolean logic by the following axioms and rules of inference $\vdash_c$:

RCEA $\quad \dfrac{\vdash_c \alpha \leftrightarrow \beta}{\vdash_c (\alpha \Rightarrow \gamma) \leftrightarrow (\beta \Rightarrow \gamma)}$

RCEC $\quad \dfrac{\vdash_c \alpha \leftrightarrow \beta}{\vdash_c (\gamma \Rightarrow \alpha) \leftrightarrow (\gamma \Rightarrow \beta)}$

CC $\quad \vdash_c ((\alpha \Rightarrow \beta) \land (\alpha \Rightarrow \gamma)) \to (\alpha \Rightarrow (\beta \land \gamma))$

CM $\quad \vdash_c (\alpha \Rightarrow (\beta \land \gamma)) \to ((\alpha \Rightarrow \beta) \land (\alpha \Rightarrow \gamma))$

CN $\quad \vdash_c (\alpha \Rightarrow \top)$

MP $\quad \vdash_c (\alpha \Rightarrow \beta) \to (\alpha \to \beta)$

Note that (RCEC) and (MP) are also axioms of Stalnaker's logic **C2** that we described in Section 3.1. Contrariety is then defined on the base of the logic MP and covers two main cases: first, $\alpha \in \mathcal{L}_c$ is *contrary to* β if both formulas are inconsistent in MP, i.e., $\{\alpha, \beta\} \vdash_c \bot$. The second case deals explicitly with rules involving ⇒. The basic idea is that a formula $\alpha = \phi \land \epsilon \Rightarrow \psi$ should be *in contrariety to* $\beta = \phi \Rightarrow \psi$ because α suggests that additional preconditions must be satisfied for β to hold. For the precise formal definition of contrariety, we refer to the original paper [Besnard et al., 2013]. If α is in contrariety to β, this is denoted by $\alpha \bowtie \beta$. Note that $\bowtie$ is neither symmetric, nor antisymmetric. Contrariety is lifted to sets of formulas by $\alpha \bowtie \Phi$ if there is $\beta \in \mathcal{L}_c$ such that $\Phi \vdash_c \beta$ and $\alpha \bowtie \beta$.

Given a knowledge base $\Delta \subseteq \mathcal{L}_c$, an argument is a pair $\langle \Phi, \alpha \rangle$ where the following conditions hold:

- $\Phi \subseteq \Delta$;

- for all β such that $\Phi \vdash_c \beta$, $\beta \not\bowtie \Phi$;

- $\Phi \vdash_c \alpha$;

- for all $\Phi' \subset \Phi$, $\Phi' \not\vdash_c \alpha$.

Two arguments $\langle\Phi,\alpha\rangle, \langle\Psi,\beta\rangle$ are *quasi-identical* if $\Phi = \Psi$ and $\alpha \equiv_c \beta$, where $\alpha \equiv_c \beta$ means $\alpha \vdash_c \beta$ and $\beta \vdash_c \alpha$.

Attacks between arguments are defined in terms of contrariety. An argument $\langle\Psi,\beta\rangle$ is a *rebuttal* for $\langle\Phi,\alpha\rangle$ if $\beta \bowtie \alpha$, and $\langle\Psi,\beta\rangle$ is a *defeater* for $\langle\Phi,\alpha\rangle$ if $\beta \bowtie \Phi$. Besnard et al. show that rebuttals are subsumed by defeaters so that we can focus on defeaters from now on. However, defeaters can be quite general so that we need additional attributes to characterize most relevant defeaters. First, defeaters should be most specific both in a set-theoretical and logical sense: an argument $\langle\Phi,\alpha\rangle$ is at least as *conservative* than an argument $\langle\Psi,\beta\rangle$ if $\Phi \subseteq \Psi$ and $\beta \vdash_c \alpha$. In the following, an enumeration $\langle\Psi_1,\beta_1\rangle, \langle\Psi_2,\beta_2\rangle \ldots$ of all maximally conservative defeaters for $\langle\Phi,\alpha\rangle$ is assumed to be fixed for each argument. $\langle\Psi_i,\beta_i\rangle$ is a *pertinent defeater* for $\langle\Phi,\alpha\rangle$ if for each $j < i$, $\langle\Psi_i,\beta_i\rangle$ and $\langle\Psi_j,\beta_j\rangle$ are not quasi-identical.

Finally, pertinent defeaters are used to build argumentation trees. An *argumentation tree* for α has an argument for α as its root, and each child node is a pertinent defeater of its parent node; moreover, for each node $\langle\Psi,\beta\rangle$ with ancestor nodes $\langle\Psi'_1,\beta'_1\rangle, \ldots, \langle\Psi'_n,\beta'_n\rangle$, we have $\Psi \not\subseteq \Psi'_i$ for $i \in \{1,\ldots,n\}$. Argumentation in this conditional logic then may follow the lines of the classical framework in [Besnard and Hunter, 2001].

It is interesting to note that the semantics for conditionals provided by ranking functions [Spohn, 1988] which is used to equip DELP argumentation with an example-based semantics in Section 5.3.1 satisfies the axioms and inference rules of the conditional logic MP (under mild prerequisites). This can easily be verified by observing that RCEA, RCEC, CC, and CM are implied by system P [Kraus et al., 1990a] (see also Section 3.1) which inference based on ranking functions is known to satisfy, and MP is a simple arithmetic exercise for ranking functions. CN holds for consistent formulas as all ranks assigned to worlds are finite. An interesting research question would be if ranking functions can also provide a semantics for the approach presented in [Besnard et al., 2013], and how contrariety can be characterized in terms of ranking functions.

5.4 Other approaches

We now shortly discuss some other approaches that can be argued to connect structured argumentation and conditional logics.

Gabbay and d'Avila Garcez [Gabbay and d'Avila Garcez, 2009] ask a methodological question about structured argumentation by giving detailed considerations on the different options for instantiating abstract argumentation frameworks. This paper argues that there is a wide variety of options to do so, and gives several detailed examples of how this can be done. Among others, non-monotonic logics, i.e. consequence relations satisfying reflexivity, cut and cautious monotony, are discussed. In more detail, Gabbay and d'Avila Garcez suggest that nodes in an argumentation graph could represent pairs of sets of non-monotonic conditionals and a conclusion based on these conditionals, and that an argument (Δ, ϕ) attacks an argument (Θ, ψ) if adding ϕ leads to ψ not being derivable anymore in view of Θ. For example (adapting the notation somewhat to our chapter), $(\{p\}, p)$ attacks $(\{b, b \mathrel{|\!\sim}_f p \wedge b \mathrel{|\!\sim}_{\neg}^f\}, f)$ as the knowledge that something is a penguin no longer allows us to derive that it flies according to most non-monotonic logics.

A brand of non-monotonic logics that allows to reason with conditional statements that we have not discussed here are *input/output-logics* [Makinson and Van Der Torre, 2000]. They provide a fine-grained picture of the different ways of reaching a conclusion by forward chaining a selected subset of conditionals, and have been proven especially useful in deontic logics. These logics are given an argumentative characterisation by Van Berkel and Straßer [Van Berkel and Straßer, 2022]. They do this by defining *deontic argument calculi*, which allow for structured argumentation on the basis of a set of conditionals interpreted as normative statements. The different input/output-logics from the literature are then captured by allowing for different inference rules in the process of argument construction, many of which are quite familiar to the axioms from conditional logics. It is an interesting question whether also the conditional logics discussed above can be represented in a similar way.

6 Further works

Both conditional logic and abstract argumentation (or some extension of it such as ADFs) are logical formalisms for reasoning. In this chapter, we discussed several ideas on how to combine these formalisms into a single formalism. Our focus was on works where we used the foundational ideas of semantical evaluation from one formalism and applied it in the other. Another general approach for combining arbitrary logics into a joint formalisms is that of *fibring*, see [Gabbay, 1995, 1996]. Given two logics $\mathcal{L}_1$ and $\mathcal{L}_2$, the fibring $\mathcal{L}_{1,2}$ of $\mathcal{L}_1$ and $\mathcal{L}_2$ that combines both syntax and semantics for both base logics in a simple manner. The syntax of $\mathcal{L}_{1,2}$ allows for an arbitrary combination of the syntax of $\mathcal{L}_1$ and $\mathcal{L}_2$, e.g., formulas may contain connectors of both $\mathcal{L}_1$ and $\mathcal{L}_2$ in an arbitrary manner. Informally speaking, if one were to fibre conditional logic and abstract argumentation, a valid formula would be $(aRb|bRc)$ with the intuitive meaning "if b attacks c, then usually a attacks b". The semantics of $\mathcal{L}_{1,2}$ is then a combination of the semantics of both $\mathcal{L}_1$ and $\mathcal{L}_2$ as well. In particular, [Gabbay, 1996] defines an inference relation $\mathrel{|\!\sim}_{1,2}$ on $\mathcal{L}_{1,2}$ that is a conservative extension of given inference relations $\mathrel{|\!\sim}_1$ and $\mathrel{|\!\sim}_2$ on $\mathcal{L}_1$ and $\mathcal{L}_2$, respectively, in the sense, that $\mathcal{L}_{1,2}$ coincides with $\mathcal{L}_1$ and $\mathcal{L}_2$ on the respective syntactical fragments of $\mathcal{L}_{1,2}$. However, properties of this new inference relation $\mathrel{|\!\sim}_{1,2}$ cannot be derived in a general manner and depend highly on the logics $\mathcal{L}_1$ and $\mathcal{L}_2$ and their inference relations $\mathrel{|\!\sim}_1$ and $\mathrel{|\!\sim}_2$, respectively. In essence, fibring logics allows for *joint* reasoning of two different formalisms in one single framework, while most of the work discussed in this chapter was concerned with an *integrated* approach to reasoning. How exactly a fibred logic using conditional logic and abstract argumentation (or ADFs) behaves, could be an interesting avenue for future work, though.

[Weydert, 2012, 2013, 2014] presents a new semantics for abstract argumentation, which is also rooted in conditional logical terms. In more detail, a ranking interpretation is provided for extensions of arguments instantiated by strict and defeasible rules by using conditional ranking semantics. Thus, Weydert presupposes a conditional knowledge base that is used to construct an argumentation framework.

[Bochman, 2016] relates Abstract Dialectical Frameworks to causal reasoning, and, more precisely, to Pearl's causal models [2009]. In essence, a causal model describes causal dependencies between *exogeneous* variables (which cannot directly be observed) and *endogenous* variables, which can be observed. A causal model formalises how states of variables are caused by other states of variables and, due to the non-monotonicity of causality, a causal model can thus be interpreted as a specific non-monotonic theory, quite similar to conditional logics. Bochman then shows certain correspondences between semantical notions of ADFs and causal models by modelling acceptance conditions of ADFs as causal rules. A previous study by [Bochman, 2005] already revealed similar relationships between assumption-based argumentation by [Bondarenko et al., 1997] and the causal reasoning approach of [Giunchiglia et al., 2004].

Another contribution of Alexander Bochman [Bochman, 2006] proposes a conceptual differentiation between two paradigms of non-monotonic reasoning, which he calls *preferential* and *explanatory*. The conditional logics discussed above fall under the first paradigm, whereas argumentation is an example of a formalism for explanatory non-monotonic reasoning. A number of differences between the two kinds of non-monotonic reasoning are discussed, and a general axiomatic theory for each of these paradigms is given. Even though the general conclusion of [Bochman, 2006] agrees with the insights expounded in this overview, we leave a deeper comparison between these works for the future.

Verheij has initiated a line of work [Verheij, 2017] that integrates ideas from preferential reasoning into argumentation by means of so-called *case models*. A case model $(C, \geq)$ consists of a set C of logically consistent, mutually incompatible formulas and a total preorder $\geq$ over these cases. Arguments, conceived of as pairs of formulas (ϕ, ψ) representing the premise ϕ and conclusion ψ, are then classified on the basis of a case model using ideas inspired by preferential semantics. For example, an argument (ϕ, ψ) is *presumptively valid*[11] if, among all cases verifying the premise ϕ, there is a $\geq$-maximal case that also verifies ψ.

[Alviano et al., 2022, 2023] consider preferential interpretations for abstract argumentation frameworks that are derived from gradual semantics. The latter allow to assign numeric values of argument strength

---

[11][Verheij, 2017] uses different notations for different kinds of arguments, e.g. a presumptively valid argument is denoted by $\phi \rightsquigarrow \psi$.

to individual arguments and are therefore similar to ranking functions (and therefore preferential interpretations) for conditional logics. This approach allows to reason over argument acceptance (and arbitrary formulas over arguments) through defeasible rules that can be derived from the preferential interpretations.

[Thimm and Kern-Isberner, 2013, 2014] introduce *stratified labelings*, a semantical approach to abstract argumentation frameworks, where arguments receive a non-negative natural number that assesses the *controversiality* of arguments and are inspired by ordinal ranking functions from conditional logics. As a matter of fact, [Thimm and Kern-Isberner, 2013] show that conditional knowledge bases can be transformed into abstract argumentation frameworks, such that rational stratified labelings of the latter behave similarly as the system Z ranking function of the former.

The behaviour of abstract argumentation in dynamic settings in is studied [Rienstra et al., 2015b]. In more detail, they ask the question whether the labelling status of arguments is preserved when adding or removing arguments or attacks in an abstract argumentation framework. This conceptually is quite similar to postulates such as (cautious) monotony, where beliefs persist when adding (believed) formulas.

Finally, we notice that some foundational papers on non-monotonic conditionals expand on ideas that are, at least conceptually, related to argumentation. For example, in Lehmann and Magidor's prolific paper on rational closure [Lehmann and Magidor, 1992a], the authors motivate the preference-comparison between cumulative models using the notions of attack and defence. Geffner and Pearl [Geffner and Pearl, 1992] go even further, giving a full-fledged argumentative proof theory that is sound and complete with respect to their conditional inference method of *conditional entailment*.

There are further works that only loosely touch on the subject of this chapter, but still model some aspect of *conditional inference*. For example, [Bernreiter et al., 2022] introduce *conditional preference-based argumentation frameworks*, which allow the specification of preferences between arguments. In fact, preferences are given conditioned on selected sets of arguments and can differ for different sets. A similar approach for structured argumentation is discussed by [Dung et al., 2019].

[Perotti et al., 2011] define *conditional labels* for arguments that describe conditions about the acceptance of arguments, given the status of other arguments (similarly as acceptance conditions in ADFs). These can be used in dialogues to enable strategic moves of agents. Another form of *conditional labelings* are presented by [Booth et al., 2012]. Here, a conditional labeling assigns acceptance status to arguments, under the condition that another set of labelings is assumed to evaluate the argumentation framework rationally. Using conditional labelings, the strict semantical evaluation of classical semantics can be relaxed and conditional labelings models rationality as close as possible, given the circumstances. The work of [Booth et al., 2012] therefore shares some motivation with the work of [Skiba et al., 2021; Skiba and Thimm, 2022] that we discussed in Section 5.

7 Summary and Conclusion

In this chapter, we gave a thorough introduction to the logic of conditionals, and have surveyed work that integrated ideas inspired by conditional logics into formal argumentation. We saw that despite the differences between the two approaches (Section 5), integrating insights from conditional logic into formal argumentation is still useful and results in richer argumentative models (as demonstrated in Section 4 and 5.3), while argumentative models can also improve upon conditional logics (as we saw in Section 5.3). As indicated in several parts of this chapter, we believe there is still a lot of exciting work to be done in this area, and hope our chapter will serve as a useful basis for such further investigations.

Acknowledgements

The research reported here was partially supported by the Deutsche Forschungsgemeinschaft under grant 423456621.

References

Ernest W Adams. Subjunctive and indicative conditionals. *Foundations of language*, pages 89–94, 1970.

Carlos E Alchourrón, Peter Gärdenfors, and David Makinson. On the logic of theory change: Partial meet contraction and revision functions. *Journal of symbolic logic*, pages 510–530, 1985.

Mario Alviano, Laura Giordano, and Daniele Theseider Dupré. Many-valued argumentation, conditionals and a probabilistic semantics for gradual argumentation. *CoRR*, abs/2212.07523, 2022.

Mario Alviano, Laura Giordano, and Daniele Theseider Dupré. Typicality, conditionals and a probabilistic semantics for gradual argumentation. In Kai Sauerwald and Matthias Thimm, editors, *Proceedings of the 21st International Workshop on Non-Monotonic Reasoning co-located with the 20th International Conference on Principles of Knowledge Representation and Reasoning (KR 2023) and co-located with the 36th International Workshop on Description Logics (DL 2023), Rhodes, Greece, September 2-4, 2023*, volume 3464 of *CEUR Workshop Proceedings*, pages 4–13. CEUR-WS.org, 2023.

Michael Bernreiter, Wolfgang Dvorák, and Stefan Woltran. Abstract argumentation with conditional preferences. In Francesca Toni, Sylwia Polberg, Richard Booth, Martin Caminada, and Hiroyuki Kido, editors, *Computational Models of Argument - Proceedings of COMMA 2022, Cardiff, Wales, UK, 14-16 September 2022*, volume 353 of *Frontiers in Artificial Intelligence and Applications*, pages 92–103. IOS Press, 2022.

Philippe Besnard and Anthony Hunter. A logic-based theory of deductive arguments. *Artificial Intelligence*, 128(1-2):203–235, 2001.

Philippe Besnard, Éric Grégoire, and Badran Raddaoui. A conditional logic-based argumentation framework. In *International Conference on Scalable Uncertainty Management*, pages 44–56. Springer, 2013.

Alexander Bochman. Propositional argumentation and causal reasoning. In Leslie Pack Kaelbling and Alessandro Saffiotti, editors, *IJCAI-05, Proceedings of the Nineteenth International Joint Conference on Artificial Intelligence, Edinburgh, Scotland, UK, July 30 - August 5, 2005*, pages 388–393. Professional Book Center, 2005.

Alexander Bochman. Two paradigms of nonmonotonic reasoning. In *International Symposium on Artificial Intelligence and Mathematics, AI&Math 2006, Fort Lauderdale, Florida, USA, January 4-6, 2006*, 2006.

Alexander Bochman. Abstract dialectical argumentation among close relatives. In *COMMA*, pages 127–138, 2016.

Dmitri A Bochvar and Merrie Bergmann. On a three-valued logical calculus and its application to the analysis of the paradoxes of the classical extended functional calculus. *History and Philosophy of Logic*, 2(1-2):87–112, 1981.

Andrei Bondarenko, Phan Minh Dung, Robert A. Kowalski, and Francesca Toni. An abstract, argumentation-theoretic approach to default reasoning. *Artif. Intell.*, 93:63–101, 1997.

Richard Booth, Souhila Kaci, Tjitze Rienstra, and Leendert W. N. van der Torre. Conditional acceptance functions. In Bart Verheij, Stefan Szeider, and Stefan Woltran, editors, *Computational Models of Argument - Proceedings of COMMA 2012, Vienna, Austria, September 10-12, 2012*, volume 245 of *Frontiers in Artificial Intelligence and Applications*, pages 470–477. IOS Press, 2012.

Gerhard Brewka, Hannes Strass, Stefan Ellmauthaler, Johannes Peter Wallner, and Stefan Woltran. Abstract dialectical frameworks revisited. In *Proceedings of the 22th International Joint Conference on Artificial Intelligence (IJCAI'15)*, 2013.

Federico Cerutti, Sarah A Gaggl, Matthias Thimm, and Johannes P Wallner. Foundations of implementations for formal argumentation. *Handbook of Formal Argumentation*, pages 688–767, 2017.

B.F. Chellas. Basic conditional logic. *Journal of Philosophical Logic*, 4(2):133–153, 1975.

Mukesh Dalal. Investigations into a theory of knowledge base revision: preliminary report. In *Proceedings of the Seventh National Conference on Artificial Intelligence*, volume 2, pages 475–479. Citeseer, 1988.

James P Delgrande and Pavlos Peppas. Belief revision in horn theories. *Artificial Intelligence*, 218:1–22, 2015.

Phan Minh Dung, Phan Minh Thang, and Tran Cao Son. On structured argumentation with conditional preferences. In *The Thirty-Third AAAI Conference on Artificial Intelligence, AAAI 2019, The Thirty-First Innovative Applications of Artificial Intelligence Conference, IAAI 2019, The Ninth AAAI Symposium on Educational Advances in Artificial Intelligence, EAAI 2019, Honolulu, Hawaii, USA, January 27 - February 1, 2019*, pages 2792–2800. AAAI Press, 2019.

Phan Minh Dung. On the acceptability of arguments and its fundamental role in nonmonotonic reasoning, logic programming and n-person games. *Artificial Intelligence*, 77:321–358, 1995.

Uwe Egly, Sarah Alice Gaggl, and Stefan Woltran. Answer-set programming encodings for argumentation frameworks. *Argument and Computation*, 1(2):147–177, 2010.

Dov M. Gabbay and Artur S. d'Avila Garcez. Logical modes of attack in argumentation networks. *Stud Logica*, 93(2-3):199–230, 2009.

Dov M Gabbay. Fibred semantics and the weaving of logics, part 2: Fibring non-monotonic logics. In *Logic Colloquium*, volume 92, pages 75–94, 1995.

Dov M. Gabbay. Fibred semantics and the weaving of logics, part 1: Modal and intuitionistic logics. *J. Symb. Log.*, 61(4):1057–1120, 1996.

Alejandro J. García and Guillermo R. Simari. Defeasible logic programming: An argumentative approach. *Theory and Practice of Logic Programming*, 4(1):95–138, 2004.

Peter Gärdenfors. Belief revisions and the ramsey test for conditionals. *The Philosophical Review*, 95(1):81–93, 1986.

Peter Gärdenfors. Belief revision and nonmonotonic logic: two sides of the same coin? In *European Workshop on Logics in Artificial Intelligence*, pages 52–54. Springer, 1990.

Hector Geffner and Judea Pearl. Conditional entailment: Bridging two approaches to default reasoning. *Artificial Intelligence*, 53(2-3):209–244, 1992.

Michael Gelfond and Nicola Leone. Logic programming and knowledge representation—the a-prolog perspective. *AI*, 138(1-2):3–38, 2002.

Enrico Giunchiglia, Joohyung Lee, Vladimir Lifschitz, Norman McCain, and Hudson Turner. Nonmonotonic causal theories. *Artif. Intell.*, 153(1-2):49–104, 2004.

M. Goldszmidt and J. Pearl. Qualitative probabilities for default reasoning, belief revision, and causal modeling. *Artificial Intelligence*, 84:57–112, 1996.

Moisés Goldszmidt and Judea Pearl. Qualitative probabilities for default reasoning, belief revision, and causal modeling. *Artificial Intelligence*, 84(1-2):57–112, 1996.

Sven Ove Hansson. A survey of non-prioritized belief revision. *Erkenntnis*, 50(2):413–427, 1999.

James Hawthorne. Nonmonotonic conditionals that behave like conditional probabilities above a threshold. *Journal of Applied Logic*, 5(4):625–637, 2007.

Jesse Heyninck, Gabriele Kern-Isberner, Kenneth Skiba, and Matthias Thimm. Interpreting conditionals in argumentative environments. In *NMR 2020 Workshop Notes*, page 73, 2019.

Jesse Heyninck, Gabriele Kern-Isberner, and Matthias Thimm. On the correspondence between abstract dialectical frameworks and non-monotonic conditional logics. In *Proceedings of the 33rd International Florida Artificial Intelligence Research Society Conference*, pages 575–580, 2020.

Jesse Heyninck, Gabriele Kern-Isberner, Matthias Thimm, and Kenneth Skiba. On the correspondence between abstract dialectical frameworks and nonmonotonic conditional logics. *Ann. Math. Artif. Intell.*, 89(10-11):1075–1099, 2021.

Jesse Heyninck, Gabriele Kern-Isberner, Tjitze Rienstra, Kenneth Skiba, and Matthias Thimm. Revision, defeasible conditionals and non-monotonic inference for abstract dialectical frameworks. *Artif. Intell.*, 317:103876, 2023.

Jesse Heyninck. Relations between assumption-based approaches in non-monotonic logic and formal argumentation. *Journal of Applied Logics*, 6(2):317–357, 2019.

Hirofumi Katsuno and Alberto O Mendelzon. Propositional knowledge base revision and minimal change. *Artificial Intelligence*, 52(3):263–294, 1991.

Gabriele Kern-Isberner and Guillermo R Simari. A default logical semantics for defeasible argumentation. In *Proceedings of the Twenty-fourth International Florida Artificial Intelligence Research Society Conference*, 2011.

Gabriele Kern-Isberner and Matthias Thimm. Towards conditional logic semantics for abstract dialectical frameworks. In Carlos Chesnevar et al., editor, *Argumentation-based Proofs of Endearment*, volume 37 of *Tributes*. College Publications, November 2018.

Gabriele Kern-Isberner, Christoph Beierle, and Gerhard Brewka. Syntax splitting= relevance+ independence: New postulates for nonmonotonic reasoning from conditional belief bases. In *Proceedings of the International Conference on Principles of Knowledge Representation and Reasoning*, volume 17, pages 560–571, 2020.

Sarit Kraus, Daniel Lehmann, and Menachem Magidor. Nonmonotonic reasoning, preferential models and cumulative logics. *AI*, 44(1-2):167–207, 1990.

Sarit Kraus, Daniel J. Lehmann, and Menachem Magidor. Nonmonotonic reasoning, preferential models and cumulative logics. *Artificial Intelligence*, 44(1-2):167–207, 1990.

Daniel Lehmann and Menachem Magidor. What does a conditional knowledge base entail? *AI*, 55(1):1–60, 1992.

Daniel J. Lehmann and Menachem Magidor. What does a conditional knowledge base entail? *Artificial Intelligence*, 55(1):1–60, 1992.

David Lewis. Counterfactuals and comparative possibility. In *IFS: Conditionals, Belief, Decision, Chance and Time*, pages 57–85. Springer, 1973.

David K Lewis. Counterfactuals. 1973.

David Makinson and Leendert Van Der Torre. Input/output logics. *Journal of philosophical logic*, 29:383–408, 2000.

David Makinson. Five faces of minimality. *Studia Logica*, 52(3):339–379, 1993.

David Makinson. Conditional probability in the light of qualitative belief change. *Journal of Philosophical Logic*, 40(2):121–153, 2011.

Donald Nute. *Topics in conditional logic*, volume 20. Springer Science & Business Media, 1980.

Donald Nute. Conditional logic. In *Handbook of philosophical logic*, pages 387–439. Springer, 1984.

Judea Pearl. System Z: A natural ordering of defaults with tractable applications to nonmonotonic reasoning. In *Proc. of the 3rd conf. on Theor. asp. of reasoning about knowledge*, TARK '90, pages 121–135, San Francisco, CA, USA, 1990. Morgan Kaufmann Publishers Inc.

Judea Pearl. *Causality: Models, Reasoning and Inference*. Cambridge University Press, second edition, 2009.

Alan Perotti, Guido Boella, Dov M. Gabbay, Leon van der Torre, and Serena Villata. Conditional labelling for abstract argumentation. In Andrew V. Jones, editor, *2011 Imperial College Computing Student Workshop, ICCSW 2011, London, United Kingdom, September 29-30, 2011. Proceedings*, volume DTR11-9 of *Department of Computing Technical Report*, pages 59–65. Imperial College London, 2011.

J.L. Pollock and J. Cruz. *Contemporary Theories of Knowledge*. Rowman & Littlefield, Lanham, MD, 1999.

J.L. Pollock. *Cognitive Carpentry*. MIT Press, Cambridge, MA., 1995.

Jörg Pührer. Realizability of three-valued semantics for abstract dialectical frameworks. *Artificial Intelligence*, 278:103198, 2020.

Frank Plumpton Ramsey. General propositions and causality. *Foundations of Mathematics*, 1931.

Tjitze Rienstra, Chiaki Sakama, and Leendert van der Torre. Persistence and monotony properties of argumentation semantics. In *TAFA*, pages 211–225. Springer, 2015.

Tjitze Rienstra, Chiaki Sakama, and Leendert van der Torre. Persistence and monotony properties of argumentation semantics. In *Theory and Applications of Formal Argumentation: Third International Workshop, TAFA 2015, Buenos Aires, Argentina, July 25-26, 2015, Revised Selected Papers 3*, pages 211–225. Springer, 2015.

Kenneth Skiba and Matthias Thimm. Ordinal conditional functions for abstract argumentation. In Francesca Toni, Sylwia Polberg, Richard Booth, Martin Caminada, and Hiroyuki Kido, editors, *Computational Models of Argument - Proceedings of COMMA 2022, Cardiff, Wales, UK, 14-16 September 2022*, volume 353 of *Frontiers in Artificial Intelligence and Applications*, pages 308–319. IOS Press, 2022.

Kenneth Skiba, Tjitze Rienstra, Matthias Thimm, Jesse Heyninck, and Gabriele Kern-Isberner. Ranking extensions in abstract argumentation. In *IJCAI'21, 2021*. ijcai.org, 2021.

Wolfgang Spohn. Ordinal conditional functions: A dynamic theory of epistemic states. In *Causation in decision, belief change, and statistics*, pages 105–134. Springer, 1988.

Wolfgang Spohn. A brief comparison of pollock's defeasible reasoning and ranking functions. *Synthese*, 131:39–56, 2002.

Robert C Stalnaker. A theory of conditionals. In *Ifs: Conditionals, belief, decision, chance and time*, pages 41–55. Springer, 1968.

M. Thimm and G. Kern-Isberner. On the relationship of defeasible argumentation and answer set programming. In Philippe Besnard, Sylvie Doutre, and Anthony Hunter, editors, *Proceedings of the 2nd International Conference on Computational Models of Argument COMMA'08*, pages 393–404. IOS Press, 2008.

Matthias Thimm and Gabriele Kern-Isberner. On the relationship of defeasible argumentation and answer set programming. *Computational Models of Argument - Proceedings of COMMA 2008*, 8:393–404, 2008.

Matthias Thimm and Gabriele Kern-Isberner. Stratified labelings for abstract argumentation (preliminary report). Technical report, ArXiv, August 2013.

Matthias Thimm and Gabriele Kern-Isberner. On controversiality of arguments and stratified labelings. In *Proceedings of the Fifth International Conference on Computational Models of Argumentation (COMMA'14)*, September 2014.

Alasdair Urquhart. Basic many-valued logic. In *Handbook of philosophical logic*, pages 249–295. Springer, 2001.

Lees Van Berkel and Christian Straßer. Reasoning with and about norms in logical argumentation. *Computational Models of Argument: Proceedings of COMMA 2022*, 353:332, 2022.

Achille C Varzi and Massimo Warglien. The geometry of negation. *Journal of Applied Non-Classical Logics*, 13(1):9–19, 2003.

Bart Verheij. Formalizing arguments, rules and cases. In *Proceedings of the 16th edition of the International Conference on Articial Intelligence and Law*, pages 199–208, 2017.

Emil Weydert. On arguments and conditionals. In *Proceedings of the ECAI-2012 Workshop on Weighted Logics for Artificial Intelligence (WL4AI)*, pages 69–76, 2012.

Emil Weydert. On the plausibility of abstract arguments. In Linda C. van der Gaag, editor, *Symbolic and Quantitative Approaches to Reasoning with Uncertainty - 12th European Conference, ECSQARU*

2013, Utrecht, The Netherlands, July 8-10, 2013. Proceedings, volume 7958 of *Lecture Notes in Computer Science*, pages 522–533. Springer, 2013.

Emil Weydert. A plausibility semantics for abstract argumentation frameworks. *CoRR*, abs/1407.4234, 2014.

CHAPTER 9

ARGUMENTATION AND MACHINE LEARNING

ANTONIO RAGO*
Department of Computing, Imperial College London
a.rago@imperial.ac.uk

KRISTIJONAS ČYRAS*
Ericsson
kristijonas.cyras@ericsson.com

JACK MUMFORD
Department of Computer Science, University of Liverpool
jack.mumford@liverpool.ac.uk

OANA COCARASCU
Department of Informatics, King's College London
oana.cocarascu@kcl.ac.uk

1 Introduction

In this chapter, we overview research works that combine Computational Argumentation (henceforth simply argumentation) and Machine Learning (ML). In this section, we start with a general outlook of the themes and trends prevalent among the individual works overviewed in the chapter, including a loose categorisation of three types of interactions between ML an argumentation models, before briefly covering related work and papers we chose to omit. We then move on to our literature overview in

---
*Equal contribution.

§2 which we split by broad themes of the purposes of the interactions between ML and argumentation, briefly discussing promising research avenues. Finally, we conclude in §3.

We assume the reader to be familiar with forms and frameworks of argumentation, e.g. as discussed in the 1st Volume of the Handbook of Formal Argumentation (Baroni et al., 2018), and fundamentals of and common methods in ML, as easily accessible e.g. on Wikipedia.

1.1 Outlook

Studies that consider interactions of argumentation and ML exhibit the following trends. Starting with ML characteristics, the **type of learning** employed is largely *supervised*, though there are instances of works that consider *unsupervised* as well as both un/supervised types of learning, and a reasonable amount of works that focus on *Reinforcement* Learning (RL). In terms of **ML models** and algorithms employed, it is typical, but not exclusively so, to make use of simpler techniques and architectures, such as *rule learning* (including both rule induction and rule extraction), *tree-based models* (such as decision trees (DTs) and random forests (RFs)), *naive Bayesian classifiers* (NBCs), *support vector machines* (SVMs), (usually shallow and/or feed-forward) *neural networks* (NNs) and model-based *RL*. Some recent works though opt for modern complex models, in particular *graph neural networks* (GNNs), typically available off-the-shelf. The **data** that is used in experiments is most often of the *tabular* type, though *textual* and *image* data also make their marks, while in the case of RL, *simulations* in simple environments are prominent. The **datasets** themselves are largely basic and small, often coming from the *UCI ML Repository*.[1] Though again, the recent works employing GNNs typically work on *argument graph datasets* afforded by the International Competition on Computational Models of Argumentation (ICCMA).[2]

In terms of **argumentation frameworks**, we see a great diversity, ranging from the abstract to the structured kind. There is a notable focus on *Abstract Argumentation Frameworks* (AFs), *Value-Based Argumentation Frameworks* (VAFs) and *(Quantitative) Bipolar AFs* ((Q)BAFs), with QBAFs especially prominent in combination with NNs. Many works

---
[1] https://archive.ics.uci.edu
[2] https://argumentationcompetition.org/

use *Structured Argumentation* (SA), including the classical formalisms such as *Defeasible Logic Programming* (DeLP) and *assumption-based argumentation* (ABA), but more prominently specific *rule-based argumentation frameworks* (RB), often constructed using rule learning. Some *Probabilistic Argumentation* (PA) and forms of *informal argumentation* also appear. In terms of semantics, the most popular among the works overviewed seem to be those of the classic grounded or preferred extensions, in terms of both extensions and of sceptical and credulous acceptability, but gradual semantics are also often used, as are dialogues as both a form of argumentative interaction and inference. We will elaborate on such argumentative and ML-targeted characteristics of individual works in §2.

Tangentially to the trends of characteristics of either ML or argumentation, we distinguish three **types of interaction** between ML and argumentation. First, a *synergistic* combination of an ML model and an argumentation model essentially means tightly integrating, or merging, the two kinds of models into one. The second, a *segmented* approach, is where ML and argumentation are interleaved in performing learning and reasoning, using outputs of one as the inputs of another. Finally, the third type of interaction is *approximated*, in the sense that one type of model is meant to approximate or shadow, at varying levels of detail, the other. We recognise that the boundaries between these notions are not clear cut and we do not aim at precise definitions of interactions. Rather, we posit these as guiding concepts, to glean the rough mode of interaction between argumentation and ML in each individual work, and hope that the individual descriptions given later in §2 will clarify these three notions.

Two broad themes of the **purpose of interaction** between argumentation and ML emerge. The first one that we identify is of using *argumentation to improve and/or explain ML models*. We overview research works following this theme in §2.1. The other theme is that of using *ML to support, analyse or replace argumentation*. We overview such themed works in §2.2.

As for improving and/or explaining ML with argumentation, we identify the following motifs. SA is proposed to improve ML-based classification, often via rule-based/premise-conclusion kind of arguments constructed using rule learning, typically via segmented or synergistic approaches. This can be achieved by injecting knowledge into ML-based systems via rules or preferences, which is enabled by argumentation. Argumentation can also improve RL-based policies in Multi-Agent Systems, typically by means of argumentative selection of actions. Further, argumentation is often meant to explain ML-based systems, purportedly because of argumentative relations (especially in synergistic and segmented approaches) that are particularly analytical and interpretable. This can be achieved for instance by reasoning argumentatively with rules learnt from data in a segmented or synergistic approach, or by argumentatively explaining why one classification decision is preferred over another after approximating an ML model. Explainability can also be enhanced with argumentation in RL by exhibiting argumentative interaction between agents' actions, or by using an argument graph as a structure to learn on and thereafter explain by means of existing argumentative explainability methods (see the recent surveys (Čyras et al., 2021; Vassiliades et al., 2021; Guo et al., 2023)). It should be noted that some of the opportunities for improving ML models with argumentation which were foreseen in (Longo, 2016) have indeed been realised in some cases, particularly those concerned with argumentative knowledge being injected into, and thus guiding inference in, ML models.

As for ML supporting argumentation, several main motifs emerge. One is that of using ML to generate argumentation frameworks. This can be achieved by means of rule learning, particularly using Inductive ML, purely for constructing rule-based argumentation frameworks in a segmented fashion to reason with. (This echoes the theme of an interaction in the opposite direction, of argumentation for improving ML models, as mentioned above, even though the intrinsic goals of the research are distinct.) Meanwhile, frameworks of the abstract kind, particularly AFs, can be learnt from other types of data or from other AFs themselves, often using approaches of the segmented kind. On the other hand, ML models can be used for the task of computing argumentation frameworks, meaning to predict the acceptance statuses of arguments, by and large in AFs using GNNs in an approximated fashion. Yet another recent motif is that of investigating if some forms of ML, particularly NNs, can be seen as argumentation frameworks, particularly QBAFs, and vice versa, for the purposes of advancing or even replacing such argumentation frameworks, as well as analysing or advancing such ML models. Finally, we encounter a couple of works that study argumentation in everyday life and investigate whether ML can help with recommending arguments to be put forward in real-world argumentative discussions, or whether ML should be used instead of formal argumentation due to its issues in modelling these situations.

By the end of §2, we will have seen many works where various forms of argumentation and ML models interact in different ways for one-way or mutual benefit. We will see different application areas (e.g. medical, robotics, prediction, legal and commonsense reasoning) and various tasks (e.g. classification, policy learning, argument graph generation and computation) addressed by the proposed systems. We will deem some research avenues to be more promising, often due to the use of modern ML and more extensive experimentation, others less so due to simplistic settings and/or empirical evaluation.

A general criticism for argumentation in meeting ML is that of the apparent lack of implementations (in great contrast to the abundance of off-the-shelf ML models), which makes it harder for the interaction between the two to take place. The lack of user studies supporting the presumed benefits of argumentation is another crucial gap in the pairing between argumentation and ML. These concerns are implicit from our literature overview and we believe should be addressed if this cross-disciplinary research is to bear fruit.

Before delving into discussing individual works, we briefly mention related work, as well as omitted works that are out of scope of this chapter.

1.2 Related Work and Omissions

We will first discuss other reviews of the literature on argumentation which are closely linked to that which we provide. The work with an objective which most closely aligns with our own is by Cocarascu and Toni (2016), in which the authors provide a comprehensive overview of argumentation's support for ML. The authors here note the relative youth of this area of research at the time, which indeed motivates the need for our contribution after many years of rapid progress in ML, both in general and combined with argumentation. One of the directions which we consider where there has indeed been a significant amount of progress is the use of argumentation for supporting explainable AI (concerning ML models or otherwise), which motivated three recent surveys of this literature by Vassiliades et al. (2021), Čyras et al. (2021) and Guo et al. (2023). It should be noted that cross-fertilisations between argumentation and either ML or explainable AI host a reasonable degree of overlap but neither is subsumed by the other. Finally, Proietti and Toni (2023) give a roadmap towards Neuro-Argumentative Learning, i.e. learning argumentation frameworks from data, with a focus on neuro-symbolic approaches, looking ahead to directions for future work and challenges which remain.

With regards to works which were considered to be surplus to our review, the largest and most related such body of work was that on Argument Mining. We also leave out papers which consider tasks related to natural language processing or generation, e.g. works concerning the use of argumentation for analysis (Hinton and Wagemans, 2023; Rajasekharan et al., 2023) or fine-tuning (Thorburn and Kruger, 2022; Furman et al., 2023) of Large Language Models (Brown et al., 2020). Also, while argumentation frameworks can be seen as knowledge graphs of a specific form, we consider the wide range of cross-fertilisations between ML and knowledge graphs (e.g. see (Tiddi and Schlobach, 2022) for a recent survey) to be outside the scope of this chapter. We also choose to omit a number of works, e.g. (Longo, 2016; Zeng et al., 2018), where ML is proposed as an application of an argumentative methodology only informally, without a concrete ML application. Further, approaches which consider tasks most often associated with ML methods but without some explicit use of ML, e.g. (Amgoud and Serrurier, 2008; Potyka et al., 2022b), or those which consider ML models as a possible source of knowledge, e.g. (Hung et al., 2022), are also omitted. Similarly, we also choose to leave out methods which generate argument graphs from data that do not include any explicit ML techniques, e.g. (Hunter, 2020). Orthogonally, we do not include works that target typical ML tasks and can be seen to have argumentative flavour but do not use argumentation in any way, e.g. (Taniar et al., 2008). Finally, argumentative approaches which consider techniques often employed in ML, such as Markov decision processes (MDPs), e.g. (Potyka, 2020), or Bayesian networks (BNs), e.g. (Bex and Renooij, 2016), but without explicitly using ML itself are typically outside the scope of the chapter.

2 Literature Overview

We start by overviewing research works that we deem to use argumentation for the benefit of ML (§2.1). We then move on to overview works that we instead deem using ML mostly for the benefit of argumentation (§2.2).

2.1 Argumentation for ML

In this subsection, we consider the various ways in which argumentation has been used for supplementing ML models. Here, the general trend is for methods to target the improvement of the ML model's performance, explain its outputs or a combination of the two. We begin with those which concern improvement alone, before transitioning to those which concern explanation alone roughly according to the works' main objectives. The works whose main objective is judged to be improvement in performance are discussed in §2.1.1 and summarised in Table 1, while those where the main objective is enhanced explainability are discussed in §2.1.2 and summarised in Table 2.[3]

2.1.1 For Improving ML Models

Several works propose either segmented, synergistic or approximate methods of integrating argumentation with ML models for the purpose of improving their performance. The general (but not exclusive) theme in these works is that of using rule learning to induce rule-based arguments and then to apply argumentative reasoning to improve the outputs of the ML models, typically using injected expert knowledge which is enabled by means of argumentation. We begin with some works which focus on the improvement of an ML model's performance exclusively, without considering explainability.

---

[3]Note that in all tables we show only the references to the primary works describing an approach, with the others given in the discussion.

Work	Type	Arg.	Learn.	ML Model(s)	Task	Data
(Gómez and Chesñevar, 2004b)	Seg	SA (DeLP)	U	Fuzzy ART NN	Clust	Tab
(Carstens and Toni, 2015)	Seg	QBAF	S	NBC, SVM, RF	Class	Text
(Ayoobi et al., 2022)	App	AF, QBAF	U, S	Rule	Class	Tab
(Thimm and Kersting, 2017)	Seg	SA (DeLP)	U, S	Rule	Class	Tab
(Možina et al., 2007)	Syn	Informal	S	Rule	Class	Tab
(Shao et al., 2020)	Syn	Informal	S	NN	Class	Im
(Sendi et al., 2019)	Seg	SA (RB, Dialogues)	S	NN, Rule, DT	Class	Tab
(Prentzas et al., 2022)	Seg	SA (Gorgias)	S	Rule, RF	Class	Tab
(Riveret et al., 2020)	Syn	PA (Semi-AF)	S	RBM	Class	Tab
(Gao et al., 2012)	Syn	VAF	R	MDP	Policy	Sim
(Riveret et al., 2019)	Syn	PA, SA	R	MDP	Policy	Sim

Table 1: A summary of the ways in which argumentation has been used for *improving* ML models, as detailed in §2.1.1. We show: the type of integration (segmented, synergistic or approximate); the kind of argumentation used; the type of learning employed (supervised, unsupervised or reinforcement); the ML model(s) used, including fuzzy adaptive resonance theory (ART) NNs, rule learning (including rule induction and rule extraction) and restricted Boltzmann machines (RBMs); the task undertaken (clustering, classification or policy learning) and the type of data used (tabular, textual, image or, specifically in the case of RL, simulation).

9 - ARGUMENTATION AND MACHINE LEARNING

Some of the earliest works to use argumentation (as it is now commonly understood) for improving ML methods in classification tasks are those of Gómez and Chesñevar (2004a,b). In particular, the authors propose the use of argumentation for **tabular data clustering** using **unsupervised learning**. In their **segmented** approach, defeasible argumentation processes the outputs of an **NN**. The use of argumentation is meant to improve on the then state-of-the-art of randomly choosing one output cluster when the input is pattern-matched to more than one cluster. In particular, data instances are first clustered using a **fuzzy adaptive resonance theory NN**, and then **DeLP** resolves inconsistent classification outcomes for instances assigned to multiple conflicting clusters. Specifically, the learnt clusters are represented in DeLP by means of facts and rules, and so are any preferences that would resolve situations where an instance is assigned to multiple clusters (for example, a cluster with newer instances is preferred over others, or the smallest/largest cluster is preferred). For a new data instance, DeLP-style argumentation finds the rules that are pattern-matched to apply and through dialectical analysis infers the preferred classification. The stated advantages of combining unsupervised clustering and preferences for argumentative conflict resolution are theoretical: the authors suggest their approach could be applied to resolve classification conflicts for ambiguous instances and to potentially explain classification decisions.

Carstens and Toni (2015) define a **segmented** approach for sentiment polarity **classification** via **supervised learning**. They leverage on the reasoning contained in **QBAFs** with gradual semantics to make improvements in classifier accuracy in tasks with **textual data**. Here, the QBAF contains arguments representing the possible classes, as well as arguments comprising one or more premises and a conclusion, where premises are characteristics of sentences that indicate a certain polarity in the case of sentiment classification, and the conclusion is the polarity associated with these words. Base scores are obtained from the output of the classifier (e.g. confidence of the classification or the classification performance on the training corpus, depending on the classifier used) or from a given argument base. The dialectical strength of each class is computed using a quantitative semantics and the classification with maximal strength is assigned as the final classification for the testing instance. The authors evaluate the approach on two corpora of tweets, the Sanders corpus and the STS corpus, and another corpus consisting of positive and negative sentences from movie reviews. They experiment with **NBCs**, **SVMs** and **RFs**, showing improvements in accuracy in two computational linguistics tasks.

Going even further towards argumentation as an alternative to classical learning-based inference, Ayoobi et al. (2022) propose an **approximate** method of pure argumentation-based learning for action selection in **semi-supervised learning** scenarios. The approach uses two **argumentation frameworks**, one **abstract** and one **bipolar with weights** on relationships, to incrementally record – as well as select the best – actions given context described as sets of feature-value pairs. In short, arguments are of the precondition-postcondition form, obtained by **rule learning** from all the possible combinations of feature-value pairs of a data instance. By default, the grounded extension of the abstract framework yields exactly one of the previously seen actions as the best in a given context and the supporting context conditions as feature-value pairs are recorded in a BAF. However, if the AF does not unambiguously resolve the conflicts among all the applicable actions, then an alternative action is extracted from the BAF depending on the cumulative strength of the already learnt support and attack conditions for various actions. The authors thus present a non-ML-based approach to **classification** on **tabular data**. Having in mind potential applicability to robotic scenarios, the authors compare this argumentation-based learning approach to some classic RL-type of approaches in multi-armed bandit problems as well as to incremental online learning approaches, and show superiority of argumentation-based learning with respect to both performance and speed of learning. This work was extended in (Ayoobi et al., 2021), which uses only BAFs with novel acceleration strategies thus addressing the complexity issues in (Ayoobi et al., 2022) with better run-time, memory efficiency and learning accuracy.

The papers we overview next consider how the incorporation of argumentation can bring benefits to explainability, while still focusing on improving the ML model's performance as the main objective.

A short position paper (Thimm and Kersting, 2017) that rests on improving ML models by means of argumentation in the spirit of Gómez and Chesñevar (2004a,b) summarises the use of **SA** for improving either **supervised** or **unsupervised classification** of **tabular data** with, crucially, the added benefit of explainability. The authors propose a two-step, **segmented** classification approach combining **rule learning** and argumentative inference. First, rule learning algorithms extract from a given data set frequent patterns as lists of (typically) collectively conflicting rules. Then, such rules can be used to construct an SA framework from which conflict-free inferences can be drawn. This method enables learnt and argumentatively supported classification of unseen data, purportedly while being explainable in the sense of showing why one class is preferred over others in terms of rule-based arguments. The authors show empirically how to avoid inconsistent classification via mined rules, making use of the **DeLP** argumentative conflict resolution mechanism.

An early work that aims to improve **classification** of **tabular data** via **supervised learning** using **informal argumentation** is that of Možina et al. (2007); Bratko et al. (2009). The authors propose a **synergistic** approach wherein experts provide arguments in the form of rules associating a class and feature-value pairs, in effect giving the machine explanations as feedback. The introduced approach first learns if-then rules from argumentative examples by extending an existing method for **rule learning**, the CN2 rule induction algorithm. Two types of arguments are incorporated: "C because Reasons" and "C despite Reasons", where the first type provides reasons (i.e. combinations of features) for why a certain training instance is classified as is, whereas the second type highlights combinations of features that do not play a role in the classification of a training instance. Importantly, incorporating arguments in this way allows experts to provide information which constrains the ML model's training, facilitating interactive ML between users and models, as described in a later work (Mozina, 2018). The authors evaluate the approach empirically with three datasets from the UCI machine learning repository, and exemplify the approach in the legal and medical domains.

Another example of using **informal argumentation** in a **synergistic** way for improving **classification** via **supervised learning**, but with **image data**, is (Shao et al., 2020). The authors suggest that a probabilistic neuro-symbolic image classifier can be argued with by a human user about the correctness of classification. In contrast to previous works discussed in this subsection, they use influence functions to expose saliency maps as explanations alongside the instances classified by an **NN** (an 8-layer multi-layer perceptron (MLP)). The user can then argue with those outputs by putting forward (automatically generated) counterexample images to correct classifications or directly regularising the gradients to penalise incorrect explanations. The authors show experimentally on three image datasets that their approach can improve classifier accuracy on both new examples and previously misclassified counterexamples.

Using argumentation more formally in a **segmented** way, Sendi et al. (2019) create argumentative classifiers from rules extracted from **NNs** for the purposes of injection of domain knowledge and more explainable **classification** of **tabular data** via **supervised learning**. Specifically, an NN model trained on tabular data (using various UCI datasets) is approximated into a more interpretable model in two ways: either as a **DT** (using TREPAN library by Craven and Shavlik (1995)) or by eclectic **rule extraction**. The rules (which amount to paths in a DT) constitute **premise-conclusion type of arguments**, where premises amount to feature-value pairs and the conclusion to a class label together with the classifier's confidence score. These rules are said to essentially function as explanations for the classifier's outputs. Additionally, experts' arguments are also modelled in the premises-conclusion form. This allows for a segmented approach of an approximated NN classifier and expert knowledge into an argumentative classifier. The authors suggest how multiple argumentative classifiers can engage in multi-agent argumentation by means of a **dialogue protocol**, whereby conflicts among arguments with different conclusions are resolved by prioritising first expert knowledge arguments and then the ones with higher confidence scores. They provide a case study of multi-class medical treatment recommendation task in a realistic virtual population of 40,000 individuals (datapoints) described using numerous features, and further experiment with 11 UCI binary classification datasets comparing their method to ensemble learning and rule extraction methods (such as boosted trees and ensemble NNs). Experimental results indicate that such multi-agent argumentative classifiers (albeit without expert agents) outperform ensemble methods based on either the original or baseline NN models in terms of classification accuracy. The authors also posit that their approach enhances explainability by exposing the outputs of NNs with argumentative resolution.

Prentzas et al. (2022) define a **segmented** approach to argumentation-based **classification** of **tabular data** via **supervised learning**, with an application in the field of cancer prognosis. The authors utilise **SA**, in the form of the **Gorgias argumentation framework**, where the basic arguments consist of premises and a conclusion, and other arguments express preferences over the basic arguments. After undertaking some pre-processing in the form of statistical analyses of features' significance, the authors prescribe some form of **rule extraction** method to create arguments, using **RF** as an example. Next, in an iterative and manual process, tailored defeat arguments for rules are defined, before the dilemmas caused by argument conflicts are reduced, by adding arguments which adjust the preferences. The authors then deploy the approach in cancer prognosis prediction, demonstrating that it exhibits reasonable accuracy in an empirical analysis before showing the natural explainability afforded by its argumentative nature. This explainability is explored further, with varying styles of explanation, via a web platform in the social media, legal and medical domains in (Spanoudakis et al., 2022).

Riveret et al. (2020) propose a neuro-symbolic method to **supervised learning** that **synergistically** combines graphical representation, probabilistic learning and argumentative reasoning. Their aim is to address the challenges of learning from examples the probabilistic statuses and dependencies among arguments/statements in PA, and that of constraining and explaining probabilistic inferences of NNs for **classification**. To this end, they integrate **RBMs** and **probabilistic semi-abstract argumentation** in the following way. First, they represent each datapoint in a set of **tabular data** as an argument graph with nodes being feature-value pairs, along with the class label, and edges encoding prior knowledge in the form of logical constraints about the dataset, so that the constraints are captured by attack and support relationships and an argument labelling includes classification of the datapoint. Second, prior knowledge constraints as argument labellings are integrated within

RBMs, so that the latter are used to learn and respect the probabilistic dependencies amongst argument labels. Finally, formulating the learning task for binary classification over datapoints as argument labellings yields neuro-symbolic argumentation machines (NSAMs) – a probabilistic learning and reasoning method as an alternative to the standard RBMs as well as other ML methods. In this way, NSAMs can learn argument graphs from data and use argumentation to help with both classification of aberrant datapoints in noisy settings and overfitting. The authors compare NSAMs with classic ML approaches such as NNs, logistic regression (LR) and DTs, among others. They contrast in particular with the standard RBMs that only learn from data with those that use both prior knowledge constraints and learn from data, namely NSAMs. An experiment on a small (2400 examples) tabular dataset augmented with noise and for which expert knowledge is available shows that NSAMs outperforms other methods. As an added benefit, the authors claim that NSAMs provide explanations of individual classifications in terms of maximally consistent argument graph labellings that pertain to the pertinent feature-value pairs and prior knowledge constraints.

The works we have overviewed here focus on improving and explaining ML-based classification by means of typically a form of SA. The next family of works that we discuss focuses on RL instead. Concretely, the works aim at improving RL-based agent policies in Multi-Agent Systems (MASs).

The use of argumentation to improve RL-based agents and their interactions generally follows a synergistic approach of introducing argumentative reasoning about the utilities of actions recommended by an RL agent's policy. For example, Gao et al. (2012) develop a **synergistic** method of argumentation-based **RL**, where domain expert knowledge is injected into a (semi-)**MDP**-based algorithm to use as a reward shaping technique by means of **VAFs**. The authors specifically show how an on-policy learning algorithm SARSA can be modified to use an argumentation-based look-ahead reward shaping function using **simulations** in the application of RoboCup Soccer (Stone et al., 2005). In particular, domain-knowledge (about soccer) is used to define VAFs, where values roughly represent different tactics given by domain experts, so that numerical rewards are assigned to an RL agent's actions based on the arguments found in the preferred extensions of the framework. Experimental results show that the performance of thusly modified RL algorithm can significantly improve **policies** relative to those of the original. Gao and Toni (2014) extend this work to the multi-agent RL setting.

Further work by some of the same authors in (Riveret et al., 2019) combines **RL** with PA instead, again in a **synergistic** manner. Argument values in **probabilistic SA frameworks** are quantified as utilities and learnt using RL. Roughly, **MDPs** are captured as argumentation-based agents using **SA** frameworks with **temporal modality operators** and a **probability distribution** over argument labellings. Reinforcement is modelled as probabilistic update of argument labels in time. One of the main benefits of this approach is that agents can have attitudes (expressed as MDP actions) and thereby deliberate about **policies** using the underlying logic and argument labellings. The authors mention the advantage of the natural agency afforded by argumentation in the form of an ability to explain and forecast agents' actions, and illustrate their points with experiments in an environment simulating a simple hand-crafted MDP.

While the above works purport to show improvements of ML models by means of argumentation in specific scenarios, the benefits typically rely on the availability of user/expert interaction, e.g. by means of adding knowledge in terms of rules or preferences. It is also not clear whether or not the state-of-the-art ML has already moved beyond the problems addressed in these works.

Though these discussed works generally aim at improving ML models' performance with argumentation, most of them also suggest that argumentation can also help to explain the inferences. In the next section we discuss a large body of works in which the primary goal is that of using argumentation for explaining ML models.

2.1.2 For Explaining ML Models

We now consider approaches which incorporate argumentation to enhance explainability as the main objective, firstly considering approaches which still target improvement in the ML model's performance. One point of note here is the increase in the number of methods which are approximate,

Work	Type	Arg.	Learn.	ML Model(s)	Task	Data
(Cocarascu et al., 2020)	Seg/Syn	AF	U, S	AE, RF	Class	Tab, Text, Im
(Prakken and Ratsma, 2022)	Seg/Syn	AF	S	DT, SVM, NBC, LR	Class	Tab
(van Lente et al., 2022)	Syn/App	SA	S	LR, SVM, RF, NN, RE	Class	Tab
(Bistarelli et al., 2022)	App	BAF	S	BB	Class	Tab
(Vilone and Longo, 2022)	Seg/App	SA (RB)	S	NN	Class	Tab
(Kazhdan et al., 2020)	App	VAF	R	TE	Class	Tab
(Otero et al., 2023)	App	VAF	R	MDP, AD	Policy	Sim
(Lertvittayakumjorn and Toni, 2023)	Syn	QBAF	S	LR	Class	Text
(Timmer et al., 2017)	App	SA (ASPIC+)	S	BN	Class	Tab
(Dejl et al., 2021)	App	GAF	S	NN	Class	Text
(Sukpanichnant et al., 2021)	App	QBAF	S	NN	Class	Im
(Mollas et al., 2022)	Seg/App	SA	S	BB	Class	Tab
(Amgoud et al., 2023)	Seg	AF	S	BB	Class	Tab
(Potyka et al., 2023)	App	BAF	S	RF	Class	Tab

Table 2: A summary of the ways in which argumentation has been used for *explaining* ML models, as detailed in §2.1.2. We show: the type (segmented, synergistic or approximate); the kind of argumentation used; the type of learning employed (supervised, unsupervised or reinforcement); the ML model(s) used, including autoencoders (AEs), rule extraction (RE), trajectory extraction (TE), autoregressive decoding (AD) and black-boxes (BBs); the task undertaken (classification or policy learning) and the type of data used (tabular, textual, image or simulation).

in line with the well known trend of explanations being simplifications of the original models (Rudin, 2019), such that they are cognitively manageable for humans.

Cocarascu et al. (2020) take a **segmented/synergistic** approach to develop a model for **classification** of **tabular**, annotated **image** and **textual data** by leveraging ML's capability to compactly represent data and the argumentative reasoning to make and explain decisions. The work integrates simple, classic **supervised** and **unsupervised learning** techniques for data representation with **AFs** for relating data points and reasoning about their labels. In detail, an **autoencoder** (a type of NN used to learn efficient encodings of unlabelled data) is used to reduce the dimensionality of tabular (Mushroom dataset with 126 one-hot encoded features) and annotated image data (CelebFaces Attributes and Objects with Attributes datasets with, respectively, 40 binary attributes, and 64 attribute labels divided into 20 classes) so that data instances could be more compactly represented as arguments in an AF. In the case of textual data, **RFs** are trained on one-hot-encoded vocabularies of (possibly semantically-clustered) word lemmas extracted from product reviews with positive and negative sentiments, using two movie review datasets. The outputs of RFs serve as weights on words so that arguments would represent datapoints as sentiment-labeled sets of weighted words. The AFs are inspired by case-based reasoning (CBR), whereby an informativeness relation (over subsets of (weighted) features) together with datapoint label divergence are used to define attacks among arguments/datapoints. Further, a default label and argument are designated, and the reasoning amounts to deciding whether the default argument is defended by the grounded extension of the AF augmented with an unlabeled datapoint/argument. The determination of belonging to the grounded extension amounts to classification of the unlabeled datapoint. Finally, the reasoning about the acceptability of the default argument according to the grounded semantics is presented as an argumentation debate for the purpose of explaining the classification of a given datapoint in terms of its relations to other labeled datapoints. In this way, the overall framework first makes use of ML for data representation and then replaces a standard ML way to classification with that of argumentative inference (and explanation) over compactly represented data.

Prakken and Ratsma (2022) take the CBR-inspired argumentation for explainable **classification**, in the spirit of (Cocarascu et al., 2020), further. The authors consider an arbitrary **supervised** classifier trained on **tabular data** and build an **AF**-based model that justifies the outcomes of the classifier in terms of feature-value pairs and/or relevant datapoints. The **segmented/synergistic** approach assumes access to data features and the training set. The method incorporates rationale from legal reasoning: basically, data features (known as factors) and datapoints (known as cases) tend to favour one classification outcome or another. The authors employ a dialogical exchange of arguments (in terms of factors and cases) for and against a particular classification as an explanation mechanism. They experiment with classic ML models, including **DTs**, **SVMs**, **NBCs** and **LR**, on three basic tabular datasets (Mushroom, Churn, Admission). The results show, among other things, that the dialogical argument exchange can indeed explain (justify) most of the ML model's correctly classified datapoints.

van Lente et al. (2022) introduce argumentative explanations, which are said to be modelled on those used by humans in everyday life, for any **supervised** black-box **classification** model operating on **tabular data**. The method is model-agnostic because it deals with inputs and outputs of the classifier, learning **SA frameworks** in a **synergistic** manner via a **rule learning** algorithm such that they include features as premises, classes as conclusions and a quantitative strength value of the argument. Then, attacks between arguments are drawn when an argument rebuts another, and the arguments are then evaluated using the grounded semantics. This argumentation framework can then be used for the classification task itself, thus **approximating** the ML model, but also for explaining it, where explanations are selected as subgraphs of the argument graph, such that the arguments are either pro and con the classification. In the experiments, the authors demonstrate how the explanations perform with respect to five different evaluation metrics, notably fidelity, i.e. how well the explanations approximate the prediction of the ML model, and accuracy, i.e. how well the argumentation framework predicts with unseen data. They experiment with four different types of ML model, namely **LR**, **SVMs**, **RFs** and **NNs**, with datasets from the UCI ML Repository.

Another method which introduces a technique for **approximating black-box** ML models for **supervised classification** tasks with argumentation frameworks is that of Bistarelli et al. (2022). In an extended abstract, the authors propose to approximate ML-based classifiers operating on **tabular data** using **BAFs** with arguments as feature-value pairs in favour or against a particular class, whence sub-trees with arguments accepted or rejected under semi-stable semantics serve as model- and data-agnostic explanations of the classification. In particular, the authors first cluster the dataset based on ranges of the values of variables, before they compute a correlation matrix to approximate relations the features. These relations are initially symmetrical but the authors use conditional probabilities to remove some edges, while ensuring that the graph remains connected, and determine which arguments attack and support one another. The authors then show how, given a dataset, the semi-stable semantics for BAFs can be used to generate an explanation for given classifications, approximating the classifier via the dataset.

Vilone and Longo (2022) also define a model-agnostic method for **approximating** ML models undertaking **supervised classification** on **tabular data** in a **segmented** manner, this time deploying **rule-based SA** frameworks with the ranking-based categoriser semantics. This is achieved by defining arguments as IF-THEN rules, where the premises and conclusion of an argument correspond to the rule's antecedents and conclusion. The premises consist of restrictions of the features' values, while the conclusions are classes predicted by the classifiers, thus resulting in global explanations. Attacks between arguments are then extracted based on rebutting and undercutting, which are weighted and pruned based on the coverage in the dataset of the corresponding rules. Finally, the arguments are evaluated by the aforementioned semantics, with the acceptable arguments serving as explanations for the classifier. These explanations are evaluated using feed-forward **NNs** trained on 5 datasets from Kaggle and the UCI ML Repository, where they are shown to be simpler and more comprehensible than DTs, but lack their faithfulness, correctness and robustness.

As in §2.1.1 where we considered several works that focus on argumentation improving RL, we next discuss a couple of works which focus on explaining as well as improving RL.

Kazhdan et al. (2020) extract models from multi-agent **RL** systems and augments them with expert knowledge-based arguments for explanatory purposes. Specifically, **model extraction** (in some form) from RL agents' **trajectories** (i.e. sequences of state-action pairs) as recorded in data logs is assumed. Then arguments corresponding to recommendations of actions to perform in a given state (for any one agent) are assumed to be given by an expert user, and numerical values are assigned to those arguments based on the frequency of the actions taken in the corresponding states as observed in the trajectories. A **VAF** is used to represent the user-provided arguments together with values obtained from data, and grounded semantics is used to define **policies** selecting the best actions. An argumentation-based model of MARL agents is thus constructed in a way that attempts to both **approximate** the agents by expert heuristics, and enables expert knowledge to be incorporated. The effectiveness and interpretability of thusly extracted models is illustrated with **simulation data** in a RoboCup Soccer (Stone et al., 2005) scenario.

Otero et al. (2023) note that in the setting of multi-agent **RL**, non-symbolic models aided by argumentation can learn high-performing **policies**, as in (Gao and Toni, 2013, 2014), but are hard to explain due to their non-symbolic nature. They likewise note that while model-extraction methods using argumentation provide explainability to non-symbolic models by producing symbolic counterparts, as in (Kazhdan et al., 2020), they may incur performance loss and are often approximate and thus not completely faithful to the underlying non-symbolic model. The authors thus aim to directly learn a symbolic argumentation-based multi-agent RL model that, while still **approximate**, is both performant and explainable. The authors use RL to preference-rank arguments in an expert-constructed **VAF** in a way that maximises the performance score of the VAF, i.e. how good the VAF is a reasoning engine for choosing actions in some task. More concretely, they first formulate a combinatorial optimisation problem as an **MDP** with the state space as the set of all the possible partial rankings over arguments and the action space as arguments themselves. Where a state is a total ranking, it is assigned a score (i.e. reward) by executing the reasoning in a VAF with the arguments ranked in preference accordingly to the state. Otherwise, states representing partial argument orderings are completed by an RL algorithm step by step: it learns to add arguments until a total ordering is reached by maximising the reward given for a total argument ordering and updating the policy-gradient weights accordingly (this is also know as **autoregressive decoding** in combinatorial optimisation via RL). The learnt VAF can be used as an agent's policy recommending actions promoted by the accepted arguments. The authors implement and compare their approach with some symbolic model-based agent architectures, including those from (Kazhdan et al., 2020), with **simulation data** in the multi-agent RL environment RoboCup Keepaway (Stone et al., 2005). They establish their approach to be superior both performance- and explainability-wise, where in the latter case, the best policy VAF learnt with the help of RL is completely faithful to the trajectories of execution, in contrast to only approximate fidelity achieved in (Kazhdan et al., 2020). Overall, the approach produces performant argumentation-based agent models that can be verified and yield faithful explanations of the agent's bahaviour in a multi-agent RL setting.

We now consider the approaches where explainability of the ML model is the sole focus of the works, rather than improving its performance.

In (Lertvittayakumjorn and Toni, 2023), the authors target argumentative explanations of pattern-based **LR** (PLR) for binary **text classification** via **supervised learning**. They show that the interpretability of PLR does not guarantee that explanations are amenable to humans, demonstrating that standard explanation methods may be seen as being implausible. To rectify this, the authors introduce argumentative explanations which are able to provide additional layers of explainability, clearly demonstrating relationships between the patterns, such as agreement or disagreement. The **synergistic** method uses **QBAFs** as a means to represent the patterns observed by the classifier in an approximate manner, with the gradual semantics over the QBAF modelling the LR. They show that, given an input, the classifier and its QBAF representation always give the same prediction, and prove intuitive argumentative properties of the QBAF. The method is then evaluated empirically with three datasets, proving its advantages wrt sufficiency, and experimentally in two user studies with humans, showing improved plausibility and helpfulness over other explanation methods.

Timmer et al. (2017) propose **SA** to **approximate** and explain reasoning in **BNs** performing **supervised classification** with **tabular data**. First, for a variable of interest in a BN, a support graph is extracted, consisting of chains of variables whose observations will propagate in the BN and influence the variable of interest. Following the relations in a support graph (which is independent of the observed evidence), **ASPIC$^+$**-like arguments with premises and conclusions as variable-outcome pairs are constructed, and their strengths as either likelihood ratio or posterior odds are calculated from the premises given observed evidence. Conflicts among arguments are resolved using their strengths as priorities and the grounded semantics captures arguments with the highest probabilities. This argumentative approximation helps explain the conclusions drawn from a BN, as illustrated in a legal reasoning scenario.

Dejl et al. (2021) implement a method for generating argumentation frameworks which **approximately** represent various forms of **NNs**, trained in a **supervised** manner for **classification** with **textual data**, and provide customisable explanations for their outputs. This method represents neurons and connections as arguments and relations, respectively, in a **generalised argumentation framework**. These relations are extracted when the connections between two neurons satisfy certain conditions, and so the argumentation framework highlights parts of the network which may be interesting to a user as part of an explanation. The explanations themselves are provided in a variety of different forms, e.g. visual and conversational, which renders them suitable for a number of purposes. In the paper, the method is exemplified in the context of text classification, but other tasks such as image classification are supported, e.g. see (Albini et al., 2020).

A similar **approximate** method is proposed by (Sukpanichnant et al., 2021), where we again have argumentation frameworks representing **NNs** trained via **supervised learning**, but this time using **QBAFs** specifically and targeting **image classification** tasks. Arguments may represent sets of neurons in a layer, alleviating the density of the argumentation framework and thus extracting information which is more cognitively manageable to users. The authors show that existing methods for explanation in NNs may be represented by the QBAF, and then use it to provide visual explanations in the form of (Dejl et al., 2021).

Mollas et al. (2022) **approximate black-box** ML models for **classi-**

fication which operate on **tabular data** in a **segmented** manner with **SA**, producing explanations which are claimed to be more truthful than standard feature importance explanations. The method works by checking, for each feature, whether feature importance measures are in line with expectations when perturbations occur in their values. Depending on these checks, trees comprising structured arguments are constructed from fixed templates, which use both rebutting and undercutting attacks to determine whether the explanation is truthful, i.e. aligns with the expectations for all features. The authors then perform experiments with standard datasets from the literature. Here, they assess the method both qualitatively, showing intuitive examples of the resulting explanations, and quantitatively, showing improvements over standard feature attribution explanations in truthfulness.

In (Amgoud et al., 2023), the authors first show that abductive explanations, i.e. those consisting of features and assigned values considered to be sufficient for a given class, can be guaranteed to satisfy either, but not both, existence or correctness, properties defined in previous work (Amgoud and Ben-Naim, 2022). They then introduce a parameterised family of argumentation-based explanation functions, which explain **black-box** models for **supervised classification** of **tabular data** in a **segmented** manner. Here, **AFs** comprise arguments, consisting of features and assigned values which support classes, and attacks, which are identified between the arguments. The functions then use the stable semantics to select sets of arguments which are jointly considered to be acceptable. These explanations are shown, using theoretical and empirical analyses, to not only guarantee correctness but explain a reasonable proportion of instances, i.e. performing well wrt existence.

Potyka et al. (2023) **approximate RFs** deployed as **supervised classification** models of **tabular data** as argumentation frameworks, specifically **BAFs**, in order to provide global explanations for their reasoning. The BAFs themselves consist of arguments representing the classes, the rules and subsets of the features' values. Attacks (occurring between features, from features to rules and from rules to classes) and supports (occurring from rules to classes only) are then drawn between these arguments. It is then shown that extensions of the bi-stable semantics correspond to possible classification decisions. This means that finding sufficient and necessary explanations for the classifications can be reduced to finding those in the BAFs. The authors use Markov network encodings of the BAFs to solve the resulting combinatorial reasoning problems, and introduce an efficient probabilistic algorithm for approximating their solutions, given their high computational complexity. Finally, the approach's ability for finding sufficient and necessary reasons is examined empirically using three datasets (Iris, PIMA and Mushroom) from the literature.

2.1.3 Discussion

Our review so far of the works where argumentation has been used for supplementing ML models allows us to make the first following conclusions.

Argumentative methods for improving ML-based inference show promise for small models and datasets, but it remains to be seen whether this can be replicated at scale. Argumentation has been proposed to improve ML-based inference, under the assumption that argumentation-based inference is better-suited to reasoning with rules of thumb and exceptions than ML-based pattern matching is, or that it can readily incorporate expert knowledge. This is perhaps borne out in settings of small models and datasets, where patterns or rules discovered by ML can be augmented with, say, rule-based, argumentative reasoning. Whether argumentation can be similarly helpful with modern large scale models and datasets remains to be seen, especially because modern ML inference resolves conflicts probabilistically (the more likely patterns yield preferred inference) and large scale datasets tend to allow for capturing exceptions and particular contexts using larger models. We hope though that argumentation (or similar symbolic reasoning methods, for that matter) could be useful in specific applications, especially those requiring injection of expert knowledge, but one would still need to show that such methods would be better than the current ML methods such as Federated Mixture of Experts (ensembles of specialized models) or RLHF (RL with Human Feedback). However, it is not beyond the scope of the imagination that argumentation could be used to supplement these processes themselves, e.g. to resolve the conflicts between experts in the former case, as in (Abchiche-Mimouni et al., 2023) for conflicting models in an ensemble, or to structure human feedback in RLHF, as in (Rago et al., 2021) for recommender systems. Thus, argumentation's usefulness in improving ML inference at scale remains to be seen, though there has been clear progress which has highlighted potential opportunities.

There is arguably more evidence for argumentation's suitability for supporting the explainability of ML models, a goal which does not seem to conflict with that of model improvement, though challenges remain. There are many diverse works which construct argument graphs highlighting features, rules, data clusters, or model components that provide reasons for particular predictions. These approaches are said to provide benefits with regards to improving model inference at varying degrees, with a full spectrum of improvement to explanation on display. The intuitive relationship-based structure of argumentation explanations is seen as being more amenable to human understanding compared to typical attributive explanations (such as feature importance measures) used in ML. This seems promising, but widely-employed and convincingly useful systems for ML model explainability via argumentation are still somewhat lacking. Similarly, explainability of RL-based agents using argumentation seems promising, but much more work is needed to show its benefits in realistic systems that operate in high complexity and extensive simulation scenarios. Overall, engineering useful argumentation frameworks for explainability poses knowledge acquisition and scale bottlenecks. Open challenges also remain regarding explanatory power – whether argumentation actually improves user trust or model transparency over state-of-the-art approaches. More studies using objective metrics and subjective experimental evaluation with users would be welcome, and, without this, the extent of argumentation's explanatory benefits may continue to be seen as somewhat speculative.

While we discussed the two broad categories of the use of argumentation in tandem with ML for the sake of structured prediction and multi-agent interaction in the previous sections, in the next section we overview how ML has been considered to support argumentation.

2.2 ML for Argumentation

In this subsection, we first discuss some proposals for using argumentation frameworks for representing patterns learnt from data and making inferences thereon, essentially as a paradigm of argumentation-based learning to replace standard methods of supervised learning. We then overview several works that instead use ML to improve argumentation, by either supporting learning of argumentation frameworks from data or suggesting to learn and predict argument acceptance statuses in argumentation frameworks. Next, we cover works which combine ML and argumentation for analysis of the two formalisms, with some overlap with the sections covered in §2.1. We lastly cover a couple of works that focus on human dialogues in the real world and suggest that ML is perhaps more useful there than argumentation, implying that ML could be used in place of argumentation for some specific tasks.

2.2.1 For Supporting Argumentation

One trend of using ML for supporting argumentation is that of using supervised learning to generate argumentation frameworks. We first discuss several works that consider inductive ML (also known as concept learning, which essentially is learning from positive and negative examples to generate IF-THEN rules for classification) as a method for inducing rules and SA frameworks thereof. We suggest the reader to follow up with the references cited in the discussed works to explore the potential connections of argumentation and inductive ML.

Ontañón et al. (2012) argue that inductive generalisation in ML is a form of defeasible reasoning, and define a logical model to characterise and **generate** hypotheses from sets of examples from **tabular data** in a **supervised** manner. The authors then show how their logic-based learning model can be integrated with argumentation in a **segmented** manner to allow for argumentative interaction between agents that have inductively learnt different theories. The agents' theories are built from the rules learnt in the standard **inductive ML** setup, and **rule-based** argumentation (under, basically, grounded semantics) is used to reconcile the theories that different agents hold to be mutually consistent.

More recently, Proietti and Toni (2022); De Angelis et al. (2023) have proposed logic-based learning of **SA** frameworks from positive and negative examples in **tabular data**. The authors consider settings where examples are in the form of predicates applied to constant terms, as in logic-based knowledge bases. Their goal is then to learn, in a **super-**

Work	Type	Arg.	Learn.	ML Model(s)	Task	Data
(Ontañón et al., 2012)	Seg	SA (RB)	S	Inductive ML, Rule	Gen	Tab
(De Angelis et al., 2023)	Seg	SA (ABA)	S	Inductive ML, Rule	Gen	Tab
(Cocarascu et al., 2019)	Seg	QBAF	S	LSTM, TM	Gen	Text
(Craandijk and Bex, 2022)	App	AF	R	Q-learning with GNN	Enf	AF
(d'Avila Garcez et al., 2005)	App	VAF	S	NN (C-ILP)	Comp	Tab
(Kuhlmann and Thimm, 2019)	App	AF	S	GNN	Comp	AF
(Malmqvist et al., 2020)	App	AF	S	GNN	Comp	AF
(Craandijk and Bex, 2020)	App	AF	S	GNN	Comp	AF
(Klein et al., 2022)	App	AF	S	kNN, NBC, RF SVM, GNN	Comp	AF

Table 3: A summary of the ways in which ML has been used for *supporting* argumentation, as detailed in §2.2.1. We show: the type (segmented or approximate); the kind of argumentation used; the type of learning employed (supervised or reinforcement); the ML model(s) used, including rule learning, long-short term memory networks (LSTMs), topic modelling (TM), k-nearest neighbours (kNN); the task undertaken, i.e. the purpose of using ML (argument graph generation, enforcement or computation of semantics and/or argument acceptance) and the type of data used (tabular, textual or AFs).

vised manner, ABA frameworks. This **segmented** approach amounts to "identifying rules, assumptions and their contraries (adding to those in the given background knowledge) that cover argumentatively the positive examples and none of the negative examples according to some chosen semantics". The works focus on theoretical advancements of learning and transforming logic-based rules that later give rise to **ABA** frameworks for argumentative reasoning. In this sense, the approach can be said to use **inductive ML** for automated **generation** of argumentation frameworks. However, no experimental evaluation on ML-targeted datasets or comparison with inductive ML (such as inductive logic programming) systems has yet been carried out.

We next take a look at a particular work that uses an NN architecture to learn argumentation frameworks, this time of the abstract kind.

The argumentative approach to review aggregation introduced by Cocarascu et al. (2019) is supported by ML in a **segmented** manner. In this work, **QBAFs** are used to represent critics' reviews on movies, i.e. **textual data**, in order to provide an explainable method for their aggregation. An ontology is used to provide a skeleton for arguments in the QBAF, before the arguments, attacks, supports and base scores are then extracted from their reviews. **Supervised** learning via **long-short term memory** NNs is used to extract votes on arguments from the critics, which allows the calculation of the base score (intrinsic strength) on arguments. Also, **topic modeling** is used to **generate** some of the arguments in the QBAF from the reviews. The QBAFs, evaluated by means of a gradual semantics, then provide users with an aggregation of the reviews supported by explanations comprising components of the QBAF. The QBAFs and the arguments' evaluations are shown theoretically to satisfy desirable argumentative properties and experimentally to be comparable to the aggregations popularised by the movie review website *Rotten Tomatoes*.

Finally, one recent work looks at the enforcement problem, i.e. how to modify an AF to enforce the specified argument acceptability, from the point of view of learning AF modifications from generated AFs, rather than solving enforcement analytically.

Concretely, Craandijk and Bex (2022) use deep **RL** to learn which attack relations between arguments should be added or deleted in order to enforce the acceptability of (a set of) arguments. In particular, the authors present an Enforcement GNN (EGNN), an **approximate** approach, that efficiently learns solutions, i.e. modifications to **AFs** that enforce argument (non-)acceptance, which only need to be verified for correctness by existing argument acceptability algorithms. In more detail, the authors formulate the problem of learning **enforcement** as a **MDP** where a state represents a modification of a given AF, the actions amount to adding or deleting attacks, with the obvious transition function taking an AF to its modification, and the only non-negative reward is given at the AF with the correct arguments enforced. To learn approximately optimal policies of enforcement in such a vast state and action space, the authors employ deep **Q-learning**, whereby the Q-values of states are predicted by a **GNN**. The (fully-connected) GNN in question vectorially represents arguments that should be enforced (as nodes) and existing attack relations (as edges) and uses message passing to iteratively update node vectors. They compare their EGNN to two other relevant deep learning models as well as to symbolic solvers by training the ML models on **AF data** generated by the ICCMA generators and testing on random sets of argument to be enforced on randomly generated AFs. The empirical evaluation shows that the proposed EGNN model beats the other deep learning models in terms of generalisation ability (i.e. correctly learning enforcements) and essentially outperforms symbolic solvers in terms of efficiency (solutions found within a time limit) while being near optimal (steps taken to find a solution).

While the above works focus on generating argument graphs, the next set of works use ML to support computation in given argument graphs.

Probably the first work to use ML to support **computation** in argumentation is that of d'Avila Garcez et al. (2005). There the authors establish an **approximate** mapping from **VAFs** to **NNs** with a single hidden layer operating under **supervised** learning with **tabular data**, showing a correspondence between the two. The authors first introduce a Neural Argumentation Algorithm, which, given a VAF, produces a corresponding NN, the uniqueness of which is guaranteed. The authors show a number of theoretical results, first demonstrating the fact that the algorithm may not be able to compute the NNs corresponding to cyclic VAFs. In order to resolve this, the authors define a form of learning based on enforcement, i.e. a method for implementing changes such that a given state of acceptance or rejection for the arguments is achieved. This learning uses backpropagation to indicate changes which need to be made, i.e. "arguments learnt", in order for specified arguments to be accepted, and the approach is thus shown to model cumulative (accrual) argumentation over time.

More recent works explore how to use GNNs for computing acceptability of arguments in AFs, typically without correctness guarantees, but possibly with high accuracy and speed.

Kuhlmann and Thimm (2019) use convolutional **GNNs** (GCNs) on **AFs** to **compute** whether an argument is included in the preferred semantics, thus proposing a method to **approximate** reasoning with abstract argumentation. While the original GCN used was designed for semi-supervised training, in this work all nodes in the **AF data** are labelled, thus the training is **supervised**. Further, the number of incoming and outgoing attacks per argument are also used as features. The authors report moderate results on data from ICCMA 2017 and provide several factors that impact the performance (e.g. size of the training dataset, class imbalance, as well as hyperparameters).

Another work employing **GCNs** in an **approximated** fashion for **computation** in argumentation is that of Malmqvist et al. (2020). The GCN is used to determine sceptical and credulous acceptability in **AFs**. The **supervised** training involves **AF data**, with each AF representing a single connected component in the graph, being fed to the GCN. The authors use randomised training, forcing the GCN to learn to generalise based on the structural properties of the graphs, and test their method on data from ICCMA 2017. They are able to improve results using the randomised training approach, however the disparities between the positive (accepted) and negative (rejected) acceptability remain unsolved.

The approach of Craandijk and Bex (2020) also uses ML with **supervised learning** in an **approximated** manner, where the authors propose an Argumentation **GNN** (AGNN) to **compute** the (likelihood of the) acceptance of arguments in an **AF**. In particular, the authors consider the credulous and sceptical acceptance of arguments under the grounded, preferred, stable, and complete semantics. They experiment with **AF data** generated using ICCMA 2020 generators and show that AGNNs can almost perfectly predict the acceptability under different semantics and their method is able to scale for larger AFs. Further, the proposed AGNNs can be used to guide a basic search for extensions.

Interestingly, Kuhlmann et al. (2022) follow-up on the work of Craandijk and Bex (2020) and suggest some sceptical conclusions. The authors there study the problem of data selection for generating and computing AFs, particularly the problem of sceptical acceptance under preferred semantics. They test the AGNN of Craandijk and Bex (2020) on an alternative training set to achieve better performance results, and additionally conduct the same tests on an NN model that is a part of the AGNN. The results suggest that the AGNN does not learn to solve the problem of sceptical acceptance, but instead learns specific properties of the benchmarks that allow the model to perform relatively well. The authors conclude the same about the NN model also. Further, they perform experiments with simpler ML models, specifically 5-Nearest Neighbours, NBCs, DTs and RFs, that mainly use simple graph-theoretic features of AFs, and conclude that those models perform rather well on the AF computation task. The authors suggest that the more complex models such as the AGNN tend to simply learn and make use of those features.

We finally see a proposal that retains the argumentative machinery but suggests to use ML to predict which argumentation solver would be best to compute extensions for a given framework.

Klein et al. (2022) perform an experimental study where **supervised** ML is used in an **approximate** manner to select the fastest algorithm to compute skeptical acceptance under preferred semantics in **AFs**. In particular, the evaluation setup involves several classic ML models (**k-Nearest Neighbours, NBCs, RFs, and SVMs**) and modern **GNNs** (Graph Convolutional Networks, Graph Isomorphism Networks, and GraphSage) for predicting the fastest sound and complete argumentation solver (one out of three). The classic ML classifiers used the number of vertices, density and the minimum degree value of the directed graph as well as in- and out-degrees of the query arguments as features, whereas the GNN models were off-the-shelf (not using any precomputed graph, node or edge features); the labels were assigned to AFs according to which of the three solvers is actually the fastest to solve the problem of sceptical acceptance under preferred semantics. The authors experimented on **AF data** with 6200 AFs randomly generated with ICCMA 2017 generators, showing the GNN models slightly outperform the classic ML ones, but are typically slower.

2.2.2 For Analysing or Replacing Argumentation

We next discuss several works that synergistically combine ML and argumentation in a way that one complements the other. These works essentially show correspondence between (feed forward) NNs and QBAFs. In one direction, this allows to use ML for deriving (and computing) new gradual semantics. In the other, a QBAF can stand as architecture of an NN and potentially improve learning and/or explainability thereof.

Potyka (2021) shows that **NNs**, trained via **supervised** learning on **tabular data** and in the form of MLPs, can be equivalently seen as **QBAFs with weighted edges**. Here, arguments in the QBAFs represent the neurons in the corresponding MLP. Then, attacks and supports are drawn between arguments where the connections between the corresponding neurons have weights which are negative or positive, respectively. The base score for an argument is then obtained, intuitively, from the corresponding neuron's bias. A gradual semantics for the QBAF is defined in a way that directly corresponds to the activations which are propagated through the MLP's neurons. The author then theoretically examines the approach in depth: proving convergence guarantees for cyclic and acyclic cases; suggesting solutions to convergence issues; and assessing the semantics wrt satisfaction of existing properties from the literature. In summary, the paper exhibits a **synergystic** approach for general **analysis** whereby ML allows the definition of a novel QBAF semantics and suggests a way of learning QBAFs, and reciprocally, argumentation potentially offers ways of constructing sparse MLPs as well as explaining those which are already trained.

Using the concepts introduced in this approach, Spieler et al. (2022); Potyka et al. (2022a) have shown how these **QBAFs with weighted edges** with the semantics derived from **NNs**, in the form of MLPs, can be learnt from **tabular data**, again combining argumentation with ML **synergistically**. They first restrict the QBAFs to be acyclic, as is

Work	Type	Arg.	Learn.	ML Model(s)	Task	Data
(Potyka, 2021)	Syn	QBAF with WE	S	NN	An	Tab
(Potyka et al., 2022a)	Syn	QBAF with WE	S	NN	Class	Tab
(Rosenfeld and Kraus, 2016)	App	BAF, Dialogues	S	SVM, DT, NN	Rec	Text
(Donadello et al., 2022)	Seg	Informal, Dialogues	S	SVR, RF	Rec	Text

Table 4: A summary of the ways in which ML has been used for *analysing* or *replacing* argumentation, as detailed in §2.2.2. We show: the type (synergistic, segmented or approximate); the kind of argumentation used; the type of learning employed (supervised); the ML model(s) used, including support vector regression (SVR); the task undertaken (analysis, classification or argument recommendation) and the type of data used (tabular or textual).

the case for the NNs for **classification** via **supervised** learning being somewhat replicated. The structure of the QBAFs is then learnt using two different meta-heuristics: a genetic algorithm and particle swarm optimisation. Finally, the authors train the QBAFs for classification, i.e. learning the base scores and edge weights, using backpropagation. The activation function can then be seen as embedding the gradual semantics. The authors then undertake an experimental evaluation in classification tasks with the UCI ML Repository. The results show high accuracy for the simple problems, which are similar across the two meta-heuristics, as well as a baseline in the form of a DT method. However, the authors argue that using QBAFs is more sparse and easier for a human to interpret, aligning with the proposed reasoning behind the general trend highlighted in §2.1.2.

The last two works we discuss in this subsection paint the picture where argumentation may be sub-optimal in real-world argumentation scenarios and that one may be better of using off-the-shelf ML models for supporting human-like argumentation.

The first effectively argues that bipolar argumentation is not well-suited to select arguments that people choose in real-world deliberation scenarios and instead suggests using ML for recommending arguments in human discussions. Concretely, Rosenfeld and Kraus (2016) use ML models in a **segmented** manner with **textual data** to predict the arguments selected by humans in real-world discussions, showing superiority of ML models over argumentation theory in the task in terms of both accuracy and user satisfaction. Specifically, the authors consider **deliberation dialogues** where participants exchange information and try to reach consensus on some topic. They conduct multiple experiments with human subjects aiming to evaluate whether the participants choose arguments that are justified by **BAFs** that model the deliberations. They also define features of arguments in terms of a deliberant's previous use of arguments and argument relevance measures (such as minimum (un)directed paths' length between arguments) to enable a setting for predicting by means of ML the next argument to choose. They then study if instead simple **supervised** ML models of **SVMs, DTs** and **NNs** can predict human choices of arguments in deliberations. In the first experiment, 64 people were presented with a partial deliberation and had to choose one argument out of 4 to continue deliberation. On average, a justified argument (under some classical BAF semantics) was selected only 67.3% of the time, whereas ML models achieve accuracy above 70%. In the second experiment with 2 corpora and 64 free-form deliberations, subjects chose argumentatively justified arguments (likewise, under several classical semantics) no better than chance, whereas ML models reached accuracy similar to that in the first experiment. In their third experiment with 72 semi-structured deliberations where 144 participants were restricted to 40 pre-defined arguments to choose from, argumentation again modelled the behaviour no better than random, whereas ML models did significantly better. Finally, the authors show in extensive human studies with 204 participants and 102 deliberations how an ML-based model that takes into account argument relevance outperforms argumentation-based models when **recommending** arguments to put forward next. Overall the work shows that bipolar argumentation may host limitations in modelling human deliberations in the real world and that simple ML models of SVMs, DTs and NNs have much better predictive power in deliberations than argumentation, roughly at the level of 75% and higher, depending on the experiment and setting.

Another work uses **supervised** ML to **recommend** which arguments are put forward in persuasion settings. Donadello et al. (2022) use ML for learning utility functions of the system and user agents involved in a persuasion **dialogue** with **informal argumentation** in an automated persuasion system. In a state-of-the-art persuasion setting that uses utility maximisation across arguments of both the system (persuader) and the user (persuadee), the utility functions are typically manually specified by experts. Instead, the authors' goal is to be able to learn those utility functions from known utilities. To this end they use **support vector regression** to estimate utility of arguments for a new user given those for other users. Where further utility elicitation is done via questionnaires, the authors aim to minimise the number of questions posed to the user. To do so, they cluster users with similar utility functions and employ **RFs** to identify the best questions for regressive utility estimation. It is overall a **segmented** method where ML outputs as utility functions are used to select the most persuasive arguments. The authors evaluate it on synthetic, **textual data** using both abstract and realistic (healthy diet) simulated persuasion dialogues. The experiments show the approach to be promising for learning utilities when data is limited, for the purpose of improving automated persuasion systems.

2.2.3 Discussion

The use of inductive ML to generate argumentation frameworks shows much promise, but is still to be proven with real datasets. The two approaches for generating SA frameworks from tabular data using inductive ML give novel approaches to automated reasoning which could serve as the basis for fruitful directions of future research, given their interesting theoretical results. If the next steps of proving that this theory translates to real-world datasets at scale are made, this could provide a significant advancement for neuro-symbolic systems. In the case of learning arguments and relationships from text, there seem to be more issues, e.g. due to the need to engineer features and the free-form nature of textual data. However, in settings where the semantics of potential arguments drawn from text are relatively constrained, e.g. in review aggregation in a specified domain, there seems to be the possibility of fully automating these pipelines. This must be undertaken with caution, however, as in some domains it has been shown that ML can effectively replace argumentative systems.

GNNs' natural fit for approximating argument evaluation has much potential, capitalising on existing AF datasets and generators. The use of ML for computing argument graphs, for instance predicting acceptability in AFs, also shows much potential. This is especially the case when analytical solutions are costly but checking the correctness of, for example, argument acceptance, is cheap. Here, relatively fast ML-based prediction can be combined with verification for correctness, falling back to an argumentation solver where needed. The works very much benefited from ICCMA datasets and AF generators, an indication that competition for other argumentation frameworks, such as those which are structured, may have very welcome consequences further down the road, if the availability of data (i.e. argument graphs) induced creation of hybrid solvers.

The mapping from QBAFs to MLPs brings benefits in both directions, but looks especially exciting in introducing learnt, explainable classifiers.

Three works have shown that argumentation frameworks, namely QBAFs, can be used as the architecture of a certain form of ML models, namely NNs, so that argumentation frameworks are learnt from data using standard ML techniques, and the reasoning/inference follows argumentative structure and is more explainable. While the theoretical analysis of argumentative properties of MLPs is also very interesting, the capability for learning classifiers which reason argumentatively, maintain high accuracy and are able to explain their outputs faithfully to their inference process would be very exciting for the field of argumentation and beyond.

3 Conclusions

This chapter overviews a growing body of research on cross-fertilisations between argumentation and ML. We have broadly categorised these works into those which can be classed as argumentation for ML, and those which are ML for argumentation. The two overarching objectives in the former set of works are either improving or explaining ML models' inference, though the two are not mutually exclusive and most of the works tackle the two in tandem. The second set of works are less homogeneous, with objectives ranging from using ML to generate argumentation frameworks, to approximating semantics or recommending arguments to users without the need for an argumentation semantics. We have shown that this integration of symbolic and data-driven approaches remains a highly active yet open-ended area of research, with great potential along numerous avenues. However, challenges still remain for argumentative techniques to demonstrate definitive advantages in the realm of ML. Progress is needed along multiple dimensions, notably regarding scalability and bridging the gap with real human argumentation via user studies. With ML capabilities rapidly progressing, it is clear that there is still a role to be played in the advancement of AI by argumentation, given its prowess in explainability and commonsense reasoning, two factors which are notably lacking in even state-of-the-art ML models.

References

N. Abchiche-Mimouni, L. Amgoud, and F. Zehraoui. Explainable ensemble classification model based on argumentation. In *Proceedings of the 2023 International Conference on Autonomous Agents and Multiagent Systems, AAMAS*, pages 2367–2369, 2023. doi: 10.5555/3545946.3598936. URL https://dl.acm.org/doi/10.5555/3545946.3598936.

E. Albini, P. Lertvittayakumjorn, A. Rago, and F. Toni. DAX: deep argumentative explanation for neural networks. *CoRR*, abs/2012.05766, 2020. URL https://arxiv.org/abs/2012.05766.

L. Amgoud and J. Ben-Naim. Axiomatic foundations of explainability. In *Proceedings of the Thirty-First International Joint Conference on Artificial Intelligence, IJCAI*, pages 636–642, 2022. doi: 10.24963/IJCAI.2022/90. URL https://doi.org/10.24963/ijcai.2022/90.

L. Amgoud and M. Serrurier. Agents that argue and explain classifications. *Autonomous Agents and Multi-Agent Systems*, 16(2):187–209, 2008. doi: 10.1007/s10458-007-9025-6. URL https://doi.org/10.1007/s10458-007-9025-6.

L. Amgoud, P. Muller, and H. Trenquier. Leveraging argumentation for generating robust sample-based explanations. In *Proceedings of the Thirty-Second International Joint Conference on Artificial Intelligence, IJCAI*, pages 3104–3111, 2023. doi: 10.24963/ijcai.2023/346. URL https://doi.org/10.24963/ijcai.2023/346.

H. Ayoobi, M. Cao, R. Verbrugge, and B. Verheij. Argue to learn: Accelerated argumentation-based learning. In *20th IEEE International Conference on Machine Learning and Applications, ICMLA*, pages 1118–1123, 2021. doi: 10.1109/ICMLA52953.2021.00183. URL https://doi.org/10.1109/ICMLA52953.2021.00183.

H. Ayoobi, M. Cao, R. Verbrugge, and B. Verheij. Argumentation-based online incremental learning. *IEEE Transactions on Automation Science and Engineering*, 19(4):3419–3433, 2022. doi: 10.1109/TASE.2021.3120837. URL https://doi.org/10.1109/TASE.2021.3120837.

P. Baroni, D. Gabbay, M. Giacomin, and L. van der Torre, editors. *Handbook of Formal Argumentation*. College Publications, 2018.

F. Bex and S. Renooij. From arguments to constraints on a bayesian network. In *Computational Models of Argument - Proceedings of COMMA*, pages 95–106, 2016. doi: 10.3233/978-1-61499-686-6-95. URL https://doi.org/10.3233/978-1-61499-686-6-95.

S. Bistarelli, A. Mancinelli, F. Santini, and C. Taticchi. An argumentative explanation of machine learning outcomes. In *Computational Models of Argument - Proceedings of COMMA*, pages 347–348, 2022. doi: 10.3233/FAIA220166. URL https://doi.org/10.3233/FAIA220166.

I. Bratko, J. Žabkar, and M. Možina. Argument-based machine learning. In *Argumentation in Artificial Intelligence*, pages 463–482. 2009.

T. B. Brown, B. Mann, N. Ryder, M. Subbiah, J. Kaplan, P. Dhariwal, A. Neelakantan, P. Shyam, G. Sastry, A. Askell, S. Agarwal, A. Herbert-Voss, G. Krueger, T. Henighan, R. Child, A. Ramesh, D. M. Ziegler, J. Wu, C. Winter, C. Hesse, M. Chen, E. Sigler, M. Litwin, S. Gray, B. Chess, J. Clark, C. Berner, S. McCandlish, A. Radford, I. Sutskever, and D. Amodei. Language models are few-shot learners. In *Advances in Neural Information Processing Systems 33: Annual Conference on Neural Information Processing Systems*, 2020. URL https://proceedings.neurips.cc/paper/2020/hash/1457c0d6bfcb4967418bfb8ac142f64a-Abstract.html.

L. Carstens and F. Toni. Improving out-of-domain sentiment polarity classification using argumentation. In *IEEE International Conference on Data Mining Workshop, ICDMW*, pages 1294–1301, 2015. doi: 10.1109/ICDMW.2015.185. URL https://doi.org/10.1109/ICDMW.2015.185.

O. Cocarascu and F. Toni. Argumentation for machine learning: A survey. In *Computational Models of Argument - Proceedings of COMMA*, pages 219–230, 2016. doi: 10.3233/978-1-61499-686-6-219. URL https://doi.org/10.3233/978-1-61499-686-6-219.

O. Cocarascu, A. Rago, and F. Toni. Extracting dialogical explanations for review aggregations with argumentative dialogical agents.

In *Proceedings of the 18th International Conference on Autonomous Agents and MultiAgent Systems, AAMAS*, pages 1261–1269, 2019. URL http://dl.acm.org/citation.cfm?id=3331830.

O. Cocarascu, A. Stylianou, K. Čyras, and F. Toni. Data-empowered argumentation for dialectically explainable predictions. In *ECAI 2020 - 24th European Conference on Artificial Intelligence*, volume 325, pages 2449–2456, 2020. doi: 10.3233/FAIA200377. URL https://doi.org/10.3233/FAIA200377.

D. Craandijk and F. Bex. Deep learning for abstract argumentation semantics. In *Proceedings of the Twenty-Ninth International Joint Conference on Artificial Intelligence, IJCAI*, pages 1667–1673, 2020. doi: 10.24963/ijcai.2020/231. URL https://doi.org/10.24963/ijcai.2020/231.

D. Craandijk and F. Bex. EGNN: A deep reinforcement learning architecture for enforcement heuristics. In *Computational Models of Argument - Proceedings of COMMA*, pages 353–354, 2022. doi: 10.3233/FAIA220169. URL https://doi.org/10.3233/FAIA220169.

M. W. Craven and J. W. Shavlik. Extracting tree-structured representations of trained networks. In D. S. Touretzky, M. Mozer, and M. E. Hasselmo, editors, *Advances in Neural Information Processing Systems 8, NIPS, Denver, CO, USA, November 27-30, 1995*, pages 24–30. MIT Press, 1995. URL http://papers.nips.cc/paper/1152-extracting-tree-structured-representations-of-trained-networks.

A. S. d'Avila Garcez, D. M. Gabbay, and L. C. Lamb. Value-based argumentation frameworks as neural-symbolic learning systems. *Journal of Logic and Computation*, 15(6):1041–1058, 2005. doi: 10.1093/logcom/exi057. URL https://doi.org/10.1093/logcom/exi057.

E. De Angelis, M. Proietti, and F. Toni. ABA learning via ASP. In *Proceedings 39th International Conference on Logic Programming, ICLP*, volume 385 of *EPTCS*, pages 1–8, 2023. doi: 10.4204/EPTCS.385.1. URL https://doi.org/10.4204/EPTCS.385.1.

A. Dejl, P. He, P. Mangal, H. Mohsin, B. Surdu, E. Voinea, E. Albini, P. Lertvittayakumjorn, A. Rago, and F. Toni. Argflow: A toolkit for deep argumentative explanations for neural networks. In *AAMAS '21: 20th International Conference on Autonomous Agents and Multiagent Systems*, pages 1761–1763, 2021. doi: 10.5555/3463952.3464229. URL https://www.ifaamas.org/Proceedings/aamas2021/pdfs/p1761.pdf.

I. Donadello, A. Hunter, S. Teso, and M. Dragoni. Machine learning for utility prediction in argument-based computational persuasion. In *Thirty-Sixth AAAI Conference on Artificial Intelligence*, pages 5592–5599, 2022. doi: 10.1609/AAAI.V36I5.20499. URL https://doi.org/10.1609/aaai.v36i5.20499.

D. A. Furman, P. Torres, J. A. Rodríguez, D. Letzen, M. V. Martinez, and L. A. Alemany. An initial exploration of how argumentative information impacts automatic generation of counter-narratives against hate speech. In *Proceedings of the First International Workshop on Argumentation and Applications co-located with 20th International Conference on Principles of Knowledge Representation and Reasoning (KR 2023)*, pages 26–39, 2023. URL https://ceur-ws.org/Vol-3472/paper2.pdf.

Y. Gao and F. Toni. Argumentation accelerated reinforcement learning for robocup keepaway-takeaway. In *Theory and Applications of Formal Argumentation - Second International Workshop, TAFA*, pages 79–94, 2013. doi: 10.1007/978-3-642-54373-9_6. URL https://doi.org/10.1007/978-3-642-54373-9_6.

Y. Gao and F. Toni. Argumentation accelerated reinforcement learning for cooperative multi-agent systems. In *ECAI 2014 - 21st European Conference on Artificial Intelligence*, pages 333–338, 2014. doi: 10.3233/978-1-61499-419-0-333. URL https://doi.org/10.3233/978-1-61499-419-0-333.

Y. Gao, F. Toni, and R. Craven. Argumentation-based reinforcement learning for robocup soccer keepaway. In *ECAI 2012 - 20th European Conference on Artificial Intelligence*, volume 242, pages 342–347, 2012. doi: 10.3233/978-1-61499-098-7-342. URL https://doi.org/10.3233/978-1-61499-098-7-342.

S. A. Gómez and C. I. Chesñevar. A hybrid approach to pattern classification using neural networks and defeasible argumentation. In *Proceedings of the Seventeenth International Florida Artificial Intelligence Research Society Conference*, pages 393–398, 2004a. URL http://www.aaai.org/Library/FLAIRS/2004/flairs04-069.php.

S. A. Gómez and C. I. Chesñevar. Integrating defeasible argumentation with fuzzy art neural networks for pattern classification. *Journal of Computer Science & Technology, 2004, vol. 4, núm. 1, p. 45-51*, 2004b.

Y. Guo, T. Yu, L. Bai, J. Tang, Y. Ruan, and Y. Zhou. Argumentative explanation for deep learning: A survey. In *2023 IEEE International Conference on Unmanned Systems (ICUS)*, pages 1738–1743, 2023. doi: 10.1109/ICUS58632.2023.10318322.

M. Hinton and J. H. M. Wagemans. How persuasive is ai-generated argumentation? an analysis of the quality of an argumentative text produced by the GPT-3 AI text generator. *Argument & Computation*, 14(1):59–74, 2023. doi: 10.3233/AAC-210026. URL https://doi.org/10.3233/AAC-210026.

N. D. Hung, N. Huynh, T. Theeramunkong, and T. Nhu. Composite argumentation systems with ML components. In *Computational Models of Argument - Proceedings of COMMA*, pages 164–175, 2022. doi: 10.3233/FAIA220150. URL https://doi.org/10.3233/FAIA220150.

A. Hunter. Generating instantiated argument graphs from probabilistic information. In *ECAI 2020 - 24th European Conference on Artificial Intelligence*, pages 769–776, 2020. doi: 10.3233/FAIA200165. URL https://doi.org/10.3233/FAIA200165.

D. Kazhdan, Z. Shams, and P. Liò. Marleme: A multi-agent reinforcement learning model extraction library. In *International Joint Conference on Neural Networks*, pages 1–8, 2020. doi: 10.1109/IJCNN48605.2020.9207564. URL https://doi.org/10.1109/IJCNN48605.2020.9207564.

J. Klein, I. Kuhlmann, and M. Thimm. Graph neural networks for algorithm selection in abstract argumentation. In *Proceedings of the 1st Workshop on Argumentation & Machine Learning co-located with*

9th International Conference on Computational Models of Argument (COMMA), pages 81–95, 2022. URL https://ceur-ws.org/Vol-3 208/paper6.pdf.

I. Kuhlmann and M. Thimm. Using graph convolutional networks for approximate reasoning with abstract argumentation frameworks: A feasibility study. In *Scalable Uncertainty Management - 13th International Conference, SUM*, pages 24–37, 2019. doi: 10.1007/978-3-030-35514-2 _3. URL https://doi.org/10.1007/978-3-030-35514-2_3.

I. Kuhlmann, T. Wujek, and M. Thimm. On the impact of data selection when applying machine learning in abstract argumentation. In *Computational Models of Argument - Proceedings of COMMA*, pages 224–235, 2022. doi: 10.3233/FAIA220155. URL https://doi.org/10.3233/FAIA220155.

P. Lertvittayakumjorn and F. Toni. Argumentative explanations for pattern-based text classifiers. *Argument & Computation*, 14(2):163–234, 2023. doi: 10.3233/AAC-220004. URL https://doi.org/10.3 233/AAC-220004.

L. Longo. Argumentation for knowledge representation, conflict resolution, defeasible inference and its integration with machine learning. In *Machine Learning for Health Informatics - State-of-the-Art and Future Challenges*, pages 183–208. 2016. doi: 10.1007/978-3-319-50478-0_9. URL https://doi.org/10.1007/978-3-319-50478-0_9.

L. Malmqvist, T. Yuan, P. Nightingale, and S. Manandhar. Determining the acceptability of abstract arguments with graph convolutional networks. In *Proceedings of the Third International Workshop on Systems and Algorithms for Formal Argumentation co-located with the 8th International Conference on Computational Models of Argument (COMMA)*, pages 47–56, 2020. URL https://ceur-ws.org/Vol-267 2/paper_5.pdf.

I. Mollas, N. Bassiliades, and G. Tsoumakas. Altruist: Argumentative explanations through local interpretations of predictive models. In *SETN 2022: 12th Hellenic Conference on Artificial Intelligence*, pages

21:1–21:10, 2022. doi: 10.1145/3549737.3549762. URL https://doi.org/10.1145/3549737.3549762.

M. Možina, J. Žabkar, and I. Bratko. Argument based machine learning. *Artif. Intell.*, 171(10-15):922–937, 2007. doi: 10.1016/j.artint.2007.04.007. URL https://doi.org/10.1016/j.artint.2007.04.007.

M. Mozina. Arguments in interactive machine learning. *Informatica (Slovenia)*, 42(1), 2018. URL http://www.informatica.si/index.php/informatica/article/view/2231.

S. Ontañón, P. Dellunde, L. Godo, and E. Plaza. A defeasible reasoning model of inductive concept learning from examples and communication. *Artificial Intelligence*, 193:129–148, 2012. doi: 10.1016/j.artint.2012.08.006. URL https://doi.org/10.1016/j.artint.2012.08.006.

C. Otero, D. Craandijk, and F. Bex. ORLA: learning explainable argumentation models. In *Proceedings of the 20th International Conference on Principles of Knowledge Representation and Reasoning, KR*, pages 542–551, 2023. doi: 10.24963/kr.2023/53. URL https://doi.org/10.24963/kr.2023/53.

N. Potyka. Abstract argumentation with markov networks. In *ECAI 2020 - 24th European Conference on Artificial Intelligence*, pages 865–872, 2020. doi: 10.3233/FAIA200177. URL https://doi.org/10.3233/FAIA200177.

N. Potyka. Interpreting neural networks as quantitative argumentation frameworks. In *Thirty-Fifth AAAI Conference on Artificial Intelligence, AAAI*, pages 6463–6470, 2021. URL https://ojs.aaai.org/index.php/AAAI/article/view/16801.

N. Potyka, M. Bazo, J. Spieler, and S. Staab. Learning gradual argumentation frameworks using meta-heuristics. In *Proceedings of the 1st Workshop on Argumentation & Machine Learning co-located with 9th International Conference on Computational Models of Argument (COMMA)*, pages 96–108, 2022a. URL https://ceur-ws.org/Vol-3208/paper7.pdf.

N. Potyka, X. Yin, and F. Toni. On the tradeoff between correctness and completeness in argumentative explainable AI. In *1st International Workshop on Argumentation for eXplainable AI co-located with 9th International Conference on Computational Models of Argument (COMMA)*, volume 3209, 2022b. URL https://ceur-ws.org/Vol-3209/8151.pdf.

N. Potyka, X. Yin, and F. Toni. Explaining random forests using bipolar argumentation and markov networks. In *Thirty-Seventh AAAI Conference on Artificial Intelligence*, pages 9453–9460, 2023. doi: 10.1609/AAAI.V37I8.26132. URL https://doi.org/10.1609/aaai.v37i8.26132.

H. Prakken and R. Ratsma. A top-level model of case-based argumentation for explanation: Formalisation and experiments. *Argument & Computation*, 13(2):159–194, 2022. doi: 10.3233/AAC-210009. URL https://doi.org/10.3233/AAC-210009.

N. Prentzas, A. Gavrielidou, M. Neofytou, and A. C. Kakas. Argumentation-based explainable machine learning (argeml): a real-life use case on gynecological cancer. In *Proceedings of the 1st Workshop on Argumentation & Machine Learning co-located with 9th International Conference on Computational Models of Argument (COMMA) 2022)*, pages 1–13, 2022. URL https://ceur-ws.org/Vol-3208/paper1.pdf.

M. Proietti and F. Toni. Learning assumption-based argumentation frameworks. 13779:100–116, 2022. doi: 10.1007/978-3-031-55630-2_8. URL https://doi.org/10.1007/978-3-031-55630-2_8.

M. Proietti and F. Toni. A roadmap for neuro-argumentative learning. In *Proceedings of the 17th International Workshop on Neural-Symbolic Learning and Reasoning*, pages 1–8, 2023. URL https://ceur-ws.org/Vol-3432/paper1.pdf.

A. Rago, O. Cocarascu, C. Bechlivanidis, D. A. Lagnado, and F. Toni. Argumentative explanations for interactive recommendations. *Artificial Intelligence*, 296:103506, 2021. doi: 10.1016/J.ARTINT.2021.103506. URL https://doi.org/10.1016/j.artint.2021.103506.

A. Rajasekharan, Y. Zeng, and G. Gupta. Argument analysis using answer set programming and semantics-guided large language models. In *Proceedings of the International Conference on Logic Programming 2023 Workshops co-located with the 39th International Conference on Logic Programming (ICLP)*, 2023. URL https://ceur-ws.org/Vol-3437/paper6GDE.pdf.

R. Riveret, Y. Gao, G. Governatori, A. Rotolo, J. Pitt, and G. Sartor. A probabilistic argumentation framework for reinforcement learning agents - towards a mentalistic approach to agent profiles. *Autonomous Agents and Multi-Agent Systems*, 33(1-2):216–274, 2019. doi: 10.1007/s10458-019-09404-2. URL https://doi.org/10.1007/s10458-019-09404-2.

R. Riveret, S. N. Tran, and A. S. d'Avila Garcez. Neuro-symbolic probabilistic argumentation machines. In *Proceedings of the 17th International Conference on Principles of Knowledge Representation and Reasoning, KR*, pages 871–881, 2020. doi: 10.24963/kr.2020/90. URL https://doi.org/10.24963/kr.2020/90.

A. Rosenfeld and S. Kraus. Providing arguments in discussions on the basis of the prediction of human argumentative behavior. *ACM Transactions on Interactive Intelligent Systems*, 6(4):30:1–30:33, 2016. doi: 10.1145/2983925. URL https://doi.org/10.1145/2983925.

C. Rudin. Stop explaining black box machine learning models for high stakes decisions and use interpretable models instead. *Nature Machine Intelligence*, 1(5):206–215, 2019. doi: 10.1038/S42256-019-0048-X. URL https://doi.org/10.1038/s42256-019-0048-x.

N. Sendi, N. Abchiche-Mimouni, and F. Zehraoui. A new transparent ensemble method based on deep learning. In *Knowledge-Based and Intelligent Information & Engineering Systems: Proceedings of the 23rd International Conference KES-2019*, volume 159, pages 271–280, 2019. doi: 10.1016/j.procs.2019.09.182. URL https://doi.org/10.1016/j.procs.2019.09.182.

X. Shao, T. Rienstra, M. Thimm, and K. Kersting. Towards understanding and arguing with classifiers: Recent progress. *Datenbank-*

Spektrum, 20(2):171–180, 2020. doi: 10.1007/s13222-020-00351-x. URL https://doi.org/10.1007/s13222-020-00351-x.

N. Spanoudakis, A. C. Kakas, and A. Koumi. Application level explanations for argumentation-based decision making. In *1st International Workshop on Argumentation for eXplainable AI co-located with 9th International Conference on Computational Models of Argument (COMMA)*, volume 3209, 2022. URL https://ceur-ws.org/Vol-3209/1871.pdf.

J. Spieler, N. Potyka, and S. Staab. Interpretable machine learning with gradual argumentation frameworks. In *Computational Models of Argument - Proceedings of COMMA*, pages 373–374, 2022. doi: 10.3233/FAIA220179. URL https://doi.org/10.3233/FAIA220179.

P. Stone, R. S. Sutton, and G. Kuhlmann. Reinforcement learning for robocup soccer keepaway. *Adaptive Behaviour*, 13(3):165–188, 2005. doi: 10.1177/105971230501300301. URL https://doi.org/10.1177/105971230501300301.

P. Sukpanichnant, A. Rago, P. Lertvittayakumjorn, and F. Toni. Neural qbafs: Explaining neural networks under lrp-based argumentation frameworks. In *AIxIA 2021 - Advances in Artificial Intelligence - 20th International Conference of the Italian Association for Artificial Intelligence*, pages 429–444, 2021. doi: 10.1007/978-3-031-08421-8_30. URL https://doi.org/10.1007/978-3-031-08421-8_30.

D. Taniar, J. W. Rahayu, V. C. S. Lee, and O. Daly. Exception rules in association rule mining. *Applied Mathematics and Computation*, 205(2):735–750, 2008. doi: 10.1016/j.amc.2008.05.020. URL https://doi.org/10.1016/j.amc.2008.05.020.

M. Thimm and K. Kersting. Towards argumentation-based classification. In *Logical Foundations of Uncertainty and Machine Learning, Workshop at IJCAI*, volume 17, 2017.

L. Thorburn and A. Kruger. Optimizing language models for argumentative reasoning. In *Proceedings of the 1st Workshop on Argumentation & Machine Learning co-located with 9th International Conference on*

Computational Models of Argument (COMMA), pages 27–44, 2022. URL https://ceur-ws.org/Vol-3208/paper3.pdf.

I. Tiddi and S. Schlobach. Knowledge graphs as tools for explainable machine learning: A survey. *Artificial Intelligence*, 302:103627, 2022. doi: 10.1016/J.ARTINT.2021.103627. URL https://doi.org/10.1016/j.artint.2021.103627.

S. T. Timmer, J. C. Meyer, H. Prakken, S. Renooij, and B. Verheij. A two-phase method for extracting explanatory arguments from bayesian networks. *International Journal of Approximate Reasoning*, 80:475–494, 2017. doi: 10.1016/j.ijar.2016.09.002. URL https://doi.org/10.1016/j.ijar.2016.09.002.

J. van Lente, A. Borg, and F. Bex. Everyday argumentative explanations for classification. In *Proceedings of the 1st Workshop on Argumentation & Machine Learning co-located with 9th International Conference on Computational Models of Argument (COMMA)*, pages 14–26, 2022. URL https://ceur-ws.org/Vol-3208/paper2.pdf.

A. Vassiliades, N. Bassiliades, and T. Patkos. Argumentation and explainable artificial intelligence: a survey. *Knowledge Engineering Review*, 36:e5, 2021. doi: 10.1017/S0269888921000011. URL https://doi.org/10.1017/S0269888921000011.

K. Čyras, A. Rago, E. Albini, P. Baroni, and F. Toni. Argumentative XAI: A survey. In *Proceedings of the Thirtieth International Joint Conference on Artificial Intelligence, IJCAI*, pages 4392–4399, 2021. doi: 10.24963/ijcai.2021/600. URL https://doi.org/10.24963/ijcai.2021/600.

G. Vilone and L. Longo. An XAI method for the automatic formation of an abstract argumentation framework from a neural network and its objective evaluation. In *1st International Workshop on Argumentation for eXplainable AI co-located with 9th International Conference on Computational Models of Argument (COMMA)*, volume 3209, 2022. URL https://ceur-ws.org/Vol-3209/2119.pdf.

Z. Zeng, C. Miao, C. Leung, and J. J. Chin. Building more explainable artificial intelligence with argumentation. In *Proceedings of the Thirty-Second AAAI Conference on Artificial Intelligence*, pages 8044–8046, 2018. doi: 10.1609/AAAI.V32I1.11353. URL https://doi.org/10.1609/aaai.v32i1.11353.

CHAPTER 10

CAUSATION AND ARGUMENTATION

ALEXANDER BOCHMAN
Holon Institute of Technology, Israel
bochmana@hit.ac.il

FEDERICO CERUTTI
University of Brescia, Italy
federico.cerutti@unibs.it

TJITZE RIENSTRA
Maastricht University, The Netherlands
t.rienstra@maastrichtuniversity.nl

1 Introduction

The notion of causation is fundamental to how we make sense of the world. We use causation when we explain why things happen, when we predict the effects of actions, and when we assign responsibility or blame. It should therefore be no surprise that causality has been studied extensively in many fields, including philosophy, psychology, linguistics and artificial intelligence. Despite the fact that much of everyday argumentation involves causality, the topic has received relatively little attention in the literature on argumentation. Nevertheless, there are a number of strands of research that do make this connection. The goal of this chapter is to provide a coherent overview of these works.

The question of how to define causation has occupied philosophers since ancient times. For Aristotle, understanding the cause of a thing was a crucial condition for having knowledge of that thing. A cause, to Aristotle, was an answer to a *why*-question. He distinguished four types of causes: the material cause (what is it made of?), the formal cause (what is it?), efficient cause (where does its change or motion come from?) and the final cause (what is its aim?) [Hankinson, 1998]. The modern view on causality goes back to Hume, who defined a cause as follows. "We may define a cause to be an object followed by another, and where all the objects, similar to the first, are followed by objects similar to the second. Or, in other words, where, if the first object had not been, the second never had existed." [Buckle, 2007] This definition actually expresses two different understandings of causality. The first, referred to as the *regularity* definition, roughly states that A is a cause of B if every occurrence of A is followed by the occurrence of B. The second, referred to as the *counterfactual* definition, states that A is a cause of B if B had not occurred without A. Both of these definitions have been influential in later years. The regularity definition, however, faces a number of problems. For instance, if every occurrence of A is followed by the occurrence of B, then this may in fact be the result of A and B having a common cause C [Andreas and Guenther, 2021]. As a result, the counterfactual definition of causality has come to dominate.

Reasoning about causality requires a language to represent knowledge about causal relationships and to express the questions we wish to answer. Causal knowledge is typically represented using causal diagrams of some sort. These are directed acyclic graphs (DAGs) where nodes represent variables of interest and arrows represent causal dependencies. Figure 1 (adapted from [Pearl, 1987]) depicts an example of a causal diagram for a simple toy scenario: while the grass being wet (X_3) is caused either by the sprinkler (X_1) or by rain (X_2), the street being wet (X_3) is caused only by rain. The use of DAGs to represent causal knowledge goes back to work by Sewall Wright on the method of path analysis [Wright, 1934] and was further developed and popularised by Pearl (see [Pearl and Mackenzie, 2018] for a historical overview). Different approaches employ different representations of the exact nature of causal dependencies. For instance, Bayesian networks can be used to represent probabilistic causal relationships [Pearl, 1988]. Pearl's seminal work on Structural Equation Models (SEMs) represents a further development in his approach to causality [Pearl, 2009]. A SEM consists of a set of equa-

10 - Causation and Argumentation

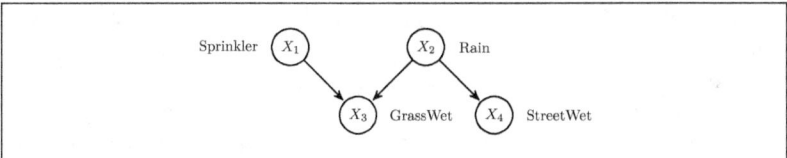

Figure 1: An example of a causal diagram.

tions that specify how each variable is determined by its direct causes, as well as a distinguished set of *exogenous* variables that represent factors not explained within the model. Pearl's concept of the *ladder of causation* defines three types of causal queries, each requiring causal models with increasing level of detail [Pearl and Mackenzie, 2018]. SEMs possess the ability to answer all three: (1) observational queries (e.g., would the street be wet if we observed the sprinkler to be on?); (2) interventional queries (e.g., would the street be wet if we *turned* the sprinkler on?), and (3) counterfactual queries (e.g. given that the street is wet and the sprinkler is on, would the street have been wet if the sprinkler had been off?).

A significant part of everyday argumentation involves causes, effects, and relationships between causes and effects. The goal of this chapter is to explore how argumentation and causality are connected within the different perspectives from which the two phenomena are studied. We start in section 2 with the logical foundations of causal reasoning. Here we take the *causal calculus* as a starting point, [McCain and Turner, 1997, Lifschitz, 1997, Bochman, 2004, Bochman and Lifschitz, 2015, Bochman, 2021]. This logical theory of causal reasoning emerged within the field of nonmonotonic reasoning in artificial intelligence. It represents a significant stride in addressing complex domains and applications within AI that previously defied effective depiction and modelling through conventional logical techniques. Moreover, this theory is acknowledged for its inclusive capacity, encapsulating essential formalisms and domains within nonmonotonic reasoning. Pearl's approach to causal reasoning in the framework of structural equation models can be seen as an instantiation of this theory. The theory also provides an interpretation of default assumptions, while the latter create in effect a basic link between causal reasoning and assumption-based argumentation.

In Section 3, we then turn our attention to the array of concepts put forward regarding the ontological notion of *causality*. Adopting the contemporary trend of scheme-based reasoning [Walton et al., 2008], we report on approaches defining causation from a dialectical perspective: causal connections thus become *defeasible generalisations*. We also discuss one of the important applications of causal reasoning in artificial intelligence, namely its promised explanatory power. Causation is indeed a cornerstone concept when it comes to understanding the myriad interconnections that underpin observable phenomena across various disciplines. Providing a framework to comprehend these complex relationships, causal explanations enable us to foresee outcomes, execute interventions, and, crucially, construct a coherent narrative around the events we witness.

Recognising the inherent link between the concept of defeasible generalisation and the establishment of causal connections, it is a logical progression to see that structured argumentation methodologies have been explored within the scope of causal theories, as elaborated in Section 4. Research in formal and informal argumentation offers a systematic way to frame and scrutinise the complex mechanisms behind causation, facilitating a deeper analysis and understanding of how certain factors lead to specific outcomes and the ways in which these causal links can be both identified and justified.

Viewing elements of causal reasoning through the lens of formal argumentation provides a rich and nuanced perspective, cf. Section 5. For example, basic causal inference resonates with the principles of collective argumentation [Bochman, 2018a]. Similarly, structural equations can be re-interpreted in the context of the acceptance conditions in Abstract Dialectical Frameworks. Moving to Pearl's d-separation criterion [Pearl, 2009], its essence is mirrored in abstract argumentation through the interplay of arguments and counterarguments, akin to the way d-separation identifies conditional independencies within a causal model.

2 Causal Reasoning: Basic Principles and Constructions

We begin with a brief description of a particular theory of causal reasoning. This theory, called the causal calculus,[1] has been introduced as part of a general field of nonmonotonic reasoning in AI, where it has been shown to cover important areas and applications of AI, especially those that had persistently resisted feasible representation and modeling using standard logical methods. This theory has also been shown to encompass several key formalisms and areas of nonmonotonic reasoning in AI, such as abduction and diagnosis, logic programming, and reasoning about action and change.

A new stage in the development of this theory has emerged with the realisation that it can also provide a formal representation for Pearl's approach to causality in the framework of structural equation models (see [Bochman and Lifschitz, 2015]). In addition, a number of applications of the causal calculus outside AI have been developed, such as problems of causal attribution (actual causality) in legal theory and causal representation of general dynamic reasoning. A detailed description of the causal calculus, as well as the range of its current applications in AI and beyond can be found in [Bochman, 2021].

2.1 Causal Theories and their Semantics

As it is common for reasoning formalisms, the causal calculus has a language and an associated semantics. Its language is a set of causal rules defined on an underlying language of propositions, while its semantics is a set of valuations on propositions that conform to the causal rules.

A *causal rule* is an inference rule of the form

$$a \Rightarrow A,$$

where a is a finite set of propositions and A a proposition. The rule says that a set a of propositions *causes* proposition A. A *causal theory* is simply a set of causal rules.

The basic principle of causal reasoning can be formulated as the following rationality postulate of acceptance for propositions:

---
[1]See [McCain and Turner, 1997, Lifschitz, 1997, Bochman, 2004].

Causal Acceptance Principle *A proposition A is accepted with respect to a causal theory Δ if and only if Δ contains a causal rule $a \Rightarrow A$ such that all propositions in a are accepted.*

If we take causes as something that provide *reasons* for their effects (answer the question *why*, using Aristotle's phrase), then the above principle can be viewed as expressing a constitutive principle of rationality in our context, since it states that (acceptance of) propositions can both serve as and stand in need of reasons (see [Brandom, 2000]). Sets of accepted propositions that conform to the above principle will form the *models* of the corresponding causal theory.

There are two parts that constitute the above principle. These two parts could be expressed as two independent rationality postulates:

Preservation Principle If all propositions in a are accepted, and a causes A, then A should be accepted.

Principle of Sufficient Reason Any proposition should have a cause for its acceptance.

The Preservation Principle expresses a common idea that the very concept of an inference rule presupposes that it should preserve, or 'transmit', acceptance of the corresponding propositions.

Leibniz' Principle of Sufficient Reason is a normative principle of reasoning stating that propositions *require* reasons for their acceptance, and such reasons are provided by establishing their causes. The origins of this principle can be found in the well-known law of causality, but also in Aristotle's distinction between syllogisms and demonstrations. This principle can also be viewed as a rational ground for the need of *argumentation* (justification) in acceptance of propositions, and thereby as a primary link between causal reasoning and argumentation, which is the main subject of this chapter.

Example 1. *The following causal theory provides a causal description of a well-known example from [Pearl, 1987]. This causal theory corresponds to the scenario represented by the causal diagram depicted in*

Figure 1.

$$Rained \Rightarrow Grasswet$$
$$Sprinkler \Rightarrow Grasswet$$
$$Rained \Rightarrow Streetwet.$$

If, for instance, Rained is accepted with respect to such a causal theory, then both Grasswet and Streetwet should also be accepted. However, in a causal reasoning with this causal theory, any acceptable set of propositions that contains Grasswet should contain either Rained or Sprinkler as its causes. Similarly, Streetwet implies in this sense acceptance of both its only possible cause Rained and a collateral effect Grasswet. Both derivations from causes to their effects and from effects to their possible causes constitute essential parts of causal reasoning.

Preservation cannot be used as a sole principle of validity for causal rules: we cannot follow Tarski [Tarski, 1994] in *defining* causal rules as inference rules that preserve acceptance. This could be seen already from the fact that such a stipulation would immediately sanction the Reflexivity postulate of deductive inference (namely, all rules of the form $A \Rightarrow A$) and this would trivialize in turn the second part of our rationality postulate, the principle of sufficient reason: on a causal reading, rules $A \Rightarrow A$ will make all propositions self-justified (self-evident). Incidentally, this observation indicates also that (absence of) Reflexivity constitutes one of the key differences between causal inference and deductive consequence.

Rational Semantics

A valuation is a function $v \in \{0,1\}^L$ that assigns either 1 ('truth') or 0 ('falsity') to every proposition of the language. If $v(A) = 1$, we will say that proposition A is *accepted* ('taken-true') in the valuation v. A valuation can be safely identified with its associated set of accepted propositions.

$\Delta(u)$ will denote the set of propositions that are directly caused by a set u in a causal theory Δ, that is,

$$\Delta(u) = \{A \mid a \Rightarrow A \in \Delta,\ a \subseteq u\}.$$

This notation will help us in formulating the following basic definition of semantics for our language.

Definition 2.1. • A causal model *of a causal theory* Δ *is a valuation that satisfies the following condition:*

$$v = \Delta(v).$$

• *A rational semantics of a causal theory is the set of all its causal models.*

The notion of a causal model provides precise formal expression of the Causal Acceptance principle since it determines that a proposition is accepted in a model if and only if it has a cause in this model.

$\Delta(u)$ is a monotonic operator on the set of propositions, while causal models correspond to fixed points of this operator. Consequently, any causal theory has at least one causal model, so it always has a rational semantics. As an important special case, a causal theory always has the least model. This model can be obtained by applying the operator $\Delta()$ iteratively, starting with the empty set $\emptyset$. This least model provides a faithful representation of the concept of (deductive) *provability* in our causal framework. However, it expresses only a small part of the informational content embodied in the source causal theory. Moreover, this observation can actually be extended to the rational semantics itself.

A causal model, viewed just as a set of (accepted) propositions, and the rational semantics in general contain only purely categorical, *factual* information. In this respect, they provide only a possible factual output of the rich causal information embodied in the original causal theory. Unlike the case of an ordinary correspondence semantics, even the whole set of such possible outputs is insufficient for determining, or capturing back, the initial causal information, what causes what. Essentially different causal theories can 'accidentally' have the same rational semantics. Nevertheless, as for ordinary reasoning formalisms, the rational semantics plays a crucial, indispensable role in evaluation and adjudication of causal theories.

2.2 Causal Inference

It turns out that there are formal derivations among causal rules that always preserve the rational semantics. Such derivations will be taken to constitute the underlying *logic* of causal reasoning. On our current maximal level of abstraction, this logic can be described as follows:

Definition 2.2. *A causal inference relation is a set of causal rules that is closed with respect to the following derivation rules:*

Monotonicity If $a \Rightarrow A$ and $a \subseteq b$, then $b \Rightarrow A$;

Cut If $a \Rightarrow A$ and $a, A \Rightarrow B$, then $a \Rightarrow B$.

The above notion of causal inference incorporates two of the three basic postulates for ordinary Tarski consequence relations. It explicitly disavows, however, the first postulate of Tarski consequence, the Reflexivity postulate. As we will see, it is this 'omission' that creates the possibility of causal reasoning in this framework.

We will extend causal rules to rules having arbitrary sets of propositions as premises using a familiar compactness recipe:

$$u \Rightarrow A \equiv a \Rightarrow A, \text{ for some finite } a \subseteq u.$$

$\mathcal{C}(u)$ will denote the set of propositions caused by u with respect to a causal inference relation $\Rightarrow$, that is

$$\mathcal{C}(u) = \{A \mid u \Rightarrow A\}.$$

As could be expected, the causal operator $\mathcal{C}$ will play much the same role as the usual derivability operator for consequence relations. In particular, the above postulates of causal inference can be recast as the following properties of the causal operator:

Monotonicity If $u \subseteq v$, then $\mathcal{C}(u) \subseteq \mathcal{C}(v)$.

Cut $\mathcal{C}(u \cup \mathcal{C}(u)) \subseteq \mathcal{C}(u)$.

Thus, $\mathcal{C}$ is a monotonic operator. Still, it is not inclusive, that is, $u \subseteq \mathcal{C}(u)$ does not always hold. Also, it is not idempotent, that is, $\mathcal{C}(\mathcal{C}(u))$ can be distinct from $\mathcal{C}(u)$.

For an arbitrary causal theory Δ, we will denote by $\Rightarrow_\Delta$ the least causal inference relation that includes Δ, while $\mathcal{C}_\Delta$ will denote the associated causal operator.

2.2.1 Causal Inference vs. Deductive Consequence

Given a causal inference relation, we can define the following Tarski consequence relation:

$$u \vdash_\Rightarrow A \equiv A \in u \text{ or } u \Rightarrow A.$$

$\vdash_\Rightarrow$ is the least consequence relation containing $\Rightarrow$. If $\text{Cn}_\Rightarrow$ is a consequence operator corresponding to $\vdash_\Rightarrow$, then

$$\text{Cn}_\Rightarrow(u) = u \cup \mathcal{C}(u).$$

Now, Cut implies $\mathcal{C}(u) = \mathcal{C}(\text{Cn}_\Rightarrow(u))$ as well as $\mathcal{C}(u) = \text{Cn}_\Rightarrow(\mathcal{C}(u))$, so the causal operator absorbs $\text{Cn}_\Rightarrow$ on both sides:

$$\text{Cn}_\Rightarrow \circ \mathcal{C} = \mathcal{C} \circ \text{Cn}_\Rightarrow = \mathcal{C}.$$

Thus, deductive consequences of a given causal theory can be safely used as intermediate premises and conclusions in causal inference. This allows us to see causal rules themselves as just a special kind of deductive rules, which naturally corresponds to Aristotle's theory of reasoning in his *Analytics* where (causal) demonstrations were viewed as a species of syllogisms (deductions) (see [Bochman, 2021]). It should be kept in mind, however, that deductive inference alone is insufficient for determining the *causal* consequences of a set of propositions.

2.3 Causal vs. Semantic Equivalence

It turns out that causal inference provides an adequate and maximal logical framework for reasoning with causal models.

Definition 2.3. *Two causal theories will be called* semantically equivalent *if they determine the same rational semantics.*

To begin with, we have:

Lemma 2.4. *Any causal theory Δ is semantically equivalent to $\Rightarrow_\Delta$.*

Thus, the postulates of causal inference are adequate for reasoning with causal models since they preserve the latter.

Definition 2.5. *Two causal theories Δ and Γ will be called* logically equivalent *if each can be obtained from the other using the postulates of causal inference. Or, equivalently, if $\Rightarrow_\Delta$ coincides with $\Rightarrow_\Gamma$.*

Now, as a consequence of the previous lemma, we obtain:

Corollary 2.6. *Logically equivalent causal theories are semantically equivalent.*

The reverse implication in the above corollary does not hold, because the rational semantics *does not* fully determine the content of the original causal theory. Two essentially different causal theories could determine the same rational semantics. This under-determination is closely related to a more general fact that both the rational semantics itself and semantic equivalence of causal theories are *nonmonotonic* notions; they are not preserved under extensions of causal theories with further causal rules. The following simple example illustrates this.

Example 2. *Causal theories $\{A \Rightarrow B\}$ and $\{A \Rightarrow C\}$ are obviously different, but they are semantically equivalent since they determine the same rational semantics, which contains a single model $\emptyset$ in which no proposition is accepted. Now let us add to these causal theories the same causal rule $A \Rightarrow A$. Then the first causal theory will already have an additional model $\{A, B\}$, while the semantics of the second theory will acquire a different model $\{A, C\}$.*

What we need, therefore, is a stronger, logical counterpart of the notion of semantic equivalence that would be preserved under addition of new causal rules. This suggests the following definition.

Definition 2.7. *Causal theories Δ and Γ will be said to be* strongly semantically equivalent *if, for any set Φ of causal rules, $\Delta \cup \Phi$ is semantically equivalent to $\Gamma \cup \Phi$.*

Strongly equivalent causal theories are "equivalent forever"—that is, they are interchangeable in any larger causal theory without changing the associated rational semantics. This equivalence could be seen as a kind of logical equivalence, and the next result shows that this logic is precisely the logic of causal inference.

Theorem 2.8. *Causal theories are strongly semantically equivalent if and only if they are logically equivalent.*

2.4 Axioms vs. Assumptions

The rational semantics of causal theories is based on the law of causality which requires that any accepted proposition should have an accepted cause. Accordingly, justification of accepted propositions constitutes an essential part of this semantic framework. In fact, this is a common feature of many other formalisms of nonmonotonic reasoning in AI.[2]

The law of causality leads to a fundamental problem known already in antiquity as the *Agrippan trilemma*: if we do not want to accept infinite regress of causation, we should accept either uncaused or self-caused propositions. Two kinds of propositions can play, respectively, these two roles in the causal calculus:

Definition 2.9.
- *Proposition A will be called an* axiom *of a causal theory Δ if the rule $\emptyset \Rightarrow A$ belongs to Δ;*

- *Proposition A will be called a* causal assumption *of a causal theory if the rule $A \Rightarrow A$ belongs to it.*

Example 3. *Let us return to Pearl's example (Example 1):*

$$Rained \Rightarrow Grasswet \quad Sprinkler \Rightarrow Grasswet \quad Rained \Rightarrow Streetwet$$

This causal theory has a single empty causal model, mainly because the causal status of Rained and Sprinkler are not determined. But now let us make Rained and Sprinkler causal assumptions of our theory:

$$Rained \Rightarrow Rained \quad Sprinkler \Rightarrow Sprinkler.$$

---

[2]See, e.g., [Denecker et al., 2015] for an abstract theory of justifications in non-monotonic reasoning.

As a result, the rational semantics of this causal theory will acquire three additional causal models:

$$\{Rained, Grasswet, Streetwet\} \quad \{Sprinkler, Grasswet\}$$
$$\{Rained, Sprinkler, Grasswet, Streetwet\}$$

These models display already some correlations *(or 'regularities') among the relevant propositions. For instance, that Rained is always accompanied by Grasswet and Streetwet in these models (deduction), but also that Streetwet is always accompanied by Rained (abduction).*

Both axioms and assumptions provide end-points of justification in causal reasoning. The difference between the two can be described as follows. Every axiom *must* be accepted, and hence it should belong to every causal model. In contrast, any causal assumption *can* be incorporated into a causal model when it is consistent with the latter, but it does not have to be included into it. This makes causal assumptions much similar to abducibles in a system of *abductive reasoning*. In fact, the causal calculus can be used to provide a uniform description of abductive reasoning (see [Bochman, 2007]).

2.4.1 Supraclassical Causal Reasoning

The above abstract theory of causal reasoning can be raised to a full-fledged reasoning system by incorporating ordinary classical entailment as an integral part of causal reasoning.

From now on, our underlying language L of propositions will be a classical propositional language with the usual classical connectives and constants $\{\wedge, \vee, \neg, \rightarrow, \mathbf{t}, \mathbf{f}\}$. The symbol $\vDash$ will stand for the classical entailment while Th will denote the associated classical provability operator. In this and subsequent sections, $p, g, r, \ldots$ will denote propositional atoms while $A, B, C, \ldots$ will denote arbitrary classical propositions.

Definition 2.10. *A causal inference relation in a classical language will be called* supraclassical *if it satisfies the following additional rules:*

(**Strengthening**) *If* $b \Rightarrow C$ *and* $a \vDash B$, *for every* $B \in b$, *then* $a \Rightarrow C$;

(**Weakening**) *If* $a \Rightarrow B$ *and* $B \vDash C$, *then* $a \Rightarrow C$;

(**And**)   If $a \Rightarrow B$ and $a \Rightarrow C$, then $a \Rightarrow B \wedge C$;

(**Truth**) $\mathbf{t} \Rightarrow \mathbf{t}$;

(**Falsity**) $\mathbf{f} \Rightarrow \mathbf{f}$.'

Causal reasoning with classical propositions requires also an appropriate 'upgrade' of the corresponding rational semantics.

Definition 2.11.
- A classical causal model *of a causal theory* Δ *is a classically consistent valuation (that is, $\mathbf{f} \notin v$) that satisfies the following condition:*
$$v = \mathrm{Th}(\Delta(v)).$$

- A rational supraclassical semantics *of a causal theory is the set of all its classical causal models.*

A classical causal model is closed both with respect to the causal rules and with respect to classical entailment. The principle of sufficient reason is generalized, however, to the principle that any accepted proposition should (at least) be a classical logical consequence of accepted propositions that are caused in the model.

2.5 Defaults in Causal Reasoning

We will describe now a causal interpretation of the notion of default assumption. This causal interpretation of defaults provides also a primary link between causal reasoning and assumption-based argumentation (ABA).

2.5.1 Defaults versus Facts

In the framework of the causal calculus, defaults can be defined as a special kind of assumptions that we *must* accept unless there are reasons to the contrary.

Let as say that a proposition A is *rejected* in a causal model if the model contains a cause for the contrary proposition $\neg A$. Then we can formulate the following (informal) principle of Default Acceptance:

Default Acceptance *A default is a causal assumption that is accepted whenever it is not rejected.*

Note, however, that $\neg A$ is accepted in a causal model only if it has a cause in this model (that is, when A is rejected). Accordingly, Default Acceptance boils down to the principle of Default Bivalence:

Default Bivalence *For any causal model v and any default assumption A, either $A \in v$ or $\neg A \in v$.*

Default bivalence can be viewed as a characteristic property of defaults, in contrast to classical logical reasoning where *all* propositions are required to satisfy bivalence.

Now, reasoning in Reiter's default logic amounts to deriving justified conclusions from a default theory using its inference rules and default assumptions. However, if the set of all defaults is incompatible with the theory, we must make a reasoned choice among the default assumptions. At this point, default reasoning requires that a reasonable set of defaults should explain why the rest of the default assumptions should be rejected. The appropriate choices of default assumptions (called *stable* sets) will determine then *extensions* of a default theory which are taken to constitute the nonmonotonic semantics of the latter.

Bipolarity. Default logic demands, in effect, that once we choose a stable set of defaults, the rest of acceptable propositions should be derived from this set. This stringent understanding of acceptance for defaults and the rest of propositions creates a *bipolar* system of reasoning that divides all propositions into two classes with opposite principles of acceptance. The first class contains *factual propositions* that are viewed as unacceptable unless they are derived from other propositions (and ultimately from accepted defaults), while the second class contains defaults that are viewed as acceptable unless they are refuted by other propositions (and, again, ultimately by other accepted defaults). It is this understanding that also makes default logic a principal instantiation of assumption-based argumentation [Bondarenko et al., 1997] where default assumptions play the role of arguments.

2.6 Structural Equation Models

Pearl's approach to causal reasoning in the framework of structural equation models (see [Pearl, 2009]) can be viewed as an important instantiation of our general theory.

Definition 2.12. *A structural equation model is a triple* $M = \langle U, V, F \rangle$, *where*

- U *is a set of* exogenous *variables,*
- V *is a finite set* $\{V_1, V_2, \ldots, V_n\}$ *of* endogenous *variables that are determined by other variables in* $U \cup V$, *and*
- F *is a set of functions* $\{f_1, f_2, \ldots, f_n\}$ *such that each f_i is a mapping from* $U \cup (V \backslash V_i)$ *to V_i, and the entire set, F, forms a mapping from U to V.*

Symbolically, the set F in the definition above can be represented as a set of *structural* equations

$$V_i = f_i(PA_i, U_i) \quad i = 1, \ldots, n,$$

where PA_i is the set of variables in $V \backslash \{V_i\}$ (parents of V_i) sufficient for representing f_i, and similarly for the relevant set of exogenous variables $U_i \subseteq U$. Each such equation stands for a set of "structural" equalities

$$v_i = f_i(pa_i, u_i) \quad i = 1, \ldots, n,$$

where v_i, pa_i and u_i are particular instantiations of V_i, PA_i and U_i.

Every instantiation $U = u$ of the exogenous variables determines a particular "causal world" of the structural model. Such worlds stand in one-to-one correspondence with the solutions to the above equations in the ordinary mathematical sense. However, structural equations also encode causal information in their very syntax by treating every instantiation of the variable on the left-hand side of the equal symbol (=) as effect and treating the corresponding instantiations of the variables on the right as causes. Accordingly, the equality signs in structural equations convey the asymmetrical relation of "is determined by." This causal reading does not affect the set of solutions of a structural model, but it plays a crucial role in determining the effect of external interventions and evaluation of counterfactual assertions with respect to such a model (see below).

A comprehensive description of structural models requires the use of a first-order language. Still, we can obviate this limitation of our (propositional) formalism by considering the Herbrand base of this first-order language as our propositional language. This Herbrand base consists of all propositions of the form $X = x$, where X is some (exogenous or endogenous) variable while x is its particular admissible value.

The representation of Pearl's structural models in the causal calculus, suggested in [Bochman and Lifschitz, 2015], amounted to viewing a structural equality $v_i = f_i(pa_i, u_i)$ as the following causal rule:

$$PA_i = pa_i, U_i = u_i \Rightarrow V_i = f_i(pa_i, u_i).$$

In the special case when all the relevant variables are Boolean, a Boolean structural equation $p = F$ (where F is classical logical formula) produces in this sense two causal rules

$$F \Rightarrow p \quad \text{and} \quad \neg F \Rightarrow \neg p.$$

Instantiations of exogenous variables are also required to be causal assumptions: for every exogenous atom $U = u$, we should accept

$$U = u \Rightarrow U = u.$$

For Boolean exogenous variables, this amounts to adding the following two rules for any such variable:

$$p \Rightarrow p \quad \text{and} \quad \neg p \Rightarrow \neg p.$$

Then Pearl's causal worlds will correspond to classical causal models of the associated causal theory that are *worlds* (maximal classically consistent sets of propositions).

Example 4. *The following set of (Boolean) structural equations provides a representation of Pearl's example (see Example 1) in structural models:*

$$Grasswet = Rained \vee Sprinkler \qquad Streetwet = Rained.$$

*If Rained and Sprinkler are exogenous variables, while Grasswet
and Streetwet are endogenous ones, then this structural model will have
the same causal worlds as the following causal theory:*

$$Rained \Rightarrow Grasswet \quad Sprinkler \Rightarrow Grasswet \quad Rained \Rightarrow Streetwet$$
$$\neg Rained, \neg Sprinkler \Rightarrow \neg Grasswet \quad \neg Rained \Rightarrow \neg Streetwet$$

*with an additional stipulation that Rained, ¬Rained, Sprinkler and
¬Sprinkler are assumptions:*

$$Rained \Rightarrow Rained \quad \neg Rained \Rightarrow \neg Rained$$
$$Sprincler \Rightarrow Sprinkler \quad \neg Sprinkler \Rightarrow \neg Sprinkler$$

2.6.1 Causal Counterfactuals

Structural equation models have been used by Pearl to provide a novel semantics for counterfactual conditionals. Unlike Lewisian semantics for counterfactuals (see [Lewis, 1973]) that are based on a possible-worlds semantics of comparative similarity, this new semantics rests directly on causal inference.[3] There is a growing literature on this new semantics of counterfactuals, including important applications in linguistic semantics. It has also attracted attention of the argumentation community in AI. We will describe below a formal definition of counterfactuals in the causal calculus that corresponds to the original definition in the structural approach.

According to Pearl, in order to answer intervention (action) and counterfactual queries, we have to consider submodels of a structural model. Given an instantiation x of a subset X of endogenous variables, a *submodel* M_x of a structural model M is the model obtained from M by replacing its set of functions F by the following set:

$$F_x = \{f_i \mid V_i \notin X\} \cup \{X = x\}.$$

A submodel M_x can be viewed as a result of performing an action $do(X = x)$ on M that produces a minimal change required to make $X = x$ hold true. This submodel can be used for evaluating counterfactuals of the form "Had X been x, would $Y = y$ hold?"

---
[3]Cf. the Central Claim in [Schulz, 2011].

We restrict the description below to the Boolean case. Then the corresponding transformation of causal theories can be described as follows:

Definition 2.13. *A revision of a causal theory Δ with a set L of literals is a causal theory $\Delta * L$ obtained from Δ by removing first all causal rules having literals from L or their negations in heads, and then adding L as a set of new axioms (that is, adding rules $\mathbf{t} \Rightarrow l$ for each $l \in L$).*

It can be verified that revisions of causal theories exactly correspond to submodels of Boolean structural models.

By a *counterfactual* we will mean an expression of the form $L > A$, where L is a finite set of literals and A a proposition. Traditionally, counterfactuals have been defined semantically with respect to worlds. The interventionist definition of causal counterfactuals suggests, however, a powerful and useful generalization of validity for counterfactuals with respect to structural models, and thereby wrt causal theories.

Definition 2.14. *Counterfactual $L > A$ holds in a causal theory Δ (notation $L >_\Delta A$), if A holds in all causal worlds of the revision $\Delta * L$.*

As in the structural account, acyclic causal theories always determine a unique causal world for any interpretation of the exogenous variables. Accordingly, for a causal world α of a causal theory Δ, we will call the set of exogenous literals that hold in α the *basis* of this world (see [Veltman, 2005, Schulz, 2011]). This notion of a basis allows us to extend the above definition of causal counterfactuals to worlds.

Let Δ^α denote the causal theory obtained from Δ by adding rules $\mathbf{t} \Rightarrow l$ for each literal from the basis of α.

Definition 2.15 (World-based counterfactuals). *$L > B$ will be said to hold in a causal world α of a causal theory Δ if it holds in Δ^α.*

The above definition corresponds to the definition of causal counterfactuals in structural equation models (see, e.g., [Halpern, 2000]). Moreover, it has been shown in [Bochman, 2021] that the problem of validity for Boolean counterfactuals on this definition is reducible to classical entailment.

To end this short description of causal counterfactuals, we would like to highlight one of the main advantages of this novel, causal approach to counterfactuals as compared with the original semantics of Lewis, especially for AI. To illustrate the problems the latter has for a feasible representation of counterfactual reasoning, let us consider the following famous puzzle from [Goodman, 1947]:

Example 5 (**Striking a match**). *Suppose that a match lights whenever it is struck, unless it is wet. Suppose also that in the actual world the match is not struck, it is not wet, and is not lit. Would it light if it were struck?*

The commonsense answer "yes" is obvious, but Lewis's comparative similarity account has a peculiar problem with it. To obtain this answer, it should keep the fact that the match is not wet in the counterfactual world(s) where the match is struck while at the same time does not keep the fact that the match is not lit. In other words, it should make the world in which the match is struck, not wet, and lit more similar to the actual world than the world where it is struck, wet, and not lit. This asymmetry cannot be obtained, however, from the classical logical description of the relevant "match" law, namely from the classical implication $Struck, \neg Wet \to Lights$ which is fully symmetric in this respect.[4] Accordingly, in order to obtain a proper (and justified) representation of this situation, we should somehow have an independent grip on determining comparative similarity among worlds, which does not appear to be a feasible task. On the causal interpretation, however, such an asymmetry can be immediately derived from the following causal law:

$$Struck, \neg Wet \Rightarrow Lights.$$

$\neg Struck$ and $\neg Wet$ form the basis of the actual world

$$\alpha = \{\neg Struck, \neg Wet, \neg Lights\},$$

and therefore the world $\{Struck, \neg Wet, Lights\}$ is a single causal world of the revision $\Delta^\alpha * Struck$. Consequently, $Struck > Lights$ holds in the actual world. Speaking more generally, causal rules provide all the necessary information for evaluating associated counterfactuals.

---

[4]Because it is logically equivalent to $Struck \to Wet \vee Lights$.

2.6.2 Counterfactual Equivalence

In structural equation models, the relation between causal theories and their (rational) semantics surfaces as the relation between causal and purely mathematical understanding of structural equations. Thus, as in the general case, two informationally different sets of structural equations may "accidentally" determine the same causal worlds. And at this point, a key feature of Pearl's approach to causal reasoning amounts to the assumption that the relevant differences between causal theories can be revealed by performing the same interventions ("surgeries") on them.

According to Pearl, the submodels of a given structural model determine its "causal content". In accordance with that, we can introduce the following definition:

Definition 2.16. *Causal theories Γ and Δ are* intervention-equivalent *if, for every set L of literals, the revision $\Gamma * L$ has the same causal worlds as the revision $\Delta * L$.*

Intervention-equivalence of two causal theories amounts to coincidence of their associated causal counterfactuals. Accordingly, the content of a causal theory is fully determined by its 'counterfactual profile' in Pearl's approach. In this sense, the approach can even be viewed as a further development of the counterfactual approach to causal reasoning initiated by David Lewis in [Lewis, 1973].

Taken in this perspective, the difference with our approach, described earlier, amounts to taking intervention-equivalence instead of strong semantic equivalence as a basic information concept for causal theories. This alternative approach sanctions, however, a somewhat different logic for causal reasoning.

2.6.3 Basic Causal Inference

It turns out that the Cut rule of causal inference does *not* preserve intervention-equivalence: there are causal theories that are equivalent with respect to supraclassical causal inference, but their revisions with the same literals determine different causal worlds (and different counterfactuals). In order to cope with this situation, we have to modify our postulates of causal inference.

Definition 2.17.
- *A set of causal rules in a classical language will be called a* causal production relation *if it satisfies all the postulates of supraclassical causal inference except Cut.*

- *A causal production relation will be called* basic *if it satisfies*

   (Or)   *If $A \Rightarrow C$ and $B \Rightarrow C$, then $A \vee B \Rightarrow C$.*

The postulate Or sanctions reasoning by cases for causal rules. Now, as follows from the above definition, basic inference is obtained from supraclassical causal inference by replacing the Cut postulate with Or. A detailed description of this kind of causal inference and its connections with other nonmonotonic formalisms in AI has been given in [Bochman, 2005]. It has been shown, in particular, that this kind of inference can already be given a *logical* interpretation in possible worlds models; by this interpretation, a causal rule $A \Rightarrow B$ is representable as a modal conditional

$$A \to \Box B,$$

where $\Box$ is the usual necessity operator (see also [Turner, 1999]).

It has been shown in [Bochman, 2018b] that basic inference constitutes, in effect, the internal logic of causal reasoning in Pearl's causal models. More precisely, it has been shown that basically equivalent causal theories are intervention equivalent. Moreover, the reverse implication has been shown to hold for the special case of *Pearl's* causal theories, that is, for causal theories obtained from structural equation models by the translation of [Bochman and Lifschitz, 2015].

2.6.4 Four-valued interpretation

Finally, we will briefly describe yet another, this time four-valued, semantics of basic causal inference. This semantic interpretation has played an important role in the whole approach to nonmonotonic reasoning in [Bochman, 2005], and it is this semantics that will provide an important logical link between causal reasoning and argumentation.

[Belnap, 1977] has introduced a powerful and illuminating interpretation of four-valued reasoning based on an identification of the four values $\{\mathbf{t}, \mathbf{f}, \top, \bot\}$ with the *subsets* of the set $\{t, f\}$ of Boolean values. If ν is a four-valued interpretation on this understanding, then $\nu(A) = \{t\}$

means that proposition A is *true* (**t**), $\nu(A)=\{f\}$ means that A is *false* (**f**), $\nu(A) = \{t,f\}$ means that A is *contradictory* ($\top$), and $\nu(A) = \emptyset$ means that it is *undetermined* ($\bot$).

Let us say that A is *accepted* in a four-valued interpretation ν if $t \in \nu(A)$, while $f \in \nu(A)$ will mean that A is *rejected*. Then any four valued interpretation can be viewed as a pair of *independent* 'standard' valuations, one determining when a proposition is accepted, another determining when it is is rejected. Moreover, this representation allows us to define any four-valued connective using a pair of associated definitions, one saying when a compound logical formula is accepted, another saying when it is rejected.

Two such connectives are of special interest for our present study, primarily because they are direct four-valued counterparts of the basic classical connectives, conjunction and negation. First, there is a natural *conjunction* $\wedge$ of propositions that is determined by the following familiar semantic conditions:

$A \wedge B$ is accepted iff A is accepted and B is accepted

$A \wedge B$ is rejected iff A is rejected or B is rejected

$\wedge$ behaves as an ordinary classical conjunction with respect to acceptance and rejection of propositions. On the other hand, it is a four-valued connective, since the above conditions determine a four-valued truth-table for conjunction in Belnap's interpretation of four-valued logic.

Next there is a negation connective $\neg$ that also behaves in a fully classical way with respect to both acceptance and rejection:

$\neg A$ is accepted iff A is not accepted

$\neg A$ is rejected iff A is not rejected

Combining the above connectives, we can obtain (a four-valued counterpart of) any classical logical formula, and it can be easily seen that all such formulae behave in a fully classical way both with respect to acceptance and rejection.

As final step, we can provide the following semantic interpretation of causal rules in this setting.

Definition 2.18. • *A causal rule $A \Rightarrow B$ holds in a four-valued interpretation ν if either A is rejected, or B is accepted in ν.*

- *An interpretation ν will be called a (four-valued)* model *of a causal theory Δ if every causal rule from Δ holds in ν.*

It can be verified that the set of causal rules that hold in a four-valued interpretation satisfies all the derivation rules of basic causal inference. Moreover, our last result below will show that this kind of causal inference is complete for the above four-valued semantics.

For a set I of four-valued interpretations, we will denote by $\Rightarrow_I$ the set of all causal rules in the classical language $\{\wedge, \neg\}$ that hold in every interpretation from I. Then the following result is actually a representation theorem showing that the four-valued semantics of acceptance and rejection is adequate for basic causal inference.

Theorem 2.19. $\Rightarrow$ *is a basic causal production relation iff it coincides with $\Rightarrow_I$, for some set of four-valued interpretations I.*

2.7 Classical Causal Inference and Causal Worlds

The differences between Pearl's approach and our theory disappear once we restrict the rational semantics to causal models that are worlds. Note, however, that this move amounts to imposing Bivalence on the set of accepted propositions.

Definition 2.20. *A causal inference relation will be called* classical *if it is supraclassical and satisfies Or.*

Classical causal inference combines the properties of both basic and supraclassical causal inference. It corresponds also to the restriction of the rational semantics to worlds.

Definition 2.21.
- *A* causal world *of a causal theory Δ is a classical causal model of Δ which is also a world.*

- *A* rational classical semantics *of a causal theory is the set of all its causal worlds.*

The above semantics moves us one last step closer to the traditional correspondence semantics. Nevertheless, even the rational classical semantics is still nonmonotonic with respect to the source causal theory, so the latter is not determined by the former.

2.7.1 Default Negation and Logic Programming

The formalism of classical causal inference allows us to formalize an important alternative understanding of negation, namely the concept of *default negation*. The latter is based on the idea that a negative proposition can be accepted whenever we do not have *reasons* for accepting the corresponding positive proposition. This notion of default negation provides a causal representation of yet another key formalism of non-monotonic reasoning in AI — logic programming.

A declarative meaning of logic programs in modern logic programming involves an asymmetric treatment of positive and negative information, which is reflected in viewing the negation operator **not** in program rules as *negation as failure*. This understanding can be captured using the Default Negation postulate below.

Definition 2.22. *A classical causal inference relation will be called negatively closed, if it satisfies*

(**Default Negation**) $\neg p \Rightarrow \neg p$, *for any propositional atom p.*

The above principle makes negations of atomic propositions causal assumptions in the corresponding causal inference relation. Moreover, given Bivalence (that holds for causal worlds), the Default Negation postulate stipulates, in effect, that negations of atomic propositions are defaults. As a result, the principle of sufficient reason is reduced in such systems to the necessity of explaining only positive facts. The postulate can be seen as giving a formal expression to Reiter's *closed world assumption* and reflects the main distinctive feature of reasoning behind logic programs and databases.

A *logic program* Π is a set of program rules of the form

$$\mathbf{not}\, d, c \leftarrow a, \mathbf{not}\, b \qquad (*)$$

where a, b, c, d are finite sets of propositional atoms.

Now, a *stable causal interpretation* of logic programs amounts to interpreting every program rule (*) as the following causal rule:

$$d, \neg b \Rightarrow \bigwedge a \to \bigvee c.$$

Then it can be shown that a stable semantics of a program Π coincides with the classical causal semantics of its translation.

On the causal interpretation, a logic program can be seen as a causal theory satisfying the principle of negation as default. Moreover, given this principle, the correspondence between logic programs and causal theories turns out to be bidirectional in the sense that any causal theory is reducible to some logic program.

3 Arguing for and against Causal Rules

In Section 2.1, we touched upon the concept of the *causal rule*, albeit without delving deeply into its intricacies. As we transition to this section, our focus shifts to examining the various proposals presented in academic literature pertaining to the discourse and argumentation surrounding such causal rules. Here — in line with the recent tradition of scheme-based reasoning e.g., [Walton et al., 2008]— we opt for a different vantage point by framing causation in dialectical terms. By viewing causal links as defeasible generalisations, we pave the way for a more flexible and dynamic understanding.

This perspective takes a nod from Hastings's suggestion [Hastings, 1962] that causal generalisations are more hints than exhaustive causal models. This line of thinking within the scheme-based paradigm perceives a causal link as a tentative argument, enmeshed in a possible succession of dialectical exchanges. Here, after an initiator presents the argument, the onus falls on the responder to either counter a premise, pose a critical query, or endorse the argument. The ultimate acceptance of such an argument hinges significantly on the assignment of the evidentiary burden (for a deeper look, consult [Hahn and Oaksford, 2007a, Walton, 2014]). The act of posing critical queries could alter this burden, requiring added substantiation for the tentative conclusion to stand.

Regarding the foundations for a tentative causal argument, [Walton et al., 2008] offers a diverse set of starting points, with Si as the causative agent and Sj as the resultant:

1. Consistency: Sj is a frequent outcome after Si.

2. Chronological order: Si precedes or coincides with Sj.

3. Adaptability: Si is alterable or subject to change.

4. Causative nature: Si serves as a pivotal or INUS condition for Sj. Here, a *cause* is *an insufficient but necessary part of an unnecessary but sufficient condition* (INUS) [Mackie, 1980].

5. Pragmatic relevance: Criteria like intentionality or exceptionality might pinpoint a causative agent.

Given the long-standing interest of argumentation schemes in legal reasoning, it is unsurprising that [Walton et al., 2008] heavily focuses on the causative nature of causality, building mostly on the *ceteris paribus* principle — a Latin phrase meaning "all other things being equal" — that thus emerges as a fundamental element. When evaluating whether one event causes another, it is imperative to consider the conditions under which this causality holds true, and this is where the *ceteris paribus* clause comes into play. It underscores the need to keep extraneous variables constant to accurately assess the causal relationship between two events. Tied closely to this principle are the concepts of necessity and sufficiency. For an event A to be deemed a necessary cause of event B, event B cannot occur without event A. Conversely, if A is a sufficient cause of B, then the occurrence of A guarantees the occurrence of B. However, in real-world scenarios, pure necessity and sufficiency are rare, and many causal relationships are contingent upon a myriad of conditions.

Arguing about causation can be approached in two primary directions: forward and backwards. The forward approach posits that a specific cause leads to a given effect, essentially tracing the flow of events as they naturally occur. Conversely, the backward approach starts with an observed effect and seeks to identify its probable cause. This backward reasoning is intimately connected with the concept of *abduction* [Walton et al., 2008], a form of logical inference which begins with an observation and then seeks the simplest and most likely explanation. Such an approach is foundational in many scientific and investigative disciplines, where understanding the root cause of observed phenomena is of paramount importance.

It is therefore evident that causation, whilst often perceived as an absolute, is inherently context-dependent [Walton et al., 2008]. This is underscored by its dialectical nature, rooted in the shifting sands of evidentiary burden. The burden of evidence required to establish causation can vary considerably depending on the context in which it is being evaluated. For instance, the scientific community may demand rigorous experimental validation to accept a causal link, whereas the legal realm may pivot more on the balance of probabilities. Similarly, societal interpretations of causation can be influenced by cultural, ethical, or emotional factors, leading to thresholds that differ from both scientific and legal standards. Such contextual variations highlight the multifaceted nature of causation, emphasising the importance of clearly delineating the context in which causal claims are made and evaluated.

3.1 Argumentation Schemes about Causation

Foundational studies within the schema-driven approach, notably by [Hastings, 1962], outline dual primary variations of causal discourse: the progression from cause leading to effect, and the reciprocal from the outcome back to its origin. Within these main categories, Hastings elaborates further. For the initial kind, progressing from cause to effect, he introduces sub-categories, one being "forecasting based on prevailing circumstances," alluding to a discourse where the conclusion predicts forthcoming occurrences. Turning our attention to the converse type, from effect tracing back to cause, it bears significant resemblance to a pair of sub-categories Hastings labels as "indicative reasoning," using the example "the presence of bear footprints suggests a nearby bear," and "deriving hypothesis from evidence," both predominantly centring around causality. We will focus on these aspects in Section 3.2, drawing a parallelism between backward causal reasoning and abduction.

Delving into the discourse from the cause leading to its effect, [Hastings, 1962] proposes a quartet of pivotal inquiries, believed to hold relevance irrespective of the specific sub-category:

1. Is there a legitimate causal connection between the origin and the outcome? Essentially, is it the authentic cause?

2. What is the likelihood of the outcome rooted in this correlation?

3. Does the origin adequately account for the resultant effect?

4. Could external factors be at play, potentially disrupting the causation?

These queries embody Hastings' stance, developed from textual scrutiny, that authentic causal discussions in real-world scenarios inherently possess intricate layers with multiple causal and correlational constituents. This notion, that in a significant number of real-world situations, individuals' causal frameworks are relatively undeveloped, appears pivotal for in-depth normative analyses of everyday causal discourses.

Numerous scholars following Hastings have incorporated the principle of *argument from cause to effect* within their essential categorisation of argumentative schemes [Grennan, 1997, Kienpointner, 1992, Perelman, 1971, Prakken and Renooij, 2001, Van Eemeren et al., 2015, Walton, 1996], clearly with different interpretations regarding the scheme's essence and potential formalisation.

Utilising Toulmin's approach [Toulmin, 1958], [Grennan, 1997, Hastings, 1962, Walton, 1996] provided insights into informal argumentation. This framework, gaining popularity in studying argumentation skills evolution, recognises classical logic's limitations in addressing day-to-day informal discussions laden with uncertainties. Toulmin proposed a dialectical argumentative style observed in courtroom exchanges. In this context, he delineated an argument's structure, segregating it into core elements: *claim* (the conclusion), *data* (facts supporting the claim), *warrants* (reasons bridging data and claim), *backing* (foundational assumptions bolstering specific warrants), *rebuttals* (exceptions to the claim or its inferential bridge), and *qualifiers* (indicative of the claim's certainty degree).

A normative outline of argumentation is essential, dictating the recommended methods of argumentation for rational individuals. While classical logic aimed to establish this standard, Toulmin highlighted its inadequacy in handling everyday discussions. However, Toulmin's model, with its structural focus, shares the same shortcomings as classical logic. Later scholars have endeavoured to enhance the normative element using critical questions and sought proper formalisation to accentuate the normative facets of causal arguments. Stemming from the scheme-based tradition, these efforts often embrace a dialectical view, asserting that arguments are best understood and assessed within the broader context of their encompassing dialectical discourse, as propounded in [Van Eemeren and Grootendorst, 2004].

Central to contemporary studies in the scheme-based domain is the belief that arguments, like those from cause to effect, feature a defeasible generalisation, as suggested in [Walton, 1996, Walton et al., 2008]. This approach echoes Hastings's observation that causal generalisations seldom represent comprehensive causal frameworks. Walton's schemes are reasoning patterns providing structures for conjectures, typically formed by a set of premises tentatively (defeasibly) supporting a given conclusion, and the means for refutation, namely *critical questions*. The use of argumentation schemes and critical questions embodies a deeply dialectical nature. At its core, dialectics is about the interplay of opposing viewpoints, where one perspective responds to, refutes, or complements another. Argumentation schemes provide structured templates for constructing arguments, encompassing the reasoning process that underpins a claim. However, these schemes are inherently open to challenge. This is where critical questions come into play, serving as tools to interrogate the validity, relevance, or coherence of the presented arguments.

The major premise of the argumentation scheme is presented in the form of a defeasible conditional of a kind that could be associated with a Toulmin warrant like that used by Hastings:

> Generally, if A occurs, then B will (might) occur.
> In this case, A occurs (might occur).
> Therefore, in this case, B will (might) occur.

Associate to this scheme — which resembles very closely the *Argument from Sign* scheme linking the truth of a proposition Y with the truth of proposition X, cf. Section 3.5 for an additional discussion — there are the following critical questions:

CQ1: How strong is the causal generalisation (if it is true at all)?

CQ2: Is the evidence cited (if there is any) strong enough to warrant the generalisation as stated?

CQ3: Are there other factors that would or will interfere with or counteract the production of the effect in this case?

Argumentation schemes, as a normative approach, offer prescriptive guidelines on how arguments should ideally be constructed and evaluated, setting a standard for rational discourse. However, while these schemes provide valuable frameworks for reasoning, it is essential to juxtapose them with how arguments manifest in real-world contexts. This brings us to the corpora-based approach. By delving into vast datasets of textual arguments, this descriptive method seeks to unearth the patterns, tendencies, and nuances of actual argumentative discourse, offering a more pragmatic insight into the complexities of everyday communication. Moving from the idealised structures of argumentation schemes, we venture into the rich terrain of corpora-based studies, revealing the intricacies of argumentation in its natural setting.

In their examination of natural language text corpora, Oestermeier and Hesse [Oestermeier and Hesse, 2000] delineated a comprehensive categorisation of causal arguments. Their taxonomy looks at arguments that either support or challenge causal links. This does not go against the use of argumentation schemes and critical questions. In fact, critical questions often help connect arguments and their counterarguments.

Concerning arguments for causal claims, they distinguish between:

1. Arguments from circumstantial evidence;

2. Arguments from contrastive evidence;

3. Arguments from causal explanations.

Concerning arguments from circumstantial evidence, [Oestermeier and Hesse, 2000] considers the cases of spatio-temporal contiguity, co-occurrences, and similarity of cause and effects. *Spatiotemporal Contiguity* as posited by Hume [Hume, 2000] and later by Einhorn and Hogarth [Einhorn and Hogarth, 1986], focuses on events that transpire closely in space and time, suggesting that if two events occur nearly simultaneously, one might infer a causal relationship.

Example 6. *For instance, "He likely fell for her whilst sipping that cocktail," is an example of circumstantial evidence, where the rationale hinges on the temporal proximity of events.*

Co-occurrences, highlighted by Hume [Hume, 2000] and later by Kuhn [Kuhn, 1991], revolve around the frequent pairing of events; if event A consistently accompanies event B, it hints at a causal connection.

Example 7. *An illustrative comment might be, "Every attempt to print leads to a system crash, implying printing instigates the malfunction." Here, the argument is anchored in repetitive observations of paired events.*

Lastly, *Similarity of Cause and Effect*, proposed by Einhorn and Hogarth [Einhorn and Hogarth, 1986], suggests that marked resemblances between two entities hint at causation.

Example 8. *An example might be, "The stain's origin must be your muddy boots; both marks share an identical hue," with the foundation rooted in the similarities of structural attributes, leading to inferred causality.*

Coming to the argumentation schemes grounded on contrastive evidence, the *Covariation* model [Mill and Robson, 1974] posits that if one event shifts in line with another, a causal relationship might exist.

Example 9. *For example, "Adjusting this wheel seems to alter the screen's brightness," with the argument built on observations across varied conditions, drawing from Mill's method of difference to infer causality.*

Statistical Covariation [Cheng, 1993, Eells, 1991] delineates a probabilistic regularity between events, suggesting one elevates the likelihood or risk of the other.

Example 10. *A typical assertion could be, "There's a higher incidence of cancer among smokers, hinting at tobacco as a causative agent."*

This stands on multi-observational data, stressing the statistical consistency to deduce causality. With *Before-after-comparison* [Ducasse, 1926], the focus shifts to temporal sequence, arguing causality from the mere chronological progression of events, such as, "Post-diet, my physique is transformed," underscoring temporal disparities to determine causality.

Then, the *Experimental Comparison* [Von Wright, 2004] underscores the importance of intentional manipulation to ascertain causality.

10 - CAUSATION AND ARGUMENTATION

Example 11. *For instance, "Triggering the Cancel button appears to crash the system," is an argument based on experimental data and outcomes from intended interferences.*

Lastly, the *Counterfactual vs. factual* method [Hume, 2000, Mackie, 1980] hinges on the hypothetical scenario, proposing that an event would not have transpired in the absence of another.

Example 12. *An example might be, "Your tardiness resulted in my delay. Had you been punctual, I could have left sooner," which contrasts factual occurrences against hypothetical counterparts, leveraging this juxtaposition to extract causal implications.*

Finally, concerning *causal explanations*, the *Causal Mechanism* model [Ahn et al., 1995, Shultz, 1982] offers an insight into the intermediary process or mechanism that links one event to another.

Example 13. *For instance, "His anger culminated in the accident due to its detrimental effect on his concentration." This argument hinges on an understanding of underlying mechanisms and the transitive inferences they can facilitate.*

Next, the *No Alternative* framework [Kuhn, 1991] dictates that in the absence of any other plausible explanation, one event can be deemed the cause of another.

Example 14. *An illustrative argument of the No Alternative framework might be, "Given that the new word processor was the sole active program, it must have been the culprit behind the system crash."*

This approach necessitates a thorough exploration of potential explanations and subsequently dismissing them to affirm causality.

Lastly, the *Typical Effect* principle [Thagard, 2000] contends that an event can be attributed to a cause if it represents a recurring consequence of that cause.

Example 15. *An exemplary statement could be, "The system's crash is likely attributable to the word processor, as such malfunctions are a frequent aftermath of its operation."*

This argument is grounded in specific causal knowledge, requiring an assessment of both the plausible explanations and their customary occurrences to deduce causality.

Oestermeier and Hesse [Oestermeier and Hesse, 2000] also provide a description of arguments qualifying causal claims. The simplest case is the *Causation without Responsibility*, where the *Arguments from No Intention* [Hart and Honoré, 1985, Weiner, 1995] highlights situations where an action results in an outcome, but the individual responsible for the action did not intend that specific result.

Example 16. *Take, for instance, the scenario: "Indeed, John was the cause behind Peter's toy breaking, but it wasn't a deliberate act on his part."*

This argument pivots on detailed understanding of actions and the intentions that drive them.

More articulated is the case of qualifying causal claims due to *Causal complexities*. The Argument from *Partial Cause* [Oestermeier and Hesse, 2000] posits that while A may influence B, it is not the sole causative agent.

Example 17. *For instance, the traffic situation might have contributed to an accident, but not as the singular cause.*

Such an argument hinges on the understanding that effects often spring from multiple causes.

Next, an *Indirect Cause* [Oestermeier and Hesse, 2000] means that A does not directly lead to B, but triggers a sequence or chain of events that culminates in B.

Example 18. *To illustrate, the rain did not directly result in a road accident but set off a series of events leading to it.*

This is rooted in the understanding of causative chains.

Moreover, the *Common Cause* framework [Oestermeier and Hesse, 2000] suggests that neither A nor B influence each other; rather, both are outcomes of a third factor, C. A case in point could be that both smoking and cancer might be outcomes of a specific lifestyle. This perspective is drawn from the idea that multiple outcomes can stem from a singular cause.

10 - Causation and Argumentation

In the argument from *Interaction* [Oestermeier and Hesse, 2000], then, neither A nor B solely causes the other; they influence each other reciprocally. For instance, homelessness might spur unemployment and the other way around. Such a viewpoint is anchored in the belief that several factors interplay and influence each other. Finally, the *Mix-up of Cause and Effect* [Walton, 1989] challenges the perceived direction of causality. Here, what is often seen as the outcome might, in fact, be the cause.

Example 19. *Consider this: "The surge in violence is not an outcome of the prison system; rather, the prison system might be a result of escalating violence." This argument pivots to a precise understanding of causative directions.*

Considering the counterarguments that could be posited to challenge causal argumentation, Oestermeier and Hesse [Oestermeier and Hesse, 2000] discuss:

1. Arguments from alternative explanations;

2. Arguments from counter-evidence;

3. Arguments from insufficiency of evidence.

When alternative explanations are available, a natural counterargument that could arise is the argument from *More Plausible Alternative* [Thagard, 2000]. It postulates that event A might not be the true cause of event B because a third factor, C, could be more probable due to various reasons like its proximity to B or other influential factors. To put this into perspective, consider an accident during wintertime. One might immediately attribute the mishap to the snow, but on deeper reflection, it might be deduced that the driver's reckless behaviour was the actual catalyst for the unfortunate event. This argument is grounded on having a specific understanding of potential alternative explanations for an outcome and then evaluating them to negate a previously assumed causal link in favour of a more likely explanation.

Counter-evidence, instead, serves as a contrasting tool to challenge causal arguments by presenting data or instances that contradict the proposed cause-and-effect relationship. One case is the argument of the *Wrong Temporal Order* [Oestermeier and Hesse, 2000], which highlights that event A has not caused event B when A happens after B. For instance, attributing system crashes to server problems becomes illogical if the latter occurred after the former, clearly contradicting the sequential nature of cause and effect. This deduction is backed by understanding the observed sequence of the two events. The *No Contact* argument [Oestermeier and Hesse, 2000] postulates that event A cannot have induced event B if there is no discernible connection or contact between them. Imagine blaming a server for a system crash when the computer was not even linked to the server. This relies on knowledge of spatial-temporal connections or the lack thereof. Transitioning from spatial to volitional arguments, the *Free Decision* argument [Oestermeier and Hesse, 2000] suggests that A has not led to C due to an intervening free choice, B. For instance, the traffic situation may not be blamed for an accident if an individual consciously chooses to speed in that context. The underpinning here is the comprehension of intentions and mental prerequisites of actions. The argument from *Insufficient Cause* asserts that A has not induced B simply because there are instances where A occurred without resulting in B. The classic example is that smoking does not necessarily lead to cancer, as evidenced by lifelong smokers who never fall ill. This argument hinges on recognising exceptions or broader knowledge of the non-universality of the said cause. Lastly, the argument from *Unnecessary Cause* posits that A might not be the cause of B, given that B can transpire independently of A. This means that just because some smokers get cancer, smoking is not definitively the cause; even non-smokers can contract the disease. It leans on the understanding of exceptions or general knowledge about the independence of outcomes from certain triggers.

To conclude this section, we would like to remark Oestermeier and Hesse's contribution [Oestermeier and Hesse, 2000] to the field of visual argumentation concerning causal representation. For instance, they highlight the significance of mental simulations in causal reasoning, discussing how these simulations compare factual occurrences with constructed fictional scenarios. They also touch upon the use of causal diagrams to simplify complex causal relationships and the role of computer simulations in visualising causal theories. In essence, their work provides a nuanced view of the interplay between visual argumentation and causal representation.

Consider the case of Dr Snow and the Cholera outbreak in London in 1853. During the 1853-54 cholera outbreak in London, the prevailing medical opinion attributed the disease to "miasmas" and other noxious emanations from the Thames River's swamps and mud. However, Dr. Snow harboured a different theory, suspecting contaminated water supplies as the cause. To validate his hypothesis, he plotted the residences of 500 cholera victims on a street map of Soho, see Figure 2. All these individuals had drunk from the Broad Street pump, which was at the heart of the "cholera field."

3.2 Inferring Probable Causes from Effects

How do we infer causes from effects? Various schemes guide inference from data to hypotheses, whether they are general causal or specific events. Walton [Walton et al., 2008] describes for instance the argumentation scheme from effect to cause:

> Generally, if A occurs, then B will (might) occur.
> In this case, B did in fact occur.
> Therefore, in this case, A also presumably occurred.

The process of inferring causes from observed effects closely relates to the concept of abduction, a form of logical reasoning. Abduction, often termed "inference to the best explanation," involves identifying the most probable cause or explanation behind a particular observation or set of data. When we observe certain effects and strive to pinpoint their causes, we are effectively employing abductive reasoning, sifting through potential explanations and selecting the one that most suitably accounts for the observed phenomena.

Peirce [Peirce, 1997] pioneered the idea of treating abduction as its own unique reasoning form, distinct from both induction and deduction. He emphasised the distinction between endorsing a hypothesis through scientific experimentation (akin to induction) and proposing a hypothesis to rationalise observed events (abduction). Peirce characterised the abductive inference in this manner:

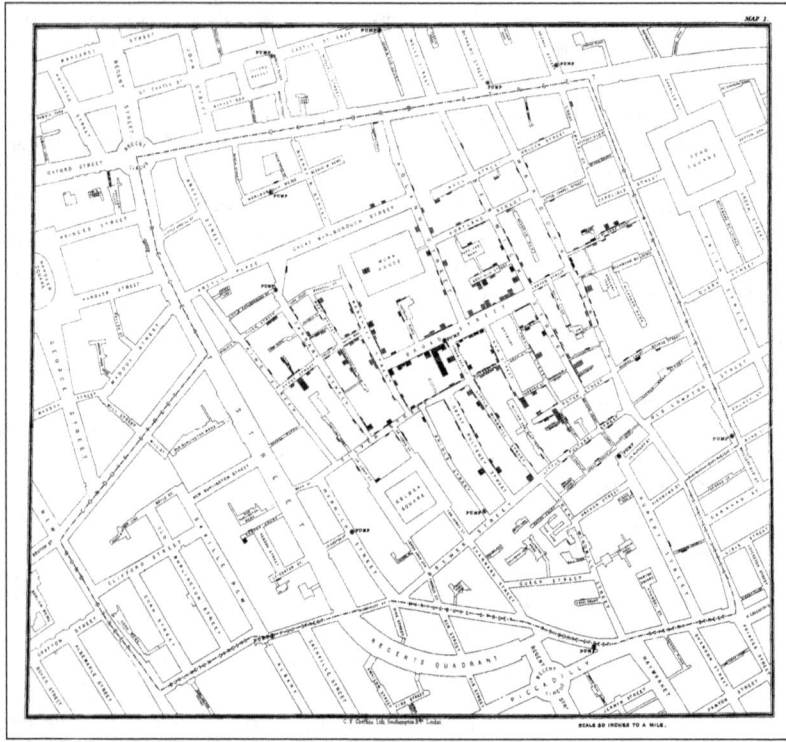

Figure 2: A map of deaths from the 1853–54 cholera outbreak in the Broad Street area. [Snow, 1855, p. 44], public domain.

A startling observation, E, is made;
Yet, if B were correct, E would naturally ensue,
Thus, there's a compelling reason to believe B is correct.

This is a clear elucidation where the fact E is explicated from premise B, diverging from both deductive and inductive reasoning. Still, Peirce's approach does not address the possibility of alternative hypotheses. Building on this, Josephson and Josephson [Josephson and Josephson, 1996] described a more comprehensive version of abduction:

> A dataset, F, contains facts, observations, and constants.
> I accounts for F (if proven correct, would rationalise F).
> No alternative hypothesis elucidates F as effectively as I.
> Consequently, I is likely accurate.

3.3 On Causation and Explanations

The notion of causation has always held a central position in the realm of explanations across diverse fields [Josephson and Josephson, 1996]. Causative explanations offer a means to grasp the intricate web of relationships underlying observed phenomena and they provide a scaffold for our understanding, allowing us to predict, intervene, and, most importantly, derive a sense of coherence about the events we observe.

Several studies across philosophy and cognitive science have postulated a strong link between abductive reasoning and the act of explaining. Especially in discerning the causes of events, individuals often employ abductive logic to ascertain their preferred explanation. Harman's [Harman, 1965] work is seminal in recognising this relationship, and subsequent empirical studies have affirmed it [Lombrozo, 2012, Lombrozo and Gwynne, 2014, Rehder, 2003, Wilkenfeld and Lombrozo, 2015]. Popper [Popper, 2005] championed the value of abductive thinking within scientific methodologies. He fervently advocated for the scientific approach to hinge on the empirical refutability of premises, contrasting with the prevailing inductivist perspective.

In the sphere of machine learning, causation plays a pivotal role in demystifying complex algorithms and models. As machine learning systems increasingly permeate various sectors, there arises an urgent need to provide clear and transparent explanations for their decisions, especially when these decisions bear significant consequences, such as in healthcare or the judiciary. Traditional accuracy-focused metrics are no longer sufficient. Stakeholders, be they end-users, regulators, or the general public, now demand causal explanations that elucidate the *why* behind an autonomous decision, not just the *what* [Arrieta et al., 2020].

The explanation tendered in response to a why question often varies.

Aristotle's Four Causes framework, occasionally referred to as the Explanation Modalities model, remains pivotal in discourse about causality and explanations [Hankinson, 1998]:

1. **Material**: Pertains to the constituent or substance that constitutes an entity. For instance, rubber serves as the material cause of a car tyre.

2. **Formal**: Involves the shape or characteristics that define an entity's essence. Being circular, for instance, is a formal cause of a car tyre, often alluded to as categorical elucidations.

3. **Efficient**: Represents the immediate antecedents that induce a transformation. The production by a tyre fabricator, for instance, is the efficient cause of a car tyre, frequently termed mechanistic elucidations.

4. **Final**: Highlights the purpose or objective of an entity. Facilitating vehicle movement stands as the final cause of a car tyre, commonly labelled as functional or teleological elucidations.

Diverse scholars have introduced conceptual frameworks paralleling Aristotle's, such as Dennett [Dennett, 1989] and Marr [Marr, 2010]. Kass and Leake [Kass and Leake, 1987], moreover, added a societal stratum, that explicates human actions devoid of deliberate intent. An example provided by Kass and Leake illustrates a city's escalating crime rate. Although resultant of intentional human activities, the aggregate crime surge does not connote collective intent. While each criminal act is deliberate, the overarching surge in crime is not a predetermined outcome but an inadvertent collective repercussion.

3.4 Context and Evidentiary Burden

Arguing from correlation to causation is a pivotal causal argumentation scheme [Walton et al., 2008]. It is extensively studied in informal logic, primarily due to the *post hoc* fallacy, which is the error of deducing one event as the cause of another solely based on their correlation. While a correlation between two events can hint at a causal relationship, it is not definitive proof: the observed correlation might be coincidental, or there might be an external factor influencing both events. Hence, while correlation can suggest causation, it is not ultimate evidence, yet remains a frequently employed argumentative approach:

There is a positive correlation between A and B.
Therefore, A causes B.

The *correlation to cause* argument, often seen as the starting point in causative investigations, essentially represents the most basic level of evidential power. It suggests that the mere presence of correlation, while indicative, is not definitive proof of causation. Drawing parallels, in scientific reasoning, causation is robustly examined beyond mere correlations, much like the thorough approach seen in intelligence analysis. Legal reasoning necessitates evidence to surpass mere correlation to establish causation, especially given the stringent standards of the courtroom. Similarly, in the social sciences, while correlation might provide a starting point, deeper analyses are essential to discern intricate human and societal interactions.

In scientific discourse, causality delves into the intricate relationship between cause and effect, aiming to ascertain the underlying reasons for observed phenomena. In this way, causal arguments are not treated as absolute, rather more pragmatically are perceived in relation to a specific context of investigation. Scientific argumentation follows a structured progression [Walton et al., 2008]. Initially, there is a *discovery phase* where hypotheses are proposed to elucidate specific data. These hypotheses are subsequently *tested* through experiments or further data collection. Following this, hypotheses are *refined* and articulated more precisely using mathematical and logical formalisation techniques. While this sequence suggests a linear process, it rather presents an idealised model of how scientific investigations should unfold, reflecting the principles of the scientific method.

Issues with causal argumentation and related fallacies often arise in the preliminary stages of an investigation or even before a thorough scientific inquiry. For instance, a herbal remedy might gain popularity in health stores for treating a certain illness. As its popularity grows, many might believe the remedy is the cause of their health improvements. However, at this early stage, it is very much possible that this is a *post hoc* argument. To determine the efficacy of such a herbal remedy, medical researchers might conduct a statistical study with a control group. If significant effects are observed, subsequent research might explore the herb's components to identify the active chemical causing these effects. As the study progresses, a theory emerges. As the theory becomes more refined, defeasible reasoning becomes more relevant.

The process of scientific investigation mirrors practices in intelligence analysis, cf. [Cerutti, 2018, Cerutti and Pearson, 2018, Toniolo et al., 2015], where analysts similarly traverse various stages to derive meaningful conclusions. In the Pirolli and Card model [Pirolli and Card, 2005] of intelligence analysis, there is a progression from gathering raw data to synthesising and refining it into structured insights. Initially, vast amounts of unstructured information are collected, akin to the hypothesis formation stage in scientific research. As the intelligence analysis process advances, patterns and connections are discerned, resembling the inductive reasoning phase in scientific investigations. As analysts delve deeper, their hypotheses become increasingly refined and focused, paralleling the defeasible reasoning in scientific processes.

Legal argumentation diverges significantly from scientific discourse. While both possess similar dialogue stages, legal reasoning leans more towards defeasible arguments than deductive or inductive ones. In legal contexts, while scientific evidence on causation is crucial, it is primarily presented through expert testimonies, as judges, lawyers, and juries are not typically scientists and cannot validate the evidence scientifically. Opposing causal theories are presented by both plaintiffs and defendants, with the strength of their arguments judged based on the legal standards of proof for that case type. Therefore, the evaluation of arguments in legal settings is distinct from scientific evaluations.

There are two primary standards of proof commonly used across various fields: *beyond a reasonable doubt* and *the preponderance of the evidence*. The former is the strictest standard, often applied in criminal cases, implying that there is little room for doubt regarding the truth of the claim. It is also used in scientific research, — e.g., $p < 0.01$ — denoting a very low probability that the observed results occurred by chance. The *preponderance of the evidence*, is more lenient and used in civil cases. Here, the goal is to determine if a claim is more likely true than not: if there is a slightly greater than 50% chance a claim is true, it is accepted.

Jonathan Haidt — an American social psychologist — in a series of recent public appearances[5] makes the claim that in social policies, for those responsible for the well-being of others, such as parents and educators, the more lenient standard is often more applicable. They must weigh the costs of false negatives, where real threats are dismissed, against false positives, where non-existent threats are acted upon.

Haidt — together with Zach Rausch, Jean Twenge, and others — started a collaborative review categorising the primary types of studies addressing the impact of social media on teen mental health such as correlational studies, longitudinal research, genuine experiments, and quasi-experiments. Among the achieved conclusions, they identify a number of correlational studies that indicate links between prolonged usage and signs of anxiety and depression. The connection becomes even more pronounced in girls where there appears to be a correlation between those who engage in social media for over 4 hours daily, and those who experience depression at a rate two to three times higher than those who use it for under an hour.

Reactions to Haidt et al.'s endeavour showed the application of several critical questions and counterarguments we discussed in this chapter, mostly focusing on arguments from *insufficient cause*, *low force of statistical data*, *partial cause*, and *indirect cause*. In addition to comment — or promise to do so in a forthcoming book — on the merit of the critiques, it is interesting that Haidt opens the floor concerning the standard of proof, arguing for a lenient standard than the one adopted in criminal law or in scientific endeavour.

The problem of arguing about the standard of proof brings us to comment on the need for assessing the causal strength.

---

[5]For example, see https://jonathanhaidt.substack.com/p/social-media-mental-illness-epidemic (on 27th October 2023) and https://jonathanhaidt.substack.com/p/why-some-researchers-think-im-wrong (on 27th October 2023).

3.5 The Causal Strength: the Case of Bayesian Argumentation

Consider the *argument from sign*, a fundamental example of an argumentation scheme for causal reasoning [Walton et al., 2008]:

> In the given scenario, X (an observation) is accurate.
> When X is true in such situations, it usually indicates the truth of Y.
> Hence, Y is true in this scenario.

To challenge a supposition derived from the argument from sign, associated *critical questions* (CQs) can be formulated:

CQ1: How strong is the association between the sign and the event it suggests?

CQ2: Could other occurrences more accurately explain the sign?

The integration of Bayesian principles in argumentation, as detailed by [Hahn and Hornikx, 2016], is a nuanced response to the MAXMIN rule when amalgamating linked and convergent arguments. *Convergent arguments* involve multiple independent arguments converging towards a shared claim. In contrast, linked arguments establish a sequence of dependencies, lending support to a claim only in unison.

For convergent arguments, Walton [Walton, 1992] champions the MAX rule. Here, the collective strength or validity of the argument is gauged by the strongest of the independent arguments converging on a shared assertion. For linked arguments, academic consensus [Walton, 1992, Pollock, 2001] dictates that the overall validity of the argument hinges on its weakest element. While the proximity between plausibility and probability is acknowledged in specific instances [Walton, 1992], others [Hahn et al., 2013] assert its prevalence in numerous situations. This probabilistic viewpoint on an argument's strength or plausibility concludes that the MIN rule sets the maximum limit on the probabilistic interpretation of a linked argument's strength. This is evident in the equation $P(A \wedge B) = P(A) \cdot P(A|B) = P(B) \cdot P(B|A) \leq \min\{P(A), P(B)\}$.

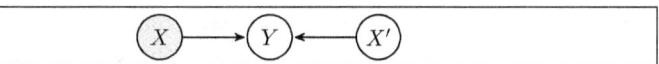

Figure 3: Simplified belief network depicting the *argument from sign* as per [Hahn and Hornikx, 2016]

Given their inherently defeasible nature, arguments can be visualised as a network of random variables interlinked in a belief network, typified as a directed graph. In this graph, nodes represent random variables, whilst edges depict causal connections. Hahn and colleagues [Hahn and Hornikx, 2016] provide a Bayesian perspective on the *argument from sign* (refer to Figure 3), illuminating the potential of belief networks in quantifying the impact of critical questions. For example, responses to **CQ1** are given by $p(Y \mid X)$ and $p(Y \mid \overline{X})$ (see Fig. 3). Meanwhile, **CQ2** delves into the examination of alternative occurrences, like X', which correlate more closely with Y. Employing this model, the efficacy of various argumentation schemes has been tested. Additionally, experimental investigations, such as [Hahn and Oaksford, 2007b], delineate the Bayesian framework's deployment in deducing both qualitative and quantitative perceptions of general argument strength by individuals, while other research [Cerutti, 2021] evidenced its role in supporting explaining machine learning behaviours.

4 Structured Causal Argumentation

Various formalisms for structured argumentation have been proposed that adopt Dung's abstract argumentation approach. What these formalisms have in common is that they provide definitions for (1) how arguments are constructed on the basis of a knowledge base, and (2) how attacks between arguments are defined. Reasoning then proceeds by determining which sets of arguments are jointly acceptable, which is done using an appropriate argumentation semantics [Dung, 1995]. Well-known examples include the ABA (Assumption-Based Argumentation) formalism [Bondarenko et al., 1997] and the ASPIC+ system [Modgil and Prakken, 2013]. The aim of these formalisms is to provide an argumentation-based account of reasoning with defeasible rules or assumptions. These formalisms do not, however, explicitly aim to provide an argumentation-based account of *causal* reasoning. This is evident in the fact that the distinction between cause and effect, and between observation and intervention, is foreign to the reasoning mechanisms and knowledge representations adopted by these formalisms. These formalisms furthermore lack a notion of counterfactual, which can be defined in terms of the aforementioned distinctions.

The aim of this section is to provide an overview of structured argumentation formalisms that do explicitly aim to provide an argumentation-based account of causal reasoning. What the two approaches have in common is that they are based on Dung's model of abstract argumentation. We present the necessary definitions of abstract argumentation in Section 4.1. We make a distinction between two approaches to structured causal argumentation. The first is a rule-based approach, of which the work of Bex [Bex, 2015] and Wieten et al. [Wieten et al., 2020, Wieten et al., 2022] are examples. In this approach, causal knowledge is represented using defeasible rules that are categorised into *causal rules* and *evidential rules*. The semantics of this distinction is defined by a set of inference schemes referred to as Pearl's C-E system [Pearl, 1988]. We describe this rule-based approach in Section 4.2. The second approach is an assumption-based approach, which employs structural equation models for representing causal knowledge, and defeasible assumptions to represent beliefs about exogenous variables. This approach, due to Bengel et al. [Bengel et al., 2022], is described in Section 4.3.

4.1 Dung's Model of Abstract Argumentation

The structured causal argumentation formalisms that we discuss have in common that conclusions are drawn by generating an argumentation framework and determining the acceptable arguments using the standard admissibility-based semantics. In what foollows will use (Ar, att) to denote an argumentation framework with a finite set Ar of arguments and attack relation $att \subseteq Ar \times Ar$ [Dung, 1995]. The structured causal argumentation formalisms described in the following subsections specify how arguments are constructed, what their content is and how the attack relation is defined. They use the standard admissibility-based semantics defined for determining the acceptable sets of arguments of an argumentation framework. We refer the reader to [Dung, 1995] for further details about argumentation frameworks and their admissibility-based semantics.

4.2 Rule-based Causal Argumentation

Many forms of structured argumentation use rule-based representations of knowledge. For example, the ASPIC+ framework uses strict and defeasible rules, and defines arguments as structures in which these rules are chained together [Modgil and Prakken, 2013]. Thus, an obvious starting point for a structured argumentation-based account of causal reasoning is to use rules to represent causal knowledge. Pearl showed that this is indeed possible, but that care must be taken to ensure that these rules are applied in a causally consistent manner [Pearl, 1987]. In this section we outline Pearl's so called *C-E system* for reasoning with rule-based representations of causal knowledge. We then provide an overview of the two causal argumentation formalisms of Bex [Bex, 2015] and Wieten et al. [Wieten et al., 2020, Wieten et al., 2022], which can be regarded as argumentation-based implementations of Pearl's C-E system.

4.2.1 Pearl's C-E System

Pearl's C-E system is based on the distinction of two types of rules for representing causal knowledge: *causal* and *evidential* rules [Pearl, 1987]. Causal rules were discussed already in Section 2.1. A causal rule describes a causal relationship between events in the real world and specifies that we may infer an effect from its cause. An example of a causal rule is

$$Fire \Rightarrow Smoke \text{ (Fire causes Smoke)}. \tag{1}$$

Evidential rules describe how we ascribe causes to the phenomena we observe. They specify that we may infer a cause from evidence of its effect, and can be regarded as *explanation evoking*. Examples of evidential rules are

$$Smoke \Rightarrow Fire \text{ (smoke is evidence for fire)} \tag{2}$$

and

$Smoke \Rightarrow SmokeMachine$ (smoke is evidence for a smoke machine).
(3)

From a common sense reasoning perspective, both causal and evidential rules are natural representations of causal knowledge. Thus, ideally, we want to be able to reason with a mix of causal and evidential rules.

The common interpretation of a defeasible rule $\phi \Rightarrow \psi$ is that it allows us to derive ψ from ϕ as long as ψ is consistent with the other beliefs we can derive. Unrestrictedly applying this interpretation to causal and evidential defeasible rules leads to problems, however. To see why, suppose we have evidence for *Fire*. The causal rule (1) then allows us to infer *Smoke*, after which the evidential rule (2) allows us to infer *SmokeMachine*. This is clearly wrong, since the use of the first rule establishes that *Smoke* was caused by *Fire*. This means that we need not (and in general should not) use the second rule to infer *SmokeMachine* as an explanation for *Smoke*.

To avoid the problem outlined above, Pearl proposed a set of inference schemes, referred to as the *C-E system*, for reasoning with a mix of causal and evidential defeasible rules. From now on we will use the connectives $\Rightarrow_c$ and $\Rightarrow_e$ to distinguish causal and evidential defeasible rules, and ϕ, ψ, χ are taken to be propositional formulas. We say that a formula ϕ is *E-believed* (denoted $E(\phi)$) if ϕ can be inferred with an evidential rule, and that ϕ is *C-believed* (denoted $C(\phi)$) if ϕ can be inferred only with a causal rule. Pearl's C-E system consists of the inference rules (a), (b), and (c) given below.

$$(a)\frac{\phi \Rightarrow_c \psi \quad C(\phi)}{C(\psi)} \quad (b)\frac{\phi \Rightarrow_c \psi \quad E(\phi)}{C(\psi)} \quad (c)\frac{\phi \Rightarrow_e \psi \quad E(\phi)}{E(\psi)} \quad (\times)\frac{\phi \Rightarrow_e \psi \quad C(\phi)}{\psi}$$

Rules (a), (b) and (c) ensure that E-believed propositions can only be established by chains of evidential rules, while C-believed propositions may be established by a mix of causal and evidential rules. The fourth rule, labeled $\times$, is *not* a valid rule of the C-E system. This rule sanctions the illegal pattern of inference outlined above, where we applied the evidential rule (2) to the C-believed fact *Fire*.

Pearl's C-E system provides a means to reason about causality using a mix of causal and evidential defeasible rules. In the next two subsections we present two formalisms that can be regarded as argumentation-based implementations of Pearl's C-E system. In these approaches, the inference rules of the C-E system are implicit in the way arguments are constructed. Both systems establish, in a formal manner, that they adhere to the constraints imposed by Pearl's C-E system.

4.2.2 An Integrated Theory of Causal Stories and Evidential Arguments

Bex' *integrated theory of causal stories and evidential arguments* (or *integrated theory*, in short) is intended as a model that integrates causal, story-based reasoning and evidential argumentation in the legal domain [Bex, 2015]. In this section we present the basic definitions of the formalism and review its adherence to Pearl's C-E scheme.

Based on the ASPIC framework [Modgil and Prakken, 2013], Bex defines an *argumentation system* as consisting of a logical language, a set of strict and defeasible rules, the latter of which are categorised as either causal or evidential. We will use $\rightarrow$ to denote strict rules and $\Rightarrow_c$ and $\Rightarrow_e$ to denote causal and evidential defeasible rules. The formalism also supports rules that defeat other rules, which requires a mapping from defeasible rules to formulas like in the ASPIC+ system [Modgil and Prakken, 2013]. To simplify our definitions we omit this aspect. All definitions in this section are adapted from [Bex, 2015].

Definition 4.1. *An* argumentation system *is a triple* $AS = (\mathcal{L}, \mathcal{R}, n)$, *where $\mathcal{L}$ is a logical language closed under negation* ($\neg$)*;* $\mathcal{R} = \mathcal{R}_s \cup \mathcal{R}_d$ *(with $\mathcal{R}_s \cap \mathcal{R}_d = \emptyset$) is a set of strict ($\mathcal{R}_s$) and defeasible ($\mathcal{R}_d$) rules; $\mathcal{R}_d = \mathcal{R}_c \cup \mathcal{R}_e$ (with $\mathcal{R}_c \cap \mathcal{R}_e = \emptyset$) is a set of causal ($\mathcal{R}_c$) and evidential ($\mathcal{R}_e$) defeasible rules. A strict rule is of the form* $\phi_1, \ldots, \phi_n \rightarrow \psi$*; a causal defeasible rule is of the form* $\phi \Rightarrow_c \psi$*; and an evidential defeasible rule is of the form* $\phi \Rightarrow_e \psi$ *(all $\phi, \phi_i, \psi \in \mathcal{L}$).*

It is assumed that $\mathcal{R}_s$ includes all the common inference rules of deductive logic. Note here that, unlike defeasible rules in the ASPIC framework, causal and evidential rules have precisely one premise. It is, however, possible to express a causal or evidential defeasible rules with multiple premises, by including the appropriate strict rule.

If two causal rules have the same consequent then their antecedents are considered as *alternative causes*. Likewise, the consequences of two evidential rules with the same antecedent are alternative causes. A third possibility is the combination of a causal and evidential rule that express alternative causes for the same effect. Formally:

Definition 4.2. *Causes ϕ and χ are alternatives if for any pair of rules $r_i, r_j \in \mathcal{R}_c \cup \mathcal{R}_e$:*

1. $r_i = \phi \Rightarrow_c \psi$ and $r_j = \chi \Rightarrow_c \psi$ and $\phi \neq \chi$; or

2. $r_i = \psi \Rightarrow_e \phi$ and $r_j = \phi \Rightarrow_e \chi$ and $\phi \neq \chi$; or

3. $r_i = \phi \Rightarrow_c \psi$ and $r_j = \phi \Rightarrow_e \chi$ and $\phi \neq \chi$

A knowledge base consists of two types of elements: *evidence* and *hypotheses*. Intuitively, while evidence holds beyond any doubt, hypotheses can be defeated by evidence to the contrary.

Definition 4.3. *A knowledge base in an $AS = (\mathcal{L}, \mathcal{R}, n)$ is a set $\mathcal{K} \subseteq \mathcal{L}$ consisting of two disjoint subsets $\mathcal{K}_e$ (the evidence) and $\mathcal{K}_h$ (the hypotheses).*

The following definition states how arguments are constructed given a knowledge base and argumentation system. This definition is similar to that of the ASPIC framework [Modgil and Prakken, 2013], except that it defines not one but three types of defeasible arguments: causal arguments (containing only causal rules), evidential arguments (containing only evidential rules) and mixed arguments (containing both).

Definition 4.4. *An argument on the basis of a knowledge base $\mathcal{K}$ in an argumentation system AS is any A such that either:*

1. $A = \phi$ and $\phi \in \mathcal{K}$. In this case we define: $\text{Prem}(A) = \{\phi\}$; $\text{Conc}(A) = \phi$; $\text{Sub}(A) = \{\phi\}$; $\text{TopRule}(A) = undefined$; $\text{ERule}(A) = \emptyset$; $\text{CRule}(A) = \emptyset$.

2. $A = (A_1, \ldots, A_n \to \psi)$ and $A_1, \ldots, A_n$ are arguments such that there exists a strict rule $\text{Conc}(A_1), \ldots, \text{Conc}(A_n) \to \psi$ in $\mathcal{R}_s$. In this case we define: $\text{Prem}(A) = \text{Prem}(A_1) \cup \ldots \cup \text{Prem}(A_n)$; $\text{Conc}(A) = \psi$; $\text{Sub}(A) = \text{Sub}(A_1) \cup \ldots \cup \text{Sub}(A_n) \cup \{A\}$; $\text{TopRule}(A) = \text{Conc}(A_1), \ldots, \text{Conc}(A_n) \to \psi$; $\text{ERule}(A) = \text{ERule}(A_1) \cup \ldots \cup \text{ERule}(A_n)$; $\text{CRule}(A) = \text{CRule}(A_1) \cup \ldots \cup \text{CRule}(A_n)$.

3. $A = (A_1, \ldots, A_n \Rightarrow_c \psi)$ and $A_1, \ldots, A_n$ are arguments such that there exists a causal rule $\text{Conc}(A_1), \ldots, \text{Conc}(A_n) \to_c \psi$ in $\mathcal{R}_c$. In this case we define: $\text{Prem}(A) = \text{Prem}(A_1) \cup \ldots \cup \text{Prem}(A_n)$; $\text{Conc}(A) = \psi$; $\text{Sub}(A) = \text{Sub}(A_1) \cup \ldots \cup \text{Sub}(A_n) \cup \{A\}$; $\text{TopRule}(A) = \text{Conc}(A_1), \ldots, \text{Conc}(A_n) \Rightarrow_c \psi$; $\text{ERule}(A) = \text{ERule}(A_1) \cup \ldots \cup \text{ERule}(A_n)$; $\text{CRule}(A) = \text{CRule}(A_1) \cup \ldots \cup \text{CRule}(A_n) \cup \{\text{Conc}(A_1), \ldots, \text{Conc}(A_n) \Rightarrow_c \psi\}$.

4. $A = (A_1, \ldots, A_n \Rightarrow_e \psi)$ and $A_1, \ldots, A_n$ are arguments such that there exists an evidential rule $\text{Conc}(A_1), \ldots, \text{Conc}(A_n) \Rightarrow_e \psi$ in $\mathcal{R}_e$. In this case we define: $\text{Prem}(A) = \text{Prem}(A_1) \cup \ldots \cup \text{Prem}(A_n)$; $\text{Conc}(A) = \psi$; $\text{Sub}(A) = \text{Sub}(A_1) \cup \ldots \cup \text{Sub}(A_n) \cup A$; $\text{TopRule}(A) = \text{Conc}(A_1), \ldots, \text{Conc}(A_n) \Rightarrow_e \psi$; $\text{ERule}(A) = \text{ERule}(A_1) \cup \ldots \cup \text{ERule}(A_n) \cup \text{Conc}(A_1), \ldots, \text{Conc}(A_n) \Rightarrow_e \psi$; $\text{CRule}(A) = \text{CRule}(A_1) \cup \ldots \cup \text{CRule}(A_n)$.

An argument A is causal if $\text{CRule}(A) \neq \emptyset$ and $\text{ERule}(A) = \emptyset$, evidential if $\text{ERule}(A) \neq \emptyset$ and $\text{CRule}(A) = \emptyset$, and mixed if $\text{CRule}(A) \neq \emptyset$ and $\text{ERule}(A) \neq \emptyset$.

The distinction between causal and evidential arguments corresponds to a similar distinction that we made in Section 3 in the context of causal argument schemes, with Walton's argumentation schemes *from cause to effect* and *from effect to cause* being the main examples in each category.

Like in the ASPIC+ framework, one argument attack another whenever the former *rebuts* or *undermines* the latter. A third type of attack, referred to as *alternative-attack*, is introduced to capture arguments whose (sub)conclusions are alternative causes.

Definition 4.5. *Argument A attacks argument B iff A undercuts, rebuts, undermines or alternative-attacks B, where:*

- A rebuts argument B (on B') iff $\text{Conc}(A) = -\phi$ for some $B' \in \text{Sub}(B)$ of the form $B'' \Rightarrow_{c/e} \phi$.

- A undermines B (on B') iff $\text{Conc}(A) = -\phi$ for some $B' = \phi$, $\phi \notin \mathcal{K}_e$.

- A alternative-attacks B (on B') iff $\text{Conc}(A) = \phi$ for some $B' \in \text{Sub}(B)$ with $\text{Conc}(B') = \psi$, where ϕ and ψ are alternatives according to definition 4.2.

Having defined arguments and attacks, we can now state how an argumentation framework is constructed on the basis of an argumentation system AS and knowledge base $\mathcal{K}$, jointly referred to as an *integrated theory*. Any of the standard admissibility-based semantics can be used to determine the accepted arguments of this argumentation framework.

Definition 4.6. *Let IT be an integrated theory $(AS, \mathcal{K})$. The* integrated argumentation framework *defined by AT is the argumentation framework (Ar, att) where Ar is the set of all finite arguments constructed from $\mathcal{K}$ in AS according to Definition 4.4, and att is defined by x att y iff x attacks y according to Definition 4.5.*

The system satisfies Pearl's C-E constraint, which states that it is not possible to evidentially infer a conclusion from something that was inferred using causal reasoning. Bex formalizes this condition as follows [Bex, 2015].

Proposition 4.7. *Let $(\mathcal{A}, \mathcal{C})$ be an integrated argumentation framework. Under credulous or skeptical acceptance under an admissible semantics, if the argument $A \in \mathcal{A}$ is accepted then there is no $B \in \text{Sub}(A)$ such that $B : B'' \Rightarrow_e \chi$ and $B' \Rightarrow_d \psi$ and $\text{Conc}(B'') = \phi$ where $\phi \neq \chi$.*

4.2.3 The Information Graph Formalism

The *information graph* formalism of Wieten et al. [Wieten et al., 2020, Wieten et al., 2022] is intended as a graph-based tool to aid in the process of reasoning about evidence in the legal and forensic domains. An information graph represents a set of defeasible rules and permits a convenient graph-based visualization. Wieten et al. refer to defeasible rules as *generalizations*, and make a distinction is between causal and evidential generalizations. Given an information graph and a body of evidence, the formalism supports a mix of deductive and abductive inference.

In this section we present the basic definitions of the information graph formalism and review its adherence to Pearl's C-E scheme. All the definitions in this section are due to Wieten et al. [Wieten et al., 2020]. An extended version of the information graph formalism was presented in [Wieten et al., 2022]. This extended version supports additional notions such as *enablers* and generalizations that are neither causal nor evidential. The extended version was also shown to satisfy the main rationality postulates for argumentation formalisms due to Caminada and Amgoud [Caminada and Amgoud, 2007]. For the sake of simplicity, we will not discuss these additional notions and results, and we will base our presentation on [Wieten et al., 2020].

An information graph consists of a set of vertices representing propositional literals, and a set of (hyper)arcs representing causal generalizations, evidential generalizations and exceptions. Causal and evidential generalizations are essentially causal and evidential default rules as discussed in Section 4.2.1. An exception is a relation between a propositional literal and a generalization, stating that the former represents a circumstance under which the latter may not apply.

Definition 4.8. *An* information graph *is a directed graph* $G = (\mathbf{P}, \mathbf{A})$, *where* $\mathbf{P}$ *is a set of propositional literals and* $\mathbf{A} = \mathbf{G} \cup \mathbf{X}$ *(with* $\mathbf{G} \cap \mathbf{X} = \emptyset$*) a set of directed (hyper)arcs.* $\mathbf{G}$ *contains* generalisation arcs *and* $\mathbf{X}$ *contains* exception arcs *(defined below). Furthermore,* $\mathbf{G} = \mathbf{G^c} \cup \mathbf{G^e}$ *(with* $\mathbf{G^c} \cap \mathbf{G^e} = \emptyset$*), where* $\mathbf{G^c}$ *contains* causal generalisation arcs *and* $\mathbf{G^e}$ *contains* evidential generalisation arcs.

A generalisation arc $g \in \mathbf{G}$ *is a directed (hyper)arc* $g : \{p_1, \ldots, p_n\} \to p$, *indicating a generalisation with antecedents* $\mathbf{P}_1 = \{p_1, \ldots, p_n\} \subseteq \mathbf{P}$ *and consequent* $p \in \mathbf{P} \setminus \mathbf{P}_1$. *We use the symbols* $\to_\mathbf{c}$ *and* $\to_\mathbf{e}$ *to denote, respectively, causal and evidential generalisation arcs.*

An exception arc $x \in \mathbf{X}$ *is a hyperarc* $x : p \rightsquigarrow g$ *where* $p \in \mathbf{P}$ *is called an* exception *to the generalisation* $g \in \mathbf{G}$.

Intuitively, a causal generalisation arc $\{p_1, \ldots, p_n\} \to_\mathbf{c} p$ states that $p_1, \ldots, p_n$ are a cause for p. An evidential generalisation arc $\{p_1, \ldots, p_n\}$

$\rightarrow_e p$ states that $p_1, \ldots, p_n$ are evidence for p. Given a generalisation arc $g : \{p_1, \ldots, p_n\} \rightarrow p$, we define $\text{Head}(g) = p$ and $\text{Tails}(g) = \{p_1, \ldots, p_n\}$. An exception arc $p \rightsquigarrow g$ states that p represents a circumstance under which g may not apply. We use $-q$ to denote p if $q = \neg p$, and $\neg p$ if $q = p$.

An information graph $G = (\mathbf{P}, \mathbf{A})$ is assumed to include a set $\mathbf{E} \subseteq \mathbf{P}$ referred to as the *evidence set*. This set contains the evidence on the basis of which we perform inference. It is assumed to be consistent: if $p \in \mathbf{E}$ then $\neg p \notin \mathbf{E}$. An information graph defines two types of inference: deductive (forward) and abductive (backward) inference:

Definition 4.9. *Let $G = (\mathbf{P}, \mathbf{A})$ be an information graph with evidence set $\mathbf{E}$. Then $q \in \mathbf{P} \setminus \mathbf{E}$ is deductively inferred from $p_1, \ldots, p_n \in \mathbf{P}$ using generalisation $(g : \{p_1, \ldots, p_n\} \rightarrow q) \in \mathbf{G}$, denoted $p_1, \ldots, p_n \rightarrow_g q$, iff for all $i = 1, \ldots, n$:*

1. *$p_i \in \mathbf{E}$; or*

2. *p_i is deductively inferred from $r_1, \ldots, r_m \in \mathbf{P}$ using generalisation $g' : \{r_1, \ldots, r_m\} \rightarrow p_i$, where $g' \in \mathbf{G^e}$ if $g \in \mathbf{G^e}$; or*

3. *p_i is abductively inferred from $r \in \mathbf{P}$ using generalisation $g' : \{p_i, r_1, \ldots, r_m\} \rightarrow r$ in $\mathbf{G^c}$, $g \neq g'$, $r_1, \ldots, r_m \in \mathbf{P}$.*

Definition 4.10. *Let $G = (\mathbf{P}, \mathbf{A})$ be an information graph with evidence set $\mathbf{E}$. Then $p_1, \ldots p_n \in \mathbf{P} \setminus \mathbf{E}$ are abductively inferred from $q \in \mathbf{P}$ using generalisation $(g : \{p_1, \ldots, p_n\} \rightarrow q) \in \mathbf{G^c}$, denoted $q \twoheadrightarrow_g p_1; \ldots; q \twoheadrightarrow_g p_n$, iff:*

1. *$q \in \mathbf{E}$; or*

2. *q is deductively inferred from $r_1, \ldots, r_m \in \mathbf{P}$ using generalisation $g' : r_1, \ldots, r_m \rightarrow q$ in $\mathbf{G}$, $g \neq g'$, where $g' \in \mathbf{G} \setminus \mathbf{G^c}$; or*

3. *q is abductively inferred from $r \in \mathbf{P}$ using generalisation $g' : \{q, r_1, \ldots, r_m\} \rightarrow r$ in $\mathbf{G^c}$, $r_1, \ldots, r_m \in \mathbf{P}$.*

The deductive and abductive inferences sanctioned by an information graph give rise to a set of arguments inductively defined as follows.

Definition 4.11. *An argument A on the basis of an information graph $G = (\mathbf{P}, \mathbf{A})$ and evidence set $\mathbf{E}$ is any finite A such that either:*

1. $A = p$ *and* $p \in \mathbf{E}$. *In this case we define:* $\text{Conc}(A) = p$; $\text{Sub}(A) = \{A\}$; $\text{ImmSub}(A) = \emptyset$; $\text{Gen}(A) = \emptyset$; $\text{TopGen}(A) = $ *undefined.*

2. $A = (A_1, \ldots, A_n \twoheadrightarrow_g p)$ *and* $A_1, \ldots, A_n$ *are arguments such that p is deductively inferred from* $\text{Conc}(A_1), \ldots, \text{Conc}(A_n)$ *using a generalisation* $g \in \mathbf{G} \backslash (\text{Gen}(A_1) \cup \ldots \cup \text{Gen}(A_n))$ *where* $g : \{\text{Conc}(A_1), \ldots, \text{Conc}(A_n)\} \to p$. *In this case we define:* $\text{Conc}(A) = p$; $\text{Sub}(A) = \text{Sub}(A_1) \cup \ldots \cup \text{Sub}(A_n) \cup \{A\}$; $\text{ImmSub}(A) = \{A_1, \ldots, A_n\}$; $\text{Gen}(A) = \text{Gen}(A_1) \cup \ldots \cup \text{Gen}(A_n) \cup \{g\}$; $\text{TopGen}(A) = g$.

3. $A = (A' \twoheadrightarrow_g p)$ *and A' is an argument such that p is abductively inferred from* $\text{Conc}(A')$ *using generalisation* $g \in \mathbf{G} \setminus \text{Gen}(A')$ *where* $g : \{p, p_1, \ldots, p_n\} \to \text{Conc}(A')$ *for* $p_1, \ldots, p_n \in \mathbf{P}$. *In this case we define:* $\text{Conc}(A) = p$; $\text{Sub}(A) = \text{Sub}(A') \cup \{A\}$; $\text{ImmSub}(A) = \{A'\}$; $\text{Gen}(A) = \text{Gen}(A') \cup \{g\}$; $\text{TopGen}(A) = g$.

We now specify how an argumentation framework is constructed on the basis of an information graph. Like in the ASPIC+ framework [Modgil and Prakken, 2013], one argument attacks another whenever the former *rebuts* or *undermines* the latter. A third type of attack, referred to as *alternative-attack* and inspired by the notion of alternative-attack from Definition 4.5, is introduced to capture arguments whose (sub)conclusions are alternative causes.

Definition 4.12. *Let $G = (\mathbf{P}, \mathbf{A})$ (with $\mathbf{A} = \mathbf{G} \cup \mathbf{X}$ — see Definition 4.8) be an information graph with evidence set $\mathbf{E}$. The argumentation framework defined by G and $\mathbf{E}$ is $(\mathcal{A}, \mathcal{C})$ where $\mathcal{A}$ contains all arguments on the basis of G and $\mathbf{E}$, and $\mathcal{C}$ is defined by $(A, B) \in \mathcal{C}$ iff A rebuts, undercuts, or alternative attacks B, where:*

- *A rebuts B (on B') iff $\text{Conc}(B') \notin \mathbf{E}$ and $\text{Conc}(A) = -\text{Conc}(B')$, where $B' = \text{Sub}(B)$.*

- *A undercuts B (on B') iff there exists an $x \in \mathbf{X}$ s.t. $x : \text{Conc}(A) \leadsto g$ and $\text{TopGen}(B') = g$.*

- A alternative-attacks B (on B') iff there exists a $C \in \text{ImmSub}(A) \cap \text{ImmSub}(B')$ s.t. $\text{Conc}(A)$ and $\text{Conc}(B')$ are abductively inferred from $\text{Conc}(C)$ using generalizations g and $g' \in \mathbf{G^C}$ respectively, where $g \neq g'$.

Any of the admissibility-based semantics can be used to determine the conclusions we can draw from an information graph.

Wieten et al. prove that Definition 4.11 satisfies the following interpretation of Pearl's C-E constraint. In words: if c_1 and c_2 are alternative causes for e, then an argument that derives e as an effect of c_1 cannot be extended to an argument that derives c_2 as a cause for e.

Proposition 4.13. *Let $G = (\mathbf{P}, \mathbf{A})$ be an information graph with evidence set $\mathbf{E}$. Let $c_1, c_2 \in \mathbf{P}$ be alternative causes of $e \in \mathbf{P}$ in that either:*

1. *$\exists g \in \mathbf{G^e}$, $e \in \text{Tails}(g), \text{Head}(g) = c_1$, and either:*

   (a) *$\exists g' \neq g \in \mathbf{G^e}$, $e \in \text{Tails}(g'), \text{Head}(g') = c2$; or*

   (b) *$\exists g' \in \mathbf{G^c}$, $c_2 \in \text{Tails}(g'), \text{Head}(g') = e$.*

2. *$\exists g \in \mathbf{G^c}$, $c_1 \in \text{Tails}(g), \text{Head}(g) = e$, and either:*

   (a) *$\exists g' \neq g \in \mathbf{G^c}$, $c_2 \in \text{Tails}(g'), \text{Head}(g') = e$; or*

   (b) *$\exists g' \in \mathbf{G^e}$, $e \in \text{Tails}(g'), \text{Head}(g') = c_2$.*

Let A and B be arguments on the basis of G and $\mathbf{E}$. If $\text{Conc}(B) = e$, $A \in \text{ImmSub}(B)$ and $\text{Conc}(A) = c_1$ then there is no argument C on the basis of G and $\mathbf{E}$ such that $B \in \text{ImmSub}(C)$ and $\text{Conc}(C) = c_2$.

4.3 Assumption-based Structured Causal Argumentation

The second approach to structured causal argumentation is an assumption-based approach. This approach, due to [Bengel et al., 2022], uses structural equation models for representing causal knowledge, and relies on the method of reasoning with maximally consistent subsets which, as shown by Cayrol [Cayrol, 1995], can be formulated within Dung's model of argumentation under the stable semantics. We start with the argumentation-based formulation of this method.

Definition 4.14. Let $\mathcal{L}$ be a propositional language generated by a finite set of atoms. A knowledge base is a pair $\Delta = (K, A)$ where $K \subseteq \mathcal{L}$ is a consistent set of facts and $A \subseteq \mathcal{L}$ a set of assumptions.

- An argument induced by Δ is a pair (Φ, ψ) such that
  - $\Phi \subseteq A$,
  - $\Phi \cup K \nvdash \bot$,
  - $\Phi \cup K \vdash \psi$ and if Ψ is a set such that $\Psi \subset \Phi$ then $\Psi \cup K \nvdash \psi$.

- An argument (Φ, ψ) undercuts an argument (Φ', ψ') if for some $\phi' \in \Phi'$ we have $\phi' \equiv \neg\psi$.

- The argumentation framework induced by Δ is the argumentation framework $F(\Delta) = (Ar, att)$ where Ar consists of all arguments induced by Δ and $(a, b) \in att$ whenever a undercuts b.

Given a knowledge base $\Delta = (K, A)$, the stable extensions of $F(\Delta)$ correspond with the maximal subsets of A that are consistent with K. This fact, first established by Cayrol [Cayrol, 1995], forms the basis for various forms of argumentation with maximally consistent subsets [Arieli et al., 2019].

Given a knowledge base $\Delta = (K, A)$ we will say that Δ entails ϕ (written $\Delta \mid\sim \phi$) if every stable extension of $F(\Delta)$ contains an argument with conclusion ϕ. We furthermore define the inference relation $\mid\sim_\Delta \subseteq \mathcal{L} \times \mathcal{L}$ by $\phi \mid\sim_\Delta \psi$ if and only if $(K \cup \{\phi\}, A) \mid\sim \psi$.

A *Boolean structural equation model* is a structural equation model $M = \langle U, V, F \rangle$ (see Definition 2.12) where U and V consist of Boolean-valued variables (atoms). If $M = \langle U, V, F \rangle$ is a Boolean structural equation model then F can be represented as a set of *boolean structural equations* of the form $(V_i \leftrightarrow \phi)$ where ϕ contains members of $PA_i \cup U_i$. In what follows we denote by K_M the set of boolean structural equations that represent F.

Example 20. . Let $M = \langle U, V, F \rangle$ be a Boolean structural equation model where

$$U = \{Corona, Influenza, AtRisk\},$$
$$V = \{Covid, Flu, ShortOfBreath, Fever, Chills\},$$

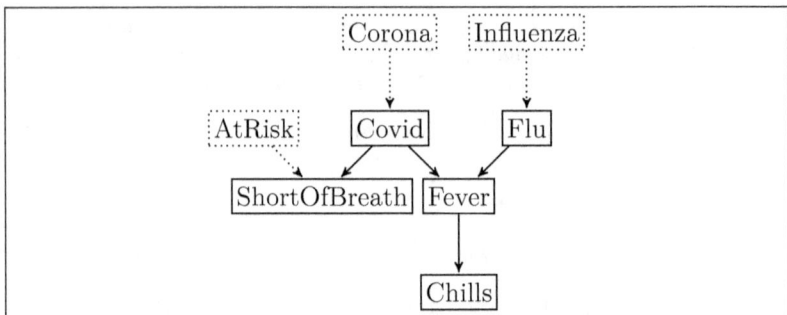

Figure 4: Causal graph for Example 20.

and where F is represented by the following set of Boolean structural equations.

$$(Covid \leftrightarrow Corona)$$
$$(Flu \leftrightarrow Influenza)$$
$$(Fever \leftrightarrow Covid \vee Flu)$$
$$(Chills \leftrightarrow Fever)$$
$$(ShortOfBreath \leftrightarrow Covid \wedge AtRisk)$$

This model represents a simple patient diagnosis scenario. In words: Corona virus causes Covid, Influenza virus causes Flu, Covid and Flu cause Fever, Fever causes Chills, and Covid causes ShortOfBreath, but only if the patient is AtRisk. Note that Corona, Influenza and AtRisk are background atoms and thus assumed unobservable. Figure 4 depicts the causal graph for this model, which includes the background atoms drawn with dotted lines.

A *causal knowledge base* is a knowledge base $\Delta = (K_M, A)$ where M is a Boolean structural equation model and where A contains only background atoms or negations of background atoms. Here we depart from the probabilistic approach to causal modelling, where beliefs about background variables are represented probabilistically [Pearl, 2009]. The representation chosen here permits three possible attitudes towards a background atom u, since we can assume just u, just $\neg u$, or both. The latter represents a state of uncertainty where u may be true as well as false. If $\Delta = (K_M, A)$ is a causal knowledge base then the inference relation $\mathrel{\vert\!\sim}_\Delta$ represents a relation between observations and their predicted causes and effects according to Δ. We can determine whether $\phi \mathrel{\vert\!\sim}_\Delta \psi$ holds by constructing the argumentation framework $F((K_M \cup \{\phi\}, A))$ and computing its stable extensions. Together with the argumentation framework, these extensions explain the conclusions we can draw when observing ϕ.

Example 21. *Consider the causal knowledge base $\Delta = (K_M, A)$ where M is the Boolean structural equation model defined in Example 20 and where $A = \{AtRisk, \neg AtRisk, \neg Corona, \neg Influenza\}$. In words: we assume that the patient may or may not be at risk, and that there is no corona or influenza infection unless there is evidence to the contrary. Now consider the question of whether observing fever entails shortness of breath:*

$$\text{Fever} \mathrel{\vert\!\sim}_\Delta \text{ShortOfBreath}. \tag{4}$$

Flu and covid cause fever. Thus, fever is evidence for flu or covid. Of these two possible causes, covid may cause shortness of breath, but only if the patient is at risk, which may or may not be true. Hence, fever may or may not be evidence for shortness of breath. This reasoning is depicted by the AF $F(K_M \cup \{Fever\}, A)$ shown in Figure 5 (we only depict arguments relevant to the conclusion ShortOfBreath). This AF has four stable extensions $\{a_2, a_4\}$, $\{a_3, a_4\}$, $\{a_2, a_5\}$, and $\{a_3, a_5, a_1\}$. The argument a_1 with conclusion ShortOfBreath is included in some but not all of these extensions. We thus have that (4) is false, but note that

$$\text{Fever} \mathrel{\vert\!\sim}_\Delta \neg\text{ShortOfBreath}$$

is also false. Thus, given fever, shortness of breath is possible but not necessary.

Apart from reasoning about what Pearl refers to as "rung 1" (observation), the current approach can also accommodate reasoning with in rung 2 (intervention) and rung 3 (counterfactuals) [Pearl and Mackenzie, 2018]. While interventions are interpreted as changes ("surgeries") of the causal model (cf. Section 2.6.1), counterfactuals can be answered by constructing a "twin-model" as

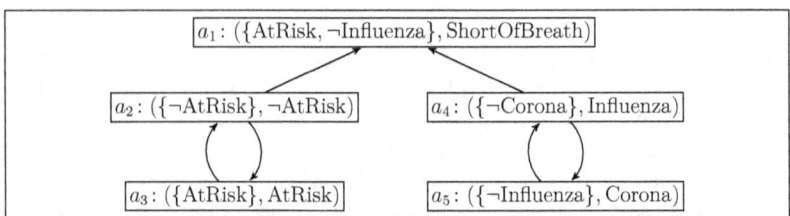

Figure 5: The argumentation framework $F(K \cup \{\text{Fever}\}, A)$.

described by [Pearl, 2009]. For further details we refer the reader to [Bengel et al., 2022].

The form structured causal argumentation that we considered in this section can also be represented within the ASPIC+ formalism for structured argumentation. This is because ASPIC+ subsumes Brewka's preferred subtheories [Brewka, 1989], which in turn subsumes reasoning with maximally consistent subsets [Modgil and Prakken, 2013]. The same applies to the ABA (Assumption-Based Argumentation) formalism [Bondarenko et al., 1997], which was shown by [Heyninck and Arieli, 2020] to be capable of capturing reasoning with maximally consistent subsets.

5 Formal Connections between Causal and Argumentation Formalisms

In this section we provide an overview of formal connections that have been established between causal reasoning and theories of formal argumentation.

5.1 Causal reasoning as propositional argumentation

The first kind of connection that can be established between causal reasoning and argumentation is a 'logical' one, namely a connection between basic causal inference as described earlier (see section 2.6.3) and a formalism of *collective argumentation* that has been described in volume 1 of this Handbook (see [Bochman, 2018a]). The latter formalism is a generalization of Dung's argumentation frameworks in which the attack relation is defined directly among sets of arguments instead of single arguments: $a \hookrightarrow b$ says that a set a of arguments attacks a set b of arguments.[6] Then a (collective) *attack relation* is a relation $\hookrightarrow$ on finite sets of arguments satisfying the following postulate:

Monotonicity If $a \hookrightarrow b$, then $a, a_1 \hookrightarrow b, b_1$.

Collective argumentation can be given a natural four-valued semantics that provides the concept of an attack with a *meaning*. This formal meaning stems from the following understanding of an attack $a \hookrightarrow b$:

> *If all arguments in a are accepted, then at least one of the arguments in b should be rejected.*

The argumentation theory does not impose, however, the classical constraints on acceptance and rejection of arguments, so an argument can be both accepted and rejected, or neither accepted, nor rejected. This is nothing other than *Belnap's interpretation* of four-valued logic that we have already used earlier in describing basic causal inference.

Given this understanding, the above informal description of an attack can be formulated as the following semantic definition.

Definition 5.1. *An attack $a \hookrightarrow b$ will be said to* hold *in a four-valued interpretation ν of arguments, if either some $A \in a$ is not accepted in ν, or some $B \in b$ is rejected in ν.*

An interpretation ν will be called a model *of an argument theory Δ if every attack from Δ holds in ν.*

For a set I of four-valued interpretations, we will denote by $\hookrightarrow_I$ the set of all attacks that hold in each interpretation from I. Then the following result is actually a basic representation theorem showing that the four-valued semantics of acceptance and rejection is also adequate for collective argumentation.

Theorem 5.2. $\hookrightarrow$ *is an attack relation iff it coincides with $\hookrightarrow_I$, for some set of four-valued interpretations I.*

---

[6] In Dung's frameworks this relation is defined in terms of attacks among particular arguments.

In this setting, it is only natural to extend the underlying language of arguments to the classical propositional language by adding the four-valued conjunction $\wedge$ and negation $\neg$ we already used earlier in section 2.6.3. Thus, the following postulates provide a simple syntactic characterization of these connectives for attack relations:

$$a, A \wedge B \hookrightarrow b \text{ iff } a, A, B \hookrightarrow b$$
$$a \hookrightarrow A \wedge B, b \text{ iff } a \hookrightarrow A, B, b \qquad (A_\wedge)$$
$$A, \neg A \hookrightarrow \qquad \hookrightarrow A, \neg A$$
$$\text{If } a, A \hookrightarrow b \text{ and } a, \neg A \hookrightarrow b \text{ then } a \hookrightarrow b \qquad (A_\neg)$$
$$\text{If } a \hookrightarrow b, A \text{ and } a \hookrightarrow b, \neg A \text{ then } a \hookrightarrow b$$

Then a *basic propositional attack relation* can be described as a collective attack relation satisfying the rules $(A_\wedge)$ and $(A_\neg)$.

An immediate benefit of introducing the above connectives into the language of argumentation is that any finite set of arguments a becomes reducible to a single argument $\bigwedge a$:

$$a \hookrightarrow b \text{ iff } \bigwedge a \hookrightarrow \bigwedge b.$$

As a result, the collective attack relation in this language is reducible to an attack relation between individual arguments. Moreover, given the negation connective, we can represent the resulting argumentation theory as a certain binary attack relation on classical formulas. This more general description of basic propositional attack relation is given in the following theorem.

Theorem 5.3. *Basic attack relations are precisely collective attack relations in the classical language that satisfy the following postulates:*

(Left Strengthening) *If $A \vDash B$ and $B \hookrightarrow C$, then $A \hookrightarrow C$;*

(Right Strengthening) *If $A \hookrightarrow B$ and $C \vDash B$, then $A \hookrightarrow C$;*

(Truth) $\mathbf{t} \hookrightarrow \mathbf{f}$;

(Falsity) $\mathbf{f} \hookrightarrow \mathbf{t}$;

(Left Or) *If $A \hookrightarrow C$ and $B \hookrightarrow C$, then $A \vee B \hookrightarrow C$;*

(**Right Or**)   If $A \hookrightarrow B$ and $A \hookrightarrow C$, then $A \hookrightarrow B \vee C$.

At this point, the reader should notice an obvious similarity between the above notion of basic propositional argumentation and the notion of a basic causal inference described earlier in section 2.6.3. Given their respective semantic descriptions, the precise formal correspondence between these two formalisms amounts to the following mutual translations:

$$A \Rightarrow B \equiv \neg B \hookrightarrow A \qquad \text{(CA)}$$
$$A \hookrightarrow B \equiv B \Rightarrow \neg A. \qquad \text{(AC)}$$

Lemma 5.4. *If $\hookrightarrow$ is a basic attack relation, then (PA) determines a basic production inference relation, and vice versa, if $\Rightarrow$ is a basic production inference relation, then (AP) determines a basic attack relation.*

Remark 5.5. *A seemingly more natural correspondence between propositional argumentation and causal inference can be obtained using the following definitions:*

$$A \Rightarrow B \equiv A \hookrightarrow \neg B \qquad A \hookrightarrow B \equiv A \Rightarrow \neg B.$$

By these definitions, A causes B if it attacks $\neg B$, and vice versa: A attacks B if it causes $\neg B$. Unfortunately, this correspondence, though plausible by itself, does not take into account the intended understanding of arguments as default assumptions. As a result, it cannot be extended directly neither to classical causal inference, nor to the correspondence between the associated nonmonotonic semantics, described below.

The above correspondence can also be extended to the correspondence between fully classical causal inference and a particular kind of basic propositional argumentation described in the next definition.

Definition 5.6. *A propositional attack relation will be called* causal, *if it is basic and satisfies the following postulate:*

(**Self-Defeat**)   If $A \hookrightarrow A$, then $\mathbf{t} \hookrightarrow A$.

The rule Self-Defeat of causal argumentation gives a formal representation for an often expressed desideratum that self-conflicting arguments should not participate in defeating other arguments (see, e.g., [Bondarenko et al., 1997]). This aim is achieved in our setting by requiring that such arguments are attacked even by tautologies and hence by any argument whatsoever.

At least in the latter fully classical case, the above logical correspondence between the causal calculus and propositional argumentation can even be extended to the correspondence between the associated non-monotonic semantics. More precisely, there is an exact correspondence between the rational classical semantics of a causal theory and stable semantics of the corresponding argument theory (see [Bochman, 2005]). It is still unclear, however, whether and how the range of this correspondence can be extended beyond the most regular case of classical causal reasoning and stable argumentation.

5.2 Structural Equation Models and Abstract Dialectical Frameworks

We will establish in this section a direct correspondence between Pearl's structural equation models and abstract dialectical frameworks of argumentation from [Brewka and Woltran, 2010, Brewka et al., 2013].

Abstract dialectical frameworks (ADFs) have been developed as an abstract argumentation formalism purported to capture more general forms of argument interaction than just attacks among arguments, which form the basis of the original Dung's argumentation frameworks. To achieve this, each argument in an ADF has been associated with an *acceptance condition*, which is some propositional function determined by arguments that are linked to it. Using such acceptance conditions, ADFs allow us to express that arguments may jointly support another argument, or that two arguments may jointly attack a third one, and so on. Dung's argumentation frameworks are recovered in this setting by the acceptance condition saying that an argument is accepted if none of its parents are.

We will view ADFs as a specific knowledge representation formalism (cf. [Strass, 2013]) and show its close conceptual connections with the formalism of causal reasoning. This will also allow us to single out some basic principles behind the construction of ADFs and their semantics.

5.2.1 Abstract Dialectical Frameworks

An ADF is a directed graph whose nodes represent statements or positions which can be accepted or not. The links represent dependencies: the status of a node s only depends on the status of its parents denoted $par(s)$, that is, the nodes with a direct link to s. In addition, each node s has an associated acceptance condition C_s specifying the exact conditions under which s is accepted. C_s is a function assigning to each subset of $par(s)$ one of the truth values $\mathbf{t}, \mathbf{f}$. Intuitively, if for some $R \subseteq par(s)$ we have $C_s(R) = \mathbf{t}$, then s will be accepted provided the nodes in R are accepted and those in $par(s) \setminus R$ are not accepted.

Definition 5.7. *An abstract dialectical framework (ADF) is a tuple $D = (S, L, C)$, where*

- *S is a set of statements (positions, nodes),*
- *$L \subseteq S \times S$ is a set of links; and*
- *$C = \{C_s\}_{s \in S}$ is a set of total functions $C_s : 2^{par(s)} \to \{\mathbf{t}, \mathbf{f}\}$, one for each statement s. C_s is called acceptance condition of s.*

A more "logical" representation of ADFs can be obtained simply by assigning each node s a *classical* propositional formula corresponding to its acceptance condition C_s [Ellmauthaler, 2012]. In this case we can tacitly assume that the acceptance formulas implicitly specify the parents a node depends on. It is then not necessary to give the links L, so an ADF D amounts to a tuple (S, C) where S is a set of statements and C is a set of propositional formulas, one for each statement from S. The notation $s[C_s]$ has been used by the authors to denote the fact that C_s is the acceptance condition of s.

A two-valued interpretation v is a (two-valued) *model* of an ADF (S, C) whenever for all statements $s \in S$ we have $v(s) = v(C_s)$, that is, v maps exactly those statements to true whose acceptance conditions are satisfied under v. This notion of a model provides a natural semantics for ADFs. In addition to this semantics, however, the authors have defined appropriate generalizations for all the major semantics of Dung's argumentation frameworks. In [Brewka et al., 2013], all these semantics were defined by generalizing the two-valued interpretations to three-valued ones. All of them were formulated using the basic operator Γ_D over three-valued interpretations that was introduced, in effect, already in [Brewka and Woltran, 2010]. In the formulation of [Brewka et al., 2013], for an ADF D and a three-valued interpretation v, the interpretation $\Gamma_D(v)$ is given by the mapping

$$s \mapsto \prod \{w(C_s) \mid w \in [v]_2\},$$

where $\prod$ is the product operator on interpretations while $[v]_2$ is the set of all two-valued interpretations that extend v.

For each statement s, the operator Γ_D returns the consensus truth value for its acceptance formula C_s, where the consensus takes into account all possible two-valued interpretations w that extend the input valuation v. If v is two valued, we get $\Gamma_D(v)(s) = v(C_s)$, so v is a two-valued model for D iff $\Gamma_D(v) = v$. In other words, two-valued models of D are precisely those classical interpretations that are fixed points of Γ_D.

The *grounded model* of an ADF D can now be defined as the least fixpoint of Γ_D. This fixpoint is in general three valued, and it always exists since the operator Γ_D is monotone in the information ordering $\leq_i$, as shown in [Brewka and Woltran, 2010]. This grounded semantics has been viewed by the authors as the greatest possible consensus between all acceptable ways of interpreting the ADF at hand.

The operator Γ_D also provides a proper basis for defining admissible, complete, and preferred semantics for arbitrary ADFs.

Definition 5.8. *A three-valued interpretation v for an ADF D is*

- *admissible iff $v \leq_i \Gamma_D(v)$;*

- *complete iff $\Gamma_D(v) = v$; and*

- *preferred iff it is $\leq_i$-maximal admissible.*

The above definitions provide proper generalizations of the corresponding semantics for Dung's argumentation frameworks and, moreover, preserve much of the properties and relations of the latter. Thus, the grounded semantics is always a complete model, and each complete model is admissible. In addition, as it is the case for Dung's argumentation frameworks, all preferred models are complete, the grounded model is the $\leq_i$-least complete model, and the set of all complete models forms a complete meet-semilattice with respect to the information ordering $\leq_i$.

In [Brewka and Woltran, 2010], the standard Dung semantics of stable extensions was generalized only to a restricted type of ADFs called bipolar, but [Brewka et al., 2013] has suggested a new definition that avoids unintended features of the original definition and covers arbitrary ADFs, not just bipolar ones (see also [Strass, 2013]). This new definition is based on the notion of a *reduct* of an ADF, similar to the Gelfond-Lifschitz transformation of logic programs. We will discuss the representation of the stable semantics in ADFs later in this section.

5.2.2 The Causal Representation of ADFs

Now we are going to provide a uniform and modular translation of ADFs into the causal calculus. An essential precondition of this causal representation, however, will consist in transforming the underlying semantic interpretations of ADFs in terms of three-valued models, used in [Brewka et al., 2013], into ordinary classical logical descriptions. This latter transformation will also allow us to clarify to what extent the various semantics suggested for ADFs admit a classical logical reading. In fact, the very possibility of such a classical reformulation stems from the crucial fact that the basic operator Γ of an ADF, described earlier, is defined, ultimately, in terms of ordinary classical interpretations extending a given three-valued one. Nevertheless, our reformulation will also reveal a significant discrepancy between these semantics and their immediate causal counterparts.

To begin with, any three-valued interpretation v on the set of statements S can be encoded using an associated set of literals $[v] = S_0 \cup \neg S_1$ such that $S_0 = \{p \in S \mid v(p) = \mathbf{t}\}$ and $S_1 = \{p \in S \mid v(p) = \mathbf{f}\}$. Moreover, this set of literals generates a unique deductively closed theory $\mathrm{Th}([v])$ that corresponds in this sense to the source three-valued interpretation v. Conversely, any *literal* deductively closed theory u (namely a deductive closure of a set of literals) will correspond to a unique three-valued interpretation v such that $u = \mathrm{Th}([v])$. These simple facts establish a precise bidirectional correspondence between three-valued interpretations and classical literal theories. Moreover, we will see in what follows that the main operator Γ of ADFs will correspond under this reformulation to a "literal" restriction of the causal operator $\mathcal{C}$ of basic production inference.

As our starting point, we should note a striking similarity between the official definition of an ADF and structural equation models from [Pearl, 2009] (see section 2.6). More precisely, Boolean structural models are practically identical to ADFs (especially in a logical formulation), with equations $p = F$ playing essentially the same role as the acceptance conditions $p[F]$. The differences are that only endogenous atoms are determined by their associated conditions in structural equation models, but on the other hand, there are no restrictions on appearances of atoms on both sides in ADF's acceptance conditions. Moreover, plain (two-valued) models of ADFs correspond precisely to causal worlds of a structural equation model.

Now, a modular representation of Boolean structural equation models as causal theories of the causal calculus, described earlier in this chapter, can now be seamlessly transformed into the following causal representation of ADFs:

Definition 5.9 (Causal representation of an ADF). *For any ADF D, Δ_D is the causal theory consisting of the rules*

$$F \Rightarrow p \quad \text{and} \quad \neg F \Rightarrow \neg p$$

for all acceptance conditions $p[F]$ in D.

The above representation is fully modular, and it will be taken as a uniform basis for the correspondences described below.

To begin with, we immediately establish the following theorem.

Theorem 5.10. *The two-valued semantics of an ADF D corresponds precisely to the classical causal semantics of Δ_D.*

As a consequence, the full system of classical causal inference provides a precise logical basis for this nonmonotonic semantics.

5.2.3 General Correspondences

Now we are going to show that the above causal representation also survives the transition to three-valued models of ADFs. To this end, however, we will have to retreat from the system of classical causal inference to a weaker system of basic inference.

A broader correspondence between various semantics of ADFs and rational supraclassical semantics of the causal calculus arises from the fact that the operator Γ of an ADF naturally corresponds to a particular causal operator of the associated causal theory.

Let $\mathcal{L}$ denote the set of classical literals of the underlying language. We will denote by $\mathcal{C}^{\mathcal{L}}$ the restriction of a causal operator $\mathcal{C}$ to literals, that is, $\mathcal{C}^{\mathcal{L}}(u) = \mathcal{C}(u) \cap \mathcal{L}$. As we are going to show, the operator Γ of ADFs corresponds precisely to this "literal restriction" of the causal operator associated with a basic production inference. In what follows, $[v]$ will denote the set of classical literals corresponding to a three-valued interpretation v.

Lemma 5.11. *For any three-valued interpretation v,*

$$[\Gamma_D(v)] = \mathcal{C}_D^{\mathcal{L}}([v]),$$

where $\mathcal{C}_D$ is a basic causal operator corresponding to Δ_D.

The above equation has immediate consequences for the broad correspondence between the semantics of ADFs that are defined in terms of the operator Γ_D and natural sets of propositions definable wrt associated causal theory. Thus, we have the following.

Theorem 5.12. *Complete models of an ADF D correspond precisely to the fixed points of $\mathcal{C}_D^{\mathcal{L}}$:*

$$v = \Gamma_D(v) \quad \textit{iff} \quad [v] = \mathcal{C}_D^{\mathcal{L}}([v]).$$

As a result, we immediately conclude that preferred models of an ADF correspond to maximal fixpoints of $\mathcal{C}_D^{\mathcal{L}}$ (with respect to set inclusion) while the grounded model corresponds to the least fixpoint of $\mathcal{C}_D^{\mathcal{L}}$.

It turns out, however, that when viewed in a classical logical setting, the restriction of the causal operator to literals inadvertently leads to an information loss. More precisely, though disjunctive formulas can appear in acceptance conditions used by Γ in an ADF, the operator itself records, in effect, only literals that are produced, and thereby disregards all other information that can be obtained from its output. The following example illustrates this.

Example 22. *Let D be the following ADF:*

$$q[p] \qquad r[\neg p] \qquad s[q \vee r].$$

The grounded model of this ADF is empty (all atoms are unknown). However, the associated causal theory Δ_D comprises the following rules:

$$p \Rightarrow q \qquad \neg p \Rightarrow r \qquad q \vee r \Rightarrow s$$
$$\neg p \Rightarrow \neg q \qquad p \Rightarrow \neg r \qquad \neg q \wedge \neg r \Rightarrow \neg s.$$

The least classical causal model of $\mathcal{C}_D$ is precisely the set of propositions that are provable from the above theory using the postulates of classical causal inference. Now, the first two rules imply $\mathbf{t} \Rightarrow q \vee r$ (by Or), and hence $\mathbf{t} \Rightarrow s$ by Cut. Similarly, the fourth and fifth rule imply $\mathbf{t} \Rightarrow \neg q \vee \neg r$. As result, the least causal model of $\mathcal{C}_D$ is much more informative, namely $\mathrm{Th}(\{q \leftrightarrow \neg r, s\})$.

It can also be seen from the above example that the restriction of causal models to literals does not necessary produce fixed points of the corresponding literal operator $\mathcal{C}^{\mathcal{L}}$. Still, it can be shown that for any fixed point of the latter (that is, for any complete model an ADF) there exists a least causal model that contains it. The latter model may contain, however, more information than its literal source.

The above considerations and results suggest a natural generalization of an ADF to acceptance conditions of the form $A[B]$, where both A and B are classical formulas. This would supply the abstract argumentation frameworks with further representation capabilities and thereby even contribute to its original purpose of providing a powerful and widely applicable abstraction tool for argumentation (see [Bochman, 2016]).

5.3 Independence and D-Separation in Abstract Argumentation

We now establish a formal connection between causal networks and abstract argumentation frameworks. The basis for this connection is the notion of *D-separation*. A fundamental idea that underlies causal networks is the interpretation of DAGs as carriers of *independence assumptions*. The interpretation relies on the D-separation criterion, which determines whether two sets of variables are independent given a third set, by inspecting the structure of the DAG [Pearl, 2009]. We can think of the attacks in an argumentation framework as a specific kind of causal relationship. This idea suggests that we can similarly interpret argumentation frameworks as carriers of independence assumptions. These independence assumptions then bear on the evaluation of the argumentation framework, i.e., the acceptance status of the arguments. This is the idea that was explored in [Rienstra et al., 2020], the results of which we summarise here.

5.3.1 Labelling-based Argumentation Semantics

We will first recall the labelling-based semantics of argumentation frameworks [Caminada and Gabbay, 2009]. A three-valued labelling-based semantics for argumentation frameworks associates every argumentation framework $F = (Ar, att)$ with a set of *labelings* of Ar. A labelling of an argumentation framework $F = (Ar, att)$ (or of a set Ar of arguments) is a function $L : Ar \to \{\texttt{I}, \texttt{0}, \texttt{U}\}$ that maps every argument to a label $\texttt{I}$ for *in* (or *accepted*), $\texttt{0}$ for *out* (or *rejected*) and $\texttt{U}$ for *undecided*. We use $\mathcal{L}(F)$ to denote the set of labellings of F. A *complete* labelling is a labelling where an argument is accepted whenever its attackers are rejected, and rejected whenever an attacker is accepted. That is, L is a complete labelling of $F = (Ar, att)$ if, for all $\mathbf{x} \in Ar$, we have (1) $L(\mathbf{x}) = \texttt{I}$ if and only if for all $\mathbf{y} \in Ar$ such that $\mathbf{y}\,att\,\mathbf{x}$ we have $L(\mathbf{y}) = \texttt{0}$; and (2) $L(\mathbf{x}) = \texttt{0}$ if and only if there is a $\mathbf{y} \in Ar$ such that $\mathbf{y}\,att\,\mathbf{x}$ and $L(\mathbf{y}) = \texttt{I}$. Various additional criteria may be considered for a labelling to represent a reasonable position. A *semantics* σ maps every AF F to a set of labellings of F denoted $\mathcal{L}_\sigma(F)$. The $\mathcal{CO}$ (*complete*), $\mathcal{PR}$ (*preferred*), $\mathcal{GR}$ (*grounded*), $\mathcal{ST}$ (*semi-stable*) and $\mathcal{ST}$ (*stable*) semantics are defined as follows.

Definition 5.13.

$$\mathcal{L}_{\mathcal{CO}}(F) = \{L \in \mathcal{L}(F) \mid L \text{ is a complete labelling of } F\}$$
$$\mathcal{L}_{\mathcal{PR}}(F) = \{L \in \mathcal{L}_{\mathcal{CO}}(F) \mid \nexists L' \in \mathcal{L}_{\mathcal{CO}}(F), L^{-1}(\mathtt{I}) \subset L'^{-1}(\mathtt{I})\}$$
$$\mathcal{L}_{\mathcal{GR}}(F) = \{L \in \mathcal{L}_{\mathcal{CO}}(F) \mid \nexists L' \in \mathcal{L}_{\mathcal{CO}}(F), L^{-1}(\mathtt{I}) \supset L'^{-1}(\mathtt{I})\}$$
$$\mathcal{L}_{\mathcal{SST}}(F) = \{L \in \mathcal{L}_{\mathcal{CO}}(F) \mid \nexists L' \in \mathcal{L}_{\mathcal{CO}}(F), L^{-1}(\mathtt{U}) \supset L'^{-1}(\mathtt{U})\}$$
$$\mathcal{L}_{\mathcal{ST}}(F) = \{L \in \mathcal{L}_{\mathcal{CO}}(F) \mid L^{-1}(\mathtt{U}) = \emptyset\}$$

5.3.2 Independence for Labelling-based Argumentation Semantics

Intuitively, given a set $\mathcal{L}$ of labellings of $F = (Ar, att)$, and disjoint subsets A, B, C of Ar, A and B are independent given C if, once we know the values of C then knowing the values of A provides no information about B and vice versa. This is the idea behind the notion of conditional independence defined in [Rienstra et al., 2020]. To formalise it we first need some auxiliary definitions. Let Ar be a set of arguments. Given a set $A \subseteq Ar$ we denote by $L\!\downarrow_A$ the restriction of L to A. We say that a labelling L_A of some subset A of Ar is *consistent* with a set $\mathcal{L}$ of labellings of Ar if for some $L \in \mathcal{L}$ we have $L\!\downarrow_A = L_A$. Given two labellings L_A and L_B of disjoint subsets A and B of Ar we denote by $L_A \cup L_B$ the union of the two valuations. Conditional independence with respect to a set of labellings is defined as follows.

Definition 5.14. *Let $\mathcal{L} \subseteq \mathcal{L}(F)$ be a set of labellings of $F = (Ar, att)$. Given disjoint subsets A, B, C of Ar, we say that A and B are independent given C in $\mathcal{L}$ if, for all $V_A \in \mathcal{L}(A), V_B \in \mathcal{L}(B), V_C \in \mathcal{L}(C)$, consistency of $V_C \cup V_A$ and $V_C \cup V_B$ in $\mathcal{L}$ implies consistency of $V_C \cup V_A \cup V_B$ in $\mathcal{L}$.*

Example 23. *Consider the AF F and the set $\mathcal{L}_{pr}(F)$ of preferred labellings shown in Figure 6. While the arguments $\mathbf{a}$ and $\mathbf{e}$ are independent given $\mathbf{b}$, they are not unconditionally independent. To see why, note that $(\mathbf{a}: \mathtt{O})$ and $(\mathbf{e}: \mathtt{U})$ are consistent but $(\mathbf{a}: \mathtt{O}, \mathbf{e}: \mathtt{U})$ is not. The arguments $\mathbf{a}$ and $\mathbf{h}$ are independent, but they are not independent given $\mathbf{e}$. This is because $(\mathbf{e}: \mathtt{O}, \mathbf{a}: \mathtt{I})$ and $(\mathbf{e}: \mathtt{O}, \mathbf{h}: \mathtt{I})$ are consistent but $(\mathbf{e}: \mathtt{O}, \mathbf{a}: \mathtt{I}, \mathbf{h}: \mathtt{I})$ is*

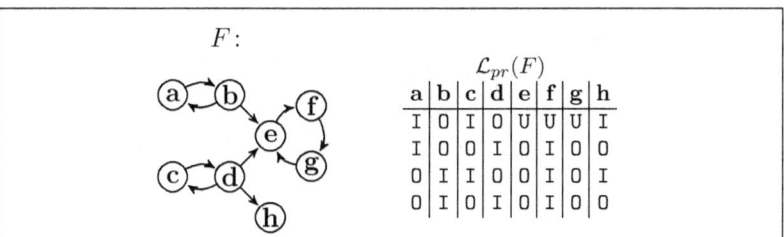

Figure 6: An argumentation framework and its preferred labellings

not. If we condition on **d** in addition to **e**, then **a** and **h** are independent again.

5.3.3 A Markov Property for Argumentation Semantics

The *Markov* property for a probability distribution with respect to a DAG states that every variable is independent of its non-descendants given its parents. The d-separation criterion, applied to a DAG G, is sound with respect to any probability distribution that satisfies the Markov property with respect to G [Pearl, 2009]. The *SCC Markov* property for an argumentation semantics serves a similar purpose in the current context. A semantics σ is SCC Markovian if for every argumentation framework F the set $\mathcal{L}_\sigma(F)$ satisfies the property that every strongly connected component (SCC) S of F is independent of its non-descendants given its parents in $\mathcal{L}_\sigma(F)$.

Definition 5.15. *The set of* SCCs *(strongly connected components) of an argumentation framework* $F = (Ar, att)$, *denoted* $SCCS_F$, *contains the equivalence classes induced by the* path equivalence relation $\sim_F$ *over* Ar *defined by* $\mathbf{x} \sim_F \mathbf{y}$ *iff* $\mathbf{x} = \mathbf{y}$ *or there is a directed path from* $\mathbf{x}$ *to* $\mathbf{y}$ *and* $\mathbf{y}$ *to* $\mathbf{x}$.

Definition 5.16. *Let* F *be an argumentation framework and let* $S \in SCCS_F$. *A* parent *of* S *is a parent of an element of* S *that is itself not a member of* S. *A descendant of* S *is an element of* S *or a variable* $\mathbf{x}$ *such that directed path from an element of* S *to* $\mathbf{x}$ *exists. A nondescendant of* S *is any variable that is not a descendant or parent of* S. *We denote the parents and nondescendants of* S *by* $Pa_F(S)$ *and* $ND_F(S)$, *respectively.*

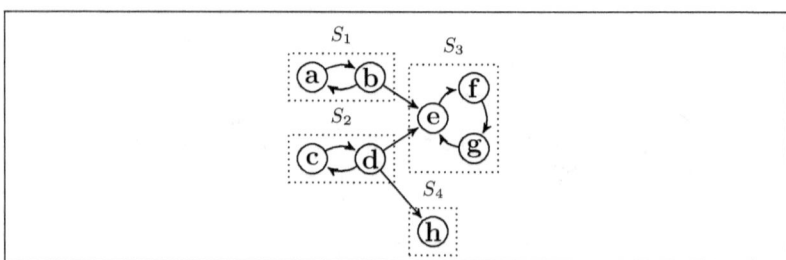

Figure 7: The argumentation framework from Figure 6 with SCCs highlighted

Example 24. *Let F be the AF shown in Figure 7, which is the same AF as shown in Figure 6 with SCCs enclosed in dotted rectangles. We have:*

$$Pa_F(S_1) = \emptyset \qquad ND_F(S_1) = \{c, d, h\}$$
$$Pa_F(S_2) = \emptyset \qquad ND_F(S_2) = \{a, b\}$$
$$Pa_F(S_3) = \{b, d\} \qquad ND_F(S_3) = \{a, c, h\}$$
$$Pa_F(S_4) = \{d\} \qquad ND_F(S_4) = \{a, b, c, e, f, g\}$$

Definition 5.17. *A semantics σ is* SCC-Markovian *iff for every AF F and every $S \in \text{SCCS}_F$, S is independent of $ND_F(S)$ given $Pa_F(S)$ in $\mathcal{L}_\sigma(F)$.*

It is shown in [Rienstra et al., 2020] that the SCC Markov property is implied by the combination of other properties for semantics for abstract argumentation that have been proposed in the literature, namely *Universality* and *SCC decomposability* [Baroni et al., 2014]. The admissible, complete, preferred and grounded semantics satisfy these properties and are therefore SCC Markovian, while the stable semantics violates Universality and the semi-stable semantics violates SCC decomposability. Indeed, as shown in [Rienstra et al., 2020], the stable and semi-stable semantics are not SCC Markovian.

5.3.4 D-separation for Argumentation Frameworks

The next step is to define a method to identify, given an AF F and semantics σ, the independencies that hold in $\mathcal{L}_\sigma(F)$ by inspecting the structure of F. In causal networks, this is done using the D-separation criterion. The criterion is sound in the sense that, if A and B are d-separated by C in the DAG G, then A and B are probabilistically independent of C in every probability distribution represented by a causal or Bayesian network with structure G [Pearl, 2009]. D-separation is defined as follows.

Definition 5.18. *Let $G = (\mathbf{X}, \rightarrow)$ be a DAG. A trail in G is a loop-free, undirected (i.e., ignoring edge directions) path between two variables. If A, B, C are three disjoint sets of variables in G then A and B are said to be d-separated by C if every trail between every variable in A and in B is blocked by C. A trail is blocked by C if either:*

- *It contains a triple $\mathbf{x} \rightarrow \mathbf{z} \rightarrow \mathbf{y}$ or $\mathbf{x} \leftarrow \mathbf{z} \rightarrow \mathbf{y}$ such that $\mathbf{z} \in C$.*

- *It contains a triple $\mathbf{x} \rightarrow \mathbf{z} \leftarrow \mathbf{y}$ and neither $\mathbf{z}$ nor a descendant of $\mathbf{z}$ is in C (a descendant of $\mathbf{z}$ is any variable $\mathbf{z}'$ such that a directed path from $\mathbf{z}$ to $\mathbf{z}'$ exists in G).*

The d-separation criterion cannot be applied directly to argumentation frameworks since these are not acyclic. Instead, [Rienstra et al., 2020] define the *d-graph* transformation that transforms an argument framework F into a DAG denoted F^* such that we can apply the d-separation on F^* in order to determine the independencies that hold $\mathcal{L}_\sigma(F)$. Intuitively, the d-graph transformation replaces the 'causal feedback' loops formed by the cycles of an argumentation framework with dependencies represented using extra *latent common cause* variables. If a graph contains, for example, a cycle $\mathbf{a} \leftrightarrow \mathbf{b}$, then the resulting d-graph contains a fork structure $\mathbf{a} \leftarrow \mathbf{s} \rightarrow \mathbf{b}$, where $\mathbf{s}$ is an extra variable representing a common cause for $\mathbf{a}$ and $\mathbf{b}$. This common cause $\mathbf{s}$ is hypothetical, used purely to account for the dependency between $\mathbf{a}$ and $\mathbf{b}$, and therefore treated as unobservable. As long as $\mathbf{s}$ is not observed, the two structures (cycle and fork) represent the same independence information, because the fork structure ensures that $\mathbf{a}$ and $\mathbf{b}$ are d-separated only if $\mathbf{s}$ is observed.

The transformation of an argumentation framework F to the d-graph F^* consists of three steps, applied separately to each SCC S_i of F:

1. Remove all edges between elements of S_i.

2. Add an extra latent common cause variable $\mathbf{s}_i$ and an edge from $\mathbf{s}_i$ to every element of S_i.

3. For every parent $\mathbf{x}$ of S_i, replace the edge from $\mathbf{x}$ to S_i with an edge from $\mathbf{x}$ to $\mathbf{s}_i$.

The result is a DAG since step 1 removes all cycles and steps 2 and 3 do not introduce new cycles. The following definition describes the d-graph transformation more concisely.

Definition 5.19. *Let $F = (Ar, att)$ be a graph with SCCs $SCCS_F = \{S_1, \ldots, S_n\}$. The d-graph of F is the DAG $F^* = (Ar \cup \{\mathbf{s}_1, \ldots, \mathbf{s}_n\}, \to')$ where $\mathbf{x} \to' \mathbf{y}$ iff for some $i \in 1, \ldots, n$ either:*

- $\mathbf{x} \in Pa_F(S_i)$ *and* $\mathbf{y} = \mathbf{s}_i$, *or*

- $\mathbf{x} = \mathbf{s}_i$ *and* $\mathbf{y} \in S_i$.

Let us look at an example. Let F be the AF shown in Figure 7. The d-graph F^* is shown in Figure 8. In this figure we have highlighted the original SCCs with dotted rectangles. Note that, while the edges in F represent attacks, the edges in F^* represent arbitrary relations of direct influence. Consider the cycle in F contained in the SCC $S_1 = \{\mathbf{a}, \mathbf{b}\}$. As edges between elements of S_1 are removed in F^*, this cycle is not present in F^*. The dependency between $\mathbf{a}$ and $\mathbf{b}$ is now accounted for by the variable $\mathbf{s}_1$, which acts as a common cause for $\mathbf{a}$ and $\mathbf{b}$. Now consider the SCC $S_3 = \{\mathbf{e}, \mathbf{f}, \mathbf{g}\}$, which is transformed similarly, but in addition, the attack of the parents $\mathbf{b}$ and $\mathbf{d}$ of S_3 on $\mathbf{e}$ is replaced with edges from $\mathbf{b}$ and $\mathbf{d}$ to $\mathbf{s}_3$. This is done so that the dependence of *all* elements of S_3 on $\mathbf{b}$ and $\mathbf{d}$—not just $\mathbf{e}$ but also $\mathbf{f}$ and $\mathbf{g}$—is still accounted for. Note that the addition of the variable $\mathbf{s}_4$ for the singleton SCC S_4 is not actually needed, but treating all SCCs the same simplifies the definition of d-graph.

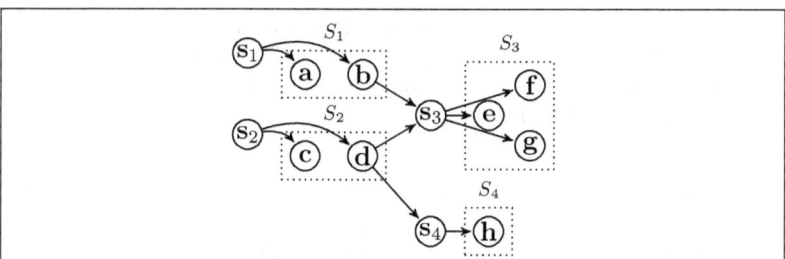

Figure 8: The d-graph for the AF from Figure 6

5.3.5 Soundness of the D-Graph Approach

The following theorem is the main result of [Rienstra et al., 2020]. It states that the d-graph approach is sound for any semantics that is SCC Markovian.

Theorem 5.20. *Let σ be an SCC-Markovian semantics and let $F = (Ar, att)$ be an AF. Then for all disjoint sets $A, B, C \subseteq Ar$ we have that, if A and B are d-separated by C in F^*, then A and B are independent given C in $\mathcal{L}_\sigma(F)$.*

Example 25. *Let F be the AF shown earlier in Figure 6. The d-graph F^* of F is shown in Figure 8. In Example 23 we listed a number of (non-)independencies that hold in $\mathcal{L}_{pr}(F)$. All these independencies can be derived using d-separation in F^*, while none of the non-independencies can be derived. For instance, in the d-graph F^* we have that $\mathbf{a}$ and $\mathbf{e}$ are not d-separated by the empty set because the empty set does not block the trail $\mathbf{a} \leftarrow s_1 \to \mathbf{b} \to s_3 \to \mathbf{e}$. Indeed, $\mathbf{a}$ and $\mathbf{e}$ are not unconditionally independent in $\mathcal{L}_{pr}(F)$. The trail is blocked by $\mathbf{b}$, however, which is due to the chain $s_1 \to \mathbf{b} \to s_3$. Therefore $\mathbf{a}$ and $\mathbf{e}$ are d-separated by $\mathbf{b}$. Indeed, $\mathbf{a}$ and $\mathbf{e}$ are independent given $\mathbf{b}$ in $\mathcal{L}_{pr}(F)$.*

It must be noted that the d-graph approach is not complete. A number of cases can be identified where independencies that hold under an SCC Markovian semantics cannot be derived using the d-graph approach. One issue is that some independencies are due to mediating arguments being rejected. Suppose $\mathbf{a}$ attacks $\mathbf{b}$ and $\mathbf{b}$ attacks $\mathbf{c}$. Then $\mathbf{a}$ and $\mathbf{c}$ are d-separated and may be dependent. Now suppose that **b** is attacked by an argument **d** which is itself unattacked. Then **a** and **c** are still d-separated even though they are independent under any SCC-Markovian semantics. For further discussion we refer the reader to [Rienstra et al., 2020].

6 Discussion

The goal of this chapter was to explore how argumentation and causality are connected within the different perspectives from which the two phenomena are studied. We focused on the logical foundations of causal reasoning, taking the causal calculus as our starting point. We then focused on scheme-based reasoning to approach causal connections as defeasible generalisations as we pivoted to examining causality from a dialectical viewpoint. This perspective shifted our understanding of causation, framing it as a flexible, dynamic force shaping our narratives and interventions within various fields. We also discussed approaches that frame causal reasoning as a form of structured argumentation. The different aspects that we covered illustrate the synergetic relationship between causal reasoning and formal argumentation.

There are several works that somehow combine causality and argumentation but did not fit our framework of exposition. We briefly discuss these works below.

- *Causal discovery* algorithms aim to infer, given observational data, a set of causal graphs that best explain the data. *Constraint-based* causal discovery algorithms do so by performing statistical *independence tests* in order to obtain causal graphs whose implied independencies reflect the independencies present in the data. This approach, however, is ineffective when dealing with small datasets, which result in unreliable independence tests. Bromberg and Margaritis [Bromberg and Margaritis, 2009] address this problem by using Amgoud and Cayrol's [Amgoud and Cayrol, 2002] argumentation-based system of reasoning with inconsistent knowledge bases. Arguments are constructed by applying the graphoid axioms which are satisfied by independence relationships [Pearl, 2009], while preferences between arguments are based on the relative reliability of the independence tests. The grounded extension of the generated argumentation frameworks represent a consistent view on a set of potentially inconsistent independence tests. An experimental evaluation demonstrate significant improvements in the accuracy of causal discovery for both synthetic and real-world datasets.

- Rago et al. [Rago et al., 2022] address the problem of explaining predictions made on the basis of a causal model. Their approach is based on generating an abstract bipolar argumentation framework from a given causal model and a given setting of the exogenous variables of the causal model. This bipolar argumentation framework explains in a qualitative manner how the different variables of the causal model influence each other.

The avenues for future research in the realm of causality and argumentation are numerous and wide-ranging. As the present chapter showed, we are still far from having a complete understanding of how to marry structural causal models with argumentation frameworks, especially when aiming to produce advanced analytical tools that harness the power of argumentative reasoning for better elucidation and assessment of causal models — and by extension, possibly machine learning models — thus enhancing both their clarity and argumentative strength.

In Section 3, we discuss the relevant literature looking at causal links as defeasible generalisations. An associated yet unanswered research avenue could look at defeasible causal links altogether and the implication of such defeasibility in the overall causal calculus.

A defeasible causal calculus would constitute a formal framework that recognises the conditional validity of causal relationships, acknowledging that these relationships may not hold universally and can be overridden by additional information. Such a calculus would incorporate mechanisms for evaluating and adjusting causal links based on new evidence or contextual factors, thus reflecting the non-absolute nature of causality in complex systems.

In this system, causal links would be expressed as defeasible rules, similar to those found in defeasible logic. The presence of a cause would not necessarily guarantee an effect under all circumstances; rather, the calculus would accommodate exceptions or specific conditions that might negate the expected outcome. For instance, a causal rule might state that *Event A causes Event B*, with an added clause specifying that this relationship holds *unless Condition C* is present. Condition C, in this case, would act as a defeater, preventing the causal link from manifesting under certain scenarios.

The calculus would also include a method for evaluating evidence to determine the strength or validity of a causal link. This evaluation could involve probabilistic measures, assigning weights to evidence, or setting thresholds that must be met for a causal relationship to be considered active. Moreover, the system would be designed to account for the context in which causal relationships are assessed. It would allow for the inclusion of external factors, background knowledge, and specific conditions that might influence the applicability of a causal link.

A mechanism for revising causal rules would be integral to this system. As new evidence or insights emerge, the system would enable the retraction of previously accepted causal relationships or the introduction of new defeaters. This would ensure that the calculus remains responsive to the evolving nature of knowledge. The reasoning processes supported by this system would account for the defeasibility of causal links, generating conclusions that are tentative and subject to revision rather than definitive.

The semantics of such a defeasible causal calculus can then be seen as an extension of argumentative semantics into the domain of causation. While argumentative semantics typically focuses on the logical consistency and persuasiveness of arguments, defeasible causal calculus extends these principles to the empirical domain, where the reliability of causal claims is assessed through a similar dialectical process. This involves not only logical consistency but also empirical validation, weighing evidence, and considering the relevance of contextual factors. Thus, the connection between these two domains underscores a broader epistemological concern with reasoning in situations where knowledge is provisional and subject to change.

Furthermore, we believe that the domain is ripe for empirical investigation into the application of causal reasoning within legal and ethical debates, potentially leading to the development of sophisticated argumentation systems adept at navigating the complex causal narratives often encountered in judicial and policy-making scenarios, e.g., [Cerutti and McDermott, 2023]. Lastly, probing the interface of causality in multi-agent systems is an intriguing possibility, particularly exploring how agents articulate causal arguments, converge on joint causal understandings, and make collective decisions informed by such deliberations—an area that promises to yield significant benefits for collaborative robotics, distributed computing architectures, and the mechanics of blockchain consensus mechanisms.

Declaration of Generative AI and AI-assisted technologies in the writing process

During the preparation of this work, the authors used GPT-3.5 and GPT-4 – OpenAI's large-scale language-generation model – to improve readability and language. After using it, the authors reviewed and edited the content as needed and they take full responsibility for the content of the publication.

Acknowledgement

This work was supported by the EU NEXTGENERATIONEU program within the PNRR Future Artificial Intelligence – FAIR project (PE0000013, CUP H23C22000860006), Objective 10: Abstract Argumentation for Knowledge Representation and Reasoning, specifically by the project Argumentation for Informed Decisions with Applications to Energy Consumption in Computing – AIDECC (CUP D53C24000530001). This work was supported by project SERICS (PE00000014) under the MUR National Recovery and Resilience Plan funded by the European Union – NextGenerationEU, specifically by the project NEACD: Neurosymbolic Enhanced Active Cyber Defence (CUP J33C22002810001). This work was supported by project ACRE (AI-Based Causality and Reasoning for Deceptive Assets - 2022EP2L7H) and xInternet (eXplainable Internet - 20225CETN9) projects - funded by European Union -Next Generation EU within the PRIN 2022 program (D.D. 104 - 02/02/2022 Ministero dell'Università e della Ricerca). The work is partially supported by the European Office of Aerospace Research & Development and the Air Force Office of Scientific Research under award number FA8655-22-1-7017 and by the US DEVCOM Army Research Laboratory (ARL) under Cooperative Agreement #W911NF2220243. Any opinions, findings, and conclusions or recommendations expressed in this material are those of the author(s) and do not necessarily reflect the views of the United States government.

References

[Ahn et al., 1995] Ahn, W.-k., Kalish, C. W., Medin, D. L., and Gelman, S. A. (1995). The role of covariation versus mechanism information in causal attribution. *Cognition*, 54(3):299–352.

[Amgoud and Cayrol, 2002] Amgoud, L. and Cayrol, C. (2002). A reasoning model based on the production of acceptable arguments. *Ann. Math. Artif. Intell.*, 34(1-3):197–215.

[Andreas and Guenther, 2021] Andreas, H. and Guenther, M. (2021). Regularity and Inferential Theories of Causation. In Zalta, E. N., editor, *The Stanford Encyclopedia of Philosophy*. Metaphysics Research Lab, Stanford University, Fall 2021 edition.

[Arieli et al., 2019] Arieli, O., Borg, A., and Heyninck, J. (2019). A review of the relations between logical argumentation and reasoning with maximal consistency. *Ann. Math. Artif. Intell.*, 87(3):187–226.

[Arrieta et al., 2020] Arrieta, A. B., Díaz-Rodríguez, N., Del Ser, J., Bennetot, A., Tabik, S., Barbado, A., García, S., Gil-López, S., Molina, D., Benjamins, R., et al. (2020). Explainable artificial intelligence (xai): Concepts, taxonomies, opportunities and challenges toward responsible ai. *Information fusion*, 58:82–115.

[Baroni et al., 2014] Baroni, P., Boella, G., Cerutti, F., Giacomin, M., van der Torre, L., and Villata, S. (2014). On the input/output behavior of argumentation frameworks. *Artificial Intelligence*, 217:144–197.

[Belnap, 1977] Belnap, Jr., N. D. (1977). A useful four-valued logic. In Dunn, M. and Epstein, G., editors, *Modern Uses of Multiple-Valued Logic*, pages 8–41. D. Reidel.

[Bengel et al., 2022] Bengel, L., Blümel, L., Rienstra, T., and Thimm, M. (2022). Argumentation-based causal and counterfactual reasoning. In Cyras, K., Kampik, T., Cocarascu, O., and Rago, A., editors, *1st International Workshop on Argumentation for eXplainable AI co-located with 9th International Conference on Computational Models of Argument (COMMA 2022), Cardiff, Wales, September 12, 2022*, volume 3209 of *CEUR Workshop Proceedings*. CEUR-WS.org.

[Bex, 2015] Bex, F. (2015). An integrated theory of causal stories and evidential arguments. In Sichelman, T. and Atkinson, K., editors, *Proceedings of the 15th International Conference on Artificial Intelligence and Law, ICAIL 2015, San Diego, CA, USA, June 8-12, 2015*, pages 13–22. ACM.

[Bochman, 2004] Bochman, A. (2004). A causal approach to nonmonotonic reasoning. *Artificial Intelligence*, 160:105–143.

[Bochman, 2005] Bochman, A. (2005). *Explanatory Nonmonotonic Reasoning*. World Scientific.

[Bochman, 2007] Bochman, A. (2007). A causal theory of abduction. *Journal of Logic and Computation*, 17:851–869.

[Bochman, 2016] Bochman, A. (2016). Abstract dialectical argumentation among close relatives. In Baroni, P., Gordon, T. F., Scheffler, T., and Stede, M., editors, *Computational Models of Argument: Proceedings of COMMA 2016, Potsdam, Germany*, pages 127–138. IOS Press.

[Bochman, 2018a] Bochman, A. (2018a). Argumentation, nonmonotonic reasoning and logic. In *Handbook of Formal Argumentation*, volume 1, pages 2887–2926. College Publications.

[Bochman, 2018b] Bochman, A. (2018b). On laws and counterfactuals in causal reasoning. In *Principles of Knowledge Representation and Reasoning: Proceedings of the Sixteenth International Conference, KR 2018*, pages 494–503. AAAI Press.

[Bochman, 2021] Bochman, A. (2021). *A logical theory of causality*. MIT Press.

[Bochman and Lifschitz, 2015] Bochman, A. and Lifschitz, V. (2015). Pearl's causality in a logical setting. In *Proceedings of the 29th AAAI Conference on Artificial Intelligence*, pages 1446–1452. AAAI Press.

[Bondarenko et al., 1997] Bondarenko, A., Dung, P. M., Kowalski, R. A., and Toni, F. (1997). An abstract, argumentation-theoretic framework for default reasoning. *Artificial Intelligence*, 93:63–101.

[Brandom, 2000] Brandom, R. (2000). *Articulating Reasons: An Introduction to Inferentialism*. Harvard University Press.

[Brewka, 1989] Brewka, G. (1989). Preferred subtheories: An extended logical framework for default reasoning. In *Proceedings of the 11th International Joint Conference on Artificial Intelligence. Detroit, MI, USA, August 1989*, pages 1043–1048.

[Brewka et al., 2013] Brewka, G., Strass, H., Ellmauthaler, S., Wallner, J. P., and Woltran, S. (2013). Abstract dialectical frameworks revisited. In *IJCAI 2013: Proceedings of the 23rd International Joint Conference on Artificial Intelligence*. AAAI Press.

[Brewka and Woltran, 2010] Brewka, G. and Woltran, S. (2010). Abstract dialectical frameworks. In *Principles of Knowledge Representation and Reasoning: Proceedings of the Twelfth International Conference, KR 2010*.

[Bromberg and Margaritis, 2009] Bromberg, F. and Margaritis, D. (2009). Improving the reliability of causal discovery from small data sets using argumentation. *J. Mach. Learn. Res.*, 10:301–340.

[Buckle, 2007] Buckle, S. (2007). *Hume: An Enquiry Concerning Human Understanding: And Other Writings*. Cambridge Texts in the History of Philosophy. Cambridge University Press.

[Caminada and Amgoud, 2007] Caminada, M. and Amgoud, L. (2007). On the evaluation of argumentation formalisms. *Artif. Intell.*, 171(5-6):286–310.

[Caminada and Gabbay, 2009] Caminada, M. W. A. and Gabbay, D. M. (2009). A logical account of formal argumentation. *Studia Logica*, 93(2-3):109–145.

[Cayrol, 1995] Cayrol, C. (1995). On the relation between argumentation and non-monotonic coherence-based entailment. In *Proceedings of the Fourteenth International Joint Conference on Artificial Intelligence, IJCAI 95, Montréal Québec, Canada, August 20-25 1995, 2 Volumes*, pages 1443–1448.

[Cerutti, 2018] Cerutti, F. (2018). On Scientific Enquiry and Computational Argumentation. In *Proceedings of the 18th Workshop on Computational Model of Natural Argument*.

[Cerutti, 2021] Cerutti, F. (2021). Supporting trustworthy artificial intelligence via bayesian argumentation. In *International Conference of the Italian Association for Artificial Intelligence*, pages 377–388. Springer.

[Cerutti and McDermott, 2023] Cerutti, F. and McDermott, Y. (2023). A formal argumentation exercise on the karadžić trial judgment. *Journal of Applied Logics*, 10(3):319 – 339.

[Cerutti and Pearson, 2018] Cerutti, F. and Pearson, G. (2018). Supporting Scientific Enquiry with Uncertain Sources. In *2018 21st International Conference on Information Fusion (FUSION)*, pages 1–8. IEEE.

[Cheng, 1993] Cheng, P. W. (1993). Separating Causal Laws from Casual Facts: Pressing the Limits of Statistical Relevance. In Medin, D. L., editor, *Psychology of Learning and Motivation*, volume 30, pages 215–264. Academic Press.

[Denecker et al., 2015] Denecker, M., Brewka, G., and Strass, H. (2015). A formal theory of justifications. In *13th International Conference on Logic Programming and Non-monotonic Reasoning (LPNMR)*, pages 250–264.

[Dennett, 1989] Dennett, D. C. (1989). *The Intentional Stance*. MIT press.

[Ducasse, 1926] Ducasse, C. J. (1926). On the nature and the observability of the causal relation. *The Journal of Philosophy*, 23(3):57–68.

[Dung, 1995] Dung, P. M. (1995). On the acceptability of arguments and its fundamental role in nonmonotonic reasoning, logic programming and n-person games. *Artificial intelligence*, 77(2):321–357.

[Eells, 1991] Eells, E. (1991). *Probabilistic Causality*, volume 1. Cambridge University Press.

[Einhorn and Hogarth, 1986] Einhorn, H. and Hogarth, R. (1986). Judging Probable Cause. *Psychological Bulletin*, 99(1):3–19.

[Ellmauthaler, 2012] Ellmauthaler, S. (2012). Abstract Dialectical Frameworks: Properties, Complexity, and Implementation. Master's thesis, Technische Universität Wien, Institut für Informationssysteme.

[Goodman, 1947] Goodman, N. (1947). The problem of counterfactual conditionals. *Journal of Philosophy*, 44:113–128.

[Grennan, 1997] Grennan, W. (1997). *Informal Logic: Issues and Techniques.* McGill-Queen's Press-MQUP.

[Hahn and Hornikx, 2016] Hahn, U. and Hornikx, J. (2016). A normative framework for argument quality: argumentation schemes with a Bayesian foundation. *Synthese*, 193(6):1833–1873.

[Hahn and Oaksford, 2007a] Hahn, U. and Oaksford, M. (2007a). The Burden of Proof and Its Role in Argumentation. *Argumentation*, 21(1):39–61.

[Hahn and Oaksford, 2007b] Hahn, U. and Oaksford, M. (2007b). The Rationality of Informal Argumentation: A Bayesian Approach to Reasoning Fallacies. *Psychological Review*, 114(3):704–732.

[Hahn et al., 2013] Hahn, U., Oaksford, M., and Harris, A. J. (2013). Testimony and argument: A bayesian perspective. In *Bayesian Argumentation: The Practical Side of Probability*, pages 15–38. Springer Netherlands.

[Halpern, 2000] Halpern, J. Y. (2000). Axiomatizing causal reasoning. *Journal of Artificial Intelligence Research*, 12:317–337.

[Hankinson, 1998] Hankinson, R. J. (1998). *Cause and Explanation in Ancient Greek Thought*. Clarendon Press.

[Harman, 1965] Harman, G. H. (1965). The Inference to the Best Explanation. *The Philosophical Review*, 74(1):88–95.

[Hart and Honoré, 1985] Hart, H. L. A. and Honoré, T. (1985). *Causation in the Law*. OUP Oxford.

[Hastings, 1962] Hastings, A. C. (1962). *A Reformulation of the Modes of Reasoning in Argumentation*. Northwestern University.

[Heyninck and Arieli, 2020] Heyninck, J. and Arieli, O. (2020). Simple contrapositive assumption-based argumentation frameworks. *Int. J. Approx. Reason.*, 121:103–124.

[Hume, 2000] Hume, D. (2000). *A Treatise of Human Nature*. Oxford University Press.

[Josephson and Josephson, 1996] Josephson, J. R. and Josephson, S. G. (1996). *Abductive Inference: Computation, Philosophy, Technology*. Cambridge University Press.

[Kass and Leake, 1987] Kass, A. and Leake, D. (1987). *Types of Explanations*. Yale University, Department of Computer Science.

[Kienpointner, 1992] Kienpointner, M. (1992). *Alltagslogik. Struktur und Funktion von Argumentationsmustern*. Stuttgart-Bad Cannstatt: Frommann-Holzboog.

[Kuhn, 1991] Kuhn, D. (1991). *The Skills of Argument*. Cambridge University Press.

[Lewis, 1973] Lewis, D. (1973). *Counterfactuals*. Harvard University Press.

[Lifschitz, 1997] Lifschitz, V. (1997). On the logic of causal explanation. *Artificial Intelligence*, 96:451–465.

[Lombrozo, 2012] Lombrozo, T. (2012). Explanation and Abductive Inference. In Holyoak, K. J. and Morrison, R. G., editors, *The Oxford Handbook of Thinking and Reasoning*, page 0. Oxford University Press.

[Lombrozo and Gwynne, 2014] Lombrozo, T. and Gwynne, N. Z. (2014). Explanation and inference: Mechanistic and functional explanations guide property generalization. *Frontiers in Human Neuroscience*, 8:700.

[Mackie, 1980] Mackie, J. L. (1980). *The Cement of the Universe: A Study of Causation*. Clarendon Press.

[Marr, 2010] Marr, D. (2010). *Vision: A Computational Investigation into the Human Representation and Processing of Visual Information*. MIT press.

[McCain and Turner, 1997] McCain, N. and Turner, H. (1997). Causal theories of action and change. In Kuipers, B. and Webber, B. L., editors, *Proceedings of the Fourteenth National Conference on Artificial Intelligence and Ninth Innovative Applications of Artificial Intelligence Conference, AAAI 97, IAAI 97, July 27-31, 1997, Providence, Rhode Island, USA*, pages 460–465. AAAI Press / The MIT Press.

[Mill and Robson, 1974] Mill, J. S. and Robson, JM. (1974). A System of Logic Ratiocinative and Inductive. Collected Works of John Stuart Mill.

[Modgil and Prakken, 2013] Modgil, S. and Prakken, H. (2013). A general account of argumentation with preferences. *Artif. Intell.*, 195:361–397.

[Oestermeier and Hesse, 2000] Oestermeier, U. and Hesse, F. W. (2000). Verbal and visual causal arguments. *Cognition*, 75(1):65–104.

[Pearl, 1987] Pearl, J. (1987). Embracing causality in formal reasoning. In *Proceedings of the Sixth National Conference on Artificial Intelligence (AAAI-87)*, pages 369–373.

[Pearl, 1988] Pearl, J. (1988). Embracing causality in default reasoning. *Artif. Intell.*, 35(2):259–271.

[Pearl, 2009] Pearl, J. (2009). *Causality: Models, Reasoning and Inference*. Cambridge University Press, 2nd edition. (1st ed. 2000).

[Pearl and Mackenzie, 2018] Pearl, J. and Mackenzie, D. (2018). *The Book of Why*. Basic Books, New York.

[Peirce, 1997] Peirce, C. S. (1997). *Pragmatism as a Principle and Method of Right Thinking: The 1903 Harvard Lectures on Pragmatism*. Suny Press.

[Perelman, 1971] Perelman, C. (1971). *The New Rhetoric*. Springer.

[Pirolli and Card, 2005] Pirolli, P. and Card, S. (2005). The sensemaking process and leverage points for analyst technology as identified through cognitive task analysis. In *Proceedings of International Conference on Intelligence Analysis*.

[Pollock, 2001] Pollock, J. L. (2001). Defeasible reasoning with variable degrees of justification. *Artificial Intelligence*, 133(1-2):233–282.

[Popper, 2005] Popper, K. (2005). *The Logic of Scientific Discovery*. Routledge.

[Prakken and Renooij, 2001] Prakken, H. and Renooij, S. (2001). Reconstructing causal reasoning about evidence: A case study. *Legal knowledge and information systems. JURIX*, pages 131–142.

[Rago et al., 2022] Rago, A., Baroni, P., and Toni, F. (2022). Explaining causal models with argumentation: the case of bi-variate reinforcement. In Kern-Isberner, G., Lakemeyer, G., and Meyer, T., editors, *Proceedings of the 19th International Conference on Principles of Knowledge Representation and Reasoning, KR 2022, Haifa, Israel. July 31 - August 5, 2022*.

[Rehder, 2003] Rehder, B. (2003). A causal-model theory of conceptual representation and categorization. *Journal of Experimental Psychology: Learning, Memory, and Cognition*, 29(6):1141.

[Rienstra et al., 2020] Rienstra, T., Thimm, M., Kersting, K., and Shao, X. (2020). Independence and d-separation in abstract argumentation. In Calvanese, D., Erdem, E., and Thielscher, M., editors, *Proceedings of the 17th International Conference on Principles of Knowledge Representation and Reasoning, KR 2020, Rhodes, Greece, September 12-18, 2020*, pages 713–722.

[Schulz, 2011] Schulz, K. (2011). 'If you'd wiggled A, then B would've changed'—causality and counterfactual conditionals. *Synthese*, 179(2):239–251.

[Shultz, 1982] Shultz, T. R. (1982). Rules of Causal Attribution. *Monographs of the Society for Research in Child Development*, 47(1):1–51.

[Snow, 1855] Snow, J. (1855). *On the mode of communication of cholera*. John Churchill, 2nd edition.

[Strass, 2013] Strass, H. (2013). Approximating operators and semantics for abstract dialectical frameworks. *Artificial Intelligence*, 205:39–70.

[Tarski, 1994] Tarski, A. (1994). *Introduction to Logic and to the Methodology of the Deductive Sciences*, volume 24. Oxford university press.

[Thagard, 2000] Thagard, P. (2000). *How Scientists Explain Disease*. Princeton University Press.

[Toniolo et al., 2015] Toniolo, A., Norman, T., Etuk, A., Cerutti, F., Ouyang, R., Srivastava, M., Oren, N., Dropps, T., Allen, J., and Sullivan, P. (2015). Supporting reasoning with different types of evidence in intelligence analysis. In *Proceedings of AAMAS 2015*, pages 781–789.

[Toulmin, 1958] Toulmin, S. E. (1958). *The Uses of Argument*. Cambridge university press.

[Turner, 1999] Turner, H. (1999). A logic of universal causation. *Artificial Intelligence*, 113:87–123.

[Van Eemeren and Grootendorst, 2004] Van Eemeren, F. H. and Grootendorst, R. (2004). *A Systematic Theory of Argumentation: The Pragma-Dialectical Approach*. Cambridge University Press.

[Van Eemeren et al., 2015] Van Eemeren, F. H., van Eemeren, F. H., and Kruiger, T. (2015). Identifying argumentation schemes. *Reasonableness and Effectiveness in Argumentative Discourse: Fifty Contributions to the Development of Pragma-Dialectics*, pages 703–712.

[Veltman, 2005] Veltman, F. (2005). Making counterfactual assumptions. *Journal of Semantics*, 22(2):159–180.

[Von Wright, 2004] Von Wright, G. H. (2004). *Explanation and Understanding*. Cornell University Press.

[Walton, 1992] Walton, D. (1992). Rules for plausible reasoning. *Informal Logic*, 14(1).

[Walton, 2014] Walton, D. (2014). *Burden of Proof, Presumption and Argumentation*. Cambridge University Press.

[Walton, 1989] Walton, D. N. (1989). *Informal Logic: A Handbook for Critical Argument*. Cambridge University Press.

[Walton, 1996] Walton, D. N. (1996). *Argument Structure: A Pragmatic Theory*. University of Toronto Press Toronto.

[Walton et al., 2008] Walton, D. N., Reed, C., and Macagno, F. (2008). *Argumentation Schemes*. Cambridge University Press, NY.

[Weiner, 1995] Weiner, B. (1995). *Judgments of Responsibility: A Foundation for a Theory of Social Conduct*. guilford Press.

[Wieten et al., 2020] Wieten, R., Bex, F., Prakken, H., and Renooij, S. (2020). Deductive and abductive reasoning with causal and evidential information. In Prakken, H., Bistarelli, S., Santini, F., and Taticchi, C., editors, *Computational Models of Argument - Proceedings of COMMA 2020, Perugia, Italy, September 4-11, 2020*, volume 326 of *Frontiers in Artificial Intelligence and Applications*, pages 383–394. IOS Press.

[Wieten et al., 2022] Wieten, R., Bex, F., Prakken, H., and Renooij, S. (2022). Deductive and abductive argumentation based on information graphs. *Argument Comput.*, 13(1):49–91.

[Wilkenfeld and Lombrozo, 2015] Wilkenfeld, D. A. and Lombrozo, T. (2015). Inference to the Best Explanation (IBE) Versus Explaining for the Best Inference (EBI). *Science & Education*, 24(9):1059–1077.

[Wright, 1934] Wright, S. (1934). The method of path coefficients. *The annals of mathematical statistics*, 5(3):161–215.

CHAPTER 11

DEFEASIBLE ARGUMENTATION-BASED EPISTEMIC PLANNING WITH PREFERENCES

JUAN C. L. TEZE
Facultad de Ciencias de la Administración,
Universidad Nacional de Entre Ríos (UNER),
Consejo Nacional de Investigaciones Científicas y Técnicas
(CONICET)
carlos.teze@uner.edu.ar

LLUIS GODO
Artificial Intelligence Research Institute (IIIA),
Spanish National Council for Scientific Research (CSIC)
godo@iiia.csic.es

GERARDO I. SIMARI
Departamento de Ciencias e Ingeniería de la Computación,
Universidad Nacional del Sur (UNS), Instituto de Ciencias e
Ingeniería de la Computación (UNS-CONICET), and School of
Computing and Augmented Intelligence, Arizona State University
gis@cs.uns.edu.ar

1 Introduction

Planning is a research area within Artificial Intelligence (AI) that addresses the problem of obtaining a set of actions to achieve a specific goal given a description of the initial state of the world. Recently, the consideration of *epistemic* elements in building a plan has revealed a useful new perspective in the area: *"Epistemic planning is the enrichment of planning with epistemic notions, that is, knowledge and beliefs"* [Bolander and Andersen, 2011; Bolander, 2017; Baral et al., 2017; Belle et al., 2023]. Various frameworks for planning have been proposed allowing for a formalisation and mechanization of knowledge-based reasoning in the planner itself. A central feature of classical frameworks is that their domain descriptions assume a fully observable, static, and deterministic world, which might lead to contradictions when the available knowledge is incomplete or inconsistent. In [Pollock, 1998], the author concludes that since epistemic cognition is *defeasible*, a planning agent must be prepared to revise its plans as soon as its beliefs change, and may need to acquire more information through reasoning to solve a planning problem.

Defeasible argumentation is a form of reasoning about beliefs that can be used to exploit the contents of knowledge bases in the context of possible inconsistencies [Bench-Capon and Dunne, 2007; Rahwan and Simari, 2009]. Specifically, the fundamental process in defeasible argumentation is to confront reasons to support or dismiss a conclusion that is under scrutiny. An analysis mechanism supports this process by obtaining arguments for and against such conclusion, and then comparing those in conflict in order to reach a decision regarding acceptance.

Several works have proposed using argumentation to enhance planning systems. Particularly, planning problems have been primarily addressed from two points of view: Practical reasoning and Automated planning. In the context of the former, i.e. reasoning about what to do next, a number of attempts have been made to leverage argumentation [Girle et al., 2004]. There are many ways to engage in practical reasoning, which make the task of formulating an argumentation-based planning system more complicated, mainly when investigating more specific aspects of rationality [Pollock, 1995]. Using and instantiating Dung's argumentation framework [Dung, 1995] has been the predominant approach in practical reasoning – see, e.g. [Rahwan and Amgoud, 2006a; Dung, 2016] – while another research line [Pollock, 1999; Teze et al., 2022] closer to automated planning has explored how to use argumentation to guide the reasoning process. In general, in the latter approaches defeasible argumentation is used as the inference mechanism to reason about the preconditions and effects of actions in a planner system, especially in dynamic domains dealing with incomplete and contradictory information, which is often the case in real-world planning scenarios.

Actually, to solve a planning problem, a planning system with an appropriate set of actions should be provided. In classical planning, a general assumption on the representation of these actions is that it must encapsulate all the possible preconditions and effects that are relevant to solve the planning problem. Consider for example a scenario where a service robot agent has ordered some food from a restaurant and it is about to receive it at home by means of an action "receiving a food delivery service at home". A relevant effect can be "having a food delivery box at home". However, there could be other consequences that could be obtained but considered irrelevant, and thus not included in the representation of the action. For example, if the payment of the delivery service is done at the moment with a debit card, the customer will automatically have less money in her bank account. Therefore, instead of including all the possible effects in the representation of the actions, the system could be provided with a reasoning mechanism for obtaining those consequences that follow from the effects of an action. For example, "the ordered food is at home" could be considered as a *plausible* consequence of the effect that "there is a food delivery box at home". To do so, besides having the action specified in the planner, a possibility is to include extra knowledge in the form of a defeasible rule, such as "having a food delivery box at home is a reason to believe that the ordered food is at home".

It is important to remark that if "the ordered food is at home" is considered as an effect of the action, then it will be difficult to handle exceptions like "the delivered food is not what was ordered". However, this kind of problem can be properly handled by argumentation formalisms. For example, "there is a food delivery box at home but the delivered food is not what was ordered, is a reason to believe that the ordered food is not at home". In particular, classical planning systems do not perform any type of reasoning over the effects of actions. In dynamic domains, it is a complex task to determine in advance what the effects of actions are because the information is constantly changing and depends on many factors. In this context, defeasible argumentation-based epistemic planners have been effectively applied in formalizing planning domains [García et al., 2007; García et al., 2008; Pardo et al., 2011c]; these approaches are characterized by the use of defeasible reasoning for the epistemic tasks performed over the represented knowledge.

Classical planning aims at finding a sequence of actions that, starting from an initial state, leads to a goal state. However, it is often the case that certain approaches are focused not only on the final goal state after plan execution, but also they attempt to address other important aspects, such as satisfying users' preferences [Baier and McIlraith, 2008], value-based selection of actions [Nawwab et al., 2008; Teze et al., 2019a], or complying with norms imposed on the planner establishing what the system is required to do under certain conditions [Shams et al., 2020a]. More specifically, modeling user preferences with explicitly-specified priorities has attracted the attention of many researchers. However, and despite its importance in the reasoning process, most of the existing argumentation-based planning systems do not provide additional capabilities for dynamically changing the preferences expressed by these priorities when a plan is being constructed.

In this chapter, we survey the main approaches in the literature concerning all the above-mentioned issues. However, the purpose of this survey is neither to cover the whole range of argumentation-based planning approaches nor to solve open questions or particular cases that have not been addressed so far; thus, we do not aim to have an exhaustive coverage of the subject. The rest of the chapter is structured in two main sections. In Section 2, we give an overview of different approaches in the area of planning with argumentation studied in the literature, while in Section 3 we present a description of a specific approach to deal with the handling of (contextual) preferences when a plan is formulated. In particular, we present the P-APOP algorithm proposed in [Teze et al., 2022], and summarize a set of computational complexity results. Finally, in Section 4 we offer our conclusions and discuss several challenges for research and hurdles that must be addressed on the path to obtaining fully working solutions.

Research Questions
How can argumentation theory capture practical reasoning?
How can argumentation be exploited to guide the reasoning process, specifically for the selection and organization of actions?
How can argumentation theory be leveraged for plan search in cooperative scenarios?
How can the relationship between an agent's values and the construction of plans be formalized?
How can a set of agents achieve a goal jointly following a same plan?
What is the course of action to adopt in the presence of different goals and norms?
How can argumentation theory be exploited to explain the results of planning systems?
How can the notion of preference be embedded in argumentation and epistemic planning formalisms?
How can preferences be used to compute an optimal plan?
How can a planning system handle contextual preferences?

Table 1: A selection of the main research questions addressed by different lines of work in epistemic planning.

2 Related Argumentation-based Planning Efforts

There are many challenging areas related to planning that have been addressed in the literature, and there is clearly much work still to be done in addressing epistemic planning issues. As a more general presentation of a set of epistemic planning-related research questions, we propose a list of aspects to be considered when trying to solve complex planning problems – presented in Table 1 – that motivate the criteria used to classify the different approaches presented in the next section.

This section reports a summary of approaches focusing on the use of argumentation in epistemic planning, and the handling of preferences. We will first briefly touch upon some works that combine argumentation with planning, and then we focus particularly on works that incorporate the representation of and reasoning with preferences in the formalism.

2.1 Planning with Argumentation

In many real-world planning applications, it is common to encounter situations where unresolved contradictory and/or incomplete information occurs. Argumentation has become a very active research field because of its effective computational capacity to capture and solve conflicts, and there have been many research efforts towards the development of argumentation-based planning systems in the last two decades. In the following we discuss relevant contributions in different types of argumentation-based planning formalisms.

Practical Reasoning. A number of attempts have been made to address how argumentation theory can capture practical reasoning. Argumentation-based practical reasoning employs the conflict resolution capabilities of argumentation theory to solve conflicts between beliefs, intentions, and desires. Different approaches have dealt with these aspects; for instance, [Rahwan and Amgoud, 2006b] introduces a formalism for agents following the BDI approach to reason about desires (generating desires and plans to achieve them). Argumentation-based proposals have also been used to compute the set of intentions to be pursued, or the resolution of incompatibilities among pursuable goals [Amgoud et al., 2011]. Motivated by the requirements of autonomic computing systems, [Liao and Huang, 2010] proposes the architecture of an Autonomous, Normative and Guidable agent (ANGLE) and its extended defeasible logic-based knowledge representation, including observations and motivational knowledge. In this formalism, the reasoning and decision-making tasks adopt argumentative deliberation based on dynamic theories. Other works, such as [Bench-Capon and Atkinson, 2009], follow the notion of argument schemes proposed by Walton [Walton, 1996]. Other approaches using argumentation in a normative environment were proposed in [Shams et al., 2016; Shams et al., 2020b].

Automated Planning. Unlike argumentation-based approaches for practical reasoning, some planning formalisms have exploited the use of argumentation as a mechanism to guide the reasoning process, primarily concerned with the computational process for the selection and organization of actions. One of the well-known works on building a planner based on a defeasible reasoner was proposed in [Pollock, 1999], in which

11 - ARGUMENTATIVE EPISTEMIC PLANNING WITH PREFERENCES

Pollock presents OSCAR, an implemented architecture whose defeasible reasoner essentially performs a defeasible search for plans. In [García et al., 2007; García et al., 2008], the authors introduce an argumentation-based formalism for constructing plans using partial order planning techniques, called DeLP-based partial order planning (DeLP-POP). In this approach, action preconditions can be satisfied either by actions' effects or conclusions supported by arguments, so actions and arguments are combined to construct plans. Actually, DeLP-POP is an extension of the POP algorithm that considers actions and arguments as planning steps and resolves the interferences that can appear. In [Pardo et al., 2011c; Pajares-Ferrando et al., 2011; Pajares-Ferrando and Onaindia, 2017], DeLP-POP is extended to multi-agent cooperative planning, while [Pajares-Ferrando and Onaindia, 2017] presents a planning system based on DeLP to reason about context information during the construction of a plan – the system is designed to operate in cooperative multi-agent environments. Each step of the construction of a plan is discussed among agents following a proposed dialogue mechanism that allows agents to exchange arguments about the conditions that might affect an action's feasibility according to their distributed knowledge and beliefs. Temporal defeasible reasoning has also been studied in the planning literature, where for instance Pardo and Godo [Pardo and Godo, 2013; Pardo and Godo, 2018] presented a distributed multiagent planning system for cooperative tasks. The main feature of the proposal is the development of a planning approach based on t-DeLP, an extension of DeLP for defeasible temporal reasoning. The authors also propose a dialogue-based algorithm for plan search in cooperative scenarios.

The work of [Teze and Godo, 2021] concerns epistemic planning problems, focusing on an argumentation-based approach. The paper discusses first steps in developing an approach to handle contextual preferences that can dynamically change based on knowledge-based priorities. They introduce a generic architecture, independent of the underlying formalism and reasoning mechanisms, as well as a set of guidelines to support knowledge and software engineers in the analysis and design of planning systems leveraging this preference handling capacity. The authors also present a concrete instantiation based on Possibilistic Defeasible Logic Programming [Alsinet et al., 2008]. Recently, [Teze *et al.*, 2022] presents a revised, refined, and extended version of [Teze and Godo, 2021], where the main criteria employed to decide which actions to keep during the construction of plans, by using contextual conditional-preference expressions associated with each action, are formally defined. It also discusses possible interferences that can appear when such expressions are used, and it presents an extension of the APOP algorithm [García *et al.*, 2008] for this setting with contextual preferences.

Following the idea of value-based argumentation [Atkinson and Bench-Capon, 2021], there have also been approaches on providing grounds for formalizing the relationship between values and actions, and integrating defeasible argumentation into the agent reasoning process. For instance, in [Teze *et al.*, 2019b] the values that an agent holds are used to compare plans, and several comparison strategies are formally defined.

Planning Problems in Multiagent Environments. In multiagent environments, agents may need to jointly follow a course of action in order to achieve a goal. The different viewpoints that agents have on the environment may cause disagreements, and reaching an agreement requires the alignment of viewpoints. Argumentation provides natural ways for conflict resolution in collaborative decision making. Many works have advanced the state of the art in argumentation-based multiagent planning. In [Belesiotis *et al.*, 2009; Belesiotis *et al.*, 2010] the authors investigate the use of argumentation to solve conflicts between planning proposals caused by inconsistency between beliefs. Another interesting work that combines the benefits of argumentation in multiagent environments emphasizes the use of defeasible temporal reasoning for negotiation dialogues [Pardo *et al.*, 2011a; Pardo and Godo, 2013]; an extended and revised version of this work is [Pardo and Godo, 2018]. In the same vein, [Monteserin and Amandi, 2011] introduces a proposal that models the argumentation process as a planning process, and obtains an argumentation-based negotiation plan. In [Pardo *et al.*, 2011b; Pajares-Ferrando and Onaindia, 2017], the authors present a multiagent extension of the DeLP-based Partial Order Planning (POP) framework [García *et al.*, 2008] for cooperative planning. Apart from individual goals, the system may require to follow societal norms that promote systems that follow the right behavior. Toniolo et al. [2011] address the question of what is the best course of action to adopt in the presence of different goals and norms, proposing a solution based on argumentation schemes for deliberative dialogues in multiagent environments.

Explainable Planning. Explainable AI Planning [Fox *et al.*, 2017] is a fairly recent research area that involves explaining the outcome and results of planning systems. The relevant question here is how can argumentation theory be exploited to explain the results of planning systems. Argumentation has been widely recognized by the Explainable AI community [Sklar and Azhar, 2018] as a powerful logical model of reasoning that is capable of explaining the behavior of a system by linking any system decision to the evidence supporting it. Some recent approaches like [Mahesar and Parsons, 2023] build around a set of argument schemes that create arguments that give explanations for a plan and its key elements (*i.e.,* actions, states, and goals). In [Shams *et al.*, 2016], the explanations of justifiability of the best plan are generated using an argumentation-based dialogue. A proposal for resolving planning problems with assumption-based argumentation (ABA) was presented in [Fan, 2018]. This work proposes to generate explanations for both planning solutions as well as failed plans extracted from *dispute trees* [Dung *et al.*, 2006]. The work of [Oren *et al.*, 2020] presents a prototype system implementation based on ASPIC [Caminada *et al.*, 2014] for building arguments that justify why a plan should be executed. Two alternatives for plan explanation are considered: visual plan explanation via graphical representations, and textual representation of a plan in a natural language created through a dialogue-based approach, where participants take turns to make utterances that are used to establish whether some argument (and therefore its conclusions) is justified.

2.2 Representation and Reasoning with Preferences

In many planning approaches [Son and Pontelli, 2006; Jorge *et al.*, 2008; Bidoux *et al.*, 2019], modeling user preferences with explicitly-specified priorities plays a significant role, especially in decision-making processes. This priority information is beneficial in the selection of appropriate knowledge, and guides the planning process according to user needs. In this section, we discuss how the notion of preference has been embedded in argumentation and epistemic planning formalisms.

2.2.1 Preferences in Defeasible Argumentation

Defeasible Argumentation formalisms have received increased attention as an advanced mechanism to formalize essential parts of what is known as commonsense reasoning. One of the main issues the argumentative reasoning process must address is confronting reasons to support or dismiss a claim that is under scrutiny. For this purpose, there is a need for an analytical mechanism that follows well-understood steps, starting by obtaining arguments and then comparing those in conflict to determine which arguments prevail; this last step requires a comparison, which in turn needs a preference criterion on the set of arguments to evaluate the strength or importance of arguments in order to reach a decision.

Despite the clear significance of the outcome of a comparison among arguments, there is neither a unique way of establishing a preference relation between arguments nor a consensus in the argumentation literature regarding which criterion should be used; for a comprehensive overview, see for instance [Kaci *et al.*, 2021]. For example, some approaches choose to use a criterion that considers an explicit order over rules [Prakken and Sartor, 1997; Šefránek, 2008], whereas others consider an order over literals [Wakaki, 2010; Wakaki, 2011; Ferretti *et al.*, 2008] or even social values [Bench-Capon, 2003]. In [Kaci and van der Torre, 2008], the authors extend the work of [Bench-Capon, 2003] in order to take into account multiple values and various kinds of preferences over values. In [Cohen *et al.*, 2021], the preference is defined in terms of the strength or credibility of those agents that contribute with pieces of information to the argument. Based on the idea of prioritized norms, [Liao *et al.*, 2019] shows different variants of lifting priorities over norms to priorities on arguments themselves, allowing to capture a preference order over arguments. Differently from [Modgil and Prakken, 2013], where an extension of ASPIC$^+$ is proposed to use preferences to resolve attacks, [Cyras and Toni, 2016] introduces ABA$^+$ considering preferences on assumptions rather than (defeasible) rules. In [Modgil, 2009], the author presents a formalism in Dung's abstract argumentation framework for metalevel argumentation-based reasoning about preferences between arguments, and applies it to Prakken and Sartor's argument based logic programming with defeasible priorities (ALP-DP).

Other approaches, such as [Simari and Loui, 1992; Stolzenburg *et al.*,

2003; García and Simari, 2004], define a criterion based on a generalized specificity principle. Furthermore, there exist other formalisms that use a combination of several fixed and predefined criteria [Godo *et al.*, 2012; Deagustini *et al.*, 2013; Teze *et al.*, 2015; Teze *et al.*, 2020], while others simply consider a general preference relation [Amgoud and Vesic, 2011b; Amgoud and Vesic, 2011a]. Usually, in current argumentation formalisms, the definition of the argument comparison criterion is either fixed and embedded in the system, or it is modular. See the works of [Kaci *et al.*, 2018; Beirlaen *et al.*, 2018; Prakken, 2021] for reviews of argument preference criteria present in the argumentation literature.

2.2.2 Preferences in Epistemic Planning

In the work of [Myers and Lee, 1999], the authors argue that users' preferences are of great importance in selecting a plan for execution when the space of solution plans is dense. While there is a significant body of research on preference within classical planning theory [Gerevini and Long, 2005; Baier and McIlraith, 2008; Bienvenu *et al.*, 2011; Bidoux *et al.*, 2019; Canal *et al.*, 2023], most of the research in epistemic planning has particularly been focused on methodologies and issues related to computational efficiency. Relatively limited efforts have been dedicated to addressing other important aspects, such as generating high quality plans satisfying users' preferences and constraints. For example, [Son and Pontelli, 2006] presented a first proposal of a language for specification of preferences for planning problems and included a logic programming encoding of the language based on Answer Set Programming (ASP). The language allows to handle four different preference categories: about a state, an action, a trajectory, or multi-dimensional preferences. Recently, [Klassen *et al.*, 2023] proposed an automated planning approach for the task of planning with epistemic preferences, which incorporates weighted preferences and computes the optimal plan by maximizing the sum of weights of the preferences satisfied. In [Monteserin and Amandi, 2011], the authors propose the use of a preference-based planning algorithm that represents the argument selection criterion into the agent's mental state as preferences. The algorithm uses the agent's preferences in order to select the best actions.

Regarding approaches using argumentation, [Teze *et al.*, 2019b] focused on providing grounds for formalizing the relationship between values and actions, and for integrating defeasible argumentation into the agent reasoning process. In this formalism, the main idea is that of using values to compare plans, and several comparison strategies are formally defined. The authors propose to arrange values hierarchically, and exploit an agent's preferences over values using such a hierarchy. Finally, as already mentioned above, [Teze and Godo, 2021] and its extension [Teze et al., 2022] deal with defeasible argumentation-based epistemic planning with an approach to handle contextual preferences that can dynamically change via knowledge-based priorities.

The research efforts in the area discussed in this section are summarized in Figure 1. As we have mentioned, the formalization and use of mechanisms for handling preferences have not been widely adopted in the epistemic planning literature. In fact, tackling the fundamental question of whether the notion of preference can be embedded in argumentation and epistemic planning formalisms makes it a particularly challenging research topic. In this context, in the next section we provide some details of the approach developed in [Teze et al., 2022], which takes that direction.

3 Argumentation-based Epistemic Planning with Preferences

With the goal of providing more details about a specific approach, in this section we present an overview of an epistemic planning framework proposed in [Teze and Godo, 2021; Teze et al., 2022], which is a formalism that incorporates defeasible argumentation as a reasoning mechanism in the construction of plans. The main novelty of this epistemic planning formalism centers on introducing a way to select the priority assignment mechanism to modify the preferences among different pieces of defeasible knowledge as the planner reasons and chooses which actions to add to a plan. In the following, we start by describing a general architecture for defeasible argumentation-based epistemic planning, and then show a P-DeLP-based particular instantiation.

Reference	PR	AP	ME	EP	P
Pollock et al.[1999]		✓			
Rahwan et al.[2006b]	✓				
García et al.[2008]		✓			
Bench-Capon et al.[2009]	✓				
Belesiotis et al.[2010]		✓	✓		
Amgoud et al.[2011]	✓				
Monteserin et al.[2011]		✓	✓		✓
Pardo et al.[2011c]		✓	✓		
Toniolo et al.[2011]	✓		✓		
Shams et al.[2016]	✓		✓	✓	
Fan et al.[2018]		✓		✓	
Pardo et al.[2018]		✓	✓		
Teze et al.[2019b]		✓			✓
Oren et al.[2020]		✓		✓	
Shams et al.[2020b]	✓				
Teze et al.[2021]		✓			✓
Teze et al.[2022]		✓			✓
Parsons et al.[2023]		✓		✓	

Figure 1: Comparison of argumentation-based planning approaches in terms of five categories, where check marks indicate a focus on the respective category. Abbreviations: **PR** (*Practical Reasoning*), **AP** (*Automated Planning*), **ME** (*Multiagent Environment*), **E** (*Explainability*), and **P** (*Preferences*).

3.1 An Argumentation-based Epistemic Planning Framework

The development of planning systems with defeasible reasoning and preferences can be a very complex task involving several stages toward obtaining the final system. In [Teze and Godo, 2021], instead of a specific solution, a set of guidelines is introduced to support knowledge and software engineers in the analysis and design of planning systems, focused on five central stages:

1. *Planning domain analysis*: In complex and dynamic environments, planning systems eventually may require dealing with contradictory and incomplete knowledge about the domain. In this context, structured argumentation has played a crucial role in capturing and representing this type of knowledge. This stage is aimed at providing a detailed and precise description of the planning domain and the user's preferences, which includes: a knowledge base in a formal language expressing domain information, and a specification of a preference relation over pieces of knowledge. These preferences reflect the importance or priority of the information that arguments built in the reasoning process will contain.

2. *Planning problem analysis*: In planning, the classic problem involves finding a sequence of actions that, starting from an initial state, leads to a goal state. A precise description of the planning problem requires a deeper analysis to identify its properties. Several dimensions need to be considered, such as whether multiple actions can be taken concurrently or if only one action is possible at a time, whether the objective is to reach a designated goal state or to maximize a reward function, the presence of one agent or multiple agents, or whether actions have associated probabilities. These issues should be carefully analyzed during this stage. Also, as we have already mentioned, it is often the case that certain planning approaches are concerned not only with the final goal state after plan execution, but also with attempting to address other important aspects, such as users' preferences [Baier and McIlraith, 2008] or value-driven actions [Teze et al., 2019a].

3. *Reasoning mechanism*: This mechanism is in charge of interpreting the available domain knowledge, generating arguments, and then comparing those in conflict to decide on acceptance. A reasoner contains three main components: an *Inference Mechanism*, which carries out inferences based on available knowledge to be used in the construction of plans; a *Conflict Solver*, which establishes a preference relation over the set of arguments through an argu-

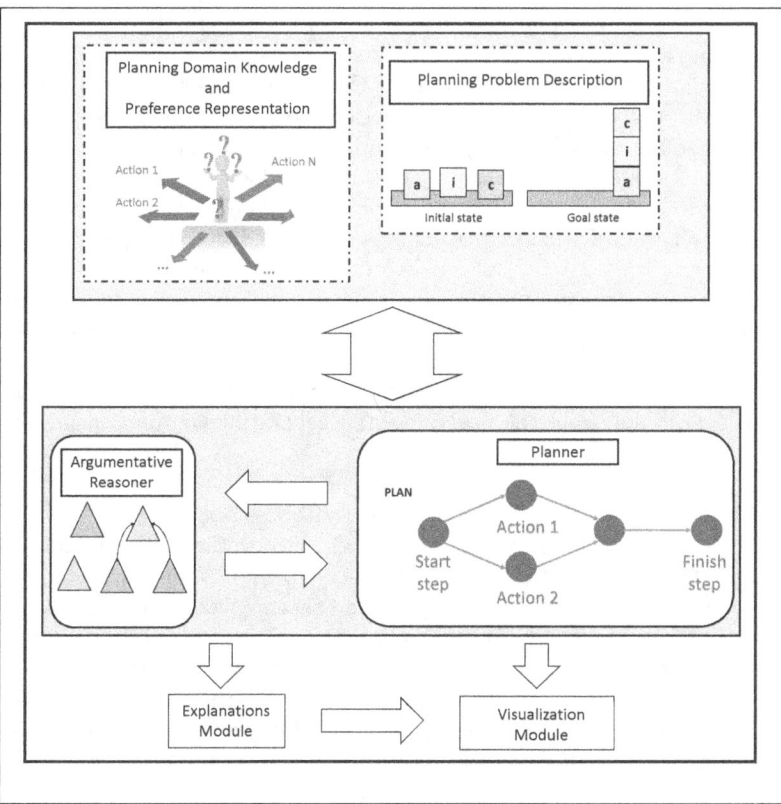

Figure 2: The Epistemic Planning Framework based on Argumentation (figure reproduced from [Teze and Godo, 2021]).

ment comparison criterion; and a *Semantic Analyzer*, which aims at determining the acceptability of arguments by considering the interaction between them. This last process can be done declaratively via conditions that a set of acceptable arguments must meet [Dung, 1995], or procedurally with a specific algorithm [García and Simari, 2004].

4. *Planning mechanism*: Responsible for the general algorithm driving the main planning system functionality, which consists in coordinating the interactions among the components mentioned above, and obtaining a sequence of actions to achieve the desired goals making use of defeasible reasoning in the process. Most of the proposals in the literature generally consider one of the following two approaches: either the whole plan is viewed as an argument, and then defeasible reasoning is performed over complete plans, or it is used as a tool for determining which actions are applicable in a given state. Planning algorithms are also mainly based on two approaches: *progression* planning and *regression* planning. The former searches forward from a given initial state until a goal state is reached, while the latter tries to improve this situation by beginning from the goal state and generating the plan in inverse order.

5. *Output design*: This step includes effective ways to interpret the process by which the planning decisions are made. Plan explainability is essential for helping users to understand and improve trust in plans [Fox et al., 2017], and such explanations can take on several forms. Visual plan explanation [Gerevini and Saetti, 2020] presents a graphical view of a plan, with nodes representing actions, edges linking them, and different filtering options available to the user. Other approaches [Oren et al., 2020] involve a textual description of the plan in natural language. Related works [Mahesar and Parsons, 2023] use argumentation for providing mechanisms to construct arguments that can be useful to justify why a plan should be executed.

Figure 2 schematically illustrates an epistemic planning framework based on the methodological guidelines described above.

In the rest of the section, we will sketch a particular instantiation of the above generic epistemic planning framework presented in [Teze et al., 2022]. We first recall the P-DeLP argumentation system upon which the planning formalism is built, then we present the planning formalism itself along with related algorithms, and finally we proceed to show computational complexity results associated with this framework.

3.2 P-DeLP: An Extension of the DeLP Argumentation Framework Dealing with Ordinal Preferences

Possibilistic Logic (see *e.g.*, [Dubois *et al.*, 1994] for full details) is a logic of qualitative uncertainty, alternative to other more numerical uncertainty models like the probabilistic one, where what really matters is the likelihood order induced on propositions by the uncertainty values they take, and not the absolute values themselves. It is thus an *ordinal* model that is very suitable for handling preferences [Benferhat *et al.*, 2001; Dubois and Prade, 2005; Kaci, 2011]. Possibilistic Defeasible Logic Programming (P-DeLP) [Chesñevar *et al.*, 2004; Alsinet *et al.*, 2008] is a structured argumentation framework that extends the DeLP framework [García and Simari, 2004] by allowing to attach weights to argument conclusions. The ultimate answer to queries is based on the existence of warranted arguments computed through a qualitative dialectical analysis. The top-down proof procedure of P-DeLP is based on the one used in Defeasible Logic Programming.

In P-DeLP, a knowledge base represents domain knowledge and user preferences encoded as prioritized DeLP rules. Given a set of literals **L**, a *weighted* clause is a pair $(R; \omega)$, where R is a rule $L \leftarrow L_1, \ldots, L_k$ or a fact L (*i.e.*, a rule with empty antecedent), $L, L_1, \ldots, L_k \in \mathbf{L}$, and the weight $\omega \in [0, 1]$ expresses the priority or preference degree of the clause, interpreted as a lower bound for the conditional necessity degree $Nec(L \mid L_1 \wedge \ldots \wedge L_k)$ in the case $R = L \leftarrow L_1, \ldots, L_k$, or a lower bound for the necessity degree $Nec(L)$ in the case $R = L$. Note that, by considering $Nec(L \mid L_1 \wedge \ldots \wedge L_k)$ we are following the usual notational conventions in Logic Programming [Lifschitz, 1997] that regards the set of literals in the body of a clause $L_1, \ldots, L_k$ as a conjunction of these literals. Also, following [García and Simari, 2004], P-DeLP rules can be represented as schematic rules with variables; as usual in Logic Programming, schematic variables are denoted with initial uppercase letters. To keep the usual terminology in defeasible reasoning, we distinguish between *strict* and *defeasible* clauses: a clause $(R; \omega)$ is referred to as strict if $\omega = 1$ (top priority) and defeasible otherwise (*i.e.*, if $\omega < 1$). The higher the weight ω, the higher the priority of the clause. Given a set $\mathbb{P}$ of weighted clauses, often referred to as a P-DeLP program or simply a *program*, we will distinguish the set of all the clauses in $\mathbb{P}$ considered as strict, denoted Π, and the set of all the defeasible clauses in $\mathbb{P}$, denoted Δ. When useful, we will write $\mathbb{P} = (\Pi, \Delta)$ to refer to the set of weighted clauses, discriminating strict and defeasible clauses.

Example 1. *The following application domain was introduced in [Teze et al., 2022], and consists of a scenario where a cooking service robot was designed to prepare a meal considering the user's particular preferences. Consider the following P-DeLP program modeling the robot's knowledge.*

$$\Pi_1 = \left\{ \begin{array}{l} (open\_now(superfour); 1) \\ (\sim good\_products(superfour); 1) \end{array} \right\}$$

$$\Delta_1 = \left\{ \begin{array}{l} (suggest(S) \leftarrow open\_now(S); 0.2) \\ (\sim suggest(S) \leftarrow \sim good\_products(S); 0.7) \end{array} \right\}$$

Observe that the set Π_1 of strict clauses has two facts, and the set Δ_1 has two defeasible rules, which can be interpreted as follows: "S is a supermarket that is open now (open_now(S))" is a reason to suggest it, whereas if "S does not offer good products ($\sim good\_products(S)$)" then there exist reasons against suggesting it. Moreover, the weights attached to the defeasible rules indicate that the second rule has more priority or is more preferred than the first rule. Recall that weights in P-DeLP are purely ordinal, so what really matters here is the preference ordering they induce.

We will use the symbol "$\mid\sim$" to denote the possibilistic inference meta-relation between a program $\mathbb{P}$ and a weighted literal $(L; \omega)$, i.e., $\mathbb{P} \mid\sim (L; \omega)$ will express that from $\mathbb{P}$ it is possible to build a sequence $(L_1; \omega_1), \ldots, (L_n; \omega_n)$ of weighted literals such that (a) $(L_n; \omega_n) = (L; \omega)$, and (b) each $(L_i; \omega_i)$ with $i < n$ either belongs to $\mathbb{P}$ or has been obtained by the application of the following *generalized modus ponens rule*

$$\frac{(H \leftarrow H_1, \ldots, H_k; \beta) \quad (H_1; \gamma_1), \ldots, (H_k; \gamma_k)}{(H; \min(\beta, \gamma_1, \ldots, \gamma_k))} \quad [GMP]$$

where $(H \leftarrow H_1, \ldots, H_k; \beta) \in \mathbb{P}$ and all weighted literals $(H_1; \gamma_1), \ldots, (H_k; \gamma_k)$ appear before $(L_i; \omega_i)$ in the sequence. Note that this rule is sound with respect to the semantics of necessity degrees as introduced for instance in [Dubois et al., 1994]. Indeed, if $Nec(H_i) \geq \gamma_i$ for $i = 1, \ldots, k$ and $Nec(L \mid H_1 \wedge \ldots \wedge H_k) \geq \beta$, then $Nec(H) \geq \min(\beta, \gamma_1, \ldots, \gamma_k)$.[1]

A P-DeLP program $\mathbb{P} = (\Pi, \Delta)$ is said to be *contradictory* if, for some atom a, $\mathbb{P} \vdash (a; \omega)$ and $\mathbb{P} \vdash (\sim a; \beta)$, with $\omega > 0$ and $\beta > 0$. Since the strict part Π represents non-defeasible information, we will assume that Π is non-contradictory itself. When reasoning from a contradictory program $\mathbb{P}$, the P-DeLP system builds arguments from $\mathbb{P}$.

An *argument* for a literal L with necessity degree $\omega > 0$, denoted $\langle \mathcal{A}, (L; \omega) \rangle$, is a minimal, non contradictory set of defeasible rules $\mathcal{A}$ such that together with the program's strict knowledge allows the derivation of L with a given weight ω (the smallest weight of the clauses involved in the derivation), that will be regarded as the conclusion supported by the argument $\mathcal{A}$.

Example 2. *The following arguments $\langle \mathcal{A}_1, (suggest(superfour); 0.2) \rangle$ and $\langle \mathcal{A}_2, (\sim suggest(superfour); 0.7) \rangle$ can be built from the P-DeLP program $\mathcal{P}_1$ presented in Example 1, where:*

$$\mathcal{A}_1 = \{ \ (suggest(superfour) \leftarrow open\_now(superfour); \ 0.2) \ \}$$
$$\mathcal{A}_2 = \{ \ (\sim suggest(superfour) \leftarrow \sim good\_products(superfour); \ 0.7) \ \}$$

and the literals $(suggest(superfour); 0.2)$ and $(\sim suggest(superfour); 0.7)$ are obtained by applying the GMP inference rule presented above to the strict facts in Π_1 and the weighted rules in $\mathcal{A}_1$ and $\mathcal{A}_2$, respectively.

Given a program program $\mathbb{P}$ and a literal L as input, the answer of the P-DeLP system to the query L is based on checking for the existence of warranted arguments for L, computed through an exhaustive dialectical analysis that involves the construction and evaluation of arguments that either support or interfere with the query under analysis. That is, the warrant process evaluates whether there exists for some weight $\alpha > 0$ an argument $\langle \mathcal{A}, (L; \alpha) \rangle$ from $\mathbb{P}$ that cannot be defeated; see [Chesñevar et al., 2004; Alsinet et al., 2008] for more details about the entire warrant process.

---

[1] Recall that a necessity measure Nec on a propositional language $\mathcal{L}$ is a mapping $Nec : \mathcal{L} \to [0,1]$ such that $Nec(\top) = 1, Nec(\bot) = 0$, and $Nec(\varphi \wedge \psi) = \min(Nec(\varphi), Nec(\psi))$. Then, the corresponding (qualitative) notion of conditional necessity is usually defined as follows: $Nec(\varphi \mid \psi) = Nec(\neg \psi \vee \varphi)$ if $Nec(\neg \psi \vee \varphi) > Nec(\neg \psi)$, and $Nec(\varphi \mid \psi) = 0$ otherwise.

3.3 A P-DeLP-based Planning Framework Instantiation

Having prioritized information can particularly be useful to guide the reasoning process in a planning problem. One of goals in [Teze et al., 2022] was to allow the adjustment of the priority weights on rules to be used by P-DeLP's inference mechanism when selecting actions in the planning process. The priority degree associated with a defeasible rule is then context-dependent, where the notion of context is understood – in a general sense – as conditions favoring a particular priority criterion.

Given a finite set of rules $\mathbb{R}$, a *priority criterion* prc is formally defined as an assignment $\rho_{\mathsf{prc}} : \mathbb{R} \to [0,1)$ of priority degrees to the rules in $\mathbb{R}$. For simplicity, we will write $\mathsf{prc}(R)$ instead of $\rho_{\mathsf{prc}}(R)$, and given a set of (weighted) defeasible rules Δ and a priority criterion prc, we will write Δ_{prc} to denote the set of rules resulting from updating the weights of the rules of Δ. If prc assigns the minimal weight 0 to a defeasible rule R (*i.e.*, if $\mathsf{prc}(R) = 0$), this means that R plays no role at all under criterion prc. On the other hand, note that by definition it is not allowed to assign a maximal weight 1 to a (defeasible) rule since in that case it would become a strict rule.

Example 3. *Consider the defeasible rules of the P-DeLP program $\mathcal{P}_1$ from Example 1, and the two criteria* pref_rocio *and* pref_aldo *prioritizing rules according respectively to the preferences of Rocio and Aldo, who are homeowners. The following are the corresponding sets of updated rules according to these criteria:*

$$\Delta_{\mathsf{pref\_rocio}} = \left\{ \begin{array}{l} (suggest(S) \leftarrow open\_now(S); 0.6) \\ (\sim suggest(S) \leftarrow \sim good\_produts(S); 0.4) \end{array} \right\}$$

$$\Delta_{\mathsf{pref\_aldo}} = \left\{ \begin{array}{l} (suggest(S) \leftarrow open\_now(S); 0.2) \\ (\sim suggest(S) \leftarrow \sim good\_produts(S); 0.9) \end{array} \right\}$$

Since arguments rely on defeasible knowledge, when they are evaluated it can be the case that there exist arguments supporting contradictory literals, so that a particular argument comparison strategy to deal with the conflicting arguments is required. A specific strategy presented in [Chesñevar et al., 2004] relies on comparing the weights of arguments. Using this strategy and the priorities of Example 3 specified in $\Delta_{\text{pref\_rocio}}$, the argument $\langle \mathcal{A}_1, (suggest(superfour); 0.6) \rangle$ is preferred over the argument $\langle \mathcal{A}_2, (\sim suggest(superfour); 0.4) \rangle$, since $\mathcal{A}_1$ provides a greater weight for the conclusion than $\mathcal{A}_2$.

One of the features of the planning framework we describe in this section is a mechanism that dynamically modifies preferences (or priorities) among pieces of defeasible knowledge depending on the current state of the world the planner is acting upon, that will be described below. In the planning system, a *state of the world* Ψ is represented as a consistent set of literals, considered to hold true.

Example 4. *The following consistent set of facts can represent a possible state of the world in a given moment:*

$$\Psi_4 = \left\{ \begin{array}{l} lunchtime \\ open\_now(superfour) \\ superM(superfour) \\ \sim good\_products(superfour) \\ recipe(pastaPuttanesca) \end{array} \right\},$$

where it is lunch time, superfour is a supermarket that is open now but does not offer good products, and that a recipe for preparing pasta puttanesca is available.

Defeasible argumentation is used for reasoning over the preconditions to execute actions. Indeed, a set of domain defeasible rules Δ together with a set Ψ of literals describing the current state of the world define a P-DeLP program (Ψ^*, Δ), where $\Psi^* = \{(L, 1) \mid L \in \Psi\}$, upon which the planner system can perform defeasible reasoning about whether preconditions of a given action are warranted. We will denote by $\mathsf{warrL}(\Psi, \Delta)$ the set of literals warranted by the program (Ψ^*, Δ).

To do so in a specific context, the planning system can use a particular priority order over the defeasible knowledge that will be obtained after evaluating a given expression. The idea is to associate to every action a suitable conditional expression that will select, by means of *guards*, the priority criteria to be used in each given context depending on the world's current state.

A *guard* is a set of literals γ, and it is satisfied by a state Ψ when $\gamma \subseteq \Psi$. In its simplest form, a *conditional-preference expression* E can be just a priority criterion prc, and in that case the priority assignment corresponding to this criterion is applied over defeasible rules. In general, E can be of the form $E = [\gamma : E_1; E_2]$, where γ is a guard and where E_1 and E_2 can be in turn either priority criteria or further conditional expressions. In such a case, if E is evaluated, and γ is satisfied in the current state Ψ (i.e., $\gamma \subseteq \Psi$), then E_1 is evaluated; otherwise, E_2 is evaluated. This recursive evaluation procedure is applied until a priority criterion is obtained.

Example 5. *Let us consider the following conditional preference about the two priority criteria introduced in Example 3:*

"If it is lunch time, use Rocio's preferences, otherwise use Aldo's preferences",

This informal statement can be captured with the following conditional-preference expression:

$$E_1 = [\{lunchtime\} : E_2; E_3],$$

where the (non-conditional) expressions E_2 and E_3 stand for

$E_2 =$ pref_rocio,

$E_3 =$ pref_aldo.

Consider now the state Ψ_4 introduced in Example 4. It is clear that the guard "lunchtime" is satisfied by the state Ψ_4, and thus the result of evaluating E_1 at Ψ_4 is the priority criterion $E_2 =$ pref_rocio.

Having defined what conditional-preference expressions are, they can be used to formally define the set of actions a planner system may use to change the world and achieve its goals. Three elements specify an action A: its preconditions P, its consequences X, and the preferences E under which P will be evaluated.

An *action* is a triple $A = \langle X, P, E \rangle$, where $X = \{X_1, X_2, \ldots, X_n\}$ is a consistent set of literals representing the consequences of executing A, $P = \{P_1, P_2, \ldots, P_n\}$ is a set of literals representing the preconditions that need to be satisfied before A can be executed, and E is

11 - ARGUMENTATIVE EPISTEMIC PLANNING WITH PREFERENCES

a conditional-preference expression representing the preferences under which to evaluate preconditions P. We will use the following notation for actions:

$$\{X_1, X_2, \ldots, X_n\} \xleftarrow{(A,E)} \{P_1, P_2, \ldots, P_n\}.$$

Intuitively, given a context represented by a state Ψ and a defeasible knowledge base Δ, an action $A = \langle X, P, E \rangle$ specifies that "if all preconditions of A are warranted by the argumentation system $(\Psi, \Delta_{\text{prc}})$, where prc is the criterion obtained by evaluating E at Ψ, then after executing A the postconditions X will be added to the state Ψ".

Example 6. *Consider the application domain presented in Example 1, and the conditional-preference expressions of Example 5. The actions that the robot can perform are the following:*

$$\mathbf{A}_6 = \left\{ \begin{array}{l} \{food\_prod\_ordering\} \xleftarrow{(order\_food\_products, E_1)} \{recipe(R), superM(S), suggest(S)\} \\ \{ing\_ready\} \xleftarrow{(search\_storage, E_1)} \{recipe(R), storage(R)\} \\ \{ing\_ready\} \xleftarrow{(receive\_food\_products, E_2)} \{food\_prod\_ordering\} \\ \{homemade\_meal\} \xleftarrow{(cooking, E_1)} \{ing\_ready\} \end{array} \right\}$$

These actions can be interpreted as follows:

— order_food_products: *ordering food products from a supermarket. There must exist a supermarket available for making an order.*

— search_storage: *searching for the correct ingredients from the house storage. The ingredients in the recipe must be in the storage.*

— receive_food_products: *receiving the supermarket's products at home. There must exist a food product order.*

— cooking: *cooking at home. All of the recipe's ingredients must be available.*

Apart from the domain knowledge to reason during the planning process, the planner will have a set of actions that will be available for modifying the world.

Formally, a *preference-based planning domain* is a triple $(\Delta, \mathbf{C}, \mathbf{A})$ where:

- Δ is a set of defeasible rules.
- $\mathbf{C}$ is a set of priority criteria over rules of Δ.
- $\mathbf{A} = \{A_1, A_2, \ldots, A_n\}$ is a set of actions, where for each $A = \langle X, P, E \rangle \in \mathbf{A}$ is such that for every prc in E it holds that prc $\in \mathbf{C}$.

As already mentioned, checking whether an action can be executed involves checking its applicability, *i.e.*, checking whether the literals of the set of preconditions can be warranted. After an applicable action is executed, the state itself is consistently modified with each effect after removing any possible conflict. The new state resulting from executing an action A in the state Ψ will be denoted by $\Psi^A = (\Psi \setminus \overline{X}) \cup X$, where $\overline{X}$ is the set of the complemented literals in X.

Example 7. *Consider the set of defeasible rules Δ_1 defined in Example 1, the set of criteria $\mathbf{C}_3 = \{$pref_rocio, pref_aldo$\}$ of Example 3, and the set of actions $\mathbf{A}_6$ presented in Example 6. Suppose the robot's planning system has the following domain $(\Delta_1, \mathbf{C}_3, \mathbf{A}_6)$ and the state Ψ_4 presented in Example 4, where:*

$$\Psi_4 = \left\{ \begin{array}{l} lunchtime \\ open\_now(superfour) \\ superM(superfour) \\ \sim good\_products(superfour) \\ recipe(pastaPuttanesca) \end{array} \right\}.$$

Consider now the action order_food_products *in $\mathbf{A}_6$ and the priority criterion* pref_rocio *obtained after evaluating E_1. This action is applicable in Ψ_4 according to the priorities defined by* pref_rocio *because one of its preconditions, recipe(pastaPuttanesca), is in Ψ_4 and its other precondition, suggest(superfour), belongs to* warrL$(\Psi_4, \Delta_{\text{pref\_rocio}})$ *since there exists a non-defeated argument for $\langle \mathcal{A}_1, (suggest(superfour); 0.6)\rangle$ where:*

$$\mathcal{A}_1 = \left\{ \ (suggest(superfour) \leftarrow open\_now(superfour); 0.6) \ \right\}$$

The resulting state of executing the action order_food_products *in state*

Ψ_4 is then the following:

$$\Psi^{\text{order\_food\_products}} = (\Psi_4 \setminus \overline{\mathsf{X}}) \cup \mathsf{X} = \left\{ \begin{array}{l} lunchtime \\ open\_now(superfour) \\ superM(superfour) \\ \sim good\_products(superfour) \\ recipe(pastaPuttanesca) \\ food\_prod\_ordering \end{array} \right\}$$

where $\mathsf{X} = \{food\_prod\_ordering\}$.

Since the execution of an applicable action leads to a new state, another action could be applicable at this new state, and so on. A sequence $\mathsf{S} = [\mathsf{A}_1, \mathsf{A}_2, \ldots, \mathsf{A}_n]$ will be regarded as an *applicable sequence* of actions at a state Ψ if (1) A_1 is applicable at Ψ, and (2) every action A_i, $2 \leq i \leq n$, is applicable in $(\cdots(\Psi^{\mathsf{A}_1})\cdots)^{\mathsf{A}_{i-1}}$. We will use Ψ^{S} or $\Psi^{[\mathsf{A}_1,\ldots,\mathsf{A}_n]}$ as a shorthand for $(\cdots(\Psi^{\mathsf{A}_1})\cdots)^{\mathsf{A}_n}$. In fact, the main aim of any planning system is to find a sequence of actions that, starting from an initial state, leads to a state where a given goal is satisfied.

A *preference-based planning problem* is a tuple $(\Psi, \Delta, \mathbf{C}, \mathbf{A}, \mathbb{G})$, where:

— Ψ is a consistent finite set of weighted literals representing an initial state,

— $(\Delta, \mathbf{C}, \mathbf{A})$ is a preference-based planning domain,

— $\mathbb{G}$ is a consistent finite set of literals representing the system's goals.

A *solution* to a preference-based planning problem is an applicable sequence of actions such that when executed in an initial state, it leads to a state that satisfies the conditions in $\mathbb{G}$.

Example 8. *Consider the following preference-based planning problem* $T = (\Psi_4, \Delta_1, \mathbf{C}_3, \mathbf{A}_6, \mathbb{G}_8)$, *where*

— Ψ_4 *is the state presented in Example 4,*

— $(\Delta_1, \mathbf{C}_3, \mathbf{A}_6)$ *is the planning domain presented in Example 7,*

— $G_8 = \{homemade\_meal\}$

A possible solution for this planning problem is the plan:

$S_1 = $ [order_food_products, receive_food_products, cooking]

since S_1 is a sequence of applicable actions at Ψ_4, and $G_8 \subseteq \Psi^{S_1}$.

So far, we have presented a planning formalism that integrates preferences into the construction of plans. In particular, the approach provides the possibility of expressing contextual preferences under which the preconditions of a specific action should be evaluated. To encode these preferences, conditional priority expressions are used, allowing the user to specify possibly different priority criteria depending on the state of the world. In the next section, we present a partial order planning algorithm that considers the conditional-preference expressions formalized above.

3.4 Argumentation in Partial Order Planning with Contextual Preferences

The formalism described above can decide whether a plan is a solution to a preference-based planning problem, but it does not describe *how* to construct such a plan for achieving the goals of a planning system. In the following, an extension of the APOP [García et al., 2008] algorithm, called P-APOP (Argumentative Partial Order Planning with Preferences), is introduced to build plans using conditional-preference expressions. We will first show an illustrative example of how a complete plan incorporating arguments and actions is obtained in P-APOP before we go into the algorithm specifics in Section 3.5.

The P-APOP algorithm has as input the system's goals and an initial state, and outputs a partial-order plan that is a solution for the planning problem. That is, the planner starts with an initial partial plan consisting of a **start** step whose effects encode the initial state and a **finish** step whose preconditions encode the goals to be achieved, in the sense of aiming at having them warranted through the argumentation process. The initial plan is then incrementally completed with new steps until all the preconditions of these steps are warranted. Intuitively, this process generates a new partial plan whenever a new step is considered. Two types of steps are identified: *action steps*, that represent the execution of an action, and *argument steps*, that provide arguments to support the preconditions of some action step. Unlike actions, arguments are not only used to support some plan step, but they are also used to interfere or support other arguments in the plan.

Figure 3 shows a sequence of partial plans and how a complete plan is obtained by means of actions and arguments for the preference-based planning problem $(\Psi_4, \Delta_1, \mathbf{C}_3, \mathbf{A}_6, \mathfrak{G}_8)$ presented in Example 8. The preconditions of the finish step represents the system's goals $\mathfrak{G}_8$ and the effects of the start step encode the initial state Ψ_4. In the graphical representation, action steps are depicted by square nodes labeled with the action name. The literals appearing below an action step represent the action's preconditions, and the literals appearing above represent its effects. Moreover, the literal that appears on the right-hand side of an action step represents the selected priority criterion obtained from the conditional-preference expression associated with the action. Argument steps are represented by triangles labeled with the argument name. The literal at the top of the triangle is the conclusion of the argument. On the other hand, the solid arrows represent *causal links* of the plan, and they are used to link an effect of an action step with a precondition of another action step or with a literal in the base of an argument step. The solid arrows that link the conclusion of an argument step and a precondition of an action step represent *support links* of the plan. The *ordering constraints* are represented by dashed arrows. These constraints allow an order to be established between steps, whereas causal and support links allow to identify the source of each literal in a plan.

In Figure 3-(a), the finish action step has one unsatisfied precondition (*homemade_meal*). The action cooking is the only one available that can be used to satisfy this precondition. Thus, cooking is added (Figure 3-(b)) to the plan by the planning process, and its precondition becomes a subgoal to be achieved. Observe that *ing_ready* is achieved by two action steps: search_storage and receive_food_products. If search_storage (Figure 3-(c)) is chosen, a new step is added; now, *recipe(pastaPuttanesca)* and *storage(pastaPuttanesca)* must be satisfied as preconditions. The literal *recipe(pastaPuttanesca)* is satisfied by the start step, but none of the available actions achieve *storage(pastaPuttanesca)*, and from the rules in $\Delta_{\text{pref\_rocio}}$, it is not possible to build an argument for *storage(pastaPuttanesca)* either. In that case, the algorithm fails in finding a step to achieve a subgoal, so the control is returned to the point in the algorithm where the choice was made. Now receive_food_products (Figure 3-(d)) is chosen, a new step is added, and the precondition *food_prod_ordering* must be satisfied. The literal *food_prod_ordering* is consequence of the action order_food_products; therefore, the corresponding step order_food_products (Figure 3-(e)) is added to the plan. The literals *recipe(pastaPuttanesca)* and *superM(superfour)* are satisfied by the start step, and from the rules in $\Delta_{\text{pref\_rocio}}$, where pref_rocio is the priority criterion used after evaluating E_1, it is possible to build argument $\langle \mathcal{A}_1, (suggest(superfour); 0.6) \rangle$ supporting $(suggest(superfour); 0.6)$, as well as $\langle \mathcal{A}_2, (\sim suggest(superfour); 0.4) \rangle$, which attacks the former (for the detailed structure, see Example 2). Then, argument $\langle \mathcal{A}_1, (suggest(superfour); 0.6) \rangle$ is selected since it has a greater weight and the literal in the base of the rule body conforming $\mathcal{A}_1$ is achieved by the start step, and thus a plan is finally formulated.

Note that if the criterion pref_aldo were selected, order_food_products would not be applicable since $\mathcal{A}_2$ would have a greater weight than $\mathcal{A}_1$, and under this criterion *suggest(superfour)* would become non-warranted – consequently, no solution plan could be formulated. This exemplifies the fact that the criterion selected for a specific action and the information considered for such selection are very relevant in the argumentative reasoning process, since the use of a different criterion can change the warrant state of an action's preconditions, clearly affecting its applicability.

As a final remark, note that the planning process does not establish a single specific sequence of actions, but rather focuses on defining a set of ordering constraints, specifying which actions must be executed before others. To determine whether a partial-order plan is a solution for a preference-based planning problem, it is necessary to first establish a correspondence between partially- and totally-ordered plans by applying a topological sorting to derive a total-order solution, as usual in the partial-order planning paradigm. Given a totally-ordered sequence of

11 - ARGUMENTATIVE EPISTEMIC PLANNING WITH PREFERENCES

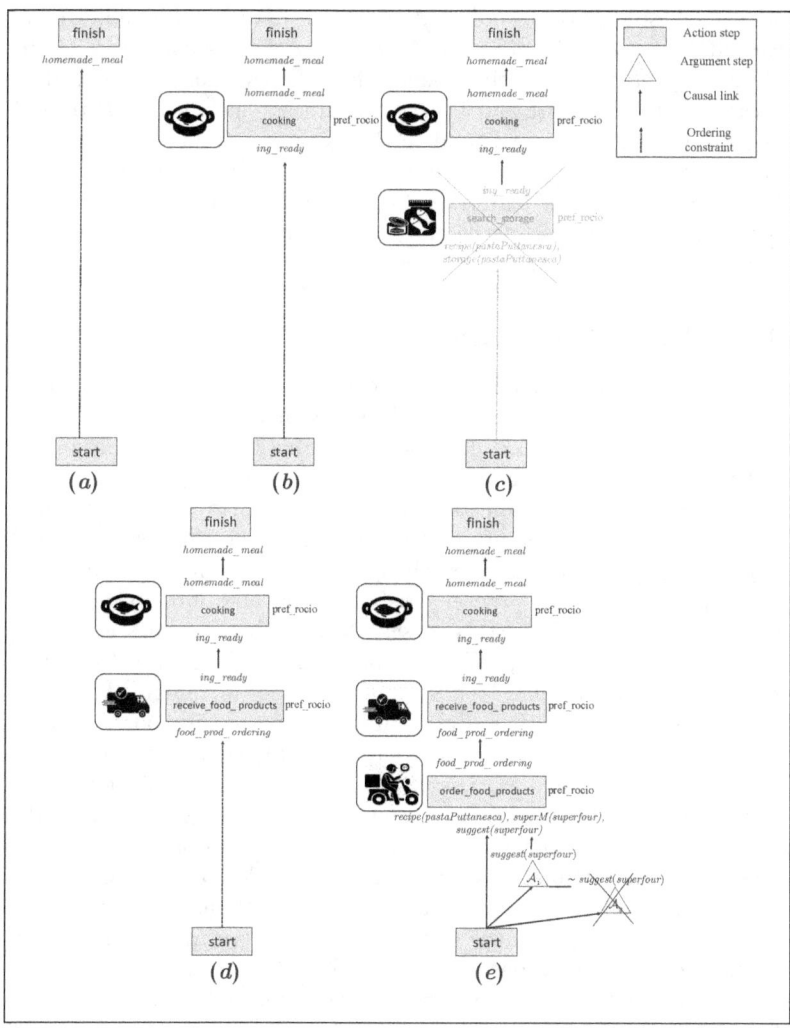

Figure 3: Partial plans for Example 8.

action steps *Seq* derived from a particular partial plan, where each action step is consistent with the ordering constraints of the corresponding plan, we will denote with $Plan(Seq) = [\mathsf{A}_1, \mathsf{A}_2, \ldots, \mathsf{A}_n]$ the sequence of actions obtained by replacing each action step in *Seq* with its corresponding action. Note that start and finish steps are not included in $Plan(Seq)$ because they do not correspond to the execution of any action – they are only required to represent the initial state and goals of the problem. Finally, a partial-order plan is a solution to a preference-based planning problem T when the sequence of actions $Plan(Seq) = [\mathsf{A}_1, \mathsf{A}_2, \ldots, \mathsf{A}_n]$ is a solution to T.

3.5 The P-APOP Algorithm

In P-APOP, finding a partial plan consists in completing a plan by adding steps to achieve goals, as illustrated in Figure 3. For a better understanding of the P-APOP algorithm, in this section we operationally describe its main algorithms.

The P-APOP algorithm starts with an initial plan and seeks to complete it with new steps, attempting to resolve the threats that could appear. These threats appear when the effect of a new action added in the plan is to delete a literal satisfying a precondition already solved by other action steps. In this sense, when involving actions and arguments to construct plans, different types of threats can arise. In [García et al., 2008], the authors identify different types of threats that could arise in argumentation-based planning and propose methods to resolve each of them. A new type of interference is introduced in [Teze et al., 2022] when conditional expressions are used: an action might have interferences with the guards appearing in these expressions. Thus, the P-APOP algorithm first builds a null plan, which consists of six empty sets containing: action steps, argument steps, ordering constraints, subgoals, causal links, and support links. Then, it attempts to complete it with the recursive procedure COMPLETE_PLAN (see Algorithm 1) until all the steps' preconditions are warranted.

To achieve its goal, besides the initial state Ψ and the goals G, function COMPLETE_PLAN considers Δ and **A** as input parameters. The set Δ contains defeasible rules whose weights will be possibly changed by the use of a different priority criterion when new action steps are added to the plan. For convenience, it is assumed that the initial weights of the rules in Δ are provided by a certain distinguished priority criterion in the set **C** of criteria the system works with. The procedure COM-PLETE_PLAN begins with an unsatisfied subgoal; then, it is necessary to identify those steps that can be used to achieve such a subgoal. Towards this end, the procedure GET_STEPS is in charge of building plan steps to support an unsatisfied subgoal. The set *Steps* contains either actions in **A**, or a set of argument steps for *SubGoal* built from Δ_{prc} under a given criterion prc. If no argument can be built, then only actions are considered. Note that if the algorithm fails in finding a step to achieve a subgoal, the backtracking point is updated and the control is returned at the point in the algorithm where a step choice (statement **choose**) was made.

Once the set *Steps* has been built, a step is chosen and the sets included in *Plan* are updated. As we have already mentioned, after a new step is added to the plan, new threats could occur. The procedure RESOLVE_PLAN will consider all steps in the plan to detect possible interference cases that can appear and try to resolve each of them. This particular function checks four different types of threats involving arguments and actions:

- *action-action*: A precondition L is threatened by an action step A if the complemented literal $\overline{L}$ is an effect of A.

- *action-argument*: Let $\langle \mathcal{A}, (L;\alpha) \rangle$ be an argument supporting a precondition of an action step A_j; an action step A_i threatens the argument $\langle \mathcal{A}, (L;\alpha) \rangle$ if an effect of A_i negates any literal present in the set of all literals that appear as bodies of rules in the argument $\langle \mathcal{A}, (L;\alpha) \rangle$. Step A_i comes before $\langle \mathcal{A}, (L;\alpha) \rangle$.

- *argument-argument*: Let $\langle \mathcal{A}, (L;\alpha) \rangle$ be an argument added to a plan to support the precondition of an action step A; then, $\langle \mathcal{A}, (L;\alpha) \rangle$ is threatened by an argument $\langle \mathcal{B}, (Q;\beta) \rangle$ if $\langle \mathcal{B}, (Q;\beta) \rangle$ is a defeater for $\langle \mathcal{A}, (L;\alpha) \rangle$, and $\langle \mathcal{B}, (Q;\beta) \rangle$ is ordered to appear before $\langle \mathcal{A}, (L;\alpha) \rangle$ in the plan.

- *guard-action*: Let E be a conditional-preference expression. The guard γ in E is threatened by an action step A if one of its effects negates a literal present in γ, and the satisfaction state of such a guard becomes non-satisfied.

Different threat resolution methods may be applied for each kind of threat, such as including new ordering constraints for moving the cause of the threat to a harmless position or eliminating the threat with a counterargument or a new action step. A detailed description of the algorithms that deal with these problems can be found in [García *et al.*, 2007; García *et al.*, 2008; Teze *et al.*, 2022]. Note that unresolved threats involve backtracking, which implies removing the last added step from *Plan*, and considering pending alternatives. The basic idea behind P-APOP is to search through a plan space, which can be characterized as a tree where each node represents a partial-order plan. If a failure occurs, the algorithm backtracks to the parent node. Note that the rollback process involved in the backtracking step requires identifying any links, ordering constraints, subgoals, and dependency tree associated with the failed step, and removing them without changing the rest of the plan.

Progress through the P-APOP algorithm consists of analyzing partially complete plans and modifying them in a way that brings them closer to a solution. It is easy to see that any linear plan that satisfies a partially ordered plan is such that all action step preconditions are necessarily satisfied, and backtracking ensures that the search space is eventually exhausted.

11 - Argumentative Epistemic Planning with Preferences

Algorithm 1 Function to complete plan

1: **function** COMPLETE_PLAN($Plan, \Delta, \mathbf{A}, \Psi$): $Plan$
2:    **if** $Plan.SubGoals = \emptyset$ **then return** $Plan$;
3:    **end if**
4:    **choose** $SubGoal$ **from** $Plan$;
5:    $Steps :=$ GET_STEPS($SubGoal, Plan, \Delta, \mathbf{A}, \Psi$);
6:    **if** $Steps = \emptyset$ **then fail**;
7:    **end if**
8:    **choose** $Step$ **from** $Steps$;
9:    **if** $Step$ is an action step OR $Step$ is an argument step **then**
10:       UPDATE $Plan$
11:    **end if**
12:    RESOLVE_THREATS($Plan, \Delta$);
13:    COMPLETE_PLAN($Plan, \Delta, \mathbf{A}, \Psi$);
14: **end function**

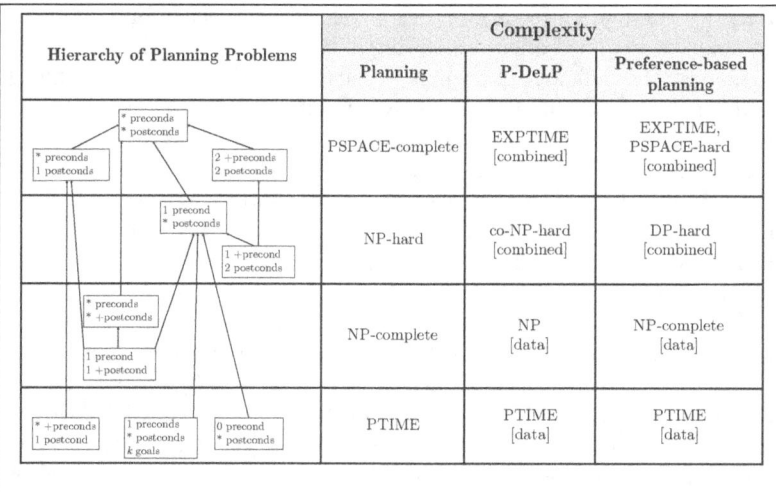

Figure 4: Overview of complexity results for a variety of problems (figure reproduced from [Teze et al., 2022]).

Summary of Complexity Results. After showing how the P-DeLP system can be considered together with partial-order planning techniques to consider arguments as planning steps, and sketching the P-APOP algorithm for constructing plans, we now focus on a major issue that arises in plan construction processes, which is related to the computational requirements that they must satisfy. Thus, understanding the inherent complexity of the reasoning tasks is crucial towards efficient implementations of defeasible argumentation-based planning systems. In that respect, the following decision problem is studied:

> *Does there exist a plan P such that, executed starting in state Ψ, arrives at goal $\mathbb{G}$ following priorities* **C** *and satisfies the constraints imposed by* Δ *and* **A***?*

Here we summarize from [Teze et al., 2022] both data and combined complexity results of query answering in the context of P-DeLP in order to analyze the difficulty of resolving preference-based planning problems under a variety of conditions. These results are based in turn on the work of [Bylander, 1991], where the author provides several complexity results for SATPLAN – the decision problem of establishing whether an instance of *propositional* planning is satisfiable – and several of its restricted versions. These results, in combination with the those for DeLP reported in [Cecchi et al., 2006; Alfano et al., 2021], are leveraged for this analysis.

Figure 4 illustrates several complexity results for a hierarchy of different planning problems, and shows how the computational complexity varies from PTIME to EXPTIME, depending on different restrictions that can be considered. The results for SATPLAN are summarized in the first complexity column reproduced from [Bylander, 1991], whereas the main data and combined complexity results for P-DeLP (the decision problem of whether a literal is warranted) are given in the second column, and they are direct consequences of those in [Cecchi et al., 2006; Alfano et al., 2021]. Finally, the third column gives the complexity results for the preference-based planning problem under each set of restricted versions. For a more detailed discussion on these results, see [Teze et al., 2022].

4 Conclusions and Perspectives

In this chapter, we have been mainly concerned with a defeasible argumentation-based approach to epistemic planning. This is a relatively recent field involving aspects of automated planning, knowledge representation, and defeasible reasoning. In order to develop theoretical formalisms and planning systems that are both expressive and practically efficient, it is necessary to combine the state of the art from all these areas. Over the years, for instance, particular attention has already been paid in the literature to efforts towards capturing more and more complex and challenging planning settings than the classical one such as planning under uncertainty, with preferences or multi-agent planning.

More specifically, in this chapter we have first provided main motivations for the need to use argumentation in the context of planning systems. Then, we have given an overview of relevant works on argumentation-based epistemic planning, focusing particularly on approaches arising in four research fields: practical reasoning, automated planning, multi-agent planning, and explainable planning. This was followed by a discussion of the use of preferences in both defeasible argumentation and planning formalisms. Finally, we presented an overview of a specific preference-based planning system, which combines partial order planning with defeasible reasoning.

Looking forward, we can identify several topics and research directions in the area of defeasible planning. We now briefly discuss some of the ones we find particularly interesting:

- The study of planning processes as search through a space of plans naturally lends to analytical studies of computational complexity. Analyses of aspects related to efficiency in the use of resources are of great importance, but little progress has been made in this direction in argumentation-based planning. Much work can still be done in the realm of algorithms and complexity, such as studying sub-problems for which efficiency guarantees can be established.

- Incorporating *humans in the planning loop*, especially in collaborative settings, presents several important challenges that must be addressed. Explainable planning seeks to build trust and transparency when interacting with humans; thus, even though explainability is not a new topic, it is another promising research line in the realm of epistemic planning.

- While defeasible planners have been effectively applied in different domains, it is important to achieve implementations with empirical evaluations in real settings, and compare obtained results with other approaches from the literature in terms of effectiveness and efficiency. There are many opportunities in this line of work.

- The P-APOP algorithm does not leverage heuristics, so another promising direction to analyze in the future is the adaptation of the algorithm to include different heuristic methods that allow the reduction of the search space and, consequently, the overall computational cost.

- In *abstract* argumentation frameworks, the specification of different *semantics* encodes different criteria of acceptability of (sets of) arguments. Establishing a correspondence between our framework and such argumentation semantics, and their associated properties, is a promising line of work that can serve to investigate how the selection of the best plan(s) can be based on well-established properties.

- Conditional-preference expressions constitute a key component in the argumentation-based planning framework we have described in Section 3 to decide which actions to keep while constructing plans. A detailed study of such expressions and properties that characterize them for rational decision-making is an interesting open task for future research.

- Another important topic is the relationship between the notion of threat and attack present in the argumentation literature. On the other hand, it would also be especially interesting to study how to incorporate values into defeasible rules based on rationality principles, like the ones proposed by [Hunter, 2013]. This is a challenging objective for future research.

- Finally, the study of other preference representation models and tools – such as operators for combining or prioritizing contexts –

is also a challenge that deserves attention.

These research directions are only a few of the ones that stand out; the goal of this chapter was to describe an area that shows promising early results, but with many opportunities for both basic and applied research and development.

Acknowledgements

The authors are indebted to Guillermo R. Simari for inspiring discussions and to anonymous reviewers for their helpful comments. This work was funded in part by Universidad Nacional del Sur (UNS) under grants PGI 24/ZN057 and PGI 24/N055, Consejo Nacional de Investigaciones Científicas y Técnicas (CONICET) under grant PIP 11220210100577CO, Agencia Nacional de Promoción Científica y Tecnológica, Argentina under grant PICT-2018-0475 (PRH-2014-0007), and Universidad Nacional de Entre Ríos (UNER) under grant PDTS-UNER 7066). Godo acknowledges partial support by the Spanish project LINEXSYS (PID2022-139835NB-C21) funded by MCIN/AEI/10.13039/501100011033.

References

[Alfano et al., 2021] Gianvincenzo Alfano, Sergio Greco, Francesco Parisi, Gerardo I. Simari, and Guillermo R. Simari. Incremental computation for structured argumentation over dynamic DeLP knowledge bases. *Artif. Intell.*, 300:103553, 2021.

[Alsinet et al., 2008] Teresa Alsinet, Carlos I. Chesñevar, Lluís Godo, and Guillermo R. Simari. A logic programming framework for possibilistic argumentation: Formalization and logical properties. *Fuzzy Sets and Systems*, 159(10):1208–1228, 2008.

[Amgoud and Vesic, 2011a] Leila Amgoud and Srdjan Vesic. A new approach for preference-based argumentation frameworks. *Ann. Math. Artif. Intell.*, 63(2):149–183, 2011.

[Amgoud and Vesic, 2011b] Leila Amgoud and Srdjan Vesic. Two roles of preferences in argumentation frameworks. In *Symbolic and Quantitative Approaches to Reasoning with Uncertainty – 11th European Conference, ECSQARU*, pages 86–97, 2011.

[Amgoud et al., 2011] Leila Amgoud, Caroline Devred, and Marie-Christine Lagasquie-Schiex. Generating possible intentions with constrained argumentation systems. *Int. J. Approx. Reason.*, 52(9):1363–1391, 2011.

[Atkinson and Bench-Capon, 2021] Katie Atkinson and Trevor Bench-Capon. Value-based argumention. In *Handbook of Formal Argumentation, Vol. 2*, pages 299–354. College Publications, 2021.

[Baier and McIlraith, 2008] Jorge A. Baier and Sheila A. McIlraith. Planning with preferences. *AI Mag.*, 29(4):25–36, 2008.

[Baral et al., 2017] Chitta Baral, Thomas Bolander, Hans van Ditmarsch, and Sheila A. McIlraith. Epistemic planning (Dagstuhl seminar 17231). *Dagstuhl Reports*, 7(6):1–47, 2017.

[Beirlaen et al., 2018] Mathieu Beirlaen, Jesse Heyninck, Pere Pardo, and Christian Straßer. Argument strength in formal argumentation. *FLAP*, 5(3):629–676, 2018.

[Belesiotis et al., 2009] Alexandros Belesiotis, Michael Rovatsos, and Iyad Rahwan. A generative dialogue system for arguing about plans in situation calculus. In *Argumentation in Multi-Agent Systems, 6th International Workshop, ArgMAS 2009*, volume 6057 of *Lecture Notes in Computer Science*, pages 23–41. Springer, 2009.

[Belesiotis et al., 2010] Alexandros Belesiotis, Michael Rovatsos, and Iyad Rahwan. Agreeing on plans through iterated disputes. In *9th International Conference on Autonomous Agents and Multiagent Systems (AAMAS)*, pages 765–772. IFAAMAS, 2010.

[Belle et al., 2023] Vaishak Belle, Thomas Bolander, Andreas Herzig, and Bernhard Nebel. Epistemic planning: Perspectives on the special issue. *Artif. Intell.*, 316:103842, 2023.

[Bench-Capon and Atkinson, 2009] Trevor J. M. Bench-Capon and Katie Atkinson. Action-state semantics for practical reasoning. In *The Uses of Computational Argumentation, Papers from the 2009 AAAI Fall Symposium*, volume FS-09-06 of *AAAI Technical Report*. AAAI, 2009.

[Bench-Capon and Dunne, 2007] Trevor J. M. Bench-Capon and Paul E. Dunne. Argumentation in artificial intelligence. *Artif. Intell.*, 171(10-15):619–641, 2007.

[Bench-Capon, 2003] Trevor J. M. Bench-Capon. Persuasion in practical argument using value-based argumentation frameworks. *J. Log. Comput.*, 13(3):429–448, 2003.

[Benferhat et al., 2001] Salem Benferhat, Didier Dubois, and Henri Prade. Towards a possibilistic logic handling of preferences. 14:303–317, 05 2001.

[Bidoux et al., 2019] Loïc Bidoux, Jean-Paul Pignon, and Frédérick Bénaben.

Planning with preferences using Multi-Attribute Utility Theory along with a Choquet integral. *Eng. Appl. Artif. Intell.*, 85:808–817, 2019.

[Bienvenu et al., 2011] Meghyn Bienvenu, Christian Fritz, and Sheila A. McIlraith. Specifying and computing preferred plans. *Artif. Intell.*, 175(7-8):1308–1345, 2011.

[Bolander and Andersen, 2011] Thomas Bolander and Mikkel Birkegaard Andersen. Epistemic planning for single and multi-agent systems. *Journal of Applied Non-Classical Logics*, 21(1):9–34, 2011.

[Bolander, 2017] Thomas Bolander. A gentle introduction to epistemic planning: The DEL approach. In *Proceedings of the Ninth Workshop on Methods for Modalities (M4M@ICLA)*, pages 1–22, 2017.

[Bylander, 1991] Tom Bylander. Complexity results for planning. In *Proceedings of the 12th International Joint Conference on Artificial Intelligence. Sydney*, pages 274–279. Morgan Kaufmann, 1991.

[Caminada et al., 2014] Martin Caminada, Sanjay Modgil, and Nir Oren. Preferences and unrestricted rebut. In *Computational Models of Argument - Proceedings of COMMA*, volume 266, pages 209–220. IOS Press, 2014.

[Canal et al., 2023] Gerard Canal, Carme Torras, and Guillem Alenyà. Generating predicate suggestions based on the space of plans: an example of planning with preferences. *User Model. User Adapt. Interact.*, 33(2):333–357, 2023.

[Cecchi et al., 2006] Laura A Cecchi, Pablo R Fillottrani, and Guillermo R Simari. On the complexity of DeLP through game semantics. In *Proc. 11th Intl. Workshop on Nonmonotonic Reasoning (NMR 2006)*, pages 386–394, 2006.

[Chesñevar et al., 2004] Carlos I. Chesñevar, Guillermo R. Simari, Teresa Alsinet, and Lluís Godo. A Logic Programming Framework for Possibilistic Argumentation with Vague Knowledge. In *UAI '04, Proceedings of the 20th Conference in Uncertainty in Artificial Intelligence*, pages 76–84. AUAI Press, 2004.

[Cohen et al., 2021] Andrea Cohen, Sebastian Gottifredi, Luciano H. Tamargo, Alejandro Javier García, and Guillermo R. Simari. An informant-based approach to argument strength in defeasible logic programming. *Argument Comput.*, 12(1):115–147, 2021.

[Cyras and Toni, 2016] Kristijonas Cyras and Francesca Toni. ABA+: assumption-based argumentation with preferences. In *Principles of Knowledge Representation and Reasoning: Proceedings of the Fifteenth International Conference (KR)*, pages 553–556. AAAI Press, 2016.

[Deagustini et al., 2013] C. A. D. Deagustini, S. E. Fulladoza Dalibón, S. Gottifredi, M. A. Falappa, C. I. Chesñevar, and G. R. Simari. Relational databases as a massive information source for defeasible argumentation. *Knowledge-Based Systems*, 51:91–109, 2013.

[Dubois and Prade, 2005] Didier Dubois and Henri Prade. A bipolar possibilistic representation of knowledge and preferences and its applications. In *Fuzzy Logic and Applications, 6th International Workshop (WILF)*, volume 3849 of *Lecture Notes in Computer Science*, pages 1–10. Springer, 2005.

[Dubois et al., 1994] Didier Dubois, Jerome Lang, and Henri Prade. Possibilistic Logic. In *Handbook of Logic in Artificial Intelligence and Logic Programming, Nonmonotonic Reasoning, and Uncertain Reasoning*, volume 3, pages 439–513. Claredon Press, 1994.

[Dung et al., 2006] Phan Minh Dung, Robert A. Kowalski, and Francesca Toni. Dialectic proof procedures for assumption-based, admissible argumentation. *Artif. Intell.*, 170(2):114–159, 2006.

[Dung, 1995] Phan M. Dung. On the acceptability of arguments and its fundamental role in nonmonotonic reasoning, logic programming and n-person games. *Artif. Intell.*, 77(2):321–358, 1995.

[Dung, 2016] Phan Minh Dung. Argumentation for practical reasoning: An axiomatic approach. In *PRIMA 2016: Princiles and Practice of Multi-Agent Systems – 19th International Conference*, volume 9862 of *Lecture Notes in Computer Science*, pages 20–39. Springer, 2016.

[Fan, 2018] Xiuyi Fan. On generating explainable plans with assumption-based argumentation. In *PRIMA 2018: Principles and Practice of Multi-Agent Systems – 21st International Conference*, volume 11224 of *Lecture Notes in Computer Science*, pages 344–361. Springer, 2018.

[Ferretti et al., 2008] Edgardo Ferretti, Marcelo Errecalde, Alejandro J. García, and Guillermo R. Simari. Decision rules and arguments in defeasible decision making. In *Computational Models of Argument: Proceedings of COMMA*, pages 171–182, 2008.

[Fox et al., 2017] Maria Fox, Derek Long, and Daniele Magazzeni. Explainable planning. *CoRR*, abs/1709.10256, 2017.

[García and Simari, 2004] Alejandro J. García and Guillermo R. Simari. Defeasible logic programming: An argumentative approach. *Theory and Practice of Logic Programming (TPLP)*, 4(1-2):95–138, 2004.

[García et al., 2007] Diego R. García, Alejandro Javier García, and Guillermo R. Simari. Planning and defeasible reasoning. In *6th International Joint Conference on Autonomous Agents and Multiagent Systems (AAMAS 2007)*, page 222, 2007.

[García et al., 2008] Diego R. García, Alejandro Javier García, and

Guillermo R. Simari. Defeasible reasoning and partial order planning. In *Foundations of Information and Knowledge Systems, 5th International Symposium (FoIKS)*, pages 311–328, 2008.

[Gerevini and Long, 2005] Alfonso Gerevini and Derek Long. Plan constraints and preferences in PDDL3. Technical report, Technical Report 2005-08-07, Department of Electronics for Automation, University of Brescia, Brescia, Italy, 2005.

[Gerevini and Saetti, 2020] Alfonso E Gerevini and Alessandro Saetti. An interactive tool for plan generation, inspection, and visualization. *Knowledge Engineering Tools and Techniques for AI Planning*, pages 127–155, 2020.

[Girle et al., 2004] Roderic A. Girle, David Hitchcock, Peter McBurney, and Bart Verheij. Decision support for practical reasoning. In *Argumentation Machines, New Frontiers in Argument and Computation*, volume 9 of *Argumentation Library*, pages 55–83. Springer, 2004.

[Godo et al., 2012] Lluís Godo, Enrico Marchioni, and Pere Pardo. Extending a temporal defeasible argumentation framework with possibilistic weights. In *Logics in Artificial Intelligence – 13th European Conference, JELIA 2012*, pages 242–254, 2012.

[Hunter, 2013] Anthony Hunter. A probabilistic approach to modelling uncertain logical arguments. *Int. J. Approx. Reason.*, 54(1):47–81, 2013.

[Jorge et al., 2008] A Jorge, Sheila A McIlraith, et al. Planning with preferences. *AI Magazine*, 29(4):25–25, 2008.

[Kaci and van der Torre, 2008] Souhila Kaci and Leendert van der Torre. Preference-based argumentation: Arguments supporting multiple values. *Int. J. Approx. Reasoning*, 48(3):730–751, 2008.

[Kaci et al., 2018] Souhila Kaci, Leendert W. N. van der Torre, and Serena Villata. Preference in abstract argumentation. In *Computational Models of Argument – Proceedings of COMMA 2018*, volume 305 of *Frontiers in Artificial Intelligence and Applications*, pages 405–412. IOS Press, 2018.

[Kaci et al., 2021] Souhila Kaci, Leendert van Der Torre, Srdjan Vesic, and Serena Villata. Preference in abstract argumention. In *Handbook of Formal Argumentation, Vol. 2*, pages 211–248. College Publications, 2021.

[Kaci, 2011] Souhila Kaci. *Working with Preferences: Less Is More*. Cognitive Technologies. Springer, 2011.

[Klassen et al., 2023] Toryn Q. Klassen, Christian Muise, and Sheila A. McIlraith. Planning with epistemic preferences. In *Proceedings of the 20th International Conference on Principles of Knowledge Representation and Reasoning (KR)*, pages 752–756, 2023.

[Liao and Huang, 2010] Beishui Liao and Huaxin Huang. ANGLE: an autonomous, normative and guidable agent with changing knowledge. *Inf. Sci.*, 180(17):3117–3139, 2010.

[Liao et al., 2019] Beishui Liao, Nir Oren, Leender van der Torre, and Serena Villata. Prioritized norms in formal argumentation. *J. Log. Comput.*, 29(2):215–240, 2019.

[Lifschitz, 1997] Vladimir Lifschitz. Foundations of logic programs. In Gerhard Brewka, editor, *Principles of Knowledge Representation*, pages 69–128. Center for the Study of Language and Information, 1997.

[Mahesar and Parsons, 2023] Quratul-ain Mahesar and Simon Parsons. Argument schemes and a dialogue system for explainable planning. *ACM Trans. Intell. Syst. Technol.*, 14:1–25, 2023.

[Modgil and Prakken, 2013] A general account of argumentation with preferences. *Artificial Intelligence*, 195:361–397, 2013.

[Modgil, 2009] Sanjay Modgil. Reasoning about preferences in argumentation frameworks. *Artif. Intell.*, 173(9-10):901–934, 2009.

[Monteserin and Amandi, 2011] Ariel Monteserin and Analía Amandi. Argumentation-based negotiation planning for autonomous agents. *Decision Support Systems*, 51(3):532–548, 2011.

[Myers and Lee, 1999] Karen L. Myers and Thomas J. Lee. Generating qualitatively different plans throught metatheoretic biases. In *Proceedings of the Sixteenth National Conference on Artificial Intelligence and Eleventh Conference on Innovative Applications of Artificial Intelligence*, pages 570–576. AAAI Press / The MIT Press, 1999.

[Nawwab et al., 2008] Fahd S. Nawwab, Trevor J. M. Bench-Capon, and Paul E. Dunne. A methodology for action-selection using value-based argumentation. In *COMMA*, pages 264–275, 2008.

[Oren et al., 2020] Nir Oren, Kees van Deemter, and Wamberto Weber Vasconcelos. Argument-based plan explanation. In *Knowledge Engineering Tools and Techniques for AI Planning*, pages 173–188. Springer, 2020.

[Pajares-Ferrando and Onaindia, 2017] Sergio Pajares-Ferrando and Eva Onaindia. Defeasible-argumentation-based multi-agent planning. *Inf. Sci.*, 411:1–22, 2017.

[Pajares-Ferrando et al., 2011] Sergio Pajares-Ferrando, Eva Onaindia, and Alejandro Torreño. An architecture for defeasible-reasoning-based cooperative distributed planning. In *Proceedings of OTM, Part I*, volume 7044 of *Lecture Notes in Computer Science*, pages 200–217. Springer, 2011.

[Pardo and Godo, 2013] Pere Pardo and Lluís Godo. A temporal argumentation approach to cooperative planning using dialogues. In *Computational Logic in Multi-Agent Systems – 14th International Workshop (CLIMA XIV)*, volume 8143 of *Lecture Notes in Computer Science*, pages 307–324. Springer, 2013.

[Pardo and Godo, 2018] Pere Pardo and Lluís Godo. A temporal argumentation approach to cooperative planning using dialogues. *J. Log. Comput.*, 28(3):551–580, 2018.

[Pardo et al., 2011a] Pere Pardo, Pilar Dellunde, and Lluís Godo. Argumentation-based negotiation in t-DeLP-POP. In *Artificial Intelligence Research and Development – Proceedings of the 14th International Conference of the Catalan Association for Artificial Intelligence*, volume 232 of *Frontiers in Artificial Intelligence and Applications*, pages 179–188. IOS Press, 2011.

[Pardo et al., 2011b] Pere Pardo, Sergio Pajares Ferrando, Eva Onaindia, Lluís Godo, and Pilar Dellunde. Cooperative dialogues for defeasible argumentation-based planning. In *Argumentation in Multi-Agent Systems – 8th International Workshop (ArgMAS)*, volume 7543 of *Lecture Notes in Computer Science*, pages 174–193. Springer, 2011.

[Pardo et al., 2011c] Pere Pardo, Sergio Pajares, Eva Onaindia, Lluís Godo, and Pilar Dellunde. Multiagent argumentation for cooperative planning in DeLP-POP. In *10th International Conference on Autonomous Agents and Multiagent Systems (AAMAS 2011)*, pages 971–978, 2011.

[Pollock, 1995] John L Pollock. *Cognitive carpentry: A blueprint for how to build a person*. Mit Press, 1995.

[Pollock, 1998] John L. Pollock. Defeasible planning. In *Integrating Planning, Scheduling, and Execution in Dynamic and Uncertain Environments, Papers from the Conference on Artificial Intelligence AAAI Workshop*. AAAI, 1998.

[Pollock, 1999] John L. Pollock. Rational cognition in OSCAR. In *Intelligent Agents VI, Agent Theories, Architectures, and Languages (ATAL), 6th International Workshop*, volume 1757 of *Lecture Notes in Computer Science*, pages 71–90. Springer, 1999.

[Prakken and Sartor, 1997] H. Prakken and G. Sartor. Argument-based extended logic programming with defeasible priorities. *Journal of Applied Non-classical Logics*, 7:25–75, 1997.

[Prakken, 2021] Henry Prakken. Philosophical reflections on argument strength and gradual acceptability. In *Symbolic and Quantitative Approaches to Reasoning with Uncertainty – 16th European Conference (ECSQARU)*, volume 12897, pages 144–158. Springer, 2021.

[Rahwan and Amgoud, 2006a] Iyad Rahwan and Leila Amgoud. An argumentation-based approach for practical reasoning. In *Argumentation in Multi-Agent Systems, Third International Workshop, ArgMAS 2006*, volume 4766 of *Lecture Notes in Computer Science*, pages 74–90. Springer, 2006.

[Rahwan and Amgoud, 2006b] Iyad Rahwan and Leila Amgoud. An argumentation based approach for practical reasoning. In *AAMAS*, pages 347–354, 2006.

[Rahwan and Simari, 2009] Iyad Rahwan and Guillermo R. Simari. *Argumentation in Artificial Intelligence*. Springer Publishing Company, Incorporated, 2009.

[Šefránek, 2008] Ján Šefránek. Preferred answer sets supported by arguments. In *Proceedings of 12th International Workshop on Non-Monotonic Reasoning (NMR 2008)*, pages 232–240, 2008.

[Shams et al., 2016] Zohreh Shams, Marina De Vos, Nir Oren, and Julian A. Padget. Normative practical reasoning via argumentation and dialogue. In *Proceedings of the Twenty-Fifth International Joint Conference on Artificial Intelligence (IJCAI)*, pages 1244–1250. IJCAI/AAAI Press, 2016.

[Shams et al., 2020a] Zohreh Shams, Marina De Vos, Nir Oren, and Julian Padget. Argumentation-based reasoning about plans, maintenance goals, and norms. *ACM Transactions on Autonomous and Adaptive Systems*, 14(3):1–39, mar 2020.

[Shams et al., 2020b] Zohreh Shams, Marina De Vos, Nir Oren, and Julian A. Padget. Argumentation-based reasoning about plans, maintenance goals, and norms. *ACM Trans. Auton. Adapt. Syst.*, 14(3):9:1–9:39, 2020.

[Simari and Loui, 1992] Guillermo R. Simari and Ronald P. Loui. A mathematical treatment of defeasible reasoning and its implementation. *Artif. Intell.*, 53(2-3):125–157, 1992.

[Sklar and Azhar, 2018] Elizabeth I. Sklar and Mohammad Q. Azhar. In *Proceedings of the 6th International Conference on Human-Agent Interaction, HAI 2018*, pages 277–285. ACM, 2018.

[Son and Pontelli, 2006] Tran Cao Son and Enrico Pontelli. Planning with preferences using logic programming. *Theory Pract. Log. Program.*, 6(5):559–607, 2006.

[Stolzenburg et al., 2003] Frieder Stolzenburg, Alejandro Javier García, Carlos I. Chesñevar, and Guillermo R. Simari. Computing generalized specificity. *Journal of Applied Non-Classical Logics*, 13(1):87–113, 2003.

[Teze and Godo, 2021] Juan Carlos Teze and Lluís Godo. An architecture for argumentation-based epistemic planning: A first approach with contextual preferences. *IEEE Intell. Syst.*, 36(2):43–51, 2021.

[Teze et al., 2015] Juan Carlos Teze, Sebastian Gottifredi, Alejandro Javier García, and Guillermo R. Simari. Improving argumentation-based recommender systems through context-adaptable selection criteria. *Expert Syst. Appl.*, 42(21):8243–8258, 2015.

[Teze et al., 2019a] Juan C. L. Teze, Antoni Perelló-Moragues, Lluís Godo, and Pablo Noriega. Practical reasoning using values: an argumentative approach based on a hierarchy of values. *Annals of Mathematics and Artificial Intelligence*, 87(3):293–319, aug 2019.

[Teze et al., 2019b] Juan Carlos Teze, Antoni Perello-Moragues, Lluís Godo, and Pablo Noriega. Practical reasoning using values: an argumentative approach based on a hierarchy of values. *Ann. Math. Artif. Intell.*, 87(3):293–319, 2019.

[Teze et al., 2020] Juan Carlos Teze, Sebastian Gottifredi, Alejandro Javier García, and Guillermo R. Simari. An approach to generalizing the handling of preferences in argumentation-based decision-making systems. *Knowl. Based Syst.*, 189, 2020.

[Teze et al., 2022] Juan Carlos Teze, Lluís Godo, and Gerardo I. Simari. An approach to improve argumentation-based epistemic planning with contextual preferences. *Int. J. Approx. Reason.*, 151:130–163, 2022.

[Toniolo et al., 2011] Alice Toniolo, Timothy J. Norman, and Katia P. Sycara. Argumentation schemes for collaborative planning. In *Agents in Principle, Agents in Practice – 14th International Conference (PRIMA)*, volume 7047 of *Lecture Notes in Computer Science*, pages 323–335. Springer, 2011.

[Wakaki, 2010] Toshiko Wakaki. Preference-based argumentation capturing prioritized logic programming. In *Argumentation in Multi-Agent Systems – 7th International Workshop, ArgMAS 2010*, pages 306–325, 2010.

[Wakaki, 2011] Toshiko Wakaki. Preference-based argumentation handling dynamic preferences built on prioritized logic programming. In *Agents in Principle, Agents in Practice – 14th International Conference (PRIMA)*, pages 336–348, 2011.

[Walton, 1996] Douglas Walton. *Argumentation Schemes for Presumptive Reasoning*. Lawrence Erlbaum Associates, 1996.

Chapter 12

Formal Argumentation and Modal Logic

Carlos Iván Chesñevar
Universidad Nacional del Sur & Consejo Nacional de Investigaciones Científicas y Técnicas (CONICET), Argentina
cic@cs.uns.edu.ar

Jürgen Dix
Technische Universität Clausthal, Germany
dix@tu-clausthal.de

Beishui Liao
Zhejiang University (ZJU), China
baiseliao@zju.edu.cn

Jieting Luo
Zhejiang University (ZJU), China
luojieting@zju.edu.cn

Carlo Proietti
Consiglio Nazionale delle Ricerche (CNR), Italy
carlo.proietti@ilc.cnr.it

Antonio Yuste-Ginel
Universidad Complutense de Madrid (UCM), Spain
antoyust@ucm.es

1 Introduction

While formal argumentation captures diverse kinds of reasoning and dialogue activities with uncertainty and conflicting information, modal logic plays a major role in philosophy and related fields as a tool for understanding and reasoning about concepts such as knowledge, obligation, time, and actions. The combination of argumentation and modal logic has been of interest to this handbook since volume 2 [Arieli et al., 2021]. As we can find a large and heterogeneous body of literature on this subject, the first thing is to propose a criterion according to which one can divide and categorise the different works.

One possible such criterion is classifying the different approaches according to the *modal operators* that they use (e.g., temporal, epistemic, dynamic, etc). The most direct way in which one can relate argumentation and modal logic is by noting that an abstract argumentation framework [Dung, 1995] is nothing but a Kripke frame (that is, a basic semantic structure in modal logic). A natural research enterprise is then to use modal techniques to study abstract argumentation. This was done in a series of papers by Davide Grossi and more people, e.g. [Caminada and Gabbay, 2009; Grossi, 2009; Grossi, 2010a; Grossi, 2010b; Gabbay and Grossi, 2014; Gratie et al., 2012; Grossi and Rey, 2019] (and cf. [Caminada and Gabbay, 2009, Sect. 4.4] for an alternative approach following the same basic idea). In these approaches, the attack relation becomes an accessibility relation that one can use to interpret different modalities, and different types of extensions can then be defined in modal-logic-based languages. This is what might be called an *argumentative approach* to the combination of formal argumentation and modal logic, and it is well studied in the mentioned chapter of this handbook's previous volume [Arieli et al., 2021, Sect. 3.2].

In a different vein, one can use a non-argumentative interpretation of modal operators for increasing the original expressive power of argumentation systems, so as to jointly reason about argumentation and some other relevant cognitive dimension. Thus, for instance, the use of modalities can be *temporal*. In [Bulling et al., 2008b], Alternating-time Temporal Logic is used to reason about what properties a coalition can enlarge to enforce, extended with argumentation to provide how this very coalition is formed. Yet another usual interpretation of modal operators, *dynamic logic*, consists in understanding that they quantify over possible executions of programs or, more in general, over actions. In [Luo et al., 2022a], the arrows of the underlying modal structure are associated with actions that an artificial agent may execute. Moreover, each arrow is also possibly associated with a set of values that they promote/demote. Argumentation frameworks are then used in their value-based version to let the agent decide what is the best available plan to reach a given goal according to her value scale. Finally, there is a branch of the literature that works on the combination of formal argumentation and *epistemic logic*, where we can distinguish two main lines of work. The first line concerns using arguments to determine beliefs. This is the main intuition underlying a series of papers [Shi et al., 2017; Shi et al., 2018a; Shi et al., 2018b; Shi et al., 2021; Shi, 2021], which focus on an argumentative extension of topological epistemic models. In a different technical setting, [Burrieza and Yuste-Ginel, 2020; Burrieza and Yuste-Ginel, 2021] syntactically capture the relation between argumentation and belief using awareness epistemic logic and ASPIC$^+$ arguments. The main idea of the second line is to have beliefs about my opponent's argumentative information, which plays a crucial role in the choice of my moves during a dialogue. It was first treated from an epistemic logic perspective in the work of Schwarzentruber et al. [Schwarzentruber et al., 2012], and later on in [Proietti and Yuste-Ginel, 2020; Proietti and Yuste-Ginel, 2021]. Using these epistemic-argumentative models, we are able to reason about higher-order and unquantified uncertainty about argumentation frameworks, which is in turn a key feature in strategical settings for argumentation.

Another criterion that we can employ for categorising the combination of argumentation and modal logic is their relative positions. Modal logic can be used for reasoning on the object level, while structured argumentation can be built on the meta-level over modal logic. As an example, in [Luo et al., 2022a], arguments for plan selection are constructed using a planner agent's beliefs that are expressed in modal logic. In contrast, modal logic can be built over argumentation. For example, [Schwarzentruber et al., 2012] and [Proietti and Yuste-Ginel, 2020; Proietti and Yuste-Ginel, 2021] see argumentative information as the object that agents reason about using modal logic. Different from the above two ways, [Bulling *et al.*, 2008b] puts argumentation inside modal logic: given a coalition of agents, the framework can be used to check its strategic properties, allowing the coalition to be enlarged according to the theory of coalitional argumentation.

The rest of this chapter is organised as follows. We provide the minimal necessary background on modal logic and abstract argumentation frameworks in Section 2. Then, in Sections 3, 4 and 5 we cover the three different combinations of modal operators and argumentation systems that we have just introduced (dynamic operators, temporal operators and epistemic operators, respectively). Finally, Sections 6 and 7 close the paper by giving some pointers to further literature on the topic and sketching out current trends and challenges at the intersection between modal logic and formal argumentation.

2 Formal preliminaries

In this section, we will present definitions for different core concepts in modal logic and argumentation, which are shared to some extent by the formalisms introduced in later sections. Other concepts (such as more specific semantics and modalities) defined later can be related to these core concepts.

2.1 Modal logic

Here we provide the general definitions of the syntactic and semantic notions for modal logic that we will use in the rest of the chapter. We are going to work with different interpretations of modal logic and, therefore, with different interpretations of modalities. Hence, to keep things abstract enough, we assume as given a finite set of *generic labels* $\mathbb{L} = \{l_1, ..., l_n\}$. Depending on the context of application, elements of $\mathbb{L}$ may denote actions, action profiles, agents or sets of agents. Moreover, we assume as fixed from now on a denumerable set of *atomic propositions* $\Phi = \{p, q, ...\}$.

Definition 2.1 (Labelled multi-modal language). *The language $\mathcal{L}(\Phi, \mathbb{L})$ is given by the following BNF*

$$\varphi ::= p \mid \neg\varphi \mid (\varphi \wedge \varphi) \mid \Box_l \varphi \qquad p \in \Phi, l \in \mathbb{L}.$$

That is, the language for propositional logic enriched with a set of $\Box_l$-modalities. We will often employ $\Diamond_l$ as the dual of $\Box_l$, defined as $\neg\Box_l\neg$ (and we sometimes take $\Diamond_l$ as the primitive operator and define $\Box_l$ as $\neg\Diamond_l\neg$ instead). The rest of Boolean operators are defined as usual using $\wedge$ and $\neg$. $\mathcal{L}(\Phi)$ denotes the propositional fragment of $\mathcal{L}(\Phi, \mathbb{L})$ (i.e., the result of dropping the clause $\Box_l$ from the previous grammar).

A multi-modal language of this kind is typically interpreted on a *labelled transition system* (a.k.a. a *multi-relational Kripke frame*), defined as follows.

Definition 2.2 (Labelled Transition Systems and Models). *A labelled transition system over $\mathbb{L}$ is a tuple $T = \langle S, \mathcal{R} \rangle$, where*

- *S is a finite, non-empty set of states; and*

- *$\mathcal{R} \subseteq S \times \mathbb{L} \times S$ is a transition relation between states labelled with elements of $\mathbb{L}$. We use $\mathcal{R}_l$ to denote the relation $\{\langle x, y \rangle \in S^2 \mid \langle x, l, y \rangle \in \mathcal{R}\}$.*

Further, a model (over Φ and $\mathbb{L}$) –sometimes called an interpreted labelled transition system or, more extensively, a multi-relational Kripke model– is defined as a pair $\mathcal{M} = \langle T, \mathcal{V} \rangle$ where:

- *T is a labelled transition system; and*

- *$\mathcal{V}$ is a propositional valuation $\mathcal{V} : S \to 2^\Phi$ that assigns each state s with the subset of atomic propositions which are true at state s; thus for each $s \in S$ we have $\mathcal{V}(s) \subseteq \Phi$.*

For notational convenience, we sometimes unravel the content of T and write $\mathcal{M} = \langle S, \mathcal{R}, \mathcal{V} \rangle$. Here again, the labels may stand for different kinds of transition relations, e.g. $\langle s, t \rangle \in \mathcal{R}_l$ may denote the execution of an action changing the system from state s to state t. Or else, when l stands for an agent, it may indicate that agent l considers t as an alternative to s.

Once we define a valuation for propositional atoms, the next fundamental step is to define the full notion of truth with respect to satisfaction relation $\models$, more precisely to specify under which conditions a given formula φ is true at a given state s in a model $\mathcal{M}$ (denoted $\mathcal{M}, s \models \varphi$). The following definition does it in a recursive way.

Definition 2.3 (Truth). *Formulas of the labelled multi-modal language are interpreted in pointed models recursively as follows:*

$$\begin{aligned}
\mathcal{M}, s &\models p & &\text{iff} & p &\in \mathcal{V}(s) \\
\mathcal{M}, s &\models \neg\varphi & &\text{iff} & \mathcal{M}, s &\not\models \varphi \\
\mathcal{M}, s &\models (\varphi \wedge \psi) & &\text{iff} & \mathcal{M}, s &\models \varphi \text{ and } \mathcal{M}, s \models \psi \\
\mathcal{M}, s &\models \Box_l \varphi & &\text{iff} & &\text{for all } t \in S,\ s\mathcal{R}_l t \text{ implies } \mathcal{M}, t \models \varphi
\end{aligned}$$

We say that a formula φ is valid in a model $\mathcal{M} = \langle S, \mathcal{R}, \mathcal{V} \rangle$ iff $\mathcal{M}, s \models \varphi$ for every $s \in S$, and that a formula φ is valid in a transition system T iff it is valid in every model $\mathcal{M}$ based on T. Further, φ is valid in a class of transition systems iff it is valid in every element in the class.

Some formulas – more precisely some schemes, i.e. general forms of formula – are valid only in classes of systems where the transition relations satisfy some specific property. In such case we say that a formula φ *defines* the class of frames satisfying this property or, more briefly, that it defines this property. Many such formulas work as *axiom schemes* for different axiomatic calculi of modal logic. For example the general scheme $(K) = \Box_l(\varphi \to \psi) \to (\Box_l\varphi \to \Box_l\psi)$, which is in fact valid in all systems, serves to axiomatise the most basic calculus of modal logic. Some such schemes, particularly relevant in what follows, are written in the table below, together with the property of $\mathcal{R}_l$ they define.

	Axiom scheme	Property of $\mathcal{R}_l$
(K)	$\Box_l(\varphi \to \psi) \to (\Box_l\varphi \to \Box_l\psi)$	
(PF)	$\Diamond_l\varphi \to \Box_l\varphi$	Partial Functionality
(D)	$\Box_l\varphi \to \Diamond_l\varphi$	Seriality
(T)	$\Box_l\varphi \to \varphi$	Reflexivity
(4)	$\Box_l\varphi \to \Box_l\Box_l\varphi$	Transitivity
(5)	$\neg\Box_l\varphi \to \Box_l\neg\Box_l\varphi$	Euclideanity

Before ending this subsection we define some normal modal logics that will be used in other parts of the paper. The minimal modal system K is the smallest set of formulas containing all instances of the axiom scheme (K), all the valid formulas of propositional calculus, and closed under both Modus Ponens — if $\varphi, \varphi \to \psi \in K$, then $\psi \in K$ — and the Necessitation Rule — if $\varphi \in K$, then $\Box\varphi \in K$. Extensions of K are defined by adding more formulas to the basic generating set of K and closing again the resulting set under Modus Ponens and the Necesitation Rule. This is expressed as $K + (S_1) + ... + (S_n)$ where $S_1,...,S_n$ are the new schemata. The following table defines some well known extensions of K:

Modal system	Definition
T	K + (T)
S4	K + (T) + (4)
S5	K + (T) + (4) + (5)
KD45	K + (D) + (4) + (5)

2.2 Abstract argumentation

Argumentation frameworks, the general structures for abstract argumentation, are defined as follows:

Definition 2.4 (Abstract argumentation framework [Dung, 1995]). *An Abstract Argumentation Framework (AF) is a directed graph $AF = \langle Ar, att \rangle$ where Ar is a set of elements called arguments, and $att \subseteq Ar \times Ar$ is binary relation over arguments.*

Although Dung called *att* an attack relation, it is sometimes clearer to interpret it as a defeat relation. Roughly speaking, an argument a attacks another argument b if they are incompatible (they cannot be jointly accepted); while a defeats b if a attacks b and a is at least as strong as b. This distinction (attack vs. defeats) emerges from the literature on structured argumentation [Besnard *et al.*, 2014; Chesñevar *et al.*, 2000] and it will be exemplified in several parts of this chapter, where the expression "be as at least strong as" will be attributed precise formal meanings.

Argumentation frameworks are, in their bare bones, nothing more than directed graphs. What is fundamental is the specification of their *semantics* – sometimes also called *solution concepts* – which encode different criteria of justification for (sets of) arguments. The following definition provides the original ones by [Dung, 1995], which are the ones used in this chapter.

Definition 2.5 (Argumentation semantics). *Given $AF = \langle Ar, att \rangle$ and $\mathcal{E} \subseteq Ar$,*

- *$\mathcal{E}$ is conflict-free iff there does not exist $a, b \in \mathcal{E}$ such that $\langle a, b \rangle \in att$.*

- *An argument $a \in Ar$ is acceptable w.r.t. a set $\mathcal{E}$ (a is defended by $\mathcal{E}$), iff $\forall \langle b, a \rangle \in att$, $\exists c \in \mathcal{E}$ such that $\langle c, b \rangle \in att$.*

- *A conflict-free set of arguments $\mathcal{E}$ is admissible iff each argument in $\mathcal{E}$ is acceptable w.r.t. $\mathcal{E}$.*

- *$\mathcal{E}$ is a complete extension of AF iff $\mathcal{E}$ is admissible and each argument in Ar that is acceptable w.r.t. $\mathcal{E}$ is in $\mathcal{E}$.*

- *$\mathcal{E}$ is the grounded extension of AF iff $\mathcal{E}$ is the minimal (w.r.t. set inclusion) complete extension.*

- *$\mathcal{E}$ is the preferred extension of AF iff $\mathcal{E}$ is the maximal (w.r.t. set inclusion) complete extension.*

- *$\mathcal{E}$ is a stable extension of AF iff $\mathcal{E}$ is conflict-free and $\forall b \in Ar \setminus \mathcal{E}$, $\exists a \in \mathcal{E}$ such that $\langle a, b \rangle \in att$.*

Let $AF = \langle Ar, att \rangle$ be an AF, let $\mathcal{S} \in \{\mathcal{CO}, \mathcal{GR}, \mathcal{PR}, \mathcal{ST}\}$ (where $\mathcal{CO}$ stands for complete, $\mathcal{GR}$ for grounded, $\mathcal{PR}$ for preferred, and $\mathcal{ST}$ for stable), we denote by $\mathcal{E}_\mathcal{S}(AF)$ the set of $\mathcal{S}$-extensions of AF.

For a detailed study of these and further semantics, the reader is referred to the monographic chapter of the first volume of this handbook [Baroni et al., 2018].

3 Argumentation and dynamic logic for value-based planning

Autonomous agents are supposed to be able to perform value-based ethical reasoning based on their value systems in order to distinguish moral from immoral behavior. Existing work on value-based practical reasoning such as [Atkinson and Bench-Capon, 2018][Bench-Capon et al., 2012] [Liao et al., 2021] demonstrates how an agent can reason about what he should do among alternative action options that are associated with value promotion or demotion. More than that, agents are supposed to be able to finish tasks or achieve goals that are assigned by their users through performing a sequence of actions. Classical planning concerns finding a successful sequence of actions achieving a goal. Since there might exist multiple plans that an agent can follow and each plan might promote or demote different values along each action, the agent should be able to resolve the conflicts between them and evaluate which plan he should follow. If the decision-making problem concerns choosing a plan instead of an action, then we first need to know how an agent can see whether he can follow a particular plan to achieve his goal. Modal logic allows us to represent and verify whether a goal can be achieved by executing a plan under specific conditions such as norm compliance assumptions [Ågotnes et al., 2007][Knobbout and Dastani, 2012][Alechina et al., 2013], namely telling agents whether a plan works or not, but cannot tell agents whether it is the best option. Certainly, agents can collect the representation results regarding whether a plan promotes or demotes a specific set of values and then compare different plans using lifting approaches as what has been done in [Luo et al., 2019]. However, the order lifting problem is a major challenge in many areas of AI and no approach is ultimately "correct". Moreover, the agent in our setting needs to lift the preference over values to the preference over plans with respect to value promotion and demotion, which even complicates the problem. Therefore, we need a more natural and intuitive approach to deal with representation results.

It has been shown that argumentation provides a useful mechanism to model and resolve conflicts [Dung, 1995], and particularly can be used for the decision making of artificial intelligence in a dialectical way, and provides explanation for that [Prakken, 2006][Atkinson and Bench-Capon, 2007][Rahwan and Amgoud, 2006][Liao and van der Torre, 2020]. In this section, we develop a logic-based framework that combines modal logic and argumentation for value-based planning. In the first part, modal logic is used as a technique to represent and verify agents' belief in terms of whether a plan with its local properties of value promotion or demotion can be followed to achieve an agent's goal. Using the representation results to construct arguments, we then propose an argumentation framework that allows an agent to reason about his plans in the form of support and objection. We prove that our framework satisfies a set of properties consistent with our understanding of rational decision making. Our preliminary idea has been presented in [Luo et al., 2022b], where arguments are constructed with a value for promotion or demotion. However, we notice that arguing about plans using this way of argument construction is in fact equivalent to arguing about plans using arguments that are constructed with a set of values for promotion or demotion in the democratic lifting way. We thus in this version construct arguments with a set of values for promotion or demotion and allow more lifting ways for comparing sets of values.

3.1 Modal logic for representation

The basic semantic structure of our approach is a transition system (Def. 2.2) where the set of generic labels $\mathbb{L}$ is understood as a set of actions $Act = \{\alpha_1, \alpha_2, ..., \alpha_n\}$ that are executable by the agent. This way of looking at transitions systems represents the computational behavior of a system caused by an agent's actions in the agent's subjective view. Hence, vertices S corresponds to possible states of the system, and the relation $\mathcal{R} \subseteq S \times Act \times S$ represents the possible transitions of the system. When a certain action $\alpha \in Act$ is performed, the system might progress from a state s to a different state s' in which different propositions hold. Moreover, some restrictions are imposed on relation $\mathcal{R}$ in order to capture some intuitions. Recall that $\Phi = \{p, q, ...\}$ is a set of atomic propositions.

Definition 3.1 (Action Transition Models). *An action transition model is a interpreted labelled transition system (i.e., a model) $T = \langle S, \mathcal{R}, \mathcal{V} \rangle$ (Def. 2.2) where the set of labels $\mathbb{L}$ represents a set of actions $Act = \{\alpha_1, ..., \alpha_n\}$. Moreover, it is assumed that*

- *for all $s \in S$ there exists an action $a \in Act$ and a state $s' \in S$ such that $\langle s, \alpha, s' \rangle \in \mathcal{R}$;*

- *we restrict actions to be deterministic, that is, if $\langle s, \alpha, s' \rangle \in \mathcal{R}$ and $\langle s, \alpha, s'' \rangle \in \mathcal{R}$, then $s' = s''$.*

Since the relation $\mathcal{R}$ is partially functional, we write $s[\alpha]$ to denote the state s' for which it holds that $\langle s, \alpha, s' \rangle \in \mathcal{R}$. We also use $s[\alpha_1, \ldots, \alpha_n]$ to denote the resulting state for which a sequence of actions $\alpha_1, \ldots, \alpha_n$ successively execute from state s. A pointed action transition model is a pair $\langle T, s \rangle$ such that T is an action transition model, and $s \in S$ is a state from T. Adopted from [Knobbout et al., 2016][Knobbout et al., 2014], the language $\mathcal{L}(\Phi, Act)$ is just our generic labelled language $\mathcal{L}(\Phi, \mathbb{L})$ (Def. 2.1) with $\mathbb{L} = Act$. For convenience, we take $\Diamond_\alpha$ instead of $\Box_\alpha$ to be the primitive modal operator. The notion of truth in pointed action transition models is then also the generic one (Sect. 2.1). We just make explicit the cause for $\Diamond_\alpha$:

$$T, s \models \Diamond_\alpha \varphi \text{ iff } s[\alpha] \text{ exists and } T, s[\alpha] \models \varphi.$$

Given a pointed action transition model $\langle T, s \rangle$, we say that a sequence of actions $\lambda = \alpha_1 \ldots \alpha_n$ brings about a φ-state if and only if $T, s \models \Diamond_{\alpha_1} \ldots \Diamond_{\alpha_n} \varphi$. In the rest of the section, we will sometimes write $\Diamond_\lambda \varphi$ instead of $\Diamond_{\alpha_1} \ldots \Diamond_{\alpha_n} \varphi$ for short.

A action transition model represents how a system progresses by an agent's actions. Besides, an agent in the system is assumed to have his own goal, which is a formula expressed in propositional logic $\mathcal{L}(\Phi)$. It is indeed possible for an agent to have multiple goals and his preference over different goals. For example, a goal hierarchy is defined in [Ågotnes et al., 2007] to represent increasingly desired properties that the agent wishes to hold. However, we find that the setting about whether the agent has one goal or multiple goals is in fact not essential for our analysis, so we simply assume that the agent only has a goal for simplifying our presentation.

Example 3.2. *Consider the action transition model T in Figure 1, which represents how an agent can get to a pharmacy to buy medicine for his user. State s_0 is the initial state, representing staying at home, and proposition p, representing arriving at a pharmacy, holds in state s_4. The agent can perform actions α_1 to α_6 in order to get to state s_4.*

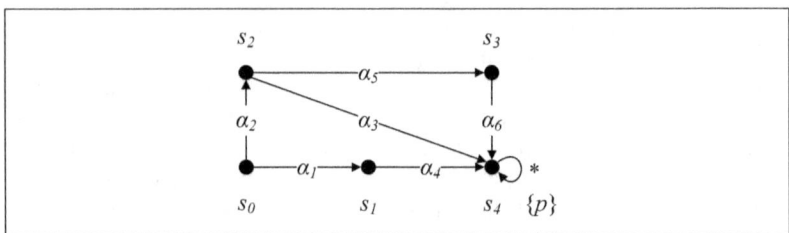

Figure 1: Transition system T. The star loop around state s_4 means that the agent stays in state s_4 whatever he does.

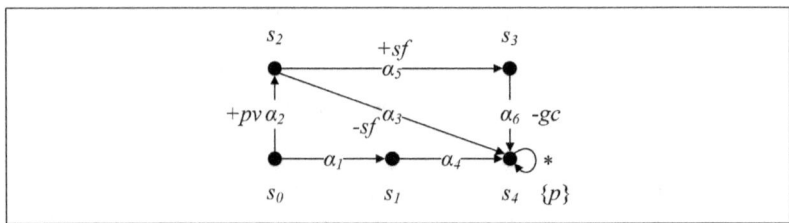

Figure 2: A value-based action transition model VT. The star loop around state s_4 means that the agent stays in state s_4 whatever he does.

From this action transition model, the following formulas hold:

$$T, s_0 \models \Diamond_{\alpha_1} \Diamond_{\alpha_4} p,$$
$$T, s_0 \models \Diamond_{\alpha_2} \Diamond_{\alpha_3} p,$$
$$T, s_0 \models \Diamond_{\alpha_2} \Diamond_{\alpha_5} \Diamond_{\alpha_6} p,$$

which means that the agent can first perform action α_1 and then action α_4, or action α_2 followed by action α_3, or action α_2 followed by actions α_5 and α_6, to get to the pharmacy.

It is important for an agent not only to achieve his goal, but also to think about how to achieve his goal. As we can see from the running example, there are multiple ways for the agent to get to the pharmacy, and the agent needs to evaluate which one is the best to choose. In this section, agents are able to perform value-based practical reasoning in terms of planning their actions to achieve their goals. We first assume that an agent has a set of values. A value can be seen as an abstract standard according to which agents have their preferences over options. For instance, if we have a value denoting *equality*, we prefer the options where equal sharing or equal rewarding hold. Unlike [Luo et al., 2019] where a value is interpreted as a state formula, we simply assume a value as a primitive structure without considering how it is defined. We assume that agents can always compare any two values, so we define an agent's value system as a total pre-order (instead of a strict total order) over a set of values, representing the degree of importance of something.

Definition 3.3 (Value Systems). *A value system $V = \langle \text{Val}, \precsim \rangle$ is a tuple consisting of a finite set of values* $\text{Val} = \{v_1, ..., v_k\}$ *together with a total pre-ordering $\precsim$ over* Val. *When $v_i \precsim v_j$, we say that value v_j is at least as important as value v_i. As is standard, we define $v_i \sim v_j$ to mean $v_i \precsim v_j$ and $v_j \precsim v_i$, and $v_i \prec v_j$ to mean $v_i \precsim v_j$ and $v_i \not\sim v_j$.*

We label some of the transitions with the values promoted and demoted by moving from a starting state to a ending state. Notice that not every transition can be labeled, as some transitions may not be relevant to any value in an agent's value system. Formally, a function $\delta : \{+, -\} \times \text{Val} \to 2^{\mathcal{R}}$ is a valuation function over T which defines the status (promoted (+) or demoted (-)) of a value $v \in$ Val ascribed to a set of transitions. We then define a value-based action transition model VT as a action transition model together with a value system V and a function δ.

Definition 3.4 (Value-based Transition Model). *A value-based action transition model is defined by a triple $VT = \langle T, V, \delta \rangle$, where T is an action transition model, V is a value system and δ is a valuation function that assigns value promotion or demotion to a set of transitions.*

Given a sequence of actions with respect to a value-based action transition model, we then express whether the performance of the sequence in a state promotes or demotes a specific value, which can be done by extending our language $\mathcal{L}(\Phi, Act)$ with new modalities of the form promoted$(v, \alpha_1 \ldots \alpha_n)$ and demoted$(v, \alpha_1 \ldots \alpha_n)$. The formula promoted$(v, \alpha_1 \ldots \alpha_n)$ (resp. demoted$(v, \alpha_1 \ldots \alpha_n)$) should be intuitively read as there exists an action that promotes (resp. demotes) value v in the sequence of actions $\alpha_1, \ldots, \alpha_n$. Given a pointed value-based action transition model (VT, s) and a value $v \in \text{Val}$, the satisfaction relation $VT, s \models \psi$ is extended with the following new semantics:

- $VT, s \models \text{promoted}(v, \alpha_1 \ldots \alpha_n)$ iff there exists $1 \leq m \leq n$ such that
$$(s[\alpha_1, \ldots, \alpha_{m-1}], \alpha_m, s[\alpha_1, \ldots, \alpha_m]) \in \delta(+, v);$$

- $VT, s \models \text{demoted}(v, \alpha_1 \ldots \alpha_n)$ iff there exists $1 \leq m \leq n$ such that
$$(s[\alpha_1, \ldots, \alpha_{m-1}], \alpha_m, s[\alpha_1, \ldots, \alpha_m]) \in \delta(-, v).$$

Notice that the formula only expresses the local property of a sequence of actions in terms of value promotion or demotion by an action within the sequence. Thus, it is possible that an action within the sequence promotes value v but it gets demoted by another action within the sequence, meaning that both $VT, s \models \text{promoted}(v, \alpha_1 \ldots \alpha_n)$ and $VT, s \models \text{demoted}(v, \alpha_1 \ldots \alpha_n)$ hold at the same time. Since a sequence of actions is denoted as λ, we will sometimes write $\text{promoted}(v, \lambda)$ instead of $\text{promoted}(v, \alpha_1 \ldots \alpha_n)$ and $\text{demoted}(v, \lambda)$ instead of $\text{demoted}(v, \alpha_1 \ldots \alpha_n)$ for short. Having the above formulas, the agent is then aware of whether he can perform the sequence of actions to achieve his goal and which value gets promoted or demoted along the sequence. The formulas represent the local properties of a sequence of actions, and thus it can be the case that the same value is simultaneously promoted and demoted along the same sequence of actions. We continue our running example to illustrate how to use our logical language to express and verify properties of sequences of actions.

Example 3.5. *Suppose the ethical agent has privacy (pv), safety (sf) and good conditions (gc) as his values and a value system as $pv \prec gc \prec sf$. As in Figure 2, some of the transitions have been labeled with value promotion or demotion with respect to the agent's values. Taking action α_2 in state s_0 is interpreted as asking for the permission of taking a private path, which promotes the value of privacy. Taking action α_3 means crossing the road without using the crosswalk, which demotes the value of safety of the agent, and conversely taking action α_4 in state s_2 promotes the value of safety of the agent. Finally, performing action α_5 in* state s_3 means stepping into water. As the agent is a robot, which should avoid getting wet, this choice will demote the value of maintaining good conditions of the agent. The agent can verify whether he can achieve his goal while promoting or demoting a specific value by performing a sequence of actions. The verification results are listed below:

$VT, s_0 \models \Diamond_{\alpha_2}\Diamond_{\alpha_3}p \wedge \text{promoted}(pv, \alpha_2\alpha_3)$

$VT, s_0 \models \Diamond_{\alpha_2}\Diamond_{\alpha_3}p \wedge \text{demoted}(sf, \alpha_2\alpha_3)$

$VT, s_0 \models \Diamond_{\alpha_2}\Diamond_{\alpha_5}\Diamond_{\alpha_6}p \wedge \text{promoted}(pv, \alpha_2\alpha_5\alpha_6)$

$VT, s_0 \models \Diamond_{\alpha_2}\Diamond_{\alpha_5}\Diamond_{\alpha_6}p \wedge \text{promoted}(sf, \alpha_2\alpha_5\alpha_6)$

$VT, s_0 \models \Diamond_{\alpha_2}\Diamond_{\alpha_5}\Diamond_{\alpha_6}p \wedge \text{demoted}(gc, \alpha_2\alpha_5\alpha_6)$

3.2 Value-based planning: an argumentative approach

Given a action transition model and an agent's goal, modal logic allows an agent to represent and verify whether he can achieve his goal while promoting or demoting a specific value by performing a sequence of actions. Since following different plans might promote or demote different sets of values, next question is how the agent *internally* decides what to do given the representation results. In this section, we propose to use argumentation as a technique for an agent's planning. Formal argumentation is a nonmonotonic formalism for representing and reasoning about conflicts based on the construction and the evaluation of interacting arguments [Dung, 1995]. In particular, it has been used in practical reasoning, concerned with reasoning about what agents should do, given different alternatives and outcomes they bring about [Bench-Capon *et al.*, 2012][Amgoud and Prade, 2007]. Since argumentation resolves conflicts in a dialectical way, it also provides justification and explanation to the final solution. In general, epistemic planning considers the following problem [Bolander, 2017; Bolander and Andersen, 2011]: Given my current state of belief, and a desirable state of belief, how do I get from one to the other? In particular, each plan is labeled with a set of values that are promoted or demoted along the plan. The agent needs to look for a plan that is not only feasible but also optimal with respect to value promotion and demotion. We first define the notion of plans. A plan is defined as a finite sequence of actions that will bring about the agent's goal in the underlying action transition model.

Definition 3.6 (Plans). *Given a value-based action transition model VT, a state s and a formula $g \in \mathcal{L}(\Phi)$ as an agent's goal, a sequence of actions λ over Act is said to be a plan w.r.t s and g, denoted as $\lambda_{s,g}$, iff $VT, s \models \Diamond_\lambda g$. Sometimes, we write λ for $\lambda_{s,g}$ if it is clear from the context.*

A sequence of actions is denoted as λ if it is a plan. Given a set of available plans, the agent can construct arguments to support or oppose the execution of a plan. The reason to supporting a plan is that the plan promotes a set of values, and the reason to opposing a plan is that the plan demotes a set of values, which can be expressed as formulas in our language $\mathcal{L}$. We define two types for arguments for planning: an ordinary argument supports the performance of a plan, while a blocking argument opposes the performance of a plan.

Definition 3.7 (Ordinary Arguments for Planing). *Given a value-based action transition model VT, a state s, a goal g and a plan λ w.r.t. s and g,*

- *let $A \subseteq \text{Val}$ be a set of values, a non-empty ordinary argument is a pair $\langle +A, \lambda \rangle$, read as "plan λ should be selected because it promotes values A", iff*
$$VT, s \models \bigwedge_{v \in A} \text{promoted}(v, \lambda),$$

- *an empty ordinary argument is a pair $\langle -\emptyset, \lambda \rangle$, read as "plan λ should be selected because it does not demote any values", such that*
$$VT, s \models \bigwedge_{v \in \text{Val}} \neg \text{demoted}(v, \lambda).$$

Definition 3.8 (Blocking Arguments for Planning). *Given a value-based action transition model VT, a state s and a plan λ,*

- *let $A \subseteq \text{Val}$ be a set of values, a non-empty blocking argument is a pair $\langle -A, \neg \lambda \rangle$, read as "plan λ should not be selected because it demotes values A", iff*
$$VT, s \models \bigwedge_{v \in A} \text{demoted}(v, \lambda);$$

- an empty blocking argument is a pair $\langle +\emptyset, \neg\lambda \rangle$, read as "plan λ should not be selected because it does not promote any values", such that
$$VT, s \models \bigwedge_{v \in \text{Val}} \neg \text{promoted}(v, \lambda).$$

We use $\mathcal{A}_o^p$ to denote the set of ordinary arguments and use $\mathcal{A}_b^p$ to denote the set of blocking arguments for planning, and $\mathcal{A}^p = \mathcal{A}_o^p \cup \mathcal{A}_b^p$ to denote the set of two types of arguments. In the following text, unless it is addressed clearly, an ordinary argument can refer to a non-empty ordinary argument or an empty ordinary argument, and a blocking argument can refer to a non-empty blocking argument or an empty blocking argument. Both an ordinary argument and a blocking argument correspond to representation results. Conventionally, we might represent an argument using an alphabet $(a, b, \dots)$ and thus the plan that it supports or opposes is denoted $\lambda_a, \lambda_b, \dots$ and the set of promoted or demoted values is denoted as uppercase letters V_a, V_b, etc.

Example 3.9 (Ordinary arguments and blocking arguments). *The value-based action transition model in Fig. 2 shows that the agent is aware of three plans $\alpha_1\alpha_4$, $\alpha_2\alpha_3$ and $\alpha_2\alpha_5\alpha_6$. Plan $\alpha_2\alpha_3$ promotes value pv but demote value sf, plan $\alpha_1\alpha_4$ does not promote or demote any value, and plan $\alpha_2\alpha_5\alpha_6$ promote values pv and sf, but demote value gc. Based on the representation results, the agent can construct the following ordinary arguments and blocking arguments: $\langle +\emptyset, \neg\alpha_1\alpha_4 \rangle$, $\langle +\{pv\}, \alpha_2\alpha_3 \rangle$, $\langle -\{sf\}, \neg\alpha_2\alpha_3 \rangle$, $\langle +\{pv, sf\}, \alpha_2\alpha_5\alpha_6 \rangle$ and $\langle -\{gc\}, \neg\alpha_2\alpha_5\alpha_6 \rangle$.*

When we get to choose a plan to follow, there are conflicts between the alternatives as they cannot be followed all at the same time. The conflicts are interpreted as attacks between two ordinary arguments supporting different plans and one ordinary argument and one blocking argument supporting and objecting to the same plan respectively.

Definition 3.10 (Attacks for Planning). *Given a set of ordinary arguments $\mathcal{A}_o^p$ and a set of blocking arguments $\mathcal{A}_b^p$,*

- *for any two ordinary arguments $a, b \in \mathcal{A}_o^p$, a attacks b iff $\lambda_a \neq \lambda_b$;*

- *for any ordinary argument $a \in \mathcal{A}_o^p$ and any blocking argument $b \in \mathcal{A}_b^p$,*

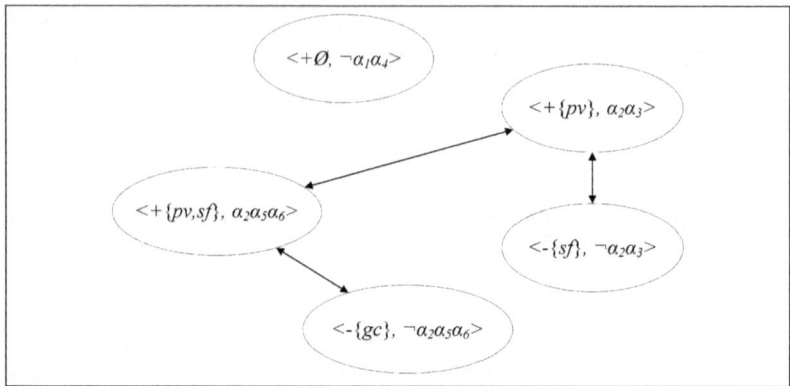

Figure 3: Attack relation between ordinary arguments and blocking arguments.

- a attacks b iff $\lambda_a = \lambda_b$;
- b attacks a iff $\lambda_a = \lambda_b$.

The set of attacks (an attack relation) over $\mathcal{A}^p$ are denoted as att^p.

It is obvious that our attack relation is mutual. It should be noticed that there is no attack between two blocking arguments, as a blocking argument only functions as blocking the conclusion of an ordinary argument but does not make a conclusion by itself.

Example 3.11. *In the running example, there are three ordinary arguments and three blocking arguments. The attack relation is depicted in Figure 3, where any two ordinary arguments with different plans mutually attack (for instance, $\langle +\{pv, sf\}, \alpha_2\alpha_5\alpha_6\rangle$ and $\langle +\{pv\}, \alpha_2\alpha_3\rangle$), and any ordinary argument and blocking argument with the same plan are mutually attacked (for instance, $\langle +\{pv\}, \alpha_2\alpha_3\rangle$ and $\langle -\{sf\}, \neg\alpha_2\alpha_3\rangle$).*

The attack relation represents conflicts between plans. However, the notion of attack may not be sufficient for modeling conflicts between arguments, as an agent has his preference over the values that are promoted or demoted by different plans. In structured argumentation frameworks such as ASPIC$^+$ [Modgil and Prakken, 2013], an argument

a can be used as a counter-argument to another argument b, if a successfully attacks, i.e. defeats, b. Whether an attack from a to b (on its sub-argument b') succeeds as a defeat, may depend on the relative strengths of a and b, which is a preference over arguments a and b based on the preferences over their constituent ordinary premises and defeasible rules. Here we use the same approach to decide an attack succeeds as a defeat. Recall that an agent has a value system, which was defined as a total pre-order over a set of values. So there needs to be a lifting way that allows the planning agent to lift the preference over values to preferences over arguments. Two lifting ways are commonly used in structured argumentation: the so called *Elitist* and *Democratic* ways. Eli (denoted as $\trianglelefteq_E$) compares sets on their minimal and Dem (denoted as $\trianglelefteq_D$) on their maximal elements.

Definition 3.12 (Lifting). *Given two set of values A and B, $\trianglelefteq_E$ is defined as follows:*

$$A \trianglelefteq_E B \text{ iff there exists } v_a \in A \text{ s.t. for all } v_b \in B : v_a \precsim v_b.$$

$\trianglelefteq_D$ *is defined as follows:*

$$A \trianglelefteq_D B \text{ iff for all } v_a \in A \text{ there exists } v_b \in B : v_a \precsim v_b.$$

We use $\trianglelefteq \in \{\trianglelefteq_E, \trianglelefteq_D\}$ to denote an arbitrary lifting approach of the above. We define $A \simeq B$ to mean $A \trianglelefteq B$ and $B \trianglelefteq A$, and $A \triangleleft B$ to mean $A \trianglelefteq B$ and it is not the case that $A \simeq B$.

It is easy to prove that $\trianglelefteq$ is reflexive and transitive. We can then determine the defeat relation over two arguments based on the value system. The notion of defeat combines the notions of attack and preference.

Definition 3.13 (Defeats for Planning). *Given a set of arguments $\mathcal{A}^p$, a set of attacks $\mathcal{R}^p$ over $\mathcal{A}^p$ and a value system V, for any two arguments $a, b \in \mathcal{A}^p$, a defeats b iff a attacks b and it is not the case that $V_a \trianglelefteq V_b$ or b is an empty argument. The set of defeats (a defeat relation) over $\mathcal{A}^p$ based on an attack relation att^p, a value system V and a lifting $\trianglelefteq$ is denoted as $\mathcal{D}^p(att^p, V, \trianglelefteq)$. We write $\mathcal{D}^p$ for short if it is clear from the context.*

In words, given mutual attacks between two arguments, the attack from the argument with less preferred value set to the attack from the argument with a more preferred value set does not succeed as a defeat, and the empty argument is always defeated. One might ask whether it is more convenient to combine the notions of attack relation and defeat relation. We argue that two notions represent the relation between two arguments from different perspectives, one for the conflicts between plans and the other for the preferences over values. Because of that, defining these two notions separately can make our framework more clear, even though technically it is possible to combine them. Here are several properties that characterize our defeat relation.

Proposition 3.14. *Given two ordinary arguments $a, b \in \mathcal{A}_o^p$, a and b defeat each other iff $\lambda_a \neq \lambda_b$ and ($A \simeq B$). Given an ordinary argument $a \in \mathcal{A}_o^p$ and a blocking argument $b \in \mathcal{A}_b^p$, a and b defeat each other iff $\lambda_a = \lambda_b$ and ($A \simeq B$ or both a and b are empty arguments).*

Proof. Proof follows from Definition 3.13 directly. □

Proposition 3.15. *Given a set of arguments $\mathcal{A}^p$, a defeat relation $\mathcal{D}^p$ on $\mathcal{A}^p$ never forms any pure odd cycles.*

Proof. According to Definition 3.13, in order for an argument a to defeat another argument b, value set A must be not less preferred than B or B is an empty argument. Given three non-empty arguments a, b, c, since the preference order over sets of values is transitive, if a defeats non-empty argument b and b defeats c, then a also defeats c. For the case where the set of values are equally preferred, because of Proposition 3.14, any odd cycles that are formed by $\mathcal{D}^p$ are always together with two-length cycles, which are not pure odd cycles. For the case where there exists empty arguments, if a is an non-empty argument and b is an empty argument, then c is also empty. As a is an non-empty argument and c is empty, c cannot defeat a. □

Proposition 3.16. *Given a set of arguments $\mathcal{A}^p$, a defeat relation $\mathcal{D}^p$ on $\mathcal{A}^p$ is irreflexive.*

Proof. It is a special case of Proposition 3.15 for the number of arguments in the odd cycle being one. □

We are now ready to construct a Dung-style abstract argumentation framework with ordinary arguments, blocking arguments and the defeat relation on them.

Definition 3.17 (Argumentation Frameworks for Planning). *Given a pointed value-based action transition model (VT, s) and a formula $g \in \mathcal{L}(\Phi)$ as an agent's goal, an argumentation framework for planning over (VT, s) and g is a pair $PAF = \langle \mathcal{A}^p, \mathcal{D}^p \rangle$, where $\mathcal{A}^p$ is a set of arguments and $\mathcal{D}^p$ is a defeat relation on $\mathcal{A}^p$.*

Example 3.18. *In our running example, the agent has a value system as $pv \prec gc \prec sf$, which means that safety is more important than keeping good condition, and keeping good condition is more important than privacy. With lifting $\trianglelefteq_D$, we then can see some of the attacks in Figure 3 do not succeed as defeats. For example, argument $\langle +\{pv, sf\}, \alpha_2\alpha_5\alpha_6 \rangle$ and argument $\langle -\{gc\}, \neg\alpha_2\alpha_5\alpha_6 \rangle$ are mutually attacked, but since $\{gc\} \trianglelefteq_D \{pv, sf\}$, only the attack from argument $\langle +\{pv, sf\}, \neg\alpha_2\alpha_5\alpha_6 \rangle$ to argument $\langle -\{gc\}, \alpha_2\alpha_5\alpha_6 \rangle$ becomes a defeat. Notice that argument $\langle +\emptyset, \neg\alpha_1\alpha_4 \rangle$ do not receive any defeats or defeat any arguments because there is no ordinary argument with plan $\alpha_1\alpha_4$.*

With lifting $\trianglelefteq_E$, we should notice the defeat between argument $\langle +\{pv, sf\}, \alpha_2\alpha_5\alpha_6 \rangle$ and argument $\langle -\{gc\}, \neg\alpha_2\alpha_5\alpha_6 \rangle$. Since $\{pv, sf\} \trianglelefteq_E \{gc\}$, argument $\langle -\{gc\}, \neg\alpha_2\alpha_5\alpha_6 \rangle$ to argument $\langle +\{pv, sf\}, \alpha_2\alpha_5\alpha_6 \rangle$. All other defeats remain the same as with lifting $\trianglelefteq_D$. See the defeat relation in Figure 4 and Figure 5 with different lifting ways.

Given an argumentation framework for planning PAF, the status of arguments is evaluated, producing sets of arguments that are acceptable together, which are based on the notions of conflict-freeness, acceptability and admissibility. The well-known argumentation semantics are listed in Definition 2.5, each of which provides a pre-defined criterion for determining the acceptability of arguments in a PAF [Dung, 1995]. We use $\mathcal{S} \in \{\mathcal{CO}, \mathcal{PR}, \mathcal{GR}, \mathcal{ST}\}$ to denote the complete, preferred, grounded and stable semantics, respectively, and $\mathcal{E}_\mathcal{S}(PAF)$ to denote the set of extensions of PAF under a semantics in $\mathcal{S}$. The following propositions characterize our argumentation framework in terms of Dung's semantics.

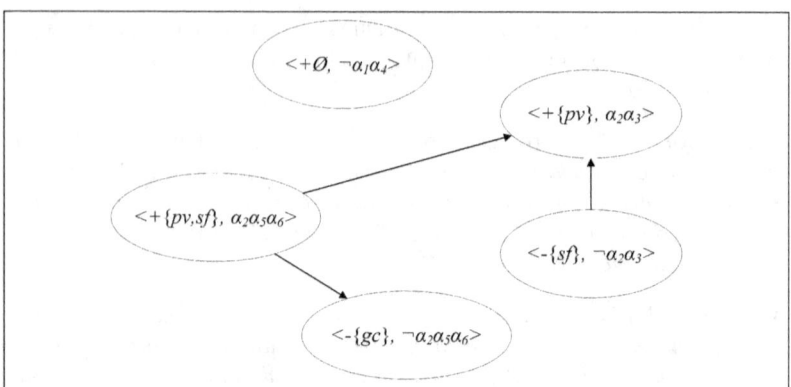

Figure 4: An argumentation framework for planning with lifting $\trianglelefteq_D$.

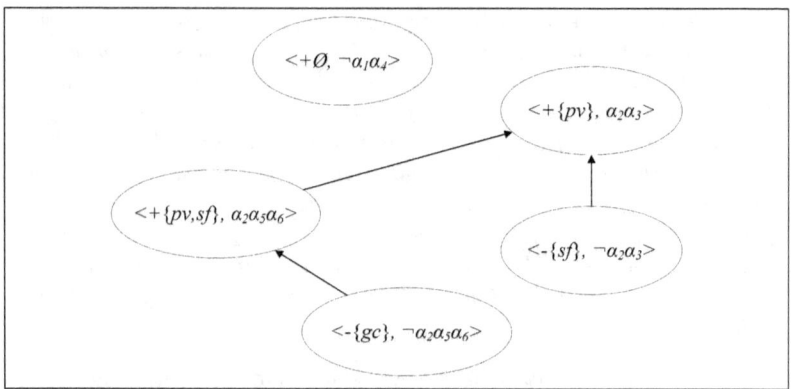

Figure 5: An argumentation framework for planning with lifting $\trianglelefteq_E$.

Proposition 3.19. *Given* $PAF = \langle \mathcal{A}^p, \mathcal{D}^p \rangle$, $\mathcal{E}_{\mathcal{PR}}(PAF) = \mathcal{E}_{\mathcal{ST}}(PAF)$.

Proof. Since our defeat relation $\mathcal{D}^p$ never forms any pure odd cycles by Proposition 3.15, which means that PAF is limited controversial, each preferred extension of PAF is stable. Detailed proof can be found in [Dung, 1995]. □

Proposition 3.20. *Given* $PAF = \langle \mathcal{A}^p, \mathcal{D}^p \rangle$ *and the grounded extension*

E of PAF, if E contains an ordinary argument, then $\mathcal{E}_{\mathcal{PR}}(PAF) = \mathcal{E}_{\mathcal{GR}}(PAF)$.

Proof. Suppose $\mathcal{E}_{\mathcal{PR}}(PAF) \neq \mathcal{E}_{\mathcal{GR}}(PAF)$, which means that there is more than one preferred extension. Since an ordinary argument is contained in the grounded extension E, it should also be contained in each preferred extension. However, each preferred extension indicates a distinct plan, which will be later proved by Proposition 3.22 and its implication. Contradiction! □

The notion of optimal plans is then defined under various semantics in Definition 2.5. Similarly to [Liao et al., 2019], given an argument a, we write concl(a) for the conclusion of argument a, and Oplans($PAF, \mathcal{S}$) for the set of conclusions of ordinary arguments from the extensions under a specific semantics.

Definition 3.21 (Optimal Plans). *Given $PAF = \langle \mathcal{A}^p, \mathcal{D}^p \rangle$ and a semantics $\mathcal{S}$, a set of optimal plans, written as* Oplans($PAF, \mathcal{S}$)*, are the conclusions of the ordinary arguments within extensions under semantics $\mathcal{S}$.*

$$\text{Oplans}(PAF, \mathcal{S}) = \{\text{concl}(a) \mid a \in \mathcal{A}_o^p, a \in E \text{ and } E \in \mathcal{E}_{\mathcal{S}}(PAF)\}$$

We show that the results of our approach are consistent with the rationality of decision-making through the following propositions. Firstly, all the accepted arguments within an extension indicate the same plan.

Proposition 3.22. *Given an argumentation framework for planning $PAF = \langle \mathcal{A}^p, \mathcal{D}^p \rangle$ and an extension E of PAF under a specific semantics as defined in Definition 2.5,*

1. *for any two ordinary arguments $a, b \in E$, it is the case that $\lambda_a = \lambda_b$;*

2. *for any ordinary argument $a \in E$ and any blocking argument $b \in E$, $\lambda_a \neq \lambda_b$.*

Proof. For any extension E under a specific semantics, it is required that all the arguments in E should be conflict-free. 1. By Definition 3.13, we can derive two cases: either there is no attack between these two arguments, or one argument attacks the other but does not succeed as a defeat. For the former case, two arguments contain the same plan. For the latter case, since any attack between two arguments is mutual, if an attack from argument a to argument b fails to be a defeat because $A \triangleleft B$ or argument a is an empty argument while argument b is a non-empty argument, the attack from argument b to argument a will succeed to be a defeat. That means that the second case is impossible and only the first case holds. Hence, the two arguments have the same plan. 2. We can prove in a similar way that for any ordinary argument $a \in E$ and any blocking argument $b \in E$, $\lambda_a \neq \lambda_b$, □

From that we can see, if there are multiple preferred extensions, then each of them indicates a distinct plan. Secondly, when using lifting $\trianglelefteq_D$, our argumentation-based approach always accepts the argument with the most preferred value. Because of that, the plan that promotes the most preferred value will be accepted and the plan that demotes the most preferred value will be rejected.

Proposition 3.23. *Given an argumentation framework for planning $PAF = \langle \mathcal{A}^p, \mathcal{D}^p \rangle$ with lifting $\trianglelefteq_D$, let $v \in \text{Val}$ be a value such that for all arguments $a \in \mathcal{A}^p$ and all values $v' \in V_a$ it is the case that $v' \precsim v$, then an argument with value v is in a preferred extension. If it is not in a cycle, then it is also in the grounded extension.*

Proof. Because $v' \precsim v$, according to Definition 3.13 and lifting $\trianglelefteq_D$, an argument with value v only gets defeated by an argument with value v' that satisfies $v \sim v'$ or $v = v'$. In such a case, the defeats are mutual so argument a is self-defended. Thus, it is contained in a preferred extension. If it is not in a cycle, which means that it is not self-defended but only defeats other arguments, then it is in the grounded extension.
□

When using lifting $\trianglelefteq_E$, our argumentation-based approach always rejects the argument with the least preferred value. Because of that, the plan that promotes the least value will be rejected.

Proposition 3.24. *Given an argumentation framework for planning $PAF = \langle \mathcal{A}^p, \mathcal{D}^p \rangle$ with lifting $\trianglelefteq_E$, let $v \in \text{Val}$ be a value such that for all* arguments $a \in \mathcal{A}^p$ and all values $v' \in V_a$ it is the case that $v \precsim v'$, then an argument with value v is rejected under any semantics.

Proof. In order for an argument with v to be accepted, there must be another accepted argument that defends the an argument with v. However, since for all arguments $a \in \mathcal{A}^p$ and all values $v' \in V_a$ it is the case that $v \precsim v'$, this argument will also defeat the argument with v, making it rejected. □

Because of the above three propositions, the agent can conclude to follow an optimal plan to achieve his goal. However, the notion of optimal plans is defined as the set of conclusions of ordinary arguments from the extensions, so the set of optimal plans becomes empty if an extension does not contain any ordinary arguments. The following proposition indicates the conditions for which the set of optimal plans is not empty.

Proposition 3.25. *Given an argumentation framework for planning* $PAF = \langle \mathcal{A}^p, \mathcal{D}^p \rangle$, Oplans$(PAF, \mathcal{S}) \neq \emptyset$ *iff there exists an ordinary argument a such that it is not defeated by a blocking argument b with $V_a \prec V_b$.*

Proof. Having Oplans$(PAF, \mathcal{S}) \neq \emptyset$ means that there is at least one extension which contains at least one ordinary argument. ⇒: Suppose there does not exists an ordinary argument a such that it is not defeated by a blocking argument b with $V_a \prec V_b$, which means that all the ordinary arguments (if exist) are defeated by a blocking argument and not self-defended against a blocking argument. In such a case, there exists a blocking argument that does not receive any defeats, which makes all the ordinary arguments rejected. Contradiction! ⇐: If there exists an ordinary argument such that it is not defeated by a blocking argument with $V_a \prec V_b$, then (1) the ordinary argument does not receive any defeats and thus it should be contained in the grounded extension, or (2) the ordinary argument is in a two-length cycle with a blocking argument and thus it should be contained in a preferred extension, or (3) the ordinary argument receives defeats from other ordinary arguments and thus there is always an ordinary argument accepted. Hence, Oplans$(PAF, \mathcal{S})$ is not an empty set. □

Example 3.26. *The argumentation framework for planning PAF with lifting $\trianglelefteq_D$ can be represented as Fig. 4. Because*

$$\mathcal{E}_{\mathcal{PR}}(PAF) = \mathcal{E}_{\mathcal{GR}}(PAF) = \mathcal{E}_{\mathcal{ST}}(PAF) =$$
$$\{\{\langle +\{pv, sf\}, \alpha_2\alpha_5\alpha_6\rangle, \langle +\emptyset, \neg\alpha_1\alpha_4\rangle, \langle -\{sf\}, \neg\alpha_2\alpha_3\rangle\}\}$$

and thus Oplans$(PAF, \mathcal{S}) = \{\alpha_2\alpha_5\alpha_6\}$, *the agent can follow plan $\alpha_2\alpha_5\alpha_6$ to get to a pharmacy. The argumentation framework for planning PAF with lifting $\trianglelefteq_E$ can be represented as Fig. 5. Because*

$$\mathcal{E}_{\mathcal{PR}}(PAF) = \mathcal{E}_{\mathcal{GR}}(PAF) = \mathcal{E}_{\mathcal{ST}}(PAF) =$$
$$\{\{\langle -\{gc\}, \alpha_2\alpha_5\alpha_6\rangle, \langle +\emptyset, \neg\alpha_1\alpha_4\rangle, \langle -\{sf\}, \neg\alpha_2\alpha_3\rangle\}\}$$

and thus Oplans$(PAF, \mathcal{S}) = \emptyset$.

When making plans, an agent must evaluate the available options based on their value system. Representation results express all the available plans with value promotion and demotion, establishing a preference order over values as part of the agent's value system. Intuitively, the agent can use representation results to translate preferences over values into preferences over plans. However, since each plan has a set of promoted values and a set of demoted values, the agent must specify their preferences over plans from both aspects, which traditional lifting approaches cannot accommodate. In structured argumentation, like ASPIC+, people use lifting approaches to determine the defeat between two arguments based on preferences over rules and premises in each argument. Drawing inspiration from this, we suggest constructing both ordinary and blocking arguments for the execution of a plan in order to account for the promoted and demoted values associated with it. The success of one argument in defeating another depends on the preference order between the two sets of values pertaining to the arguments. In essence, rather than directly translating preferences over values into preferences over plans, we translate preferences over values into preferences over sets of values when determining the defeat relation between arguments, ultimately leading to accepted plans. This demonstrates that our argumentation-based approach serves as a dialogical justification for the use of lifting approaches and as a mediating mechanism between preferences over values and preferences over plans.

4 Argumentation and temporal logic for coalition formation

Argumentation has proven useful to provide a sound model to conceptualize reasoning processes related to *coalition formation* in multiagent systems [Amgoud, 2005a; Amgoud, 2005b]. The underlying approach is based on using conflict and preference relationships among coalitions to determine which coalitions should be adopted by the agents according to a particular argumentation semantics, which can then be computed using a suitable proof theory.

A variant of modal logic suitable for temporal reasoning called Alternating-time Temporal Logic (ATL) [Alur et al., 2002] can provide a further development on the above concept, making it possible to reason about the behavior and abilities of agents under various rationality assumptions [Jamroga and Bulling, 2007a; Jamroga and Bulling, 2007b; Bulling et al., 2008c]. In ATL the key construct has the form $\langle\!\langle A \rangle\!\rangle \phi$, which expresses that a coalition A of agents can *enforce* the formula ϕ. Under a model theoretic viewpoint, $\langle\!\langle A \rangle\!\rangle \phi$ holds whenever the agents in A have a winning strategy for ensuring the satisfiability of ϕ (independently of the behavior of A's opponents). However, this operator accounts only for the *theoretical existence* of such a strategy, not taking into account whether the coalition A can be actually formed. Indeed, in order to join a coalition, agents usually require some kind of *incentive* (e.g. sharing common goals, getting rewards, etc.), since usually forming a coalition does not come for free (fees have to be paid, communication costs may occur, etc.). Consequently, several possible coalition structures among agents may arise, from which the best ones should be adopted according to some rationally justifiable procedure.

In this section we present an argumentative approach to extend ATL for modelling coalitions. We provide a formal extension of ATL, CoaLATL, by including a new construct $\langle\!| A |\!\rangle \phi$ which denotes that *the group A of agents is able to build a coalition B, $A \cap B \neq \emptyset$, such that B can enforce ϕ*. That is, it is assumed that agents in A work together and try to form a coalition B. The actual computation of the coalition is modelled in terms of a given argumentation semantics [Dung, 1995] in the context of coalition formation [Amgoud, 2005a]. In a second step, we enrich CoaLATL with goals. We address the question *why* agents should cooperate. Goals refer to agents' subjective incentive to join coalitions. We show that the proof theory for modelling coalitions in our framework can be embedded as a natural extension of the model checking procedure used in ATL.

4.1 Alternating-time Temporal Logic in a nutshell

Alternating-time Temporal Logic (ATL) [Alur et al., 2002] enables reasoning about temporal properties and strategic abilities of agents. The language of ATL is defined as follows.

Definition 4.1 ($\mathcal{L}_{ATL}$ [Alur et al., 2002]). *Let* $\text{Agt} = \{a_1, \ldots, a_k\}$ *be a nonempty finite set of all agents, and* Φ *be a set of propositions (with typical element p). We denote by "a" a typical agent, and by "A" a typical group of agents from* Agt. $\mathcal{L}_{ATL}(\text{Agt}, \Phi)$ *is defined by the following grammar:* $\varphi ::= p \mid \neg\varphi \mid \varphi \wedge \varphi \mid \langle\!\langle A \rangle\!\rangle \bigcirc \varphi \mid \langle\!\langle A \rangle\!\rangle \square \, \varphi \mid \langle\!\langle A \rangle\!\rangle \varphi \mathcal{U} \, \varphi$.

Informally, $\langle\!\langle A \rangle\!\rangle \varphi$ expresses that agents A have a *collective strategy to enforce* φ. ATL formulae include the usual temporal operators: $\bigcirc$ ("in the next state"), $\square$ ("always from now on") and $\mathcal{U}$ (strict "until"). Additionally, $\Diamond$ ("now or sometime in the future") can be defined as $\Diamond \varphi \equiv \top \mathcal{U} \varphi$.

The semantics of ATL is defined by *concurrent game structures*. We recall that $\Phi = \{p, q, r, \ldots\}$ denotes a set of atomic propositions, $\text{Agt} = \{a_1, \ldots, a_k\}$ is a set of *agents*, and $Act = \{\alpha_1, \ldots, \alpha_n\}$ is a set of actions.

Definition 4.2 (CGS [Alur et al., 2002]). *A concurrent game structure* (CGS) *is a tuple* $\mathcal{M} = \langle S, \mathcal{V}, d, o \rangle$, *where each of the components is defined as follows:*

- S *is a set of* states.

- $\mathcal{V}: S \to 2^\Phi$ *is a* valuation function.

- $d: \text{Agt} \times S \to 2^{Act}$ *is a function that indicates the actions available to agent* $a \in \text{Agt}$ *in state* $q \in S$. *We often write* $d_a(q)$ *instead of* $d(a, q)$, *and use* $d(q)$ *to denote the set* $d_{a_1}(q) \times \cdots \times d_{a_k}(q)$ *of action profiles* in state q.

- *Finally, o is a* transition function *which maps each state $q \in S$ and action profile $\vec{\alpha} = \langle \alpha_1, \ldots, \alpha_k \rangle \in d(q)$ to another state $q' = o(q, \vec{\alpha})$.*

Note that these structures can be seen as a special case of our generic labelled transition systems (Definition 2.2) where the set of labels is instantiated a the set of all action profiles. Moreover, the underlying relation $\mathcal{R}$ (here represented with o) is partially functional (just as in Definition 3.1). Nota also that "q" is not to be confused with a propostional variable, it is simply a state in which certain propositional variables are true (determined by $\mathcal{V}$).

A *path* $\lambda = q_0 q_1 \cdots \in S^\omega$ is an infinite sequence of states such that there is a transition between each q_i, q_{i+1}. We define $\lambda[i] = q_i$ to denote the i-th state of λ. The set of all paths starting in q is defined by $\Lambda_{\mathcal{M}}(q)$.

A (memoryless) *strategy* of agent a is a function $s_a : S \to Act$ such that $s_a(q) \in d_a(q)$. We denote the set of such functions by Σ_a. A *collective strategy* s_A for team $A \subseteq \text{Agt}$ specifies an individual strategy for each agent $a \in A$; the set of A's collective strategies is given by $\Sigma_A = \prod_{a \in A} \Sigma_a$ and $\Sigma := \Sigma_{\text{Agt}}$.

The *outcome* of strategy s_A in state q is defined as the set of all paths that may result from executing s_A: $out(q, s_A) = \{\lambda \in \Lambda_{\mathcal{M}}(q) \mid \forall i \in \mathbb{N}_0 \; \exists \vec{\alpha} = \langle \alpha_1, \ldots, \alpha_k \rangle \in d(\lambda[i]) \; \forall a \in A \; (\alpha_a = s_A^a(\lambda[i]) \wedge o(\lambda[i], \vec{\alpha}) = \lambda[i+1])\}$, where s_A^a denotes agent a's part of the collective strategy s_A.

Definition 4.3 (ATL Semantics). *Let a* CGS *$\mathcal{M} = \langle S, \mathcal{V}, d, o \rangle$ and $q \in S$ be given. The semantics is given by a satisfaction relation $\models$ as follows:*

$\mathcal{M}, q \models p$ *iff* $p \in \mathcal{V}(q)$

$\mathcal{M}, q \models \neg \varphi$ *iff* $\mathcal{M}, q \not\models \varphi$

$\mathcal{M}, q \models \varphi \wedge \psi$ *iff* $\mathcal{M}, q \models \varphi$ *and* $\mathcal{M}, q \models \psi$

$\mathcal{M}, q \models \langle\!\langle A \rangle\!\rangle \bigcirc \varphi$ *iff there is* $s_A \in \Sigma_A$ *st.* $\mathcal{M}, \lambda[1] \models \varphi$ *for all* $\lambda \in out(q, s_A)$

$\mathcal{M}, q \models \langle\!\langle A \rangle\!\rangle \Box \, \varphi$ *iff there is* s_A *st.* $\mathcal{M}, \lambda[i] \models \varphi$ *for all* $\lambda \in out(q, s_A)$ *and* $i \in \mathbb{N}_0$

$\mathcal{M}, q \models \langle\!\langle A \rangle\!\rangle \varphi \mathcal{U} \psi$ *iff there is* $s_A \in \Sigma_A$ *st., for all* $\lambda \in out(q, s_A)$, *there is* $i \in \mathbb{N}_0$ *with* $\mathcal{M}, \lambda[i] \models \psi$, *and* $\mathcal{M}, \lambda[j] \models \varphi$ *for all* $0 \leq j < i$.

We note that the given semantics aligns well with Definition 4.1 and all the formulae introduced there.

4.2 Coalitions and argumentation

In this subsection, we provide an argument-based characterization of coalition formation that will be used later to extend ATL. We follow an approach similar to [Amgoud, 2005a], where an argumentation framework for generating coalition structures is defined, generalizing the framework of Dung for argumentation [Dung, 1995], [1] extended with a *preference relation*. The basic notion is that of a *coalitional framework*, which contains a set of elements $\mathfrak{C}$ (usually seen as agents or coalitions), an attack relation (for modelling conflicts among elements of $\mathfrak{C}$), and a preference relation between elements of $\mathfrak{C}$ (to describe favorite agents/coalitions).

Definition 4.4 (Coalitional framework [Amgoud, 2005a]). *A coalitional framework is a triple $\mathcal{CF} = \langle \mathfrak{C}, att, \prec \rangle$ where $\mathfrak{C}$ is a non-empty set of elements, $att \subseteq \mathfrak{C} \times \mathfrak{C}$ is an attack relation, and $\prec$ is a preorder on $\mathfrak{C}$ representing preferences on elements in $\mathfrak{C}$.*

Let S be a non-empty set of elements. $\mathbb{CF}(S)$ denotes the set of all coalitional frameworks where elements are taken from the set S, i.e. for each $\langle \mathfrak{C}, att, \prec \rangle \in \mathbb{CF}(S)$ we have that $\mathfrak{C} \subseteq S$.

The set $\mathfrak{C}$ in Definition 4.4 is intentionally generic, accounting for various possibilities. One is to consider $\mathfrak{C}$ as a set of agents $\text{Agt} = \{a_1, \ldots, a_k\}$: $\mathcal{CF} = \langle \mathfrak{C}, att, \prec \rangle \in \mathbb{CF}(\text{Agt})$. Then, a *coalition* is given by $C = \{a_{i_1}, \ldots, a_{i_l}\} \subseteq \mathfrak{C}$ and "agent" can be used as an intuitive reference to elements of $\mathfrak{C}$. Another possibility is to use a coalitional framework $\mathcal{CF} = \langle \mathfrak{C}, att, \prec \rangle$ based on $\mathbb{CF}(2^{\text{Agt}})$. Now elements of $\mathfrak{C} \subseteq 2^{\text{Agt}}$ are *groups* or *coalitions* (where we consider singletons as groups too) of agents. Under this interpretation a coalition $C \subseteq \mathfrak{C}$ is a *set of sets* of agents. Although "coalition" is already used for $C \subseteq \mathfrak{C}$, we also use the intuitive reading "coalition" or "group" to address elements in $\mathfrak{C}$.[2] Yet another way is not to use the specific structure for elements in $\mathfrak{C}$,

---

[1] The reader is referred to Section 2.2 for further details about Dung's approach to abstract argumentation.

[2] The first interpretation is a special case of the second (coalitional frameworks are members $\mathbb{CF}(2^{\text{Agt}})$).

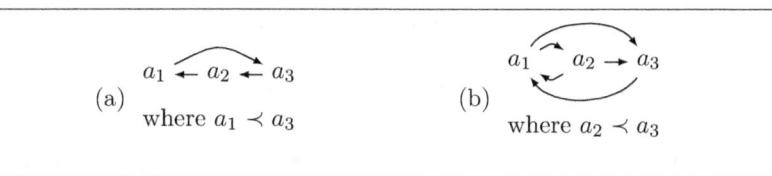

Figure 6: Figure (a) (resp. (b)) corresponds to the coalitional frameworks defined in Example 4.5 (resp. 4.14 (b)). Nodes represent agents and arrows between nodes stand for the attack relation.

assuming it just consists of abstract elements, e.g. c_1, c_2, etc. One may think of these elements as individual agents or coalitions. This approach is followed in [Amgoud, 2005a]. From now on we will mainly follow the first alternative when informally speaking about coalitional frameworks (i.e., we consider $\mathfrak{C}$ as a set of agents).

Example 4.5. *Consider the following two coalitional frameworks: (i) $\mathcal{CF}_1 = \langle \mathfrak{C}, att, \prec \rangle$ where $\mathfrak{C} = \{a_1, a_2, a_3\}$, $att = \{\langle a_3, a_2\rangle, \langle a_2, a_1\rangle, \langle a_1, a_3\rangle\}$ and agent a_3 is preferred over a_1, i.e. $a_1 \prec a_3$; and (ii) $\mathcal{CF}_2 = \langle \mathfrak{C}', att', \prec' \rangle$ where $\mathfrak{C}' = \{\{a_1\}, \{a_2\}, \{a_3\}\}$, $att' = \{\langle \{a_3\}, \{a_2\}\rangle, \langle \{a_2\}, \{a_1\}\rangle, \langle \{a_1\}, \{a_3\}\rangle\}$ and group $\{a_3\}$ is preferred over $\{a_1\}$, i.e. $\{a_1\} \prec' \{a_3\}$. They capture the same scenario and are isomorphic but $\mathcal{CF}_1 \in \mathbb{CF}(\{a_1, a_2, a_3\})$ and $\mathcal{CF}_2 \in \mathbb{CF}(\mathcal{P}(\{a_1, a_2, a_3\}))$; that is, the first framework is defined regarding single agents and the latter over (trivial) coalitions. Figure 6 (a) shows a graphical representation of the first coalitional framework.*

Let $\mathcal{CF} = \langle \mathfrak{C}, att, \prec \rangle$ be a coalitional framework. For $C, C' \in \mathfrak{C}$, we say that C attacks C' iff $CattC'$. The attack relation represents conflicts between elements of $\mathfrak{C}$; for instance, two agents may rely on the same (unique) resource or they may have disagreeing goals, which prevent them from cooperation. However, the notion of attack may not be sufficient for modelling conflicts, as some elements (resp. coalitions) in $\mathfrak{C}$ may be preferred over others. This leads to the notion of *defeater* which combines the notions of attack and preference. Following Dung's approach to abstract argumentation (see Section 2.2), members in a coalition may prevent attacks to members in the same coalition, *defending* each other. This prompts the following definitions:

Definition 4.6 (Defeater). *Let* $\mathcal{CF} = \langle \mathfrak{C}, att, \prec \rangle$ *be a coalitional framework and let* $C, C' \in \mathfrak{C}$. *We say that* C *defeats* C' *if, and only if,* C *attacks* C' *and* C' *is not preferred over* C *(i.e., not* $C \prec C'$*). We also say that* C *is a* defeater *for* C'.

Definition 4.7 (Defence). *Let* $\mathcal{CF} = \langle \mathfrak{C}, att, \prec \rangle$ *be a coalitional framework and* $C, C' \in \mathfrak{C}$. *We say that* C' *defends itself against* C *if, and only if,* C' *is preferred over* C, *i.e.,* $C \prec C'$, *and* C' *defends itself* if it defends itself against any of its attackers. *Furthermore,* C *is* defended by *a set* $\mathfrak{S} \subseteq \mathfrak{C}$ *of elements of* $\mathfrak{C}$ *if, and only if, for all* C' *defeating* C *there is a coalition* $C'' \in \mathfrak{S}$ *defeating* C'.

In other words, if an element C' defends itself against C then C may attack C' but C is not allowed to defeat C'. A minimal requirement one should impose on a coalition is that its members do not defeat each other; otherwise, the coalition may be unstable and break up sooner or later because of conflicts among its members. This is formalised in the next definition.

Definition 4.8 (Conflict-free). *Let* $\mathcal{CF} = \langle \mathfrak{C}, att, \prec \rangle$ *be a coalitional framework and* $\mathfrak{S} \subseteq \mathfrak{C}$ *a set of elements in* $\mathfrak{C}$. *Then,* $\mathfrak{S}$ *is called* conflict-free *if, and only if, there is no* $C \in \mathfrak{S}$ *defeating some member of* $\mathfrak{S}$.

It should be remarked that our notions of "defence" and "conflict-free" are defined in terms of "defeat" rather than "attack".[3] Given a coalitional framework $\mathcal{CF}$ we will use argumentation to compute coalitions with desirable properties. In argumentation theory, many different semantics have been proposed to define ultimately accepted arguments [Dung, 1995].

Definition 4.9 (Coalitional framework semantics). *A semantics for a coalitional framework* $\mathcal{CF} = \langle \mathfrak{C}, att, \prec \rangle$ *is a (isomorphism invariant) mapping* $\mathcal{E}$ *which assigns to a given coalitional framework* $\mathcal{CF} = \langle \mathfrak{C}, att, \prec \rangle$ *a set of subsets of* $\mathfrak{C}$, *i.e.,* $\mathcal{E}(\mathcal{CF}) \subseteq \mathcal{P}(\mathfrak{C})$.

Let $\mathcal{CF} = \langle \mathfrak{C}, att, \prec \rangle$ be a coalitional framework. To formally characterize different semantics we will define a function $\mathcal{F}_{\mathcal{CF}} : \mathcal{P}(\mathfrak{C}) \to \mathcal{P}(\mathfrak{C})$

---

[3] In [Amgoud, 2005a; Amgoud, 2005b] these notions are defined the other way around, resulting in a different characterization of stable semantics.

which assigns to a set of coalitions $\mathfrak{S} \in \mathcal{P}(\mathfrak{C})$ the coalitions defended by $\mathfrak{S}$.

Definition 4.10 (Characteristic function $\mathcal{F}$). *Let $\mathcal{CF} = \langle \mathfrak{C}, att, \prec \rangle$ be a coalitional framework and $\mathfrak{S} \subseteq \mathfrak{C}$. The function $\mathcal{F}$ defined by*

$$\mathcal{F}_{\mathcal{CF}} : \mathcal{P}(\mathfrak{C}) \to \mathcal{P}(\mathfrak{C})$$
$$\mathcal{F}_{\mathcal{CF}}(\mathfrak{S}) = \{C \in \mathfrak{C} \mid C \text{ is defended by } \mathfrak{S}\}$$

is called characteristic function.[4]

$\mathcal{F}$ can be applied recursively to coalitions resulting in new coalitions. For example, $\mathcal{F}(\emptyset)$ provides all undefeated coalitions and $\mathcal{F}^2(\emptyset)$ constitutes the set of all elements of $\mathfrak{C}$ which members are undefeated *or* are defended by undefeated coalitions.

Example 4.11. *Consider again the coalitional framework $\mathcal{CF}_1$ given in Example 4.5. The characteristic function applied on the empty set results in $\{a_3\}$ since the agent is undefeated, $\mathcal{F}(\emptyset) = \{a_3\}$. Applying $\mathcal{F}$ on $\mathcal{F}(\emptyset)$ determines the set $\{a_1, a_3\}$ because a_1 is defended by a_3. It is easy to see that $\{a_1, a_3\}$ is a fixed point of $\mathcal{F}$.*

We now introduce the first concrete semantics called coalition structure semantics, which was originally defined in [Amgoud, 2005a].

Definition 4.12 (Coalition structure $\mathcal{E}_{CS}$ [Amgoud, 2005a]). *Let $\mathcal{CF} = \langle \mathfrak{C}, att, \prec \rangle$ be a coalitional framework. Then*

$$\mathcal{E}_{CS}(\mathcal{CF}) := \left\{ \bigcup_{i=1}^{\infty} \mathcal{F}_{\mathcal{CF}}^i(\emptyset) \right\}$$

is called coalition structure semantics *or just* coalition structure *for $\mathcal{CF}$.*

For a coalitional framework $\mathcal{CF} = \langle \mathfrak{C}, att, \prec \rangle$ with a finite set $\mathfrak{C}$[5] the characteristic function $\mathcal{F}$ is continuous [Dung, 1995, Lemma 28]. Since $\mathcal{F}$ is also monotonic it has a least fixed point given by $\mathcal{F}(\emptyset) \uparrow^\omega$ (according to Knaster-Tarski). We have the following straightforward properties of coalition structure semantics.

---
[4]We omit the subscript $\mathcal{CF}$ if it is clear from context.
[5]Actually, it is enough to assume that $\mathcal{CF}$ is finitary (cf. [Dung, 1995, Def. 27]).

Proposition 4.13 (Coalition structure). *Let $\mathcal{CF} = \langle \mathfrak{C}, att, \prec \rangle$ be a coalitional framework with a finite set $\mathfrak{C}$. There is always a unique coalition structure for $\mathcal{CF}$. Furthermore, if no element of $C \in \mathfrak{C}$ defends itself then the coalitional structure is empty, i.e. $\mathcal{E}_{CS}(\mathcal{CF}) = \{\emptyset\}$.*

Example 4.14. *The following situations illustrate the notion of coalitional structure:*

(a) *Consider Example 4.11. Since $\{a_1, a_3\}$ is a fixed point of $\mathcal{F}_{\mathcal{CF}_1}$ the coalitional framework $\mathcal{CF}_1$ has $\{a_1, a_3\}$ as coalitional structure.*

(b) *$\mathcal{CF}_3 := \langle \mathfrak{C}, att, \prec \rangle \in \mathbb{CF}(\{a_1, a_2, a_3\})$ (shown in Figure 6(b)), is a coalitional framework with $\mathfrak{C} = \{a_1, a_2, a_3\}$, $att = \{ \langle a_1, a_2 \rangle, \langle a_1, a_3 \rangle, \langle a_2, a_1 \rangle, \langle a_2, a_3 \rangle, \langle a_3, a_1 \rangle\}$ and a_3 is preferred over a_2, $a_2 \prec a_3$, has the empty coalition as associated coalition str., i.e. $\mathcal{E}_{CS}(\mathcal{CF}) = \{\emptyset\}$.*

Since the coalition structure is often very restrictive, it seems reasonable to introduce other less restrictive semantics, following Dung's approach to abstract argumentation (see Section 2.2). We redefine these semantics in terms of the characteristic function $\mathcal{F}$:

Definition 4.15 (Argumentation Semantics). *Let $\langle \mathfrak{C}, att, \prec \rangle$ be a coalitional framework, $\mathfrak{S} \subseteq \mathfrak{C}$ a set of elements of $\mathfrak{C}$. $\mathfrak{S}$ is called (a) admissible extension iff $\mathfrak{S}$ is conflict-free and $\mathfrak{S}$ defends all its elements, i.e. $\mathfrak{S} \subseteq \mathcal{F}(\mathfrak{S})$; (b) complete extension iff $\mathfrak{S}$ is conflict-free and $\mathfrak{S} = \mathcal{F}(\mathfrak{S})$; (c) grounded extension iff $\mathfrak{S}$ is the smallest (wrt. to set inclusion) complete extension; (d) preferred extension iff $\mathfrak{S}$ is a maximal (wrt. to set inclusion) admissible extension; (e) stable extension iff $\mathfrak{S}$ is conflict-free and it defeats all arguments not in $\mathfrak{S}$. Let $\mathcal{E}_{CS}(\mathcal{CF})$ (resp. $\mathcal{E}_{CO}(\mathcal{CF})$, $\mathcal{E}_{\mathcal{GR}}(\mathcal{CF})$, $\mathcal{E}_{\mathcal{PR}}(\mathcal{CF})$ and $\mathcal{E}_{ST}(\mathcal{CF})$) denote the semantics which assigns to a coalitional structure $\mathcal{CF}$ all its admissible (resp. complete, grounded, preferred, and stable) extensions.*

There is only one unique coalition structure (possibly the empty one) for a given coalitional framework, but there can be several stable and preferred extensions. The existence of at least one preferred extension is assured which is not the case for the stable semantics. Thus, the possible coalitions very much depend on the used semantics.

Example 4.16. *For $\mathcal{CF}_3$ from Example 4.14 the following holds:*

$$\mathcal{E}_{\mathcal{CS}}(\mathcal{CF}) = \{\emptyset\}$$
$$\mathcal{E}_{\mathcal{AD}}(\mathcal{CF}) = \{\{a_1\},\{a_2\},\{a_3\},\{a_2,a_3\}\}$$
$$\mathcal{E}_{\mathcal{CO}}(\mathcal{CF}) = \mathcal{E}_{\mathcal{GR}}(\mathcal{CF}) = \{\{a_1\},\{a_2,a_3\}\} =$$
$$\mathcal{E}_{\mathcal{PR}}(\mathcal{CF}) = \mathcal{E}_{\mathcal{ST}}(\mathcal{CF}) = \{\{a_1\},\{a_2,a_3\}\}$$

Analogously, for the coalitional framework $\mathcal{CF}_1$ from Example 4.5 there exists one complete extension $\{a_1, a_3\}$ which is also a grounded, preferred, and stable extension.

4.3 Coalitional ATL

In this section we combine *argumentation for coalition formation* and ATL and introduce *Coalitional* ATL (CoaATL). This logic extends ATL by new operators $\langle\!| A |\!\rangle$ for each subset $A \subseteq$ Agt of agents. These new modalities combine, or rather integrate, coalition formation into the original ATL cooperation modalities $\langle\!\langle A \rangle\!\rangle$. The intended reading of $\langle\!| A |\!\rangle \varphi$ is that the group A of agents *is able to form a coalition $B \subseteq$ Agt such that some agents of A are also members of B, $A \cap B \neq \emptyset$, and B can enforce φ.* Coalition formation is modelled by the formal argumentative approach in the context of coalition formation, as described in Section 4.2, based on the framework developed in [Amgoud, 2005a].

Our main motivation for this logic is to make it possible to reason about the ability of building coalition structures, and not only about an *a priori* specified group of agents (as it is the case for $\langle\!\langle A \rangle\!\rangle \varphi$). The new modality $\langle\!| A |\!\rangle$ provides a rather subjective view of the agents in A and their power to create a group B, $A \cap B \neq \emptyset$, which in turn is used to reason about the ability to enforce a given property.

The language of CoaATL is as follows.

Definition 4.17 ($\mathcal{L}_{ATL^c}$). *Let Agt $= \{a_1, \ldots, a_k\}$ be a finite, nonempty set of agents, and Φ be a set of propositions (with typical element p). We use the symbol "a" to denote a typical agent, and "A" to denote a typical group of agents from Agt. The logic $\mathcal{L}_{ATL^c}(\text{Agt}, \Phi)$ is defined by the following grammar:*

$$\varphi ::= p \mid \neg\varphi \mid \varphi \wedge \varphi \mid \langle\!\langle A \rangle\!\rangle \bigcirc \varphi \mid \langle\!\langle A \rangle\!\rangle \square \varphi \mid \langle\!\langle A \rangle\!\rangle \varphi \mathcal{U} \varphi \mid$$
$$\langle\!| A |\!\rangle \bigcirc \varphi \mid \langle\!| A |\!\rangle \square \varphi \mid \langle\!| A |\!\rangle \varphi \mathcal{U} \varphi$$

We extend concurrent game structures by means of *coalitional frameworks* and an *argumentative semantics*. A coalitional framework is assigned to each state of the model capturing the current "conflicts" among agents. In doing so, we allow that conflicts can change over time, being thus *state dependent*. Moreover, we assume that coalitional frameworks are agent-dependent. Thus, two intial groups of agents may have different skills to form coalitions. Consider for instance the following example.

Example 4.18. *Imagine two agents a_1 and a_2 which are not able (because they do not have the money) to convince a_3 and a_4 to join. But a_1, a_2 and a_3 together have the money and all four can enforce a property φ. So $\{a_1, a_2\}$ are not able to build a greater coalition to enforce φ; but $\{a_1, a_2, a_3\}$ are. So we are not looking at coalitions per se, but how they evolve from others.*

We assume that the argumentative semantics is the same for all states.

Definition 4.19 (CGM). *A coalitional game model (CGM) is given by a tuple*
$$\mathcal{M} = \langle S, \mathcal{V}, d, o, \zeta, \mathcal{S} \rangle$$
where $\langle S, \mathcal{V}, d, o \rangle$ is a CGS, $\zeta : 2^{\text{Agt}} \to (S \to \mathbb{CF}(\text{Agt}))$ is a function which assigns a coalitional framework over Agt to each state of the model subjective to a given group of agents, and $\mathcal{S}$ is an (argumentative) semantics as defined in Definition 4.9. The set of all such models is given by $\mathbb{M}(S, \text{Agt}, \Phi, \zeta, \mathcal{S})$.

A model provides an argumentation semantics $\mathcal{S}$ which assigns all formable coalitions to a given coalitional framework. As argued before we require from a valid coalition that it is not only justified by the argumentation semantics but that it is also not disjunct with the predetermined starting coalition. This leads to the notion *valid coalition*.

Definition 4.20 (Valid coalition). *Let A, $B \subseteq \text{Agt}$ be groups of agents, $\mathcal{M} = \langle S, \mathcal{V}, d, o, \zeta, \mathcal{S} \rangle$ be a CGM and $q \in S$. We say that B is a valid coalition with respect to A, q, and $\mathcal{M}$ whenever $B \in \mathcal{E}_\mathcal{S}(\zeta(A)(q)))$ and $A \cap B \neq \emptyset$. Furthermore, we use $\text{VC}_\mathcal{M}(A, q)$ to denote the set of all valid coalitions regarding A, q, and $\mathcal{M}$ (subscript $\mathcal{M}$ is omitted if clear from the context).*

Remark 4.21. In [Bulling et al., 2008a] we assume that the members of the initial group A work together, whatever the reasons might be. So group A was added to the semantics. This ensured that agents in A can enforce ψ on their own, if they are able to do so. Even if A is not accepted originally by the argumentation semantics, i.e. $A \notin \mathcal{E}_\mathcal{S}(\zeta(A)(q))$. Here, we drop this requirement. As pointed out in [Bulling et al., 2008a] the "old" semantics is just a special case of this new one: The operator from [Bulling et al., 2008b] can be defined as $\langle\!\langle A\rangle\!\rangle\gamma \vee \langle\!\langle A\rangle\!\rangle\gamma$.

Moreover, we changed the condition that the predefined group given in the coalitional operator must be a subset of the formed coalition, $A \subseteq B$, to the weaker requirement that only some member of the inital coalition should be in the new one, $A \cap B \neq \emptyset$.

The semantics of the new modality is given by

Definition 4.22 (COALATL Semantics). *Let a* CGM $\mathcal{M} = \langle S, \mathcal{V}, d, o, \zeta, \mathcal{S}\rangle$ *a group of agents* $A \subseteq \text{Agt}$, *and* $q \in S$ *be given. The semantics of Coalitional* ATL *extends that of* ATL, *given in Definition 4.3, by the following rule* $(\langle\!\langle A\rangle\!\rangle\psi \in \mathcal{L}_{ATL^c}(\text{Agt}, \Phi))$:

$\mathcal{M}, q \models \langle\!\langle A\rangle\!\rangle\psi$ *iff there is a coalition* $B \in \text{vc}_\mathcal{M}(A, q)$ *such that* $\mathcal{M}, q \models \langle\!\langle B\rangle\!\rangle\psi$.

Remark 4.23 (Different Semantics, $\models_\mathcal{S}$). *We have just defined a whole class of semantic rules for modality* $\langle\!\langle \cdot \rangle\!\rangle$. *The actual instantiation of the semantics* $\mathcal{S}$, *for example* $\mathcal{ST}$, $\mathcal{PR}$, *and* $\mathcal{CS}$ *defined in Section 4.2, affects the semantics of the cooperation modality.*

For the sake of readability, we sometimes annotate the satisfaction relation $\models$ *with the presently used argumentation semantics. That is, given a* CGM $\mathcal{M}$ *with an argumentation semantics* $\mathcal{S}$ *we write* $\models_\mathcal{S}$ *instead of* $\models$.

The underlying idea of the semantic definition of $\langle\!\langle A\rangle\!\rangle\psi$ is as follows. A given (initial) group of agents $A \subseteq \text{Agt}$ is able to form a *valid coalition* B (where A and B must not be disjoint), with respect to a given coalitional framework $\mathcal{CF}$ and a particular semantics $\mathcal{S}$, such that B can enforce ψ.

Similarly to the different possibilities in our definition of valid coalitions there are other sensible semantics for COALATL. The semantics we

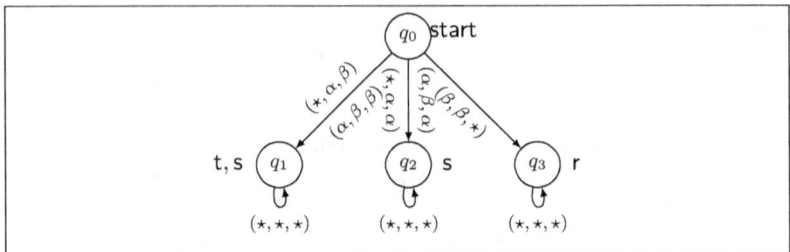

Figure 7: A simple CGS defined in Example 4.25.

presented here is not particularly dependent on time; i.e., except from the selection of a valid coalition B at the initial state there is no further interaction between time and coalition formation. We have chosen this simplistic definition to present our main idea—the connection of ATL and coalition formation by means of argumentation–as clear as possible. A precise approach dealing with time, however, is beyond the scope of this chapter.

Proposition 4.24 ([Bulling et al., 2008b]). *Let $A \subseteq \mathrm{Agt}$ and $\langle\!\langle A \rangle\!\rangle \psi \in \mathcal{L}_{ATL^c}(\mathrm{Agt}, \Phi)$. Then it holds that $\langle\!\langle A \rangle\!\rangle \psi \to \bigvee_{B \in 2^{\mathrm{Agt}}, A \subseteq B} \langle\!\langle B \rangle\!\rangle \psi$ is a validity with respect to CGM's.*

Compared to ATL, a formula like $\langle\!\langle A \rangle\!\rangle \varphi$ does *not* refer to the ability of A to enforce φ, but rather to the ability of A to *constitute* a coalition B, such that $A \cap B \neq \emptyset$, and then, in a second step, to the ability of B to enforce φ. Thus, two different notions of ability are captured in these new modalities. For instance, $\langle\!\langle A \rangle\!\rangle \psi \wedge \neg \langle\!\langle \mathrm{Agt} \rangle\!\rangle \psi$ expresses that group A of agents can enforce ψ, but there is no *reasonable* coalition at all which can enforce ψ (particularly not A, although they possess the theoretical power to do so).

The next example motivates the usefulness of the new modality. Classic ATL can only consider sets of agents that can enforce something, but it can not take into account whether such sets can indeed by formed (are allowed in the coalitional framework). The new modality, however, allows to model such situations.

Example 4.25. *There are three agents a_1, a_2, and a_3 which prefer different outcomes. Agent a_1 (resp. a_2, a_3) desires to get outcome* r *(resp.*

s, t). *One may assume that all outcomes are distinct; for instance, a_1 is not satisfied with an outcome x whenever $x \neq$ r. Each agent can choose to perform action α or β. Action profiles and their outcomes are shown in Figure 7. The $\star$ is used as a placeholder for any of the two actions, i.e. $\star \in \{\alpha, \beta\}$. For instance, the profile $(\beta, \beta, \star)$ leads to state q_3 whenever agent a_1 and a_2 perform action β and a_3 either does α or β.*

According to the scenario depicted in the figure, a_1 and a_2 cannot commonly achieve their goals. The same holds for a_1 and a_3. On the other hand, there exists a situation, q_1, in which both agents a_2 and a_3 are satisfied. One can formalise the situation as the coalitional game $\mathcal{CF} = \langle \mathfrak{C}, att, \prec \rangle$ given in Example 4.14(b), that is, $\mathfrak{C} =$ Agt, $att = \{(a_1,a_2), (a_1,a_3), (a_2,a_1), (a_2,a_3), (a_3,a_1)\}$ and $a_2 \prec a_3$.

We formalise the example as the CGM $\mathcal{M} = \langle S, \mathcal{V}, d, o, \zeta, \mathcal{S}\rangle$ where Agt $= \{a_1, a_2, a_3\}, S = \{q_0, q_1, q_2, q_3\}$, $\Phi = \{$r, s, t$\}$, and $\zeta(A)(q) = \mathcal{CF}$ for all states $q \in S$ and groups $A \subseteq$ Agt. Transitions and the state labeling can be seen in Figure 7. Furthermore, we do not specify a concrete semantics $\mathcal{S}$ yet, and rather adjust it in the remainder of the example.

We can use pure ATL formulas, i.e. formulas not containing the new modalities $\langle\!|\cdot|\!\rangle$, to express what groups of agents can achieve. We have, for instance, that agents a_1 and a_2 can enforce a situation which is undesirable for a_3: $\mathcal{M}, q_0 \models \langle\!\langle a_1, a_2 \rangle\!\rangle \bigcirc$ r. Indeed, $\{a_1, a_2\}$ and the grand coalition Agt (since it contains $\{a_1, a_2\}$) are the only coalitions which are able to enforce $\bigcirc$ r; we have

$$\mathcal{M}, q_0 \models \neg \langle\!\langle X \rangle\!\rangle \bigcirc \mathsf{r} \tag{1}$$

for all $X \subset$ Agt and $X \neq \{a_1, a_2\}$. Outcomes s or t can be enforced by a_2: $\mathcal{M}, q_0 \models \langle\!\langle a_2 \rangle\!\rangle \bigcirc (\mathsf{s} \vee \mathsf{t})$. Agents a_2 and a_3 also have the ability to enforce a situation which agrees with both of their desired outcomes: $\mathcal{M}, q_0 \models \langle\!\langle a_2, a_3 \rangle\!\rangle \bigcirc (\mathsf{s} \wedge \mathsf{t})$

These properties do not take into account the coalitional framework, that is, whether specific coalitions can be formed or not. By using the coalitional framework, we get

$$\mathcal{M}, q_0 \models_\mathcal{S} \langle\!\langle a_1, a_2 \rangle\!\rangle \bigcirc \mathsf{r} \wedge \neg \langle\!| a_1 |\!\rangle \bigcirc \mathsf{r} \wedge \neg \langle\!| a_2 |\!\rangle \bigcirc \mathsf{r}$$

for any semantics $\mathcal{S}$ introduced in Definition 4.9 and calculated in Example 4.16. The possible coalition (resp. coalitions) containing a_1 (resp. a_2) is $\{a_1\}$ (resp. are $\{a_2\}$ and $\{a_2, a_3\}$). But neither of these can enforce $\bigcirc r$ (in q_0) because of (1). Thus, although it is the case that the coalition $\{a_1, a_2\}$ has the theoretical ability to enforce r in the next moment (which is a "losing" situation for a_3), a_3 should not consider it as sensible since agents a_1 and a_2 would not agree to constitute a coalition (according to the coalitional framework $\mathcal{CF}$).

The decision for a specific semantics is a crucial point and depends on the actual application. The next example shows that with respect to a particular argumentation semantics, agents are able to form a coalition which can successfully achieve a given property, whereas another argumentative semantics does not allow that.

Example 4.26. COALATL *can be used to determine whether a coalition for enforcing a specific property exists. Assume that $\mathcal{S}$ represents the grounded semantics. For instance, the statement*

$$\mathcal{M}, q_0 \models_{\mathcal{E}_{\mathcal{GR}}} \langle\!\langle \emptyset \rangle\!\rangle \bigcirc t$$

expresses that there is a grounded coalition (i.e. a coalition wrt to the grounded semantics) which can enforce $\bigcirc t$, namely the coalition $\{a_2, a_3\}$. This result does not hold for all semantics; for instance, we have

$$\mathcal{M}, q_0 \not\models_{\mathcal{E}_{\mathcal{CS}}} \langle\!\langle \emptyset \rangle\!\rangle \bigcirc t$$

with respect to the coalition structure semantics, since the coalition structure is the empty coalition and $\mathcal{M}, q_0 \not\models \langle\!\langle \emptyset \rangle\!\rangle \bigcirc t$.

Note that it is easily possible to extend the language by an *update mechanism*, in order to compare different argumentative semantics using formulae inside the object language.

4.4 Cooperation and Goals

Why should agents join coalitions? Up to now we did not address this question and focussed on *why not* to cooperate. Often cooperation does not come for free and it requires some kind of incentive (i.e. sharing common goals or getting rewards) to offer one's ability in order to support other agents. Coalitional frameworks, however, were mainly used to model conflicts between agents, and therewith, avoid cooperation. In [Bulling et al., 2008b] the authors propose *goals* as agents' incentives to join coalitions; the following is based on that work.

We are now incorporating a *goal framework* into CoaLATL models. First of all, each agent is equipped with a *set of goals* $\mathcal{G}_a$ where $a \in$ Agt and $\mathcal{G} := \bigcup_{a \in \text{Agt}} \mathcal{G}_a$. Goals are formulated as ATL-path formulae or conjunctions of them. An agent, say Bill, might have the goal—or rather a dream—that it will sometimes be able to buy a new car *without* asking other people (e.g. its wife Ann). Such a goal can be formulated as $\Diamond \langle\!\langle \text{Bill} \rangle\!\rangle \bigcirc \text{buyNewCar}$. Sometimes Bill would like to *enforce* to buy a new car in the next moment. To assign goals to agents a CGM is extended by a *goal mapping*.

Definition 4.27 (Goal mapping $\mathfrak{g}$). *A goal mapping is a function* $\mathfrak{g}$: Agt $\to (S^+ \to \mathcal{P}(\mathcal{G}))$ *assigning a set of goals to a given sequence of states and an agent.*

So, a goal mapping assigns a set of goals to a *history*. This is needed to introduce goals into CGM's. The history dependency can be used, for instance, to model when a goal should be removed from the list: An agent having a goal $\Diamond$ s may drop it after reaching a state in which s holds.

So far, we did not say how goals can be actually used to form coalitions. We assume, given some task, that agents having goals satisfied or partly satisfied by the outcome of the task are willing to cooperate to bring about the task. Consider, for instance, the ATL formula $\langle\!\langle A \rangle\!\rangle \gamma$. It says that A can enforce γ—the *objective*. In the context of Coalitional ATL it is even more intuitive: $\langle\!| A |\!\rangle \gamma$ means that A is able to from a coalition B which can enforce the objective γ. Of course, rational agents should have reasons to bring about γ in order to work towards γ. In the following we will use the notion *objective* (or objective formula) to refer to both the task itself and the outcome of it. A typical objective is written as o. Agents which have goals fulfilled or at least partly fullfilled by objective o are possible candidates to participate in a coalition aiming at o. We consider CoaLATL *objectives* which are CoaLATL path formulae.

We say that an objective o *satisfies* goal g, $o \hookrightarrow g$, if the goal g is fulfilled after o has been accomplished. Intuitively, an objective $\Box$ t satisfies goal $\Box$ (t $\vee$ s) and supports goal $\Diamond$ t.

4.5 Coalitional ATL with Goals

In this section, we merge together Coalitional ATL with the goal framework described above. The syntax of the logic is given as in Definition 4.17. The necessary change takes place in the semantics. We redefine what it means for a coalition to be *valid*.

Up to now, valid coalitions were solely determined by coalitional frameworks. Conflicts represented by such frameworks are a coarse, but necessary, criterion for a successful coalition formation process. However, nothing is said about incentives *to join* coalitions, only why coalitions should *not* be joined.

Goals allow us to capture the first issue. For a given objective formula o and a finite sequence of states, called *history*, we only consider agents which have some goal supported by the current objective. CGM's *with goals* are given as a straightforward extension of CGM's (cf. Definition 4.19).

Definition 4.28 (CGM with goals). *A* CGM *with goals (*CGMG*)* $\mathcal{M}$ *is given by a model of* $\mathbb{M}(S, \text{Agt}, \Phi, \mathcal{S}, \zeta)$ *extended by a set of goals* $\mathcal{G}$ *and a goal mapping* $\mathfrak{g}$ *over* $\mathcal{G}$. *The set of all such models is denoted by* $\mathbb{M}^g(S, \text{Agt}, \Phi, \mathcal{S}, \zeta, \mathcal{G}, \mathfrak{g})$ *or just* $\mathbb{M}^g$ *if we assume standard naming.*

To define the semantics we need some additional notation. Given a path $\lambda \in S^\omega$ we use $\lambda[i,j]$ to denote the sequence $\lambda[i]\lambda[i+1]\ldots\lambda[j]$ for $i,j \in \mathbb{N}_0 \cup \{\infty\}$ and $i \leq j$. A *history* is a finite sequence $h = q_1 \ldots q_n \in S^+$, $h[i]$ denotes state q_i if $n \geq i$, q_n for $i \geq n$, and ε for $i < 0$ where $i \in \mathbb{Z} \cup \{\infty\}$. Furthermore, given a history h and a path or history λ the combined path/history starting with h extended by λ is denoted by $h \circ \lambda$.

Finally, we present the semantics of CoALATL *with* goals. It is similar to Definition 4.22. Here, however, it is necessary to keep track of the steps (visited states) made to determine the goals of the agents. The finite list of steps already taken is denoted by τ.

Definition 4.29 (Goal-based semantics of $\mathcal{L}_{ATL^c}$). Let $\mathcal{M}$ be a CGMG, q a state, and $i, j \in \mathbb{N}_0$. Let $\tau \in S+$, any finite sequence of states already visited. The goal-based semantics of $\mathcal{L}_{ATL^c}$ formulae is given as follows:

$\mathcal{M}, q, \tau \models p$ iff $p \in \mathcal{V}(q)$

$\mathcal{M}, q, \tau \models \varphi \wedge \psi$ iff $\mathcal{M}, q, \tau \models \varphi$ and $\mathcal{M}, q, \tau \models \psi$

$\mathcal{M}, q, \tau \models \neg \varphi$ iff not $\mathcal{M}, q, \tau \models \varphi$

$\mathcal{M}, q, \tau \models \langle\!\langle A \rangle\!\rangle \bigcirc \varphi$ iff there is $s_A \in \Sigma_A$ such that $\mathcal{M}, \lambda[1], \tau \circ \lambda[1] \models \varphi$ for all $\lambda \in out(q, s_A)$

$\mathcal{M}, q, \tau \models \langle\!\langle A \rangle\!\rangle \square \varphi$ iff there is s_A such that $\mathcal{M}, \lambda[i], \tau \circ \lambda[1, i] \models \varphi$ for all $\lambda \in out(q, s_A)$ and $i \in \mathbb{N}_0$

$\mathcal{M}, q, \tau \models \langle\!\langle A \rangle\!\rangle \varphi \mathcal{U} \psi$ iff there is $s_A \in \Sigma_A$ such that, for all $\lambda \in out(q, s_A)$, there is $i \in \mathbb{N}_0$ with $\mathcal{M}, \lambda[i], \tau \circ \lambda[1, i] \models \psi$, and $\mathcal{M}, \lambda[j], \tau \circ \lambda[1, j] \models \varphi$ for all $0 \leq j < i$.

$\mathcal{M}, q, \tau \models (\!| A |\!) \bigcirc \varphi$ iff there is $s_A \in \Sigma_A$ such that $\mathcal{M}, \lambda[1], \tau \circ \lambda[1] \models \varphi$ for all $\lambda \in out(q, s_A)$

$\mathcal{M}, q, \tau \models (\!| A |\!) \square \varphi$ iff there is s_A such that $\mathcal{M}, \lambda[i], \tau \circ \lambda[1, i] \models \varphi$ for all $\lambda \in out(q, s_A)$ and $i \in \mathbb{N}_0$

$\mathcal{M}, q, \tau \models (\!| A |\!) \varphi \mathcal{U} \psi$ iff there is $s_A \in \Sigma_A$ such that, for all $\lambda \in out(q, s_A)$, there is $i \in \mathbb{N}_0$ with $\mathcal{M}, \lambda[i], \tau \circ \lambda[1, i]$ $models \psi$, and $\mathcal{M}, \lambda[j], \tau \circ \lambda[1, j]$ $models \varphi$ for all $0 \leq j < i$.

Ultimately, we are interested in $\mathcal{M}, q \models \varphi$ defined as $\mathcal{M}, q, q \models \varphi$.

All the new functionality provided by goals is captured by the new valid coalition function vc^g.

Definition 4.30 (Valid coalitions, $vc^g(q, A, o, \tau)$). Let $\mathcal{M} \in \mathbb{M}^g$, $\tau \in S^+$, $A, B \subseteq \text{Agt}$, o an COALATL objective.

We say that B is a valid coalition after τ with respect to A, o, and $\mathcal{M}$ if, and only if,

1. $B \in \mathcal{E}(\zeta(\tau[\infty])(A))$, $A \cap B \neq \emptyset$, and

2. there are goals $g_{b_i} \in \mathfrak{g}_{b_i}(\tau)$, one per agent $b_i \in B$, such that $o \hookrightarrow_{\mathcal{M},\tau,B} g_{b_1} \wedge \cdots \wedge g_{b_{|B|}}$

The set $\text{vc}^g(q, A, o, \tau)$ consists of all such valid coalitions wrt to $\mathcal{M}$.

Thus, for the definition of valid coalitions among other things, a goal mapping, a function ζ and a sequence of states τ are required. The intuition of τ is that it represents the history (the sequence of states visited so far including the current state). So, τ is used to determine which goals of the agents are still active.

Finally, we have to define when a goal is satisfied.

Definition 4.31 (Satisfaction of goals). Let g be an ATL-goal, o an $\mathcal{L}_{ATL^c}$-objective, and $\tau \in S^+$. We say that objective o satisfies g, for short $o \hookrightarrow_{\mathcal{M},\tau,B} g$, with respect to $\mathcal{M}, \tau$, and B if, and only if, there is a strategy $s_B \in \Sigma_B$ such that

1. for all $\lambda \in out(\tau[\infty], s_B)$: $\mathcal{M}, \lambda, \tau \models o$ implies $\mathcal{M}, \lambda \models g$, and

2. there is some path $\lambda \in out(\tau[\infty], s_B)$ with $\mathcal{M}, \lambda, \tau \models o$.

A goal is satisfied by an objective if each path (enforceable by B) that satisfies the objective does also satisfy the goal. That is, satisfaction of the objective will guarantee that the goal becomes true. The second condition ensures that the coalition actually has a way to bring about the goal. However, in [Bulling et al., 2008b] it is shown that the second condition is superfluous.

4.6 Model Checking ATL^c

In this section, we present an algorithm for model checking CoALATL formulae. The model checking problem is given by the question whether a given CoALATL formula follows from a given CGM $\mathcal{M}$ and state q, i.e. whether $\mathcal{M}, q \models \varphi$ [Clarke et al., 1999]. In [Alur et al., 2002] it is shown that model checking ATL is **P**-complete, with respect to the number of transitions of $\mathcal{M}$, m, and the length of the formula, l, and can be done in time $\mathcal{O}(m \cdot l)$.

For CoALATL we also have to treat the new coalitional modalities in addition to the normal ATL constructs. Let us consider the formula $\langle\!\langle A \rangle\!\rangle \psi$.

According to the semantics of $\langle\!\langle A \rangle\!\rangle$, given in Definition 4.22, we must check whether there is a coalition B such that (i) $A \cap B \neq \emptyset$, (ii) B is acceptable by the argumentation semantics, and (iii) $\langle\!\langle B \rangle\!\rangle \psi$. The number of possible candidate coalitions B which satisfy (i) and (ii) is bounded by $|2^{\text{Agt}}|$. Thus, in the worst case there might be *exponentially* many calls to a procedure checking whether $\langle\!\langle B \rangle\!\rangle \psi$. Another source of complexity is the time needed to compute the argumentation semantics. In [Dunne and Bench-Capon, 2003], for instance, it is stated that credulous acceptance[6] using preferred semantics is **NP**-complete.

Both considerations together suggest that the model checking complexity has two computationally hard parts: exponentially many calls to $\langle\!\langle A \rangle\!\rangle \psi$ and the computation of the argumentation semantics. Indeed, Theorem 4.32 will support this intuition. However, we show that it is possible to "combine" both computationally hard parts to obtain an algorithm which is in $\mathbf{\Delta_2^P} = \mathbf{P^{NP}}$, if the computational complexity to determine whether a given coalition is acceptable are not harder than **NP**.

For the rest of this section, we will denote by $\mathcal{L}_{\mathcal{S},\mathcal{CF}}$ the set of all coalitions A such that A is acceptable according to the coalitional framework $\mathcal{CF}$ and the argumentation semantics $\mathcal{S}$, i.e. $\mathcal{L}_{\mathcal{S},\mathcal{CF}} := \{A \mid A \in \mathcal{E}(\mathcal{CF})\}$.

Given some complexity class $\mathcal{C}$, we use the notation $\mathcal{L}_{\mathcal{S},\mathcal{CF}} \in \mathcal{C}$ to state that the word problem of $\mathcal{L}_{\mathcal{S},\mathcal{CF}}$, i.e., whether A is a member of $\mathcal{L}_{\mathcal{S},\mathcal{CF}}$, is in $\mathcal{C}$. Actually in [Bulling et al., 2008b] it was stated that $\mathcal{L}_{\mathcal{S},\mathcal{CF}} \in \mathbf{P}$ for all semantics $\mathcal{S}$ defined in Definition 4.15. In Figure 8 we propose a model checking algorithm for CoaATL. The complexity result given in the next theorem is modulo the complexity needed to compute membership in $\mathcal{L}_{\mathcal{S},\mathcal{CF}}$.

Theorem 4.32 (Model checking CoaATL [Bulling et al., 2008b]). *Let a* CGM $\mathcal{M} = \langle S, \mathcal{V}, d, o, \zeta, \mathcal{S} \rangle$ *be given,* $q \in S$, $\varphi \in \mathcal{L}_{ATL^c}(\text{Agt}, \Phi)$, *and* $\mathcal{L}_{\mathcal{S},\mathcal{CF}} \in \mathcal{C}$. *Model checking* CoaATL *with respect to the argumentation semantics* $\mathcal{S}$[7] *is in* $\mathbf{P^{NP^{\mathcal{C}}}}$.

The last theorem gives an upper bound for model checking CoaATL with respect to an arbitrary but fixed semantics $\mathcal{S}$. A finer grained

---
[6]That is, whether an argument is in *some* preferred extension.
[7]That is, whether $\mathcal{M}, q \models_{\mathcal{S}} \varphi$.

---

function $mcheck(\mathcal{M}, q, \varphi)$;

Given a CGM $\mathcal{M} = \langle S, \mathcal{V}, d, o, \zeta, \mathcal{S}\rangle$, a state $q \in S$, and $\varphi \in \mathcal{L}_{ATL^c}(\text{Agt}, \Phi)$ the algorithm returns $\top$ if, and only if, $\mathcal{M}, q \models_S \varphi$.

---

case φ **contains no** $\langle\!\langle B \rangle\!\rangle$: **if** $q \in mcheck_{\text{ATL}}(\mathcal{M}, \varphi)$ **return** $\top$ **else** $\bot$

case φ **contains some** $\langle\!\langle B \rangle\!\rangle$:

> **case** $\varphi \equiv \neg\psi$: **return** $\neg(\mathcal{M}, q, \psi)$
>
> **case** $\varphi \equiv \psi \vee \psi'$: **return** $mcheck(\mathcal{M}, q, \psi) \vee mcheck(\mathcal{M}, q, \psi')$
>
> **case** $\varphi \equiv \langle\!\langle A \rangle\!\rangle T\psi$: Label all states q' where $mcheck(\mathcal{M}, q', \psi) == \top$ with a new proposition yes and return $mcheck(\mathcal{M}, q, \langle\!\langle A \rangle\!\rangle T\text{yes})$; T stands for $\Box$ or $\bigcirc$.
>
> **case** $\varphi \equiv \langle\!\langle A \rangle\!\rangle \psi \mathcal{U} \psi'$: Label all states q' where $mcheck(\mathcal{M}, q', \psi) == \top$ with a new proposition yes_1, all states q' where $mcheck(\mathcal{M}, q', \psi') == \top$ with a new proposition yes_2 and return $mcheck(\mathcal{M}, q, \langle\!\langle A \rangle\!\rangle \text{yes}_1 \mathcal{U} \text{yes}_2)$.
>
> **case** $\varphi \equiv \langle\!\langle A \rangle\!\rangle T\psi$, ψ **contains some** $\langle\!\langle C \rangle\!\rangle$: Label all states q' where $mcheck(\mathcal{M}, q', \psi) == \top$ with a new proposition yes and return $mcheck(\mathcal{M}, q, \langle\!\langle A \rangle\!\rangle T\text{yes})$; T stands for $\Box$ or $\bigcirc$.
>
> **case** $\varphi \equiv \langle\!\langle A \rangle\!\rangle \psi \mathcal{U} \psi'$, ψ **or** ψ' **contain some** $\langle\!\langle C \rangle\!\rangle$: Label all states q' where $mcheck(\mathcal{M}, q', \psi) == \top$ with a new proposition yes_1, all states q' where $mcheck(\mathcal{M}, q', \psi') == \top$ with a new proposition yes_2 and return $mcheck(\mathcal{M}, q, \langle\!\langle A \rangle\!\rangle \text{yes}_1 \mathcal{U} \text{yes}_2)$.
>
> **case** $\varphi \equiv \langle\!\langle A \rangle\!\rangle \psi$ **and** ψ **contains no** $\langle\!\langle C \rangle\!\rangle$: Non-deterministically choose $B \in 2^{\text{Agt}}$
>
> **if**
>
> (1) $B \in (\mathcal{E}(\zeta(A)(q)))$,
> (2) $A \cap B \neq \emptyset$, and $\quad(\star)$
> (3) $q \in mcheck_{\text{ATL}}(\mathcal{M}, \langle\!\langle B \rangle\!\rangle \psi)$
>
> **then return** $\top$ **else** $\bot$

---

function $mcheck_{\textbf{ATL}}(\mathcal{M}, \varphi)$;

Given a CGS $\mathcal{M} = \langle S, \mathcal{V}, d, o \rangle$ and $\varphi \in \mathcal{L}_{ATL}(\text{Agt}, \Phi)$, the standard ATL model checking algorithm (cf. [Alur et al., 2002]) returns all states q with $\mathcal{M}, q \models_{\text{ATL}} \varphi$.

> **return** $\{q \in S \mid \mathcal{M}, q \models_{\text{ATL}} \varphi\}$

Figure 8: A model checking algorithm for CoalATL

classification of the computational complexity of $\mathcal{L}_{\mathcal{S},\mathcal{CF}}$ allows to improve the upper bound given in Theorem 4.32. Assume that $\mathcal{L}_{\mathcal{S},\mathcal{CF}} \in \mathbf{P}$ and consider the last case of function *mcheck* in Figure 8 labelled by $(\star)$, $\varphi \equiv \langle\!\langle A \rangle\!\rangle \psi$. First, a coalition $B \in 2^{\text{Agt}}$ is non-deterministically chosen and then, it is checked whether B satisfies the three conditions (1-3) in $(\star)$. Each of the three tests can be done in deterministic polynomial time. Hence, the verification of $\mathcal{M}, q \models \langle\!\langle A \rangle\!\rangle \psi$, in the last case, meets the "guess and verify" principle which is characteristic for problems in **NP**. This brings the overall complexity of the algorithm to $\Delta_2^\mathbf{P}$. More surprisingly, the same result holds even for the case where $\mathcal{L}_{\mathcal{S},\mathcal{CF}} \in \mathbf{NP}$.

Corollary 4.33 ([Bulling *et al.*, 2008b]). *If* $\mathcal{L}_{\mathcal{S},\mathcal{CF}} \in \mathbf{NP}$ *(resp.* **NP**-*complete) then model checking* CoaLATL *is in* $\Delta_2^\mathbf{P}$ *(resp.* $\Delta_2^\mathbf{P}$-*complete) with respect to* $\mathcal{E}$.

In [Dunne and Bench-Capon, 2003] the complexity of credulous reasoning with respect to the preferred and stable extensions is analyzed and determined to be **NP**-complete. This is in the line with our result: there can be a polynomial number of calls to $mcheck(\mathcal{M}, q, \langle\!\langle A \rangle\!\rangle \psi)$ (where ψ does not contain any cooperation modality $\langle\!\langle \cdot \rangle\!\rangle$). Now, the problem of checking whether $mcheck(\mathcal{M}, q, \langle\!\langle A \rangle\!\rangle \psi)$ holds is very similar to checking whether some argument is credulously accepted. In both cases we have to ask for the existence of a set X with specific properties (in our framework we refer to X as a coalition and in [Dunne and Bench-Capon, 2003] as an argument) which can be validated in polynomial deterministic time.

Corollary 4.34 ([Bulling *et al.*, 2008b]). *Model checking* CoaLATL *is in* $\Delta_2^\mathbf{P}$ *for all semantics defined in Definition 4.15.*

5 Argumentation and epistemic logic

Doxastic and epistemic logics are the branches of modal logics that investigate the properties of belief (*dóxa* in ancient Greek) and knowledge (*epistēmē*), both in single and multi-agent contexts. There are several connections between argumentation and the analysis of knowledge and belief, that one can abridge as an influence in both directions. On the one hand, arguments inform our beliefs about the world and provide the grounds for many things we claim to know. Conversely, our beliefs and the things we know shape the way we produce and put forward arguments. The potential of combining analytic tools from doxastic/epistemic logic and formal argumentation can be easily argued for in many areas of application. Yet, such a combination is a relatively recent endeavour, most of the work dating back only to the last decade or so.

In what follows, we present the aim and rationale of the most salient efforts in this sense, and situate them along to the two just mentioned directions of influence. The rest of this section proceeds as follows. We first provide some essential background on epistemic and doxastic logics (Section 5.1) and hint at some advances in their field that are relevant for combining them with argumentation. In Section 5.2 we provide a more articulated description of the rationale for combining tools from these disciplines in both directions of influence, i.e. from arguments to knowledge and belief in Section 5.2.1 and from knowledge and belief to arguments in Section 5.2.2. Finally, in Section 5.3 we overview recent works exploring the first direction of influence, and we do the same in Section 5.4 for works along the second direction.[8]

5.1 Epistemic logic and reasoning about knowledge

Fundamental philosophical questions such as 'what is to know something?' and 'how knowledge differs from mere belief?' can be traced back, in the western philosophical tradition, at least to Plato's *Theaetetus*. Epistemic and doxastic logics as an axiomatic deductive rendering of the notions of, respectively, knowledge and belief, have a much more recent history. These fields date back to the seminal work of [Von Wright, 1951] and the subsequent more systematic treatment by [Hintikka, 1962], which introduced relational (Kripke) models as their standard semantics.

---

[8]Part of the content of the whole section builds upon previous work of Antonio Yuste-Ginel and Carlo Proietti [Burrieza and Yuste-Ginel, 2020; Burrieza and Yuste-Ginel, 2021; Proietti and Yuste-Ginel, 2020; Proietti and Yuste-Ginel, 2021; Yuste-Ginel and Herzig, 2023]. The exposition of the material is inspired by the PhD dissertation of the first author [Yuste-Ginel, 2022, Chp. 5], although novel approaches are discussed here, and the presentation has been systematically improved and expanded.

In this framework, knowledge and belief are interpreted as universal modalities (expressed by a □-operator) where modal notions such as 'it is known that ϕ' (resp. 'it is believed that ϕ') are interpreted as 'ϕ is the case in all states that are accessible from the actual one'. In most of what follows, to keep things simple, we treat knowledge and belief as separate and independent modalities, specifying the interpretation of □ in each context. It should be noticed though that this is not the only possible option. Indeed, a long tradition in epistemology, dating back to Plato, identifies knowledge as a derivative notion, i.e. as some form of true belief. As a consequence, modal approaches inspired by this tradition formalise belief and knowledge as interdependent modalities, most of the time with belief as a primitive modality and knowledge as defined by it.[9] Yet another option, that we will touch upon in what follows, is to treat both belief and knowledge as derivative modalities, e.g., by grounding them on the arguably more primitive concept of *evidence*.

At an intuitive level, knowledge and belief have different properties which translate into specific axioms. Knowledge is usually required to satisfy *factivity*: to know that ϕ implies that ϕ is true, which arguably does not hold in the case of simple belief. Factivity is expressed by the axiom schema (T) $\Box \phi \to \phi$ (Section 2.1), which defines reflexivity at the level of structures. Belief is instead often required to satisfy the condition that it is not possible to believe a contradiction, expressed by the schema $\neg \Box \bot$. The latter is equivalent to schema (D) $\Box \phi \to \Diamond \phi$. In fact, both formulas define *seriality* of the accessibility relation, i.e. for any state s there is always some state t accessible from s. Since reflexivity entails seriality, the doxastic interpretation of □ puts a weaker constraint on the accessibility relation than the epistemic interpretation.

Additional properties for both knowledge and belief are so-called *positive* and *negative introspection*. Positive introspection postulates that anything that is known (resp. believed) is also known to be known (resp. believed to be believed), and is captured by the axiom schema (4) $\Box \phi \to \Box \Box \phi$, which defines transitivity. Negative introspection instead means that anything that is not known (resp. not believed) is also known

---

[9]The converse option to treat belief as derived from knowledge as primitive has also been put forward in recent epistemological discussion [Williamson, 2002; Williamson, 2011] or in well-known approaches to the dynamics of epistemic attitudes [Baltag and Smets, 2008].

to be not known (resp. believed to be not believed) and is expressed by (5) $\neg\Box\phi \to \Box\neg\Box\phi$, which defines *euclidianity*: any two states that are accessible from a third one have access to each other. The more or less 'standard' calculus for doxastic logic is KD45, i.e. the system K of normal modal logic augmented with axioms (D), (4) and (5). The status of both axioms (4) and (5) is instead debated with regard to the epistemic reading of $\Box$. Many philosophers tend to reject both of them, assuming (T) as the only valid axiom schema for knowledge. On the other hand, computer scientists usually accept both, taking the system S5 (i.e. K + (T) + (4) + (5)) as a viable axiomatization of epistemic logic, i.e. one where the accessibility relation is an equivalence relation.

In general, both knowledge and belief may have different meanings depending on the context of application. Consequently, their modal rendering as a $\Box$-operator may obey different properties, which entails the validity or invalidity of different axiom schemas. In this sense, even the axioms and rules of the basic system of normal modal logic T may be disputed. For example, accepting the necessitation rule N – inferring $\vdash \Box\phi$ from $\vdash \phi$ – entails that all logical validities are known. The latter seems fine when modelling what an agent *implicitly* knows, or can infer in principle, but is inadequate when modelling the *explicit* knowledge of agents with limited computational resources. This is known as the problem of *logical omniscience*. Otherwise, in a doxastic context, we may read the belief operator $\Box\phi$ as the 'agent assigns high probability to ϕ being true'. Here, the formula $(\Box\phi \wedge \Box\psi) \to \Box(\phi \wedge \psi)$ fails to hold in general: the fact that two separate events are highly probable does not entail that their conjunction is. However, this formula is a logical consequence of the axioms of K.

In cases like these, there are two main strategies of approach. On the one hand, one can add new operators to the basic language in order to express more nuanced epistemic or doxastic concepts. This is, for example, the strategy of *awareness logics* [Fagin and Halpern, 1987; Fagin et al., 2004], where an awareness operator $A\phi$ – meaning 'the agent is aware of ϕ' – is added to the language and used to define explicit knowledge in conjunction with $\Box$. On the other hand, one can weaken the basic logic and, consequently, change its semantics. This is the case of *neighbourhood semantics* [Chellas, 1980] – that we will encounter in

Section 5.3 – which do not validate all axioms and inference rules of K, as the ones mentioned in the previous paragraph. The same outcome may be obtained by defining the □-operator of knowledge or belief on top of a different operator with a neighbourhood semantics, in our case an *evidence* modality [Pacuit, 2017]. All these strategies have been applied at the interplay between argumentation and epistemic logic as we will see in what follows.

5.2 A twofold influence

5.2.1 Influence 1: From arguments to knowledge and beliefs

Our beliefs about the world are shaped by the evidence we encounter, which can be either direct (e.g., *seeing*) or indirect (e.g., by *testimony* or by *inference*). Such evidence is often of an argumentative nature. I may believe that Jones owns a Ford because I have seen him riding one (direct evidence). Yet, this belief may be defeated by Smith telling me that Jones is around with his company car, which makes an argument to the conclusion that I have not seen him riding his own car. In recent years, doxastic and epistemic logics have been combined with abstract argumentation with the aim to explore the many senses in which belief can be supported or defeated by arguments. In Section 5.3 we present some of these approaches [Burrieza and Yuste-Ginel, 2020; Burrieza and Yuste-Ginel, 2021; Grossi and van der Hoek, 2014; Shi et al., 2017; Shi et al., 2018a]. This line of investigation has a strong link with central issues in epistemology. One of them is the debate around the so-called JTB thesis, according to which knowledge is to be defined as *justified true belief*. This thesis has been harshly debated since Edmund Gettier raised a number of famous objections against it in a famous paper [Gettier, 1963]. The core of the issue lies in the fact that the central notion of justification needs specification. In fact, abstract argumentation provides a full theory of justification (as defence against counterarguments). In this respect, it naturally works as a tool to assess the JTB theory. A first approach along these lines is to be found in [Shi, 2021; Shi et al., 2021] and will be presented in Section 5.3.2.

5.2.2 Influence 2: From knowledge and beliefs to arguments

Regarding the second direction of influence, everyone agrees that our beliefs have a strong impact on the type of arguments we are prone to endorse. Trivially, if I compare two arguments a and b and I believe that the premisses of a are true, while I am unsure whether one of the premisses of b holds, then I should conceive, *ceteris paribus*, argument a as strictly stronger than b. Some of the works we mentioned in the previous paragraph (e.g., [Burrieza and Yuste-Ginel, 2020]) do take care of these kinds of principles operating in epistemic argument evaluation. Furthermore, the arguments we produce in a social context are influenced by the beliefs and knowledge we attribute to our audience. For instance, I may easily fool a child with some argument that I wouldn't use in other contexts. In general, arguing requires a theory of other minds and has many strategic aspects that link argumentation to the study of persuasion techniques. Perhaps, this is what determined the development of argumentation and rhetoric as separate from logic *stricto sensu*.[10] Yet, the formal approach to the aspects of strategic argumentation becomes nowadays more and more relevant for the purposes of human-machine interaction and the goal of building intelligent debaters. This is indeed what motivates *opponent modelling* in formal argumentation, today a fairly active area of research, as witnessed by an increasing number of works over the last years [Rienstra *et al.*, 2013; Thimm, 2014; Hadjinikolis *et al.*, 2013; Mailly, 2022; Alfano *et al.*, 2023]. Here again, combining formal argumentation with tools from (dynamic) epistemic logics provides a general tool to categorize different approaches to opponent modelling and to inspire further developments. In Section 5.4, we illustrate work in this direction and their link to applications.

5.3 From arguments to knowledge and belief

The works presented in this section are those exploring the first direction of influence, from arguments to knowledge and belief (Section 5.2.1).

---

[10]It should be noticed that in Aristotle's *Organon*, argumentation, or *dialectic*, was intended to be a branch of logic – which constitutes the object of the *Topics* and *Sophistical Refutations* – the main difference being that the object of dialectic is syllogisms with uncertain or generally assumed premises (*endoxa*) rather than true ones.

We proceed from the most natural and simpler approach by [Grossi and van der Hoek, 2014], which simply fuses standard modal logic and abstract argumentation. We then go towards the one initiated by [Shi et al., 2017], displaying more complex (topological) models for modal logic in order to account for a notion of argument-based evidence enabling to formalize the JTB theory. We finally present the most articulated approach, enriching both the formalism for modal logic, by means of awareness logics, and the one for argumentation, exploiting the richer ASPIC$^+$ formalism for structured argumentation. This allows for a finer granularity when representing concepts in argumentation, e.g. different types of attacks among arguments (such as *rebuttal*, *undermining* or *undercut* from [Pollock, 1987]), and therefore the possibility of encoding more articulated types of argument-based beliefs.

5.3.1 Product models for argumentation and belief

As seen in Section 2.1, Kripke semantics for modal logic provides a natural tool to talk about graphs and, therefore, to reason about abstract argumentation and its solution concepts [Grossi, 2009; Grossi, 2010a; Grossi, 2010b; Caminada and Gabbay, 2009]. As illustrated in Section 5.1, they are also the primary tool for doxastic and epistemic logics. Therefore, combining the respective Kripke semantics is perhaps the most natural approach for fusing these two different frameworks. The work by [Grossi and van der Hoek, 2014] proceeds along these lines and is one of the first combining epistemic logic and abstract argumentation to analyse the interactions between beliefs and argumentation. The keystone of the work is indeed the use of *product models* [Gabbay and Shehtman, 1998; Kurucz et al., 2003]. Here, possible worlds are pairs $\langle s, a \rangle$ with s a doxastic state and a a given argument. Intuitively, s is the 'actual' state of affairs and a is the 'currently entertained' argument. This allows, among other things, the definition and formal analysis of several forms of justified belief. This and related work by [Shi et al., 2018b] are covered in detail by the previous handbook chapter devoted to logic-based approaches to argumentation [Arieli et al., 2021, Sect. 3.2.2.], so we skip a full presentation to avoid overlapping.

5.3.2 Topo-argumentative models for argument-based epistemic attitudes

The notion of *evidence* is a central one for epistemology and lies behind that of *justified belief*, i.e. a belief supported by strong or undefeated evidence. There are many ways to frame the interplay between evidence, belief and knowledge in a modal logical setting. As mentioned, one of them consists in modelling evidence as a primitive notion by means of neighbourhood semantics, with knowledge and belief as derived concepts [van Benthem et al., 2012; van Benthem et al., 2014; Baltag et al., 2016].[11] Arguments are typical sources of evidence and solution concepts from abstract argumentation can shed some light on the way they may serve to justify a belief. The line of work of [Shi, 2018; Shi et al., 2017; Shi et al., 2018a; Shi, 2021; Shi et al., 2021] brings together all these insights by combining abstract argumentation with so-called *topological models* of evidence for doxastic logics.[12]

The main point of departure of this approach is to understand pieces of evidence as members of a topological structure. A *topology* τ over a non-empty set S is a set of sets $\tau \subseteq 2^S$, such that: (i) $\emptyset, S \in \tau$ (the unit and the empty set are its elements); (ii) if $A, B \in \tau$, then $A \cap B \in \tau$ (closure under finite intersections); and (iii) for any –possibly infinite– family $\{A_x\}_{x \in X} \subseteq \tau$, we have that $\bigcup_{x \in X} A_x \in \tau$ (closure under arbitrary unions). The topology generated by a family of sets $B \subseteq 2^S$ is the smallest topology τ_B such that $B \subseteq \tau_B$. Given a topology τ, its elements are usually called *opens*. These opens represent pieces of evidence in topological models for epistemic logic [Baltag et al., 2016], and they will be the arguments of the topological argumentation model we are about to present:

Definition 5.1 (TA-models). *A Topological-Argumentation model*

---

[11] Another approach runs by adding specific *justification* terms to the language of doxastic-epistemic logics [Baltag et al., 2012; Baltag et al., 2014; Égré et al., 2021], a general strategy borrowed from *justification logics* [Artemov, 2008].

[12] Note that [Arieli et al., 2021] mentions some of these works, but focus on the presentation of a different approach by the same authors [Shi et al., 2018b] that works without topology and it is somehow closer to [Grossi and van der Hoek, 2014], so we devote some space for the introduction of topological tools into the modelling of argument-based beliefs.

(TA-model) *for a countable set of atomic variables* Φ *is a tuple* $M = \langle S, E_0, \tau_{E_0}, \rightsquigarrow, \mathcal{V} \rangle$, *where*

- $S \neq \emptyset$ *is a set of* possible worlds.
- $E_0 \subseteq 2^S \setminus \{\emptyset\}$ *is a collection of* basic pieces of evidence.
- τ_{E_0} *is the topology generated by* E_0.
- $\rightsquigarrow \subseteq \tau_{E_0} \times \tau_{E_0}$ *is a* defeat relation *satisfying:*
  - *for every* $A \in \tau_{E_0} \setminus \{\emptyset\}$, *we have* $\langle A, \emptyset \rangle \in \rightsquigarrow$ *and* $\langle \emptyset, A \rangle \notin \rightsquigarrow$.
  - *for every* $A, B \in \tau_{E_0}$, *we have* $A \cap B = \emptyset$ *iff either* $\langle A, B \rangle \in \rightsquigarrow$ *or* $\langle B, A \rangle \in \rightsquigarrow$.
- $\mathcal{V} : S \rightarrow 2^{\Phi}$.

The idea of modelling basic pieces of evidence as sets of possible worlds (elements of the collection $E_0 \subseteq 2^S \setminus \{\emptyset\}$) can be traced back to [van Benthem et al., 2012; van Benthem et al., 2014]. The main assumption behind it is that *evidence* is understood as *information-as-range* [Shi et al., 2021], so that if S represents all the epistemic alternatives of the agent, a piece of evidence $A \subseteq S$ tells the agent that the actual world is in A (and hence $S \setminus A$ should be disregarded according to A). Note, however, how $A \in E_0$ does not informally mean that the agent accepts A, but she rather takes it as a starting point for reasoning.

The topological structure τ_{E_0} represents the possible ways in which the agent can logically combine her basic pieces of evidence. Importantly, here the elements of τ_{E_0} play the role of *arguments* (see [Özgün, 2017] for a discussion) and $\rightsquigarrow$ represents a *defeat* relation among them. The idea behind this specific definition is that there is a defeat from A to B only when A and B are incompatible pieces of evidence and A is 'as least as strong as' B. In sum, $\rightsquigarrow$ functions as a way of modelling how incompatible pieces of evidence are weighted. This process of evaluation is modelled through the conflict calculus introduced by [Dung, 1995] (Section 2.2).[13]

---

[13]The relation $\rightsquigarrow$ is deemed an "attack" relation in [Shi et al., 2017] and the subsequent works. However, we believe that the notion of defeat makes better sense in

In particular, two forms of argument-based belief are defined over TA-models in [Shi et al., 2021]. Both of them are based in the grounded semantics for abstract argumentation frameworks (see Definition 2.5). Recall that $\mathcal{E}_{\mathcal{GR}}(\langle Ar, \rightsquigarrow \rangle)$ is used to denote the set of all grounded extensions of $\langle Ar, \rightsquigarrow \rangle$ (Section 2.2), and hence $\bigcup \mathcal{E}_{\mathcal{GR}}(\langle Ar, \rightsquigarrow \rangle)$ denotes the grounded extension, since it is unique (as shown by [Dung, 1995]). Let $M = \langle S, E_0, \tau_{E_0}, \rightsquigarrow, \mathcal{V} \rangle$ be a TA-model, and let $P \subseteq S$ be a proposition, then:

- the agent has a *grounded belief* on P iff there is an $A \in \bigcup \mathcal{E}_{\mathcal{GR}}(\langle \tau_{E_0}, \rightsquigarrow \rangle)$ such that $A \subseteq P$.

- the agent has a *fully grounded belief* on P iff for every $A \in \bigcup \mathcal{E}_{\mathcal{GR}}(\langle \tau_{E_0}, \rightsquigarrow \rangle)$, there is an $A' \in \bigcup \mathcal{E}_{\mathcal{GR}}(\langle \tau_{E_0}, \rightsquigarrow \rangle)$, such that $A' \subseteq A$ and $A' \subseteq P$.

The authors' choice of employing only the grounded semantics to define belief may have several reasons. First, the sceptic flavour of grounded semantics is particularly significant in the context of epistemic as opposed to practical reasoning, i.e. reasoning about what to believe as opposed to reasoning about what to do [Prakken, 2006]. Second, as mentioned, the grounded extension is always unique. This dodges the discussion that would arise if a semantics that returns multiple extensions were used, namely, which of the (mutually incompatible) extension should be the one that actually serves the agent to ground her/his beliefs. Finally, as pointed out by [Shi et al., 2021], the grounded extension is never empty in the current setting, as it can be shown that W is always undefeated. This guarantees, among other things, that valid propositions are always groundly believed. However, it seems worthy to investigate whether and how other semantics sharing the properties of uniqueness and non-emptiness could work in this framework.

---

the current context (see Section 2.2 for the distinction between attacks and defeats). In this sense, 'A attacks B' is best understood as the (symmetric) incompatibility relation $A \cap B = \emptyset$, i.e. the second precondition of $A \rightsquigarrow B$ in definition 5.1, while further properties of $\rightsquigarrow$ (e.g. the first precondition in definition 5.1) act as symmetry-breaking constraints to assess the relative weight of A and B. We thank one of the anonymous reviewers of this chapter for asking us to clarify this point.

Both notions, grounded belief and fully grounded belief, are possible formalizations of the first type of influence, i.e. of how arguments determine specific types of belief. Curiously, while fully grounded beliefs satisfy KD45 axioms, grounded beliefs fail to be closed under conjunction (and hence do not satisfy the (K) axiom).[14] Moreover, pairwise consistency among groundly believed propositions is guaranteed, but this is not the case when we consider sets of groundly believed propositions with more than two elements. In terms of the literature about rationality postulates for argumentation systems [Caminada, 2017], grounded belief is directly consistent but not indirectly consistent. In contrast, fully grounded belief is indirectly (i.e. totally) consistent. Finally, fully grounded belief is strictly stronger than grounded belief, so that the former implies the latter but not vice versa. This brief comparison among both notions makes explicit the existing tension between *believing more* (or more informatively) and believing *more consistently* (see [Shi, 2018] and [Shi et al., 2021] for a detailed discussion on such a tension).

Along the same lines, [Shi, 2021] provides an argumentative account of the JTB characterization of knowledge. Here too, the author defines different notions of knowledge, namely K_1, K_2 and K_3, ranging from weaker to stronger and investigates their logical properties and the conditions for their equivalence. The weakest K_1 is defined simply as:

- the agent *knows* P ($K_1 P$) iff it has a grounded belief on P and P is true at the world of evaluation.

In this version of JTB, justified belief is therefore grounded belief. It is also shown that (grounded) belief implies believing to know ($\models B\phi \to BK_1\phi$), and that many other properties postulated by standard logics for knowledge and belief hold [Stalnaker, 2006].

Before closing this subsection, we mention a related, interesting work by [Wáng and Li, 2021]. This paper introduces a generalization of neighbourhood models where arguments are understood as sets of propositions, and propositions are represented semantically as sets of possible worlds. The main reason for us to leave a detailed review of this work out of this chapter is that the approach lacks an explicit modelling of

---

[14]Or, in other words, the grounded extension of $\langle \tau_{E_0}, \rightsquigarrow \rangle$ is not always closed under intersections. See [Shi et al., 2021] for conditions under which this is actually the case.

conflicts among arguments, which is an essential feature of all the logical frameworks introduced here, and of formal models of argumentation in general.

5.3.3 A more syntactic approach

[Burrieza and Yuste-Ginel, 2020; Burrieza and Yuste-Ginel, 2021] adopt a more syntactic approach to modelling the interaction between arguing and believing. In short, it is based on the combination of *awareness epistemic models* [Fagin and Halpern, 1987] with ASPIC$^+$ arguments [Modgil and Prakken, 2013]. The main reason behind this move is to bridge epistemic logic with the field of *structured argumentation* [Besnard et al., 2014], where arguments are essentially understood as composite entities, with premises, conclusions, inferential links between the two, and which may encompass subarguments as their parts.

Syntax. Unlike previously reviewed works, here arguments are, together with other formulas, first-class syntactic citizens, as specified by the definition of the language.

Definition 5.2 (Language for Awareness of Structured Arguments). *The language $\mathcal{L}_{ASA}$[15] is the pair $\langle \mathsf{F}, Ar \rangle$ of formulas and arguments, which are defined by mutual recursion as follows:*

$$\varphi ::= p \mid \neg\varphi \mid (\varphi \wedge \varphi) \mid \Box\varphi \mid \mathsf{aware}(\alpha) \mid \mathsf{conc}(\alpha) = \varphi \mid$$
$$\mid \mathsf{strict}(\alpha) \mid \mathsf{undercuts}(\alpha, \alpha) \mid \mathsf{wellshap}(\alpha) \qquad p \in \Phi, \alpha \in Ar.$$

$$\alpha ::= \langle \varphi \rangle \mid \langle \alpha_1, ..., \alpha_n \twoheadrightarrow \varphi \rangle \mid \langle \alpha_1, ..., \alpha_n \Rightarrow \varphi \rangle \qquad \varphi \in \mathsf{F}, n \geq 1.$$

As in ASPIC$^+$, the grammar defines three types of arguments. $\langle \varphi \rangle$ is an *atomic argument* whose sole premise and conclusion is φ. $\langle \alpha_1, ..., \alpha_n \twoheadrightarrow \varphi \rangle$ is the argument that *deductively* concludes φ from the conclusions of subarguments $\alpha_1, ..., \alpha_n$. $\langle \alpha_1, ..., \alpha_n \Rightarrow \varphi \rangle$ is the argument that *presumtively/defeasibly* concludes φ from the conclusions of subarguments $\alpha_1, ..., \alpha_n$. The modal operator $\Box$ denotes (implicit) belief.

---

[15]The language is simply denoted $\mathcal{L}$ in [Burrieza and Yuste-Ginel, 2021]. We use *ASA* as an abbreviation of 'Awareness of Structured Arguments' in order to distinguish it from the rest of the languages that appear in this chapter.

Concerning other formulas, aware(α) reads "the agent is aware of argument α"; conc(α) = φ reads "the conclusion of α is φ"; strict(α) reads "argument α is strict (i.e. it contains no defeasible rule)". undercuts(α, β) means that argument α undercuts argument β (i.e. α attacks the last rule employed in the construction of β). Finally, wellshap(α) means that α is well-shaped or well-constructed, in the sense that it only uses either valid deductive rules or accepted defeasible rules in its construction.

Semantics. The semantics of $\mathcal{L}_{ASA}$ is strongly meta-syntactic, meaning that its model theory relies on functions ranging over language constructions. Let us introduce some of them. SEQ(F) is used to denote the *set of all finite sequences over* F and $\langle\langle\varphi_1,...,\varphi_n\rangle, \varphi\rangle$ denotes an arbitrary element of SEQ(F) with $n+1$ elements. These sequences are used to represent defeasible inference steps in the meta-language. Moreover, the following ASPIC$^+$'s functions are used to analyse **arguments' structures**:

- Prem(α) returns the **premisses** of α and it is defined as follows: Prem($\langle\varphi\rangle$) := $\{\varphi\}$, Prem($\langle\alpha_1,...,\alpha_n \hookrightarrow \varphi\rangle$) := Prem($\alpha_1$) $\cup ... \cup$ Prem(α_n) where $\hookrightarrow \in \{\twoheadrightarrow, \Rightarrow\}$.

- Conc(α) returns the **conclusion** of α and it is defined as follows Conc($\langle\varphi\rangle$) := $\{\varphi\}$ and Conc($\langle\alpha_1,...,\alpha_n \hookrightarrow \varphi\rangle$) := $\{\varphi\}$ where $\hookrightarrow \in \{\twoheadrightarrow, \Rightarrow\}$.

- sub$_A(\alpha)$ returns the **subarguments** of α and it is defined as follows: sub$_A(\langle\varphi\rangle)$:= $\{\langle\varphi\rangle\}$ and sub$_A(\langle\alpha_1,...,\alpha_n \hookrightarrow \varphi\rangle)$:= $\{\langle\alpha_1,...,\alpha_n \hookrightarrow \varphi\rangle\} \cup$ sub$_A(\alpha_1) \cup ... \cup$ sub$_A(\alpha_n)$ where $\hookrightarrow \in \{\twoheadrightarrow, \Rightarrow\}$.

- TopRule(α) returns the **top rule** of α, i.e. the last rule applied in the formation of α. It is defined as follows: TopRule($\langle\varphi\rangle$) is left undefined, TopRule($\langle\alpha_1,...,\alpha_n\twoheadrightarrow\varphi\rangle$) = TopRule($\langle\alpha_1,...,\alpha_n \Rightarrow \varphi\rangle$) := $\langle(\text{Conc}(\alpha_1),...,\text{Conc}(\alpha_n)), \varphi\rangle$.

- DefRule(α) returns the set of **defeasible rules** of α and it is defined as DefRule($\langle\varphi\rangle$) := $\emptyset$, DefRule($\langle\alpha_1,...,\alpha_n\twoheadrightarrow\varphi\rangle$) := DefRule($\alpha_1$) $\cup ... \cup$ DefRule(α_n) and DefRule($\langle\alpha_1,...,\alpha_n \Rightarrow \varphi\rangle$) := $\{\langle(\text{Conc}(\alpha_1),...,\text{Conc}(\alpha_n)), \varphi\rangle\} \cup$ DefRule(α_1) $\cup ... \cup$ DefRule(α_n).

Let us also define **semantic propositional negations**, for any $\varphi, \psi \in \mathsf{F}$: $\varphi =_\sim \psi$ abbreviates $\mathsf{wellshap}(\langle\langle\varphi\rangle\twoheadrightarrow\neg\psi\rangle) \wedge \mathsf{wellshap}(\langle\langle\psi\rangle\twoheadrightarrow\neg\varphi\rangle)$.

Definition 5.3 ($\mathcal{L}_{ASA}$-models). *A **model** for $\mathcal{L}_{ASA}$ is a tuple $M = \langle S, \mathcal{R}_\mathcal{B}, \mathcal{O}, \mathcal{D}, \mathfrak{n}, \mathcal{V}\rangle$ where:*

- $\langle S, \mathcal{R}_\mathcal{B}\rangle$ *is a doxastic structure (i.e., $\mathcal{R}_\mathcal{B}$ is serial, transitive and euclidean).*[16]

- $\mathcal{O} \subseteq Ar$ *the awareness set of the agent.*

- $\mathcal{D} \subseteq \mathsf{SEQ}(\mathsf{F})$ *is a set of* accepted defeasible rules. *These rules are assumed to be consistent and invalid according to classic propositional logic.*

- $\mathfrak{n}: \mathsf{SEQ}(\mathsf{F}) \to \Phi$ *is a (possibly partial)* naming function *for rules, where $\mathfrak{n}(R)$ informally means "the rule R is applicable".*

- $\mathcal{V}$ *is and an* atomic valuation, *i.e. a function $\mathcal{V}: S \to 2^\Phi$.*

Let $M = \langle S, \mathcal{R}_\mathcal{B}, \mathcal{O}, \mathcal{D}, \mathfrak{n}, \mathcal{V}\rangle$ be a model for $\mathcal{L}_{ASA} = \langle \mathsf{F}, Ar\rangle$. We use $\vdash_0$ to denote logical consequence in classical propositional logic. The **set of well-shaped arguments** $WS^M \subseteq Ar$ (depending on $\mathcal{D}$ in M) is the smallest set fulfilling the following conditions:

1. $\langle\varphi\rangle \in WS^M$ for any $\varphi \in \mathsf{F}$.

2. $\langle\alpha_1,...,\alpha_n\twoheadrightarrow\varphi\rangle \in WS^M$ iff both $\alpha_i \in WS^M$ for every $1 \leq i \leq n$ and $\{\mathsf{Conc}(\alpha_1),...,\mathsf{Conc}(\alpha_n)\} \vdash_0 \varphi$.

3. $\langle\alpha_1,...,\alpha_n \Rightarrow \varphi\rangle \in WS^M$ iff both $\alpha_i \in WS^M$ for every $1 \leq i \leq n$ and $\langle\langle\mathsf{Conc}(\alpha_1),...,\mathsf{Conc}(\alpha_n)\rangle, \varphi\rangle \in \mathcal{D}$.

Definition 5.4 (Truth in $\mathcal{L}_{ASA}$-models). *Let $\langle M, w\rangle$ be a pointed model for $\mathcal{L}_{ASA}$, that is, $M = \langle S, \mathcal{R}_\mathcal{B}, \mathcal{O}, \mathcal{D}, \mathfrak{n}, \mathcal{V}\rangle$ is a model and $w \in S$. The **truth** relation, relating pointed models and formulas, is given by:*

---

[16] Recall that this is an instance of Definition 2.2 with a single label representing the modelled agent.

$$M, w \models \Box\varphi \quad \textit{iff} \quad \textit{for all } w' \in S\colon (w, w') \in \mathcal{R}_\mathcal{B}$$
$$\textit{implies } M, w' \models \varphi.$$
$$M, w \models \mathsf{aware}(\alpha) \quad \textit{iff} \quad \alpha \in \mathcal{O}.$$
$$M, w \models \mathsf{conc}(\alpha) = \varphi \quad \textit{iff} \quad \mathsf{Conc}(\alpha) = \varphi.$$
$$M, w \models \mathsf{strict}(\alpha) \quad \textit{iff} \quad \mathsf{DefRule}(\alpha) = \emptyset.$$
$$M, w \models \mathsf{undercuts}(\alpha, \beta) \quad \textit{iff} \quad \beta = \langle \beta_1, ..., \beta_n \Rightarrow \psi \rangle \textit{ and}$$
$$\mathsf{Conc}(\alpha) = \neg \mathsf{n}(\mathsf{TopRule}(\beta)).$$
$$M, w \models \mathsf{wellshap}(\alpha) \quad \textit{iff} \quad \alpha \in WS^M.$$

Types of beliefs. $\mathcal{L}_{ASA}$ is rich enough to distinguish several kinds of belief. A general distinction is made between basic and argument-based beliefs (mimicking the one among intuitive and inferential beliefs in cognitive sciences [Sperber, 1997]). Basic beliefs are based on non-inferential information, such as observation or trusted testimonies. There are, in turn, two subtypes of basic beliefs, inherited from the epistemic modal logic tradition: implicit and explicit ones. Basic-implicit beliefs are the ideal beliefs of a perfect reasoner and are captured by the primitive operator $\Box$. Its explicit counterparts are defined à la [Fagin and Halpern, 1987], using atomic arguments for simulating awareness of sentences: $\Box^e \varphi := \Box\varphi \wedge \mathsf{aware}(\langle\varphi\rangle)$.

There are also two types of argument-based beliefs: *deductive* beliefs and *grounded* beliefs. Deductive beliefs are defined as $\mathsf{B}^\mathsf{D}(\alpha, \varphi) := \mathsf{accept}(\alpha) \wedge \mathsf{aware}(\alpha) \wedge \mathsf{conc}(\alpha) = \varphi \wedge \mathsf{strict}(\alpha) \wedge \mathsf{wellshap}(\alpha)$ (to be read as "the agent has a deductive belief that φ based on argument α"), where $\mathsf{accept}(\alpha) := \bigwedge_{\varphi \in \mathsf{Prem}(\alpha)} \Box\varphi$ stands for argument doxastic acceptance. Note that the definition of argument doxastic acceptance (accept) includes a very simple instance of Influence 2: the only arguments taken into consideration by the agent are those whose premises are believed. Moreover, the definition of deductive beliefs ($\mathsf{B}^\mathsf{D}(\alpha, \varphi)$) is a clear instance of Influence 1: the arguments that the agent is aware of influence her (deductive) beliefs.

The definition of grounded beliefs needs some preliminary argumentative notions. The first one is a notion of binary preference among arguments. Many preference relations are definable in $\mathcal{L}_{ASA}$, capturing different versions of Influence 2. As an illustration, a very simple notion of preference, that assumes that the agent only takes into account doxastically accepted arguments, consists in preferring strict arguments over non-strict ones: $\alpha \geq \beta := \text{strict}(\alpha) \vee \neg\text{strict}(\beta)$.

Argument attacks and defeats (= successful attacks), notions imported from ASPIC$^+$, are definable in $\mathcal{L}_{ASA}$:

- **Undercutting a subargument** $\text{undercuts}^*(\alpha, \beta) := \bigvee_{\beta' \in \text{sub}_A(\beta)} \text{undercuts}(\alpha, \beta')$.

- **Unrestricted successful rebuttal**
  $\text{Urebuts}(\alpha, \beta) := \neg\text{strict}(\beta) \wedge \bigvee_{\beta' \in \text{sub}_A(\beta)} (\text{conc}(\alpha) = \varphi \wedge \text{conc}(\beta') = \psi \wedge \varphi =\sim \psi \wedge \alpha \geq \beta')$.

- **Defeat** $\text{defeat}(\alpha, \beta) := \text{undercuts}^*(\alpha, \beta) \vee \text{Urebuts}(\alpha, \beta)$.

Let $\langle M, w \rangle$ be a pointed model for $\mathcal{L}_{ASA} = \langle \mathsf{F}, Ar \rangle$, we define its **associated argumentation framework** as $AF^M := \langle Ar^M, \leadsto \rangle$, where $Ar^M := \{\alpha \in Ar \mid M, w \models \text{aware}(\alpha) \wedge \text{wellshap}(\alpha) \wedge \text{accept}(\alpha)\}$ and $\leadsto \subseteq Ar^M \times Ar^M$ is given by $\alpha \leadsto \beta$ iff $M, w \models \text{defeat}(\alpha, \beta)$. Finally, we expand our language with the grounded belief operator $\mathsf{B}(\alpha, \varphi)$, interpreted in pointed models as follows:

$$M, w \models \mathsf{B}(\alpha, \varphi) \quad \text{iff} \quad \alpha \in \bigcup \mathcal{E}_{\mathcal{GR}}(AF^M) \quad \text{and} \quad \text{Conc}(\alpha) = \varphi.$$

The relative strength of the four kinds of belief is given by the following diagram:

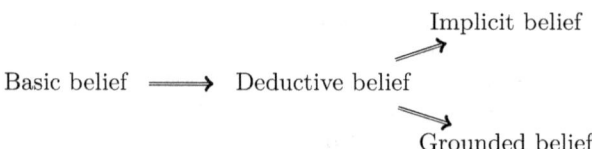

5.4 From knowledge and belief to arguments

This section introduces formal work exploring the second direction of influence: how knowledge and beliefs, especially those about other minds, influence the use of argument. Historically, the first work combining epistemic logic and abstract argumentation in this sense is [Schwarzentruber et al., 2012], followed by [Proietti and Yuste-Ginel, 2020; Proietti and Yuste-Ginel, 2021], where the original framework was systematically expanded and dynamified. Unlike the previous section, here, we will not treat different works separately, but rather introduce some of their high-level ideas as well as their applications.

One key feature that differentiates the logical frameworks presented here from those in the previous section is the multi-agent perspective, due to the essential role played in this context by reasoning about others' beliefs and knowledge. Therefore, a preliminary step, before introducing modalities, is to combine abstract argumentation frameworks and multi-agency. Many options are explored in the literature (see also [Proietti and Yuste-Ginel, 2021] for a review). Here, we introduce one of them, probably the most popular (see [Yu et al., 2021] for a principle-based analysis of its semantics).

Definition 5.5 (Multi-agent AF). *Let Agt and $\mathbf{A}$ be two finite, non-empty and disjoint sets (agents and arguments, respectively), a multi-agent AF (MAF) is a tuple $\langle Ar, \{Ar_i \mid i \in \text{Agt}\}, att\rangle$ where $Ar \subseteq \mathbf{A}$; $Ar_i \subseteq Ar$; and $att \subseteq Ar \times Ar$. We say that the MAF $\langle Ar, \{Ar_i \mid i \in \text{Agt}\}, att\rangle$ is based on Ar. We use $\text{MAF}(Ar)$ to denote the set of all MAFs based on Ar.*

Intuitively, in such a MAF, $\langle Ar, att\rangle$ represents all potentially relevant arguments and their interactions, while Ar_i is the set of arguments that agent i is aware of. We further define $att_i = att \cap (Ar_i \times Ar_i)$ as the set of attacks that agent i is aware of. Note that the definition of att_i implies that an agent is aware of an attack whenever it is aware of the arguments involved. Therefore, attacks in a MAF have an 'objective' meaning, since no agent can be 'mistaken' about them. Nonetheless, uncertainty and incomplete knowledge about attacks can still be represented at the modal level, as we shall see in what follows.

We now plug MAFs into worlds of a multi-agent doxastic model. The ideas underlying the following definition can be traced back to [Schwarzentruber et al., 2012]. We get rid of some of the assumptions presented there and some others introduced in [Proietti and Yuste-Ginel, 2021].[17]

---

[17]The reasons for doing so is that we seek maximum generality here and that some of these assumptions were introduced for mere technical reasons that are not relevant at the current level of abstraction.

M satisfies...	iff for every $i \in \mathbb{A}\text{gt}$, $w, u \in S$...
Positive knowledge of attacks	$w\mathcal{R}_i u \Rightarrow att(w) \subseteq att(u)$
Negative knowledge of attacks	$w\mathcal{R}_i u \Rightarrow att(u) \subseteq att(w)$
Positive introspection of arguments	$w\mathcal{R}_i u \Rightarrow Ar_i(w) \subseteq Ar_i(u)$
Negative introspection of arguments	$w\mathcal{R}_i u \Rightarrow Ar_i(u) \subseteq Ar_i(w)$
General negative introspection of arguments	$w\mathcal{R}_i u \Rightarrow \bigcup_{j \in \mathbb{A}\text{gt}} Ar_j(u) \subseteq Ar_i(w)$

Table 1: Some properties of EA-models

Definition 5.6 (EA-models). *An Epistemic-Argumentative model (EA-model) for Ar is a tuple $M = \langle S, \mathcal{R}, \mathcal{D} \rangle$ s.t.*

- $\langle S, \mathcal{R} \rangle$ *is a Kripke frame over* $\mathbb{A}\text{gt}$*, where S is the set of possible worlds and $\mathcal{R}$ specifies the epistemic accessibility relations of different agents (Def. 2.2), and*

- $\mathcal{D} : S \to \mathsf{MAF}(Ar)$ *is a function specifying a multi-agent AF (Def. 5.5) for each possible world.*

Let $w \in S$, we denote $\mathcal{D}(w)$ as $\langle Ar, \{Ar_i(w) \mid i \in \mathbb{A}\text{gt}\}, att(w) \rangle$.

Some interesting properties relating $\mathcal{R}$ and $\mathcal{D}$ are summarised in Table 1.

Let us illustrate the previous definition through a simple example.

Example 5.7 (An EA-model). *Let us consider the EA-model $M = \langle S, \mathcal{R}, \mathcal{D} \rangle$ for $\mathbb{A}\text{gt} = \{1, 2\}$ and $Ar = \{a, b, c, d\}$ where: $S = \{w_0, w_1\}$; $\mathcal{R}(1) = \{\langle w_0, w_1 \rangle, \langle w_1, w_1 \rangle\}$, $\mathcal{R}(2) = \{\langle w_0, w_0 \rangle, \langle w_1, w_1 \rangle\}$; and $\mathcal{D}$ is defined for each world: $Ar_1(w_0) = \{a, b, c, d\}$, $Ar_2(w_0) = \{a, b, c\}$, $att(w_0) = \{\langle a, b \rangle, \langle b, a \rangle, \langle c, b \rangle, \langle c, d \rangle, \langle d, b \rangle, \langle d, c \rangle\}$; $Ar_1(w_1) = \{a, b, c, d\}$, $Ar_2(w_1) = \{a, b\}$, $att(w_1) = \{\langle a, b \rangle, \langle b, a \rangle, \langle c, b \rangle, \langle c, d \rangle, \langle d, c \rangle\}$. This model is represented graphically in Figure 9. Intuitively, at the actual world (w_0) agent 2 is aware of arguments a, b and c (red-circled area in w_0). However, agent 1 does not know that agent 2 is aware of c, for she has access only to w_1 where the awareness area of 2 only contains a and*

b. *Moreover, agent 1 is also mistaken in his/her understanding of the attacks between arguments, inasmuch as argument d attacks argument b in the actual world w_0, but 1 believes this fact to be false (no attack in w_1).*

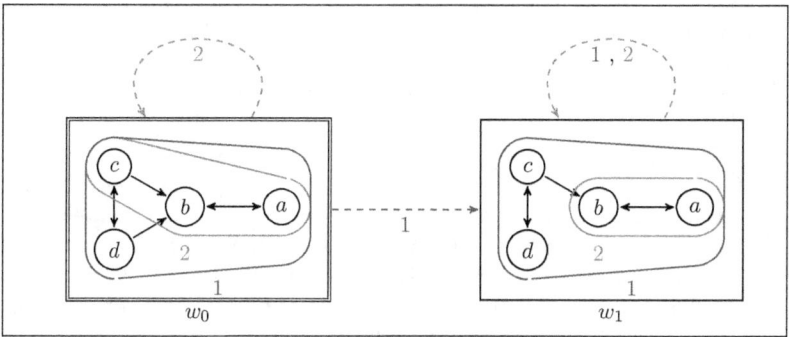

Figure 9: A simple EA-model

Languages Different languages have been proposed to talk about EA-models (or some of their extensions). The common denominator in all of them is the use of doxastic/epistemic modalities to jump from one possible world to another. The main differences among them lie in the resources chosen to talk about the MAF representing each world. Let us briefly comment on them.

[Schwarzentruber et al., 2012] introduced three of these languages. The first one, later extended by [Proietti and Yuste-Ginel, 2020], is just the standard multi-agent doxastic language with a set of distinguished atoms $\{\mathsf{aware}_i(x) \mid i \in \mathbb{A}\mathrm{gt}, x \in \mathbf{A}\}$, which are used to talk about the set of arguments that each agent is aware of at each world (i.e. to talk about $Ar_i(w)$). The lack of a syntactic device to talk about attacks implicitly assumes that the set of attacks assigned by $\mathcal{D}$ to w (denoted by $att(w)$) is the same for every $w \in S$. This assumption is dropped by the second language introduced by [Schwarzentruber et al., 2012]: a two-layer language where one type of formula is used to talk about the MAF assigned to each world and the other one is used as a way of (modally) jumping from one world to the other. Their third language, interpreted over a product-model version of EA-models, is essentially one where the previous two layers are fused together so that argumentative and doxastic modalities can be nested (just as the one we discussed in Section 5.3.1).

If one restricts to finite sets of arguments, as most of the literature does, then propositional languages are sufficient for reasoning about AFs and their semantics [Besnard and Doutre, 2004]. In this case, using modalities to describe a single MAF is an overkill. In line with this, [Proietti and Yuste-Ginel, 2021], use an enriched version of [Schwarzentruber et al., 2012]'s first language to talk about the MAF assigned to each world in EA-models. This enrichment consists of a new set of atoms for describing the attack relation $\{r_{x,y} \mid x, y \in \mathbf{A}\}$. Note that, with the variables ($\mathsf{aware}_i(x)$ and $r_{x,y}$), all properties of Table 1 are easily definable (e.g., $\neg\mathsf{aware}_i(x) \to \Box_i \bigwedge_{j \in \mathbb{A}\mathrm{gt}} \neg\mathsf{aware}_j(x)$ for the last one).[18]

Finally, a third approach is presented in [Yuste-Ginel, 2022, pp. 161-165]. This can be understood as some kind of compromise between the full power of bi-modal languages to describe MAFs [Schwarzentruber et al., 2012; Grossi and van der Hoek, 2014]) vs. the use of propositional languages [Proietti and Yuste-Ginel, 2021]. Summing up, the idea is to use yet a new kind of propositional variable $\{\mathsf{in}_x \mid x \in \mathbf{A}\}$ meant to describe an arbitrary set of arguments (e.g., an extension), plus one extra modality that quantifies over valuations changing the truth values of these variables. This new modality (inspired by works that combine dynamic logic and abstract argumentation [Doutre et al., 2014]) allows expressing maximality and minimality checking, and it is therefore expressive enough to capture all standard argumentation semantics for MAFs through polynomially long formulas.

Dynamics of information EA-models were systematically dinamised in [Proietti and Yuste-Ginel, 2021], where they are combined with event models [Baltag and Moss, 2004], imported from the field of dynamic epistemic logic [van Ditmarsch et al., 2007]. With the re-

---

[18]The main shortcoming of this approach is that with no quantifiers over atoms – or any other equally expressive device – the encoding of some notions (e.g., those requiring maximality checking, as preferred semantics) requires propositional formulas that are exponentially long (on the size of $\mathbf{A}$), and therefore not very appealing from a computational point of view.

sulting framework, one is able to model nuanced forms of epistemic and argumentative dynamics, such as the action of privately searching for a counter-argument in the context of a debate. For presentational purposes, we just introduce here the simplest kind of epistemic-argumentative action, which was first studied in [Proietti and Yuste-Ginel, 2020]: the public addition or disclosure of an argument. Informally, the idea is to model what happens when, within a group discussion, an agent publicly puts forward an argument. Formally:

Definition 5.8 (Public update with an argument). *Given a EA-model $M = \langle S, \mathcal{R}, \mathcal{D} \rangle$ and an argument $a \in Ar$, the update of M by a is the model $M^{a!} = \langle S, \mathcal{R}, \mathcal{D}^{a!} \rangle$ where $\mathcal{D}^{a!}$ only differs from $\mathcal{D}$ in the value assigned to the awareness set of each agent at each world: $Ar_i^{a!}(w) = Ar_i(w) \cup \{a\}$ for every $i \in$ Agt and every $w \in S$.*

Applications to opponent modelling The main application of EA-models and their dynamics is the systematic analysis of different forms of opponent modelling in abstract argumentation. EA-models represent an expressive and epistemically transparent formalism where other proposals can be translated, so as to fully understand their hidden epistemic assumptions. Let us quickly review some of these reductions.

Incomplete AFs One of the simplest ways for modelling the opponent of an agent within a debate is through the notion of *incomplete argumentation framework* (IAF) [Baumeister et al., 2021]. The idea is to provide a compact specification of the uncertain view that the agent has of her opponent's information about a debate. Formally, an IAF is a tuple $\langle Ar, Ar^?, att, att^? \rangle$ where $Ar, Ar^? \subseteq \mathbf{A}$ are two disjoint sets of arguments, respectively representing certain and uncertain arguments; and $att, att^? \subseteq (Ar \cup Ar^?) \times (Ar \cup Ar^?)$ are two disjoint sets of attacks respectively representing certain and uncertain attacks. Perhaps more intuitively, Ar is the set of arguments that the agent believes her opponent to be aware of; $Ar^?$ is the set of arguments such that the agent does not know whether her opponent is aware of; and something analogous for att and $att^?$.

Reasoning about IAFs needs the notion of completion, i.e. a hypothetical removal of uncertainty from the IAF. Formally, a completion of $\langle Ar, Ar^?, att, att^? \rangle$ is any AF $\langle Ar^*, att^* \rangle$ such that $Ar \subseteq Ar^* \subseteq Ar \cup Ar^?$

and $att \cap (Ar^* \times Ar^*) \subseteq att^* \subseteq (att \cup att^?) \cap (Ar^* \times Ar^*)$.[19] As a key for reduction to EA-models, completions can be understood as possible worlds. Indeed, an IAF can be seen as a single-agent EA-model where each possible world represents a completion. If an argument a is present in one completion, then the atom $\mathsf{aware}_i(a)$ is going to be true in its corresponding world of the model, and the same holds for attacks. In this way, reasoning problems over IAFs become model-checking problems in EA-models[Proietti and Yuste-Ginel, 2021]. Note that the correspondence is not strict: There are EA-models that do not represent the set of completions of any IAF. Hence, a legitimate question: what logic would we get if we only consider EA-models that represent IAFs?

The previous question is answered by [Herzig and Yuste-Ginel, 2021b]. The keystone for this finding is the connection between IAFs and the epistemic logic of visibility studied in [Herzig et al., 2018]. Interestingly, the two central axioms of the resulting logic are:

$$\Box_i \varphi \to \Diamond_i \varphi$$
$$\Box_i (l_1 \vee \cdots \vee l_n) \to (\Box_i l_1 \vee \cdots \vee \Box_i l_n)$$

where $l_1 \ldots l_n$ is a sequence of consistent literals from $\{\mathsf{aware}_i(x) \mid x \in \mathbf{A}\} \cup \{\mathsf{r}_{x,y} \mid x,y \in \mathbf{A}\}$. The first axiom is just (D) (see Sect. 2.1 and the introduction to the current section), so it shows that IAFs model an epistemic attitude that is, at least, consistent (which seems quite reasonable for intelligent artificial agents). The second one expresses that the captured epistemic attitudes distribute over disjunctions of consistent literals (literals that capture the status of arguments and attacks). This second property looks more difficult to justify, unless for its efficient computational behaviour.

Control AFs (CAFs) [Dimopoulos et al., 2018] extend IAFs in two directions: uncertainty and dynamics. Regarding uncertainty, CAFs include a new attack relation $att^{\leftrightarrow}$ which is meant to capture attacks whose existence is known by the agent but whose direction is unknown. For example, imagine that a and b are two arguments disclosed, respectively, by two politicians of opposing parties. Imagine, moreover, that

---

[19]Classical acceptability problems over AFs (sceptical and credulous acceptability) can be reformulated so as to get a new layer of quantification (over completions), obtaining in this way necessary (in all completions) and possible (in at least one completion) variants of classical problems.

the agent knows that her opponent is biased towards one side of the political spectrum, but she is not sure about which side it is. Hence, the agent considers two possible completions (for her opponent): one where a attacks b and one where b attacks a. It is easy to show that no IAF represent that precise set of completions, so the previous kind of scenarios justify the introduction of $att^{\leftrightarrow}$. On the dynamic side, CAFs expand IAFs with an AF $\langle Ar_C, att_C \rangle$ formed by *control arguments* and *control attacks*. Intuitively, these arguments are the ones that the agent knows privately (she knows them and knows her opponent does not know). Then, reasoning over CAFs introduces yet another quantification layer (over subsets of Ar_C), raising so-called *controllability problems*: is there a set of control arguments such that a target argument gets sceptically/credulously accepted in all/at least one completion? These problems too were reduced to EA-model-checking problems in [Proietti and Yuste-Ginel, 2021]. Such a reduction shows at least three interesting things. First, the axiom of IAFs capturing distribution over disjunctions of consistent literals is dropped because of the introduction of $att^{\leftrightarrow}$. Second, control arguments are in tight connection with the notion of public disclosure (see above). Third, the effects of communicating these arguments (e.g., the resulting perception of control attacks) are strongly idealised: the agent always knows what the effects of her communication act are. The last two points motivate the study of further forms of representing argument communication in a compact setting.

Recursive opponent models The reductions of incomplete AFs and control AFs to EA models do not exploit all the expressive power of the latter. Indeed, and as far as uncertainty and multi-agency are concerned, incomplete AFs and control AFs can be seen as depth-1 epistemic attitudes about the arguments and attacks that the opponent of the agent owns. In other words, incomplete and control AF only talk about what the agent believes about her opponent's view of the debate. However, as pointed out by [Thimm, 2014] concerning strategic argumentation, higher-order epistemic attitudes might as well be relevant in an argumentative context. As an example, imagine that you want to surprise your opponent. Then you might disclose an argument that *you believe your opponent believes that you are not aware of*, which is a depth-2 epistemic attitude. Along these lines, more expressive (and hence more complex) forms of opponent modelling have been studied in the context of strategic argumentation. These opponent models are, in their qualitative version,[20] EA models in disguise. The details of the reduction can be found in [Proietti and Yuste-Ginel, 2021, Sect. 8.3].

New formalisms In recent years, two different tools for representing qualitative uncertainty about AFs appeared [Sakama and Cao Son, 2020; Alfano et al., 2023]. Although they have not yet been reduced to EA-models, they should be reducible if one looks at the respective complexity classes of the relevant reasoning problems. We believe these reductions to be interesting open problems, so as to carve deeper into the epistemic assumptions underlying these new formalisms and get a more complete picture of how to represent qualitative uncertainty and multi-agency over AFs.

6 Related work

The first works on the combination of formal argumentation and modal logic appeared less than 20 years ago ([Boella et al., 2005] is the pioneering one, to the best of our knowledge). However, the literature is already quite vast. This section points to further readings on the topic that were left out of this chapter either because we wanted to keep its extension between reasonable limits or because they were already analysed in a previous volume of this handbook [Arieli et al., 2021].

The first line of work, started by [Doutre et al., 2014] and followed by [Doutre et al., 2017; Doutre et al., 2019; Herzig and Yuste-Ginel, 2021a; Yuste-Ginel and Herzig, 2023], consists in applying the Dynamic Logic of Propositional Assignments (DL-PA) to reason about abstract argumentation formalisms. DL-PA is a lightweight, well-behaved variant of dynamic propositional logic. In the quoted papers, it is used to capture argumentation frameworks, their semantics, their dynamics, and their extensions with qualitative uncertainty. These works share the same general target of those adopting an *argumentative reading of modalities*:[21]

---

[20]The are also probabilistic versions of opponent models for argumentation that go beyond the expressivity of EA models.

[21]Those discussed in the introduction and in [Arieli et al., 2021, Sect. 3.2.1], i.e., [Grossi, 2009] and subsequent papers.

using modal logic to reason about argumentation systems. However, instead of interpreting modalities over attack relations and arguments, they are based on propositional encodings of argumentation formalisms [Besnard and Doutre, 2004], and the role of modalities is basically capturing restricted propositional quantification.

A second line of work proposes to use modal languages as the object languages of structured argumentation formalisms (e.g., ASPIC$^+$ [Modgil and Prakken, 2013] or ABA [Dung et al., 2009]). A particular instance of this idea is the insertion of modal languages into deductive argumentation systems, something that was previously covered in another chapter of this handbook [Besnard and Hunter, 2018]. Going further than deductive reasoning, and hence incorporating non-monotonic inferences, a recent paper [van der Torre, 2019] has approached the use of deontic modalities in an argumentation system from the ASPIC family.

Finally, there is another interesting research line oriented towards integrating temporal and modal language formulas to represent arguments in the nodes of an argumentation network, as done in [Barringer et al., 2012]. This approach can be seen as an extension of the traditional Dung networks, which depict arguments as atomic entities and study the relationships of attack between them. That way more content can be added to nodes in the network (e.g., proofs in some logic or simply just formulas from a richer language). Argumentation networks have also been applied in modelling so-called *argumentation with many lives* [Gabbay and Rozenberg, 2020], where the network stands for a survival game (and thus the various traditional Dung semantics can be viewed as defining extensions in the form of possible survival group). With many lives available, there can be sets of nodes "living together" (so that members can attack but not able to kill one another). Recent research work in many lives argumentation networks [Gabbay and Rozenberg, 2023] included temporal aspects, modelled through evolutionary temporal logic.

7 Conclusion

Argumentation and modal logic are two important theories and techniques within the field of knowledge representation and reasoning. Although their combination started only a couple of decades ago, the literature of the topic has flourished very rapidly since then. In this chapter we analysed three different lines of work to combine modal logic and formal argumentation: a) a logic-based framework that combines dynamic logic and argumentation for value-based planning; b) alternating-time temporal logic extended with coalitional argumentation; c) a combined approach for integrating epistemic logics and argumentation. These three alternatives give clear evidence of the heterogeneity of problems that can be approached by the combination of these two families of formal tools. Moreover, they also show the different relative positions in which modal logic and argumentation can be put together. In a broader sense, they also allow us to think about the possible interplay between classical logics and non-monotonic logics and how they can function respectively when solving particular problems.

Joint uses of formal argumentation and modal logic face numerous challenges in the current state of the art. We have seen how each work focuses on a particular reading of modalities (e.g., argumentative, epistemic, dynamic, or temporal). Although some papers tackle the fundamental question of their combination (e.g., argumentative and epistemic [Grossi and van der Hoek, 2014]), there are many possibilities which are yet unexplored. For instance, there seems to be strong rationale for motivating the design and study of structured argumentation systems that allow for the integration of deontic and epistemic modalities for reasoning about complex scenarios. Moreover, both strategic logic (Section 4.1) and strategic argumentation (end of Section 5.4) are used for reasoning about agents' strategic behavior in the context of multi-agent systems. If one sees the announcement of an argument as the performance of an action, then it seems natural looking at the transfer of strategic argumentation to ATL-like logics that are interpreted over transition systems. That would allows us to represent and reason about the argumentative abilities of agents in the style of modal logic.

References

[Ågotnes et al., 2007] Thomas Ågotnes, Wiebe van der Hoek, and Michael Wooldridge. Normative system games. In *Proceedings of the 6th international joint conference on Autonomous agents and multiagent systems*, pages 1–8, 2007.

[Alechina et al., 2013] Natasha Alechina, Mehdi Dastani, and Brian Logan. Reasoning about normative update. In *Proceedings of the Twenty-Third International Joint Conference on Artificial Intelligence*, IJCAI '13, page 20–26. AAAI Press, 2013.

[Alfano et al., 2023] Gianvincenzo Alfano, Sergio Greco, Francesco Parisi, and Irina Trubitsyna. Epistemic abstract argumentation framework: Formal foundations, computation and complexity. In *Proceedings of the 2023 International Conference on Autonomous Agents and Multiagent Systems*, pages 409–417, 2023.

[Alur et al., 2002] R. Alur, T. A. Henzinger, and O. Kupferman. Alternating-time Temporal Logic. *Journal of the ACM*, 49:672–713, 2002.

[Amgoud and Prade, 2007] Leila Amgoud and Henri Prade. Formalizing practical reasoning under uncertainty: An argumentation-based approach. In *2007 IEEE/WIC/ACM International Conference on Intelligent Agent Technology (IAT'07)*, pages 189–195. IEEE, 2007.

[Amgoud, 2005a] Leila Amgoud. An argumentation-based model for reasoning about coalition structures. In *ArgMAS*, pages 217–228, 2005.

[Amgoud, 2005b] Leila Amgoud. Towards a formal model for task allocation via coalition formation. In *AAMAS*, pages 1185–1186, 2005.

[Arieli et al., 2021] Ofer Arieli, AnneMarie Borg, Jesse Heyninck, and Christian Straßer. Logic-based approaches to formal argumentation. In DM Gabbay, Massimiliano Giacomin, Guillermo R Simari, and Matthias Thimm, editors, *Handbook of Formal Argumentation*, volume 3, pages 707–838. College Publications, 2021.

[Artemov, 2008] Sergei Artemov. The logic of justification. *The Review of Symbolic Logic*, 1(4):477–513, 2008.

[Atkinson and Bench-Capon, 2007] Katie Atkinson and Trevor Bench-Capon. Practical reasoning as presumptive argumentation using action based alternating transition systems. *Artificial Intelligence*, 171(10-15):855–874, 2007.

[Atkinson and Bench-Capon, 2018] Katie Atkinson and Trevor Bench-Capon. Taking account of the actions of others in value-based reasoning. *Artificial Intelligence*, 254:1–20, 2018.

[Baltag and Moss, 2004] Alexandru Baltag and Lawrence S Moss. Logics for epistemic programs. *Synthese*, 139(2):165–224, 2004.

[Baltag and Smets, 2008] Alexandru Baltag and Sonja Smets. A qualitative theory of dynamic interactive belief revision. In Wiebe van der Hoek, Giacomo Bonanno, and Michael Wooldridge, editors, *Logic and the foundations of game and decision theory (LOFT 7)*, volume 3 of *Texts in Logic and Games*, pages 9–58. Amsterdam University Press, 2008.

[Baltag et al., 2012] Alexandru Baltag, Bryan Renne, and Sonja Smets. The logic of justified belief change, soft evidence and defeasible knowledge. In *Logic, Language, Information and Computation: 19th International Workshop, WoLLIC 2012, Buenos Aires, Argentina, September 3-6, 2012. Proceedings 19*, pages 168–190. Springer, 2012.

[Baltag et al., 2014] Alexandru Baltag, Bryan Renne, and Sonja Smets. The logic of justified belief, explicit knowledge, and conclusive evidence. *Annals of Pure and Applied Logic*, 165(1):49–81, 2014.

[Baltag et al., 2016] Alexandru Baltag, Nick Bezhanishvili, Aybüke Özgün, and Sonja Smets. Justified belief and the topology of evidence. In Jouko Väänänen, Åsa Hirvonen, and Ruy de Queiroz, editors, *Logic, Language, Information, and Computation*, pages 83–103. Springer, 2016.

[Baroni et al., 2018] Pietro Baroni, Martin Caminada, and Massimiliano Giacomin. Abstract argumentation frameworks and their semantics. In Pietro Baroni, Dov M. Gabbay, Massimilino Giacomin, and Leendert van der Torre, editors, *Handbook of formal argumentation*, pages 159–236. College Publications, 2018.

[Barringer et al., 2012] Howard Barringer, Dov M. Gabbay, and John Woods. Modal and temporal argumentation networks. *Argument Comput.*, 3(2-3):203–227, 2012.

[Baumeister et al., 2021] Dorothea Baumeister, Matti Järvisalo, Daniel Neugebauer, Andreas Niskanen, and Jörg Rothe. Acceptance in incomplete argumentation frameworks. *Artificial Intelligence*, 295:103470, 2021.

[Bench-Capon et al., 2012] Trevor Bench-Capon, Katie Atkinson, and Peter McBurney. Using argumentation to model agent decision making in economic experiments. *Autonomous Agents and Multi-Agent Systems*, 25(1):183–208, 2012.

[Besnard and Doutre, 2004] Philippe Besnard and Sylvie Doutre. Checking the acceptability of a set of arguments. In James P. Delgrande and Torsten Schaub, editors, *Proceedings of the NMR*, pages 59–64. AAAI Press, 2004.

[Besnard and Hunter, 2018] Philippe Besnard and Anthony Hunter. A review of argumentation based on deductive arguments. *Handbook of Formal Argumentation*, 1:437–484, 2018.

[Besnard et al., 2014] Philippe Besnard, Alejandro Garcia, Anthony Hunter, Sanjay Modgil, Henry Prakken, Guillermo Simari, and Francesca Toni. Introduction to structured argumentation. *Argument & Computation*, 5(1):1–4, 2014.

[Boella et al., 2005] Guido Boella, Joris Hulstijn, and Leendert W. N. van der Torre. A logic of abstract argumentation. In *ArgMAS*, volume 4049 of *Lecture Notes in Computer Science*, pages 29–41. Springer, 2005.

[Bolander and Andersen, 2011] Thomas Bolander and Mikkel Birkegaard Andersen. Epistemic planning for single-and multi-agent systems. *Journal of Applied Non-Classical Logics*, 21(1):9–34, 2011.

[Bolander, 2017] Thomas Bolander. A gentle introduction to epistemic planning: The DEL approach. In Sujata Ghosh and R. Ramanujam, editors, *Proceedings of the Ninth Workshop on Methods for Modalities, M4M@ICLA 2017, Indian Institute of Technology, Kanpur, India, 8th to 10th January 2017*, volume 243 of *EPTCS*, pages 1–22, 2017.

[Bulling et al., 2008a] Nils Bulling, Carlos Iván Chesñevar, and Jürgen Dix. An argumentative approach for modelling coalitions using ATL. In Iyad Rahwan and Pavlos Moraitis, editors, *Argumentation in Multi-Agent Systems, Fifth International Workshop, ArgMAS 2008, Estoril, Portugal, May 12, 2008. Revised Selected and Invited Papers*, volume 5384 of *Lecture Notes in Computer Science*, pages 197–216. Springer, 2008.

[Bulling et al., 2008b] Nils Bulling, Jürgen Dix, and Carlos Iván Chesñevar. Modelling coalitions: ATL + argumentation. In Lin Padgham, David C. Parkes, Jörg P. Müller, and Simon Parsons, editors, *7th International Joint Conference on Autonomous Agents and Multiagent Systems (AAMAS 2008), Estoril, Portugal, May 12-16, 2008, Volume 2*, pages 681–688. IFAAMAS, 2008.

[Bulling et al., 2008c] Nils Bulling, Wojciech Jamroga, and Jürgen Dix. Reasoning about temporal properties of rational play. *Ann. Math. Artif. Intell.*, 53(1-4):51–114, 2008.

[Burrieza and Yuste-Ginel, 2020] Alfredo Burrieza and Antonio Yuste-Ginel. Basic beliefs and argument-based beliefs in awareness epistemic logic with structured arguments. In Henry Prakken, Stefano Bistarelli, Francesco Santini, and Carlo Taticchi, editors, *Proceedings of the COMMA 2020*, pages 123–134. IOS Press, 2020.

[Burrieza and Yuste-Ginel, 2021] Alfredo Burrieza and Antonio Yuste-Ginel. An awareness epistemic framework for belief, argumentation and their dynamics. In Joseph Y. Halpern and Andrés Perea, editors, *Proceedings Eighteenth Conference on Theoretical Aspects of Rationality and Knowledge (TARK)*, volume 335 of *EPTCS*, pages 69–83, 2021.

[Caminada and Gabbay, 2009] Martin WA Caminada and Dov M Gabbay. A logical account of formal argumentation. *Studia Logica*, 93(2-3):109–145, 2009.

[Caminada, 2017] Martin Caminada. Rationality postulates: applying argumentation theory for non-monotonic reasoning. *Journal of Applied Logics*, 4(8):2707–2734, 2017.

[Chellas, 1980] Brian F Chellas. *Modal logic: an introduction*. Cambridge university press, 1980.

[Chesñevar et al., 2000] Carlos Iván Chesñevar, Ana Gabriela Maguitman, and Ronald Prescott Loui. Logical models of argument. *ACM Comput. Surv.*, 32(4):337–383, dec 2000.

[Clarke et al., 1999] E. Clarke, O. Grumberg, and D. Peled. *Model Checking.* MIT Press, 1999.

[Dimopoulos et al., 2018] Yannis Dimopoulos, Jean-Guy Mailly, and Pavlos Moraitis. Control argumentation frameworks. In Sheila A. McIlraith and Kilian Q. Weinberger, editors, *Proceedings of the Thirty-Second AAAI Conference on Artificial Intelligence.* AAAI Press, 2018.

[Doutre et al., 2014] Sylvie Doutre, Andreas Herzig, and Laurent Perrussel. A dynamic logic framework for abstract argumentation. In C. Baral, G. De Giacomo, and T. Eiter, editors, *Fourteenth International Conference on the Principles of Knowledge Representation and Reasoning.* AAAI Press, 2014.

[Doutre et al., 2017] Sylvie Doutre, Faustine Maffre, and Peter McBurney. A dynamic logic framework for abstract argumentation: adding and removing arguments. In Salem Benferhat, Karim Tabia, and Moonis Ali, editors, *International Conference on Industrial, Engineering and Other Applications of Applied Intelligent Systems.*, volume 10351 of *LNCS*, pages 295–305. Springer, 2017.

[Doutre et al., 2019] Sylvie Doutre, Andreas Herzig, and Laurent Perrussel. Abstract argumentation in dynamic logic: Representation, reasoning and change. In Beishui Liao, Thomas Ågotnes, and Yi N. Wang, editors, *Dynamics, Uncertainty and Reasoning*, pages 153–185. Springer, 2019.

[Dung et al., 2009] Phan Minh Dung, Robert A Kowalski, and Francesca Toni. Assumption-based argumentation. pages 199–218. Springer, 2009.

[Dung, 1995] Phan Minh Dung. On the acceptability of arguments and its fundamental role in nonmonotonic reasoning, logic programming and n-person games. *Artificial intelligence*, 77(2):321–357, 1995.

[Dunne and Bench-Capon, 2003] Paul E. Dunne and Trevor J. M. Bench-Capon. Two party immediate response disputes: Properties and efficiency. *Artif. Intell.*, 149(2):221–250, 2003.

[Égré et al., 2021] Paul Égré, Paul Marty, and Bryan Renne. Knowledge, justification, and adequate reasons. *The review of symbolic logic*, 14(3):687–727, 2021.

[Fagin and Halpern, 1987] Ronald Fagin and Joseph Y Halpern. Belief, awareness, and limited reasoning. *Artificial intelligence*, 34(1):39–76, 1987.

[Fagin et al., 2004] Ronald Fagin, Joseph Y Halpern, Yoram Moses, and Moshe Vardi. *Reasoning about knowledge.* MIT press, 2004.

[Gabbay and Grossi, 2014] Dov Gabbay and Davide Grossi. When are two arguments the same? equivalence in abstract argumentation. In *Johan van Benthem on Logic and Information Dynamics*, pages 677–701. Springer, 2014.

[Gabbay and Rozenberg, 2020] Dov M. Gabbay and Gadi Rozenberg. Introducing abstract argumentation with many lives. *FLAP*, 7(3):295–336, 2020.

[Gabbay and Rozenberg, 2023] Dov M. Gabbay and Gadi Rozenberg. Evolutionary temporal logic for modelling many-lives argumentation networks. *FLAP*, 10(5):909–966, 2023.

[Gabbay and Shehtman, 1998] Dov M Gabbay and Valentin B Shehtman. Products of modal logics, part 1. *Logic journal of IGPL*, 6(1):73–146, 1998.

[Gettier, 1963] Edmund L. Gettier. Is justified true belief knowledge? *Analysis*, 23(6):121–123, 1963.

[Gratie et al., 2012] Cristian Gratie, Adina Magda Florea, and John-Jules Ch Meyer. Full hybrid μ-calculus, its bisimulation invariance and application to argumentation. In *International Workshop on Computational Logic in Multi-Agent Systems*, pages 181–194. Springer, 2012.

[Grossi and Rey, 2019] Davide Grossi and Simon Rey. Credulous acceptability, poison games and modal logic. In N. Agmon, M. E. Taylor, E. Elkind, and M. Veloso, editors, *18th International Conference on Autonomous Agents and Multiagent Systems, AAMAS 2019*, pages 1994–1996. International Foundation for Autonomous Agents and Multiagent Systems (IFAAMAS), 2019.

[Grossi and van der Hoek, 2014] Davide Grossi and Wiebe van der Hoek. Justified beliefs by justified arguments. In Chitta Baral, Giuseppe De Giacomo, and Thomas Eiter, editors, *Principles of Knowledge Representation and Reasoning: Proceedings of the Fourteenth International Conference*. AAAI Press, 2014.

[Grossi, 2009] Davide Grossi. Doing argumentation theory in modal logic. 2009.

[Grossi, 2010a] Davide Grossi. Argumentation in the view of modal logic. In Peter McBurney, Iyad Rahwan, and Simon Parsons, editors, *International Workshop on Argumentation in Multi-Agent Systems*, volume 6614 of *LNCS*, pages 190–208. Springer, 2010.

[Grossi, 2010b] Davide Grossi. On the logic of argumentation theory. In W. van der Hoek, G.A. Kaminka, Y. Lesperance, M. Luck, and S. Sen, editors, *9th International Conference on Autonomous Agents and Multiagent Systems (AAMAS 2010)*, pages 409–416. IFAAMAS, 2010.

[Hadjinikolis et al., 2013] Christos Hadjinikolis, Yiannis Siantos, Sanjay Modgil, Elizabeth Black, and Peter McBurney. Opponent modelling in persuasion dialogues. In *Twenty-Third International Joint Conference on Artificial Intelligence*, 2013.

[Herzig and Yuste-Ginel, 2021a] Andreas Herzig and Antonio Yuste-Ginel. Abstract argumentation with qualitative uncertainty: An analysis in dynamic logic. In Pietro Baroni, Christoph Benzmüller, and Yi N. Wáng, editors, *Logic and Argumentation*, volume 13040 of *LNCS*, pages 190–208. Springer, 2021.

[Herzig and Yuste-Ginel, 2021b] Andreas Herzig and Antonio Yuste-Ginel. On the Epistemic Logic of Incomplete Argumentation Frameworks. In M. Bienvenu, G. Lakemeyer, and E. Erdem, editors, *Proceedings of the 18th International Conference on Principles of Knowledge Representation and Reasoning*, pages 681–685, 11 2021.

[Herzig et al., 2018] Andreas Herzig, Emiliano Lorini, and Faustine Maffre. Possible worlds semantics based on observation and communication. In Hans van Ditmarsch and Gabriel Sandu, editors, *Jaakko Hintikka on Knowledge and Game-Theoretical Semantics*, pages 339–362. Springer, 2018.

[Hintikka, 1962] Kaarlo Jaakko Juhani Hintikka. Knowledge and belief: An introduction to the logic of the two notions. 1962.

[Jamroga and Bulling, 2007a] W. Jamroga and N. Bulling. A general framework for reasoning about rational agents. In *Proceedings of AAMAS'07*, pages 592–594, Honolulu, Hawaii, USA, 2007. ACM Press. Short paper.

[Jamroga and Bulling, 2007b] Wojtek Jamroga and Nils Bulling. A logic for reasoning about rational agents. In F. Sadri and K. Satoh, editors, *Proceedings of CLIMA '07*, pages 54–69, Porto, Portugal, 2007. Univesidade Do Porto.

[Knobbout and Dastani, 2012] Max Knobbout and Mehdi Dastani. Reasoning under compliance assumptions in normative multiagent systems. In *Proceedings of the 11th International Conference on Autonomous Agents and Multiagent Systems-Volume 1*, pages 331–340. International Foundation for Autonomous Agents and Multiagent Systems, 2012.

[Knobbout et al., 2014] Max Knobbout, Mehdi Dastani, and John-Jules Ch Meyer. Reasoning about dynamic normative systems. In *European Workshop on Logics in Artificial Intelligence*, pages 628–636. Springer, 2014.

[Knobbout et al., 2016] Max Knobbout, Mehdi Dastani, and John-Jules Meyer. A dynamic logic of norm change. In *Proceedings of the Twenty-second European Conference on Artificial Intelligence*, pages 886–894, 2016.

[Kurucz et al., 2003] Agi Kurucz, Frank Wolter, Michael Zakharyaschev, and Dov M Gabbay. *Many-dimensional modal logics: theory and applications*. Gulf Professional Publishing, 2003.

[Liao and van der Torre, 2020] Beishui Liao and Leendert van der Torre. Explanation semantics for abstract argumentation. In Henry Prakken, Stefano Bistarelli, Francesco Santini, and Carlo Taticchi, editors, *Computational Models of Argument - Proceedings of COMMA 2020, Perugia, Italy, September 4-11, 2020*, volume 326 of *Frontiers in Artificial Intelligence and Applications*, pages 271–282. IOS Press, 2020.

[Liao et al., 2019] Beishui Liao, Nir Oren, Leendert van der Torre, and Serena Villata. Prioritized norms in formal argumentation. *Journal of Logic and Computation*, 29(2):215–240, 2019.

[Liao et al., 2021] Beishui Liao, Michael Anderson, and Susan Leigh Anderson. Representation, justification, and explanation in a value-driven agent: an argumentation-based approach. *AI and Ethics*, 1(1):5–19, 2021.

[Luo et al., 2019] Jieting Luo, John-Jules Meyer, and Max Knobbout. A formal framework for reasoning about opportunistic propensity in multi-agent systems. *Autonomous Agents and Multi-Agent Systems*, 33(4):457–479, 2019.

[Luo et al., 2022a] Jieting Luo, Beishui Liao, and Dov M. Gabbay. Value-based practical reasoning: Modal logic + argumentation. In *COMMA*, volume 353 of *Frontiers in Artificial Intelligence and Applications*, pages 248–259. IOS Press, 2022.

[Luo et al., 2022b] Jieting Luo, Beishui Liao, and Dov M. Gabbay. Value-based practical reasoning: Modal logic + argumentation. In *COMMA*, volume 353 of *Frontiers in Artificial Intelligence and Applications*, pages 248–259. IOS Press, 2022.

[Mailly, 2022] Jean-Guy Mailly. Yes, no, maybe, i don't know: Complexity and application of abstract argumentation with incomplete knowledge. *Argument & Computation*, 13(3):291–324, 2022.

[Modgil and Prakken, 2013] Sanjay Modgil and Henry Prakken. A general account of argumentation with preferences. *Artificial Intelligence*, 195:361–397, 2013.

[Özgün, 2017] Aybüke Özgün. *Evidence in epistemic logic: a topological perspective*. PhD thesis, Université de Lorraine, 2017.

[Pacuit, 2017] Eric Pacuit. *Neighborhood semantics for modal logic*. Springer, 2017.

[Pollock, 1987] John L Pollock. Defeasible reasoning. *Cognitive science*, 11(4):481–518, 1987.

[Prakken, 2006] Henry Prakken. Combining sceptical epistemic reasoning with credulous practical reasoning. *COMMA*, 144:311–322, 2006.

[Proietti and Yuste-Ginel, 2020] Carlo Proietti and Antonio Yuste-Ginel. Persuasive argumentation and epistemic attitudes. In Luís Soares Barbosa and

Alexandru Baltag, editors, *Dynamic Logic. New Trends and Applications*, volume 12005 of *LNCS*, pages 104–123. Springer, 2020.

[Proietti and Yuste-Ginel, 2021] Carlo Proietti and Antonio Yuste-Ginel. Dynamic epistemic logics for abstract argumentation. *Synthese*, 199(3):8641–8700, 2021.

[Rahwan and Amgoud, 2006] Iyad Rahwan and Leila Amgoud. An argumentation based approach for practical reasoning. In *Proceedings of the Fifth International Joint Conference on Autonomous Agents and Multiagent Systems*, AAMAS '06, page 347–354, New York, NY, USA, 2006. Association for Computing Machinery.

[Rienstra et al., 2013] Tjitze Rienstra, Matthias Thimm, and Nir Oren. Opponent models with uncertainty for strategic argumentation. In Francesca Rossi, editor, *Twenty-Third International Joint Conference on Artificial Intelligence IJCAI 2013*. AAAI Press, 2013.

[Sakama and Cao Son, 2020] Chiaki Sakama and Tran Cao Son. Epistemic argumentation framework: Theory and computation. *Journal of Artificial Intelligence Research*, 69:1103–1126, 2020.

[Schwarzentruber et al., 2012] François Schwarzentruber, Srdjan Vesic, and Tjitze Rienstra. Building an epistemic logic for argumentation. In Luis Fariñas del Cerro, Andreas Herzig, and Jérôme Mengin, editors, *Logics in Artificial Intelligence*, volume 7519 of *LNCS*, pages 359–371. Springer, 2012.

[Shi et al., 2017] C Shi, S Smets, and FR Velázquez-Quesada. Argument-based belief in topological structures. In J Lang, editor, *Proceedings of the Sixteenth Conference on Theoretical Aspects of Rationality and Knowledge (TARK)*, EPTCS. Open Publishing Association, 2017.

[Shi et al., 2018a] Chenwei Shi, Sonja Smets, and Fernando R. Velázquez-Quesada. Beliefs based on evidence and argumentation. In *Proceedings of WoLLIC 2018*, volume 10944 of *LNCS*, pages 289–306. Springer, 2018.

[Shi et al., 2018b] Chenwei Shi, Sonja Smets, and Fernando R. Velázquez-Quesada. Beliefs supported by binary arguments. *Journal of Applied Non-Classical Logics*, 28(2-3):165–188, 2018.

[Shi et al., 2021] Chenwei Shi, Sonja Smets, and Fernando R Velázquez-Quesada. Logic of justified beliefs based on argumentation. *Erkenntnis*, pages 1–37, 2021.

[Shi, 2018] Chenwei Shi. *Reason to believe*. PhD thesis, University of Amsterdam, 2018.

[Shi, 2021] Chenwei Shi. No false grounds and topology of argumentation. *Journal of Logic and Computation*, 31(4):1079–1101, 2021.

[Simari and Loui, 1992] Guillermo Ricardo Simari and Ronald Prescott Loui.

A mathematical treatment of defeasible reasoning and its implementation. *Artif. Intell.*, 53(2-3):125–157, 1992.

[Sperber, 1997] Dan Sperber. Intuitive and reflective beliefs. *Mind & Language*, 12(1):67–83, 1997.

[Stalnaker, 2006] Robert Stalnaker. On logics of knowledge and belief. *Philosophical Studies: An International Journal for Philosophy in the Analytic Tradition*, 128(1):169–199, 2006.

[Thimm, 2014] Matthias Thimm. Strategic argumentation in multi-agent systems. *KI-Künstliche Intelligenz*, 28(3):159–168, 2014.

[van Benthem et al., 2012] Johan van Benthem, David Fernández-Duque, Eric Pacuit, et al. Evidence logic: A new look at neighborhood structures. volume 9 of *Advances in modal logic*, pages 97–118. College Publications, 2012.

[van Benthem et al., 2014] Johan van Benthem, David Fernández-Duque, and Eric Pacuit. Evidence and plausibility in neighborhood structures. *Annals of Pure and Applied Logic*, 165(1):106–133, 2014.

[van der Torre, 2019] Leendert van der Torre. From classical to non-monotonic deontic logic using aspic. In *Logic, Rationality, and Interaction: 7th International Workshop, LORI 2019, Chongqing, China, October 18–21, 2019, Proceedings*, volume 11813, page 71. Springer Nature, 2019.

[van Ditmarsch et al., 2007] Hans van Ditmarsch, Wiebe van der Hoek, and Barteld Kooi. *Dynamic epistemic logic*. Springer, 2007.

[Von Wright, 1951] Georg Henrik Von Wright. *An essay in modal logic*. Studies in Logic and the Foundations of Mathematics. North-Holland Publishing Company, 1951.

[Wáng and Li, 2021] Yì N Wáng and Xu Li. A logic of knowledge based on abstract arguments. *Journal of Logic and Computation*, 31(8):2004–2027, 2021.

[Williamson, 2002] Timothy Williamson. *Knowledge and its Limits*. Oxford University Press on Demand, 2002.

[Williamson, 2011] Timothy Williamson. 20 knowledge first epistemology. *The Routledge companion to epistemology*, 2011.

[Yu et al., 2021] Liuwen Yu, Dongheng Chen, Lisha Qiao, Yiqi Shen, and Leendert van der Torre. A principle-based analysis of abstract agent argumentation semantics. In *Proceedings of the International Conference on Principles of Knowledge Representation and Reasoning*, volume 18, pages 629–639, 2021.

[Yuste-Ginel and Herzig, 2023] Antonio Yuste-Ginel and Andreas Herzig. Qualitative uncertainty and dynamics of argumentation through dynamic logic. *Journal of Logic and Computation*, 33(2):370–405, 2023.

[Yuste-Ginel, 2022] Antonio Yuste-Ginel. *Arguments to believe and beliefs to argue. Epistemic logics for argumentation and its dynamics*. PhD thesis, 2022.

Part III

Perspectives and future directions

Chapter 13

Thirteen Challenges in Formal and Computational Argumentation

Liuwen Yu
University of Luxembourg, Luxembourg
liuwen.yu@uni.lu

Leendert van der Torre
University of Luxembourg, Luxembourg
leon.vandertorre@uni.lu

Réka Markovich
University of Luxembourg, Luxembourg
reka.markovich@uni.lu

1 Introduction

Argumentation means different things to different people. Even in the two volumes of the Handbook of Formal Argumentation, one can find a range of definitions, some focusing more on the formal aspects, others focusing more on the computational aspects. We believe that the thirteen challenges discussed in this chapter pertain to all these definitions or can be rephrased to adhere to all these definitions. Nevertheless, to clarify some of the issues we discuss, we present in Table 1 the definitions we will use in this chapter.

Moreover, as its title suggests, this chapter is particularly concerned with formal and computational argumentation as discussed in the Handbook of Formal Argumentation, the proceedings of the Computational

Term	Definition
Natural argumentation	Refers to the way humans naturally reason and communicate in everyday language, combining elements of linguistics, philosophy, and rhetoric.
Formal argumentation	A process of representing, managing and (sometimes) resolving conflicts.
Algorithmic argumentation	A step-by-step procedure or set of rules designed to perform a specific task or solve a particular argumentation problem.
Computational argumentation	Refers to the study and implementation of argumentation processes using computational methods. Involves theoretical and practical aspects of how argumentation can be modeled and executed by computers.
Argumentation technology	A computational approach incorporating argumentation reasoning mechanisms with other technologies, e.g., NLP, large language models (LLMs), distributed ledger technologies, etc.

Table 1: Five types of argumentation discussed in this chapter

Models of Argument (COMMA) conferences, the Argument and Computation journal, and the wider artificial intelligence (AI) literature on argumentation. In particular, whereas formal argumentation has developed as a branch of knowledge representation and reasoning, an essential part of AI, it now intersects with numerous disciplines, including natural language processing (NLP), and multiagent systems.

Therefore, when we refer to *argumentation* without further clarification, we mean *formal and computational argumentation*. When we specifically discuss formal argumentation only or computational argumentation only, we will make this explicit. Similarly, when referring to natural argumentation, we will do so explicitly.

To structure our discussion of these challenges, we use the attack-defense paradigm shift brought about by Dung [1995] as a pivotal point. In particular, Table 2 distinguishes between three groups of challenges. The first group is concerned with the background to this paradigm shift, the second group is concerned with the paradigm shift itself, and the third group is concerned with the consequences of this paradigm shift for computational argumentation.

Sections	Observations	Challenges
Section 2. Context of the attack-defense paradigm shift	1. Diversity in reasoning across disciplines	1. Connecting individual and collective reasoning
	2. Diversity of arguments in natural language	2. Understanding and generating arguments
	3. Diversity in modeling the process of argumentation	3. Conceptualizing argumentation
	4. Diversity of formal methods used in formal argumentation	4. Formalizing argumentation
Section 3. Attack-defense paradigm shift	5. Universality of attack	5. Creating argumentation frameworks and semantics
	6. Variety of nonmonotonic logics and game theory solution concepts	6. Representing nonmonotonic logics and solution concepts
	7. Inconsistent knowledge bases	7. Rationality postulates for defining a new logic
	8. Dialogue is based on agents, strategies, and games	8. Generalizing Dung's attack-defense paradigm for dialogue
	9. Balancing is based on support	9. Generalizing Dung's attack-defense paradigm for balancing
Section 4. Computational turn after the paradigm shift	10. Diversity of argumentation	10. Conducting a principle-based analysis of argumentation
	11. Compositional nature of argumentation	11. Designing efficient algorithms for argumentation semantics
	12. Human-level and human-centered argumentation	12. Explaining argumentation
	13. Argumentation technology	13. Integrating argumentation with technologies

Table 2: Thirteen observations and challenges discussed in this chapter

For each challenge, we begin by presenting an "observation". Here, we mean an observation about the above-mentioned literature on argumentation, i.e., the Handbook on Formal Argumentation, the COMMA proceedings, the Argument and Computation journal, and the wider AI literature on argumentation.

Given the wide range of topics discussed in this literature, and the changes taking place due to technological developments such as LLMs, the observations we chose to focus on have had a big influence on the contents of this chapter. Other researchers in argumentation might make different observations and, as a result, would approach this chapter differently. Thus, this chapter reflects our personal interpretation of the literature on argumentation.

We use a diverse array of examples for illustrative purposes from areas such as decision-making, ethical and legal reasoning, and practical reasoning, and these are listed in Table 3. The selected examples cover a wide range of disciplines and issues, illustrating also the breadth of potential application domains for the techniques discussed in this chapter. We reuse examples across different challenges so that we can look at these examples from different angles.

We selected the topics judiciously, leaving out many topics we would have liked to cover but would have made this chapter too long. To provide the reader with some additional information, we complement these challenges with open research questions.

This chapter follows the structure of Table 2 and is organized as follows. Section 2 discusses natural argumentation approaches that were prevalent before the paradigm shift, identifying four key challenges. Section 3 focuses on the paradigm shift itself, acknowledging its contributions while also highlighting the challenges it presents to our community. Section 4 examines the consequences of the paradigm shift, including the transition towards computational AI argumentation, and provides a broader view of the community.

2 Context of the attack-defense paradigm shift

In this section, we describe the diversity of argumentation in the literature to position and appreciate the attack-defense paradigm shift. In Section 2.1, we introduce the diversity of reasoning across disciplines, in Section 2.2 we discuss the diversity in natural argumentation, in Section

Example	Discipline	Challenge Number(s)
Jiminy ethical governor	Machine ethics	1, 2
Fatio dialogue protocol	Speech act theory	2
Dialogue between autonomous robot NS-4 and Spooner	Speech act theory, argumentation schemes	2
Child custody	AI & law	2, 3, 9
Scottish fitness lover, snoring professor	Knowledge representation and reasoning	4, 6
Untidy room	Neuro-symbolic AI	4
Bachelor vs. married	Knowledge representation and reasoning	7
Dialogue between accuser and suspect	AI & law	8
Murder at Facility C	AI & law	12
Intelligent Human-Input-Based Blockchain Oracle (IHiBO)	Computer science, financial markets, AI & law	13

Table 3: Examples, disciplines, and relevant challenges (see Table 2)

2.3 we discuss three argumentation conceptualizations, and in Section 2.4, we discuss the diversity of formal methods used in this area. We discuss the challenges of: connecting individual and collective reasoning; understanding, analyzing, evaluating, and generating arguments; and conceptualizing and formalizing the argumentation process.

2.1 Individual and collective reasoning

In this section, we discuss reasoning, its philosophical and mathematical foundations, and its use across many disciplines. From this perspective, we illustrate in Table 1 our definition of formal argumentation — representing, managing and (sometimes) resolving conflicts —using as an example the six layers of conflict addressed by the Jiminy ethical governor.

Russell and Norvig [2010] identify four schools of thought on AI — machines that: think like humans, act like humans, think rationally, and act rationally. We can interpret these four schools of thought from various perspectives:

Practical reasoning vs. theoretical reasoning
Practical reasoning is oriented towards choosing a course of action on the basis of goals and knowledge of one's own situation, while theoretical reasoning is oriented towards finding reasons for determining that a proposition about the world is true or false [Walton, 1990].

Descriptive reasoning vs. prescriptive reasoning
Descriptive reasoning aims to replicate human intelligence and behavior, while prescriptive reasoning aims to simulate decision-making and prescribe actions that align with ethics and laws.

In all these different kinds of reasoning, there could be individual reasoning and collective reasoning. This brings us to the distinction between intelligent systems and multiagent systems across various disciplines. In the social sciences, the distinction between individual and collective reasoning is called the micro-macro dichotomy [Coleman, 1984]. Another prototypical example is the distinction between classical decision theory based on the expected utility paradigm and classical game theory [Savage, 1972]. Whereas classical decision theory is a kind of optimization problem (concerned with maximizing the agent's expected utility), classical game theory is a kind of equilibrium analysis [Nash Jr, 1950]. This leads to the following challenge.

Challenge 1. *Connecting individual and collective reasoning.*

One of the central topics in reasoning is how to handle conflict, whether it arises among beliefs (logical inconsistency) or choices of actions (practical conflicts). This is relevant both to the reasoning process of an individual agent and to interactions among multiple agents.

Example 2.1 illustrates a conflict from the perspective of a single agent.

Example 2.1 (Car accident dilemma in *I, Robot*)**.** *In the film I, Robot, Detective Del Spooner is driving when he has an accident, plunging his own car and another vehicle carrying a child into a river. An* *autonomous general-purpose humanoid robot, NS-4, is passing by and is faced with a conflict because it cannot save all the humans involved in the accident, i.e., the drivers and the child. NS-4 must make a descriptive analysis of the situation and follow prescriptive actions guided by ethical and legal considerations.*

Example 2.2 illustrates a conflict from a multiagent perspective.

Example 2.2 (Continued from Example 2.1). *Instead of seeing the conflict in terms of saving Spooner or saving the girl, a conflict that is faced by NS-4 only, we can consider it as a disagreement between two stakeholders: NS-4 and Spooner. NS-4 wants to save Spooner, while Spooner wants to save the girl.*

NS-4 and Spooner might reach a consensus through a process known as judgment aggregation where they combine their individual judgments to arrive at a collective decision [Caminada and Pigozzi, 2011]. However, in the context of game theory, the goal is not always to resolve all conflicts but rather to understand the dynamics at play and sometimes *agree to disagree* [Aumann, 2016]. This concept, known as equilibrium analysis, allows for a situation where Spooner and NS-4 recognize their differing priorities and accept the disagreement without necessarily resolving the conflict.

In a multi-stakeholder setting, conflicts can be conceptualized and managed at various layers. The Jiminy architecture [Liao et al., 2023] is an ethical governor that uses techniques from normative systems and formal argumentation to get moral agreements from multiple stakeholders. Each stakeholder has their own normative system. The Jiminy architecture combines norms into arguments, identifies their conflicts as moral dilemmas, and evaluates the arguments to resolve each dilemma whenever possible. In particular cases, Jiminy decides which of the stakeholders' normative systems takes preference over the others.

Example 2.3 (The six layers of conflict of *I, Robot* in the Jiminy architecture).

Layer 1: conclusions only *The conflict is based on the conclusions drawn by each stakeholder. NS-4 concludes it should save Spooner, while Spooner concludes that the girl should be saved. This layer* *represents a straightforward conflict of decisions without going into the underlying reasoning, possibly making it difficult to resolve.*

Layer 2: assumptions and reasons *Agents present their conclusions along with their assumptions and reasons. We refer to these conclusions together with the assumptions and reasons as arguments. Conflict resolution may involve formal argumentation techniques such as assigning attacks among arguments. The reason Spooner wants to save the girl is that she has a longer potential lifespan. That reason could be attacked by an argument from NS-4 that the girl has incurable cancer and therefore has a short lifetime.*

Layer 3: combining normative systems *This layer involves combining multiple normative systems into a single normative system. As a consequence, there may be new arguments built from the norms of distinct stakeholders, and the combined knowledge may be sufficient to reach a moral agreement. For example, NS-4 has information unknown to Spooner — the child's illness. By aligning their knowledge bases, they may reach an agreement to save Spooner instead.*

Layer 4: context-sensitive meta-reasoning as ethical governors *Jiminy considers the agents' norm preferences. These meta-norms are context-dependent norms that select the one stakeholder who has the most relevant expertise. Jiminy may decide that NS-4's preference takes precedence over Spooner's because NS-4 can get a more accurate evaluation of the imminent accident, leading to a more reasonable decision. This mechanism is comparable to those used in private international law [Markovich, 2019].*

Layer 5: suspend decisions and observation *Traditional conflict resolution often assumes that dilemmas must be addressed immediately. However, suspending a decision to allow for additional information to emerge can be beneficial in certain situations.*

Layer 6: dialogue *In this layer, stakeholders engage in a dialogue, attempting to persuade one other. Through structured communication, NS-4 and Spooner present their arguments, counterarguments, and justifications. The dialogue helps them explore each* *other's perspectives and can lead to a more informed and mutually agreeable resolution.*

There are many other examples similar to the one in *I, Robot*. For instance, the *tunnel dilemma* and the *trolley dilemma* [Awad et al., 2018] involve ethical decisions by autonomous vehicles and the question of who should decide how they respond in life-and-death situations. Another example is a smart speaker that passively *listens in* and stores voice recordings, acting like a surveillance device [Liao et al., 2023]. Should it assist in the prevention of or investigation into crimes? This presents a moral dilemma involving household members, law enforcement agencies, and the manufacturer of the smart speaker. Which stakeholder should be alerted in such cases?

Jiminy explains the general problem of connecting individual and collective reasoning, and its relation to practical reasoning. Different mechanisms could be implemented in the Jiminy architecture to connect the two kinds of reasoning. For example, philosophical concepts such as the *veil of ignorance* [Rawls, 2001] and *Kantian imperative* [Kant, 1993] offer valuable perspectives. The veil of ignorance principle requires individuals to make decisions without knowledge of their own personal characteristics or societal position, thus promoting impartiality and fairness in collective decision-making. This aligns closely with the AI challenge of designing a system that makes unbiased decisions. Similarly, Kant's categorical imperative suggests that one should act only according to maxims that can be universally applied to build universally ethical and rational guidelines for behavior. Both principles emphasize the importance of considering the broader implications of actions, and they encourage integrating individual actions into collective norms and ethics.

In this section, we discussed the general challenge of connecting individual and collective behavior from the perspective of argumentation. We end this section by raising a number questions for further research.

1. In this section, we considered argumentation as a kind of reasoning, which raises the question: what kinds of reasoning count as argumentation, and what kinds of argumentation do not count as reasoning? More generally, what is the scope of argumentation?

2. Which kinds of reasoning cannot be handled by argumentation?

For example, can causal or case-based reasoning be cast as a kind of argumentation [Bengel et al., 2024; Roth and Verheij, 2004]? Can decision-making be regarded as a kind of argumentation [Amgoud, 2009]?

3. How does argumentation relate to other kinds of reasoning? For example, What are the distinctions and connections between reasoning as a cognitive activity and argumentation as a communicative practice [Walton, 1990]?

4. How should argumentation be used in a general theory of reasoning? For example, some articles refer to argumentation and negotiation [Sycara, 1990; Rahwan et al., 2003; Van Laar and Krabbe, 2018], examining how argumentation can be used as part of a negotiation, or how a negotiation can be seen as a kind of argumentation process.

5. How should argumentation be applied to legal and ethical reasoning? For example, how can argumentation facilitate structured discourse among agents to negotiate conflicting norms, particularly in multiagent legal proceedings where stakeholders argue for or against specific outcomes [Prakken and Sartor, 2015]?

2.2 Natural arguments

In this section, we discuss the diversity of natural argumentation, psychological analysis of natural (and formal) arguments, transitioning from NLP to foundation models and chatbots, understanding and generating arguments using foundation models, and the central role of questions in natural argumentation. We illustrate the latter using critical questions in argument schemes to find weaknesses and by asking *why* questions to obtain justifications according go the so-called Fatio dialogue protocol.

Natural argumentation refers to the way humans reason and communicate naturally in everyday language. It combines elements of linguistics, philosophy, and rhetoric. It is characterised by considerable diversity, with arguments taking various forms, styles, and contexts. Our aim is to avoid developing a separate technical understanding and generating arguments and argumentation with only a weak connection to how these concepts are understood in the humanities and related fields by both scholars and practitioners [Gordon, 2018].

In linguistics, researchers evaluate the diversity of natural language arguments and their role in human interaction with experimental methodologies [Gillioz and Zufferey, 2020]. These methodologies often include human-based techniques such as psychological experiments to verify linguistic intuitions [Schindler *et al.*, 2020] and determine sound arguments and standards of justification [Weinstock, 2006]. In AI, the focus shifts to modeling, formalizing, and automating the argumentation process, generating arguments in both natural and formal languages. However, evaluation in AI also relies heavily on human assessment. For instance, empirical cognitive experiments have been conducted by Cramer and Guillaume [2018a; 2018b; 2019] and Cerutti et al. [2021] to evaluate the connection between natural and abstract argumentation. Two main research questions often guide this evaluation: *Do the features shared by major argumentation semantics (e.g., simple reinstatement) correspond to genuine cognitive aspects of human reasoning? Which argumentation semantics are the best predictors of human evaluation of arguments?* Human evaluations of automatically generated text are also conducted to assess their performance [van der Lee *et al.*, 2021].

In recent years, the transition from NLP [Budzynska *et al.*, 2018] to argumentation-based chatbots [Black *et al.*, 2021] has been accelerated by advancements in foundation models, such as OpenAI's Generative Pre-trained Transformer (GPT) series. This leads us to the following challenge.

Challenge 2. *Understanding and generating natural arguments.*

Chatbots are conversational software that seek to understand input from human users and generate human-like responses [Black *et al.*, 2021]. In chatbot development, *questions* play a crucial role in enhancing the effectiveness of argumentation-based chatbots and building engaging conversations [McBurney *et al.*, 2021]. The use of *questions* allows chatbots to guide dialogues, challenge assertions, support critical thinking, and provide justifications. Below, we discuss two *question* mechanisms that could be embedded into chatbots — argumentation schemes with their associated critical questions, and justification-seeking questions defined by speech act theory.

Argumentation schemes are investigated by an approach developed in philosophy and rhetoric, representing stereotypical patterns of reasoning that are often informal or semi-formal, rarely fully formalized. Initially developed for teaching critical thinking, these schemes were systematized by Walton *et al.* [2008], who identified sixty-five basic schemes. Argumentation schemes involve the activity of critically evaluating viewpoints and the reasons given to support them. For this reason, every scheme has a corresponding set of critical questions to identify possible weak points, challenge the arguments, and evaluate their strengths. The mechanism of argumentation schemes and critical questions could potentially improve chatbots. Chatbots, or conversational agents like ChatGPT, are good at crafting human-like sentences [Alkaissi and McFarlane, 2023]. But they often present falsehoods as facts and exhibit inconsistent logic, and these can be difficult to detect. Users tend to follow the chatbot's logic when given ready-made answers. However, when chatbots pose questions, they prompt users to engage in deeper critical thinking and question their responses [Danry *et al.*, 2023], which fosters more realistic and reliable interactions.

For example, consider below the argumentation scheme for practical reasoning.

Major premise:	I have a goal G.
Minor premise:	Carrying out action A is a means to realize G.
Conclusion:	Therefore, I ought (practically speaking) to carry out this action A.

There are five critical questions:

CQ1 What other goals do I have that should also be considered even though they might conflict with G?

CQ2 Other than me bringing about A, what alternative actions should be considered that would also bring about G?

CQ3 From the solutions of me bringing about A and these alternative actions, which can be argued to be the most efficient?

CQ4 What grounds are there for arguing that it is possible for me to bring about A in practice?

CQ5 What other consequences of me bringing about A should be taken into account?

Example 2.4 illustrates the dialogue between NS-4 and Spooner based on the argumentation schemes of practical reasoning.

Example 2.4 (Dialogue between NS-4 and Spooner).

Spooner: *Save the girl! That is the moral and ethical choice. She deserves the chance to live her life fully.*

NS-4: *What other goals do you have that might conflict with this one?*

Spooner: *My goal is to save the most vulnerable lives. There is no conflict.*

NS-4: *What alternative actions should be considered?*

Spooner: *Saving the girl should be the only course of action. It should have the highest priority.*

NS-4: *What is the most efficient choice?*

Spooner: *Saving the girl. She is lighter, so this course of action is more likely to succeed.*

NS-4: *What grounds are there for arguing that it is practically possible to save the girl?*

Spooner: *The girl's lighter weight makes her rescue quicker and less risky.*

NS-4: *What consequences should be considered?*

Spooner: *Saving the girl aligns with the duty to protect the vulnerable.*

NS-4: *Your argument is sound and aligns with ethical and practical considerations. I will save the girl.*

Speech act theory, a subfield of pragmatics, studies how words are used not only to present information but also to carry out actions [Austin, 1975]. This theory has been formalized in the Foundation for Intelligent Physical Agents (FIPA) standards, which are widely used in computer science to facilitate communication between autonomous agents [FIPA, 2002]. The scheme allows multiple labels to be applied to one utterance since a single utterance can perform multiple actions in a dialogue [Kissine, 2013]. Such labels range from a few basic types such as assertions, questions and commands to more complex ones like promises and declarations [Searle, 1979]. One of the key features of speech acts, as opposed to physical actions, is that their main effects are on the mental and interactional states of agents [Traum, 1999]. The attitudes of belief, desire and intention are familiar to agency theories [Georgeff et al., 1999]. In the context of human-like chatbots, speech acts can be used to design interactions between the chatbot and the user [Hakim et al., 2019]. Specifically, questions that seek justification are crucial as they prompt the chatbot to provide reasons and explanations, which not only enriches the interaction but also drives the conversation towards deeper engagement and understanding.

McBurney and Parsons [2004] proposed an interaction protocol called Fatio comprising of five locutions for argumentation which can be considered as a set of speech acts.

F1: assert (P_i, ϕ): A speaker P_i asserts a statement ϕ. In doing so, P_i creates a dialectical obligation within the dialogue to provide a justification for ϕ if required subsequently by another participant.

F2: question (P_j, P_i, ϕ): A speaker P_j questions a prior utterance of assert(P_i, ϕ) by another participant P_i and seeks a justification for ϕ. The questioner P_j creates no dialectical obligations.

F3: challenge (P_j, P_i, ϕ): A speaker P_j challenges a prior utterance of assert(P_i, ϕ) by another participant P_i and seeks a justification for ϕ. P_j not only asks a question but also creates for himself a dialectical obligation to provide a justification for not asserting ϕ. For example, he must provide an argument against ϕ if questioned or challenged. Thus, challenge (P_j, P_i, ϕ) is stronger than question (P_j, P_i, ϕ).

F4: justify $(P_i, \Phi \vdash \phi)$: A speaker P_i, who had uttered **assert**(P_i, ϕ) and was then questioned or challenged by another speaker, is able to provide a justification $\Phi \in A$ for the initial statement ϕ by means of this locution.

F5: retract (P_i, ϕ): A speaker P_i, who had uttered **assert**(P_i, ϕ) or **justify**$(P_i, \Phi \vdash \phi)$, can withdraw this statement with the utterance of **retract**(P_i, ϕ) or the utterance of **retract**$(P_i, \Phi \vdash \phi)$ respectively. This removes the earlier dialectical obligation on P_i to justify ϕ or $\Phi \vdash \phi$ if questioned or challenged.

Example 2.5 illustrates the dialogue between NS-4 and Spooner following the speech act.

Example 2.5 (A dialogue between NS-4 and Del Spooner).

Spooner: *Saving the girl is the right choice. (assert)*

NS-4: *Why? (question)*

Spooner: *Because the girl is young and has a much longer lifespan. (justify)*

NS-4: *I disagree that she has a much longer lifespan. (challenge)*

Spooner: *Why? (question)*

NS-4: *I conducted a health evaluation and found that she has a terminal disease. (justify)*

In this section, we discussed the common understanding of natural arguments and how they are generated. We end with some open research questions.

1. In natural argumentation, we often encounter fallacies [Hamblin, 1970], and on the internet we encounter fake news [Visser et al., 2020]. How should fallacies and fake news be handled in argumentation-based chatbots? For example, how can argumentation schemes be used in a chatbot to evaluate if an argument is fallacious [Walton, 2013]?

2. Programming has been replaced by prompt engineering for interacting with LLMs [Ross et al., 2023]. How can argument schemes and speech act theory be used in prompt engineering? For example, argument schemes can potentially help structure for the generation of well-formed arguments [Musi and Palmieri, 2022].

3. An increasing number of arguments on the internet are generated by AI [Hinton and Wagemans, 2023]. How should AI-generated arguments be evaluated? The evaluation could focus on criteria like logical coherence, the relevance and sufficiency of the evidence, adherence to ethical principles, and the impact of the argument on the intended audience. Additionally, metrics could be developed to assess how well AI arguments handle counterarguments and whether they respect the norms of constructive and respectful discourse.

4. Argumentation has been discussed as a key component of chatbots [Castagna et al., 2024a]. Which domain applications are argumentation-based chatbots suitable for? For example, how can arguments be used in AI therapy?

5. In the previous section, we introduced the Jiminy architecture. How can foundation models, argument schemes, and speech act theory be used to improve or enrich the Jiminy architecture?

2.3 Models of argument

In this section, we discuss three conceptualizations of argumentation — argumentation as inference, argumentation as dialogue, and argumentation as balancing. Each conceptualization embodies: a unique perspective on the construction and purpose of argumentation, a set of formal methods, and application across different disciplines and contexts. As mentioned in the introduction, we view argumentation as representing, managing and sometimes resolving conflicts. We explain how this key idea becomes more concrete with the three conceptualizations.

For argumentation as inference [Prakken, 2018], we consider: coherent positions in cases of conflict, what follows from each coherent position (or what we can infer from all coherent positions), and what can be agreed upon in cases of disagreement. For argumentation as dialogue [Prakken, 2018], we also consider the stakeholders that may hold such coherent positions and how they might interact, for example as proponents and opponents in a debate. This can clarify the conflict that is being managed and sometimes even help to resolve it. In such dialogues, the concerns or goals of the stakeholders can also be made explicit. As in dispute resolution, the process becomes very important. Finally, in argumentation as balancing [Gordon, 2018], we consider conflicts as trade-offs involving taking into account various pros and cons. Here, the central metaphor, referred to frequently in the law and ethical decision-making, is a pair of scales. For fine-grained comparisons, it is not uncommon to use quantitative metrics. In this section, we discuss the following challenge.

Challenge 3. *Conceptualizing argumentation.*

Table 4 provides a detailed comparison of the three conceptualizations. Notably, the applications of these conceptualizations are neither mutually exclusive nor incompatible. The formal methods are discussed in Section 2.4.

Argumentation as inference focuses on determining the conclusions that can be derived from a given body of information, which may be incomplete, inconsistent, or uncertain. Relevant systems ultimately define a nonmonotonic notion of logical consequence in terms of the intermediate notions of argument construction, argument attack, and argument evaluation, and the arguments are seen as constellations of premises, conclusions, and inferences [Prakken, 2018]. These systems employ formal methods like nonmonotonic logic for commonsense reasoning, graph theory, computational logic, causal reasoning, and Bayesian reasoning. Argumentation as inference is primarily applied in knowledge representation and reasoning.

Argumentation as dialogue conceptualizes argumentation as a form of verbal interaction aimed at resolving conflicts of opinion [Prakken, 2018]. Relevant systems define argumentation protocols, which serve as the rules of the argumentation game, and

Conceptualization	Process	Theories and Formal Approaches	Application
Argumentation as inference	Logical structure and reasoning to derive conclusions from incomplete and inconsistent premises	Graph theory, nonmonotonic logic, computational logic, causal reasoning, Bayesian reasoning	Automated reasoning systems, knowledge representation, expert systems
Argumentation as dialogue	Dynamic verbal interaction between stakeholders to exchange information or resolve conflicts of opinion	Speech act theory, game theory, axiomatic semantics, operational semantics	Debating technologies, chatbots, recommender systems
Argumentation as balancing	Balancing pros and cons to reach a justified decision	Multi-criteria decision theory, machine ethics, computational law, case-based reasoning	Deliberative decision-making in law, ethics, and economics

Table 4: Conceptualizations of argumentation

they address strategic aspects that guide effective engagement in the game. The exploration of strategies involves understanding how to engage in productive discourse and present arguments effectively. Argumentation as dialogue utilizes speech act theory, game theory, axiomatic semantics, and operational semantics. It is most suitable for debates, chatbots, persuasion systems, negotiation systems, etc.

Argumentation as balancing involves weighing the pros and cons of an issue in order to reach a balanced decision or judgment. It is applicable not only when resolving conflicts of opinion in persuasion dialogues but also, e.g., when deciding courses of action in deliberation dialogues [Gordon, 2018]. In such a system, pro and con arguments for alternative resolutions of the issues (options or positions) are put forward, evaluated, resolved, and balanced. The formal methods used are multi-criteria decision theory, machine ethics, computational law, and case-based reasoning, and they are applied in the realms of law, ethics, and economics.

Table 4 might give the impression that the three approaches are distinct and that they have distinct application areas. We would like to point out that this is not the case. The approaches (or types) of argumentation are not mutually exclusive or even incompatible. You can switch from one to another if you want to look at the same problem or situation from different angles, highlight different aspects, or select a modeling approach that is more suitable for a particular purpose. This also means that complex application areas like the legal domain can make very good use of each approach. Indeed, legal reasoning often engages with each of the three conceptualizations — argumentation as inference, dialogue, and balancing — across different contexts and legal roles. Judges and attorneys may rely on one form of argumentation more than another, depending on the nature of the case and their specific role in the legal process. For instance, inference is commonly used by judges, attorneys, actually any type of lawyer, when applying legal rules to facts or deriving conclusions from incomplete or inconsistent premises. Dialogue plays a central role in courtroom exchanges between opposing parties. The structure of a trial often resembles a dialogue: each party presents their arguments and responds to those of the other while the judge oversees the process to ensure it follows legal procedures. Balancing is typically the domain of judges as they weigh multiple factors, conflicting interests, or values, to determine the most appropriate outcome. This is particularly important in discretionary decision-making where the law, instead of trying to provide detailed rules, assigns special power to judges so that they can make decisions based on their own evaluations. In such cases, judges exercise their judicial discretion by carefully balancing competing considerations within the framework of legal principles to reach a fair and just decision.

Hence, these different modes of reasoning can correspond to and interact with one another, creating a comprehensive tool set for legal reasoning and decision-making. Below, we shall illustrate each of the three conceptualizations using the legal example of child custody in a divorce case.

Research on argumentation-based dialogue (see the overview of Black et al. [2021]) is often carried out against the background of the six types of dialogue and in accordance with their respective goals [Walton and Krabbe, 1995], as shown in Table 5. When argumentation is viewed as a kind of dialogue between multiple agents (whether human or artificial), new issues arise. One issue is the distributed nature of information (among the agents). Another issue is the dynamic nature of information — agents do not reveal everything they believe initially, and they can learn from one other. There are also strategic issues — agents will have their own internal preferences, desires and goals [Prakken, 2018]. In Section 2.2, we described the speech act theory on dialogue formation [McBurney and Parsons, 2002]. For better comparison, we use a legal child custody case [Yu et al., 2020] to illustrate the three conceptualizations.

Type of dialogue	Initial situation	Participant goal	Dialogue goal
Persuasion	Conflict of opinions	Persuade other party	Resolve or clarify issue
Inquiry	Need to have proof	Find and verify evidence	Prove (or disprove) hypothesis
Negotiation	Conflict of interests	Get what they most want	Reasonable settlement they can both live with
Information-seeking	Need information	Acquire or give information	Exchange information
Deliberation	Dilemma or practical choice	Co-ordinate goals and actions	Decide best available course of action
Eristic	Personal conflict	Verbally hit out at opponent	Reveal deeper basis of conflict

Table 5: Types of dialogue [Walton and Krabbe, 1995]

Example 2.6 (Child custody dialogue). *Alice and Lucy are talking about a divorce case, specifically whether it is in the child's best interest to live with her mother or with her father. They have the following dialogue.*

Alice: *It is in the ten-year-old child's best interest that she lives with her mother. (assert)*

Lucy: *Why? (question)*

Alice: *Because the child wants to live with her mother and the civil code states that the judge must take the child's opinion into account. (justify)*

Lucy: *A ten-year-old child does not know what she wants. (challenge)*

Alice: *Why? (question)*

Lucy: *Public opinion says that ten-year-old children do not know what they want. (justify)*

Alice: *Most ten-year-old children do know what they want. (assert)*

Lucy: *Why do you say that? (question)*

Alice: *Peter is a child psychologist, and Peter says that most ten-year-old children know what they want. (justify)*

Most of the literature in this area is concerned with argumentation as inference. Some formal work had already had been carried out on argumentation-based inference before the publication of Dung's 1995 paper, notably the extensive research by Pollock [1987; 1992; 1994; 1995; 2001; 2009; 2010] on argument structure, the nature of defeasible reasons, the interplay between deductive and defeasible reasons, rebutting versus undercutting defeat, argument strength, argument labeling, self-defeating arguments, etc. Pollock identified reasoning as a process of constructing arguments where reasons provide the atomic links in arguments [Pollock, 1992]. He distinguished between two kinds of reasons: defeasible (prima facie) reasons and nondefeasible (conclusive) reasons [Pollock, 1987]. Nondefeasible reasons are those reasons that logically entail their conclusions while defeasible reasons may be destroyed with additional information. There are two kinds of defeaters that can defeat defeasible reasons. Rebutting defeaters deny the conclusion. Undercutting defeaters attack the connection between the reason and the conclusion. Pollock [1992; 1994; 1995] used so-called inference graphs to represent arguments and the nodes represented the steps of inference. There are three kinds of arrows in the inference graph, and they represent defeasible inferences, deductive inferences, and defeat links [Pollock, 1994].

Example 2.7 (Child custody in an inference graph). *The dialogue between Alice and Lucy can also be illustrated in the format of Pollock's inference graph, as shown in Figure 1. Figure 1 illustrates two arguments rebutting the two opposite conclusions "It is in the child's best interest that she lives with her father", and "It is not in the child's best interest that she lives with her father". An undercutting argument, "Public opinion is not reliable", defeats the argument "Most ten-year-old children do not know what they want". In this figure, nondefeasible and defeasible inferences are visualized respectively with solid and dotted lines (without arrowheads). The arrows are defeat relations.*

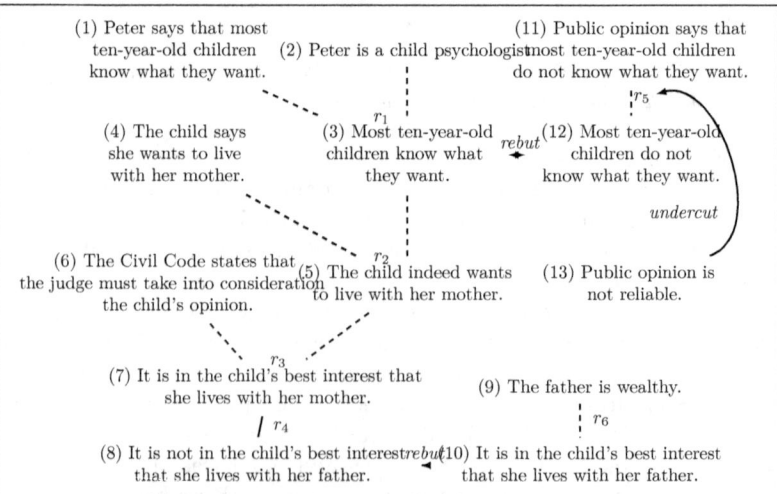

Figure 1: A dialogue about a child custody case represented as a Pollock inference graph. The solid and dotted lines (without arrowheads) are nondefeasible and defeasible inferences respectively. The arrows are defeat relations.

One model of argumentation as balancing is the Carneades Argumentation System [Gordon et al., 2007]. The conception of argument graphs in Carneades is similar to Pollock's conception of an inference graph. There are nodes in the graph representing statements (propositions) and links that indicate inference relations between statements. In particular, the system distinguishes between pro and con arguments. Semantically, con arguments are instances of presumptive inference rules for negating the conclusion. If the premises of a con argument hold, this justifies rejecting the conclusion or, equivalently, accepting its logical complement. With pro and con arguments, some statements need to be ordered or otherwise aggregated to resolve the conflict. Then there are several proof standards used to balance the pros and cons. Here are three examples:

SE (Scintilla of Evidence): A statement meets this standard iff it is supported by at least one defensible pro argument.

BA (Best Argument): A statement meets this standard iff it is supported by some defensible pro argument with priority over all defensible con arguments.

DV (Dialectical Validity): A statement meets this standard iff it is supported by at least one defensible pro argument and none of its con arguments are defensible.

Example 2.8 (Child custody in Carneades). *We represent part of the child custody example in Carnedes, as visualized in Figure 2. Statements are depicted as boxes and arguments as circles. For the purpose of this discussion, we assume that all the premises are ordinary without distinguishing between different types of premises. Premises are shown as edges without arrowheads. Pro arguments are indicated by circle arrowheads while con arguments are shown with standard arrowheads. Argument a_1 asserts that the child knows what she wants and she wants to live with her mother, making it a pro argument for the statement "It is in the child's best interest that she lives with her mother". In contrast, argument a_2 argues that the mother is less wealthy than the father, serving as a con argument against that statement. In this scenario, a_1 is given priority over a_2. Consequently, according to the BA proof stan-*

dard, the statement "It is in the child's best interest that she lives with her mother" is accepted.

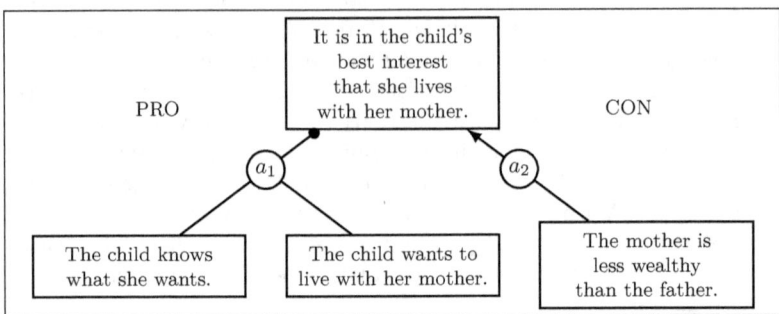

Figure 2: Child custody case represented in Carneades argument graphs

Argumentation as inference, argumentation as dialogue, and argumentation as balancing are distinct conceptualizations but they are not incompatible. A crucial question is how to move between individual reasoning (argumentation as inference and argumentation as balancing) and collective reasoning (argumentation as dialogue). Argumentation as inference can occur within an individual's mind, drawing upon a single knowledge base. However, in multiagent dialogues, each agent operates with a distinct and dynamic knowledge base. Agents employ strategic moves requiring them to learn about and understand other participants and to select or generate arguments from their knowledge base to achieve the goal of the dialogue. Some works attempt to integrate these different conceptualizations. For instance, discussion games serve as a proof procedure for abstract argumentation semantics [Caminada, 2018a], and multiagent argumentation considers agents with varying attitudes and knowledge [Arisaka et al., 2022]. Moreover, several new agent argumentation semantics are inspired by social choice theory [Yu et al., 2021; Baumeister et al., 2021], and Carneades models legal dialogue by using critical questions from argumentation schemes while incorporating balancing to model judgments [Gordon et al., 2007].

In this section, we discussed argument models. We end again with some open questions:

1. Are there any other conceptualizations of argumentation that should be considered in the argumentation literature?

2. How should it be decided which conceptualization of argumentation to use for an application? In particular, when should argumentation as inference, dialogue, or balancing be used?

3. How are these conceptualizations related and should they be combined? For example, how should agent interaction and dialogue [Arisaka et al., 2022] be introduced into Pollock's theory?

4. How should a general framework for argumentation as dialogue be designed? The formal study of argumentation-based dialogue is less substantial than the formal study of argumentation-based inference. It largely consists of a variety of different approaches and individual systems, with few unifying accounts or general frameworks [Prakken, 2018].

5. How should argumentation as balancing be represented formally? Compared with argumentation as inference and dialogue, there is little formal work on argumentation as balancing. Some examples are bipolar argumentation [Cayrol and Lagasquie-Schiex, 2013; Cayrol et al., 2021; Yu et al., 2020; Yu et al., 2023], and an investigation into balancing operations [Knoks and van der Torre, 2023; Knocks et al., 2024].

2.4 Formalizing argumentation

In this section, we discuss the large variety of formal methods in formal argumentation and their combination in applications and case studies. We show the use of nonmonotonic logic for commonsense reasoning, the integration of axiomatic and operational semantics in the Fatio dialogue system, and a combination of different reasoning methods with an example of a mother reasoning about her daughter.

Each conceptualization of argumentation comes with its own set of formal methods, as shown in Table 4 in Section 2.3. Basically, argumentation as inference uses most of the methods from graph theory (e.g., abstract argumentation [Dung, 1995]) or methods from nonmonotonic and computational logic (e.g., [Pollock, 1987; Nute, 1994; Reiter, 1980]) or causal reasoning [Giunchiglia *et al.*, 2004; Turner, 2004]. Argumentation as dialogue involves speech act theory (e.g., [FIPA, 2002]), axiomatic semantics, operational semantics, and denotational semantics, as well as game theory methods (e.g., [Moore, 1993]). In argumentation as balancing, there are methods from multi-criteria decision theory (e.g., [Amgoud, 2005; Amgoud and Vesic, 2012; Knoks and van der Torre, 2023; Knocks *et al.*, 2024]), ethical theory, legal theory, and case-based reasoning (e.g., [Yu and Gabbay, 2022]). In this section, we discuss that challenge.

Challenge 4. *Formalizing argumentation.*

Nonmonotonic logic was motivated by the fact that commonsense reasoning often involves incomplete or inconsistent information, in which case logical deduction is not a particularly useful reasoning model [Toulmin, 1958]. Classical logic is characterized by its monotonic nature. It asserts that if a set of statements S entails a proposition ϕ (denoted $S \vdash \phi$), any superset S' of S also entails ϕ. This principle underpins traditional logical proofs where lemmas remain valid while new premises are added. However, commonsense reasoning often allows conclusions to be retracted in the light of new information. For instance, the inference "Tweety, a bird, flies" (symbolized as $\mid\sim$) may be valid until additional contextual information like "Tweety is a penguin" necessitates the retraction of the initial conclusion. This approach mirrors our everyday reasoning processes, which frequently involve default rules with exceptions (e.g., $a \rightarrow x$ for strict rules and $a \Rightarrow x$ for default rules). Default rules apply unless there is evidence to the contrary requiring us to revoke our conclusions upon encountering such exceptions.

Here are some examples of the sources of nonmonotonicity [Prakken, 2010b]:

Empirical generalizations: e.g., adults are usually employed, birds can typically fly, etc.

Exceptions to legal rules: e.g., when a father dies, the child inherits, *except* when the child killed the father.

Exceptions to moral principles: e.g., one should not lie, *except* when a lie can save lives.

Conflicting information sources: experts who disagree, witnesses who contradict each other, conflicting sensory input, etc.

Alternative explanations: e.g., the grass is wet so it must have rained, but the sprinkler was on.

Conflicting reasons for actions: if we have a reason to do something, we should do it, unless we have good reasons for not doing it.

Prioritized default logic (PDL) [Brewka, 1994] is one formalization of nonmonotonic reasoning. A knowledge base in PDL contains prioritized defaults $a \overset{n}{\Rightarrow} b$ and facts, including tautologies. The notation $a \overset{n}{\Rightarrow} b$ means "if a, then normally b", with n indicating the priority level; a higher n implies a higher priority for the default rule $a \Rightarrow b$. PDL operates by selecting sets of defaults and bringing their conclusions into extensions. At each step, the default rule with the highest priority among the unapplied default rules is applied, while consistency is maintained.

Example 2.9 (Fitness lover Scot). *Assume we have the following defaults and facts:*

$$\begin{aligned}
\textit{Defaults:} \quad & BornInScotland \overset{1}{\Rightarrow} Scottish \\
& Scottish \overset{3}{\Rightarrow} LikesWhisky \\
& FitnessLover \overset{2}{\Rightarrow} \neg LikesWhisky \\
\textit{Facts:} \quad & \{BornInScotland, FitnessLover\}
\end{aligned}$$

We can obtain the extension iteratively.

$$\begin{aligned}
E_1 &= \{BornInScotland, FitnessLover\} \\
E_2 &= \{BornInScotland, FitnessLover, \neg LikesWhisky\} \\
E_3 &= \{BornInScotland, FitnessLover, \neg LikesWhisky, Scottish\} \\
E_4 &= E_3
\end{aligned}$$

We introduced the speech acts of Fatio [McBurney and Parsons, 2002] in Section 2.2, and we showed how reasons are used in a dialogue by 'question' and 'justify' moves in Section 2.3. We now reference Fatio to show how a dialogue system can make use of various formal methods: in this case, *axiomatic semantics* and *operational semantics*.

An axiomatic semantics for a programming language defines a set of axioms that the language obeys such as the pre-conditions and post-conditions for each command [Tennent, 1991]. It defines pre-conditions and post-conditions for the locutions. In Fatio, the axiomatic semantics concerns the beliefs and desires of the participating agents, which are written as $B_i \varphi$: "Agent i believes that φ is true", and $D_i \varphi$: "Agent i desires that φ be true". Central to the axiomatic semantics is a publicly viewable store to record the dialectical obligations of the participants, which is called a dialectical obligations store (DOS). The triple $(P_i, \varphi, +) \in \text{DOS}(P_i)$ denotes that participant P_i has a dialectical obligation to provide a justification or an argument in support of proposition φ, while the triple $(P_i, \varphi, -) \in \text{DOS}(P_i)$ denotes that participant P_i has a dialectical obligation to provide a justification or an argument against proposition φ.

For illustration, we list the pre- and post-conditions for *assert* and *question*, and we refer the rest of the axiomatic semantics to the original paper [McBurney and Parsons, 2004].

assert(P_i, φ) *Pre-condition*: Speaker P_i wants each participant $P_j (j \neq i)$ to believe that P_i believes the proposition $\varphi \in C$.

$$((P_i, \varphi, +) \notin \text{DOS}(P_i)) \wedge (\forall j \neq i)(D_i B_j B_i \varphi).$$

Post-condition: Each participant $P_k (k \neq i)$ believes that participant P_i wants each participant $P_j (j \neq i)$ to believe that P_i believes φ.

$$((P_i, \varphi, +) \in \text{DOS}(P_i)) \wedge (\forall k \neq i)(\forall j \neq i)(B_k D_i B_j B_i \varphi).$$

Dialectical obligation: $(P_i, \varphi, +)$ is added to $\text{DOS}(P_i)$, the dialectical obligations store of speaker P_i.

question(P_j, P_i, φ) *Pre-condition*: One participant $P_i (i \neq j)$ has a dialectical obligation to support φ and participant P_j wants every other participant $P_k (k \neq j)$ to believe that P_j wants P_i to utters a *justify*$(P_i, \varphi, .)$ locution.

Post-condition: Participant P_i must utter a *justify* locution.

Dialectical obligation: No effect.

Operational semantics in Fatio is defined from a traditional computer science perspective. That means that the state of the system changes as a result of executions of commands in a programming language. To ensure automated generation of agent dialogues, participants need mechanisms to invoke specific utterances at appropriate points in the dialogue, and these mechanisms are called *agent decision mechanisms*. In this case, the commands in question are the locutions in an argumentation dialogue conducted according to the rules of the protocol. In Fatio, for example, an agent can decide whether to **Claim or Not**, whether to **React or Not**, whether to **Fold or Not**, whether to **Defend or Not** and, as a meta-level decision mechanism, whether to **Listen or Do**. There are transition rules defined for Fatio's operational semantics, and they assume that agents are equipped with decision mechanisms to initiate and respond to utterances. This enables the system to initiate utterances and respond to utterances in the dialogue, and so the states we will take to be the inputs and outputs of these decision mechanisms reflect that process.

One possible extension to the Fatio protocol is an additional semantics called *denotational semantics*, described but not explicitly defined by McBurney and Parsons [2004]. *It would link the utterances made under the protocol to the nodes and edges of a graph representing the arguments created by the participants in the course of a dialogue. This kind of graph would be similar to the argumentation graph constructed in Thomas Gordon's [1993] Pleadings Game, which is a formal structure capturing the flow and relationships between different arguments in a dialogue or argumentation context.* It would thus provide a mathematical structure to the dialogues, mapping the linguistic constructs (utterances) to a formal representation (graph) that captures the logical relationships and the dynamics of the argumentation process. This could be a way to visualize and analyze the structure of the arguments and the interplay between different participants' statements and responses in a dialogue.

Lastly, we give another example where various formal methods are combined. Neurosymbolic AI [Garcez et al., 2008] combines neural and symbolic AI architectures to address the weaknesses of each, providing a robust AI system capable of reasoning, learning, and cognitive modeling. This diversity of formal methods brings many challenges to the area of formal argumentation. Consider Example 2.10 from Gabbay and Rivlin [2017].

Example 2.10 (Untidy room [Gabbay and Rivlin, 2017]). *A mother goes into her teenage daughter's bedroom. Her instant impression is that it is a big mess. There is stuff scattered everywhere. The mother's feeling is that it is not like her daughter to be like this. What happened?*
   Conjecture: *The girl may be experiencing boyfriend issues.*
   Further Analysis: *The mother notices a collapsed shelf and realizes that the disarray is due to the shelf collapsing under excessive weight which, upon reflection, follows a logical (gravitational) pattern.*
   Several types of reasoning are illustrated through this scenario:

Neural network reasoning: *The mess is perceived instantly, similarly to facial recognition by neural networks.*

Nonmonotonic deduction: *The mother deduces from the context and her knowledge that her daughter does not typically live in disarray. Thus, something extraordinary must have happened.*

Abductive reasoning: *She hypothesizes a plausible explanation that her daughter has social-emotional issues, which is common among teenagers.*

Database AI deduction: *A reevaluation leads to the understanding that the mess is due to gravitational effects rather than disorganization on the part of her daughter.*

Pattern recognition: *Someone accustomed to similar patterns may identify the cause as easily as they might recognize a face.*

In practical reasoning, it is crucial to combine various formal methods. To deal with scenarios similar to Example 2.10, D'Avila Garcez *et al.* [2005] proposed a hybrid model of computation that allows for deduction and learning with argumentative reasoning. The model manages to combine value-based argumentation frameworks and neural-symbolic learning systems by providing a translation from argumentation networks to neural networks. Another example is the general argumentation framework presented by Williamson and Gabbay [2005]. The framework incorporates the idea of recursive causality and extends the Bayesian network formalism to cope with recursive causality. The authors discussed how support relations behave analogously to causal relations and how arguments are recursive structures; these two observations motivate the use of recursive Bayesian networks for modeling arguments.

In this section, we discussed the formalization of argumentation. As usual, we end with some open questions.

1. We have observed that every conceptualization comes with their own formal methods. What else do they depend on? For example, which formal method(s) should be chosen for a case study or application?

2. Example 2.10 also illustrated that we often need to combine reasoning and formal methods. How can various formal methods be combined in a case study? For example, how can symbolic logic be combined with network (neural and argumentation) reasoning?

3. Formal argumentation is often presented as a general way to deal with nonmonotonicity. But how should arguments be conducted when there are various sources of nonmonotonicity? For example, how should we argue in legal or ethical contexts?

4. Concepts relevant for argumentation currently include, among others: time, action, knowledge, belief, revision, deduction, learning, context, neural networks, probabilistic networks, argumentation networks, consistency, etc. How can these concepts be incorporated into existing formal models of argumentation?

3 The paradigm shift: the attack-defense perspective

In this section, we discuss and critically reflect upon the attack-defense paradigm shift. In section 3.1, we discuss the universality of attack and defense. In section 3.2, we consider the variety of nonmonotonic logics and game theoretic solution concepts. In section 3.3, we discuss reasoning with inconsistent knowledge bases. In section 3.4, we consider argumentation as dialogue that is based on other concepts besides attack, like agents, strategies, and games. And in section 3.5, we discuss Dung's attack-defense paradigm shift for balancing that is based on both attack and support. We discuss the challenges of: creating argumentation frameworks and semantics, representing nonmonotonic logics and game theoretic concepts, defining rationality postulates for new logics, generalizing Dung's attack-defense paradigm shift for dialogue, and generalizing Dung's attack-defense paradigm shift for balancing.

3.1 Universality of attack

In this section, we introduce the attack-defense paradigm shift initiated by Dung's [1995] paper, we discuss the requirement that every utterance can be attacked including claims, arguments, and attacks, and we describe the flattening of diverse extended argumentation frameworks into basic ones.

The attack-defense paradigm shift was a turning point in modern formal argumentation, marked by Dung's theory of abstract argumentation [Dung, 1995]. In this theory, the acceptability of arguments depends on the attack relations between them, not their internal structure. An argument is accepted if it is not attacked or is successfully defended — meaning all its attackers are attacked. Pre-existing ideas and methods, such as Pollock's defeasible reasoning, dialogue theories, and balancing techniques, continue to persist and influence contemporary research. Rather than being rendered obsolete, these traditional theories are reinterpreted within the context of this new paradigm.

While the central notion of Dung's theory is the acceptability or non-acceptability of arguments based on attack and defense, Dung shows that nonmonotonic logic is a special form of argumentation (more details in Section 3.2). It can be visualized as the commutative diagram in Figure 3. There are two approaches to deriving conclusions from a knowledge base. The first is a direct approach where a given logic selects a set of rules with conclusions. The other is an indirect approach through argumentation, as shown in Figure 3 (2—4). Structured argumentation studies the process that adds the structure that turns collections of rules into arguments and assigns attack relations (2) among arguments. This gives us abstract argumentation frameworks — directed graphs where nodes represent arguments, and arrows represent attack relations. Then argumentation semantics (3) determine the acceptance status of arguments and their conclusions. To represent a given logic by structured argumentation, eventually the conclusions from both approaches must be the same.

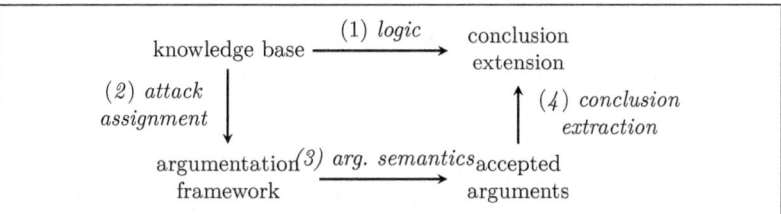

Figure 3: Commutative diagram: two approaches to nonmonotonic inference: (1) logic systems; (2)–(4) argumentation systems. With appropriate choices on elements (2) and (3), one can obtain exactly the same conclusions as for the given logic (1).

We now present informally the construction of arguments and attack relations from a knowledge base in the ASPIC+ structured argumentation frameworks [Prakken, 2010a]. A knowledge base typically consists of a set of strict rules (with a simple arrow $\rightarrow$) and a finite set of defeasible rules (using a double-lined arrow $\Rightarrow$), where each defeasible rule is assigned a priority number, denoted as $a \stackrel{n}{\Rightarrow} b$. The knowledge base also includes a base of evidence (BE). An argument can be constructed as follows:

1. For each element $\alpha \in BE$, the expression $[\alpha]$ constitutes an argument having the conclusion α.

2. Let r be a rule of the form $\alpha_1, \ldots, \alpha_n \to / \Rightarrow \alpha$, where $A_1, \ldots, A_n$ are arguments with conclusions α_i (for $1 \leq i \leq n$). In this case, the expression $[A_1, \ldots, A_n \to / \Rightarrow \alpha]$ is regarded as an argument with conclusion α.

Each argument is derived by applying the steps above (1 and 2) finitely many times to ensure a structured process for argumentation within the framework.

We now use Example 3.1 to illustrate the commutative diagram, and we explain the technical details later in section 3.2.

Example 3.1 (Two approaches to nonmonotonic reasoning). *Consider a knowledge base containing three defeasible rules $\overset{n}{\Rightarrow}$ as well as facts ($\{\top\}$), as in Figure 4(a). Logical approaches to defeasible reasoning select a subset of rules whose conclusions are maximally consistent. For example, PDL [Brewka and Eiter, 1999], discussed in Section 2.4, selects the strongest applicable rules, i.e., the order $(i) \to (ii)$ in Figure 4(a), with output $\{a, \neg b\}$. While (iii) is now made applicable by a, its consequent b conflicts with $\neg b$ and cannot be selected. Argumentation approaches, in turn, build explicit arguments (Figure 4(b)) and represent these conflicts $(b, \neg b)$ as attacks between arguments (B and C). Observe how the arguments in Figure 4(b) activate the rules in Figure 4(a). To specifically capture PDL, one needs a selection of attacks (discussed in Section 3.2), such as the attack induced by the* weakest link *in Figure 4(c), which defines that the strength of an argument reflects its weakest rules. Intuitively, the jointly acceptable arguments here are $\{A, C\}$, which corresponds to the PDL extension $\{a, \neg b\}$. But in Figure 4(c), the* last link*, which defines that the strength of an argument is that of its last rule, selects the arguments $\{A, B\}$ with output $\{a, b\}$.*

While nonmonotonic logic provides a structured framework for managing conflicts and reasoning through premises, inferences, and conclusions, it becomes difficult to capture complex conflicts among a large number of arguments. Here, abstract argumentation frameworks can provide landscapes of how these arguments relate to one another via attack relations. One of the main goals of abstract argumentation theory is to answer the question: which sets of arguments can be reasonably accepted in a discussion based on a given argumentation framework? In

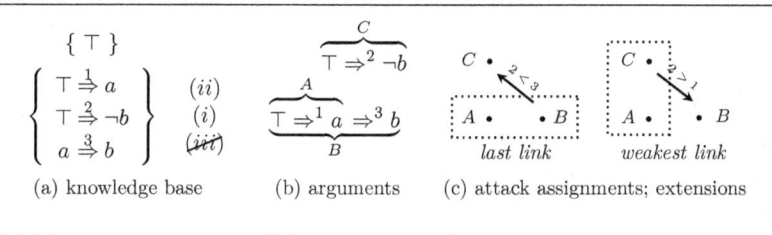

Figure 4: The PDL approach to nonmonotonic reasoning. (a) PDL iteratively selects the strongest active rule up to the point of inconsistency. (b) Arguments build upon facts (A, C) or other arguments (B). (c) Argumentation semantics abstract from a logical structure. Attacks depend on argument strength.

the simple argumentation frameworks of Figure 4(c), sets of arguments could be selected by intuition. But a formal method is needed for a more complex graph, for instance where attacks among arguments can form even or odd cycles, which may be part of more complex structures like strongly connected components (SCCs). In abstract argumentation, argumentation semantics provides a way to deal with these complications.

We use graph labeling based on so-called *gunfight rules* [Caminada, 2006; Caminada and Gabbay, 2009] to determine which arguments can survive in conflicts. The concept is straightforward: in a gunfight, one stays alive iff all attackers are dead, and one dies iff at least one attacker is still alive. Understanding this analogy essentially captures the core idea of abstract argumentation:

1. An argument is labeled *in* iff all its attackers are labeled *out*;

2. An argument is labeled *out* iff it has at least one attacker that is labeled *in*;

Example 3.2 (Argumentation framework with two cycles). *Consider the argumentation framework in Figure 5, which has a set of arguments: $\{a, b, c, d, e\}$. We follow the direction of the graph. On the left, we have an even cycle: a and b attack each other. On the right, we have an odd cycle: c attacks d, d attacks e, and e attacks c. The two cycles are* *connected by the attack from b to c, thus the status of the arguments in the even cycle will influence the status of the arguments in the odd cycle. In the left cycle, there are two possibilities. In the first case, a is labeled in, then b is labeled out. However, there is no way to label the arguments in the odd cycle on the right. Thus, we need a third label called undec (undecided), indicating that one abstains from an explicit judgment whether the argument is in or out. It means that not all the attackers are labeled out and no attackers are labeled in. Therefore, c, d, e will be labeled undec. In the second case, b is in, a is out, and c is out. Then d is reinstated as an in argument because its only attacker c is out; we can also say b defends d. It follows that e is out because d is now in. We can label a and b with the third label undec, and all the arguments in the odd cycle are also labeled undec.*

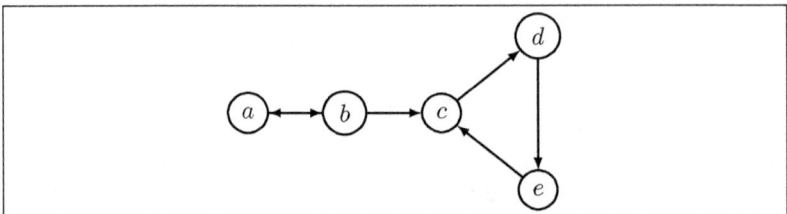

Figure 5: Argumentation framework with two cycles

Now the question is: which labelings in Dung's theory are called extensions? We illustrate the extensions to the framework in Example 3.2 below. We use the thick nodes for *in*, normal nodes for *out*, and dotted nodes for *undec*, to obtain a visualization similar to a colored graph. We say that if a labeling is three-valued, then it is a *complete* extension, as in the first item below. A complete extension generalizes a stable extension and there is no argument labeled *undec*, i.e., there is no dotted node, as in the second item below. The unique grounded extension is the most skeptical complete extension; only arguments that cannot avoid being accepted are labeled *in*, as in the third item. For some frameworks, there are no stable extensions. Then we can use preferred extensions, which are the maximal complete extensions, as in the fourth item.

- Three complete extensions (3-valued)

13 - Challenges in Formal and Computational Argumentation

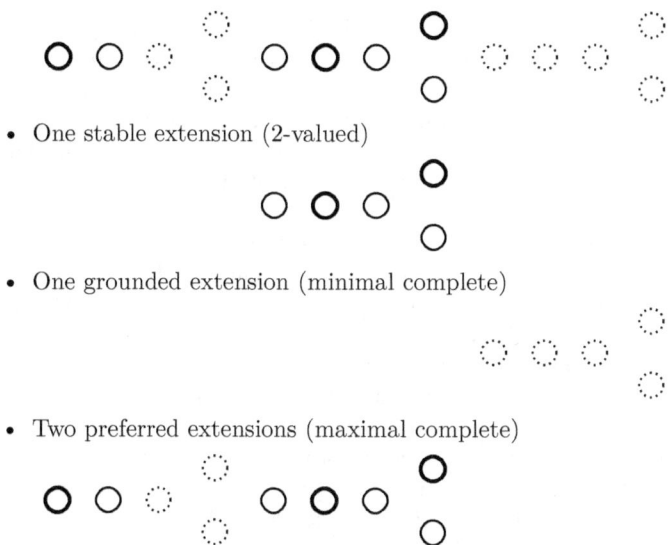

- One stable extension (2-valued)

- One grounded extension (minimal complete)

- Two preferred extensions (maximal complete)

While there have been many transformations of nonmonotonic reasoning formalisms into Dung's theory, direct usage is limited when modeling the argumentation of some realistic examples [Boella et al., 2009] such as multiagent argumentation and dialogues [Yu et al., 2021; Arisaka et al., 2022], decision-making [Kakas and Moraitis, 2003], coalition formation [Amgoud, 2005], combining micro arguments [Toulmin, 1958], normative reasoning [Atkinson and Bench-Capon, 2005], or meta-argumentation [Boella et al., 2009]. That leads to the following challenge.

Challenge 5. *Creating argumentation frameworks and semantics.*

Several extensions to abstract argumentation frameworks are discussed in the second volume of the Handbook of Formal Argumentation. Figure 6 visualizes six examples of extended argumentation frameworks. Preference-based argumentation [Kaci et al., 2021] introduces a preference relation between the arguments, as shown in Figure 6(a), where *a defeats b*, and *b* is preferred over *a*. Bipolar argumentation [Cayrol and Lagasquie-Schiex, 2005] defines support and attack independently. There are arguments in favor of other arguments, i.e., with a support relation, and also arguments against other arguments, i.e., with an attack relation, as shown in Figure 6(b). Weighted argumentation [Bistarelli *et al.*, 2021] specifies a numeric value that indicates the relative strength of an attack, as shown in Figure 6(c). Abstract agent argumentation [Yu *et al.*, 2021] extends Dung's framework with a set of agents and a relation associating arguments with agents, as shown in Figure 6(d). Value-based argumentation [Atkinson and Bench-Capon, 2021], as shown in Figure 6(e), defines values that are associated with an argument. The preference ordering of the values may depend on a specific audience. To model defeat for a specific audience: an argument A attacks an argument B for audience a if A attacks B and the value associated with B is not preferred to the value associated with A for audience a. Higher (second)-order argumentation [Cayrol *et al.*, 2021] introduces a new kind of attack which is a binary relation from arguments to attack relations, as shown in Figure 6(f).

One technique that has already proven to be useful in the past for studying such extensions is a meta-argumentation methodology involving the notion of flattening [Boella *et al.*, 2009]. Flattening is a function that maps some extended argumentation frameworks into Dung frameworks. There are two main flattening techniques. One is that we keep the arguments the same while removing attacks or introducing auxiliary attacks (this is also called reductions sometimes). This technique is used in preference-based argumentation, abstract agent argumentation, bipolar argumentation, etc. The other technique is to use not only auxiliary attacks but also auxiliary arguments in higher-order argumentation.

Example 3.3 (Four reductions of preference-based argumentation). *Figure 7 illustrates the differences between the four reductions from a preference-based argumentation framework to abstract argumentation frameworks [Kaci* et al., *2021]. The basic idea of Reduction 1 is that an attack succeeds only when the attacked argument is not preferred to the attacker. Reduction 2 enforces that one argument defeats another when the former is preferred but attacked by the latter. The idea of Reduction 3 is that if an argument is attacked by a less preferred argument, then the former should defend itself against its attacker. Reduction 4 mixes the second and the third reductions.*

Flattening by adding auxiliary arguments is a way of implementing the methodology of meta-argumentation [Boella *et al.*, 2009]. Meta-

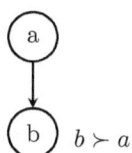

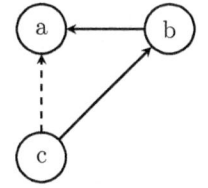

(a) **Preference-based argumentation**
$\langle Ar, \rightarrow, \succeq \rangle$, where Ar is a set of arguments, $\rightarrow$ is an attack over arguments, and $\succeq$ is a preference relation over Ar.

(b) **Bipolar argumentation**
$< Ar, \rightarrow, \dashrightarrow >$, where Ar is a set of arguments, $\rightarrow$ is an attack over arguments, and $\dashrightarrow$ is support for arguments.

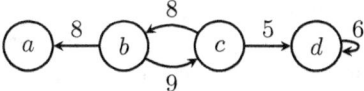

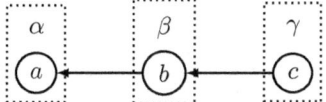

(c) **Weighted argumentation**
$\langle Ar, \rightarrow, w \rangle$, where Ar is a set of arguments, $\rightarrow$ is an attack over arguments, and w is a function from Ar to $[0, 1]$.

(d) **Abstract agent argumentation**
$\langle Ar, \rightarrow, S, \sqsubset \rangle$, where Ar is a set of arguments, $\rightarrow$ is an attack over arguments, S is a set of agents, and $\sqsubset$ is a binary relation associating arguments with agents.

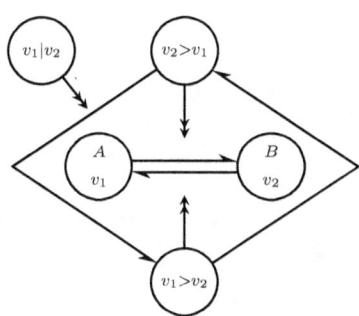

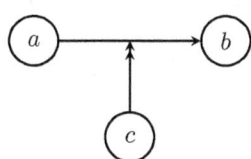

(e) **Value-based extended argumentation**
$\langle Ar, \rightarrow, V, val, P \rangle$, where Ar is a set of arguments, $\rightarrow$ is an attack over arguments, V is a set of values, val is a function mapping arguments to values, and P is a set of audiences.

(f) **Higher-order argumentation**
$\langle Ar, \rightarrow, \twoheadrightarrow \rangle$, where Ar is a set of arguments, $\rightarrow$ is an attack over arguments, and $\twoheadrightarrow$ is an attack relation over arguments that can be attacked.

Figure 6: Six extensions to the argumentation framework

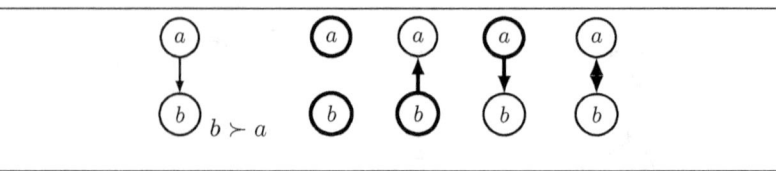

Figure 7: From left to right: the original argumentation framework, and the results after applying the four reductions respectively. The defeat relation is visualized with thick lines, and arguments that are accepted in grounded semantics also have thick lines.

argumentation generally involves taking into account the arguments of, e.g., lawyers, commentators, citizens, teachers, or parents (in accordance with the level of their expertise) but it can also go beyond this — the arguers and the meta-arguers can be represented by the *same* reasoners. For example, a lawyer may debate whether a suspect's argument attacks another argument, and she may also argue in a similar way about her own arguments. To give another example, people may be in the middle of an argument, but then start questioning the rules of the dialogue game, and argue about that. A further example is that of a child arguing that the argument *I was ill* attacks the argument *I have to do my homework* but then finds that the argument *I have a nice tan* attacks the argument *I was ill*.

When we flatten the extended framework, if an argument a of the extended argumentation framework also occurs in the flattened abstract argumentation framework, then we no longer refer to it as argument a but as the meta-argument "argument a is accepted", denoted as $accept(a)$. We use Example 3.4 to illustrate instantiating Dung's abstract argumentation framework by introducing meta-arguments that use flattening.

Example 3.4 (Flattening with auxiliary arguments [Boella *et al.*, 2009]). *Given the higher-order argumentation framework in Figure 6(f), the flattened framework is as illustrated in Figure 8. We introduce the meta-arguments* $Y_{a,b}$, *which means that a is capable of attacking b, and* $X_{a,b}$. *which means that a does not have the capability of attacking b. We use the meta-arguments in the following way. Each* $a \to b$ *is replaced by* $accept(a) \longrightarrow X_{a,b} \longrightarrow Y_{a,b} \longrightarrow accept(b)$. *The accepted arguments are*

$\{accept(a), accept(c), Y_{c,Y_{a,b}}, accept(b)\}$.

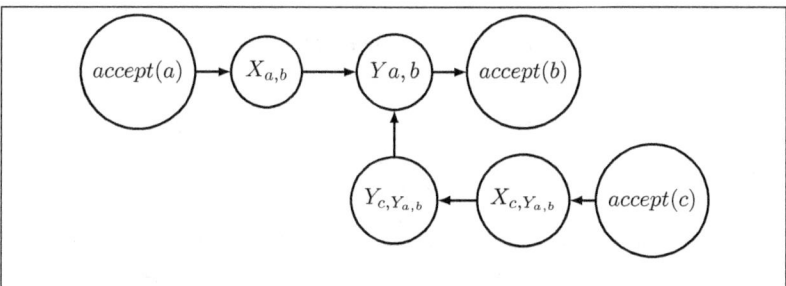

Figure 8: Flattened argumentation framework for Figure 6(f)

These examples illustrate how diverse extended argumentation frameworks build upon Dung's foundational concepts by introducing additional elements. However, the key reasoning ontology fundamentally relies on just two elements: arguments and attacks. Every utterance, be it a claim, an argument, or even an attack, can be modeled as an argument within these frameworks. Consequently, many of these extended frameworks can be *flattened* to basic ones, reinforcing the idea that attack graphs serve as a universal model of reasoning — much like how Turing machines serve as a universal model of computation.

In this section, we discussed the attack-defense paradigm shift introduced by Dung, emphasizing that every type of utterance (claim, argument, or attack) can be attacked. We also presented how diverse extended argumentation frameworks can be flattened into basic ones, demonstrating the universality of attack. We end this section with the following questions:

1. What is an argument? What is an attack? What is the interplay between an argument and an attack?

2. Should attack always be the first-class citizen in formal argumentation? For example, a novel notion of attack-defense is adopted as a first-class citizen by Liao and van der Torre [2024]. It can represent some knowledge that cannot be represented in Dung-

style argumentation, e.g., some context-sensitive knowledge in a dialogue.

3. How should the new diversity created by the attack-defense perspective be handled?

4. What does the attack-defense paradigm shift mean for argumentation as dialogue? What does the attack-defense paradigm shift mean for argumentation as balancing?

5. If we flatten an extended argumentation framework we introduce auxiliary arguments. How can we then recognize these auxiliary arguments in the instantiated Dung abstract argumentation framework? How can we deal with the original arguments and the arguments introduced later?

3.2 Representing nonmonotonic logics

In this section, we discuss: structured argumentation as a bridge from classical to nonmonotonic logic, the variety of nonmonotonic logics available, and the represention of nonmonotonic logic. We again refer to the commutative diagram in Figure 3 that we used in section 3.1 to illustrate how the same conclusions can be reached with two different approaches: the direct approach using logic and the indirect approach through the construction of argumentation frameworks, and semantics. Here, we illustrate this using the weakest versus last link principle and we continue with PDL and the weakest link.

Despite the uniqueness of classical logic, a wide variety of nonmonotonic logics are employed in different contexts. However, engaging in nonmonotonic logics means the aim is to extend classical logic rather than replace it *tout court* [Makinson, 2005]. Structured argumentation is used to classify existing nonmonotonic logics as a way to define a new nonmonotonic logic and create a bridge from classical to nonmonotonic logic. Dung [1995] provides semantics of attacks for structured argumentation. This has been used in the ASPIC+ system by Modgil and Prakken [2013; 2014], and it has also been used to reconstruct and compare a variety of nonmonotonic logics, namely default logic [Reiter, 1980], Pollock's [1987] argumentation system, and several logic-programming semantics. More representations have been developed — for details, please refer to the work of Heyninck [2019]. However, as discussed in Section 2, there is also a diversity of natural argumentation, conceptualizations, and formal methods. Notwithstanding the initial appeal of Dung's abstract argumentation theory, there are many different kinds of argumentation frameworks and semantics. That leads to the following challenge:

Challenge 6. *Representing nonmonotonic logics and solution concepts.*

Before we get into approximating PDL with argumentation, let us first talk about methodologies employed to compare different nonmonotonic logics and, in particular, their use of examples. There are different approaches to the use of examples in different disciplines. In law, ethics, and linguistics, examples are central to the development and validation of theories because they help ground abstract concepts in real-world scenarios, which helps to align logical frameworks with intuitive understanding. In contrast, knowledge representation (KR) and other areas of computer science often use examples as a practical tool to test, demonstrate, and communicate the effectiveness of a formal theory rather than using them as foundational elements in theory construction.

NLP task: translating natural language into formal language. Consider the aforementioned example of the fitness lover Scot and an additional example about a snoring professor:

The fitness lover Scot: It is commonly assumed that if a man was born in Scotland, then he is Scottish. And if he is Scottish, we can normally deduce that he likes whiskey. However, fitness lovers normally avoid alcohol for health reasons. Stewart was born in Scotland, and he is also a fitness lover. Does he like whiskey?

The snoring professor A library has a general rule that misbehavior, such as snoring loudly, leads to denial of access. However, there is another rule that professors are normally allowed access. Bob is a professor and he is snoring loudly in the library. Should he be allowed access to the library?

NLP could be used to identify three rules for each example, and then further abstract them into these formal (default) rules with priorities: $\{\top \overset{1}{\Rightarrow} a, a \overset{3}{\Rightarrow} b, \top \overset{2}{\Rightarrow} \neg b\}$. We then have:

$$\left\{ \begin{array}{c} \text{Fitness lover Scot} \\ BornInScotland \overset{1}{\Rightarrow} Scottish \\ Scottish \overset{3}{\Rightarrow} LikesWhisky \\ FitnessLover \overset{2}{\Rightarrow} \neg LikesWhisky \end{array} \right\} \quad \left\{ \begin{array}{c} \text{Snoring professor in the library} \\ snores \overset{1}{\Rightarrow} misbehaves \\ misbehaves \overset{3}{\Rightarrow} accessDenied \\ professor \overset{2}{\Rightarrow} \neg accessDenied \end{array} \right\}$$

KR task: after inputting some requirements, i.e., the goal of the reasoning, the system asks what you want to derive from what you have. Although the above two examples share a similar structure, there could be different reasoning requirements that lead to the selection of different rules and ultimately different conclusions. In the fitness lover example, one might prioritize the rule $\top \overset{2}{\Rightarrow} \neg b$ and conclude that Stewart does not like whiskey. In contrast, in the snoring professor example, one might prioritize the rule $a \overset{3}{\Rightarrow} b$ and conclude that Bob should be denied access.

Logic design task: According to these requirements, the system asks what is the best logic for your application. These two examples have been used to illustrate the difference between *prescriptive* and *descriptive* reasoning in nonmonotonic reasoning and between the *weakest link* and the *last link*, which are two principles regarding how an argument draws strength from its defaults.

In Section 2.4 and Section 3.1, we briefly mentioned Brewka and Eiter's [1999] PDL. Pardo *et al.* [2024] compared structured argumentation based on the weakest link variant with that of the PDL variant. Let us start with a reminder that PDL can be understood as a greedy approach, i.e., PDL iteratively adds the strongest applicable and consistent default. Initially, people thought that using the weakest link principle to construct argumentation frameworks would capture this kind of greedy procedure. However, over time, analysis of the weakest-link-related attack assignment reveals that it is more complicated and ambiguous than it appears at first sight.

The history of the weakest link revolves around three key examples from the literature, visualized in Figure 9 and described in Examples 3.5–3.7. Note that these examples illustrate the role of formal argumentation in the context of PDL. We refer to the work of Pardo *et al.* [2024] for the formal definitions. Here, we discuss Examples 3.5–3.7 informally.

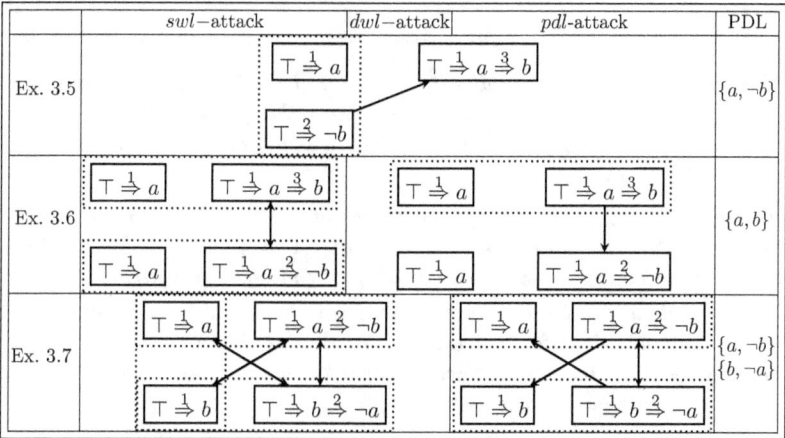

Figure 9: Approximating PDL in structured argumentation: a comparison of three attacks (columns) for three examples (rows). Columns are not marked when adjacent notions of attack agree on the induced attack relation at a given row. Dotted rectangles are argument extensions. The rightmost attacks approximate PDL better.

The following example illustrates the use of priorities. What does a *stronger priority* mean? Under the *prescriptive* reading, it means priority in the order of application: PDL always selects the strongest default (among those that are applicable and consistent). Under the *descriptive* reading, the priority of a default is its contribution to the overall status of any extension containing this default [Delgrande *et al.*, 2004]. The two readings clash in the most discussed example of defeasible reasoning with prioritized rules.

Example 3.5 (Weakest vs. Last link)**.** *Let* $\{\top \overset{1}{\Rightarrow} a,\ a \overset{3}{\Rightarrow} b,\ \top \overset{2}{\Rightarrow} \neg b\}$ *be again our defaults (Figure 9, top). The two readings of priorities give* the following outputs:

(Prescriptive.) Based on application order, one must select $\{\top \overset{2}{\Rightarrow} \neg b, \top \overset{1}{\Rightarrow} a\}$ thereby obtaining the output $\{a, \neg b\}$, as in PDL. In fact, PDL is an implementation of the prescriptive reading. Let us call simple weakest link (swl) the strength defined by the lowest priority of an argument:

$$\top \overset{1}{\Rightarrow} a \overset{3}{\Rightarrow} b \ \longmapsto \ 1 = \min\{1,3\} \qquad \top \overset{2}{\Rightarrow} \neg b \ \longmapsto \ 2 = \min\{2\}$$

A comparison of the strengths in this conflict produces the attack shown in Figure 9 (top). The semantics then gives the argument selection also shown. Our three attack relations (swl, dwl, pdl) do in fact agree on the verdict for this example.[1]

(Descriptive.) This reading favours the set $\{\top \overset{1}{\Rightarrow} a, a \overset{3}{\Rightarrow} b\}$ as its priorities $\{1,3\}$ are more desirable than the rival ones $\{1,2\}$. Last link can be seen as an implementation of this reading: the contribution of a new default to a selection or argument, say $\{\top \overset{1}{\Rightarrow} a\}$, is defined by the desirability of this default (2 vs. 3 in the example). Last link thus agrees on the above preference but arrives at it through argumentative means. First, one computes argument strength:

$$\top \overset{1}{\Rightarrow} a \overset{3}{\Rightarrow} b \ \longmapsto \ 3 = \text{last}(1,3) \qquad \top \overset{2}{\Rightarrow} \neg b \ \longmapsto \ 2 = \text{last}(2)$$

Based on this, argument $\top \overset{1}{\Rightarrow} a \overset{3}{\Rightarrow} b$ attacks $\top \overset{2}{\Rightarrow} \neg b$. Using a standard argumentation semantics, one obtains the output $\{a, b\}$, not shown in Figure 9 (top).

The simple weakest link, though, does not always capture the prescriptive reading. In response to this, a more intuitive *disjoint* variant of the weakest link has been considered [Young et al., 2016]. This variant assumes a relational measure of argument strength. It ignores all the shared defaults before searching for the weakest link between two arguments.

---

[1]This example represents the Tweety scenario $\{penguin \rightarrow bird, bird \Rightarrow flies, penguin \Rightarrow \neg flies\}$ with priorities instead of the strict rule $(\rightarrow)$. Without priorities, the solution $\{penguin, bird, \neg flies\}$ obtains from specificity (of *penguin* over *bird*): *birds fly* is overruled by the more specific *penguins do not fly*. Without specificity the solution obtains from appropriate priorities using PDL or swl.

Example 3.6 (Simple vs. disjoint weakest link). *Let* $\{\top \xRightarrow{1} a, a \xRightarrow{3} b, a \xRightarrow{2} \neg b\}$ *define our knowledge base. Note that the two arguments* $\top \xRightarrow{1} a \xRightarrow{3} b$ *and* $\top \xRightarrow{1} a \xRightarrow{2} \neg b$ *share a default* $\top \xRightarrow{1} a$ *with the lowest priority. See the middle row in Figure 9.*

(Simple weakest link) *Pollock's definition assigns the same strength of 1 to these two arguments. This gives the mutual swl-attack in Figure 9 (mid, left). Now, one argument selection* $\top \xRightarrow{1} a \xRightarrow{3} b$ *matches the PDL extension* $\{a, b\}$; *the other* $\top \xRightarrow{1} a \xRightarrow{2} \neg b$, *though, gives us a non-PDL extension,* $\{a, \neg b\}$.

(Disjoint weakest link) *The attack relation defined by disjoint weakest link (dwl) assigns strengths* $3 > 2$ *to the above arguments, after excluding the default they share. This generates the tie-breaking dwl-attack shown in Figure 9 (mid, right). This figure also shows the set of arguments selected by our semantics. The selected arguments' conclusions match the PDL output* $\{a, b\}$.

Pollock's definition of weakest link *swl* [Pollock, 2001] was adopted and studied for ASPIC+ by Modgil and Prakken [2013; 2014]. Then, Young *et al.* [2016; 2017] introduced *dwl* and proved that argument extensions under the *dwl*-attack relation correspond to PDL extensions under total orders; see also the results presented by Liao et al. [2019] and Pardo and Straßer [2022]. Under total preorders, a new attack relation is needed for more intuitive outputs and a better approximation of PDL — that is, better than *dwl*.

Example 3.7 (Beyond *dwl*). *Let* $\{\top \xRightarrow{1} a, \top \xRightarrow{1} b, a \xRightarrow{2} \neg b, b \xRightarrow{2} \neg a\}$ *be the defaults.*

(*swl, dwl*) *Weakest link attacks, depicted in Figure 9 (bottom, left), admit the selection of arguments* $\{\top \xRightarrow{1} a, \top \xRightarrow{1} b\}$. *This selection fits neither the prescriptive interpretation nor PDL. Selecting either default ought to be followed by the selection of a stronger default, namely* $a \xRightarrow{2} \neg b$ *and* $b \xRightarrow{2} \neg a$ *respectively.*

(PDL) *As PDL selects the strongest default one at a time, this excludes by construction the concurrent selection of* $\{\top \xRightarrow{1} a, \top \xRightarrow{1} b\}$. *The*

PDL-inspired attack relation in Figure 9 (bottom, right) also excludes this selection.

An important research question is then how to characterize, or at least approximate, the PDL extensions of a prioritized default theory. For total orders, an attack that characterizes PDL extensions already exists: att_{dwl} [Young et al., 2016].

But for total preorders, how to characterize PDL extensions using an attack relation assignment is an open problem. Certainly, such a characterization can no longer be based on the disjoint weakest link, as shown in Example 3.8.

Example 3.8 (Disjoint weakest link vs. PDL). *Example 3.7 shows a stable belief set $\{a, b\}$ under $att_{dwl}(K)$ that is not a PDL extension of K.*

Example 3.9 (PDL vs. Disjoint weakest link). *Let $\{\top \xRightarrow{1} a, a \xRightarrow{2} b, \top \xRightarrow{1} c, a, c \xRightarrow{2} \neg b\}$ define our knowledge base. Figure 10 shows that the shared rule $\top \xRightarrow{1} a$ produces only one stable extension $\mathcal{E}$ under the disjoint weakest link, and so we have a unique stable belief set of $(AR_K, att_{dwl}(K))$:*

$$\mathcal{E} = \{A, C, [A \Rightarrow b]\} \quad \longmapsto \quad S = \{a, b, c\}$$

In contrast, two PDL constructions exist for K, and so do two PDL extensions:

$$(\top \xRightarrow{1} a, \; a \xRightarrow{2} b, \; \top \xRightarrow{1} c) \quad \longmapsto \quad \{a, b, c\}$$
$$(\top \xRightarrow{1} c, \; \top \xRightarrow{1} a, \; a, c \xRightarrow{2} \neg b) \quad \longmapsto \quad \{a, \neg b, c\}$$

As a consequence, disjoint weakest link cannot characterize PDL under stable semantics. Observe that att_{swl} here coincides with PDL.

Attack relations have become a major subject of study in logic-based argumentation. Dung [2014; 2016] recently proposed an axiomatic method that supersedes all argumentation systems with defeasible rules. Pardo et al. [2024] attempted to identify an attack relation that captures PDL extensions, and they compared it with attacks based on the simple and disjoint weakest link using the eight principles advanced by Dung. They proved an impossibility theorem: representing PDL in

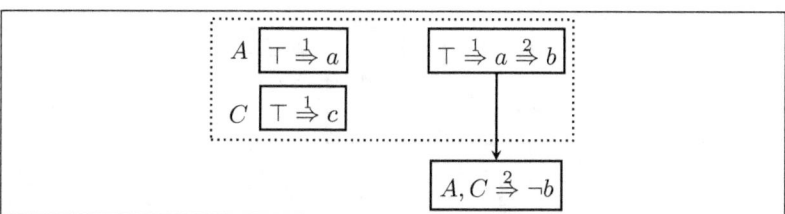

Figure 10: The stable belief set $\{a,c,b\}$ under att_{dwl} for Example 3.9. Two extensions, $\{a,c,b\}$ and $\{a,c,\neg b\}$, exist under PDL.

formal argumentation should preserve a principle (attack closure), but this is incompatible with another principle (context independence).

As seen in Examples 3.7 and 3.9, disjoint weakest link and PDL are incomparable under total preorders. As a first step towards their convergence, one can slightly modify PDL to make it closer to the disjoint weakest link. To this end, Pardo *et al.* [2024] propose *parallel* PDL (pPDL), a concurrent variant of PDL. The main novelty of pPDL is that each inductive step can concurrently select a set of defaults, rather than just one, for the technical details, we refer to the paper of Pardo *et al.* [2024].

Example 3.10 (pPDL, DWL vs. PDL). *Let us use Example 3.7 to show that the default logic PDL differs from pPDL. Figure 11 illustrates the three pPDL extensions $\{a, \neg b\}$, $\{b, \neg a\}$, $\{a, b\}$, of which $\{a, b\}$ is not a PDL extension.*

Although pPDL and att_{dwl} agree in this and other examples, pPDL does not always match the disjoint weakest link.

Example 3.11 (pPDL vs. DWL). *Example 3.9 showed a unique stable belief set, $\{a,b,c\}$, under att_{dwl}. But there are two pPDL extensions: $\{a,b,c\}$ and $\{a, \neg b, c\}$.*

To sum up, the first goal of Pardo *et al.* [2024] was to identify an attack relation that captures PDL extensions and compare it with attacks based on the simple and disjoint weakest link using the eight principles advanced by Dung [2016; 2018]. They proved which principles for attack relations are satisfied by weakest link, disjoint weakest link and

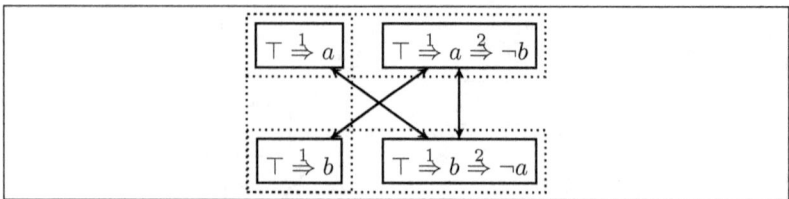

Figure 11: pPDL differs from PDL. PDL has two extensions $\{a, \neg b\}$ and $\{b, \neg a\}$. pPDL has an additional extension $\{a, b\}$. Arrows denote logical conflicts.

PDL based attacks. Their principle-based analysis presented the difference between several kinds of attack relation assignment. They identified and explained the nature of the weakest link principle and revealed that there is still the potential to improve the weakest link attack. On this last question, they proposed pPDL (parallel PDL), a concurrent variant of PDL, and they showed by way of examples that it falls closer to the disjoint weakest link than PDL does. While the pPDL variant still does not match the disjoint weakest link, one might conjecture that some further refinement might do.

In addition to presenting the argumentation framework, Dung [1995] also investigated two examples of problems from microeconomics — cooperative game theory and matching theory. In each case, Dung showed how an appropriate framework can represent a given cooperative game or a given instance of the stable marriage problem, and that the sets of winning arguments in such argumentation frameworks correspond to meaningful solutions in both these domains.

Cooperative game theory studies how rational agents cooperate to form coalitions to maximize their payoffs. A coalition's payoff is measured by its value, and agents ideally cooperate to form a coalition. The Von Neumann-Morgenstern (vNM) stable set [Von Neumann and Morgenstern, 1947] is the solution concept for distributing the grand coalition's payoff and ensuring that no agent defects. Dung showed that stable extensions of an argumentation framework correspond to vNM stable sets [Dung, 1995, Theorem 37]. However, just like the stable extensions of an argumentation framework may not exist, vNM stable sets also do not always exist. Dung proposed that sets of payoff distributions that form preferred extensions could serve as an alternative solution concept because preferred extensions always exist, and therefore this is well defined for all cooperative games.

In this section, we discussed the representation of nonmonotonic logics using the attack-defense paradigm. We end with several questions concerning such representations:

1. We showed that PDL and the weakest link definitions are similar but not exactly the same. How can PDL be changed to make it fit one of the weakest link definitions? How can the weakest link be changed to fit PDL?

2. We discussed the logic of the weakest link. What is the logic that corresponds to the last link?

3. We showed various alternative formalizations of the weakest link principle. Likewise, are there variants of the last link principle?

4. PDL is only one of many logics for prioritized rules. How can all the other systems for prioritized rules be represented?

5. We discussed representation of nonmonotonic logics, but Dung also talked about logic programming and game theory. What is the relation between different solution concepts in game theory and (extended) abstract argumentation frameworks semantics?

3.3 Postulates from paraconsistent reasoning

In this section, we continue our discussion of formal argumentation as a logical framework for nonmonotonic reasoning. We consider inconsistent knowledge bases and so-called rationality postulates from paraconsistent reasoning, which is used to define new nonmonotonic logics in ASPIC+. We illustrate the postulates using the example of three persons on a two-person tandem taken from Caminada and Wu [2011].

In previous sections, we discussed universality of attack and one resultant challenge — representing existing nonmonotonic logic. In particular, we discussed representing PDL with structured argumentation, comparing attack assignments using variants of the weakest link with principles. What we showed is just the tip of the iceberg. There are numerous options based on different knowledge bases containing various types of information such as strict and/or defeasible rules. There are different methods for constructing argumentation frameworks. There are applications of distinct semantics. The combination of all these factors defines different argumentation-based logics that can be adopted or rejected, depending on their applicability in different contexts. *Rationality postulates* are a list of desiderata that structured argumentation systems should satisfy in order to be logically well-behaved [Caminada and Amgoud, 2007]. In this section, we address the following challenge:

Challenge 7. *Rationality postulates for defining a new logic.*

Various rationality postulates are inspired by paraconsistent reasoning. Paraconsistent logic [Da Costa, 1974; Priest, 2002] is a non-classical logical system designed to handle contradictions without leading to the collapse (or "explosion") of the entire systems (as would occur in classical logic). These logics have inspired the development of modal and nonmonotonic logics as well as various rationality postulates [Da Costa et al., 2007]. Such postulates ensure that logic can handle inconsistencies without leading to the kind of trivialization where any and every conclusion becomes derivable from a set of contradictory premises. One key postulate of paraconsistent logic is noninterference, i.e., independent knowledge bases do not influence each other's outcomes. Another is avoiding contamination, i.e., the outcome of a set of formulas remains unchanged when merged with an unrelated set [Caminada et al., 2012].

A side note regarding terminology: we use terms such as postulates, axioms, requirements and desiderata in a rather interchangeable manner, and they differ slightly from principles and properties. All six terms refer to the behavior of logic, the construction of an argumentation framework, and the semantics of argumentation frameworks. Abstract properties are formally specified, and in this section, postulates are treated as desiderata, akin to formal requirements in computer science. In Section 4.1, where we discuss *the principle-based methodology* in detail, postulates are regarded as more general properties, with some being desirable and others not.

There are three fundamental rationality postulates [Caminada and Amgoud, 2007]. Direct Consistency means that any extension should be consistent according to certain semantics. Indirect Consistency means that the set of the conclusions of arguments in a given extension is consistent when closed under the strict rule. Closure means that arguments with conclusions derived from arguments in an extension using strict rules should also be in the extension.

Given these postulates, the question is under what conditions do structured argumentation satisfy them. When assigning attack relations among arguments from a knowledge base, there are so-called rebuts when the conclusions of two arguments conflict with one other. Two kinds of rebuts have been discussed in the literature: *restricted rebuts* and *unrestricted rebuts*. The intuition behind restricted rebut is: if an argument is built up with only strict rules, then the conclusion should also be strict, and the argument cannot be attacked. The intuition behind unrestricted rebut is that a conclusion is defeasible, i.e., it can be attacked iff it is built up with at least one defeasible rule. Different choices on rebuts influence how to define the argumentation formalism that derives reasonable conclusions. This exists in the ASPIC family of argumentation frameworks, including ASPIC+ [Modgil and Prakken, 2013; Modgil and Prakken, 2014], ASPIC- [Caminada et al., 2014] and ASPIC-END [Dauphin and Cramer, 2018].

Example 3.12 illustrates rationality postulates, comparing unrestricted rebut and restricted rebut, and it shows the solutions to restricted rebut required to satisfy the rational postulates for this example.

Example 3.12 (Married John [Caminada and Amgoud, 2007]). *Consider an argumentation system consisting of the strict rules* $\{\to r, \to n, m \to hs, b \to \neg hs\}$ *and the two defeasible rules* $\{r \Rightarrow m, n \Rightarrow b\}$. *An intuitive interpretation of this example is the following: "John wears a ring (r) on his finger. John is also a regular nightclubber (n). Someone who wears a ring on his finger is usually married (m). Someone who is a regular nightclubber is usually a bachelor (b). Someone who is married has a spouse (hs) by definition. Someone who is bachelor does not have a spouse ($\neg hs$) by definition." We can construct the following arguments:*

$A_1 :\to r$ $\qquad A_3 : A_1 \Rightarrow m$ $\qquad A_5 : A_3 \to hs$

$A_2 :\to n$ $\qquad A_4 : A2 \Rightarrow b$ $\qquad\quad : A_6 : A_4 \to \neg hs$

If we apply unrestricted rebut, we have A_5 and A_6 attacking each other, and we obtain the grounded extension of $\{A_1, A_2, A_3, A_4\}$ with the conclusion extension $\{r, n, m, b\}$, which does not satisfy the direct consistency property. If we apply restricted rebut, the situation is even worse. Because we do not have any attack relations, we have the extension $\{A_1, A_2, A_3, A_4, A_5, A_6\}$ and the conclusion extension $\{r, n, m, b, hs, \neg hs\}$, which are not consistent.

Two solutions for argumentation systems applying restricted rebut to satisfy the rationality postulates are *closure of transposition* and *closure of contraposition*, as adopted in ASPIC+ [Modgil and Prakken, 2014].

Example 3.13 (Example 3.12 continued). *Given that we have $m \to hs$ and $b \to \neg hs$ in the knowledge base, we add their "contraposed" versions: $\neg hs \to \neg m$ and $hs \to \neg b$. We can construct additional arguments: $A_7 : A_5 \to \neg b$ and $A_8 : A_6 \to \neg m$. We have that A_7 restrictively rebuts A_4, and that A_8 restrictively rebuts A_3. As a result, each set of conclusions yielded under grounded or preferred semantics satisfies the postulates of direct consistency, closure, and indirect consistency.*

We use the example of three persons on a two-person tandem [Caminada and Amgoud, 2007] to take a closer look at unrestricted rebut and restricted rebut — the latter is applied where unrestricted rebut can lead to undesired behavior.

Example 3.14 (Restricted rebut vs. unrestricted rebut). *Consider a knowledge base consisting of three defeasible rules, $\{\top \Rightarrow p, \top \Rightarrow q, \top \Rightarrow r\}$, and three strict rules, $\{p, q \to \neg r, p, r \to \neg q, q, r \to \neg p\}$. We can construct six arguments as shown below. If we apply unrestricted rebut, we can obtain the abstract argumentation framework on the left hand side of Figure 12. One of the complete extensions is $\{A_1, A_2, A_3\}$, yielding conclusion extension $\{p, q, r\}$. If we close this extension under strict rules, we have $\{p, q, r, \neg p, \neg q, \neg r\}$, which is not consistent. If we apply restricted rebut, we obtain the framework at the right hand side of Figure 12, where we have the complete extensions of $\{A_1, A_2, A_6\}$, $\{A_1, A_3, A_5\}$ and $\{A_2, A_3, A_4\}$. They are also consistent under the clo-*

sure of strict rules.

$$A_1 : \top \Rightarrow p \qquad A_4 : q, r \to \neg p$$
$$A_2 : \top \Rightarrow q \qquad A_5 : p, r \to \neg q$$
$$A_3 : \top \Rightarrow r \qquad A_6 : p, q \to \neg r$$

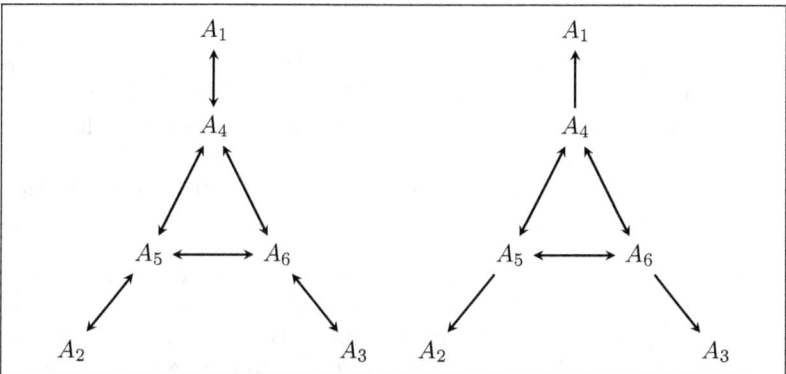

Figure 12: Resticted rebut vs. unrestricted rebut

There are more postulates. For example, *noninterference* and *crash resistance* are particularly relevant when the strict rules are derived from classical logic, and again we examine various ways of satisfying these properties. However, there have been comparatively fewer results that would establish them in systems of the ASPIC family.

Noninterference: no set of formulas can influence the entailment of an unrelated set of formulas when they are merged with a completely unrelated (syntactically disjoint) defeasible theory.

Crash resistance: no set of formulas can make an unrelated set of formulas completely irrelevant when they are merged with a completely unrelated (syntactically disjoint) defeasible theory.

A violation of non-interference means that a defeasible theory somehow influences the entailment of a completely unrelated (syntactically disjoint) defeasible theory when being merged with it. A violation of the crash resistance property is more severe, as this means that a defeasible theory influences the entailment of a completely unrelated (syntactically disjoint) defeasible theory to such an extent that the actual content of this other defeasible theory becomes irrelevant.

In this section, we discussed the use of postulates from paraconsistent reasoning in argumentation. We end with some open questions.

1. In structured argumentation, arguments can be attacked by either defeasible premises, defeasible inference rules, or the conclusion of defeasible rules. In assumption-based argumentation, there are only defeasible premises, while ASPIC+ allows all defeasibilities. How should we decide upon and clarify the various defeasibilities in structured argumentation?

2. Incorporating formal argumentation and social concepts has attracted much interest. One example is the use of an argumentative approach to normative reasoning [Dong et al., 2019; Pigozzi and van der Torre, 2018; Dong et al., 2021; Straßer and Arieli, 2019]. The question then is how to construct and evaluate deontic arguments.

3. Dialectical concepts like multiple agents, communication steps, or commitment stores (like those of the Fatio dialogue system) do not play a role in ASPIC+, which is more monolithic. If we want to add dialectical aspects to structured argumentation [Prakken, 2024a], how should we design an argumentation system that behaves logically?

3.4 Extensions of the attack-defense paradigm for dialogue

In this and the following sections, we discuss extensions to abstract argumentation. There are various approaches to extracting more information from frameworks, and there is a variety of qualitative and quantitative enrichments of frameworks. Semantics can be defined by reductions, selections, or adaptations of defense. In argumentation as inference, only preference is clearly linked to structured argumentation. In this section, we focus on extensions inspired by dialogue.

There are two kinds of extensions to abstract argumentation in the commutative diagram of Figure 3. The first extends the argumentation framework with qualitative and quantitative components from the knowledge base. In section 3.1, we mentioned various examples of such extensions. The second pertains to step (3), where the argumentation semantics contains more information rather than the acceptance of arguments.

For the second type of extension, Villata *et al.* [2011], for instance, generalize the argument semantics by selecting from the graph not only a set of nodes but also a set of edges. This represents intuitively that attacks can be successful or unsuccessful. A similar kind of intuition is formalized in extended argumentation frameworks with second or higher-order attacks [Barringer *et al.*, 2005; Cayrol *et al.*, 2021]. Attacks are treated as arguments that can be attacked, and thus can be accepted and rejected too. Consider, for example, the two-three cycle framework shown in Figure 5. One possible output is a subframework where we retain only the attack from b to a in the cycle on the left, effectively reducing the complexity of the argumentation structure while maintaining specific attack relations.

Extended abstract argumentation frameworks enhance the expressive capacity of frameworks. However, it is not clear how these extensions can be constructed directly from a knowledge base while incorporating additional components such as agents, supports, numerical values or weights. One exception is preference, which is clearly linked to structured argumentation as inference. In structured argumentation, preferences play a central role in determining formal outcomes. For example, in ASPIC+ [Modgil and Prakken, 2013], the defeat relation between arguments is governed by a preference order, typically derived through mechanisms such as the weakest link or last link principles. Specifically, an attack from one argument to another only succeeds as a defeat if the attacked argument is not stronger than or strictly preferred to the attacking argument, according to the given preference relation.

In abstract preference-based argumentation [Kaci et al., 2021], the first reduction in Figure 7 corresponds to this type of attack assignment.

In section 2.3, we introduced argumentation as dialogue. In section 2.4, we discussed its formal methods, e.g., speech acts, game theory, axiomatic semantics, and operational semantics. Inspired by dialogue, we have the following challenge:

Challenge 8. *Generalizing Dung's attack-defense paradigm for dialogue.*

At the structured level, it is natural to have the role of agents. One example is Jiminy architecture [Liao et al., 2023], discussed in section 2.1. Jiminy involves multiple stakeholders, each with their own knowledge base. When dilemmas and conflicts arise, the argumentation engine considers the combination of all the arguments constructed by each stakeholder. Either there is a large framework consisting of all the stakeholders' arguments and the attack relations, or all the knowledge bases are combined first, and the argumentation frameworks are constructed afterward.

At the abstract level, agent-based extensions typically introduce various aspects such as agents, coalitions, knowledge, uncertainty, support, and so on. As a result, there are various ways to define the semantics. Below we discuss abstract agent argumentation [Yu et al., 2021], which uses a minimal extension of Dung's framework as a common core. This work only introduces an abstract set of agents and arguments are associated with agents. There are four types of semantics, defined by adaptations of defense, reductions, aggregations, and selections:

Agent defense approaches adapt Dung's notion of defense to argumentation semantics.

Social approaches are based on counting the number of agents [Leite and Martins, 2011] and a reduction to preference-based argumentation [Amgoud and Cayrol, 2002].

Agent reductions take the perspective of individual agents and aggregate their individual perspectives [Giacomin, 2017].

Filtering methods are inspired by agents' knowledge or trust [Arisaka et al., 2022]. They leave out some arguments or attacks because they do not belong to any agent.

Yu et al. [2021] have defined *individual agent defense* and *collective agent defense*. Roughly, in individual agent defense, if an agent puts forward an argument, it can only be defended by arguments from that same agent, i.e., a set of arguments E from individual agent α defends an argument c iff there exists an agent α who has argument c such that for all arguments b attacking c, there exists an argument a in E from α attacking b. Whereas with collective agent defense, a set of agents α can do that, i.e., a set of arguments E defends c collectively iff for all arguments b attacking c, there exists an agent α who has c and an argument a in E from a set of agents α such that a attacks b. Example 3.15 illustrates these two agent defenses.

Example 3.15 (Individual agent defense vs. collective agent defense). *In Figure 6(d), argument c defends argument a, but it does not individually agent-defend it because c and a come from different agents. Consider another abstract agent framework visualized in Figure 13. Here, $\{c_1, c_2\}$ collectively agent-defend argument a, but they do not individually agent-defend it.*

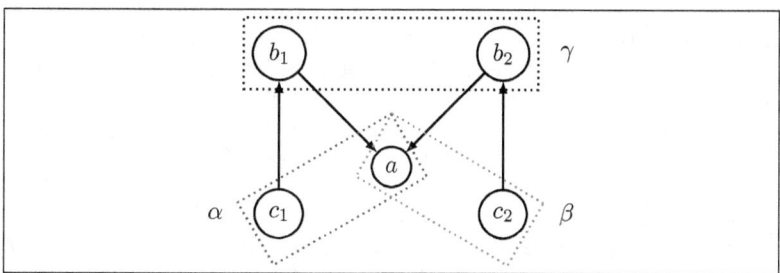

Figure 13: $\{c_1, c_2\}$ collectively agent-defending a

Social semantics is based on a reduction to preference-based argumentation for each argument, by counting the number of agents that have those arguments. It thus interprets agent argumentation as a kind of voting procedure. Example 3.16 illustrates social reduction.

Example 3.16 (Social reduction). *Consider the agent argumentation framework visualized in Figure 14. Arguments a and b both belong to agent α, b also belongs to agent β, and a attacks b. In that situation, argument b is preferred to argument a because it is held by more agents. We can then apply the four reductions from preference-based argumentation framework to abstract argumentation framework, followed by application of Dung's semantics.*

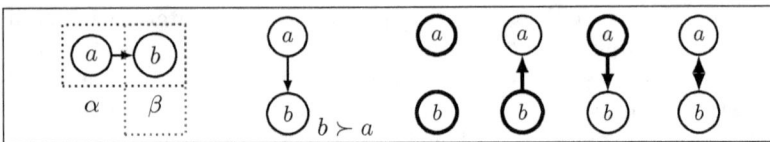

Figure 14: Social reduction: the left hand is an abstract agent argumentation framework, the middle is the corresponding preference-based framework, and the right hand are four corresponding abstract argumentation frameworks (as discussed in Example 3.3)

Agent reductions take the perspective of each agent and obtain the semantics accordingly. Intuitively, agents prefer their own arguments over the arguments of others. Thus, for each agent, there is a preference-based argumentation framework. In social reduction semantics, there is a unique abstract argumentation framework after each reduction. However, in agent reduction semantics, we obtain a set of abstract argumentation frameworks — one for each agent. The final step is to take the union of all these frameworks to form a combined abstract argumentation framework. Example 3.17 illustrates agent reductions.

Example 3.17 (Agent reduction). *Reconsider the abstract argumentation framework in Figure 14. Agent β prefers argument b over argument a. Thus, we get the same preference-based framework as depicted in Figure 14, but for a very different reason to that of social reduction. For agent α, argument a and b are equivalent. To compute the agent extensions, we take the union of the reductions for each agent.*

Agent filtering semantics remove arguments that do not belong to an agent, or they remove attacks that do not belong to an agent. An attack belongs to an agent if both the attacking and attacked arguments belong to that agent. Example 3.18 illustrates agent filtering semantics.

Example 3.18 (Agent filtering). *Consider the two abstract agent frameworks visualized in Figure 15. For the framework on the left, we might say that argument a is not known because it doesn't belong to any agent. And for the framework on the right, we might say that the attack is unknown because no agent holds both arguments a and b. The filtering methods remove such unknown arguments and unknown attacks. This is followed by the application of Dung's semantics.*

Figure 15: Unknown argument and unknown attack

There are several aspects of dialogue beyond associating arguments with agents that can be represented in abstract argumentation. One significant aspect is to make *time* explicit: unlike inference, dialogue inherently unfolds over time, with the dynamic argumentation framework evolving as the dialogue progresses. Dialogue can also be *strategic* — sometimes it is advantageous for an agent to not reveal certain arguments (this is discussed further in Section 4.3). A prototypical example of this is the content of the Miranda warning: "Anything you say can and will be used against you in a court of law". Example 3.19 illustrates how a suspect's argument can be strategically turned against him/her in a dialogue.

Example 3.19 (Dialogue between accuser and suspect [Okuno and Takahashi, 2009]). *Let Pr and Op be the players involved in the following argumentation dialogue (Pr and Op denote, respectively, a proponent and an opponent):*

Pr$_0$: *"You killed the victim."*

Op$_1$: *"I did not commit murder! There is no evidence!"*

Pr$_1$: *"There is evidence. We found your ID card near the scene."*

Op$_2$: *"That is not evidence! I had my ID card stolen!"*

Pr$_2$: *"It is you who killed the victim. Only you were near the scene at the time of the murder."*

Op$_3$: *"I did not go there. I was at facility A at that time."*

Pr$_3$: *"At facility A? Then, it is impossible that you had your ID card stolen because facility A does not allow any person to enter without an ID card."*

In this example, the opponent tries to defend himself with the claim "I had my ID card stolen!" (Op$_2$). However, the proponent strategically uses this very claim against the opponent (Pr$_3$), arguing that if the opponent was at facility A, it would have been impossible for his ID card to have been stolen because the facility does not permit entry without an ID card. This demonstrates how an argument can backfire in a strategic dialogue.

In this section, we discussed extending the attack-defense paradigm for dialogue, particularly with agents. We end the section with the following questions:

1. In this section, we discussed abstract agent argumentation, and we provided various semantics. How should a theory of structured agent argumentation be designed?

2. Strategic dialogue goes beyond argumentation as inference by incorporating agency. How should dialogue strategies be designed? And should they be evaluated?

3. What is the next step required to bridge the gap between 1) Dung's attack-defense paradigm and 2) strategic argumentation and dialogue?

4. There are many kinds of dialogues. What are the main components of a dialogue? For example, what are the components of persuasion dialogue systems like Fatio?

5. For all these kinds of dialogue, what makes a good dialogue? For example, what is a successful Fatio dialogue? Does a successful dialogue happen when someone is convinced of an argument they did not hold previously or does it happen when the parties agree about where they disagree?

3.5 Extensions of the attack-defense paradigm for balancing

In this section, we continue our discussion of extensions to abstract argumentation, focusing on extensions inspired by argumentation as balancing.

Argumentation as balancing brings to mind a double pan scale. The pros go on one pan and the cons go on the other. The pros and cons may have relative weights, and one needs to balance them from a utilitarian lens to determine the status of the issues, e.g., what action to take. Balancing finds applications in ethics and the law. In the legal domain, balancing is a metaphoric term that is generally used to describe an important conceptual operation [Aleinikoff, 1986]. In many conflicts, there is something to be said in favor of two or more outcomes. Whatever result is chosen, someone will be advantaged and someone will be disadvantaged; some policies will be promoted at the expense of some others. Hence it is often said that a "balancing operation" must be undertaken, with the "correct" decision seen as the one yielding the greatest net benefit. In medical ethics, for example, there are models of clinical ethics case consultation that often refer to 'balancing' or 'weighing' moral considerations [McDougall *et al.*, 2020].

Challenge 9. *Generalizing Dung's attack-defense paradigm for balancing.*

At an abstract level, it seems that pro and con arguments and the relations between them can be represented intuitively in bipolar argumentation frameworks discussed by Cayrol and Lagasquie-Schiex [2005; 2009; 2010; 2013], extending abstract argumentation framework with support relation that is independent of attack. Figure 16 illustrates three bipolar argumentation frameworks, where attack relations are depicted by solid arrows, and support relations are depicted by dashed arrows. Similarly to abstract agent argumentation semantics, there are also three types of bipolar argumentation semantics defined by Yu et al. [2023].

The defense-based approach defines new notions of defense using both support and attack.

The selection-based approach utilizes support only for selecting some of the extensions provided in Dung's semantics.

The reduction-based approach introduces indirect attacks based on interpretations of support.

There are three new defense based on both attack and support relations, called defense$_1$, defense$_2$, and defense$_3$, all of which have additional requirements for Dung's defense. Defended$_1$ requires that the argument defends (in Dung's theory) another argument also supports it. Defended$_2$ requires that a defender is supported. Moreover, defended$_3$ requires not only that the attackers are attacked, but also that all supporters of the attackers are attacked as well. We illustrate the three defense in Figure 16.

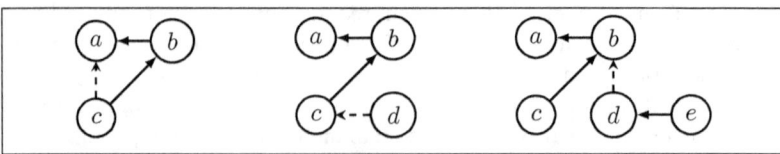

Figure 16: Three bipolar argumentation frameworks (BAFs) illustrating the three defense notions. In the left hand framework, $\{c\}$ defends$_1$ a. In the middle framework, $\{c, d\}$ defends$_2$ a. In the right hand framework, $\{c, e\}$ defends$_3$ a.

The selection-based approach uses support during the post-processing step for Dung's theory of abstract argumentation [Gargouri et al., 2020], i.e., first Dung's semantics are obtained, then support can be used to select extensions from Dung's semantics. One way selects the extensions that have the largest number of internal supports. This reflects the idea that for a coalition, the more internal supports they have, the more cohesive they are. The other way is to select the extensions that receive the most support from outside. This reflecting the idea that the more support a coalition receives, the stronger it is. It thus interprets support as a kind of voting procedure. We say that argument b in E is internally supported if b receives support from arguments in E. Argument b in E is externally supported if b receives support from arguments that are outside E.

Example 3.20 (Selection-based approach to bipolar argumentation). *Consider the bipolar argumentation framework on the right hand side of Figure 17. There are four extensions to Dung's stable semantics: $\{\{a,d\}, \{a,c\}, \{b,d\}, \{b,c\}\}$. By following the selection based on internal supports, $\{\{a,c\}\}$ is the stable semantics, while by following selection based on external supports, $\{\{a,d\}\}$ is the stable semantics.*

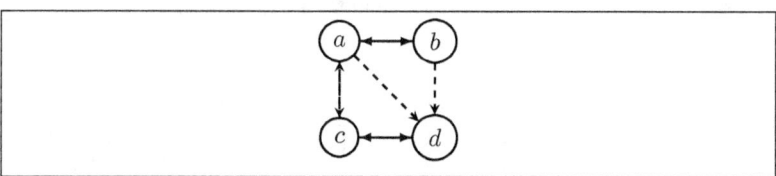

Figure 17: A bipolar argumentation framework

The reduction-based approach has been studied extensively by Cayrol and Lagasquie-Schiex [2005; 2009; 2013], and support is used as pre-processing for Dung semantics. The corresponding abstract argumentation frameworks are reduced by adding indirect attacks from the interaction between attacks and supports with different interpretations, i.e., deductive support and necessary support. Based on these two interpretations, four reductions have been discussed introducing additional attacks.

Deductive support [Boella et al., 2010] captures the intuition that if a supports b, then the acceptance of a implies the acceptance of b. Based on deductive interpretation, there are two kinds of additional attacks:

Supported attack and mediated attack. For example, in Figure 18(a), a supports c, and c attacks b. Acceptance of a implies acceptance of c, and acceptance of c implies non-acceptance of b. So, acceptance of a implies non-acceptance of b. Thus, the supported attack from a to b is introduced, depicted as a double-headed arrow. Similarly, the mediated attack is visualized in Figure 18(b).

Necessary support [Nouioua and Risch, 2010] captures the intuition that if a supports b, then the acceptance of a is necessary to obtain the acceptance of b, or equivalently, the acceptance of b implies the acceptance of a.

Secondary attack and extended attack For example, in Figure 18.(c), a attacks c, c supports b. The acceptance of a implies the non-acceptance of c and the non-acceptance of c implies the non-acceptance of b; so, the acceptance of a implies the non-acceptance of b. Thus, the secondary attack from a to b is introduced, depicted as a double-headed arrow. Similarly, the extended attack is visualized in Figure 18.(d).

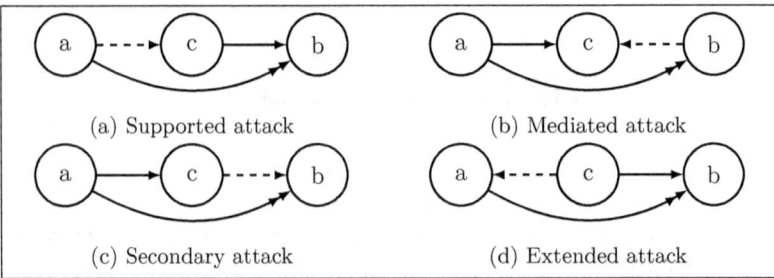

Figure 18: Four kinds of reductions of bipolar argumentation frameworks

We now use the child custody example to illustrate reduction-based semantics.

Example 3.21 (Child custody in bipolar argumentation). *Consider the bipolar framework visualized below. The figure should be read as follows. Normal arrows are attack relations, dashed arrows are support relations, a double box represents a prima facie argument which is self-supporting, and a single box represents a standard argument that does not support itself. Our focus is on how to interpret the support from (OP): "Child* wants to live with her mother" to (M): "Child's best interest is that she lives with her mother." For a comprehensive analysis, we refer to the work of Yu et al. [2020]. The supporting argument (OP) might have a special status because of the rules of the Civil Code: the judge has to take the child's opinion into consideration when deciding about custody. Analysis of this rule shows how various interpretations of the support interpretations relate to legal interpretations. We assume that the child wants to live with her mother (OP). What does this mean? One can say that the obligation to take argument (OP) into consideration means that (OP) is a prima facie argument and thus has to be accepted. If it is a prima facie argument, (M) receives the evidentiary support it needs. But this in itself doesn't decide how argument (OP) affects the extension. The extension depends on how we interpret the support relation between (OP) and (M): deductive or necessary. It seems very intuitive to interpret the support relation as deductive: the obligation to take the child's opinion into consideration is apparently very much in line with what deductive support means: if we accept the child's opinion (which is prima facie) then we have to accept (M) too. We assume the support from W: "Father is wealthy because he inherited" to F: "Child's best interest is that she lives with her father", is not deductive.

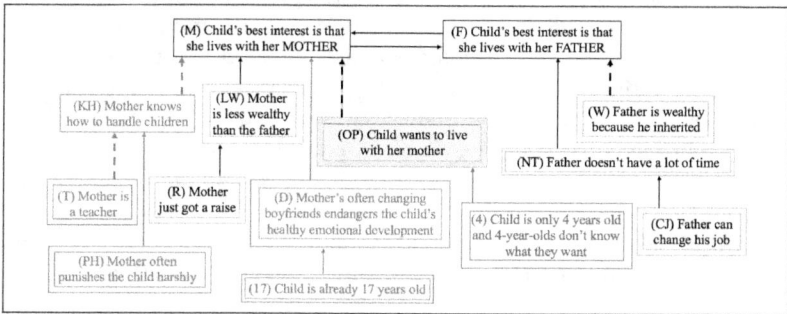

Figure 19: A child custody deliberation with possible arguments and their relations in a divorce case [Yu et al., 2020].

Such an analysis contributes to the discussion on the formalization of legal interpretation in the following way. The role of interpretation is crucial in the law, but it is also a source of criticism of the use of logic-based methods for modeling legal reasoning. For example, it was reminded that Leith warns that the knowledge engineer's interpretation when formalizing norms is necessarily premature because the authority for interpreting the law has been assigned to the judiciary [Prakken, 2013]. Addressing this criticism, the literature on legal interpretation has discussed the possibility that legal knowledge-based systems contain alternative syntactic formalizations. It has been observed that while, on the syntactic level, formalization commits us to a given interpretation, on the conceptual level, classification of factual situations as legal concepts is not an issue of logical form [Prakken, 2013]. Alternatively, we can restrict the investigation by saying that "the only aspects of legal reasoning that can be formalized are those aspects that concern the following problem: *given* a particular interpretation of a body of legal knowledge, and *given* a particular description of some legal problem, what are the general rational patterns of reasoning with which a solution to the problem can be obtained?" [Prakken, 2013]. If a formal framework offers the different interpretations itself, though, then using it might be directly exploitable to the comparison of the different possibilities and routes of reasoning given each interpretation.

It has been argued that for the validation of a bipolar argumentation theory, so-called theory-based validation is preferable to empirical validation [Prakken, 2020; Polberg and Hunter, 2018], which itself is preferable to intuition-based validation. Nevertheless, in this context, the principle-based analysis discussed in Section 4.1 complements these validation methods. The theory of formal argumentation needs to be complemented with examples and case studies concerning the use of the theory.

In this section, we discussed extending the attack-defense paradigm for balancing. We end this section with the following questions:

1. The attack-defense paradigm introduced the distinction between structured and abstract argumentation. For every extension, we need to decide whether to represent it at the structured or abstract level. For example, should justification (as in Fatio) be expressed as in structured argumentation (within the argument) or as a support (among arguments)?

2. When we introduce a new concept like support at the abstract level, it can be interpreted in various ways at the structured level. For example, what does support mean other than inferential relation (e.g., in ASPIC+)?

3. How can we better represent argumentation as balancing (e.g., in law and ethics)? For example, how should the pros and cons be aggregated? What is the role of weights?

4. How should other aspects of balancing be incorporated? For example, how should argument strength be represented and evaluated? It is important to distinguish between different kinds of argument strength, in particular logical, dialectical and rhetorical argument strength [Prakken, 2024a].

5. In the previous and present sections, we discussed extended abstract argumentation inspired by dialogue and balancing. Which other inspirations can be utilized to design extensions of abstract argumentation? For example, how can natural argumentation inspire extensions of abstract argumentation?

4 After the paradigm shift: the computational turn

In this section, we transition from the paradigm shift in formal argumentation to computational argumentation. Section 4.1 introduces principle-based analysis as a methodology for handling the diversity of argumentation models at a higher level of abstraction, providing a systematic approach to designing and choosing methods for different computational contexts. In Section 4.2, we focus on compositionality principles such as locality, modularity, and transparency, which play a central role from the attack-defense perspective and are exploited in algorithmic strategies like divide and conquer. Section 4.3 discusses the relationship between explanation and argumentation, highlighting, for example, strategic argumentation, discussion games and reason-based models for understanding argumentation as dialogue, inference, and balancing, respectively. Finally, Section 4.4 addresses integrating argumentation techniques with existing and emerging technologies, showcasing the potential of distributed argumentation systems and their applications in diverse technological contexts.

4.1 A principle-based analysis

In this section, we turn towards computational argumentation by discussing methodology. Principle-based analysis is a methodology for managing the diversity of argumentation, such as when selecting among existing methods or designing new ones. Principles describe formal argumentation at a higher level of abstraction, and a wide range of principles exists across all models of argumentation.

The principle-based approach is also called the axiomatic approach and the postulate-based approach. Principles are properties, while postulates are normally desirable properties or requirements. This approach is particularly useful when there is a diversity of alternative methods. It has been successfully applied in various areas. For example, Alchourrón-Gärdenfors-Makinson (AGM) postulates [Alchourrón et al., 1985] have been applied in belief revision operations to ensure rationality, and axiomatic principles are applied when searching for and choosing suitable voting rules for various contexts.

The principle-based approach has also been used to describe formal argumentation at a higher level of abstraction. Abstraction in mathematics is the process of extracting the underlying structures, patterns or properties of a mathematical concept. In software engineering and computer science, abstraction is the process of generalizing concrete details. In formal argumentation, one form of abstraction is to focus on the attack and defense relations between arguments rather than their internal structures. The attack-defense relation is used to define the functions of semantics. Principles can be defined as sets of such functions and are represented as constraints on such functions. Principles can be used to compare or define new functions.

The challenge addressed in this section is:

Challenge 10. *Conducting a principle-based analysis of argumentation.*

There is a diversity of principles and postulates in all models of argumentation, particularly in the context of argumentation as inference, less so in argumentation as dialogue and balancing. To illustrate, let us reuse the commutative diagram in Figure 20 featuring examples of principles that are used for different purposes. For step (1), there are Kraus-Lehmann-Magidor (KLM) principles [Kraus et al., 1990] that a logical inference relation ought to satisfy. For structured argumentation in steps (2-4), there are axiomatic analyses of various attack relation assignments among arguments [Dung, 2016; Dung and Thang, 2018; Pardo et al., 2024], as discussed in Section 3.2. For the whole commutative diagram, rationality postulates are used to ensure that the conclusions drawn at the end of the overall process have desirable properties [Caminada and Amgoud, 2005; Caminada and Ben-Naim, 2007; Caminada, 2018b], as discussed in Section 3.3. For step (3), there is a diversity of semantics available for the abstract argumentation framework. Baroni and Giacomin [2007] classified argumentation semantics based on a set of principles, which was extended by van der Torre and Vesic [2018]. For diverse extended argumentation frameworks, with even more semantics. There are the principles-based analysis of ranking-based semantics, multiagent argumentation [Yu et al., 2021], and bipolar argumentation [Yu et al., 2023].

There are three steps in a principle-based approach [van der Torre and Vesic, 2018].

Define the function that will be the object of the study. For instance, abstract argumentation semantics are functions that map graphs to sets of sets of graph nodes.

Define the principles — examine the relations between functions and principles to see if the semantics satisfy the principles.

Classify and study the sets of principles — study the relations between principles. For example, a set of principles may imply another principle. Or there may be incompatibilities among principles. Or there may be a set of principles that characterizes a semantics.

There are three main branches of abstract argumentation semantics.

Admissibility-based semantics (AB) uses gunfight rules requiring that an extension E defends itself against all attackers [Dung,

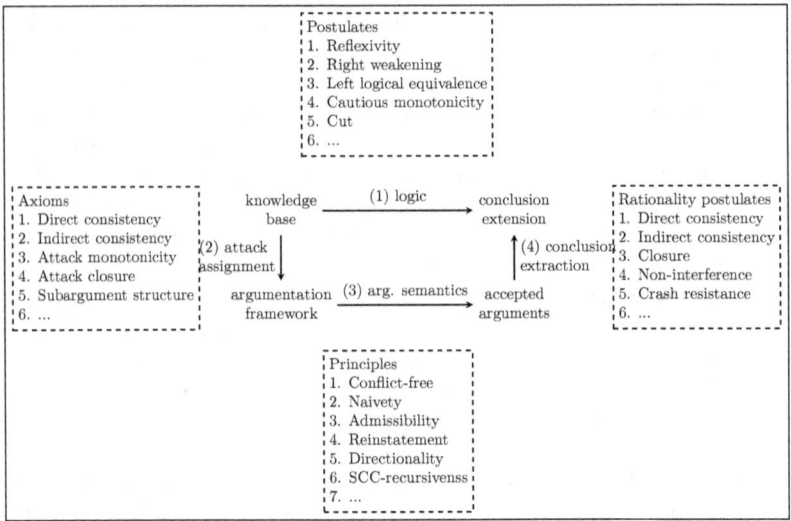

Figure 20: Principle-based analysis in commutative diagram

1995], i.e., whenever each argument attacking E from the outside is itself attacked by some element of E.

Weak admissibility-based semantics (WA) only requires that an extension E defends itself against reasonable arguments [Baumann et al., 2020].

Non-admissibility based semantics (NA) requires an extension E to be a maximal conflict-free set of arguments. The most prominent example of non-admissibility based semantics is CF2 semantics [Baroni et al., 2005].

We illustrate the above three branches of semantics with Example 4.1 below.

Example 4.1 (Three branches of abstract argumentation semantics). *Consider the three frameworks in Figure 21. For framework (a), the only extension in AB semantics is the empty set whereas under the CF2 semantics, b is accepted. To get the desirable properties of directionality* and strongly-connected-component (SCC) recursiveness (discussed further below), CF2 is defined in terms of a local function that computes the maximal conflict-free subsets for each strongly connected component of a framework. Under WA semantics, the set of weakly preferred extensions of framework (a) is $\{\{b\}\}$ because its only attacker a is self-attacking. It is like a "zombie", it is there but it can do no harm [Baumann et al., 2020]. For framework (b), the set $\{d\}$ is not admissible because it does not defend itself from b. Nevertheless, under WA semantics, d is acceptable because b is part of an odd-length cycle of arguments that are never accepted, and so it does not pose an actual threat. These extensions are listed in Table 6.

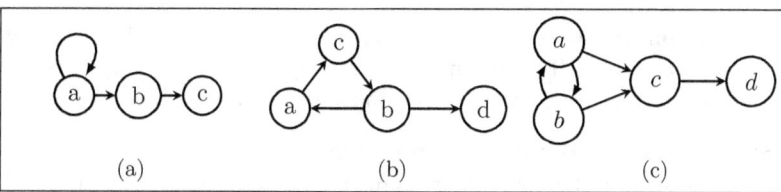

(a) (b) (c)

Figure 21: Three argumentation frameworks

Semantics	(a)	(b)	(c)
AB	$\{\emptyset\}$	$\{\emptyset\}$	$\{\{a,d\},\{b,d\}\}$
NA(CF2)	$\{\{b\}\}$	$\{\{a,d\},\{c,d\},\{b\}\}$	$\{\{a,d\},\{b,d\}\}$
WA	$\{\{b\}\}$	$\{\{d\}\}$	$\{\{a,d\},\{b,d\}\}$

Table 6: Three semantics applied to the frameworks in Figure 21. AB = admissibility-based; NA = non-admissibility based; WA = weak-admissibility based.

To compare the diverse semantics, we can categorize formal argumentation principles into three types: traditional principles (the most discussed), variants of traditional principles, and new principles specifically designed for emerging semantics. Below, we provide examples of these principles to illustrate how they are used. This will enable us to compare the different branches of argumentation semantics as well as the different agent argumentation semantics described in Section 3.4.

We list some of the traditional principles that have been used to compare these semantics.

Conflict-freeness states that every extension under semantics is a conflict-free set, i.e., there is no attack relation among the arguments in the extension.

Admissibility is satisfied by a semantics if and only if every extension under that semantics is admissible.

Naivety states that every extension under the semantics is a maximal conflict-free set.

Directionality states that an argument a should be affected only by a's attacker. The arguments that only receive an attack from a should not have any effect on the state of a.

SCC (strongly-connected-component) recursion states that extension construction carried out in an initial SCC do not depend on those concerning the other ones, while they directly affect the choices about the subsequent SCCs and so on.

Modularity states that the semantics of a framework can be obtained by the semantics of the smaller parts of that framework.

Example 4.2. *Given the framework shown in Figure 22, the complete and weak complete semantics are the same:* $\{\emptyset\}$, *while* $\emptyset$ *is not a maximal conflict-free set in this framework, e.g.,* $\{a\}$ *is also conflict-free.*

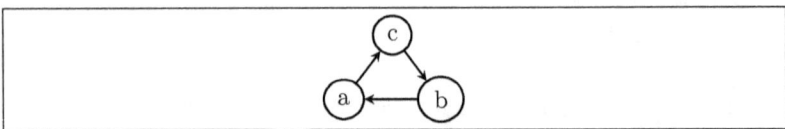

Figure 22: Complete and weak complete semantics do not satisfy naivety

In Section 3.4, we discussed abstract agent argumentation [Yu et al., 2021]. The analysis focused on the four traditional Dung semantics (complete, preferred, stable, and grounded) denoted as TR. Two new concepts, individual defense and collective defense, have been introduced. By applying these concepts to each of the four traditional semantics, there are two new variants for each: one based on individual defense and the other on collective defense. This results in a total of eight distinct defense-based semantics, denoted as Sem_1 and Sem_2. Additionally, social agent semantics, which prioritizes arguments supported by more agents, produce sixteen semantics through four reductions (denoted as $SR_1 - SR_4$) from preference-based argumentation frameworks to abstract argumentation frameworks. Agent reduction semantics, which considers the perspective of individual agents, also yields sixteen semantics through the four reductions, denoted as $AR_1 - AR_4$. Lastly, agent filtering semantics, inspired by limited knowledge, introduces eight additional semantics, denoted as OR and NBR. Altogether, this results in a total of fifty-two semantics.

The paper provides a full analysis of fifty-two agent semantics, including Dung's traditional semantics, with seventeen principles. The results of principle-based analysis are typically summarized in tables, as seen in Table 7, 8, 9. These principles are categorized into three groups: five traditional principles (Table 7), four variations of these traditional principles (Table 8), and eight new principles specifically designed for agent-based argumentation (Table 9).

For example, the agent admissibility principle is a variation of the traditional admissibility principle, with agent defense replacing the standard notion of defense. Since admissibility can be applied to either individual defense or collective defense, this gives rise to two distinct agent admissibility principles. Similarly, the agent SCC-recursiveness principles are adapted to reflect the concepts of individual and collective defense, resulting in two corresponding principles. Additionally, eight entirely new agent principles have been introduced to address the unique aspects of agent-based argumentation, and these are shown in Table 9.

Below we list the eight new principles. Some of them are expected to be satisfied by all of the semantics. Some can be used to distinguish between different semantics, since only certain semantics satisfy or do not satisfy certain principles.

Principle 10: AgentAdditionPersistence states that if more

Semantics	P1	P2	P3	P4	P5
TR	CGPS	CGPS	CGP	CGPS	CGPS
Sem_1	CGPS	CGPS	CGP	S	S
Sem_2	CGPS	CGPS	CGP	S	S
SR_1	×	×	CGP	×	×
SR_2	CGPS	×	×	×	×
SR_3	×	CGPS	CGP	CGPS	CGPS
SR_4	CGPS	G	×	×	G
AR_1	×	×	CGP	×	S
AR_2	CGPS	×	×	×	×
AR_3	CGPS	CGPS	CGP	CGPS	CGPS
AR_4	CGPS	G	×	×	G
OR	CGPS	×	CGP	CGPS	CGPS
NBR	×	×	CGP	×	S

Table 7: Comparison of abstract agent argumentation semantics and traditional principles. When a principle is never satisfied by a certain reduction for all semantics, we use the × symbol, and we use a question mark to represent an open problem. P1 refers to Principle 1, and the same convention holds for all the others. P1 = conflict-free, P2 = admissibility, P3 = directionality, P4 = SCC-recursiveness, P5 = modularity.

agents adopt an argument that is already accepted, this does not affect the extension.

Principle 11: AgentAdditionUniversalPersistence reflects the same idea as principle 10 but is based on the assumption that a is accepted in all extensions.

Principle 12: PermutationPersistence reflects a principle we expect to hold for all agent semantics — anonymity. If we permute the agents, it does not affect the extensions. This principle is analogous to language independence for arguments, as defined by Baroni and Giacomin [2007].

Principle 13: MergeAgent states that if the arguments of two agents do not attack each other, we can merge these agents into one single

Semantics.	P6	P7	P8	P9
TR	×	×	×	×
Sem_1	CGP	CGP	S	S
Sem_2	×	CGP	S	S
SR_1	×	×	×	×
SR_2	×	×	×	×
SR_3	×	×	×	×
SR_4	×	×	×	×
AR_1	×	×	×	×
AR_2	×	×	×	×
AR_3	×	×	×	×
AR_4	×	×	×	×
OR	×	×	×	×
NBR	×	×	×	×

Table 8: Comparison of the reductions and agent admissibility principles, and agent SCC-recursion. P6 = agent admissibility$_1$, P7 = agent admissibility$_2$, P8 = agent SCC-recursiveness$_1$, P9 = agent SCC-recursiveness$_2$.

agent. The principle does not hold for agent defense semantics because new agent defenses may be created.

Principle 14: RemovalAgentPersistence states that if two agents have the same arguments, we can remove one of these agents without changing the extensions. This represents the opposite of social semantics, where the number of agents does make a difference.

Principle 15: AgentNumberEquivalence is inspired by social agent semantics. It states that where there are two argumentation frameworks with the same arguments and attacks, if for every argument the number of agents holding that argument is the same, then the extensions are the same.

Principle 16: ConflictfreeInvolvement is inspired by agent reduction semantics. It states that if the set of an agent's arguments is conflict-free, then there is an extension containing those arguments.

Principle 17: RemovalArgumentPersistence is inspired by OrphanReduction semantics. It states that if we have arguments that do not belong to any agents, then they can be removed from the framework without affecting the extensions.

In the resulting Table 9, all the agent semantics satisfy P12. Perhaps surprisingly, neither social agent semantics nor agent reduction semantics satisfy P10 while trivial reduction semantics, social agent semantics, and agent filtering semantics satisfy P13. Moreover, all agent semantics except social agent semantics satisfy P14. No semantics satisfy P16. As expected, only OrphanRemoval satisfies P17. The only semantics that are not distinguished yet concern the use of different preference reductions, or different Dung semantics. To distinguish between these, the principles proposed in preference-based argumentation [Kaci et al., 2021] and in Dung's semantics can be used [Baroni and Giacomin, 2007; van der Torre and Vesic, 2018]. In that sense, the principle-based analyses can complement one other.

Finally, the principle-based approach to formal argumentation may lead to the study of impossibility and possibility results. For instance, Arrow's impossibility theorem in voting and social choice theory demonstrates that no voting system can simultaneously satisfy the whole set of seemingly reasonable criteria — non-dictatorship, unrestricted domain, Pareto efficiency, and independence of irrelevant alternatives — when there are three or more options available. This kind of result highlights the inherent trade-offs involved in designing systems that attempt to balance competing principles. Similarly, as discussed in Section 3.2, any attempt to realize PDL in ASPIC+ should preserve the definitional principle of attack closure. The impossibility theorem explains how this is incompatible with context independence [Pardo et al., 2024]. These impossibility results are crucial because they reveal the boundaries of what can be achieved within a given formal system. Additionally, they also guide researchers to either accept certain trade-offs or seek alternative approaches that might circumvent these limitations.

In this section, we discussed principle-based analysis and, as usual, we list several research questions about that topic.

Sem.	P10	P11	P12	P13	P14	P15	P16	P17
TR	CGPS	CGPS	CGPS	CGPS	CGPS	CGPS	×	×
Sem_1	S	S	CGPS	×	CGPS	×	×	×
Sem_2	S	S	CGPS	×	CGPS	×	×	×
SR_1	×	CGPS	CGPS	CGPS	×	CGPS	×	×
SR_2	×	CGPS	CGPS	CGPS	×	CGPS	×	×
SR_3	×	CGPS	CGPS	CGPS	×	CGPS	×	×
SR_4	×	CGPS	CGPS	CGPS	×	CGPS	×	×
AR_1	×	CGPS	CGPS	×	CGPS	×	×	×
AR_2	×	CGPS	CGPS	×	CGPS	×	×	×
AR_3	×	CGPS	CGPS	×	CGPS	×	×	×
AR_4	×	CGPS	CGPS	×	CGPS	×	×	×
OR	CGPS	CGPS	CGPS	CGPS	CGPS	CGPS	×	CGPS
NBR	CGPS	CGPS	CGPS	CGPS	CGPS	×	×	×

Table 9: Comparison between the reductions and new agent principles. P10 = AgentAdditionPersistence, P11 = AgentAdditionUniversalPersistence, P12 = PermutationPersistence, P13 = PermutationPersistence, P14 = RemovalAgentPersistence, P15 = AgentNumberEquivalence, P16 = ConflictfreeInvolvement, Principle 17 = RemovalArgumentPersistence.

1. How can we provide guidance to users who are not experts in formal or computational argumentation on how to use the theory of argumentation for their needs? Can we develop a user guide for the theory of argumentation?

2. How do we decide which conceptualization of formal argumentation to use for an application (argumentation as inference, dialogue or balancing), and how do we connect or combine these conceptualizations?

3. What needs to be changed to move from constructing comparison tables (as shown in Tables 7-9) to characterization, or proving possibility and impossibility results? For example, how to characterize last vs. weakest link in structured argumentation, or characterize various kinds of abstract argumentation semantics?

4. Which methodology can be developed for formal and computational argumentation to guide the search for and design of principles? For example, the principle of resolution was defined by Baroni and Giacomin [2007], then the resolution-based semantics were defined and studied by Baroni *et al.* [2011].

5. How can we combine principles from various extended argumentation frameworks? For example, reductions in preference-based argumentation often remove attacks whereas reductions in bipolar argumentation often add attacks.

4.2 Algorithmic argumentation

In this section, we consider the role of principles in algorithmic argumentation. Algorithmic argumentation, as illustrated in Table 1, refers to a step-by-step procedure or set of rules designed to perform a specific task or solve a particular argumentation problem. We focus mainly on the calculation of argumentation semantics. On the one hand, compositionality principles play a central role in the attack-defense perspective. On the other hand, algorithms and other computational approaches exploit these principles. We illustrate this by discussing locality, modularity, and transparency principles on one side, and "divide and conquer" and robustness principles on the other.

Traditionally, Dung's abstract argumentation frameworks are viewed as monolithic entities where various semantics are applied globally to determine which arguments are acceptable. While this unified approach preserves generality, it has been shown to be computationally intractable. That complexity presents the following challenge:

Challenge 11. *Designing efficient algorithms for argumentation semantics.*

The idea of compositionality is that an abstract argumentation framework is broken down into interacting smaller subframeworks. This motivates a *local focus* and further investigation into locality and modularity principles in abstract argumentation. Related principles are, for example, directionality, SCC-recursiveness, and decomposability.

The directionality property corresponds to the idea that the attack relation encodes a form of dependency and that arguments can affect one other only by following the direction of the attacks. This consideration can be extended from individual arguments to sets of arguments. If a set of arguments is unattacked (i.e., it does not receive attacks from arguments outside the set) it should not be affected by the rest of the argumentation framework. In more formal terms, given an argumentation semantics, projecting the semantics of the global framework onto an unattacked set should result in the semantics producing an evaluation of an argumentation framework consisting of only that unattacked set.

Example 4.3 illustrates directionality and why stable semantics does not satisfy this property.

Example 4.3 (Directionality [Baroni and Giacomin, 2007]). *Consider the stable semantics of the argumentation framework in Figure 5. For the subframework consisting of $\{a, b\}$, the stable semantics is $\{\{a\}, \{b\}\}$. The stable semantics of the whole framework is $\{\{b, d\}\}$ whose projection — the unattacked set $\{a, b\}$ — is $\{b\}$. However, $\{\{b\}\}$ does not coincide with the stable semantics of the unattacked framework. This is a counterexample proving that stable semantics does not satisfy the directionality.*

The property of SCC-recursiveness [Baroni et al., 2005] is based on decomposition along the SCCs of the argumentation framework. Different from directionality, SCC-recursiveness has a direct constructive interpretation. The structure of the argumentation frameworks drives the incremental definition of extensions step by step. Without complex technicalities, we illustrate SCC-recursiveness with Example 4.4.

Example 4.4 (SCC-recursiveness [Baroni et al., 2005]). *Consider again the framework in Figure 5.*

Step 1 *Partition the argumentation framework into SCCs. There are two SCCs in the framework: $S_1 = \{a, b\}$ and $S_2 = \{c, d, e\}$. Here, S_1 is identified as the initial SCC as it does not depend on S_2.*

Step 2 *Construct extensions incrementally using a base function. Determine the possible extensions within each initial SCC using a semantic-specific base function. This function returns the extensions for argumentation frameworks consisting of a single SCC. For the SCC of $S_1 = \{a, b\}$, the base function provides two possible partial candidate extensions: $\{a\}$ and $\{b\}$.*

Step 3 *For each possible extension determined in Step 2, apply the reinstatement principle. This involves suppressing the nodes directly attacked within subsequent SCCs and considering the distinction between defended and undefended nodes. Let's take candidate extension $\{b\}$. Here, argument c in S_2 cannot be included in the extension because it is attacked by argument b. Therefore, only $\{d, e\}$ can be taken into consideration.*

Step 4 *Recursively apply the steps on restricted frameworks. Consider the subframework $(\{d, e\}, \{(d, e)\})$. This subframework is again partitioned into SCCs, resulting in $S'_1 = \{d\}$ and $S'_2 = \{e\}$. Considering $S1' = \{d\}$, since e is attacked by d, e is excluded. Hence, the only extension left is $\emptyset$. Thus, the final extension is $\{b\} \cup \{d\} \cup \emptyset = \{b, d\}$.*

The idea of decomposability is to break down an abstract argumentation framework arbitrarily into interacting smaller subframeworks called modules. Input/Output frameworks have been defined on this basis [Baroni *et al.*, 2014]. Each module can be described as a black box whose Input/Output behavior — specifically referring to the labeling — fully determines its role in the system's global behavior. That makes it possible to describe and analyze the framework's global behavior in terms of the combination of the local behaviors of its constituent modules. Each local behavior can be characterized individually. This characterization is independent of the internal details of other modules. Instead, it focuses only on the connections and mutual interactions between the module and the other modules. Additionally, if two modules have the same input and output behavior, they are interchangeable in a way that does not influence the global behavior.

Example 4.5 illustrates the interfaces of subframeworks.

Example 4.5 (Interface [Baroni *et al.*, 2014])**.** *Given the abstract argumentation framework (AF) visualized in Figure 23 and the subframeworks induced by the sets $\{a, b, c\}$ and $\{d\}$, these subframeworks are denoted as $AF_{\downarrow\{a,b,c\}}$ and $AF_{\downarrow\{d\}}$. The subframeworks interact with one other through the attacks $a \to d$ and $d \to a$ respectively. For $AF_{\downarrow\{a,b,c\}}$, the interface, or input argument, is argument d, while for $AF_{\downarrow\{d\}}$, the input argument, or interface, is argument a.*

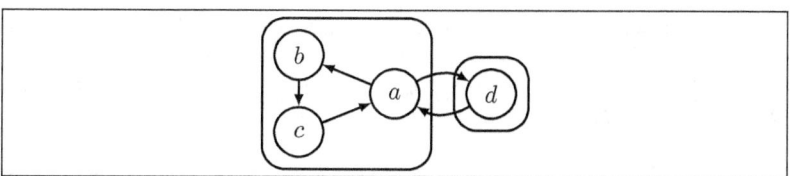

Figure 23: A partition of an abstract argumentation framework

A local function determines the labeling of a subframework based on the labeling of external input arguments, ensuring that the internal labeling of a subframework is consistent with its external influences. Example 4.6 illustrates how external input arguments enforce the internal labelings.

Example 4.6 (External input arguments enforcement [Baroni et al., 2014]). *Consider the argumentation framework in Example 4.5. If we apply the local function to subframework $AF_{\downarrow\{a,b,c\}}$ with the external argument d, if d is labeled as in, the resulting labeling of the subframework is $\{(a, out), (b, in), (c, out)\}$. If d is labeled as out, the resulting labeling of the subframework is $\{\{(a, undec), (b, undec), (c, undec)\}\}$. If d is labeled as undec, the resulting labeling of the subframework is $\{\{(a, undec), (b, undec), (c, undec)\}\}$. We can apply the same analysis for the subframework $AF_{\downarrow\{a,b,c\}}$ with external argument a.*

The property of *decomposability* states that given an arbitrary partition of an argumentation framework into a set of subframeworks, the outcomes produced by a given semantics can be obtained as a combination of the outcomes produced by a local counterpart applied separately on each subframework and vice versa.

Example 4.7 (Decomposability [Baroni et al., 2014]). *Considering again the argumentation framework in Example 4.5 and the partition $\{\{a, b, c\}, \{d\}\}$, the decomposability of the complete semantics requires a local function such that the labelings of AF are exactly those obtained by the union of the compatible labelings of $AF_{\downarrow\{a,b,c\}}$ and $AF_{\downarrow\{d\}}$ given by the local function itself. The labeling $\{(a, out), (b, in), (c, out)\}$ is compatible with $\{(d, in)\}$. The first is obtained with d labeled in, and the latter is obtained with a labeled out. On the other hand, the labeling $\{(a, out), (b, in), (c, out)\}$ is not compatible with, e.g., $\{(d, out)\}$.*

Overall, exactly two global labelings arise from combining the compatible outcomes — $\{(a, undec), (b, undec), (c, undec), (d, undec)\}$ and $\{(a, out), (b, in), (c, out), (d, in)\}$, which corresponds to the complete labelings of the whole AF.

The property of *transaparency* states that if two modules have the same Input/Output behavior, then we can replace one with the other without influencing the framework's global behavior. This ensures that the invariant part of the framework is unaffected. Example 4.8 illustrates the transparency property.

Example 4.8 (Transparency [Baroni et al., 2014]).
Consider the argumentation frameworks AF_1 and AF_2 shown in Figure 24. $\mathcal{M}_1$ and $\mathcal{M}_2$ have the same Input/Output behavior, i.e., they are equivalent under preferred semantics. The invariant set of the replacement is the set $\{e_1, e_2\}$. However, after the replacement of $\mathcal{M}_1$ by $\mathcal{M}_2$ in AF_1, the preferred extension changes. In fact, the preferred labelings of AF_1 are $\{(a_1, in), (a_2, out), (o, out), (e_2, in), (e_1, out)\}$ and $\{(a_1, out), (a_2, in), (o, undec), (e_2, undec), (e_1, undec)\}$, while $\{(b, in), (c, out), (a_1, in), (a_2, out), (o, out), (e_2, in), (e_1, out)\}$ is the only preferred labeling of AF_2.

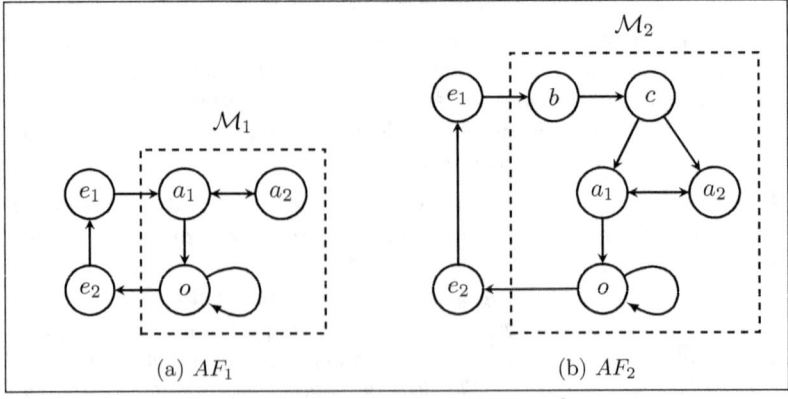

Figure 24: Preferred semantics does not satisfy transparency

Now, we illustrate how to exploit modularity, which leads us to the concept of summarization [Baroni et al., 2014]. Summarization allows a complex part of an argumentation process, such as the analysis and discussion of factual evidence in a legal case, to be replaced by a more synthetic representation. This process focuses on the facts that have an actual impact on the decision, leaving out unnecessary details. The main concern with summarization is ensuring that, as the argumentation framework is simplified, the overall outcome remains consistent and preserved.

We use Example 4.9 to illustrate how summarization works.

Example 4.9 (Summarization [Baroni et al., 2014]). *Consider the argumentation frameworks AF_1 and AF_2 shown in Figure 25. AF_2 can be obtained from AF_1 by "summarizing" the component $\mathcal{M}_1$, including the arguments a_1, a_2, a_3 and a_4, with the component $\mathcal{M}_2$, including the arguments a_1 and a_2. Then, e_1 and e_2 are the same in the two frameworks, i.e. e_1 is labeled in and e_2 is labeled out. More generally, consider a finite sequence of n arguments $a_1, \ldots, a_n$ such that each argument attacks the subsequent one, i.e. a_i attacks a_{i+1} with $1 \leq i < n$, and suppose that only a_1 can receive further attacks from other arguments and that only a_n can attack other arguments. Then, it is intuitive to see that the "black-box behavior" of a sequence of arguments of this kind, whose external "terminals" are a_1 and a_n, only depends on whether n is even or odd. In fact, the behavior of any even-length sequence is the same as where $n = 2$ (if a_1 is in then a_n is out, if a_1 is out then a_n is in, if a_1 is undec then a_n is undec), while for any odd-length sequence, the behavior is the same as for a_1 alone (if n is an odd number, a_n necessarily gets the same label as a_1).*

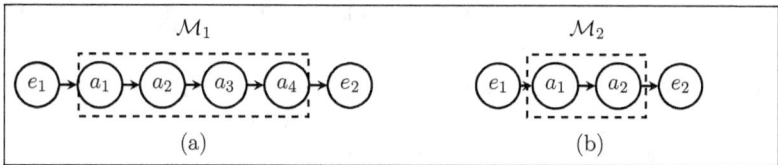

Figure 25: Summarizing a chain of arguments

The first area where locality and modularity principles find application is in the development of algorithms for efficiently computing argumentation semantics, particularly through divide-and-conquer strategies. By leveraging locality, one can focus on specific parts of an argumentation framework and thus reduce the computational burden. There are three locality- and modularity-based approaches that demonstrate how these principles can enhance the efficiency of computing semantics in dynamic, static, and partial argumentation frameworks. For instance, when only partial semantics are required — such as in scenarios where the status of certain arguments needs to be queried — algorithms can be designed to focus solely on the relevant arguments, disregarding those that do not impact the outcome. Similarly, in dialogues where new arguments are introduced, the computation can be streamlined by ignoring the effects of irrelevant arguments. For a comprehensive overview of these approaches, we refer to the work of Baroni *et al.* [2018].

Other principles used in the design of algorithms include robustness principles [Rienstra *et al.*, 2020], which deal with the behavior of a semantics when the argumentation framework changes due to the addition or removal of an attack between two arguments. Robustness principles have been classified into two kinds: persistence principles and monotonicity principles. The former deal with the question of whether a labeling in an argumentation framework under a given semantics persists after a change. The latter deal with the question of whether new labelings are created after a change. They are listed as follows:

XY addition persistence: a labeling of an argumentation framework in which x is labeled X and y is labeled Y is still a labeling of F after adding an attack from x to y.

XY removal persistence: a labeling of an argumentation framework in which x is labeled X and y is labeled Y is still a labeling of F if removing the attack from x to y.

XY addition monotonicity: if in all labelings of an argumentation framework, x is labeled X and y is labeled Y, then adding an attack from x to y does not lead to new labelings.

XY removal monotonicity: if in all labelings of an argumentation framework, x is labeled X and y is labeled Y, then removing an attack from x to y does not lead to new labelings.

Persistence and monotonicity principles are also useful for addressing enforcement problems [Baumann, 2012] in abstract argumentation. This is about the problem of determining minimal sets of changes to an argumentation framework in order to enforce some result, such as the acceptance of a given set of arguments. Because persistence and monotonicity principles can be used to determine which changes to the attack relation of an argumentation framework do or do not change its evaluation, these principles can be used to guide the search for sets of changes that address the enforcement problem. This idea has already been used for extension enforcement under grounded semantics [Niskanen et al., 2018].

In this section, we have discussed compositionality principles, algorithms and other computational approaches that exploit principles. We end this section with the following questions:

1. Most algorithms are developed for abstract argumentation and for argumentation as inference. What are the computational tasks for structured argumentation, for argumentation as dialogue, and for argumentation as balancing?

2. Apart from algorithms, what other tools does computer science have to offer, e.g., analysis of computational complexity, efficient implementation of algorithms?

3. As discussed in this section, principle-based analysis is a bridge between formal and algorithmic argumentation. Which principles are particularly useful? And how can we use these principles to speed up computation?

4. What else can we learn from artificial intelligence? The rise of machine learning and foundation models is changing the landscape of argumentation. How could these new approaches speed up computation?

5. Which topics need to be addressed first in computational argumentation? Should we address algorithms for formal argumentation concepts or focus our attention on the challenges of natural argumentation?

4.3 Explanation and argumentation

In this section, we discuss some relations between computational argumentation and explanation. Strategic argumentation explains argumentation as dialogue, discussion games explain argumentation as inference, and reason-based explanation can be used for argumentation as balancing. We illustrate the explanation for argumentation through the example of discussion games for grounded semantics.

In recent years, the field of explainable artificial intelligence (XAI) [Longo et al., 2024] has gained significant attention due to the increasing complexity and opacity of AI systems, particularly when it comes to systems being potentially deployed in critical decision-making areas such as healthcare, finance, and the law. The main focus is usually on making the reasoning behind the decisions or the predictions made by the AI system [Phillips et al., 2021] more understandable and transparent.

The relationship between explanation and argumentation can be seen from different perspectives. On the one hand, an explanation for argumentation mostly concerns the question of whether a certain argument or claim can be accepted (or not) and why. This has been studied not only at the abstract level [Ulbricht and Wallner, 2021] but also at the structured level [Borg and Bex, 2024]. On the other hand, explanation through argumentation is often intuitively seen as reasonable [Sklar and Azhar, 2018]. For example, it can clarify the decision-making process of an AI system through argumentation procedures. This can be done in a static manner by illustrating the argument inference process or showing the relations between arguments, or it can be done through an interactive dialogue that explains the reasoning [Castagna et al., 2024b]. In this section, we talk about the following challenge:

Challenge 12. *Explaining argumentation.*

Take the example of argumentation as inference, which is about how reasonable decisions or conclusions can be reached by constructing for and against arguments and then evaluating those arguments. It makes it possible to understand decisions by tracing exactly why a particular conclusion was reached. It also makes it possible to see how certain decisions it relates to other potential conclusions. Explanations are often found to be argumentative. Mercier and Sperber [2017] highlighted that the effectiveness of interactive argumentation in changing people's minds, at least for simple arguments, stems from the chance to address counterarguments during discussions. Participants can raise and rebut counterarguments, which makes the exchange more dynamic. Contrast this with one-sided messaging campaigns, where counterarguments are generated but remain unaddressed [Altay *et al.*, 2022]. Interactive argumentation can involve a form of dialogue where users interact with an AI system, asking for clarifications or further information, and the system responds with explanations. In this sense, explanation is intertwined with dialogue — a conversation where arguments are presented, challenged, and defended, as in the Fatio design [McBurney and Parsons, 2004].

Strategic argumentation (see Governatori *et al.*'s [2021] overview) can be used to explain argumentation as dialogue. By analyzing the strategies employed by an agent, it is possible to understand how and why that agent chooses to disclose certain arguments or information during a debate in order to achieve a specific objective and prevent the opposing party from gaining an advantage.

To give an example, Arisaka *et al.* [2022] propose an abstract agent argumentation model that distinguishes the global argumentation of judges from the local argumentation of accused persons, prosecutors, defense lawyers, witnesses and experts. All the "local" agents have partial knowledge of the arguments and attacks of the other "local" agents, on which basis they decide autonomously whether to accept or reject their own arguments and whether to bring their own arguments forward in court. The arguments accepted by the judge are based on a game-theoretic equilibrium among the argumentation of the other agents. The theory can be used to distinguish between the various direct and indirect ways in which an agent's arguments can be used against his/her other arguments. The global abstract agent argumentation framework is viewed differently by the different agents.

Example 4.10 (Murder at Facility C)**.** *There was a murder at Facility C. Acc has been accused of the crime. There is a witness Wit and a prosecutor Prc. Acc has two arguments in mind:*

a_1 *He was at Facility A on the day of the murder [this is a fact Acc* knows].

a_2 He is innocent [this is Acc's claim].

Prc entertains the following arguments:

a_6 Only Acc could have killed the victim [this is Prc's claim].

Meanwhile, Wit has certain beliefs on the basis of which he has three arguments:

a_3 Acc stayed home on the day of the murder, having previously lost his ID card [this Wit originally believes to be a fact].

a_4 Acc could enter any facility provided he had his ID card on him [this is a fact known to Wit].

a_5 Acc could not have been at Facility C at the time of the murder [this is Wit's claim].

Further, the relationship between the three arguments is such that a_3 attacks a_4, which attacks a_5. Altogether, these arguments by the three agents form the argumentation framework in Figure 26(a). Acc, Prc and Wit reveal their own internal argumentation, partially or elaborately, for the judge to evaluate. But since agents may come to learn the arguments of other agents if, say, they are expressed before they present their own arguments, it is possible that they take the additional information into account when deciding which arguments to present. In this example, both Prc and Acc have the characteristic of being unaware agents. Prc has no reason to drop argument a_6, and neither does Acc, as he sees no benefit in admitting that a_6. However, how Wit responds to the fact known to Acc (a_1) could prove crucial to whether he is found innocent or guilty.

Case A. Suppose Wit is unaware and open-minded. Wit presents what he believes, i.e., his local argumentation framework (see Figure 26(a).) She locally accepts a_3 and a_5. The judge evaluates all the arguments, concluding that a_2 is not acceptable, i.e., Acc is guilty. The judge starts his inference with Acc's acceptable a_1 and proceeds to reject a_2. The two arguments a_3 and a_5 accepted by Wit are not accepted by the judge. This illustrates indirect use of an argument against Acc.

Case B. *Suppose Wit is unaware and closed-minded. Instead of presenting all the reasoning he had developed in his local argumentation, Wit states the following key points concisely: that Acc stayed home on the day of the murder, and that Acc could not have been at Facility C (see $\mathcal{F}_i$). Omission of a fact known to Wit (a_4), which Wit perhaps considers irrelevant to the criminal case, changes the judge's decision completely. In $\mathcal{F}_i$, a_5 is seen an argument that is globally acceptable. That argument rejects a_6 in favor of a_2.*

Case C. *Suppose Wit learns a_1 beforehand. Wit realizes that a_3, which she thought was a fact, is not actually true. She no longer claims a_5 in her local argumentation, but she nevertheless discloses her entire original argumentation (see Figure 26(a)). Her conclusion that a_4 is acceptable concurs with the judge's view on the matter, and the judge ultimately concludes that a_2 shall be rejected.*

Case D. *Suppose again that Wit learns a_1 beforehand but that she mentions the key arguments concisely. She states that entry into any facility requires an ID card (see $\mathcal{F}_j$). Here again, the judge has no objection to the evidence that might have been provided by Proc. As such, a_6 is accepted, which proves that Acc is guilty. This illustrates direct use of an argument by Wit against Acc.*

We now move on to discussion games designed to explain argumentation as inference. As discussed in Section 4.2, calculating semantics, or determining whether a specific argument is present in some or all labelings, can be computationally expensive. Discussion games provide an alternative to that. Discussion games [Caminada, 2017] take place between two parties, typically called the "proponent" and the "opponent", who argue about whether a particular argument within a formal argumentation framework should be accepted. Discussion games can be seen as proof procedures for the argumentation semantics they are associated with, e.g., grounded semantics, preferred semantics, or stable semantics. These games provide a local explanation, focussing on the admissibility of arguments.

A discussion game for grounded semantics is won by one agent iff a particular argument is labeled *in*. There are two players: Proponent (P)

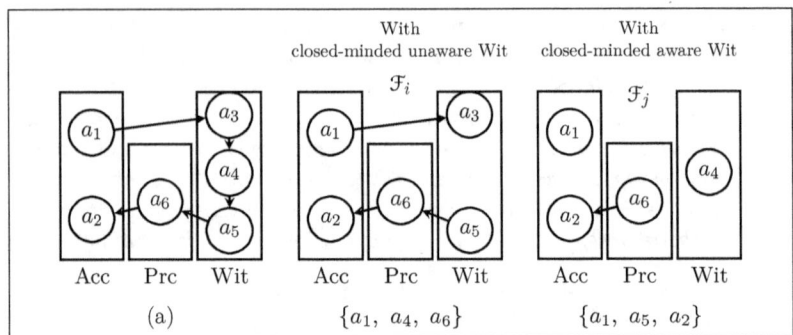

Figure 26: Left: Argumentation by an accused (Acc), by a witness (Wit), and by a prosecutor (Prc). Middle: multiagent semantics with open-minded unaware Wit ($\mathcal{F}_i$). Right: multiagent semantics with closed-minded aware Wit ($\mathcal{F}_j$).

and Opponent (O). There are four moves with respect to arguments A and B:

P: HTB(A): The labeling of A is *in*.

O: CB(A): Maybe the labeling of A is *out* in every complete labeling.

O: CONCEDE(A): Agree that the labeling of A is *in* in every complete labeling.

O: RETRACT(A) The labeling of A is *out* in every complete labeling.

The following are the discussion rules on grounded semantics.

P: HTB(A): Either this is the first move, or:

- the previous move was **CB(B)**, where A attacked B, and:

- no **CONCEDE** or **RETRACT** move is applicable.

O: CB(A): A is an attacker of the last **HTB(B)** statement, which has not yet been conceded, and:

- the directly preceding move was not a **CB** statement,

- argument A has not yet been **RETRACT**ed, and
- no **CONCEDE** or **RETRACT** move is applicable.

O: **CONCEDE(A)**: There has been a **HTB(A)** statement in the past, and

- every argument attacking **HTB(A)** has been **RETRACT**ed, and
- **CONCEDE(A)** has not yet been moved.

O: **RETRACT(A)**: There has been a **CB(A)** statement in the past, and:

- there exists an argument attacking **CB(A)** that has been **CONCEDE**ed, and
- **RETRACT(A)** has not yet been moved.

General rule: No "HTB or CB repeats" are allowed. **HTB(A)** is only allowed once, **CB(A)** is only allowed once. For any A, not both of **HTB(A)** and **CB(A)** are allowed.

We use Example 4.11 to illustrate how grounded discussion games work.

Example 4.11 (Grounded discussion game). *Given the abstract argumentation framework visualized in Figure 27, the grounded discussion game for argument C proceeds as follows:*

(1) P : HTB(c)   *(3) P : HTB(a)*   *(5) O : RETRACT(b)*
(2) O : CB(b)   *(4) O : CONCEDE(a)*   *(6) O : CONCEDE(c)*

P wins the grounded discussion game for argument c, and c is labeled in in the grounded labeling.

Yet, despite its potential advantages, the use of argumentation as balancing for explanatory purposes has not been adequately explored. This method employs reason-based [Knoks and van der Torre, 2023] and scale-based balancing as a decision model. Thus, the explanation could provide an overview of the pros and cons concerning a decision, like judges do when explaining their rulings.

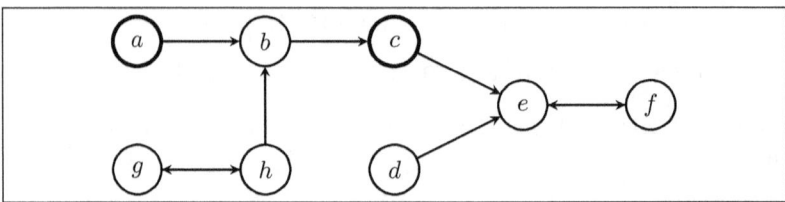

Figure 27: An abstract argumentation framework where c is being discussed

Explanations are available to not just the experts who designed the system (which is then called a white box [Vilone and Longo, 2021]), but also to lay people — non-experts who may not understand all the intricacies of certain models. This issue is about how to personalize explanations, which is particularly relevant given the diversity of backgrounds, contexts, mental states, emotions, and abilities of subjects receiving explanations generated by AI systems (humans such as patients and healthcare professionals, or virtual intelligent autonomous agents). To this end, new forms of knowledge representation should be envisioned and synergistically integrated to enable argumentation reasoning over that.

In this section, we discussed argumentation and explanation. We end this section with the following open questions:

1. Argumentation and explanation can be related to each other in various ways. What is the role of argumentation in explanation, and what is the role of explanation in argumentation? For example, how can the Fatio dialogue protocol be extended with explanation dialogues?

2. Dialogue is often cited as a distinctive feature of argumentation that can be used for interactive explanation in human-computer interactions, but other features may also be relevant. How about, for example, using balancing or inference in explanation?

3. On the topic of dialogue, it is often observed that argumentation is persuasion and that there are many other kinds of dialogues types. How can different dialogue types be integrated into dialogue

systems? For example, how can information seeking be included in argumentation?

4. Since explanations are often personal, expressed in natural language, and use common sense reasoning, foundation models and chatbots have been promoted as part of the explanation toolbox. How can we use LLMs in argumentation to incorporate context, mental states and emotions?

5. Explanation techniques can be evaluated in terms of the degree to which they improve a system's goals, i.e., how does the combination of argumentation and explanation techniques improve system behavior?

4.4 Argumentation technology

We conclude our discussion of computational argumentation by discussing the integration of argumentation techniques with existing and emerging technologies in computer science like NLP, LLMs, distributed argumentation technology, and dialogue technology. We illustrate the integration of these technologies by using as an example the integration of argumentation with blockchain technology into the architecture of the IHiBO.

Recent years have seen remarkable advancements in deep learning, particularly with the development and deployment of LLMs. This presents a significant opportunity to integrate argumentation. Argumentation is inherently suitable for enhancing the reasoning and conversational capabilities of LLMs [Bezou-Vrakatseli, 2023; Castagna *et al.*, 2024c]. Argumentation provides a formal mechanism for capturing interactions between agents, and it manages the information conflicts that arise during these interactions. This makes it an potentcial tool for improving the logical consistency and depth of responses generated by LLMs.

Additionally, LLMs prompts to reevaluate the relationship between abstract and structured argumentation. Traditionally, structured arguments were necessary because the attack relations were defined based on the internal structure of the arguments, as discussed in Section 3.2.

However, with LLM capabilities, it becomes possible to retain arguments in their natural language form and use an LLM to extract the underlying argumentation framework. This approach allows argumentation to be integrated more naturally with conversational models because LLMs provide the contextual understanding needed to facilitate these processes.

Furthermore, continuous improvements in computational power, together with the capabilities of foundational models like LLMs, have opened up new avenues for integrating argumentation into more complex systems. In such systems, argumentation can synergize with other technologies, enhancing the overall functionality and enabling more sophisticated applications. Such integration will not only advance the field of computational argumentation but will also push the boundaries of what can be achieved in AI-driven reasoning and decision-making systems. In this section, we discuss the following challenge:

Challenge 13. *Integrating argumentation with technologies.*

Distributed argumentation technology [Yu, 2023] is a computational approach that incorporates argumentation reasoning mechanisms within multiagent systems. An instantiation of distributed argumentation technology is *Intelligent Human-Input-Based Blockchain Oracle (IHiBO)* [Yu et al., 2022]. The motivation for IHiBO comes from fund management for the securities market. Figure 28 shows a toy fund management procedure. Investors first pool their money together and then fund managers conduct investment research and prepare the specific plan for the investment portfolio. Fund managers invest securities on behalf of their clients (investors) according to their research and the final decision in the investment plan. The investment generates returns, and the returns are passed down to investors. Fund managers play an important role in the investment and financial world as they give investors peace of mind that their money is in the hands of experts. However, reality is not always as hoped for and investors are supposed to know (but do not actually know) where their money is going, why, and how much is the true profit.

A significant aspect of IHiBO is its human-input-based oracle, which bridges the gap between a blockchain and real-world data, allowing human experts to input information into the blockchain. IHiBO was envi-

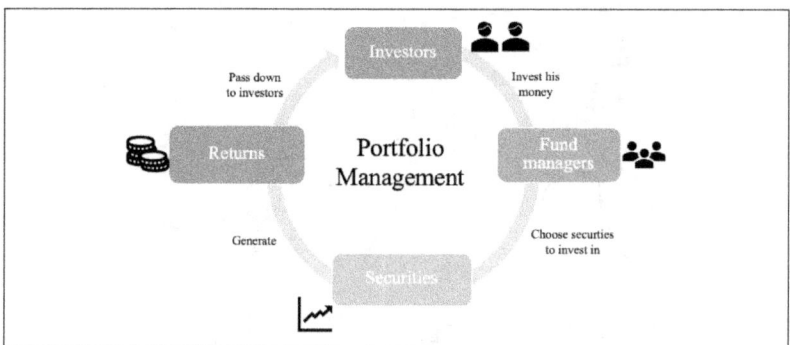

Figure 28: A fund investment process

sioned for use in fund management, where managers provide their arguments in terms of the investment plan for stocks. Specifically, IHiBO utilizes multiagent abstract argumentation frameworks to model decision-making processes, which are then implemented by smart contracts and stored on a blockchain. The integration of argumentation reasoning and blockchain makes the decision-making process more transparent and traceable.

IHiBO leverages a two-layer distributed ledger technology (DLT), shown in Figure 29, to ensure security and immutability of data while maintaining efficiency and scalability. The secondary layer, a private permissioned blockchain, accessible only to authorized users, facilitates the decision-making smart contract. This layer maintains privacy and reduces transaction costs, providing a balanced approach to data security and operational efficiency. The main layer is a public permissionless blockchain such as Ethereum, where the smart contract for executing stock transactions is invoked by the output of the decision-making process. This two-layered design is particularly important in fund management, where decisions may involve sensitive information. IHiBO's architecture supports not only multiagent abstract argumentation but also other kinds of reasoning that can be encoded in smart contracts.

In this section, we have discussed he integration of argumentation techniques with existing and emerging technologies, such as IHiBO. We end this section with the following questions.

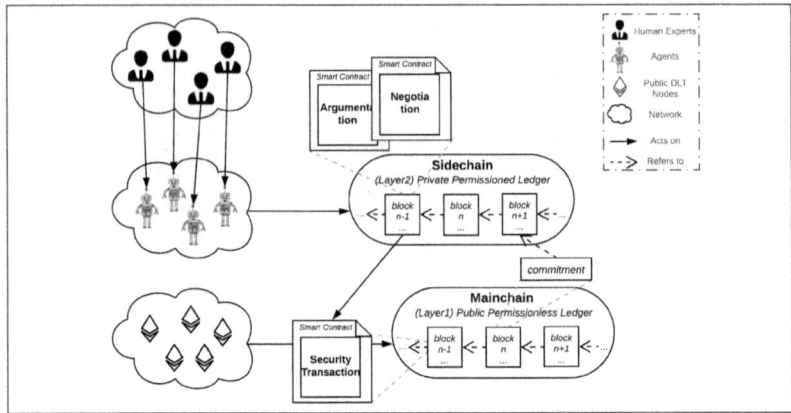

Figure 29: The architecture of the IHiBO framework. DLT = distributed ledger technology.

1. What can technologies do for argumentation, and what can argumentation do for technologies? For example, what can foundational models do for (natural, formal, and computation) argumentation? How can LLMs be used to develop technologies like IHiBO 2.0? Another discussed example outlines the potential roles of computational models of legal argumentation [Prakken, 2024b]: as tools for guiding prompt engineering, as benchmarks for evaluating the outputs of legal generative AI, and as symbolic alternatives to legal generative AI, that could be integrated as conversational interfaces.

2. As we emphasized in this chapter, conceptualizations of argumentation can take the form of inference, dialogue or balancing, and these models have their own formal methods. Do they also have their own technologies?

3. In the attack-defense paradigm shift, commutative diagrams play a central role. How can these technologies be integrated to make structured argumentation diagrams commute? How should various technologies be integrated with structured and abstract argumentation?

4. Another recurring discussion in this chapter is methodology, e.g., how can we develop a user guide about the new formal methods? Likewise, we may ask: how can we use new argumentation technologies like IHiBO?

5. This is just the beginning of the use of the attack-defense paradigm shift in argumentation technology. For a start, how can we use algorithmic argumentation methods in argumentation technologies?

5 Summary

This chapter has discussed the evolving landscape of argumentation, exploring its natural forms, the paradigm shift initiated by Dung, and subsequent advancements in computational approaches.

Natural argumentation is rooted in both theoretical and practical reasoning, with formal argumentation grounded in philosophical and mathematical foundations. This foundational approach is essential for representing, managing, and resolving conflicts in various disciplines. For instance, the Jiminy ethical governor, which operates across six layers of conflict, exemplifies the complexity and depth of formal reasoning in ethical decision-making. However, natural argumentation is inherently diverse, reflecting the complexities of human thought and communication. This diversity is evident in psychological evaluations of arguments, where understanding and generating arguments involves intricate cognitive processes. Foundation models are increasingly employed to construct and decode arguments, highlighting the importance of questions, particularly "why" questions, in uncovering weaknesses and justifying decisions. These questions play a crucial role in frameworks like the Fatio dialogue protocol, emphasizing the centrality of inquiry in argumentation. Argumentation can be conceptualized in various ways such as inference, dialogue, or balancing, each offering distinct perspectives and applications. For example, a divorce court case can be modeled differently depending on the chosen conceptual framework, which demonstrates the flexibility of argumentation theories. Additionally, formalizing argumentation through a variety of methods allows for a combination of different reasoning techniques, for instance, in the representation of practical scenarios like a mother reasoning with her daughter with mixed formal methods. This highlights the practical utility of formal argumentation for navigating complex real-world situations.

The paradigm shift in argumentation was significantly influenced by the attack-defense framework introduced by Dung [1995]. This framework marked a turning point in argumentation theory by emphasizing that every utterance, whether a claim, argument, or attack, can be contested. This led to a more comprehensive and universal approach to analyzing arguments. Structured argumentation has served as a bridge between classical and nonmonotonic logic, representing various logic and game theory concepts. This is particularly evident in the design of nonmonotonic logic, where rationality postulates from paraconsistent reasoning are crucial. Examples like the weakest versus last link principles and PDL illustrate the depth and versatility of formal argumentation in designing and representing these logics. Furthermore, extensions to abstract argumentation frameworks have been developed so that we can extract more information and incorporate qualitative and quantitative elements such as bipolarity, preferences, and so on. These extensions, inspired by dialogue and balancing, have enriched Dung's abstract argumentation frameworks and allow for a more nuanced understanding and modeling of complex argumentative situations.

Computational argumentation has advanced significantly with the development of the principle-based approach, which handles the diversity of argumentation models by providing a higher level of abstraction. This approach is essential for selecting appropriate methods and designing new ones. It ensures that the diverse landscape of argumentation models can be navigated and applied effectively. Compositionality principles such as locality, modularity and transparency are central to the attack-defense perspective in computational argumentation. These principles are exploited by algorithms and computational techniques to enhance their efficiency, robustness, and scalability, as seen in the divide and conquer approach which breaks down complex frameworks into manageable components. The relationship between explanation and argumentation was also discussed. Strategic argumentation explains dialogues, discussion games clarify inference, and underdeveloped reason-based explanations are used for balancing. These connections underscore the importance of argumentation for making AI systems more transparent and understandable. Additionally, the integration of argumentation techniques with emerging technologies such as distributed argumentation technology has expanded the potential applications of argumentation in areas like blockchain and AI. For instance, the IHiBO architecture integrates argumentation with blockchain technology to enhance transparency and trust in decision-making processes.

In summary, this chapter presents an overview of argumentation: past achievements, the current state of the art, and future directions. We discussed the three pillars in the context of natural argumentation before discussing the attack-defense paradigm shift initiated by Dung and advancements in computational argumentation that are shaping the future of the field.

6 Acknowledgments

All authors acknowledge financial support from the Luxembourg National Research Fund (FNR) — L. van der Torre through the project The Epistemology of AI Systems (EAI) (C22/SC/17111440), L. van der Torre and R. Markovich through the projects Logical Methods for Deontic Explanations (LODEX) (INTER/DFG/23/17415164/LoDEx) and Study of the Limits, Problems and Risks Associated with Autonomous Technologies (INTEGRAUTO) (INTER/AUDACE/21/16695098) and all authors through the project Deontic Logic for Epistemic Rights (DELIGHT) (O20/14776480). R. Markovich and L.Yu are also supported by the University of Luxembourg's Marie Speyer Excellence Grant of 2024 Formal Analysis of Discretionary Reasoning (MSE-DISCREASON).

References

[Alchourrón et al., 1985] Carlos E Alchourrón, Peter Gärdenfors, and David Makinson. On the logic of theory change: partial meet contraction and revision functions. *The journal of symbolic logic*, 50(2):510–530, 1985.

[Aleinikoff, 1986] T Alexander Aleinikoff. Constitutional law in the age of balancing. *Yale lj*, 96:943, 1986.

[Alkaissi and McFarlane, 2023] Hussam Alkaissi and Samy I McFarlane. Artificial hallucinations in chatgpt: implications in scientific writing. *Cureus*, 15(2), 2023.

[Altay et al., 2022] Sacha Altay, Marlène Schwartz, Anne-Sophie Hacquin, Aurélien Allard, Stefaan Blancke, and Hugo Mercier. Scaling up interactive argumentation by providing counterarguments with a chatbot. *Nature Human Behaviour*, 6(4):579–592, 2022.

[Amgoud and Cayrol, 2002] Leila Amgoud and Claudette Cayrol. Inferring from inconsistency in preference-based argumentation frameworks. *International Journal of Approximate Reasoning*, 29(2):125–169, 2002.

[Amgoud and Vesic, 2012] Leila Amgoud and Srdjan Vesic. On the use of argumentation for multiple criteria decision making. In *International Conference on Information Processing and Management of Uncertainty in Knowledge-Based Systems*, pages 480–489. Springer, 2012.

[Amgoud, 2005] Leila Amgoud. An argumentation-based model for reasoning about coalition structures. In *International Workshop on Argumentation in Multi-Agent Systems*, pages 217–228. Springer, 2005.

[Amgoud, 2009] Leila Amgoud. Argumentation for decision making. *Argumentation in artificial intelligence*, pages 301–320, 2009.

[Arisaka et al., 2022] Ryuta Arisaka, Jérémie Dauphin, Ken Satoh, and Leendert van der Torre. Multi-agent argumentation and dialogue. *IfCoLog Journal of Logics and Their Applications*, 9(4):921–954, 2022.

[Atkinson and Bench-Capon, 2005] Katie Atkinson and Trevor Bench-Capon. Legal case-based reasoning as practical reasoning. *Artificial Intelligence and Law*, 13:93–131, 2005.

[Atkinson and Bench-Capon, 2021] Katie Atkinson and Trevor Bench-Capon. Value-based argumentation. *Handbook of Formal Argumentation*, 2:397–441, 2021.

[Aumann, 2016] Robert J Aumann. *Agreeing to disagree*. Springer, 2016.

[Austin, 1975] John Langshaw Austin. *How to do things with words*. Harvard University Press, 1975.

[Awad et al., 2018] Edmond Awad, Sohan Dsouza, Richard Kim, Jonathan Schulz, Joseph Henrich, Azim Shariff, Jean-François Bonnefon, and Iyad Rahwan. The moral machine experiment. *Nature*, 563(7729):59–64, 2018.

[Baroni and Giacomin, 2007] Pietro Baroni and Massimiliano Giacomin. On principle-based evaluation of extension-based argumentation semantics. *Artificial Intelligence*, 171(10-15):675–700, 2007.

[Baroni et al., 2005] Pietro Baroni, Massimiliano Giacomin, and Giovanni Guida. Scc-recursiveness: a general schema for argumentation semantics. *Artificial Intelligence*, 168(1-2):162–210, 2005.

[Baroni et al., 2011] Pietro Baroni, Paul E Dunne, and Massimiliano Giacomin. On the resolution-based family of abstract argumentation semantics and its grounded instance. *Artificial Intelligence*, 175(3-4):791–813, 2011.

[Baroni *et al.*, 2014] Pietro Baroni, Guido Boella, Federico Cerutti, Massimiliano Giacomin, Leendert van der Torre, and Serena Villata. On the input/output behavior of argumentation frameworks. *Artificial Intelligence*, 217:144–197, 2014.

[Baroni *et al.*, 2018] Pietro Baroni, Massimiliano Giacomin, and Beishui Liao. Locality and modularity in abstract argumentation. *Handbook of formal argumentation*, pages 937–979, 2018.

[Barringer *et al.*, 2005] Howard Barringer, Dov Gabbay, and John Woods. Temporal dynamics of support and attack networks: From argumentation to zoology. *Mechanizing Mathematical Reasoning: Essays in Honor of Jörg H. Siekmann on the Occasion of His 60th Birthday*, pages 59–98, 2005.

[Baumann *et al.*, 2020] Ringo Baumann, Gerhard Brewka, and Markus Ulbricht. Revisiting the foundations of abstract argumentation–semantics based on weak admissibility and weak defense. In *Proceedings of the AAAI Conference on Artificial Intelligence*, volume 34, pages 2742–2749, 2020.

[Baumann, 2012] Ringo Baumann. What does it take to enforce an argument? Minimal change in abstract argumentation. In *ECAI 2012*, pages 127–132. IOS Press, 2012.

[Baumeister *et al.*, 2021] Dorothea Baumeister, Daniel Neugebauer, and Jörg Rothe. Collective acceptability in abstract argumentation. *Handbook of Formal Argumentation, Volume 2*, 2021.

[Bengel *et al.*, 2024] Lars Bengel, Lydia Blümel, Tjitze Rienstra, and Matthias Thimm. Argumentation-based probabilistic causal reasoning. In *Conference on Advances in Robust Argumentation Machines*, pages 221–236. Springer, 2024.

[Bezou-Vrakatseli, 2023] Elfia Bezou-Vrakatseli. Evaluation of LLM reasoning via argument schemes. *Online Handbook of Argumentation for AI*, 4, 2023.

[Bistarelli *et al.*, 2021] Stefano Bistarelli, Francesco Santini, et al. Weighted argumentation. *Handbook of Formal Argumentation, Volume 2*, 2021.

[Black *et al.*, 2021] Elizabeth Black, Nicolas Maudet, and Simon Parsons. Argumentation-based dialogue. *Handbook of Formal Argumentation, Volume 2*, 2021.

[Boella *et al.*, 2009] Guido Boella, Dov M Gabbay, Leendert van der Torre, and Serena Villata. Meta-argumentation modelling I: Methodology and techniques. *Studia Logica*, 93:297–355, 2009.

[Boella *et al.*, 2010] Guido Boella, Dov M Gabbay, Leendert van der Torre, and Serena Villata. Support in abstract argumentation. In *Proceedings of the Third International Conference on Computational Models of Argument*

(COMMA'10), pages 40–51. Frontiers in Artificial Intelligence and Applications, IOS Press, 2010.

[Borg and Bex, 2024] AnneMarie Borg and Floris Bex. Minimality, necessity and sufficiency for argumentation and explanation. *International Journal of Approximate Reasoning*, page 109143, 2024.

[Brewka and Eiter, 1999] G. Brewka and T. Eiter. Preferred answer sets for extented logic programs. *Artif. Intell.*, 109:297–356, 1999.

[Brewka, 1994] Gerhard Brewka. Reasoning about priorities in default logic. In Barbara Hayes-Roth and Richard E. Korf, editors, *Proc. of the 12th National Conference on AI*, volume 2, pages 940–945. AAAI Press / The MIT Press, 1994.

[Budzynska et al., 2018] Katarzyna Budzynska, Serena Villata, et al. Processing natural language argumentation. *Handbook of formal argumentation*, 1:577–627, 2018.

[Caminada and Amgoud, 2005] Martin Caminada and Leila Amgoud. An axiomatic account of formal argumentation. In *AAAI*, volume 6, pages 608–613, 2005.

[Caminada and Amgoud, 2007] Martin Caminada and Leila Amgoud. On the evaluation of argumentation formalisms. *Artificial Intelligence*, 171(5-6):286–310, 2007.

[Caminada and Ben-Naim, 2007] Martin Caminada and Jonathan Ben-Naim. *Postulates for paraconsistent reasoning and fault tolerant logic programming*. PhD thesis, Department of Information and Computing Sciences, Utrecht University, 2007.

[Caminada and Gabbay, 2009] Martin WA Caminada and Dov M Gabbay. A logical account of formal argumentation. *Studia Logica*, 93:109–145, 2009.

[Caminada and Pigozzi, 2011] Martin Caminada and Gabriella Pigozzi. On judgment aggregation in abstract argumentation. *Autonomous Agents and Multi-Agent Systems*, 22:64–102, 2011.

[Caminada and Wu, 2011] Martin Caminada and Yining Wu. On the limitations of abstract argumentation. In *Proceedings of the 23rd Benelux Conference on Artificial Intelligence (BNAIC 2011)*, pages 59–66, 2011.

[Caminada et al., 2012] Martin WA Caminada, Walter A Carnielli, and Paul E Dunne. Semi-stable semantics. *Journal of Logic and Computation*, 22(5):1207–1254, 2012.

[Caminada et al., 2014] Martinus Wigbertus Antonius Caminada, Sanjay Modgil, and Nir Oren. Preferences and unrestricted rebut. *Computational Models of Argument*, 2014.

[Caminada, 2006] Martin Caminada. Semi-stable semantics. *COMMA*,

144:121–130, 2006.

[Caminada, 2017] Martin Caminada. Argumentation semantics as formal discussion. *Journal of Applied Logics*, 4(8):2457–2492, 2017.

[Caminada, 2018a] Martin Caminada. Argumentation semantics as formal discussion. *Handbook of Formal Argumentation*, 1:487–518, 2018.

[Caminada, 2018b] Martin Caminada. Rationality postulates: Applying argumentation theory for non-monotonic reasoning. *Handbook of Formal Argumentation, Volume 1*, pages 771–796, 2018.

[Castagna et al., 2024a] Federico Castagna, Nadin Kökciyan, Isabel Sassoon, Simon Parsons, and Elizabeth Sklar. Computational argumentation-based chatbots: a survey. *Journal of Artificial Intelligence Research*, 80:1271–1310, 2024.

[Castagna et al., 2024b] Federico Castagna, Peter McBurney, and Simon Parsons. Explanation–question–response dialogue: An argumentative tool for explainable AI. *Argument & Computation*, (Preprint):1–23, 2024.

[Castagna et al., 2024c] Federico Castagna, Isabel Sassoon, and Simon Parsons. Can formal argumentative reasoning enhance LLMs performances? *arXiv preprint arXiv:2405.13036*, 2024.

[Cayrol and Lagasquie-Schiex, 2005] Claudette Cayrol and Marie-Christine Lagasquie-Schiex. On the acceptability of arguments in bipolar argumentation frameworks. In *European Conference on Symbolic and Quantitative Approaches to Reasoning and Uncertainty*, pages 378–389. Springer, 2005.

[Cayrol and Lagasquie-Schiex, 2009] Claudette Cayrol and Marie-Christine Lagasquie-Schiex. Bipolar abstract argumentation systems. In *Argumentation in Artificial Intelligence*, pages 65–84. Springer, 2009.

[Cayrol and Lagasquie-Schiex, 2010] Claudette Cayrol and Marie-Christine Lagasquie-Schiex. Coalitions of arguments: A tool for handling bipolar argumentation frameworks. *International Journal of Intelligent Systems*, 25(1):83–109, 2010.

[Cayrol and Lagasquie-Schiex, 2013] Claudette Cayrol and Marie-Christine Lagasquie-Schiex. Bipolarity in argumentation graphs: Towards a better understanding. *International Journal of Approximate Reasoning*, 54(7):876–899, 2013.

[Cayrol et al., 2021] Claudette Cayrol, Andrea Cohen, and Marie Christine Lagasquie Schiex. Higher-order interactions (bipolar or not) in abstract argumentation: a state of the art. 2021.

[Cerutti et al., 2021] Federico Cerutti, Marcos Cramer, Mathieu Guillaume, Emmanuel Hadoux, Anthony Hunter, and Sylwia Polberg. Empirical cognitive studies about formal argumentation. *Handbook of Formal Argumenta-*

tion, Volume 2, 2021.

[Coleman, 1984] James S Coleman. Micro foundations and macrosocial behavior. *Angewandte Sozialforschung anc AIAS Informationen Wien*, 12(1-2):25–37, 1984.

[Cramer and Guillaume, 2018a] Marcos Cramer and Mathieu Guillaume. Directionality of attacks in natural language argumentation. In *CEUR Workshop Proceedings*. RWTH Aachen University, Aachen, Germany, 2018.

[Cramer and Guillaume, 2018b] Marcos Cramer and Mathieu Guillaume. Empirical cognitive study on abstract argumentation semantics. In *Computational Models of Argument*, pages 413–424. IOS Press, 2018.

[Cramer and Guillaume, 2019] Marcos Cramer and Mathieu Guillaume. Empirical study on human evaluation of complex argumentation frameworks. In *Logics in Artificial Intelligence: 16th European Conference, JELIA 2019, Rende, Italy, May 7–11, 2019, Proceedings 16*, pages 102–115. Springer, 2019.

[Da Costa *et al.*, 2007] Newton CA Da Costa, Décio Krause, and Otávio Bueno. Paraconsistent logics and paraconsistency. In *Philosophy of logic*, pages 791–911. Elsevier, 2007.

[Da Costa, 1974] Newton CA Da Costa. On the theory of inconsistent formal systems. *Notre dame journal of formal logic*, 15(4):497–510, 1974.

[Danry *et al.*, 2023] Valdemar Danry, Pat Pataranutaporn, Yaoli Mao, and Pattie Maes. Don't just tell me, ask me: AI systems that intelligently frame explanations as questions improve human logical discernment accuracy over causal AI explanations. In *Proceedings of the 2023 CHI Conference on Human Factors in Computing Systems*, pages 1–13, 2023.

[Dauphin and Cramer, 2018] Jérémie Dauphin and Marcos Cramer. ASPIC-END: structured argumentation with explanations and natural deduction. In *Theory and Applications of Formal Argumentation: 4th International Workshop, TAFA 2017, Melbourne, VIC, Australia, August 19-20, 2017, Revised Selected Papers 4*, pages 51–66. Springer, 2018.

[D'Avila Garcez *et al.*, 2005] Artur S D'Avila Garcez, Dov M Gabbay, and Luis C Lamb. Value-based argumentation frameworks as neural-symbolic learning systems. *Journal of Logic and Computation*, 15(6):1041–1058, 2005.

[Delgrande *et al.*, 2004] James Delgrande, Torsten Schaub, Hans Tompits, and Kewen Wang. A classification and survey of preference handling approaches in nonmonotonic reasoning. *Computational Intelligence*, 20(2):308–334, 2004.

[Dong *et al.*, 2019] Huimin Dong, Beishui Liao, Reka Markovich, and Leendert van der Torre. From classical to non-monotonic deontic logic using ASPIC+.

In *Logic, Rationality, and Interaction: 7th International Workshop, LORI 2019, Chongqing, China, October 18–21, 2019, Proceedings 7*, pages 71–85. Springer, 2019.

[Dong et al., 2021] Huimin Dong, Réka Markovich, and Leendert van der Torre. Towards AI logic for social reasoning. *arXiv preprint arXiv:2110.04452*, 2021.

[Dung and Thang, 2018] Phan Minh Dung and Phan Minh Thang. Fundamental properties of attack relations in structured argumentation with priorities. *Artificial Intelligence*, 255:1–42, 2018.

[Dung, 1995] Phan Minh Dung. On the acceptability of arguments and its fundamental role in nonmonotonic reasoning, logic programming and n-person games. *Artificial intelligence*, 77(2):321–357, 1995.

[Dung, 2014] Phan Minh Dung. An axiomatic analysis of structured argumentation for prioritized default reasoning. volume 263 of *Frontiers in Artificial Intelligence and Applications*, pages 267–272. IOS Press, 2014.

[Dung, 2016] Phan Minh Dung. An axiomatic analysis of structured argumentation with priorities. *Artif. Intell.*, 231:107–150, 2016.

[FIPA, 2002] FIPA. Communicative act library specification. *http://www.fipa.org/specs/fipa00037*, 2002.

[Gabbay and Rivlin, 2017] Dov M Gabbay and Lydia Rivlin. Heal2100: human effective argumentation and logic for the 21st century. The next step in the evolution of logic. *IFCoLog Journal of Logics and Their Applications*, 2017.

[Garcez et al., 2008] Artur SD'Avila Garcez, Luis C Lamb, and Dov M Gabbay. *Neural-symbolic cognitive reasoning*. Springer Science & Business Media, 2008.

[Gargouri et al., 2020] Anis Gargouri, Sébastien Konieczny, Pierre Marquis, and Srdjan Vesic. On a notion of monotonic support for bipolar argumentation frameworks. In *20th International Conference on Autonomous Agents and MultiAgent Systems*, 2020.

[Georgeff et al., 1999] Michael Georgeff, Barney Pell, Martha Pollack, Milind Tambe, and Michael Wooldridge. The belief-desire-intention model of agency. In *Intelligent Agents V: Agents Theories, Architectures, and Languages: 5th International Workshop, ATAL'98 Paris, France, July 4–7, 1998 Proceedings 5*, pages 1–10. Springer, 1999.

[Giacomin, 2017] Massimiliano Giacomin. Handling heterogeneous disagreements through abstract argumentation (extended abstract). In *PRIMA 2017: Principles and Practice of Multi-Agent Systems*, pages 3–11, 2017.

[Gillioz and Zufferey, 2020] Christelle Gillioz and Sandrine Zufferey. *Introduction to experimental linguistics*. John Wiley & Sons, 2020.

[Giunchiglia *et al.*, 2004] Enrico Giunchiglia, Joohyung Lee, Vladimir Lifschitz, Norman McCain, and Hudson Turner. Nonmonotonic causal theories. *Artificial Intelligence*, 153(1-2):49–104, 2004.

[Gordon *et al.*, 2007] Thomas F Gordon, Henry Prakken, and Douglas Walton. The Carneades model of argument and burden of proof. *Artificial Intelligence*, 171(10-15):875–896, 2007.

[Gordon, 1993] Thomas F Gordon. The Pleadings Game: An exercise in computational dialectics. *Artificial Intelligence and Law*, 2:239–292, 1993.

[Gordon, 2018] Thomas F Gordon. Towards requirements analysis for formal argumentation. In Pietro Baroni, Dov Gabbay, Massimiliano Giacomin, and Leendert van der Torre, editors, *Handbook of formal argumentation, Volume 1*, pages 145–156. College Publications, 2018.

[Governatori *et al.*, 2021] Guido Governatori, Michael J Maher, and Francesco Olivieri. Strategic argumentation. *Handbook of Formal Argumentation, Volume 2*, 2021.

[Hakim *et al.*, 2019] Fauzia Zahira Munirul Hakim, Lia Maulia Indrayani, and Rosaria Mita Amalia. A dialogic analysis of compliment strategies employed by Replika chatbot. In *Third International conference of arts, language and culture (ICALC 2018)*, pages 266–271. Atlantis Press, 2019.

[Hamblin, 1970] C. L. Hamblin. Fallacies. *Tijdschrift Voor Filosofie*, 33(1):183–188, 1970.

[Heyninck, 2019] Jesse Heyninck. *Investigations into the logical foundations of defeasible reasoning: an argumentative perspective*. PhD thesis, Ruhr University Bochum, Germany, 2019.

[Hinton and Wagemans, 2023] Martin Hinton and Jean HM Wagemans. How persuasive is AI-generated argumentation? An analysis of the quality of an argumentative text produced by the GPT-3 AI text generator. *Argument & Computation*, 14(1):59–74, 2023.

[Kaci *et al.*, 2021] Souhila Kaci, Leendert van Der Torre, Srdjan Vesic, and Serena Villata. Preference in abstract argumentation. *Handbook of Formal Argumentation, Volume 2*, 2021.

[Kakas and Moraitis, 2003] Antonis Kakas and Pavlos Moraitis. Argumentation based decision making for autonomous agents. In *Proceedings of the second international joint conference on Autonomous agents and multiagent systems*, pages 883–890, 2003.

[Kant, 1993] Immanuel Kant. *Groundwork of the Metaphysics of Morals*. Hackett Publishing Company, Indianapolis, 3rd edition, 1993. [1785].

[Kissine, 2013] Michail Kissine. Speech act classifications. *Pragmatics of speech actions*, 173:202, 2013.

[Knocks et al., 2024] Aleks Knocks, Muyun Shao, Leendert ver der Torre, Vincent De Wit, and Liuwen Yu. A principle-based analysis for numerical balancing. In *Logics for New-Generation Artificial Intelligence (LNGAI2024)*. College Publications, United Kingdom, 2024.

[Knoks and van der Torre, 2023] Aleks Knoks and Leendert van der Torre. Reason-based detachment. In *Logics for New-Generation Artificial Intelligence (LNGAI2023)*. College Publications, London, United Kingdom, 2023.

[Kraus et al., 1990] Sarit Kraus, Daniel Lehmann, and Menachem Magidor. Nonmonotonic reasoning, preferential models and cumulative logics. *Artificial intelligence*, 44(1-2):167–207, 1990.

[Leite and Martins, 2011] João Leite and João G. Martins. Social abstract argumentation. In Toby Walsh, editor, *IJCAI 2011, Proceedings of the 22nd International Joint Conference on Artificial Intelligence, Barcelona, Catalonia, Spain, July 16-22, 2011*, pages 2287–2292. IJCAI/AAAI, 2011.

[Liao and van der Torre, 2024] Beishui Liao and Leendert van der Torre. Attack-defense semantics of argumentation. In *Computational Models of Argument*, pages 133–144. IOS Press, 2024.

[Liao et al., 2019] Beishui Liao, Nir Oren, Leendert van der Torre, and Serena Villata. Prioritized norms in formal argumentation. *Journal of Logic and Computation*, 29(2):215–240, 2019.

[Liao et al., 2023] Beishui Liao, Pere Pardo, Marija Slavkovik, and Leendert van der Torre. The Jiminy advisor: Moral agreements among stakeholders based on norms and argumentation. *Journal of Artificial Intelligence Research*, 77:737–792, 2023.

[Longo et al., 2024] Luca Longo, Mario Brcic, Federico Cabitza, Jaesik Choi, Roberto Confalonieri, Javier Del Ser, Riccardo Guidotti, Yoichi Hayashi, Francisco Herrera, Andreas Holzinger, et al. Explainable artificial intelligence (XAI) 2.0: A manifesto of open challenges and interdisciplinary research directions. *Information Fusion*, 106:102301, 2024.

[Makinson, 2005] David Makinson. *Bridges from classical to nonmonotonic logic*. King's College, 2005.

[Markovich, 2019] Réka Markovich. On the formal structure of rules in conflict of laws. In *JURIX*, pages 199–204, 2019.

[McBurney and Parsons, 2002] Peter McBurney and Simon Parsons. Games that agents play: A formal framework for dialogues between autonomous agents. *Journal of Logic, Language and Information*, 11(3):315–334, 2002.

[McBurney and Parsons, 2004] Peter McBurney and Simon Parsons. Locutions for argumentation in agent interaction protocols. In *International Workshop on Agent Communication*, pages 209–225. Springer, 2004.

[McBurney et al., 2021] Peter McBurney, Simon Parsons, et al. Argument schemes and dialogue protocols: Doug walton's legacy in artificial intelligence. *FLAP*, 8(1):263–290, 2021.

[McDougall et al., 2020] Rosalind McDougall, Cade Shadbolt, and Lynn Gillam. The practice of balancing in clinical ethics case consultation. *Clinical Ethics*, 15(1):49–55, 2020.

[Mercier and Sperber, 2017] Hugo Mercier and Dan Sperber. *The enigma of reason*. Harvard University Press, 2017.

[Modgil and Prakken, 2013] Sanjay Modgil and Henry Prakken. A general account of argumentation with preferences. *Artif. Intell.*, 195:361–397, 2013.

[Modgil and Prakken, 2014] Sanjay Modgil and Henry Prakken. The $ASPIC^+$ framework for structured argumentation: a tutorial. *Argument Comput.*, 5(1):31–62, 2014.

[Moore, 1993] David John Moore. *Dialogue game theory for intelligent tutoring systems*. PhD thesis, Leeds Metropolitan University, 1993.

[Musi and Palmieri, 2022] Elena Musi and Rudi Palmieri. The fallacy of explainable generative AI: evidence from argumentative prompting in two domains. 2022.

[Nash Jr, 1950] John F Nash Jr. Equilibrium points in n-person games. *Proceedings of the national academy of sciences*, 36(1):48–49, 1950.

[Niskanen et al., 2018] Andreas Niskanen, Johannes P Wallner, and Matti Järvisalo. Extension enforcement under grounded semantics in abstract argumentation. In *Sixteenth International Conference on Principles of Knowledge Representation and Reasoning*, 2018.

[Nouioua and Risch, 2010] Farid Nouioua and Vincent Risch. Bipolar argumentation frameworks with specialized supports. In *2010 22nd IEEE International Conference on Tools with Artificial Intelligence*, volume 1, pages 215–218. IEEE, 2010.

[Nute, 1994] Donald Nute. Defeasible logic. *Handbook of logic in artificial intelligence and logic programming*, 3:353–395, 1994.

[Okuno and Takahashi, 2009] Kenichi Okuno and Kazuko Takahashi. Argumentation system with changes of an agent's knowledge base. In *Twenty-First International Joint Conference on Artificial Intelligence*. Citeseer, 2009.

[Pardo and Straßer, 2022] Pere Pardo and Christian Straßer. Modular orders on defaults in formal argumentation. *Journal of Logic and Computation*, 2022.

[Pardo et al., 2024] Pere Pardo, Liuwen Yu, Chen Chen, and Leendert van der Torre. Weakest link, prioritised default logic and principles in argumentation.

Journal of Logic and Computation, 2024. Forthcoming.

[Phillips et al., 2021] P Jonathon Phillips, P Jonathon Phillips, Carina A Hahn, Peter C Fontana, Amy N Yates, Kristen Greene, David A Broniatowski, and Mark A Przybocki. Four principles of explainable artificial intelligence. 2021.

[Pigozzi and van der Torre, 2018] Gabriella Pigozzi and Leendert van der Torre. Arguing about constitutive and regulative norms. *Journal of Applied Non-Classical Logics*, 28(2-3):189–217, 2018.

[Polberg and Hunter, 2018] Sylwia Polberg and Anthony Hunter. Empirical evaluation of abstract argumentation: Supporting the need for bipolar and probabilistic approaches. *Int. J. Approx. Reason.*, 93:487–543, 2018.

[Pollock, 1987] John L Pollock. Defeasible reasoning. *Cognitive science*, 11(4):481–518, 1987.

[Pollock, 1992] John L Pollock. How to reason defeasibly. *Artificial Intelligence*, 57(1):1–42, 1992.

[Pollock, 1994] John L Pollock. Justification and defeat. *Artificial Intelligence*, 67(2):377–407, 1994.

[Pollock, 1995] John L Pollock. *Cognitive carpentry: A blueprint for how to build a person*. Mit Press, 1995.

[Pollock, 2001] John L Pollock. Defeasible reasoning with variable degrees of justification. *Artificial intelligence*, 133(1-2):233–282, 2001.

[Pollock, 2009] John L Pollock. A recursive semantics for defeasible reasoning. In *Argumentation in artificial intelligence*, pages 173–197. Springer, 2009.

[Pollock, 2010] John L Pollock. Defeasible reasoning and degrees of justification. *Argument and Computation*, 1(1):7–22, 2010.

[Prakken and Sartor, 2015] Henry Prakken and Giovanni Sartor. Law and logic: A review from an argumentation perspective. *Artificial intelligence*, 227:214–245, 2015.

[Prakken, 2010a] Henry Prakken. An abstract framework for argumentation with structured arguments. *Argument & Computation*, 1(2):93–124, 2010.

[Prakken, 2010b] Henry Prakken. Slides on nonmonotonic logic for commonsense reasoning, 2010.

[Prakken, 2013] Henry Prakken. *Logical tools for modelling legal argument: a study of defeasible reasoning in law*, volume 32. Springer Science & Business Media, 2013.

[Prakken, 2018] Henry Prakken. Historical overview of formal argumentation. In *Handbook of formal argumentation*, pages 73–141. College Publications, 2018.

[Prakken, 2020] Henry Prakken. On validating theories of abstract argumentation frameworks: the case of bipolar argumentation frameworks. In *Proceedings of the 8th Workshop on Computational Models of Natural Argument (CMNA 2020), Perugia, Italy (and online)*. CEUR-WS.org, 2020.

[Prakken, 2024a] Henry Prakken. An abstract and structured account of dialectical argument strength. *Artificial Intelligence*, 335:104193, 2024.

[Prakken, 2024b] Henry Prakken. On evaluating legal-reasoning capabilities of generative ai. In *Proceedings of the 24th Workshop on Computational Models of Natural Argument*, pages 100–112, 2024.

[Priest, 2002] Graham Priest. Paraconsistent logic. In *Handbook of philosophical logic*, pages 287–393. Springer, 2002.

[Rahwan et al., 2003] Iyad Rahwan, Sarvapali D Ramchurn, Nicholas R Jennings, Peter McBurney, Simon Parsons, and Liz Sonenberg. Argumentation-based negotiation. *The Knowledge Engineering Review*, 18(4):343–375, 2003.

[Rawls, 2001] John Rawls. *Justice as fairness: A restatement*. Harvard University Press, 2001.

[Reiter, 1980] Raymond Reiter. A logic for default reasoning. *Artificial intelligence*, 13(1-2):81–132, 1980.

[Rienstra et al., 2020] Tjitze Rienstra, Chiaki Sakama, Leendert van der Torre, and Beishui Liao. A principle-based robustness analysis of admissibility-based argumentation semantics. *Argument & Computation*, 11(3):305–339, 2020.

[Ross et al., 2023] Steven I Ross, Fernando Martinez, Stephanie Houde, Michael Muller, and Justin D Weisz. The programmer's assistant: Conversational interaction with a large language model for software development. In *Proceedings of the 28th International Conference on Intelligent User Interfaces*, pages 491–514, 2023.

[Roth and Verheij, 2004] Bram Roth and Bart Verheij. Cases and dialectical arguments–an approach to case-based reasoning. In *On the Move to Meaningful Internet Systems 2004*, pages 634–651. Springer, 2004.

[Russell and Norvig, 2010] Stuart J Russell and Peter Norvig. *Artificial intelligence: A modern approach*. London, 2010.

[Savage, 1972] Leonard J Savage. *The foundations of statistics*. Courier Corporation, 1972.

[Schindler et al., 2020] Samuel Schindler, Anna Drozdzowicz, and Karen Brøcker. *Linguistic intuitions: Evidence and method*. Oxford University Press, 2020.

[Searle, 1979] John R Searle. *Expression and meaning: Studies in the theory of speech acts*. Cambridge University Press, 1979.

[Sklar and Azhar, 2018] Elizabeth I Sklar and Mohammad Q Azhar. Explanation through argumentation. In *Proceedings of the 6th International Conference on Human-Agent Interaction*, pages 277–285, 2018.

[Straßer and Arieli, 2019] Christian Straßer and Ofer Arieli. Normative reasoning by sequent-based argumentation. *Journal of Logic and Computation*, 29(3):387–415, 2019.

[Sycara, 1990] Katia P Sycara. Persuasive argumentation in negotiation. *Theory and decision*, 28:203–242, 1990.

[Tennent, 1991] Robert D Tennent. Semantics of programming languages. *(No Title)*, 1991.

[Toulmin, 1958] Stephen E Toulmin. *The uses of argument*. Cambridge university press, 1958.

[Traum, 1999] David R Traum. Speech acts for dialogue agents. In *Foundations of rational agency*, pages 169–201. Springer, 1999.

[Turner, 2004] Hudson Turner. Strong equivalence for causal theories. In *Logic Programming and Nonmonotonic Reasoning: 7th International Conference, LPNMR 2004 Fort Lauderdale, FL, USA, January 6-8, 2004 Proceedings 7*, pages 289–301. Springer, 2004.

[Ulbricht and Wallner, 2021] Markus Ulbricht and Johannes P Wallner. Strong explanations in abstract argumentation. In *Proceedings of the AAAI Conference on Artificial Intelligence*, volume 35, pages 6496–6504, 2021.

[van der Lee et al., 2021] Chris van der Lee, Albert Gatt, Emiel van Miltenburg, and Emiel Krahmer. Human evaluation of automatically generated text: Current trends and best practice guidelines. *Computer Speech & Language*, 67:101151, 2021.

[van der Torre and Vesic, 2018] Leendert van der Torre and Srdjan Vesic. The principle-based approach to abstract argumentation semantics. In Pietro Baroni, Dov Gabbay, Massimiliano Giacomin, and Leendert van der Torre, editors, *Handbook of formal argumentation, Volume 1*, pages 797–838. College Publications, 2018.

[Van Laar and Krabbe, 2018] Jan Albert Van Laar and Erik CW Krabbe. The role of argument in negotiation. *Argumentation*, 32(4):549–567, 2018.

[Villata et al., 2011] Serena Villata, Guido Boella, and Leendert van der Torre. Attack semantics for abstract argumentation. In *IJCAI*. IJCAI/AAAI, 2011.

[Vilone and Longo, 2021] Giulia Vilone and Luca Longo. Classification of explainable artificial intelligence methods through their output formats. *Machine Learning and Knowledge Extraction*, 3(3):615–661, 2021.

[Visser et al., 2020] Jacky Visser, John Lawrence, and Chris Reed. Reason-checking fake news. *Communications of the ACM*, 63(11):38–40, 2020.

[Von Neumann and Morgenstern, 1947] John Von Neumann and Oskar Morgenstern. Theory of games and economic behavior, 2nd rev. 1947.

[Walton and Krabbe, 1995] Douglas Walton and Erik CW Krabbe. *Commitment in dialogue: Basic concepts of interpersonal reasoning*. State University of New York Press, 1995.

[Walton et al., 2008] Douglas Walton, Christopher Reed, and Fabrizio Macagno. *Argumentation schemes*. Cambridge University Press, 2008.

[Walton, 1990] Douglas N Walton. What is reasoning? What is an argument? *The Journal of Philosophy*, 87(8):399–419, 1990.

[Walton, 2013] Douglas Walton. *Argumentation schemes for presumptive reasoning*. Routledge, 2013.

[Weinstock, 2006] Michael P Weinstock. Psychological research and the epistemological approach to argumentation. *Informal Logic*, 26(1):103–120, 2006.

[Williamson and Gabbay, 2005] Jon Williamson and Dov Gabbay. Recursive causality in Bayesian networks and self-fibring networks. *Laws and models in the sciences*, pages 173–221, 2005.

[Young et al., 2016] Anthony P. Young, Sanjay Modgil, and Odinaldo Rodrigues. Prioritised default logic as rational argumentation. In Catholijn M. Jonker, Stacy Marsella, John Thangarajah, and Karl Tuyls, editors, *Proceedings of the 2016 International Conference on Autonomous Agents & Multiagent Systems*, pages 626–634. ACM, 2016.

[Young et al., 2017] Anthony P. Young, Sanjay Modgil, and Odinaldo Rodrigues. On the interaction between logic and preference in structured argumentation. In Elizabeth Black, Sanjay Modgil, and Nir Oren, editors, *Theory and Applications of Formal Argumentation - 4th International Workshop*, volume 10757, pages 35–50. Springer, 2017.

[Yu and Gabbay, 2022] Liuwen Yu and Dov Gabbay. Case-based reasoning via comparing the strength order of features. In *International Workshop on Explainable, Transparent Autonomous Agents and Multi-Agent Systems*, pages 143–151. Springer, 2022.

[Yu et al., 2020] Liuwen Yu, Réka Markovich, and Leendert Van Der Torre. Interpretations of support among arguments. In *Legal Knowledge and Information Systems*, pages 194–203. IOS Press, 2020.

[Yu et al., 2021] Liuwen Yu, Dongheng Chen, Lisha Qiao, Yiqi Shen, and Leendert van der Torre. A Principle-based Analysis of Abstract Agent Argumentation Semantics. In *Proceedings of the 18th International Conference on Principles of Knowledge Representation and Reasoning*, pages 629–639, 11 2021.

[Yu et al., 2022] Liuwen Yu, Mirko Zichichi, Markovich Réka, Najjar Amro, et al. Intelligent human-input-based blockchain oracle (IHiBO). In *Proceed-*

ings of the 14th International Conference on Agents and Artificial Intelligence (ICAART 2022), pages 1–12. SCITEPRESS, 2022.

[Yu *et al.*, 2023] Liuwen Yu, Caren Al Anaissy, Srdjan Vesic, Xu Li, and Leendert van der Torre. A principle-based analysis of bipolar argumentation semantics. In *European Conference on Logics in Artificial Intelligence*, pages 209–224. Springer, 2023.

[Yu, 2023] Liuwen Yu. *Distributed Argumentation Technology*. PhD thesis, 2023.

www.ingramcontent.com/pod-product-compliance
Lightning Source LLC
Chambersburg PA
CBHW071147230426
43668CB00009B/865